Encyclopedia of Gender and Information Technology

Eileen M. Trauth
The Pennsylvania State University, USA

Volume II
H–Z

IDEA GROUP REFERENCE
Hershey · London · Melbourne · Singapore

Acquisitions Editor:	Michelle Potter
Development Editor:	Kristin Roth
Senior Managing Editor:	Jennifer Neidig
Managing Editor:	Sara Reed
Copy Editors:	Julie LeBlanc, Shanelle Ramelb, April Schmidt, and Larissa Vinci
Typesetters:	Sharon Berger and Diane Huskinson
Support Staff:	Lauren Kenes
Cover Design:	Lisa Tosheff
Printed at:	Yurchak Printing Inc.

Published in the United States of America by
Idea Group Reference (an imprint of Idea Group Inc.)
701 E. Chocolate Avenue, Suite 200
Hershey PA 17033
Tel: 717-533-8845
Fax: 717-533-8661
E-mail: cust@idea-group.com
Web site: http://www.idea-group-ref.com

and in the United Kingdom by
Idea Group Reference (an imprint of Idea Group Inc.)
3 Henrietta Street
Covent Garden
London WC2E 8LU
Tel: 44 20 7240 0856
Fax: 44 20 7379 0609
Web site: http://www.eurospanonline.com

Library of Congress Cataloging-in-Publication Data

Encyclopedia of gender and information technology / Eileen Trauth, editor.
    v. <1 > cm.
  Summary: "This two volume set includes 213 entries with over 4,700 references to additional works on gender and information technology"--Provided by publisher.
  Includes bibliographical references and index.
  ISBN 1-59140-815-6 (hardcover) -- ISBN 1-59140-816-4 (ebook)
  1. Sex role in the work environment--Encyclopedias. 2. Women in computer science--Encyclopedias. 3. Information technology--Bibliography. 4. Computer science literature. I. Trauth, Eileen Moore.
  HD6060.6.E53 2006
  331.4'81004--dc22
                                2006005789

British Cataloguing in Publication Data
A Cataloguing in Publication record for this book is available from the British Library.

All work contributed to this encyclopedia set is new, previously-unpublished material. The views expressed in this encyclopedia set are those of the authors, but not necessarily of the publisher.

Editorial Advisory Board

List of Contributors

Contents
by Volume

VOLUME II

Foreword

It is commonly assumed that we are now living in a new knowledge economy, a post-industrial society. Groundbreaking developments in digitalisation and biotechnologies have led cyber gurus to assert that everything in the digital future will be different. Industrial technology may have had a patriarchal character, but digital technologies, based on brain rather than brawn, on networks rather than hierarchy, herald a new relationship between women and machines. Perhaps the link between technology and male privilege is finally being severed.

The realities of women's lives belie these simple generalisations. Certainly we are living in a very different world than that of even 30 years ago, and digital technologies increasingly mediate every area of our lives. This makes it even more imperative that we examine the extent to which existing societal patterns of gender inequality are transformed or reproduced in a new technological guise.

So, I am delighted that Eileen Trauth has taken on the mammoth task of bringing what is now a vast literature on gender and information technology together in these two volumes. In recent years, there has been a great deal of feminist writing on information and communication technologies. Studies on the Internet, cyberspace, and computer science, for example, have strengthened our analysis of how technology as a culture is implicated in the construction of masculinity. It demonstrates that the symbolic representation of technology remains sharply gendered. At the same time, feminist theory increasingly works from the basis that neither masculinity nor femininity are fixed, unitary categories but rather they are constructed in relation to each other. As a result, we now have a much more complex understanding of gender, of technology, and of the mutually-shaping relationship between them.

Cutting through the hype, this encyclopedia examines the relationship between gender and technology in all the major spheres of our lives. The theoretical approach that it adopts is much needed in an age when deep-seated technological determinism underpins much of the current debate on subjects as diverse as the ecological crisis, food safety, and genetic engineering. This determinist view represents technology as a separate sphere, developing independently of society, and can lead to pessimism about the possibilities for change. The rich collection of articles collected here certainly points in the opposite direction, providing a multiplicity of examples of activities, campaigns, and programs that are reshaping the landscape of women's relationships with machines.

The encyclopedia has many strengths, but there are two in particular that I would like to mention. The first is its interdisciplinary nature. The strength of gender theory over the last three decades has been its ability to cut across traditional disciplinary divides that not only made women invisible, but produced narrow, limited forms of knowledge. The most exciting developments in social science have been in areas that crosscut these old boundaries, and this encyclopedia exemplifies the fruitfulness of this approach. The authors cover a huge range of disciplines and approaches, and we are the richer for it.

The other notable feature of the book is that it uniquely contains contributions by authors from all over the world. Contemporary feminist studies of technology are characterised by more sensitivity to "the politics of difference" than some of the earlier literature. There is a much clearer realisation that gender, that what it is to be a woman, is experienced everywhere through such mediations as race, age, class, sexual orientation, history, and colonialism. Information technologies clearly have very different implications for Third World and First World women, within and between regions and countries. To date, this literature has been dispersed and hard to find. Now, given the comprehensive scope of this encyclopdia, we will be able to compare women's situation in different contexts, and our analysis will be enriched as a result.

I highly recommend these two volumes. They will be invaluable not only for academics and students, but for policy makers, educators, and the interested general reader. The authors represent a wealth of expertise and are at the leading edge of research in this area. If we are to transform gender power relations in the 21st century, we need a fully-rounded understanding of the relationship between gender and information technology. Eileen Trauth can be congratulated for providing us with just that.

Judy Wajcman
Author of Feminism Confronts Technology and TechnoFeminism,
Research School of Social Sciences, Australian National University
January 2006

Preface

As information technology has spread to all corners of the world and to all aspects of personal and work life, so too, has grown an interest in understanding more about the diverse characteristics of those who use, develop, and are affected by information technology (IT). This is a significant development because understanding the diverse characteristics of both developers and users has ramifications for the way in which work is done, user requirements for systems are understood, and interaction with computer-based tools is accomplished. One important aspect of this human diversity is gender. What has accompanied this increased interest in the role of gender diversity in understanding IT development and use, in recent years, has been heightened research interest in the influence of gender on information systems and technology.

As a result, a large and diffuse body of research related to the role of gender in human interactions with information technology has emerged in recent years. This body of research spans a number of disciplines including: information science and technology; information systems; computer science; engineering; education; women's studies; gender studies; labor studies; human resource management; and science, technology, and society. The focus of this research has been on issues such as similarities and differences between women's and men's use of information technology, variation in relationship to IT among members of each gender group, the effect of gender combined with other diversity characteristics (such as race or ethnicity) on IT use, and the underrepresentation of women in the IT profession. The audience for this research includes parents, educators, managers, policy makers, and other researchers. However, because this literature is located in so many different disciplines a number of problems have arisen. First, for students and scholars, it is difficult to locate the corpus of relevant gender research literature when one wants to learn about or is engaged in gender and IT research. Second, it is difficult for educators, policy makers, managers, and other consumers of this literature to find the relevant material. Consequently, there is a need to bring this research literature together into a single reference source.

The idea for a compilation of research on the topic of gender and information technology originated in 2001. While attending a panel on women and IT at an information systems conference in Australia, several of us who are engaged in gender and IT research bemoaned the lack of coherence in the research. The problem, as we saw it, was that too little research is informed by the existing literature. Instead, too much research appears to be informed by anecdotal data or personal bias. Hence, current research on gender and IT is making less of a contribution to cumulative knowledge about this topic and less of an impact in addressing the issues than it could. At that gathering, we began to talk about the need for a book that would bring together this disparate body of research literature. The opportunity to produce such a book came three years later when Dr. Mehdi Khosrow-Pour of Idea Group Inc. asked me to edit an encyclopedia on the topic of gender and IT. In accepting this invitation, my goal for this book was: (1) to bring together the research literature from all the different disciplines that are producing research about gender and information technology; (2) to bring together the gender and IT research from around the world; and (3) to produce a comprehensive resource that could be the first source to which people would turn to learn about the current state of research on gender and information technology.

What has resulted is this two-volume *Encyclopedia of Gender and Information Technology*. It is an international compilation of research on the topic of gender and information technology, representing a broad range of perspectives. Contributions to this important publication have been made by scholars throughout the world with notable research portfolios and expertise, as well as by emerging investigators. This encyclopedia provides comprehensive coverage and definitions of the most important issues, concepts, trends, and research devoted to the topic of gender and IT. It contains more than 200 articles highlighting this state-of-the-art research. These articles are written by scholars from around the world who are engaged with research into the influence of gender on the development and use of information technology as well as the impact of information technology on men and women. These articles include extensive bibliographies that, taken together, represent an exhaustive reference source for both the interested reader and the scholar engaged in research in the area of gender and IT.

In order to ensure that this encyclopedia has both geographical and disciplinary breadth, an international Advisory Board was established. The members of this Advisory Board introduce this volume by presenting overviews of their research programs in order to illustrate the ways in which the topic of gender and IT is being addressed in different countries.

To assist readers in navigating and identifying needed information, this two-volume encyclopedia has been organized by listing all entries in alphabetical order by title throughout the two volumes, and by including the title in the "Table of Contents" in the beginning of each volume. This important new publication is being distributed worldwide among academic and professional institutions and will be instrumental in providing researchers, scholars, students, and professionals with access to the latest knowledge related to research on women and men with respect to information technology.

Eileen M. Trauth, PhD
Editor-in-Chief

Acknowledgments

As editor of this encyclopedia, I would like to acknowledge the help of those individuals involved in the various phases of this project, without whose support this encyclopedia could not have been satisfactorily completed. I would like to begin by thanking Dr. Mehdi Khosrow-Pour of Idea Group Inc. for inviting me to undertake a project that I had wanted to do for several years. I would also like to thank the staff members Renée Davies and Michelle Potter at Idea Group Inc. who worked closely with me on the production of this encyclopedia.

I would like to thank the members of the Advisory Board—Alison Adam, Jane Margolis, Helen Richardson, and Liisa von Hellens—for their help in recruiting contributions to this encyclopedia and for writing the introductory articles for it. I would also like to thank the authors and the reviewers for their contributions to this two-volume encyclopedia. Without the interest and commitment of individuals around the world who are engaged in gender and IT research, this encyclopedia would not have been possible.

I would like to extend my deepest gratitude to my colleagues in the College of Information Sciences and Technology at The Pennsylvania State University who made this encyclopedia a reality. They helped me shepherd each article through the process from solicitation and proposal, to submission, review, and revision, to the final product. This encyclopedia could not have been accomplished without their diligence and dedication. Three research assistants—Haiyan Huang and Allison Morgan, working under the direction of Jeria Quesenberry—learned, firsthand, about the trials and tribulations of academic publishing as they developed the encyclopedia Web site, managed the article submission system, tracked down reviews, reminded authors, followed-up on revisions, and collated the finished product. David Hall, Associate Dean for Research, provided resources when necessary to complete this project, our Information Technology Office, and in particular, Steve Murgas and Ed Putt, who helped us with the project Web site and submission system, and Tracy Ray from our Research Office saw the final manuscript to completion.

Finally, I would like to thank Kathy Driehaus for her understanding and support, and for patiently listening to me talk about this project.

Eileen M. Trauth, PhD
State College, Pennsylvania, USA
November 2005

About the Editor

Eileen M. Trauth, PhD, is a professor of information sciences and technology and director of the Center for the Information Society at The Pennsylvania State University, USA. Her research is concerned with societal, cultural, and organizational influences on information technology, information technology work, and the information technology workforce. Her investigation of socio-cultural influences on the emergence of Ireland's information economy is published in her book, *The Culture of an Information Economy: Influences and Impacts in the Republic of Ireland*. She is currently engaged in a multi-country study of women in the information technology workforce in Australia, New Zealand, Ireland, and the U.S. Dr. Trauth has published nine books and over 100 research papers on her work. She is an associate editor of *Information and Organization* and serves on the editorial boards of several international journals. Dr. Trauth received her PhD in information science from the University of Pittsburgh.

Health Portals and Menu–Driven Identities

Lynette Kvasny
The Pennsylvania State University, USA

Jennifer Warren
The Pennsylvania State University, USA

INTRODUCTION

In this article, we make a case for research which examines the cultural inclusiveness and salience of health portals. We make our case from the standpoint of African-American women. While healthcare should be a ubiquitous social good, health disparities exist among various demographic groups. In fact, health disparities have been placed on the U.S. disease prevention and health promotion agenda. *Healthy People 2010* is an initiative sponsored by policy makers, researchers, medical centers, managed care organizations, and advocacy groups across the country. Although there is no consensus regarding what a health disparity is, sponsors agree that "racial and ethnic minorities experience multiple barriers to accessing healthcare, including not having health insurance, not having a usual source of care, location of providers, lack of transportation, lack of child care, and other factors. A growing body of evidence shows that racial and ethnic disparities in health outcomes, healthcare access, and quality of care exist even when insurance, income, and other access-related factors are controlled."[1]

In addition to healthcare, African American women have less access to the internet. Even at equivalent income levels, African Americans are less likely than either whites or English speaking Hispanics to go online. Demographically, the composition of populations not online has not changed dramatically since 2000. Overall, 60% of the total U.S. population is online with African Americans making up 11% of the total U.S. population, 8% of the online population, and 14% of the offline population. However, when looking at those who are offline, African Americans are more likely than offline whites or Hispanics to believe that they will eventually go online (Lenhart, 2003).

Although online health information is available from multiple sources, we focus solely on those health portals sponsored by the U.S. government. We made this choice based upon some early interviews with physicians and managers at a healthcare facility which serves predominantly African American clients. We learned that most clients exhibited a low degree of trust in information provided by pharmaceutical companies and other sources which seemed too commercial. Instead, clients searched for information from recognizable sources, and tended to use portals and search pages like Yahoo and Google. We found that portals sponsored by U.S. government agencies were received positively by clients. Also, portals like healthfinder.gov and cdc.gov are highly regarded by the Medical Library Association[2]. Moreover, the government is entrusted to uphold values of democracy and social justice therefore the health information that they provide should be accessible to a demographically diverse audience.

To gain insights into the cultural inclusiveness and salience of health portals, we use Nakumura's notion of menu-driven identities. For Nakumara (2002), the internet is a discursive place in which identity is enacted. She uses the term "menu-driven identities" to signify the ways in which content providers represent identities through the design of the interface and the personalization of content, and users perform their identity as they engage with the content. In what follows, we discuss health disparities and the promise of the internet in redressing inequities. Next, we further explain the ways in which users perform identity and health portals represent identities. We do this by theorizing about the health portals as mediating two-way communication between users and information providers. We conclude with directions for future research.

BACKGROUND

Health portals hold promise as an informational source for improving the health of historically underserved populations. This promise is extremely exciting given the state of health disparities in the U.S. We know from prior studies that health provider bias, stereotyping, prejudice, and clinical uncertainty may contribute to disparities along the lines of gender, class, race, and ethnicity (Balsa & McGuire, 2003). For instance, in a study by Bird and Bogart (2001), 63% of survey participants indicated that they had experienced discrimination in their interactions with their healthcare provider because of their race or ethnicity. Similarly, African Americans interviewees reported perceived discriminatory experiences such as inferior treatment, negative attitudes, being treated as if they were unintelligent, being ignored, inappropriate allegations, and racist remarks (Hobson, 2001). These negative experiences may profoundly impact attitudes towards receiving care, and willingness to comply with physician recommendations. For example, Hobson (2001) found that nearly 27% of African American survey respondents reported that, as a result of a discriminatory event, they were more hesitant to seek health services. Others avoided the healthcare facility (25.6%), avoided the provider (23.1%), avoided the personnel involved (10.3%), stopped using specific services (15.4%), or used service less frequently (7.7%).

Computer mediated communication may help minorities, women and other underserved groups to receive healthcare information in a more hospitable climate. The popularity of the internet as a medium for health communication is evidenced in two ways. First, the number of health-related Web sites has dramatically increased from a mere 15,000 sites in 1999 (Rice, 2001) to 100,000 as of 2003 (Cates, 2003). Secondly, although these sources vary in quality and relevance, the number of people seeking online health information rose to 97 million in 2001 from 60 million in 1999 (Rimal & Adkins, 2003). In a 2002 national survey (see Figure 2), researchers found that 73 million people in the U.S. or 62% of internet users have gone online to search for health information. On a typical day, about 6 million Americans go online for medical advice. This exceeds the number of Americans who actually visit health pro-

fessionals according to figures provided by the American Medical Association (Fox & Rainie, 2002). And while 42% of Americans say they don't use the internet, many of them either have been internet users at one time or have a once-removed relationship with the internet through family or household members. In fact, some exploit workarounds that allow them to use the internet by having email sent and received by online family members and by having others in their home do online searches for information they want (Lenhart, 2003). Women are more likely than men to say their latest search was at least in part for someone else—62% compared to 50% of men. Women are also more likely than men to seek healthcare and health information both online and offline (Fox & Fallows, 2003).

Identity

As an increasing number of Americans obtain health-related information online, it is important to consider that the internet is not race and gender neutral. Rather, it is a discursive space in which identities can be represented, performed, swapped, bought, sold, and stolen. Users can create profiles to personalize their experience, and create avatars which serve as visual representations of the body in cyberspace. But while spaces for fluid subjectivity abound, the internet often fails to accommodate minority cultural identities (Kolko, Nakamura, & Rodman, 2000; Kvasny, forthcoming).

Identities are inextricable from communication and are enacted in messages (Hecht, 1993). These enactments transmit and exchange values, beliefs, and norms, which may or may not affirm individuals' or groups' understandings of their own identities (Jackson, Warren, Pitts, & Wilson, under review). Identities also act as interpretative frames in the communication process (Hecht, 1993). Messages are filtered through and made sense of in relation to how individuals perceive themselves. If health messages communicate an identity, which is in conflict with how African American women perceive themselves, then the information may be viewed as unusable and we have done little to combat health disparities. As with all communication, messages that diverge from the identities of minority populations are unlikely to prove effective. Hence, health information must be situated within the target audi-

Figure 1. Menu-driven identities

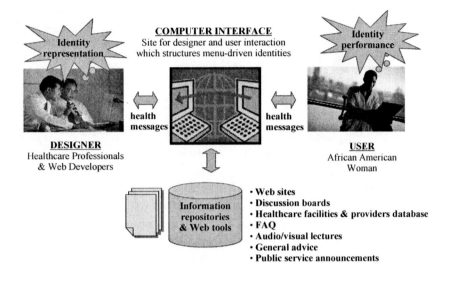

ences' sphere of experiences and understandings, or they may go unheeded.

When users search and consume information online, they perform identity. In fact, identity is the first thing that you do (create a profile or user account) before you can perform any activity. Identity performance is often a practical necessity constructed by designers who create the interface. Nakumara (2002) uses the term "architecture of belief" to signify how designers, through their choice of keywords, images, and use of language, create interfaces which represent the identities of some idealized user population(s). The interface reflects the cultural imagination of the designer, and performs familiar versions of race, gender, sexual orientation, and class. The relation between the user and the interface has been termed "menu-driven identity" performance and representation (see Figure 1) (Nakumara, 2002).

Thus, contrary to the popular notion that physical characteristics are erased online, the fact that Web users must reveal aspects of their identity suggests that bodies are often "outed" in cyberspace. For instance, we often have to define our race, gender, age, weight, marital status, and other identity factors when using health portals. In a cultural sphere such as health, these aspects of identity are crucially important factors to be considered, not superficial characteristics to be erased. Consequently, identity

is not entirely fluid and physical bodies remain important even though they are largely hidden.

Health Portals

Health portals are made spaces in which identities are enacted through computer mediating conversations between users and designers. The interface serves as a site for the production, representation, distribution, and reception of texts (health messages) which convey meanings that affects relations of power. The distinctive rhetorical conditions of the speaker, utterance, audience, and reception are created as the designer and user co-create a communicative situation in cyberspace (see Figure 2).

In what follows, we demonstrate how menu-driven identities are produced on two government Web sites. The first example is taken from the U.S. Department of Health and Human Services healthfinder portal—http://www.healthfinder.gov/justforyou/ (see Figure 3). The navigation scheme tends to reflect the unspoken biases and categories imposed by the more privileged actor (the government sponsor) in the communicative situation. Users must select an identity which is limited to two gender categories, four age categories, four ethnic/racial categories, and five roles.

Figure 2. Communicative situations

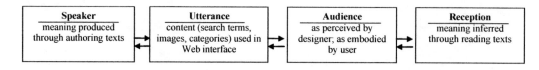

Speaker	Utterance	Audience	Reception
meaning produced through authoring texts	content (search terms, images, categories) used in Web interface	as perceived by designer; as embodied by user	meaning inferred through reading texts

For African American women, predefined categories force the performance of race and gender in ways that marginalize, and in some cases deny, their existence. This occurs when there is limited space for identity expression because the categories reproduce the limited number of choices based on historical labels and ideologies around race, gender, age, and role. Health portals, therefore, become another discursive field in which African American women are rendered invisible because they are assumed to be only African American or female—multiple selections are not allowed because there are no categories which capture both race and gender. Other underserved groups, such as gays and lesbians, are rendered completely invisible because they have no category. White is also omitted from the choices, but this is because whiteness is assumed as the default category and simply goes without saying. These familiar versions of identity are scripted and ascribed by designers when they create interfaces based upon these types of simplified categorizations. Notice how the author "selected very spe-

Figure 3. Healthfinder[3]

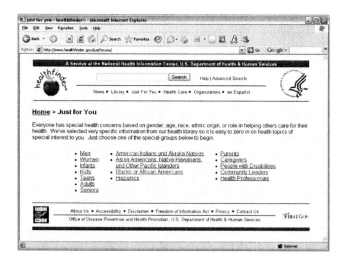

cific information from our library so that it is easy to zero in on health topics of special interest to you. Just choose one of the special groups below."

This example demonstrates how Web portals may serve as platforms for reproducing simplified discourse around difference. Category schemes are not simply passive tools through which labeling takes place, but rather are the outcomes of practices of meaning making. Designers of classification schemes constantly have to decide what categories are important, and in doing so, they develop an economy of knowledge that articulates omissions and inclusions, and ensures that only relevant features are classified (Bowker & Star, 1999). In doing so, health portals limit choices and the full participation of people who exist at the margins because users can only take the paths prescribed by the interface.

A second Web site taken from the Center for Disease Control, Office of Minority Health—http://www.cdc.gov/omh/Populations/populations.htm (see Figure 4)—demonstrates the instability of categories. For instance, race is an unstable signifier, which is socially constructed in dissimilar ways in various cultures. Blacks in the U.S. include all (including mixed race) people who trace their ancestry to Africa, but Blacks in South Africa don't include people of mixed African and European ancestry. In Britain, Black includes people with ancestry to non-African parts of the former empire such as Pakistan and China (Kolko, Nakamura, & Rodman, 2000). Racial signifiers are also unstable because they change over time. In the U.S., for example, people of African ancestry have been labeled Negro, Black, and African American.

In this health portal, the categories have been expanded to include white. Multicultural Americans are included even though it is difficult to make generalizations about health conditions because little research exists about this group. On the surface, it looks as though African American is treated as an

Figure 4. The center for disease control

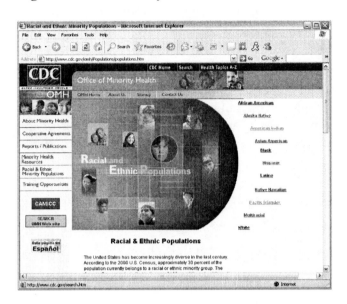

ethnic marker distinct from Black, a racial category which would include people throughout the African Diaspora. However, when selected, each category is linked to the same Web page. Hispanic and Latino are also listed as two distinct categories which, when clicked, send the user to a single Web page.

FUTURE TRENDS

Given the increasingly diverse populations of internet users and the growing usage of eHealth resources, we pose several important questions for future research:

1. How are underserved groups socially constructed on health portals?
2. What do these social constructions suggest about inclusiveness and health disparities?
3. What types of information do members of underserved groups seek, and why?
4. How well do existing government health portals serve the needs of underserved population?
5. How might we design more inclusive information resources?

We also suggest that existing health portals could better serve the needs of underserved groups. While we cannot account for the accuracy of the information provided through these portals, we did observe a wealth of information that is pertinent to various demographic groups. The challenge that we raise in this article is how best to tailor user interfaces to improve the online experiences of culturally diverse users. A tailored user interface would include:

*   A mixture of media formats including texts, audio, video, and slide presentations. Images should be representative of diverse people. Single page checklists, fact sheets, and brochures may be especially useful for women who are obtaining information for other family members and friends.
*   Texts that are comprehensible by low-literacy users and availability in languages other than English.
*   The ability to select multiple identities. This could be done with checklists or a menu structure that enables users to drill down through several demographic categories. For instance, a middle aged African American women could select from the *gender, race/ethnicity, age* categories to refine her information.
*   Normative categories such as white and heterosexual should be listed explicitly.
*   The ability to declare identities in ways that go beyond demographics. For instance status such as smoker, diabetic, HIV positive, and cancer survivor are important components of identity that influence health care needs and outcomes
*   Spaces such as chat rooms and forums for users to act as speakers and authors in the communicative situation.

CONCLUSION

The internet offers a space where African American women and other underserved groups can become empowered health consumers who access health information on their own terms (Ferguson, 1997). However, it is important to understand the extent to which online health information is inclusive of diverse users. Providing computers and internet access, and showing underserved groups the value proposition of online health resources are simply not enough. We must also consider the cultural salience of content and inclusiveness of the interface design.

REFERENCES

Balsa A. I., & McGuire T. G. (2003). Prejudice, clinical uncertainty, and stereotyping as sources of health disparities. *Journal of Health Economics*, *22*, 89-116.

Bird S. T., & Bogart L. M. (2001). Perceived race-based and socioeconomic status (SES)-based discrimination in interactions with health care providers. *Ethnicity and Disease*, *11*, 554-563.

Bowker, G. C., & Star, S. L. (1999). *Sorting things out: Classification and its consequences*. Cambridge, MA: MIT Press.

Cates, L. K. (2003). Meet online ethical expectations: New internet guidelines assist in monitoring information dissemination. *Advance News Magazine*. Retrieved November 19, 2003, from http://www.advanceforaud.com/common/editorial/editorial.aspx?CC=10095&CP=1

Ferguson, T. (1997). Health online and the empowered medical consumer. *Journal on Quality Improvement*, *23*(5), 251-257.

Fox, S., & Fallow, D. (2003, July 16). *Internet health resources: Health searches and email have become more commonplace, but there is room for improvement in searches and overall internet access*. Washington, DC: Pew Internet and American Life. Retrieved January 2005, from http://www.pewinternet.org/PPF/r/95/report_display.asp

Fox, S., & Rainie, L. (2002, May 5). *Vital decisions: How internet users decide what information to trust when they or their loved ones are sick*. Report. Washington, DC: Pew Internet and American Life. Retrieved January 2005, from http://www.pewinternet.org/PPF/r/59/report_display.asp

Hecht, M. L. (1993). 2002—A research odyssey: Toward the development of a communication theory of identity. *Communication Monographs*, *60*, 76-82.

Hecht, M. L., Jackson II, R. L., & Ribeau, S. A. (2003). *African American communication: Exploring identity and culture*. Mahwah, NJ: Lawrence Earlbaum Associates.

Hecht, M. L., Warren, J. R., Jung, E., & Krieger, J. (2004). The communication theory of identity: Development, theoretical perspective, and future directions. In W. Gudykunst (Ed.), *Theorizing about intercultural communication* (pp. 257-278). Thousand Oaks, CA: Sage Publications.

Hobson, W. D. (2001) *Racial discrimination in a health care interview project*. Special Report. Seattle, WA: Public Health Seattle and King County.

Jackson, R. L., Warren, J. R., Pitts, M. J., & Wilson, K. B. (under review). It is not my responsibility to teach culture! White graduate teaching assistants negotiating identity and pedagogy. In L. Cooks (Ed.), *Whiteness, pedagogy, and performance*. Landham, MD: Lexington Books.

Kolko, B., Nakamura, L., & Rodman, G. (2000). *Race in Cyberspace*. New York: Routledge.

Kvasny, L. (forthcoming). *Understanding the digital divide from the standpoint of the "other"*. The Data Base Advances in Information Systems.

Lenhart, A. (2003). *The ever-shifting internet population: A new look at Internet access and the digital divide*. Washington, DC: Pew Internet and American Life Project. Retrieved January 2005, from http://www.pewinternet.org/pdfs/PIP_Shifting_Net_Pop_Report.pdf

Nakumara, L. (2002). *Cybertypes: Race, ethnicity, and identity on the Internet*. New York: Routledge.

Rice, R. E. (2001). The Internet and health communication. In R. E. Rice &. J. E. Katz (Eds.), *The Internet and health communication* (pp. 5-46). Thousand Oaks, CA: Sage Publishers.

Rimal, R. N., & Adkins, D. A. (2003). Using computers to narrowcast health messages: The role of audience segmentation, targeting, and tailoring in health promotion. In T. L. Thompson, A. M. Dorsey, K. I. Miller, & R. Parrott (Eds.), *Handbook of health communication* (pp. 497-514). Mahwah, NJ: Lawrence Erlbaum and Associates.

KEY TERMS

Digital Divide: The term "digital divide" describes the fact that there exist people who do and people who don't have access to—and the capability to use—modern information technology, such as the telephone, television, or the internet. Access to these resources follows along demographic lines such as gender, income, race, ethnicity, geography, and age.

Health Disparity: Some demographic groups, such as racial and ethnic minorities, experience multiple barriers to accessing healthcare, including not having health insurance, not having a usual source of care, lack of transportation, lack of childcare. A growing body of evidence shows that health disparities persist even when insurance, income, and other access-related factors are controlled.

Health Portal: A Web site often sponsored by a large institution, which provides extensive information and contains links to a wide range of health and medical information on the internet.

Identity: Hecht (1993) identifies four frames through which identity is communicated. *Personal identity* involves self-concept, and develops through socially ascribed meanings and behaviors which are learned as one is socialized in a society (e.g., Black women stereotyped as mammy—the nurturer; jezebel—the seductress; sapphire—the wisecracking emasculating women; the welfare queen—the lazy, economically unstable mother of many bad kids). *Enacted identity* focuses on how messages express identity (e.g., I'm a black woman). *Relational identities* are those formed through one's relationships (e.g., I'm a mother). *Communal identities* are those shared by groups of people in some particular community (e.g., I belong to the Penn State community).

Menu-Driven Identity: This term signifies the ways in which content providers represent identities through the design of selection-oriented interfaces that are used to personalize content. Conversely, the term signifies the ways in which users perform their identity as they navigate within the choices that are both enabled and constrained by the interface.

ENDNOTES

[1]   http://www.healthypeople.gov/

[2]   http://www.mlanet.org/resources/medspeak/topten.html

[3]   This Web page in this screenshot served as the healthfinder start page when we conducted our analysis. This healthfinder Web site has been vastly improved, and the Web page used in our analysis is still accessible by clicking "Just for You" on the current start page.

A Historical Perspective of Australian Women in Computing

Annemieke Craig
Deakin University, Australia

INTRODUCTION

Women's participation in the Australian workforce has been increasing since the mid-1950s. In 1954, women made up 23% of the total labour force (Office for Women [OFW], 2004), but by 2004, they accounted for 44.5% (Australian Bureau of Statistics [ABS], 2004).

Over the same period, there was growth in new employment opportunities in the emerging computer industry. However, this industry did not manage to attract equal numbers of women and men, and currently women account for about one fifth of the Australian ICT workforce (Maslog-Levis, 2005). Women are paid less than men in similar positions in this sector and are less likely to hold senior management positions (Byrne & Staehr, 2003).

Gender imbalance in employment is not unique to computing. Australia's workforce is more gender segregated than that of most other industrialised countries (Gray, 2003). Over half of all female employees are employed in the clerical, sales, and service groups of occupations, and these are areas where there are substantially less men (ABS, 2000). Men dominate the trades, production, and transport occupations.

When does gender imbalance become a concern? Common sense would suggest that it has become a problem when gender imbalance has a detrimental effect on some sections of society.

The computing profession is an area where gender imbalance is of concern. New technologies bring about changes that have the potential to affect all society, and we "would be most likely to achieve maximum benefit if each significant section of society was represented in the planning decisions" (Ryan, 1994, p. 548). Without diversity in the ICT workforce, "we limit the set of life experiences that are applied, and as a result, we pay an opportunity cost, a cost in products not built, in designs not considered, in constraints not understood, in processes not invented" (Wulf, 1998).

Unless more women are employed in the areas of ICT design and development, these products and services are unlikely to meet the needs and desires of approximately half the population. Women need to be actively involved in all levels of these new technologies that have such immense potential for social change.

BACKGROUND

The ideas, curiosity, and advanced thinking that led to the creation of the computer evolved over time, with many people from around the world making a contribution.

Australia moved into the modern computing era with the development of the CSIR Mk1 in the late 1940s (Pearcey, 1994). This machine, later renamed CSIRAC, was the first computer in Australia, and it was arguably the fourth or fifth electronic stored-program computer ever developed in the world (Jones & Broomham, 1994). It provided computing service until well into the 1960s. By then, there were 34 computers in the country, and this number increased to 348 by 1965 and then to just over 3,000 machines by 1975 (Thornton & Stanley, 1978). By the year 2000, over 4 million computers were in homes around Australia, with over half (56%) of households having a home computer. In 2003, 66% of Australian households had access to a computer at home, with 53% of these households having access to the Internet (ABS, 2005).

In less than 50 years the industry had developed so rapidly that more than 264,400 people were employed in selected information-technology-related occupations in Australia (IIETF, 1993). Within this time frame, however, computers and computer work had also become stereotyped as more appropriate for males (Game & Pringle, 1984), and now only 23.6% of the current IT workforce are women. Women have tended to be highly represented in the less skilled, less sophisticated areas of data entry and computer operation, while men have made up the majority of the higher level, higher status, higher paid computer workforce (Davis, 1986).

EDUCATING THE COMPUTER PROFESSIONAL

Following the emergence of the new discipline of computing, the universities of Melbourne, Sydney, and New South Wales (NSW) first introduced courses on programming and the application of computers in 1956. By 1985, every Australian university had a computer department covering a wide range of curriculum areas. Two main approaches to the teaching of university computing emerged: computer-science and information-systems courses. In high schools, computing education was initiated in the mid-1960s "by a handful of enthusiastic maths and science teachers, mostly male" (Sale, 1994, p. 155).

The 1980s and early 1990s provided a period of enormous change in the Australian education system. High-school retention rates more than doubled during the 1980s, increasing from 34.5% in 1980 to 76.6% in 1993 (ABS, 1992). In 1987, for the first time, school retention rates for girls exceeded those of boys. Changing community expectations, a depressed teenage labour market, and government policies encouraging students to complete their secondary education were all contributing factors (Williams, Long, Carpenter, & Hayden, 1993). The result was an increased number of young people, particularly women, eligible to enter higher education. The first 130 years of Australian higher education saw women students outnumbered by men. Whereas female students made up only 45% of the student population in 1981, by 2001, women made up 57% of the 206,834 students commencing an undergraduate qualification (Office of the Status of Women [OSW], 2002).

It was in the mid-1980s that the lack of female students in computing in secondary and tertiary education as well as in the profession began to emerge as an issue and was finally recognised by academics, the industry, and politicians (see, for example, Kay, Lublin, Poiner, & Prosser, 1989; Symons, 1984). In 1990, the federal government attempted to improve the situation by setting targets to increase the proportion of women in university information-technology courses to 40% (see DEET, 1990).

During the following decade, many initiatives were created to encourage women to undertake undergraduate courses in information technology. Initiatives ranged from mentor programs to the production of videos, special classes for female students, curriculum changes to create a more inclusive curriculum, computer camps, and so on (see, for example, Clayton & Lynch, 2002; Craig, Fisher, Scollary, & Singh, 1998; Greenhill, Von Hellens, Nielsen, & Pringle, 1997).

Undergraduate female enrollments in computer science and information systems peaked at 27.2% during the early 1990s (Lang, 2003). In 2004, approximately 20% of commencing tertiary information-technology students were female (DEST, http://www.dest.gov.au/NR/rdonlyres/79212D3A-218C-D6D-9F28-7C59A9FE5F75/2464/01_Commencing_Students.xls#lTbl03>!Al). Australian women in the ICT profession are not only a small minority, but appear to be a decreasing one with a consequential reduction in diversity and creativity within the profession.

WHAT KEEPS WOMEN AWAY FROM IT?

A variety of factors impact Australian girls' decisions to not study computing.

RELATIVE COMPLEXITY
In my parents' lounge room after Christmas dinner, I am talking to my brother the computer programmer.
He is explaining to me the principles of cyberspace.
"It is only relatively complex," he says finally, peeling the icing off his fruitcake, "It is mainly a system of binaries, permutations of zero and one.

So, the data may be stored as, say, zero, zero, one, one, one, zero, zero, one."
My mother sighs.
She is next to us, half-listening.
She is knitting a fair-isle sweater.
"I'll never understand how you get your brain around it, " she says.
"It's beyond me," she says, and turns half her attention back to her fair-isle pattern: Purl, purl plain, plain, plain, plain, purl, purl. (Cate Kennedy, as cited in Senjen & Guthrey, 1996, p. 13)

Many women underestimate their ability to master complex procedures. Generations of Australian women have been able to knit complicated patterns, yet the skills involved in mastering the intricacies of this work are often unrecognised and undervalued. The language involved is distinctive to the craft work and understood in context, yet, the language of computers seems beyond comprehension to many women and can be alienating for those unfamiliar with the terminology.

With the rapidly changing nature of the industry there is a lack of information and understanding of what a career in the IT industry involves. The public image of the profession is not seen as one that involves creativity, problem solving, or working with people to help others, nor is it seen as requiring lateral thinking and good communication skills.

The impression is often that computing is word processing and spreadsheets, or made up of the Internet with students frequently unaware of the "vast body of underlying principles that make up the discipline" (Edwards & Kay, 2001, p. 334). Perceptions of IT being an industry dominated by math and science, and being technical, isolating, and lacking in teamwork and social interaction prevail (Multimedia Victoria, 2001). The enduring image of the computing industry is of young males, very technical and "nerdy," who spend most of their time working by themselves in front of a computer terminal. This is seen as unappealing and boring by many women.

In the 1980s and 1990s, high-school computing classroom experiences have often been associated with mathematics and competition to get access to scarce resources, and computing was perceived as a male domain. This influenced the decision of students, particularly females, not to pursue information

technology (Cameron, Edwards, Grant, & Kearns, 2000). In the 21st century, computer classrooms are well equipped with many more resources and are much more user-friendly places. Yet, many students still frequently describe them as being too boring (Multimedia Victoria, 2001) and are not inspired to continue to a computing career. Many of the best IT students may be more knowledgeable than the teachers, are obsessive about technology, and "live" IT. This is not seen in other areas of the curriculum and can be a discouragement to other students to continue with computing (Multimedia Victoria).

The dot-com and telecommunications crash in 2000, and increasing offshoring and outsourcing have all contributed to the impression that IT does not provide a stable career (Morton, 2005). For women who do venture into the profession, there are few role models in the industry and a lack of suitable mentors. Many also find the nature of the work makes it difficult to balance work and family responsibilities (Newmarch, Taylor-Steele, & Cumpston, 2000). The industry is also seen as a "boys' club" with few women succeeding into management levels.

Generally, people are not consciously trying to discourage women away from computing. However, as Spertus (1991, p. 75) points out,

... people's behaviour is often subconsciously influenced by stereotypes that they may not even realise they have. While perhaps it is comforting to know that no conspiracy exists against female computer scientists, it also means that the problem is harder to fight.

REDRESSING THE IMBALANCE

One approach that has been identified in the literature to redress the imbalance of women in the ICT profession is that of "women-in computing." This is where change is expected to come about via policies and strategies to create equal access to education and employment (Pringle, Nielsen, Von Hellens, Greenhill, & Parfitt, 2000). This has been described as the "add women and stir" approach, which may have limited potential as it locates the problem in women rather than in the gendered culture of the ICT workplace (Adam, Howcroft, & Richardson, 2002). A second approach, then, is "women in-computing." Here, change is expected to come about from reshaping the industry to accommodate women by altering the masculine practices of ICT work so that women can enter into such work without the loss of identity or integrity. This approach has a broader focus on the nature of techno-logical work—the source of the inequality—and how it can be made more inclusive of women.

Adopting both approaches, however, can be considered as mutually reinforcing. If more female students can be encouraged to enroll in computing courses, or participate in the industry, then there will be a greater presence that will help to transform parts of the culture. Changes in the culture itself will then encourage more females to participate. Webb and Young (2005) have suggested that it is now timely to consider adopting a different research approach, feminist epistemology, which would offer greater insights into the factors involved in the imbalance of women in the ICT profession.

The lack of women participants in the ICT profession is not unique to Australia. There are many other countries, for example, the USA, United Kingdom, Germany, the Netherlands, and South Africa, where the problem also exists (Galpin, 2002). Yet the ICT profession is gender neutral in other areas of the world such as Malaysia, Singapore, and Hong Kong. A conclusion that can be drawn from this is that it is not inherent ability that is stopping equity within ICTs, but it is much more likely to be cultural and societal influences. Teague (1997) warns that since the environmental and behavioural factors related to the imbalance of females in the ICT field are now essentially part of Western society, it may take generations to change the imbalance.

Gender shifts are possible, however. In Australia, for example, in the early 1970s less than 15% of veterinary graduates were female. Yet almost all in the veterinary profession would agree that gender is not an important attribute to be a successful vet. Currently, more than 50% of all enrollments in veterinary science are female. During this period, the image of veterinary science has undergone a remarkable shift. A very popular TV show, *A Country Practice*, introduced a female vet who was

charming, clever, and attractive. Another show, *All Creatures Great and Small*, portrayed their two male vets in a "highly positive, socially friendly and supportive, and people oriented way" (Byrne, 1994). Similarly, an increased interest in forensic-science education has been attributed to the popularity of television shows such as *CSI* (Selinger-Morris, 2005). Popular movies such as *Hackers*, *Sneakers*, *War Games*, *Antitrust*, *The Net*, and *Swordfish*, however, do not present an image of computing that is appealing to many women.

FUTURE TRENDS

The statistics indicate that in Australia, the numbers of women in computing education and the profession are not increasing. However, in 2005 it is possible to be optimistic. The issue appears to have once again been noticed by the education sector, the corporate sector, professional societies, and the government. An indication of the growing awareness can be seen by the following initiatives.

- The federal government announced plans for the Women in ICT Summit to be held in September of 2005.
- The Victorian Women in ICT Network was created as a response by the Victorian government and industry to the impact of the low participation of women on the skills base of the ICT industry.
- The Australian Computer Society (ACS) established a women's board, ACS-W.
- Many of the existing ICT networks for women that have been formed throughout Australia will work more closely together in the future. In May of 2005, the Australian Women in IT and Science Entity (AWISE) was created as a national communications umbrella that formed a collaborative voice connecting many of these ICT networks (for example, GIDGITS, Women are I.T., Women in Technology, FITT, WIIT).

These initiatives need to be the catalyst for real change.

CONCLUSION

Many programs have been introduced to try to address the complex range of factors that affect young women's choice to commence and successfully complete a degree in computing, and to equip them with the necessary skills and contacts to obtain jobs and career recognition in the industry. Due to the ever-changing nature of technology itself, it would be simplistic to expect that the problems have remained the same over time and that the strategies that have worked in the past will be the same ones producing results in the future.

Yet, it is necessary that we continue to work to redress the imbalance in the computer workforce. Women can bring a different perspective and alternative skills to the computing profession that will help create better systems for all. The entire computing industry within Australia needs to recognise the importance of the issue, and the political will needs to be found to make a lasting difference.

REFERENCES

Adam, A., Howcroft, D., & Richardson, H. (2002). Guest editorial. *Information Technology and People, 152*(1), 94-97.

Australian Bureau of Statistics (ABS). (1992). *Schools Australia 1991* (Cat. No. 4220). ABS, Canberra. Australia.

Australian Bureau of Statistics (ABS). (2000). *Work: Paid work trends in women's employment.* ABS, Canberra. Australia.

Australian Bureau of Statistics (ABS). (2004). *Labour force* (Cat. No. 6202.0.55.001). ABS, Canberra. Australia.

Australian Bureau of Statistics (ABS). (2005). *Year book Australia* (Cat. No. 1301.0). ABS, Canberra. Australia.

Byrne, E. (1994). *Critical filters, hidden helix: Policies for advancing women technology.* Brisbane: University of Queensland, Women in Science Engineering Technology Advisory Group.

Byrne, G., & Staehr, L. (2003). The participation and remuneration of women in the Australian IT industry: An exploration of recent census data. *AusWIT 2003: Participation, Progress and Potential,* 109-116.

Cameron, B., Edwards, J., Grant, J., & Kearns, P. (2000). Participation in information technology & telecommunications in education and training. *DETYA Project,* 99-102.

Clayton, D., & Lynch, T. (2002). Ten years of strategies to increase participation of women in computing programs. The Central Queensland University experience: 1999-2001. *Inroads SIGCSE Bulletin, 34*(2), 89-93.

Craig, A., Fisher, J., Scollary, A., & Singh, M. (1998). Closing the gap: Women education and information technology courses in Australia. *Journal of Systems Software, 40,* 7-15.

Davis, M. (1986). Do women get a fair DP? *Professional Computing,* 3-4.

DEET. (1990). *A fair chance for all: Higher education that's within everyone's reach.* Canberra, Australia: Australian Government Publishing Service.

Edwards, J., & Kay, J. (2001). A sorry tale: A study of women's participation in IT higher education in Australia. *Journal of Research and Practice in Information Technology, 33*(4).

Galpin, V. (2002). Women in computing around the world. *Inroads SIGCSE Bulletin, 34*(2), 94-100.

Game, A., & Pringle, R. (1984). *Gender at work.* Sydney, Australia: George Allen & Unwin.

Gray, D. (2003). *EOWW: Gender segregation in the Australian workforce.* James Cook University.

Greenhill, A., Von Hellens, L., Nielsen, S., & Pringle, R. (1997). Australian women in IT education: Multiple meanings and multiculturalism. In A. F. Grundy et al. (Eds.), *Women, work and computerization.* Berlin, Germany: Springer-Verlag.

IIETF. (1993). *The supply of people skilled in information technology: A statistical profile.* Deakin, Australian Capital Territory: Australian Information Industry Association.

Jones, B., & Broomham, R. (1994). The impact of computers on society. In J. Bennett et al. (Eds.), *Computing in Australia: The development of a profession* (pp. 183-192). Marrickville, New South Wales: Hale & Iremonger.

Kay, J., Lublin, J., Poiner, G., & Prosser, M. (1989). Not even well begun: Women in computing courses. *Higher Education, 18*(5), 511-527.

Lang, C. (2003). Flaws and gaps in the women in computing literature. *AusWIT 2003: Participation, Progress and Potential*, 39-46.

Maslog-Levis, K. (2005). Women not taking ICT courses: Academic. *ZDNet Australia.*

Morton, E. (2005). *Beyond the barriers: What women want in IT.* Retrieved September, 2005, from http://BuilderAU.com.au.

Multimedia Victoria. (2001). *Reality bytes.* Melbourne, Australia: State Government of Victoria.

Newmarch, E., Taylor-Steele, S., & Cumpston, A. (2000). *Women in IT: What are the barriers?* DETYA.

Office of the Status of Women (OSW). (2002). *Women 2002.* Australia: Comm of Australia.

Office for Women (OFW). (2004). *Women in Australia.* Retrieved December, 2004, from http://ofw.facs.gov.au/publications/wia/chapter4.html

Pearcey, T. (1994). Australia enters the computer age. In J. Bennett et al. (Eds.), *Computing in Australia: The development of a profession* (pp. 15-32). Marrickville, New South Wales: Hale & Iremonger.

Pringle, R., Nielsen, S., Von Hellens, L., Greenhill, A., & Parfitt, L. (2000). *Net gains.* Paper presented at IFIP Women, Work and Computerization, Vancouver, Canada.

Ryan, C. (1994). Raising girls' awareness of computing careers. In *Proceedings of the 5th IFIP International Conference on Women, Work and Computerization.*

Sale, A. (1994). Computer science teaching in Australia. In J. Bennett et al. (Eds.), *Computing in Australia: The development of a profession* (pp. 151-154). Marrickville, New South Wales: Hale & Iremonger.

Selinger-Morris, S. (2005, May 25). Rise of the pop jobs. *Sydney Morning Herald.*

Senjen, R., & Guthrey, J. (1996). *The Internet for women.* Melbourne, Australia: Spinifex.

Spertus, E. (1991). *Why are there so few female computer scientists?* MA: MIT Artificial Intelligence Laboratory.

Symons, L. (1984). *Girls and computing.* South Australia: Angle Park Computing Centre.

Teague, J. (1997). A structured review of reasons for the under representation of women in computing. *Second Australasian SIGCSE Conference.*

Thornton, B., & Stanley, P. (1978). *Foundation for Australian Resources report on computers in Australia.* The New South Wales Institute of Technology, Faculty of Mathematical and Computing Sciences.

Webb, P., & Young, J. (2005). Perhaps it's time for a fresh approach to ICT gender research? *Journal of Research and Practice in Information Technology, 37*(2), 147-160.

Williams, T., Long, M., Carpenter, P., & Hayden, M. (1993). *Entering higher education in the 1980s.* Australia: DEET, Australian Government Publishing Service.

Wulf, W. A. (1998). Diversity in engineering. *The Bridge, 28*(4).

KEY TERMS

CSIR Mark 1: Australia's first automatic electronic computer. It was developed by the Council of Scientific and Industrial Research in Sydney. Today, it can be found in the Melbourne Museum.

Gender Segregation: When the percentage of the share of employment for a gender in an individual area is well above or below the corresponding share of total employment for that gender.

ICT Industry: The information and communication technology industry.

Nerdy: A term used to describe a person who is very interested in and focused on computers, but who is felt to be socially inept.

Tertiary Education: Coming after primary and secondary education, tertiary education refers to university- and college-level education.

History of Feminist Approaches to Technology Studies

Jennifer Brayton
Ryerson University, Canada

INTRODUCTION

While women have historically engaged with technological practices and processes as designers, producers, users and consumers, technology itself has been socially constructed as a masculine domain and inherent to male gender identity. As a result, women have not been recognized as technological participants, nor have they had their contributions validated. To understand this exclusion, different feminist approaches have been historically utilized to help situate the framing of technology as a masculine domain that is organized by the social structures of patriarchy, capitalism, and social stratification. Feminist approaches have been used to deconstruct the defining of technology as masculine, to illuminate the historical ways in which women have been part of technological fields, and to give evidence of the pleasure and empowerment women can feel with technology.

BACKGROUND

There are two principle approaches that are typically expressed when examining the nature and role of technology in society. The first, technological instrumentalism, suggests that technology is merely a neutral tool to be used as the human agent deems necessary. For good or for bad, it is people who dictate the utilization of technology (Mowshowitz, 1985). By contrast, technological determinism suggests that technology is the underlying cause of change in society. People in society are viewed as having no control or choice about how or whether to utilize technology (MacKenzie & Wajcman, 1999). In both approaches, technology is depicted as evolutionary and self directed and on a path of change that is forever expanding with knowledge.

What is absent in both is the recognition that social and historical contexts can determine the formation and cultural meaning of technology. Technology is not a ready-made tool; it is a tool made by people. Technology is infused with the dominant beliefs, attitudes, and ideologies of the society in which it has been conceptualized and developed. Living in a world where patriarchy, capitalism, and inequality structure social relationships, technology comes from, and is shaped by, these prevalent relations (Cockburn & Furst-Dilic, 1994). To understand the meaning of technology is to go beyond the physical hardware to include the complex human activities that are also technology.

The contemporary association of science to masculinity and nature to femininity arguably arose during the Scientific Revolution where the Enlightenment philosophers inherited sexist attitudes from Aristotelian philosophy that defined women as passive and intellectually inferior to men (Arnold & Faulkner, 1985). Francis Bacon called for a science and scientific method that would permit the discovery and conquest of the secrets of nature. Nature, located and identified as female, would be penetrated, conquered, and transformed (Easlea, 1983). The rise of gender roles and identities as binary opposites came to be organized through the Enlightenment period. Masculinity has become linked with rationality, hierarchy, dominance, strength, independence, power, control, aggression, ambition, and logic. Femininity has thus become connected to the oppositional attributes: irrationality, community, submissiveness, weakness, dependence, family, intuition, and softness. More typically, these characteristics are reduced to essentialist divisions that locate male/female as mind/body and technology/nature. Thus, through the Scientific Revolution, masculine identity became connected to science/technology and intelligence, and conversely, femininity was defined in terms of its non-connection. As a result, the artificial linkage between masculinity and technology became socially enshrined and now is culturally accepted without question (Murray, 1993).

Wajcman (1991) and Cockburn (1985) take as their starting point a different historical point of analysis: the move to capitalism and the creation of a sexual division of labor. Both authors make the similar claim that the sexual division of labor under patriarchy and capitalism has given rise to particular gender roles and values that locate women in the private home sphere with children, and men in the public work sphere with technology.

Cockburn (1985) focuses specifically on the transformation towards gender-specific work from feudalism to capitalism. With this transformation, men were given more social power and status compared to women under capitalism because they had knowledge and technical skills that were necessary for work place productivity. The rise of guilds as a center for trade skills limited women's access to skill development or the manufacturing of tools. Women's work became restricted to the domestic realm, with women responsible for food, childcare, and domestic duties. However, Cockburn's analysis of technology and masculinity is limited as it only attends to the technologies utilized in the capitalist public and paid workplace. She seems to be falling into the trap of defining technology as being only the technologies of production—those valued under capitalist and patriarchal structural relations.

While Cockburn (1985) focuses exclusively upon the technologies of production, Wajcman (1991) broadens her areas of exploration to include reproductive, domestic, and architectural technologies. In her research, she argues that the very definition of technology has been shaped by patriarchal and capitalist relations that only value productive workplace technologies. By focusing on a broader range of technologies, Wajcman exposes how technology exists as masculine identity which is caught up in the domination of women. This does not imply that all women lack technological skills or that all men have technical expertise, but that women and men are both structurally located through norms of hegemonic gender roles. As a result, women's use of technologies for work has been overlooked, and technologies used by women outside of the workplace, such as reproductive, domestic and leisure

technologies, have been ignored as they are not viewed as being technologies. Wajcman outlines multiple and often intersecting social processes that lead to women's alienation from new information and communication technologies (ICTs). This includes access to technological equipment, gender biases in education and the family, and the design of computer programs and games.

A parallel historical analysis of the social construction of technology as a masculine activity and identity suggests this gendered affiliation arose as a result of the masculinization of the military and engineering. Hacker (1989) argues that historically, the military, as an institution, arose from fraternal interest groups, where men lived with their families and passed along their name through their children. This social structure inherently relocated the women as outsiders to the group dynamics. In exchange for food, among other forms of labor, men promised to protect the community of women, children, and the elderly. Male military labor was valued more highly than female community-oriented labor, as only the men were paid. Engineering arose as a discipline designed to train men to be the technical staff and administrators of the military. The first military academies were engineering schools, focused upon teaching students technical skills and occupations. Since engineering was a technical branch of the military, technology became associated with the military, masculine identity, dominance, and power. In turn, as engineers graduated, they moved these internalized values into their workplaces—back into the military, or out into the capitalist labor market. Consequently, women in contemporary society still experience significant difficulties in being accepted members of technology-based fields, such as engineering and the military.

INTERSECTIONS OF WOMEN AND TECHNOLOGY

While the dominant cultural ideology connects technology with masculinity, women have always engaged with technologies and technological practices in their every day lives. Historians and scholars have documented many of the ways in which women have been the ignored and marginalized creators, developers, and users of ICTs. Thomas Jepsen's (2000) studies on the history of women and telegraphy highlight the active role women played as telegraph operators in the mid 1800s. Women entered the field at its inception, and participated in this early telecommunications industry as managers and as operators. Similarly, women have been prominent in other information and communications fields such as the U.S. telephone industry, where young, single women were hired in the late 1870s. Women inventors have held patents on a variety of early domestic and communication technologies, including boilers and cooking devices, typewriters for the blind, rotary washers, and submarine telescopes (Herring, 1999). The history of women in the computer industry is becoming well documented, with the contributions of Ada Lovelace, Grace Hopper, Alice Burks and other women being recognized as culturally significant. Women have always been engaged with technology, yet their historical contributions have been ignored as a result of the gender encoding of technology as inherently masculine.

Feminist Reticence to Technologies

While feminist scholars have been critical of the cultural framing of technology as a masculine domain, this is not to suggest women as a whole are against technology. Rather, these scholars have been fundamental in illuminating the myriad of ways by which this social construction of technology as masculine has negatively impacted upon women. For some women, the masculinization of technology has resulted in female technophobia, and the early lack of interest by women in adopting computer technologies. Feminist responses to advances in technological fields of the 1960s and 1970s was largely negative due to the cultural construction of technology as inherently masculine and patriarchal (Stabile, 1994). A female resistance and reticence towards new technologies such as home and work computer systems, and new technological spaces such as the Internet, has been the result (Turkle, 1988).

In examining the impact of new information technologies on women's work places in Canada, Heather Menzies (1981) clearly documents how female employees were more negatively impacted with the introduction of new workplace technologies. The introduction of computers and new ICTs resulted in an increase in part-time work, isolation, unemployment, and telework from home. These practices are still evident today, where women's experiences in new ICT fields are often times negative as a result of lower salaries, barriers to management promotion, competing demands on time due to family commitments, and an increased pressure to telecommute (Kome, 2003).

Early feminist Internet writers such as Dale Spender (1996) initially identified cyberspace as problematic for women–a place where they would be excluded, flamed into silence, cyberstalked, and inundated with pornography. Men shaped and determined the early culture of the Internet to reflect their own interests, creating virtual spaces that marginalized or ignored women's equal participation as producers. Spender argues that women have been stereotyped as the consumers of online information, not creators or producers. Men are viewed as the decision makers in cyberspace. Many of these early feminist concerns over the shaping of the Internet as a space hostile to women continue to exist today, especially as the WWW becomes more capitalist and corporate, and pornography continues to grow as a dominant Internet industry (Sutton, 2003).

As the Internet was accepted as part of the new social landscape in the mid 1990s, women began to challenge the societal belief that these new information and communication spaces were only of interest to men. Women existed and continue to exist in cyberspace, and actively participate in using the Internet for work, activism, and networking. It is the capitalist and patriarchal construction of cyberspace as a male space that has made women's participation invisible and marginalized, and their online identities limited to consumptive practices. The barriers that may discourage or turn women off the Internet are culturally created and are thus open to resistance, and reformation. Technophilia becomes a different female response to the empowering potentials offered to women by ICTs.

Technology for Female Empowerment and Pleasure

The potential for feminine pleasure in the blurring of gender and technology boundaries is perhaps articulated best by Donna Haraway (1991). Through feminist cyborg theory, she offers up a new vision of technology and gender in which women and men can break free of the binary divisions allocated to gender, and take pleasure in rejecting wholeness and unity linked to identity. In feminist cyborg theory, the cyborg represents the potential for the empowerment of women, the rejection of female technophobia, and the dismissal of technology as an exclusive masculine territory. The female cyborg symbolizes a rejection of the traditional binary allocation of technology to masculine identity, positively reworked from a feminist position. The female cyborg, as an icon and identity, represents the blurring of boundaries that have been used to distinguish between male/female, and technology/nature. This permits the reinterpretation of differences in society and the creation of new identities from multiplicities where women are able to be fused with technology. As a sign of female empowerment, the female cyborg represents what women need, namely an ease with technology that will allow them to participate, take pleasure, and have control.

Haraway's (1991) theoretical construction of women empowered and taking pleasure in technology is taken up by late 1990s cyberfeminists. Cyberfeminism, as a new concept and feminist theory, moves beyond feminist concerns over the social construction of technology as masculine. Instead, recognizing that ICTs and computers are embedded features of the contemporary global landscape that cannot be rejected, cyberfeminism highlights a fundamental paradigm shift for how gender and technology can be conceptualized. In this model, women can be empowered by, and take pleasure in, ownership and management of new ICTs (Wilding, 2001).

This is not to suggest that women need to accept the dominant paradigm in order to be accepted into technological fields. What is empowering within cyberfeminism is the understanding that the stereotyping of technology as being a masculine domain and practice must necessarily fall apart across time. Women who have grown up with new information technologies have easily accepted technology by its everyday presence in contemporary society (RosieX, 1995). Women should make use of new technologies, reject the dominant societal belief that technology is only that which is productive, contribute to the global formation and shaping of new technologies

and technological spaces, and take pleasure in the overall process (Plant, 1997).

Encouraging women to enter cyberspace is not the main objective of cyberfeminism. E-commerce has been strong in encouraging women to go online in order to access a new market population of product consumers. Commercial portals are now targeting women as a distinct consumer group, and many women-oriented portals like Women.com have arisen. Yet these types of women's portals are not feminist or empowering to women, and most offer the type of traditional gender-based content found in women-oriented magazines–advice on beauty, fashion, romance, and heterosexual intimacy. Cyberfeminism is about women becoming active agents in contributing to the formation of knowledge through technological spaces such as the Internet.

While feminist cyborg theory and cyberfeminism offer a contemporary feminist challenge to the traditional construction of technology as inherently masculine, they are not without their limits for assisting women in overcoming female technophobia or reticence. Both have been challenged for being abstract and not connected to the socio-cultural inequalities that result in differing levels of access and interest by women to technologies and cyberspace. Not all women have access to the Internet and new ICTs, given differences in education, economic power, existing global infrastructures, and the dominance of English within cyberspace (United Nations Division for the Advancement of Women, 2002). In addition, many women who now utilize information technologies and personal computers for labor practices exist in a world where these technologies intensify their daily workloads.

FUTURE TRENDS

While women globally continue to engage with technologies in all aspects of their lives, the on-going masculinization of technology is still evident and has significant impact upon women's experiences with technology. Feminists still continue gender-based research on a wide range of technological practices, domains and careers, especially as new ICTs become more heavily embedded in contemporary society. Some recent issues that are being explored include women bloggers and life writing, access issues for women with disabilities, globalization and women's technological labour, the Internet and the trafficking of women, women's underemployment in video game industries, the sexual exploitation of women online, the rise of cybercottage industries, the potentials for feminist cyberactivism, and using new media and technologies for teaching women's studies.

CONCLUSION

What is becoming increasingly clear is that technology is simultaneously organized by the socially stratified systems of patriarchy and capitalism. This is not a process occurring in isolation. Technology exists in particular formations because of its intersection with power and culture. As a result, technology does not function in the existing world as a value-free artifact, but rather it exists as an object that is value-laden. The implications of this gendered encoding of technology are culturally significant, whereby technology-based disciplines such as engineering, the sciences, mathematics, computer sciences, and information technology management have been historically dominated by an ideological belief system that assumes men are the developers, producers, and primary users of technological products and processes. As a result, women's historical contributions have been marginalized, and women have not been equal participants in the development, production or use of new information technologies, and have had to fight for inclusion in technology-based fields.

REFERENCES

Arnold, A., & Faulkner, W. (1985). Smothered by invention: The masculinity of technology. In W. Faulkner & E. Arnold (Eds.), *Smothered by invention: Technology in women's lives* (pp. 18-50). London: Pluto Press Ltd.

Cockburn, C. (1985). *Machinery of dominance: Women, men, and technical know-how*. London: Pluto Press Ltd.

Cockburn, C., & Furst-Dilic, R. (1994). Introduction: Looking for the gender/technology relation. In C. Cockburn & R. Furst-Dilic (Eds.), *Bringing technology home* (pp. 1-21). Buckingham: Open University Press.

Easlea, B. (1983). *Fathering the unthinkable; Masculinity, scientists and the nuclear arms race*. London: Pluto Press Ltd.

Hacker, S. (1989). *Pleasure, power, and technology*. Boston: Unwin Hyman.

Haraway, D. (1991). *Simians, cyborgs, and women: The reinvention of nature*. New York: Routledge, Chapman, and Hall, Inc.

Herring, S. D. (1999). *Women in the history of technology: Women inventors*. University of Alabama in Huntsville, Women's Studies. Retrieved June 24, 2005, from http://www.uah.edu/colleges/liberal/womensstudies/inventor.html

Jepsen, T. (2000). *My sisters telegraphic: Women in the telegraph office, 1846-1950*. Athens: Ohio University Press.

Kome, P. (2003). *Cybercottage industries: Internet and women's work*. Retrieved June 25, 2005, from http://www.womenspace.ca/policy/research_work_paper.html

MacKenzie, D., & Wajcman, J. (1999). Introductory essay: The social shaping of technology. In D. MacKenzie & J. Wajcman (Eds.), *The social shaping of technology* (2nd ed.) (pp. 3-27). Buckingham: Open University Press.

Menzies, H. (1981). *Women and the chip: Case studies of the effects of informatics on employment in Canada*. Montreal: The Institute for Research on Public Policy.

Mowshowitz, A. (1985). On the social relations of computers. *Human Systems Management*, 5, 99-110. Retrieved from http://www-cs.engr.ccny.cuny.edu/~abbe/pub.html

Murray, F. (1993). A separate reality: Science, technology, and masculinity. In E. Green, J. Owen, & D. Pain (Eds.), *Gendered by design: Informa-*

H

tion technology and office systems (pp. 64-80). London: Taylor & Francis.

Plant, S. (1997). *Zeros and ones: Digital women and the new technoculture*. New York: Doubleday Books.

RosieX. (1995). Interview with Dr. Sadie Plant. *Geekgirl Magazine, 1*. Retrieved February 15, 1999, from http://www.geekgirl.com.au/geekgirl/001stick/sadie/sadie.html

Spender, D. (1996). *Nattering on the net: Women, power, and Cyberspace*. Toronto: Garamond Press Ltd.

Stabile, C. (1994). *Feminism and the technological fix*. Manchester: Manchester University Press.

Sutton, J. (2003). *International: UNCSW action issues*. Retrieved June 25, 2005, from http://www.womenspace.ca/policy/inter_action_issues.html

Turkle, S. (1988). Computational reticence: Why women fear the intimate machine. In C. Kramarae (Ed.), *Technology and women's voices: Keeping in touch* (pp. 41-61). London: Routledge & Kegan Paul.

United Nations Division for the Advancement of Women. (2002). *Information and communication technologies and their impact on and use as an instrument for the advancement and empowerment of women*. Retrieved June 22, 2005, from: http://www.un.org/womenwatch/daw/egm/ict2002/reports/EGMFinalReport.pdf

Wajcman, J. (1991). *Feminism confronts technology*. PA: Pennsylvania State University Press.

Wilding, F. (2001). Where is the feminism in cyberfeminism? In H. Robinson (Ed.), *Feminist art theory* (pp. 396-404). Oxford: Blackwell Publishing.

KEY TERMS

Cyberfeminism: A feminist approach that arose in the 1990s to challenge and recontextualize the functions of new information and communication technologies and the Internet for women. It favors female pleasure and empowerment through feminist technological practices, and explores the intersections between gender identity, the body, culture, language, and technology.

Cyborg Theory: A theoretical framework upon which to examine the contemporary cultural intersections of human bodies, technologies, and identities. The female cyborg symbolically represents the disruption of dominant binary divisions that links women with nature and men with technology.

Gender: Socio-cultural values that are differently assigned to the physical bodies of men and women, whereby masculine attributes are typically linked to the male body and feminine attributes are typically with the female body.

Technological Determinism: An approach to understanding the role of technology in society that claims technology exists outside of society, has its own built-in functions and goals, and is the primary cause of social change.

Technological Instrumentalism: An approach to understanding the role of technology in society which conceptualizes technology as a device or tool where the user makes choices surrounding and determining its usage.

Technology: Technology is typically understood as physical objects that exist usually in the form of machines or tools. However, technology also includes human beliefs, values, activity, creativity, energy, and knowledge.

Technophilia: An excitement and eagerness for new technologies and technological practices.

Technophobia: A fear of technological innovation, computer systems, or technological cultures.

How Gender Dynamics Affect Teleworkers' Performance in Malaysia

Chong Sheau Ching
eHomemakers, Malaysia

Usha Krishnan
eHomemakers, Malaysia

INTRODUCTION

eHomemakers (http://www.ehomemakers.net), also known as Mothers for Mothers when it was first formed in 1998 in Malaysia, is a network of mothers and working-at-home persons from multiethnic communities. They are of various ages and are involved in networking activities to develop and promote the concept of working at home. The network believes that through ICT, homemakers, especially mothers, can earn an income without having to leave their homes or sacrifice their family responsibilities in the Malaysian social context. ICTs allow women to balance home and work life, thus enabling them to have the best of both worlds (Yip, 2000). In 2003, eHomemakers was a testing partner of the Association of Progressive Communications' (APC, n.d.) worldwide Gender Evaluation Methodology, and an evaluation plan titled "How Gender Dynamics Affect Teleworkers' Performance in Malaysia" was written.

eHomemakers aims to use the evaluation results to promote the creation of telecommuting opportunities and the establishment of virtual offices to the Malaysian government and the corporate sector. The findings are especially valuable to organisations that have tried telecommuting unsuccessfully. They also serve as a performance guide for teleworkers and would-be teleworkers who juggle child care, household chores, and paid work at the same time. The study was also used as a guide for eHomemakers to launch a special national campaign to advocate the promotion of teleworking for women.

BACKGROUND

Objectives of the Study

The main objective was to explore how women's family lives and home situations affect teleworking and work performance. The evaluation team conducted group discussions and interviews to identify the following:

1. Barriers and challenges faced by women who work from home
2. Ways in which working from home has impacted the women's lives and their families
3. Optimum home-office situations of a group of virtual office members
4. Characteristics and skills needed by a woman to be able to benefit fully from working from home

Research Questions Analysed

The following research questions were analysed.

1. How ICTs and gender issues affect telecommuting
2. How (if at all) ICT can affect the efficiency and productivity of a teleworker
3. How a teleworker can use ICT as a tool to balance home life with work life, and still be efficient and productive in her work performance
4. What conditions enable women to be efficient teleworkers

Methodology

The study took 4 months to complete and involved 70 respondents. The study team used a triangular methodology: home visits, focus-group discussions (FGDs), and a questionnaire survey. Respondents of the study included members of eHomemakers' virtual team (VT), staff members who worked from home, and eHomemakers members. The selection of respondents was made based on one key criterion: They were mothers who have been multitasking and working at home for less than 3 years. The justification for the criterion is that mothers who have just become teleworkers need time to adjust to their new lifestyle and so they face more barriers and challenges than mothers who have worked at home for more than 3 years. Except for the VT members who had fixed fees per month, all respondents worked on a freelance basis with their home-based consultancies against the backdrop of an unsupportive social and business environment.

Only two fathers who telework were included in the study as a control as it is not socially acceptable still, for men and women, to telework.

RESULTS OF THE STUDY

Reasons for Getting into Telework

All the VT members had experienced working from a physical office. The majority of the women gave up their jobs to become mothers, full-time homemakers, or home-based workers. A few got into home-based work as a result of retrenchment. Most were married while one was a single mother.

A similar pattern is reported by Amyot (1997) for a Canadian survey. The respondents, predominantly women, chose teleworking as an alternative to working full time in a regular office. Edwards and Field-Hendrey (2002) found that teleworking was a viable alternative because of the greater flexibility it afforded to people who have responsibilities at home like caring for children, or aged or disabled persons.

Benefits of Teleworking

All the respondents agreed that the biggest benefit was flexible time management. The female respondents made it clear that their first priority is their families and that most of them had left their former careers to raise their children.

Some of the women mentioned an increased sense of confidence that comes from having their own income and not relying on their husbands for their expenses. They also emphasised that they enjoyed being involved in something outside the realm of their husbands and children. They felt that it is important for women to have interests beyond the home and that teleworking improves their overall relationship with their husbands and children. Their ICT skills had also greatly improved because, unlike in the office, they had to learn how to troubleshoot minor computer problems on their own.

One respondent related how she was able to negotiate with her husband to start taking on some of the household tasks. Prior to her home-based business career, she was expected to do everything at home even though she was working full time outside while her husband claimed he was too tired from work to help out at home. Now, she used the same reason (being too tired because she is working) to get her husband to do some of the tasks himself.

Other benefits cited include not having to deal with traffic jams, saving time, not having to deal with office politics, and not having to worry about office wardrobe and how you look.

Other studies have also cited the advantages of teleworking. They range from flexibility in working hours to better quality of life and increased job satisfaction (Abu Hassan Asaari & Karia, 2001), as well as a lifestyle that allows workers to concentrate on the household and thus combine paid and unpaid work in the same workplace (Osnowitz, 2005).

Factors that Affect Home-Based Work

Perception of Home-Based Work

One of the barriers encountered was the negative perception about home-based work by family members and peers. Home-based work is often not considered a real job, and consequently, family members often interrupted the respondents' work and assumed that they were available for a chat, to run errands, or to do household work because they are home based.

For the male respondents in this study, there was the added pressure of not appearing as breadwinners of the household. One respondent mentioned the difficulty of being a "modern husband" and that the Malaysian culture expects husbands to work from physical offices as the sole (or at least the primary) breadwinners.

These views, however, were expressed primarily during the first few months of starting home-based work. Over time, by explaining what they did and how they worked from home, the respondents let their families and communities know that teleworking was just as serious and as valid as office-based work. As a result, the two male respondents in particular experienced less interruptions from family members.

Support from Family Members

The majority of respondents reported that support from family members was critical in the success of working from home. Children and spouses must understand that the teleworker should not be distracted when working. Respondents with very young children made arrangements with other family members (their mothers, aunts, sisters) to take care of the children during their work hours. Most of the respondents had household help, but they preferred their kids to be looked after by family members. They also said that when starting up the home office, support from family members was very critical in the first year. For example, one respondent reported that her husband would not allow her a dedicated workspace.

On the whole, most respondents claimed that they had various forms of support from their spouses. Some spouses offered technical and work-related support. Other husbands took care of children when their wives were working, especially during weekends. A female respondent related how she had become less critical of her husband's contribution to household tasks. She had since learned to let go of her ways of doing things at home, allowing her husband to do some of the household work even if it did not meet her standards.

Lack of Technical Support

Respondents who used computers in their home-based work experienced difficulties from the lack of technical support at home. When they had technical problems, they either paid for repairs or called on VT

members for help. Some called on their husbands or their children to provide the support they needed. According to one respondent, repair services are expensive, and home-based work is more expensive than office-based work where free technical support is available. They did not confirm whether they took the trouble to learn basic computer-maintenance work.

Labour Policies in Malaysia

Respondents raised the issue that Malaysian labour laws should recognize teleworking as legitimate work and should include the same benefits and support given to government office-based workers. Presently, teleworkers do not get any low-cost insurance or tax benefits.

Cost of ICTs in Malaysia

The focus groups were interested in learning more ICT skills to improve their teleworking opportunities. However, aside from a lack of affordable training for women, the high cost of ICTs was of concern to homemakers who were just starting out with no or little support from their families.

Management Issues

Most of the respondents interviewed talked about the need for the better management of VTs. They felt that there is little opportunity for management to monitor or verify the work of the team and that management's no-nonsense mind-set is crucial in ensuring that teleworkers are accountable professionally. This type of work arrangement calls for management to be very clear and focused on what it expects from the staff. The respondents also mentioned the importance of transparency from management in terms of payment schemes, decision making, and performance evaluation criteria.

RECOMMENDATIONS FOR CREATING AN ENABLING ENVIRONMENT FOR TELEWORK

Based on the findings, the following recommendations were made.

Increased ICT Access

As Malaysia positions itself to be a regional ICT hub, teleworking should be considered a work option for the increasingly educated women in the workforce who want freedom to define their choices. Also, outsourcing to teleworkers can be a reality if conducive policies are in place. Affordable ICT tools and universal access should be ensured (Mitter, Jin, Hoon, Wong, Abdullah, Rasiah, et al., 2001). ICT access is not so much of a problem in the capital, Kuala Lumpur, and the satellite areas; however, the costs are still too high. ICT access in other areas is much higher.

A person going into telework requires, at the minimum, a PC (personal computer), telephone, printer, and Internet access (Yip, 2000). As such, not everyone can afford to set up a home office. Although the respondents saw the purchasing of these tools as necessary investments, most Malaysian women do not have such means. While the obvious solution is to lower the costs to individual homes, alternative solutions such as loan schemes and affordable community Internet-access centers could be explored.

Training and Skill Development

Affordable ICT training and working from home are necessary to develop teleworking survival skills and to ease the difficulty of entry. Peregrine Wood (2000) confirms that rapid changes in technology require women to continuously learn new skills to increase their capacity to adapt.

Professional Management

The management of home-based workers must be professional, but not totally simulating the management of normal office work. Paradigm shifts in management and work culture are necessary (Mitter et al., 2001). Alternative management plans for home-based workers should take into account the multiple roles of women (and men) working from home.

A VT needs an efficient communication system and well-tested office accountability procedures that allow for transparency in decision making and performance evaluation criteria.

Effective monitoring practices ensure the accountability of teleworkers on fixed or retainer fees. One of the benefits of teleworking is time flexibility, which allows the workers to attend to their family roles and responsibilities. Instead of focusing on time spent doing a specific task, home-based management schemes must be specifically output driven with clear tasks, deliverables, and deadlines. Such management schemes must also make full use of available technologies to ensure transparency and accountability.

Changes in National Labour Policies

The International Labour Organisation (ILO, 2005) recognizes the substantial contribution to national economies from women entrepreneurs in both the formal and informal economies. In view of this, current labour policies in Malaysia must take into consideration the multiple roles of teleworking mothers and offer all home-based workers the benefits they give office employees (Abu Hassan Asaari & Karia, 2001). Any new policies on home-based work must ensure the protection of workers' rights.

DISCUSSION AND CONCLUSION

Gender Roles

Given that the respondents have been working from home only for a few years, and teleworking in Malaysia is in its early stages, conclusive findings about how teleworking challenges traditional male and female roles in the home is still premature. The long-term effects of teleworking on women's lives and gender relations in the family cannot be drawn without the same study being conducted on a larger number of respondents. However, the present findings shed light, for the first time, on the relationship between teleworking and the gender roles and inequalities in Malaysian families.

At first glance, teleworking can be an ideal solution for women to fulfill their multiple roles (Mahmood, 2002). The rapid ICT development in

Malaysia offers more opportunities for women, especially mothers, to work from home, decreasing the stress of juggling both family responsibilities and income earning (Tan, 2000). However, the long-term socioeconomic effects and implications of teleworking on gender roles should be studied (Suzan, Sixsmith, Sullivan, Hootsmans, & Clason, 2001) to determine how it should be promoted with the least negative effects within the Malaysian family context.

Although the respondents claim to be empowered with increased confidence, how their empowerment affects their relationships with their husbands and family members should be explored further as Malaysian culture is still quite patriarchal. Does teleworking truly challenge and change existing gender roles and inequalities?

Given that Malaysian women continue to fulfill traditional gender roles despite education advancement, home-based work can clearly address practical gender needs without necessarily challenging socially (and internally) accepted roles of women and men in the home (Hulten, 2001). Home-based work can become an ideal compromise for women so they can continue to fulfill their roles as mothers and homemakers. The respondents in this study gave up their careers and accepted their roles in the family. They believed in the importance of the mother as the main caregiver in a traditional family situation.

But what of the males in the family? Does having a wife who works from home further excuse them from being more involved in household work and family roles?

The two fathers interviewed confirmed that they are taking on more household work and are more active in raising their children as a result of working from home. If so, teleworking does challenge existing gender roles in the family, and it might be better to promote it among fathers. To have them physically present in the home can result in them being more involved in household tasks and management, thereby changing the division of labour in the family. In the Asian context, however, where extended families are still prevalent and household help is affordable, teleworking fathers may not necessarily attend to household work. They can easily leave household work to other (female) members of the family or to (female) household help. In this case, having men working from home will not challenge gender roles.

Promotion of Teleworking

According to the Malaysian Department of Statistics, currently over 50% of Malaysia's female population are not in the labour force although their literacy and secondary-school graduation rate slightly surpassed that of the male population (Buku Tahunan Perangkaan [Year Book of Statistics], 2002). In 2000, women comprised only 46.7% of the total Malaysian workforce. The figure has hovered around there since the 1990s due to the lack of opportunity for women to take care of children and work at the same time. Malaysian culture still values a mother's care more than paternal involvement in child care (Kulasegaran, 1999).

Teleworking brings paid work into the home, and it needs to be further advocated to the Malaysian government. The first step to promote teleworking is to validate teleworking as real and professional work in the same way that household work must be recognized as real work (Suzan et al., 2001). Giving economic value to unpaid housework will reduce the general perception that telework is a working mother's only other option besides homemaking. Teleworking should be advocated as a means to liberate mothers from traditional means of work and propel them into the information age.

Major barriers for telework exist. Local employers are already reluctant to hire women of childbearing age. Hence, promoting teleworking to them requires further studies on effective monitoring processes and measurement of mother teleworkers' productivity. Cost-benefit studies with elements of workers' rights protection will also be needed to convince employers of the benefit of engaging mother teleworkers.

However, promoting teleworking as a viable solution for women should be done carefully not only to overcome misconception of telework, but also to overcome the perception that teleworking reverses women's advancement. Women activists have fought to get out of their homes into offices to achieve equality with men in the workplace, and to live outside of traditional homemaking. Promoting teleworking specifically for women challenges their feminist stand.

At the individual level, mother teleworkers need to be aware that self-monitoring and discipline are the keys for teleworking success. More awareness training and promotion of eHomemakers' teleworking-community network will be essential to spur the movement onward through essential information dissemination, especially on gender roles.

CONCLUDING REMARKS

The findings show that indicators and benchmarks in terms of changes in gender relations as a result of teleworking must be developed, and that the evaluation of teleworking from a gender perspective must be continuous.

If teleworking promotes gender equality in the family, both women and men must challenge traditional gender roles and stereotypes, and work on true equality between husbands and wives in all aspects of family life: in decision making, in household work, and in family responsibilities. If this equality is not embedded alongside the promotion of teleworking, all teleworking can do for women at best is to provide an opportunity for women to balance their gender-based roles and responsibilities better, and at worst, to be used to justify women's multiple burdens.

REFERENCES

Abu Hassan Asaari, M. S., & Karia, N. (2001). Factors towards telecommuting: An exploratory study. *Malaysian Management Review (MMR), 36*(1), 13-23.

Amyot, D. J. (1997). Work-family conflict and home-based work. *Masters Abstracts International, 35*(2), 454.

Association of Progressive Communications (APC). (n.d.). *Gender evaluation methodology.* Retrieved July 14, 2005, from http://www.apcwomen.org/gem/

Buku Tahunan Perangkaan (Year Book of Statistics). (2002). *The Malaysian Department of Statistics.* Kuala Lumpur, Malaysia: Jabatan Perangkaan.

Edwards, L. N., & Field-Hendrey, E. (2002). Home-based work and women's labour force decisions. *Journal of Labour Economics, 20*(1), 170-200.

Hulten, K. (2001). The computer on the kitchen table: A study of women teleworking in their homes. *Dissertation Abstracts International C, 62*(1), 34.

International Labour Organisation (ILO). (2005). *The knowledge wedge: Developing the knowledge base on women entrepreneurs.* Retrieved July 14, 2005, from http://www.ilo.org/dyn/empent/empent.portal?p_docid=SWEKNOWLEDGE&p_prog=S&p_subprog=WE

Kulasegaran, A. (1999). *Women's and children's rights—and the protection offered by domestic law.* Paper presented at the 12th Commonwealth Law Conference, Kuala Lumpur, Malaysia.

Mahmood, A. N. (2002). Work and home boundaries: Sociospatial analysis of women's live-work environments. *Dissertation Abstracts International A, 63*(03), 792.

Mitter, S., Jin, K. K., Hoon, C. S., Wong, D., Abdullah, M. C., Rasiah, R., et al. (2001). Towards an enabling environment: Recommendations. In C. Ng (Ed.), *Teleworking development in Malaysia* (pp. 130-139). Penang, Malaysia: Southbound Sdn Bhd.

Osnowitz, D. (2005). Managing time in domestic space: Home-based contractors and household work. *Gender and Society, 19*(1), 83-103.

Suzan, L., Sixsmith, J., Sullivan, C., Hootsmans, H., & Clason, C. (2001). *When work comes home: Managing stress in teleworkers' families.* Poster presented at the 63rd Annual Conference of the National Council on Family Relations, New York.

Tan, A. A. L. (2000). *A quantum leap, ace-slimp project consultant.* Kuala Lumpur, Malaysia: Ace-Slimp Project Consultant.

Wood, P. (2000). *Putting Beijing online: Women working in information and communication technologies. Experiences from the APC Women's Networking Support Programme.* Manila, Philippines: APC Women's Networking Support Programme.

Yip, T. M. K. (2000). Home sweet office. In S. C. Chong (Ed.), *Working @ home: A guidebook for working women and homemakers* (pp. 137-141). Selangor, Malaysia: Corpcom Services Sdn Bhd, P.J.

KEY TERMS

Home-Based Work: Income-generating work that involves teleworking or running a home-based business.

Information and Communication Technology (ICT): Information and communication technology is the technology required for information processing. In particular, it is the use of electronic computers and computer software to convert, store, protect, process, transmit, and retrieve information from anywhere, anytime.

Teleworker: A teleworker is a physical person working from a distance whose work involves using information and communication technologies.

Teleworking: Teleworking means working from a distance through the use of telecommunication technologies. It also means employment at home while communicating with the workplace by phone, fax, or modem. The word telecommuting is synonymous with teleworking.

Virtual Team: Remotely situated individuals affiliated with a common organisation, purpose, or project who conduct their joint effort via electronic communication.

H

ICT and Gender Inequality in the Middle East

Ahmed El Gody
Modern Sciences and Arts University, Egypt

INTRODUCTION

Information communication technologies (ICT) have become an effective force for accelerating political, economic, and social development, decreasing poverty, and fostering trade and knowledge; however the uneven distribution, usage, and implementation of ICT resulted in what is known as the "digital divide" between those who have access to and utilization of information resources and those who do not (Internet.com, 2004).

The Middle East, with the exception of Israel, is the least ICT connected area worldwide with only 1.4% of the global share (less than half of the world average of 5.2%). ICT adoption and access in the Arab world are far from adequate; only 6% of the Arab world population uses the Internet, while the penetration rate of personal computers is 2.4%, and less than 4 % of the Arab population has access to a ground telephone line (Ajeeb, 2006; NUA, 2005).

The trend of globalization forced Arab countries to realize the power of ICT as one of the most important factors in achieving sustainable growth. During the past decade, genuine efforts have been implemented by Arab governments to utilize ICT; as of May 2005, every country in the Arab world (as seen in Table 1)—except Iraq and Libya—has a clear strategy or at least a plan for promoting ICT (Dutta & Coury, 2003).

In her book, *Technology Strategies for Putting Arab Countries on the Cyber Map*, Reem Hunaidi (2002) stated that despite Arab world efforts to utilize ICT, Arabs are still far from bridging the digital divide. Hunaidi stated that the Arab world is still scoring low on the Digital Access Index (as seen in Table 2), adding that bridging the digital divide requires commitment from all development stakeholders, not only Arab governments.

The Hunaidi study concluded that development should start within the Arab society through liberating Arab human capabilities, especially those of women questioning how a society can compete in an increasingly globalized world if half of its people remain marginalized (Hunaidi, 2002).

The UNDP 2004 report on human development in the Arab world added to Hunaidi's question

Table 1. ICT in the agenda of the Arab world

Country	ICT Strategy Spelled Out	ICT Implementation Plan Articulated	Operational ICT-Dedicated Research Facilities	Plan of ICT Dedicated Research Facilities	Operational Technopole Initiative	Plan of Technopole Initiative	Existence of Technology Incubator	Planned Technology Incubator
Bahrain	✓	✓	✓			✓	✓	✓
Kuwait	✓		✓	✓		✓		✓
Oman				✓				✓
Qatar	✓							✓
Saudi A.	✓	✓	✓	✓	✓	✓		✓
UAE	✓	✓	✓	✓	✓	✓	✓	✓
Algeria		✓	✓	✓		✓		✓
Egypt	✓	✓	✓	✓	✓	✓	✓	✓
Jordan	✓	✓	✓	✓	✓	✓	✓	✓
Lebanon	✓	✓	✓	✓		✓		✓
Morocco	✓	✓	✓	✓	✓	✓	✓	✓
Syria				✓				
Tunisia	✓	✓	✓	✓		✓		✓

Source: Dutta & Coury, 2003

Table 2. Digital Access Index (DAI)

UAE	0.65
Bahrain	0.58
Qatar	0.55
Lebanon	0.48
Jordan	0.45
KSA	0.44
Oman	0.43
Libya	0.42
Tunisia	0.41
Egypt	0.40
Palestine	0.38
Algeria	0.37
Morocco	0.33
Syria	0.28
Yemen	0.18
Sudan	0.15

Source: ITU, 2004

stating that the first step in human ICT development is to bridge the gender divide within the Arab world and make use of the latent 50% of the Arab population.

The Arab world has the lowest Gender Empowerment Measure (GEM) worldwide next to Sub-Saharan Africa. Nancy Hafkin and Nancy Tagger (2001), in their study "Gender, Information Technology, and Developing Countries", stated that the degree of gender bias can be vividly seen across the Arab region. Figures indicate that Arab users constitute 4% of Internet users in comparison to 22% of users in Asia, 25% in Europe, 38% in Latin America, and 50% in the United States.

Hafkin and Tagger (2001) concluded that several challenges of socio-cultural, political, economic, and education disparities need to be addressed towards advancing Arab women's active participation in the new networked information society.

BACKGROUND

ICT Diffusion in the Arab World

The Arab world is generally known as laggard in adopting and utilizing new technologies, and ICT are no exception. The Internet first arrived in the Arab world in 1992 when Egypt established a 9.6k network connection through France. Next, several Arab states started joining the new networked world; however, the pace of ICT diffusion in Arab states was slow for various reasons (El Gody, 2003; Nour, 2002). To many Arab states, like Libya and Sudan, ICT are seen as the new arm of colonization; to others like Saudi Arabia the question of morality and culture perseverance hindered full adoption of the new technologies; to the rest, the fear of Internet liberal power on the authoritative regime stood against ICT adoption, as in Syria and Tunisia. That is why Arab countries took several measures to control ICT

Table 3. ICT control in the Arab world

Country	Laws & Regulations	Content Filtering	Tapping & Surveillance	Pricing & Taxation	Infrastructure/ Telecom Control	HW/SW Manipulation	Self Censorship
UAE		✓			✓	✓	
Bahrain		✓	✓	✓	✓	✓	
Kuwait		✓		✓	✓	✓	
Lebanon	✓	✓		✓	✓		✓
Qatar		✓	✓	✓	✓	✓	
Saudi A.		✓	✓	✓	✓	✓	✓
Syria	✓	✓	✓		✓	✓	
Jordan	✓		✓	✓	✓		✓
Oman		✓		✓	✓		✓
Libya		✓	✓		✓	✓	
Algeria		✓	✓		✓		✓
Egypt	✓				✓		✓
Tunisia	✓	✓	✓		✓		✓
Yemen		✓	✓	✓	✓	✓	✓
Palestine		✓	✓		✓		✓
Sudan		✓		✓	✓	✓	✓

Source: El Gody, 2003

Table 4. Internet/PC development in the Arab world

Year	Number of Internet Users	Number of PCs
1994		400,000
1996	640,000	900,000
1998	1,000,000	1,800,000
2000	1,800,000	3,400,000
2002	3,700,000	6,100,000
2004	5,200,000	11,300,000
2005	9,000,000 **Gender Distribution (%)** Male: 96, Female: 4 **Age Distribution** U-15: 7 16-21: 42 22-35: 31 36-50: 16 50+: 4	18,000,000 Place / % Home: 29 Work: 58 Cyber Café: 12.7 PC/Internet connection
Exp. 2007	25,000,000	42,000,000

Source: El Gody, 2003

(as seen in Table 3), ranging from imposing laws and regulations, telecommunication infrastructure control, content filtering, hardware/software manipulation, and tapping and surveillance (El Gody, 2003).

After a slow start, realizing its power, ICT diffusion has increased significantly (as illustrated in Table 4). A forecast by the Ajeeb Research Unit (Ajeeb Internet Surveys, 2006) estimates that the Arab world will experience a further increase in ICT demand, accelerating from about 9 million users in 2005 to about 25 million by the end of 2007 (Nour, 2002).

However, the level of ICT adoption in the Arab world cannot be assumed homogeneous (as seen in Table 5), as Arab countries and societies differ greatly in educational standards, financial strength, and willingness to innovate (Hafkin & Tagger, 2001). The level of political acceptance of the new medium also varies. The United Arab Emirates has the highest penetration rate in the Arab world with 24% of the population having access to ICT: Bahrain and Kuwait are a distant second and third, with penetration rates of 18% and 15%, respectively. Sudan, on the other hand, has the lowest Internet penetration rate of 0.05% (Nour, 2002).

Diab Hassan (2003), in his article "ICT Capacity Building", states that ICT in the Arab world grew by 250% between August 2001 and January 2004. While Gulf State countries like UAE, Qatar, and Kuwait possess the financial strength and state-of-the-art technologies to promote ICT infrastructure, the number of ICT users is growing more slowly in some countries, like Algeria and Sudan, which have fewer economic capacities (Hassan, 2003; Nour, 2002).

Table 5. ICT in the Arab world

Country	GDP	Population Using Internet	% of Population Using Internet	# of ISPs	# of Cyber Cafés	% of Population with Access to Telephone Lines	% of Population with Access to Mobile Lines
UAE	22000	1,200,000	24	1	200	48.8	39.6
Bahrain	16,000	200,000	18	1	140	23.5	21
Qatar	10,000	110,000	13	1	100	21.1	19
Kuwait	15,000	460,000	15	3	300	21.8	33
Saudi A.	10,500	1,400,000	3	44	2600	48.8	39.3
Oman	7,700	120,000	4	1	150	7.7	18
Yemen	1,000	25,000	0.3	1	100	1.1	1.6
Palestine	1,900	110,000	0.4	1	50	11	9.2
Jordan	3,500	300,000	4	7	3500	7.8	4
Lebanon	5,000	400,000	7	26	650	19.3	16.4
Syria	3,100	80,000	0.3	1	300	7.8	2
Egypt	3,600	3,300,000	4	67	6000	21	4
Sudan	1,000	25,000	0.05	1	70	1.1	1.6
Libya	8,900	80,000	0.5	1	400	7.25	4
Algeria	5,500	100,000	0.8	1	100	7.8	1.2
Tunisia	6,500	500,000	3	1	600	6.7	1.8
Morocco	3,500	500,000	1	1	2150	9.4	3
% Gulf to rest of Arab World *	76%	45%		83	43	38	67	76

*Note: * Percentage of penetration of mobile lines in Arab countries*
Sources: NUA, 2006; Ajeeb, 2006; Arab Advisors Group, 2003; Nour, 2002

Also, despite the recent overall positive growth trend, the market for ICT is still limited in most Arab countries, and this is apparent in the low demand, limited supply, and restricted ICT spending and investment. The Arab world embraces more than 300 million people, but as Table 5 indicates, the average share of the population having access to main telephone lines, mobile phones, or the Internet are 18.1%, 16.7%, and 5%, respectively, which are low in comparison to the rest of the world. Moreover, the average Arab countries' supply, as indicated by the average number of Internet service providers (ISPs), is very low as well (Hassan, 2003; Nour, 2002).

This limited supply is attributed to inadequate investment and infrastructure. ICT spending, ICT variables, ICT per GDP, and ICT per capita in the Arab world are minimal ($850 million) in comparison to the world average ($3.2 billion). Moreover, software-to-hardware spending ratios are lagging far behind the world's average (Dutta & Coury, 2003; Hafkin & Tagger, 2001).

From the discussion it is clear that ICT diffusion in the Arab world is still characterized by a market concentration in the richer Gulf countries and the wide digital gap between them and other Arab countries in terms of demand, supply, price, and services. That is why analysts believe that despite the recent growth in the ICT sector it still has a very limited effect in bridging the digital divide because developments are made only on the micro level, not yet touching the core issue, human development.

That is why, as previously mentioned, the first step towards narrowing the digital divide is by addressing developing Arab society. That is why the issue of women using ICT is important in today's world, to empower the Arab world.

FUTURE TRENDS

Women's Usage of ICT: Barriers and Means of Empowerment

Discussing women's empowerment means mainly discussing bridging the internal digital divide. Most women in Arab countries are in the deepest part of the divide. This divide is caused by different factors that need to be "empowered" (Egyptian Ministry of Information and Communication Technology, 2004).

Illiteracy and Education Empowerment

The single most important factor in hindering women's ability to take advantage of ICT opportunities is illiteracy. One out of two women in the Arab World is illiterate, making Arab women's illiteracy lowest next to Sub Saharan Africa. The illiteracy rate ranges from 16% in Morocco to 74% in Yemen (El Gody, 2004; El Zu3abi, 2003). Accounting for a higher English illiteracy rate, 80%, and computer illiteracy, we can see why ICT penetration is low among Arab women (El Gody, 2004).

During the past decade, Arab governments focused on providing access to basic education for girls; however, ICT needs to be integrated into the educational programs. This should help improve both the quality and reach of basic education to women (Wheeler, 1998).

Culture Empowerment

Arab culture does not accept new technologies or their diffusion easily within its system. The fast spread of ICTs made Arab governments worry about the outcome and their effect on the rigid Arab culture that is highly motivated by religious ideals (El Gody, 2003).

Religion is a major factor in shaping Arab culture. For centuries, male-dominated culture leans on religious traditions as an excuse for women disempowerment (Mianai, 1981). However, Hassan (2003) stated that Islam has nothing to do with women's disempowerment, discussing that both the Koran and the teachings of Prophet Mohamed emphasized that "acquiring knowledge is an ongoing duty on each Muslim from the cradle to the grave" and that "the quest for knowledge and science is obligatory for every Muslim man and woman." Hassan concluded that Arab authoritative governments are using religion as a tool to further women's disempowerment.

When ICT arrived, most Arab governments feared the liberalizing power of the ICT; for that reason, Arab governments used their media arms to hinder ICT diffusion within their societies. Discovering ICT benefits, Arab governments started to shift gears, increasing ICT culture awareness among societies calling for more women participation. Arab women created ICT societies that developed into the ICT Arab Regional Women Task Force (ICT/ARW-TF) (Abdel Latif, 2004).

Political Empowerment

ICT is a powerful tool in improving governance and strengthening democracy. It can be particularly useful for giving a voice to women in Arab developing countries that have frequently been labeled isolated, invisible, and silent (Hafkin & Tagger, 2001).

ICT can help empower Arab women's political participation, especially in changing women's image from the silent partner in the development process. ICT, especially Internet technology, can be used as a tool for women to create virtual networking groups to discuss socio-political issues, strengthen women's participation in the political process, increase women's access to government services, improve the performance of elected women officials, and disseminate political knowledge (Egyptian Ministry of Communications and Information Technology, 2004).

Economic Empowerment

ICT can assist women's economic development, even in rural areas. ICTs can improve women's activities in "farming, rural trade, business, and industry in a variety of ways ... for instance, female farmers could greatly increase productivity with access to information on improved agricultural inputs, weather, markets, new production techniques, and farming technologies" (Hafkin & Tagger, 2001).

Infrastructure Empowerment

One of the major problems hindering ICT development is the poor Arab telecommunication infrastructure that hinders the widespread reach of ICTs, especially in rural areas where more than 60% of Arab women live (El Gody, 2003).

Therefore, increasing women's access to ICTs involves increasing the availability of communication in areas where women live. Extension of infrastructure, particularly wireless and satellite commu-

nications, is crucial to this process. In addition, access efforts should focus on the establishment of common use facilities such as telecenters, community phone shops, and other public places convenient and accessible to women (Abdel Latif, 2004).

CONCLUSION

While most Arab countries are succeeding in narrowing the digital divide on the technical level by increasing ICT investments, the internal social divide, especially gender inequality, still blocks efforts to bridge the digital divide. Various challenges are still affecting women's active participation. That is why Arab stakeholders—governments, NGOs, international communities, and donors—need to unify their efforts in bridging the internal gap.

Recommendations: Towards Women's Active Participation

Assuring Women Access to the Information Society

It is crucial to increase women's access to ICT issues and allow them to be aware of the potential impact on women. This can be attained through introducing "tailored" programs for promoting awareness of the role of women in information society and circulating them among all countries of the region which will benefit from the experiences, activities, and initiative of specialized organizations.

It is also mandatory to motivate and encourage local and regional mass media to actively participate in spreading ICT awareness among women, especially in remote areas. Producing audio-visual media techniques is also important to facilitate women's access to information society.

Increasing Arab Government Participation

Arab governments need to include the issue of women and ICT in their national strategic plans. This can be attained through setting up committees or assigning a national body to follow up the issue of women and ICT at the national level. Arab governments need to adopt clear policies to promote and develop the information society and assure women's access and active participation.

Arab Women's Participation in ICT Decision Making

Arab governments need to be committed to the WSIS Declaration of Principles issued in 2003 that states women are "an integral part and a fundament element of the information society which should enable women to fully participate in all processes of decision-making." Arab governments and national bodies need to associate women's organizations in making national ICT policies.

Call Arab Countries to Initiate Programs to Educate Women on ICT Issues

Arab governments need to prepare ICT training programs aimed at women, with a special focus on marginalized population sectors, to promote ICT skills. Arab countries need to use existing regional initiative training programs aimed at training Arab women. This should be followed by an increase of funds allocated for research and development in the Arab region to maximize women's usage of ICT.

Creating More ICT Job Opportunities in the Information Society

All stakeholders—government, private sector, and NGOs—need to increase job opportunities to women in the field of ICT, giving more room for female creativity, design, and production.

Creation of Regional and International Cooperation Network

Arab women, especially in rural and remote areas, need to create regional and international "gateways" to discuss issues mainly dealing with empowering ICT usage and current challenges.

Private Sector Partnership

Arab businessmen and women are urged to support gender ICT business initiatives, especially small and medium projects, increasing cooperation among them,

and to connect their businesses with their counterparts worldwide.

The Role of NGOs in Women's Access to Information Society

Local and grassroots NGOs need to foster activities in the field of empowering women and ICT, encouraging regional cooperation and information exchange.

Promoting Arab ICT Programming

Since language is one of the major problems hindering the spread of ICT among Arab societies, Arab governments need to promote Arabic content Web sites, especially those useful to women.

REFERENCES

Abdel Latif, A. (2004). *Gender and citizenship in the Arab region*. Retrieved January 2, 2006, from http://web.idrc.ca/es/ev-59307-201-1-DO_TOPIC.html

Ajeeb Internet Surveys. (2006). *Internet users.* Sakhr Solutions. Retrieved February 21, 2006, from http://ajeeb.sakhr.com/

Al-Zu'bi R. (2003). *From access to effective use: A suggested model for ensuring disadvantaged Arab women's engagement with ICTs.* Retrieved January 2, 2006, from http://www.siyanda.org/static/al-zu3bi_effective use.ppt#256,1

Arab Advisors Group. (2003). *International Internet bandwidth in the Arab world: Bandwidth starved until 2005.* Retrieved February 21, 2006, from http://arabadvisors.com/Pressers/presser-030901.htm

Cockburn, C., & Ruza, D. (1994). *Bringing technology home: Gender and technology in a changing Europe.* Buckingham: Open University Press.

Dutta, S., & Coury, M. (2003). *ICT challenges for the Arab world.* Retrieved January 2, 2006, from http://topics.developmentgateway.org/knowledge/rc/filedownload.do~itemId=290684

Egyptian Ministry of Communications and Information Technology. (2004). *Cairo initiatives on women* and *ICTs*. Retrieved January 2, 2006, from http://www.ituarabic.org/womenandICT/Recommendations-eng-rev1.doc

El Gody, A. (2003). *Internet censorship in the Arab world*. Cairo: Arab U.S. Association for Communication Educators (AUSACE.)

El Gody, A. (2004). *ICT and Arab media: The post third Gulf War*. Unpublished paper presented to the Arab U.S. Association for Communication Educators (AUSACE).

Hafkin, N., & Tagger, N. (2001). *Gender, information technology, and developing countries: An analytic study*. Retrieved January 2, 2006 from http://www.usaid.gov/wid/pubs/ hafnoph.pdf

Hassan, D. (2003). *ICT capacity building*. Retrieved February 21, 2006, from http://www.efore see.info/conferences andevents/malta2003/presentations/diab.ppt

Hunaidi, R. (2001). *Information technology strategies for putting Arab countries on the cybermap*. Dubai, UAE: Gulf Information Technology Exhibition (GITEX).

Internet.com. (2004). *Global digital divide still very much in existence*. Retrieved January 2, 2006, from http://www.Internet.com

Mianai, N. (1981). *Women in Islam: Tradition and transition in the Middle East*. London: John Murray.

Nour, S. (2002). *ICT opportunities and challenges for development in the Arab world* (WIDER Discussion Paper No. 2002/83). Retrieved January 2, 2006, from http://www.wider.unu.edu/publications/dps/dps2002/dp2002-83.pdf

NUA. (2006). *How many online? NUA Web Publishing Solutions*. Retrieved February 21, 2006, from http://www.nua.com/surveys/how_many _online/index.html

Sakr, N. (2004). *Breaking down the barriers in the Arab media*. Retrieved January 2, 2006, from http://www.cmfmena.org/magazine/features/ Nieman_Sakr_Winter01.pdf

Sethuraman, S. V. (1998). *Gender, informality and poverty: A global review*. Retrieved February 21, 2006, from http://www.wiego.org

UNDP. (2004). *Arab human development report 2004: Towards freedom in the Arab world*. Geneva: Author. Retrieved January 2, 2006, from http://cfapp2.undp.org/rbas/ahdr2.cfm?menu=12

Wheeler, D. (1998). In praise of the virtual life: New communications technologies, human rights, development, and the defense of Middle Eastern cultural space. In B. Benjamin, B. Parades-Holt, & J. Slaughter (Eds.), *MONITORS: A journal of human rights and technology*. Retrieved from http://www. cwrl.utexas.edu/~monitors/1.1/wheeler/index.html

KEY TERMS

Digital Access Index (DAI): Value by access level for countries, where 1 is the highest DAI level. According to the International Telecommunication Union (ITU), most of the Arab world falls under the middle and lower access unit. The average of the Arab world is 0.40, which is considered middle to lower class.

Digital Divide: A social cultural issue which refers to the socioeconomic gap between communities that have access to computers, the Internet, and telecommunications and those which do not.

Gender Empowerment Measure (GEM): A composite index measuring gender inequality in three basic dimensions of empowerment—economic participation and decision making, political participation and decision making, and power over economic resources.

ICT/ARIE-TF: A regional Non-Governmental Organization (NGO) created for voicing Arab women ICT initiatives at the World Summit for Information Society. The NGO aims to play a role in promoting ICT among women in the Arab world.

Internal Digital Divide: A term which refers to the gaps that exist between subgroups within the same community or society due to differing levels of literacy, technical skills, or gender disparities.

Middle East: By the term Middle East, the author is focusing on Arab countries that are included in the Middle East region, with the exception of Turkey, Israel, and Iran.

Women Technology Empowerment: The term appeared for the first time in the UN's *World Survey on the Role of Women in Development* (1986) and was defined as a process that "entails much more than awareness of alternatives, women's rights and the nature of requirements. It involves the breakdown of powerful sex stereotyping, which prevents women from demanding their rights from positions of authority" (Sethuraman, 1998, p. 92).

ICT Sector Characteristics in Finland

Iiris Aaltio
Lappeenranta University of Technology, Finland

Pia Heilmann
Lappeenranta University of Technology, Finland

INTRODUCTION

The ICT sector is a newcomer in the Finnish economy. The pace of growth in the Finnish electronics industry was extraordinary over the 1990s. It led to an industrial restructuring in which knowledge replaced capital, raw materials, and energy as the dominant factor in production (Ali-Yrkkö, 2001). The Finnish ICT company Nokia is a world leader in mobile communications. Nokia connects people to each other and the information that matters to them with easy-to-use and innovative products like mobile phones, devices, and solutions for imaging, games, media, and business. The net sales of Nokia totaled •29.3 billion in 2004. Nokia provides equipment, solutions, and services for network operators and corporations. The company has 15 manufacturing facilities in nine countries, and research and development in 12 countries. At the end of 2004, Nokia employed approximately 55,500 people. Nokia is a broadly held company with listings on four major exchanges (http://www.nokia.com). While Nokia's role in the Finnish economy is considerable, there is a large number of other actors in the ICT sector: hundreds of small and medium-sized, fast-growing companies networking and cooperating with Nokia. The strong ICT sector is largely the outcome of mutually enforcing, dynamic cluster relations, which were intensified during the 1990s. ICT managers are mainly engaged in developing software. The work is largely connected to projects in which suitable applications are developed for customers' needs. Applications are usually designed through interaction with customer representatives and software developers (Heilmann, 2004). The customers and the users of ICT in Finland are both women and men, but the majority of the workforce consists of men.

At the customer's side, there are many female ICT professionals. We can have a meeting where there are more women than men, and these women are really capable.

This article considers, first, background information about the ICT sector. Then information-technology companies are analyzed as sites for women's work. Future trends and needs of research are examined next, and finally the conclusion is presented.

BACKGROUND

While Finland is highly dependent on its two main business sectors, forestry and ICT, the attractiveness of the sectors, career development in the businesses, and their future developments are interesting fields of study. The second sector (ICT) is a newcomer, and the first one (forestry) has a long history. ICT has meant for many small economies like that of Finland a possibility for growth and development. The infrastructure in Finland is also recognized to be highly supportive for sector development because of good expertise backgrounds, schooling, adult learning, and state support. Does it attract women and men equally, and what will be its attractiveness in the future? Do women and men advance in the field in similar ways?

This study is comprised of two primary forms of research: literature on the ICT sector's development and empirical interview data gathered in 2002. The interviews were held with 15 ICT managers in Finland, including 2 females and 13 males. The average age of the ICT managers was 34.13 years. The managers were highly educated; 73.3% of all the managers had an academic degree, mainly that of a master of science in technology. Twenty percent of ICT managers were undergraduates in a technical university (Heilmann, 2004).

When compared with the paper business sector (producing pulp, paper, and paperboard), an older Finnish business sector with a long tradition, typical of the ICT sector seems to be the importance of networks, team working, and togetherness between workmates, even during leisure time (Heilmann, 2004). These characteristics also guide work advancement and have an impact on career development.

MAIN THRUST OF THE ARTICLE

The ICT sector can be characterized as a cluster. Clusters are used to describe networks of organizations in which competitive advantage grows from the dynamic interaction between the actors. Cluster relations cross the boundaries of sectors and spur innovation and upgrading through spillovers and knowledge transfer. A cluster can also be defined as a "network of networks," which has economic importance at the macro level (Ali-Yrkkö, Paija, Reilly, & Ylä-Anttila, 2000). The network dynamics cause positive effects on companies' competitiveness. The information and communication cluster, based on competence and technical development, has been able to offer new job opportunities, even if the time for the most rapid growth seems to be over.

In Finland, the main areas of the ICT cluster are the manufacturing of communications equipment and service provision. These areas have increased their share in the information and communication cluster (Hernesniemi, Kylmäläinen, Mäkelä, Rantala, Rautkylä-Willey, & Valtakari, 2001). Around the key industries there are industries that are considered to harbour special potential in enhancing the competitive advantage of the system through innovative applications on ICT, or though the improvement of its functional preconditions (Paija, 2001). The growth of the ICT cluster is not only connected to the growth of the markets in question, however. It is also connected to the general rise of the technical level in production and society (see Koski, Rouvinen, & Ylä-Anttila, 2001).

Careers in the Finnish ICT Sector

Because the ICT industry went through a very dynamic expansion during the 1990s, there was an especially big demand for young ICT professionals who had not only the necessary technical skills, but who could also understand the needs of customers within the new economic environment (Ruohonen, Kultanen, Lahtonen, Liikanen, Rytkönen, & Kasvio, 2002). Universities and research institutes have been successful in producing competent human resources and world-class research and development to support the development of the cluster. The supplier industries, particularly the electronics industry, in turn, have become highly specialized over the last decade to meet the needs of the key activities of the sector. The venture-capital market, as an example of associated services, has emerged as a new and important source of funding that has greatly enhanced preconditions for growth in the cluster (Paija, 2001).

The concept of career has been changing. Career progression has typically meant vertical advancement within one or more organizations, but nowadays it often describes lateral movements within an organization or from one company to another (Stroh & Reilly, 1999). Women should cope with the changing career environment, and within ICT, this means an increasing importance of the role of professional networks in career advancement.

In the future, the software sector will grow from a "nerdy" business into a professional business. Diverse skills relating to internationalization, especially experience in business management associated with international trade; language skills; negotiation skills; and knowledge of different cultures and administrative bureaucracy will be in great demand. In terms of personal skills, visionary capabilities, the ability to perceive matters in their entirety and to concentrate on essentials, communication skills, project and teamwork skills, adaptability, the ability to manage change, creativity, and courage will be emphasized. Strategic expertise will focus especially on network-related capabilities and on understanding the changes brought by the new economy and value chains within the digital economy. Eclectic scientific knowledge, the ability to integrate and master international networks and teams consisting of persons with diverse skills, creativity, and the ability to visualise and innovate are needed (for future developments in the field, see Hernesniemi et al., 2001; http://www.etla.fi).

The need for wider and multifaceted competence in the software business is increasing. In addition to software-based technical and product competence, there are business- and marketing-competence needs. Knowledge of law, international competencies, and an understanding of the meaning of production based on customer needs are all essential. There is also a need for comprehensive understanding of new challenges and opportunities created by the network and digital economy. Basic technical competence in the area of programming (e.g., skills of C++ and Java-programming languages) will remain important. In addition to these competencies, the demand for general and personal competencies will increase. In addition to personal learning, the strategic learning of the organization and strategy management connected to it will increase (Rautkylä-Willey & Valtakari, 2001).

Finnish ICT Sector as a Site of Women's Work

The ICT sector is a sector of young men; in over half of the Finnish ICT companies, the majority of employees are males under 35. Only in one out of four Finnish companies generally is the personnel as young as this (Kandolin & Huuhtanen, 2002; see also Heilmann, 2004).

The ICT managers mentioned in the doctoral dissertation of Heilmann (2004) were mainly engaged in developing software. The work was connected to projects in which suitable applications were developed for customers' needs. Applications were designed through interaction with customer representatives and software developers.

According to Heilmann (2004), it seems that the manager should be self-assured in managing his or her workload and working time. Interesting tasks involved with ICT work may cause an individual to get carried away and can easily steal too much time from his or her life. He or she should be aware and conscious of how to divide the hours of the day between work and leisure. Many managers have recognized the need for rest and have managed to adjust their work schedules. In recent writings, the need for rest and play in the ICT sector has been emphasized (see, e.g., Kivimäki-Kuitunen, 2000). It is obvious that younger workers will want to commit extra hours to work in order to demonstrate their abilities and competence, and ascertain future employment and career possibilities in the organization. In the interviews, both managers and employees agree upon the importance of rest and recreation and also call for flexibility. When private life needs more time from the manager, then he or she is allowed direct time toward the family. On the other hand, the family gives way to business when needed. It seems to be the birth of children that makes the division of work and leisure more clear. It is not possible to work long days anymore if one wishes to stabilize life (Kivimäki-Kuitunen). In a networking work environment, this is a challenge, especially for those women with family duties.

Within this sector, production can be defined as a purely masculine area. ICT-sector professions require mathematical skills, and girls usually choose something other than mathematics to study at school. There are ever fewer female students in technical universities. This is the main reason for the scarcity of females among the technical professions, but in the interviews of male ICT managers, there were found also attitudes that favoured males and showed suspicions about the competence of women (Heilmann, 2004).

In the interviews, it was still stated that gender does not seem to have great importance in developing software; what is more important is how capable the person is. Women are welcomed into the ICT sector by men: The representation of both sexes in the working place was said to have a good effect on the working climate. Concurrently, the managers mentioned there was a lack of female workers and managers in the ICT business (Heilmann, 2004).

Women in the ICT sector seemed to work mostly on supportive assignments like testing and documentation, or as assistants (Heilmann, 2004). There are signs that women in the ICT business have accepted the masculine world and made themselves "good guys." Even if the individuality of workmates was emphasized, in the interviews there was often a clear separation between women and men of the workforce:

Women work in documentation. Only one of them works in production. I see no difficulties. They are very nice girls. (ICT manager, male)

It seems that gender matters do not have much importance in work situations. Anyhow, women notice the existence of the "glass ceiling" in their career development. The career progression of women usually stops at middle management. However, the men managers did not consider the concept of the glass ceiling in the interviews (Heilmann, 2004).

Networks in the field seem to be gendered with the dominant role of men. This is clearly recognized by the female interviewees, even if they seem to have found their own ways to cope with the situation.

I feel comfortable working with men here. I have worked a lot with them. They have taken my job positively. I think there exists some kind of glass ceiling for women, however. It doesn't bother me. I have a nice job. But there exist many men's affairs when men do things together. Other department managers are taken along more easily because they are male. I don't know if they think that it would be difficult if a woman goes along. Boys can't talk boys' business then. Actually, I notice it only when I start thinking; it doesn't bother me every day. (ICT manager, female)

In my first job the male colleagues helped me a lot. It was very nice. Both in personal relations and work occasions they have treated me very well. From customers' side and from everywhere else it is the same. I have been privileged to be in these work situations as a woman. Sometimes, in the situations where male outsiders are present and where I have not been before, males talk like there is not any female present. They do not pay any attention to female members of the group. But now when I am already older and gained competence, they listen. (ICT manager, female)

I don't think gender matters at all. Almost a half of my team members are female. When you look at the organization chart you see more males in upper levels of this organization. It would be nice to have there more females. (ICT manager, female)

Also, the male managers of ICT are aware of the gendered nature of their work environment.

There could be more women working here. I don't know; maybe it comes from different hobbies. Boys and girls do different things. Boys have always been interested in machines, computers, and so on. (ICT manager, male)

Secretaries have always been female, but I see that this workplace is equal for both sexes. (ICT manager, male)

This gender structure works here. The secretary has always been female, but there are five or six other women working in our organization as well. Mostly they work in documentation; only one works in production. There isn't any harm of them; they are nice girls. Of course, when you have to arrange some company events with a sauna, you must divide the personnel into males and females. But it happens everywhere; there is no trouble. But from a professional point of view I don't see any trouble. (ICT manager, male)

In the interviews, male managers raise female gender as a problem in terms of its minority, not so much professionally. However, the female workforce is not recognized as a source for creativity among male ICT professionals. Diversity is emphasized as a value as such in the working environment. Also, the customers in the ICT sector are both women and men.

FUTURE TRENDS AND NEEDS OF RESEARCH

Information professionals are rapidly increasing not only in numbers, but also in respect to modern companies' needs for development, which places new demands on compensation and career-development policies. Both women's and men's work are valuable, partly because customers' needs are gender dependent. Customers' needs should be understood better and studied if there is also a gender gap to be filled.

This work environment is not so one-sidedly male because on the customer side, there are also many women. Women are often even in the majority in meetings. (ICT manager, male)

Companies are still facing a clearly exceptional labour-market situation in which there is a continuous shortage of competent information professionals. The attractive labour market and biased compensation structure risks the commitment of professionals and leads to high turnover rates. As a result of this, not only company attractiveness and the working climate need to be developed, but also exceptional recruitment methods need to be introduced. Due to the pace of technical advancement, the developmental needs of information professionals are on a scale of their own. With the aging workforce, this challenges the human-resource development and career planning of the company. Because of scarce human resources, heavy workloads, and developmental needs, work exhaustion prevails, which in turn calls for the application of new and flexible working practices (Holm, Lähteenmäki, Salmela, Suomi, Suominen, & Viljanen, 2002). The combining of work and private life with family responsibilities should be taken as a serious challenge of ICT companies' human-resource management. Women and also young, family-building-aged men can benefit from this kind of support.

The Finnish ICT sector is very network based, and this is a special challenge for women who are a minority with not so many contacts in the field. Networks are often gendered by nature, involving a lot of unprofessional meetings. This evidently edges out women professionals and might hinder their career advancement within the workplace. The meaning of networks should be studied as well, from the angle of how both sexes can take advantage of them and what this means to the development of the field in general. The meaning of soft work is increasing in the field, and the sector will acquire multiskilled competence in the future.

CONCLUSION

ICT companies seem to be very masculine environments. Among men, there exists some kind of suspicion toward women's competence in the area of information and communication technology.

Women's tasks are usually connected to lower level or supportive assignments, and their career progression may stop at the glass ceiling. Concurrently, the existence of both sexes in ICT companies is appreciated, but its importance for the well-being of companies is not recognized.

This is a masculine environment. There could be more girls here. It forces boys to shape up when a girl sits at the same table. I haven't met any good female software developers, but I know they exist. It is possible for girls to develop software, though it is a technical area. (ICT manager, male)

It looks as if there is a double gaining of expertise for women of the ICT sector. The first is in the school years when technical competence is acquired. Basic competence is based on mathematical skills and education first in the secondary school, and after that in technical universities. So the ICT sector becomes a masculine world from the beginning of education. After the gained formal education, competence is developed toward more specific know-how expertise inside the ICT companies and by lateral movement between the companies. Work traditions, company cultures, and networks should take better account of the gendered nature that they carry and invite competent and highly motivated women and men to enter and advance in the field. In the future, technical skills will not be enough for competence, and women can therefore enter at least into the first expertise level more easily than before.

REFERENCES

Ali-Yrkkö, J. (2001). *Nokia's network: Gaining competitiveness from co-operation* (ETLA B174 Series). Helsinki, Finland: Taloustieto.

Ali-Yrkkö, J., Paija, L., Reilly, C., & Ylä-Anttila, P. (2000). *Nokia: A big company in a small country* (ETLA B162 Series). Helsinki, Finland: Taloustieto.

Heilmann, P. (2004). *Careers of managers: Comparison between ICT and paper business sectors.* Unpublished doctoral dissertation, Acta Universitatis Lappeenrantaensis, Lappeenranta, Finland.

Hernesniemi, H., Kylmäläinen, P., Mäkelä, P., Rantala, O., Rautkylä-Willey, R., & Valtakari, M. (2001). *Suomen avainklusterit ja niiden tulevaisuus: Tuotanto, työllisyys ja osaaminen* [The key clusters of Finnish economy: Future, production, employment and competence]. Finland: Työministeriö.

Holm, J., Lähteenmäki, S., Salmela, H., Suomi, R., Suominen, A., & Viljanen, M. (2002). Best practices of ICT workforce management: A comparable research initiative in Finland. *Journal of European Industrial Training, 26*(7), 333-341.

Kandolin, I., & Huuhtanen, P. (2002). Työajat suomalaisissa IT-yrityksissä [The working hours in the Finnish IT companies]. In M. Härmä & T. Nupponen (Eds.), *Työn muutos ja hyvinvointi tietoyhteiskunnassa* [The change of work and wellbeing in the information society] (pp. 81-92). Helsinki, Finland: Sitran Raportteja.

Kivimäki-Kuitunen, A. (2000). *Work, rest and play: Matkaevästä Nokian nuorilta esimiehiltä* [From Nokia's young managers]. Tampere, Finland: Tammer-Paino.

Koski, H., Rouvinen, P., & Ylä-Anttila, P. (2001). *Uuden talouden loppu* [The end of the new economy?]*?* Helsinki, Finland: ETLA, Sitra, Yliopistopaino.

Lammi, M. (2000). *Metsäklusteri suomen taloudessa* [The forest cluster in the Finnish economy] (Series B161). Helsinki, Finland: Elinkeinoelämän Tutkimuslaitos.

Paija, L. (2001). The ICT cluster in Finland: Can we explain it? In L. Paija (Ed.), *Finnish ICT cluster in the digital economy* (ETLA B176 Series) (pp. 56-88). Helsinki, Finland: Taloustieto.

Rautkylä-Willey, R., & Valtakari, M. (2001). Kolmen avainklusterin tulevaisuudenkuvat ja osaamistarpeet [The visions and competency needs of the three key clusters]. In H. Hernesniemi, P. Kymäläinen, P. Mäkelä, O. Rantala, R. Rautkylä-Willey, & M. Valtakari (Eds.), *Suomen avainklusterit ja niiden tulevaisuus* [The Finnish key clusters and their future] (pp. 77-120). Helsinki, Finland: ESR-Julkaisut.

Ruohonen, M., Kultanen, T., Lahtonen, M., Liikanen, H., Rytkönen, T., & Kasvio, A. (2002). Emerging knowledge work and management cultures in ICT industry: Preliminary findings. In M. Härmä & T. Nupponen (Eds.), *The change of work and wellbeing in the information society* (pp. 7-22). Helsinki, Finland: Sitran Raportteja.

Stroh, L. K., & Reilly, A. (1999). Gender and careers: Present experiences and emerging trends. In G. N. Powell (Ed.), *Gender and work* (pp. 307-324). Thousand Oaks, CA: Sage.

FURTHER READING

ETLA. (n.d.). http://www.etla.fi/finnish/research/publications/searchengine/pdf/abstract/b179fin.pdf (Retrieved November 13, 2002)

Nokia. (n.d.). http://www.nokia.com (Retrieved August 15, 2005)

KEY TERMS

Gender: Gender refers to the cultural construction of femininity and maleness.

ICT Business Sector or Software Sector: Produces software applications. It hires young employees and is vulnerable to changing economic conditions.

ICT Cluster: ICT-cluster relations cross boundaries, but it mainly includes the manufacturing of communications equipment and service provision. It can also be defined as "a network of network managers". It involves producing software, but also involves supporting related industries, associated services, and buyers and appliers.

Manager: A person working in the middle level of an organization. He or she plans, organizes, motivates, directs, and controls.

Network Relations: The social, economic, and often informal relations that construct a business sector or a cluster.

ICT Usage in Sub-Saharan Africa

Vashti Galpin
University of the Witwatersrand, South Africa

INTRODUCTION

Given the circumstances of women's lives in *sub-Saharan Africa*, it may appear that *information and communication technologies* (*ICTs*) are only for wealthy, well-educated, urbanized women with time to use them, and that they are irrelevant for other women in sub-Saharan Africa. However, this is not the case: women see ICTs as providing opportunities for change, by giving them access to the information which will help improve their circumstances, as the abundant research shows (Hafkin & Taggart, 2001; Huyer & Mitter, 2003; Morna & Khan, 2000; Pacific Institute of Women's Health [PIWH], 2002; Rathgeber & Adera, 2000).

This article presents an overview of women as ICT users in sub-Saharan Africa, covering the challenges and the success stories. Since there is a large body of literature covering this area, only a representative subset is surveyed. The focus here is usage. Information technology (IT) professionals and more technological topics are considered elsewhere in this volume. Much of the literature about usage in developing countries takes a broad definition of ICTs because of the lack of the latest technologies. For example, Holmes (2004) includes computers, the Internet, mobile phones and wireless technologies as well as telephone, radio, television, print media, listening groups, and community theatre. This article will consider all electronic technologies, from computers and networking to radio and television.

When considering ICTs and developing countries, the *digital divide* is often mentioned. This term is sometimes used specifically to refer to the Internet; for example, see DiMaggio, Hargittai, Neuman, and Robinson (2001). In line with the broad definition of ICTs given above, in this article, the term *digital divide* will be used to refer to inequality in access to ICTs and ability to use them. There are multiple divides: men vs. women, urban vs. rural, rich vs. poor, young vs. old, developed vs. developing. When considering developing countries, there is an underlying information divide—people do not have access to information sources they require, electronic or otherwise, due to poverty and lack of infrastructure. This is the real problem that needs to be solved—ICTs are a means to this end.

BACKGROUND

Sub-Saharan Africa has a population of 641 million where only 35% of the population lives in urban areas, and almost half of the total population is under 15 years of age (United Nations Development Programme (UNDP), 2004). 32% of the population is undernourished, 323 million people live on less than $1 per day, and it is estimated that 8% of the population is HIV-positive; the UNDP Human Development Index for sub-Saharan Africa has decreased during the 1990s, showing the effect of the HIV/AIDS epidemic (UNDP, 2004).

All of the countries in sub-Saharan Africa are classified as developing. Thirty-one are classified as *least developed countries* by the United Nations (UN). There is a large need for development—a legacy of the history of slavery and colonialism which has affected the region. Progress has been made; for example, the adult literacy rate is 63% compared to the youth literacy rate of 77% (UNDP, 2004) which indicates that access to education is increasing. Clearly there are differences between countries; Mauritius, South Africa, and Nigeria, for example, have less poverty, although they have large wealth disparities within their populations.

In terms of technological infrastructure, there are only 15 landlines, 39 cellular subscribers and 10 Internet users per 1,000 people (UNDP, 2004). Rural areas are less likely to have electricity than urban areas, so battery, solar-powered, or wind-up radios are prevalent.

Status of Women in Sub-Saharan Africa

Women's lives in sub-Saharan Africa are influenced by strong societal opinions about their roles including an expectation that they will focus on the home, and they have less access to education and health than men do (Huyer, 1997; Momo, 2000). The female adult literacy rate is 54% compared to the male adult literacy rate of 70%; and the female youth literacy rate is 72% compared to the male youth literacy rate of 81% (UNESCO, 2004). The gross enrolment ratio at primary level is 78% for girls, and 91% for boys; at secondary level, 24% and 30% (UNESCO, 2004).

The contribution of women in terms of housework, child-rearing, subsistence farming and community management is not valued in a cash economy, and hence overlooked (Huyer, 1997). This contribution is important, and women often want to use ICTs and other information sources to improve the conditions of their families and communities. However, because of women's multiple roles, time is limited, hence the time taken to seek out information must balance with the gain achieved (Huyer, 1997).

ACCESS TO ICT

Access to ICTs is low. In 2000, women made up 12% of Internet users in Senegal, 32% in Uganda, 38% in Zambia and 51% in South Africa, but the number of Internet users in these populations is small, hence few women have access (Hafkin & Taggart, 2001)

Some of the obstacles to use of ICTs by women are low levels of literacy and education, lack of materials in local languages, lack of time, inconvenient opening times for public ICTs, cost, safety issues, sociocultural expectations about women's roles and movement in public areas, and lack of skills in using ICTs (Hafkin & Taggart, 2001). These barriers do not just occur because of poverty; they are amplified by the second-class status of women (Hafkin & Taggart, 2001).

Education is crucial; Amolo Ng'weno notes that a high school education is required to use the Internet effectively (Carnevali, 2002). With low rates of secondary education, access alone is not sufficient. The HIV/AIDS epidemic is also an issue; girl children are more likely to be removed from school to care for sick relatives, and teacher numbers are decreasing significantly (Isaacs, 2002).

Education

Few schools in Africa have access to ICTs, and to date little research has been done. Isaacs (2002) highlights the fact that many *SchoolNet* projects do not consider gender, and this may affect how successful they are for girls. The introduction of ICTs in education may be both positive and negative. In Africa, women are a substantial number of those studying by (non-computer-based) distance learning because learning can be fitted around domestic activities, and Derbyshire (2003) suggests that the introduction of computer-based distance learning could impact the number of women studying if they have no or limited access to computers, although it could also increase their interaction with other learners.

The World Links program was found to have positive aspects for girls. This program placed computers with Internet access in schools in Senegal, Mauritania, Uganda, and Ghana. Gadio (2001) reports that in all countries, teachers felt that girls gained more academic benefit from usage because of their focus on academic material. The girls also reported increased self-confidence, and the opportunity to obtain information about health and sexuality that is not available otherwise. However, in Uganda and Ghana, girls had less access due to after-school chores and social prohibition on running which prevented them from getting to the labs before boys. In comparison to boys, no girls took part in maintenance of the labs, even at single-sex schools, although at one school, the girls used their skills to teach primary-level children about computers, which appears to indicate a lack of interest in maintenance rather than lack of confidence with computers.

Public Access

Telecenters have been proposed and implemented as a way of achieving access to telephones and other ICTs. Additionally, in many cities in Africa, commercial Internet cafes are becoming common (Levey & Young, 2002; Mbarika, Jensen, & Meso, 2002).

Cost is an issue, and unless telecenters are heavily subsidized, they are unlikely to be affordable for all (Morna & Khan, 2000).

Telecenters are not always successful for women since it is often assumed that users will find their own way once the technology is provided (Rathgeber, 2002). Urban women were more likely to use telecenters, and usage typically did not include fax, Internet, or e-mail (Rathgeber, 2002). Rural women found telecenters problematic due to their cost, inconvenient opening times, plus perceptions that they are for men and have no material for people who are illiterate (Women'sNet/Dimitra, 2004).

In Uganda, 29% of users were women, in Mozambique 35% and in Mali 23% (Rathgeber, 2002); telecenters in Nigeria, Ghana, Ethiopia, Uganda, and Mozambique had fewer women users than men (Johnson, 2003). In contrast, the telecenter in Gaseleka in South Africa has a regular user community which is more than 60% female, the centre is managed by two women, and a basic computer literacy course had a majority of women graduates (Benjamin, 2001), showing that telecenters can be accessible to women. Some forms of public access are focused on women; for example, Isis-WICCE (http://www.isis.or.ug) opened an Internet cafe in Kampala, Uganda, which is a resource for women run by female staff, and which charges less than commercial operations (Oriang', 2002; PIWH, 2002).

Business

Many large companies in sub-Saharan Africa use ICTs including the Internet daily, but smaller businesses and entrepreneurs often do not have the resources. African business women want information about trade policies, fair trade, and ways to increase income, as well as ways to market products and communicate with purchasers (Huyer, 1997).

UNIFEM (United Nations Development Fund for Women) (http://www.unifem.org) is involved in programs about ICTs and business, including the WINNER Network (http://www.winner-tips.org) which is operational in Zimbabwe and imparts ICT skills to help participants in marketing their goods, locally and internationally. In addition, the Digital Diaspora Network links up Africans with ICT skills who live outside Africa with women's *non-governmental*

organizations (*NGOs*) and business organizations for technology and skills transfer. This project will also provide business role models for African women (Carnevali, 2002). The South African Department of Trade and Industry has developed the Technology for Women in Business program (http://www.twib.co.za) to assist women in developing technology-based businesses. There are numerous other programs.

Telecenters are a source of business information, although research has shown that they are used by women mostly for social or family purposes (Rathgeber, 2002). An example of the material that can be provided is a Ugandan CD-ROM on earning money for illiterate and semi-literate women involved in subsistence farming who found it useful in making decisions (Mijumbi, 2002). The women of Twendelee Handicrafts in Nairobi, Kenya use Internet cafes to access Web pages about knitting (Oriang', 2002). Businesses also employ women in positions where computer use is required; particularly secretarial or data entry jobs (Hafkin & Taggart, 2001). With the trend for outsourcing, jobs involving ICTs such as call centre operators and data capture, are moving to Africa for cost reasons; for example, all environmental tickets issued in New York are captured in Ghana (Huyer & Mitter, 2003). A concern is the sustainability of these jobs, as they tend to be repeatedly relocated to where costs are lowest.

Development

ICTs play an important role in development, although there is often a perception that they should have lower priority than basic services such as clean water and education (Rathgeber, 2000). At a workshop on empowerment of rural women (Women'sNet/Dimitra, 2004), rural women from Africa emphasized the ability of ICTs to provide information that would enable them to improve their lives. Radio is the preferred medium because of its accessibility; telecenters were seen as less accessible and a solution proposed was to take material to women in their homes (Women'sNet/Dimitra, 2004).

An example of the role ICTs can play in development is the Pacific Institute for Women's Health project to improve communication by women's NGOs in Zimbabwe, Uganda and Zambia by providing e-mail and Internet access, and opening Internet cafes for women (PIWH, 2002). This fits with Kole's three-sector model of African society: the modern, non-modern and information sectors (Kole, 2003). NGOs that use ICTs can operate at the intersection of the three sectors, often enabling community-based organisations in the non-modern sector to use ICTs.

APC-Women has developed GEM (Gender Evaluation Methodology) (http://www.apcwomen. org/gem) as an online tool to evaluate the gender impact of ICT development initiatives. As noted earlier, there is a specific gender dimension to poverty, and the analysis provided by GEM allows for assessment of how a specific ICT project has improved women's lives.

FUTURE TRENDS

There are many important recommendations about the way forward: from the general, such as full and equal education for women at all levels and in all disciplines, improved healthcare and the ability to exercise human rights; to the specific such as ensuring that telecenters are accessible to women (Johnson, 2003; Momo, 2000; Rathgeber, 2002), equal access to computers in schools (Derbyshire, 2003; Isaacs, 2002), evaluation of the success of ICT projects in terms of gender, and appropriate use of technology for local conditions (Kole, 2003). The World Summit on the Information Society (WSIS) Gender Caucus (http://www.genderwsis.org) has identified key principles for the Information Society including gender as a fundamental issue, equal participation in decision-making, a combination of old and new ICTs, appropriate design of ICTs, and evaluation of the impact of ICTs on women. Marcelle (2001) notes that to achieve an African women's cyberspace, relevant content must be generated by African women, technology must be managed by African women, and interaction must occur between user and producer.

Gender should be part of national and international ICT policy (Hafkin & Taggart, 2001; Huyer, 1997; Marcelle, 2000; Morna & Khan, 2000) otherwise ICT projects may only serve part of the community. Groups such as APC-Women-Africa are involved in lobbying around these issues.

CONCLUSION

Development and reduction of poverty is one of the motivations for the use of ICTs in developing countries such as those in sub-Saharan Africa. ICTs are not a panacea—they can have both positive and negative effects, hence it is not sufficient just to provide ICTs though provision is one of the steps. An understanding of women's information needs, as well as of their lives is necessary to ensure the technology is provided in a way that is accessible, useable, and effective. Women in sub-Saharan Africa are seldom passive consumers of technology, but active participants looking for information and tools to improve their own conditions as well as those of their families and communities.

REFERENCES

Benjamin, P. (2001). The Gaseleka Telecentre, Northern Province, South Africa. In C. Latchem & D. Walker (Eds.), *Telecentres: Case studies and key issues* (pp. 75-84). Vancouver: The Commonwealth of Learning. Retrieved January 17, 2005, from http://www.col.org/telecentres/Telecentres_complete.pdf

Carnevali, I. (2002). *Challenging the digital divide in Africa: An interview with Amolo Ng'weno.* Retrieved September 20, 2004, from http://www.africansocieties.org/n3/eng_dic2002/amoloint.htm

Derbyshire, H. (2003). *Gender issues in the use of computers in education in Africa.* Imfundo: Partnership for IT in Education. Retrieved January 21, 2005, from http://imfundo.digitalbrain.com/imfundo/web/learn/documents/Gender%20Report.pdf

DiMaggio, P., Hargittai, E., Neuman, W., & Robinson, J. (2001). Social implications of the Internet. *Annual Review of Sociology, 27,* 307-336.

Gadio, C. (2001). *Exploring the gender impact of World Links.* World Links. Retrieved January 21, 2005, from http://world-links.org/english/html/genderstudy.html

Hafkin, N., & Taggart, N. (2001). *Gender, information technology, and developing countries:*

An analytic study. LearnLink/Office of Women in Development, USAID. Retrieved January 21, 2005, from http://learnlink.aed.org/Publications/Gender _Book/Home.htm

Holmes, R. (2004). *Advancing rural women's empowerment: Information and communication technologies (ICTs) in the service of good governance, democratic practice, and development for rural women in Africa*. Women'sNet. Retrieved January 21, 2005, from http:// womensnet.org.za/dimitra_conference/Empower ing_Rural_Women.doc

Huyer, S. (1997). *Supporting women's use of information technologies for sustainable development*. IDRC/Acacia. Retrieved January 21, 2005, from http://web.idrc.ca/en/ev-10939-201-1-DO_ TOPIC.html

Huyer, S., & Mitter, S. (2003). *ICTs, globalisation, and poverty reduction: Gender dimensions of the knowledge society*. IDRC/Gender Advisory Board. Retrieved January 19, 2005, from http://web.idrc.ca/ en/ev-60511-201-1-DO_TOPIC.html

Isaacs, S. (2002). *IT's hot for girls: ICTs as an instrument in advancing girls' and women's capabilities in school education in Africa*. Division for Advancement of Women, United Nations. Retrieved January 19, 2005 from http://www.un.org/ womenwatch/daw/egm/ict2002/reports/ Paper%20by%20Isaaks2.PDF

Johnson, K. (2003). *Telecenters and the gender dimension: An examination of how engendered telecenters are diffused in Africa*. MA Thesis, Georgetown University. Retrieved January 21, 2005, from http://cct.georgetown.edu/thesis/Kelby Johnson.pdf

Kanfi, S., & Tulus, F. (1998). *Telecentres*. IDRC/ Acacia. Retrieved February 1, 2005, from http:// web.idrc.ca/en/ev-6646-201-1-DO_TOPIC.html

Kole, E. (2003). *Digital divide or information revolution?* i4d/Information for Development. Retrieved January 21, 2005, from http:// www.i4donline.net/issue/sept-oct2003/digital _full.htm

Levey, L., & Young, S. (2002). *Rowing upstream: Snapshots of pioneers of the information age in Africa*. Johannesburg, South Africa: Sharp Sharp Media.

Marcelle, G. (2000). Getting gender into African ICT policy: A strategic view. In E. Rathgeber & E. Adera (Eds.), *Gender and the information revolution in Africa* (Chap. 3). IDRC. Retrieved January 22, 2005, from http://web.idrc.ca/en/ev-9409-201-1-DO_TOPIC.html

Marcelle, G. (2001). Creating an African women's cyberspace. *The Southern African Journal of Information and Communication, 1*(1). Retrieved January 31, 2005, from http://link.wits.ac.za/journal/ j-01-gm.htm

Mbarika, V., Jensen, M., & Meso, P. (2002). Cyberspace across sub-Saharan Africa. *Communications of the ACM, 45*(12), 17-21.

Mijumbi, R. (2002). A case on the use of a locally-developed CD-ROM by rural women in Uganda. In *UN/INSTRAW virtual seminar series on gender and ICTs*. Retrieved January 21, 2005, from http:// www.un-instraw.org/en/docs/gender_and_ict/ Mijumbi_summary.pdf

Momo, R. (2000). Expanding women's access to ICTs in Africa. In E. Rathgeber & E. Adera (Eds.), *Gender and the information revolution in Africa* (Chap. 6). IDRC. Retrieved January 22, 2005, from http://web.idrc.ca/en/ev-9409-201-1-DO_TOPIC. html

Morna, C., & Khan, Z. (2000). *Net gains: African women take stock of information and communication technologies*. APC-Africa-Women/ FEMNET. Retrieved January 21, 2005, from http:// www.apcafricawomen.org/netgains.htm

Oriang', L. (2002). Using ICT in unexpected places. In L. Levey & S. Young (Eds.), *Rowing upstream: Snapshots of pioneers of the information age in Africa* (pp. 21-40). Johannesburg, South Africa: Sharp Sharp Media.

Pacific Institute for Women's Health (PIWH). (2002). *Women connect! The power of communications to improve women's lives*. Retrieved January 21, 2005, from http://www.piwh.org/pdfs/wc2002.pdf

Rathgeber, E. (2000). Women, men, and ICTs in Africa: Why gender is an issue. In E. Rathgeber & E. Adera (Eds.), *Gender and the information revolution in Africa* (Chap. 2). IDRC. Retrieved January 22, 2005, from http://web.idrc.ca/en/ev-9409-201-1-DO_TOPIC.html

Rathgeber, E. (2002). *Gender and telecentres: What have we learned?* World Bank Group. Retrieved January 21, 2005, from http://site resources.worldbank.org/INTGENDER/Resources/16Gender.ppt

Rathgeber, E., & Adera, E. (2000). *Gender and the information revolution in Africa*. IDRC. Retrieved January 22, 2005, from http://web.idrc.ca/en/ev-9409-201-1-DO_TOPIC.html

United Nations Development Programme (UNDP). (2004). *Human development report 2004*. Retrieved January 19, 2005, from http://hdr.undp.org/reports/global/2004/

UNESCO (2004). *Education for all: Global monitoring report 2005*. Retrieved January 27, 2005, from http://www.efareport.unesco.org

Women'sNet/Dimitra (2004). *Report back on the e-consultation: ICTs for the advancement of rural women's empowerment*. Women'sNet/Dimitra. Retrieved January 23, 2005, from http://womens net.org.za/dimitra_conference/e-conference_reportback.doc

KEY TERMS

Digital Divide: A term describing the gap between those who have access to and can use technology effectively and those that cannot. There are narrower definitions that focus on the use of the Internet, and broader definitions that include all ICTs.

Information and Communication Technologies (ICTs): Electronic means of communicating and conveying information, covering media such as radio and television, computer and computer net-

working technology and telecommunications. This is a broad definition of the term.

Least Developed Countries (LDCs): Countries identified by the United Nations as having a low GDP per capita (less than $750), low levels of health and education, and economic vulnerability. Thirty-one of sub-Saharan Africa's countries are classified as LDCs: Angola, Benin, Burkina Faso, Burundi, Cape Verde, Central African Republic, Chad, Comoros, Democratic Republic of the Congo, Equatorial Guinea, Eritrea, Ethiopia, Gambia, Guinea, Guinea-Bissau, Lesotho, Liberia, Madagascar, Malawi, Mali, Mauritania, Mozambique, Niger, Rwanda, São Tomé and Principe, Senegal, Sierra Leone, Togo, Uganda, United Republic of Tanzania, Zambia (UNDP, 2004). The sub-Saharan countries that are not classified as LDCs are: Botswana, Cameroon, Congo, Côte d'Ivoire, Gabon, Ghana, Kenya, Mauritius, Namibia, Nigeria, Seychelles, South Africa, Swaziland and Zimbabwe (UNDP, 2004).

Non-Governmental Organizations (NGOs): An NGO is a non-profit organization focused on particular issues for the public good. It can operate locally, regionally, nationally or internationally, and it may provide services, lobby government or perform monitoring. Some of the NGOs involved with women and ICTs in sub-Saharan Africa are Women'sNet, South Africa; Isis-WICCE, Uganda; APC-Africa-Women, Zimbabwe Women's Resource Centre and Network, ENDA-SYNFEV and FEMNET.

SchoolNet: An organization focusing on access to ICTs within schools, network and Internet connectivity access, as well as content development. Most SchoolNet projects are small scale and donor funded (Isaacs, 2002). SchoolNetAfrica is an NGO supporting country-based SchoolNets in Africa (http://www.schoolnetafrica.net).

Sub-Saharan Africa: The area of Africa south of the Sahara. In terms of the United Nations definition, this covers all countries on the Africa continent excluding Algeria, Djibouti, Egypt, Libyan Arab Jamahiriya, Morocco, Somalia, Sudan and Tunisia, and covers the island states of Comoros, Madagascar, Mauritius, São Tomé and Principe, and

Seychelles, giving a total of 45 countries (UNDP, 2004). All countries in sub-Saharan Africa are classified as developing countries.

Telecenter: "[A] location which facilitates and encourages the provision of a wide variety of public and private information-based goods and services, and which supports local economic or social development" (Kanfi & Tulus, 1998, para. 2). A distinction can be made between telecenters which are community focused, and commercial Internet cafes or cybercafes.

ICTs for Economic Empowerment in South India

Shoba Arun
Manchester Metropolitan University, UK

Richard Heeks
University of Manchester, UK

Sharon Morgan
University of Manchester, UK

INTRODUCTION

The role of new technologies, particularly information and communications technology (ICTs) in the global society is central to both contemporary social theory and understanding transformations that are characteristics of the information society and post modernity. The emphasis on technological determinism is useful in tracing social and economic changes at large, but the economic and social shaping of technology is often illustrative of wider social relations, with local considerations. Recently, studies have demonstrated how technology is socially-contextualised, with gender differential barriers to access and use of ICTs by men and women (Hafkin & Taggart, 2001). This article argues that ICTs as a form of new technology are socially deterministic, albeit context dependent, need to take into account the role of social actors and interactions, which is often ignored in the blind pursuit of market forces.

The article is structured as follows: the Background section examines some of the debates relating to gender and ICTs; then the Main Thrust section proceeds to examine the ICT context in southern India through a case study of the Kudumbashree project and some conclusions are provided in the last section.

BACKGROUND

Gender and ICTs in a Global Society

There have been a number of perspectives in understanding the social role of technology. The social construction of the technology is useful in analysing users as agents of technology and change, where a social (and logical) perspective is applied to scientific knowledge (Kline & Pinch, 1999). Often social groups, which are dynamic in nature, play a crucial role in the development of technological artefacts. The theoretical perspective on Actor Network theory developed by Strum and Latour (1999) reveals the mutual constitution of technology and society showing how these are not fixed social structures but are continuously and actively negotiating and renegotiating relative roles. The role of the social actors both at the macro and micro-level is important as well the social link in relations between various social actors. These social actors in an ethno-methodological fashion according to Garfinkel's social interactionism, are transformed from "cultural dopes" to active social actors (Mackenzie & Wajcman, 1999).

There has been a burgeoning body of literature in relation to the role of gender relations in technological change, demonstrating the masculine nature of technology and capitalist domination. Research in industrialised countries has noted that technical fields and IS systems are highly gendered (Wajcman, 2004; Webster, 2005) highlighting the social shaping, or constructivist, theory of technology. In looking at the socio-cultural influences on the professional development and working lives of women IT professionals, Trauth (2002), rejects the essentialist view of women and their relationship to IT that has been put forth in the information systems literature arguing, instead, the primacy of societal and structural influences.

Relatively little has been researched or written about the specifics of gender relations in the ICT sector in developing countries, although there are emerging studies in relation to call centres or the Software Industry (e.g., Arun & Arun, 2001) and it is this knowledge gap that this article partly addresses. It has been shown here that to analyse the deeper issues affecting women's engagement with ICTs we need to take a wider scope, such as the "gender & technology as socially defined" and "experience of daily life" approaches. It is clear from this overview that a gender perspective may take several forms varying from those focussing primarily on the individual as the means to bringing about change, to those taking a wider scope and attempting to transform the society and culture in which women are living. The Association for Progressive Communications (APC) brings the importance of women's involvement in the "definition, design, and development of new technologies" (APC-WNSP, 2002). The gender evaluation methodology (GEM) is grounded in the view that any gender analysis should (APC-WNSP, 2002) focuses on both self and social change: addressing the relationship between the ICT initiative and the way the "self" (individual, organisation, and/or community involved) operates and also the relationship between the ICT initiative and the broader context (social, political, economic, and cultural).

The experiences in one South Indian state—Kerala[1]—that can be seen as a microcosm of the ICT experience in India with a booming software sector as well as a number of innovative state interventions in ICTs, including *Kudumbashree*, a women-led poverty reduction programme that has made use of ICTs to enable the development of ICT-based enterprises run by cooperatives of poor women. In order to do this, a qualitative and case study based approach was undertaken in July-December 2004, with empirical research based on women as workers, women as entrepreneurs, women as social and economic agents in households and communities and discussion with key informants.

MAIN THRUST OF THE ARTICLE

Gender Shaping of ICTs in South India

The software and services component of the ICT sector has emerged as one of the fastest growing

industrial segments, increasing from U.S. $170-million worth of output in 1991-92 to U.S. $8.8 billion in 2003-04 (Arun, Heeks, & Morgan, 2004). The state of Kerala in South India has formulated ICT policies, through increasing human capital, creating infrastructure and innovative strategies to use ICTs as panacea for poverty alleviation.

The Kudumbashree initiative is a state interventionist poverty eradication strategy, which strongly gender-focused. *Kudumbashree*—which means "prosperity of the family"—is an initiative of the Kerala State Poverty Eradication Mission (SPEM) was launched in 1999 as a women-oriented, participatory, and integrated approach to fight poverty (Government of Kerala, 2003). Thus, use of ICTs could help gain economic empowerment of poor women and households to tap into the broader range of relations between ICTs and enhanced social and economic development. Neighbourhood Help Groups (NHG) or *ayalkootams*—a type of cooperative of ten women from poor families (based on non-monetary indicators) are formed as the basis for each Kudumbashree unit. These micro-enterprise units include a range of activities from food processing, cleaning, handicrafts, but its most innovative aspects has been its use of ICTs to form the basis for some of its enterprises.

In all, 1,206 Kudumbashree units are now operational in a range of sectors (Kudumbashree, 2004) owned, managed and operated by women from poor families. Out of these, there are three types of ICT-based enterprise comprising of 45 IT training units which provide IT training to schools; 56 data entry and digitisation units which mainly create local digital content for public (and to a lesser extent private) sector organisations; and 5 hardware assembly/maintenance units.

A profile of these ICT enterprises is provided in Table 1, which illustrates that the main ICT activities relate to data entry, hardware assembly and servicing as well as IT training. The selected units are located in different region, from the main state capital of Thiruvanathapuram, in the southern region of the state, to both urban and per-urban areas in the northern region of the state. In all, there has been a sizeable employment impact, with the ICT units creating jobs for nearly 2,000 women; with nearly U.S. $50 is being earned by each member (Arun et al., 2004). The average number of members within

Table 1. An overview of Kudumbashree ICT micro-enterprises

	Hardware Assembly	Data Entry	IT Training
Name	InfoShree Systems	Technoworld Digitals	Divine Computers Vidyasree
Start Year and Location	2003, North Kerala (peri-urban)	1999, South Kerala (State Capital)	2002, North Kerala (Urban)
Activities	Hardware: assembly & service Data Entry & DTP work: Training	Data Entry--Digitalisation of Public Records; Training: Students	IT Training to High school students
Source of Finance	Bank Loan, Subsidy, & Group Contribution	Bank loan, Subsidy, & Group contribution	Bank loan, Subsidy, & Group contribution
Employment & Human Capital	10 group members, (Electronics and computing): Four employees (all men)	10 members, (PGDCA, DCA); one Male supervisor: 52 casual workers	six members (DTP, PGDCA and CTTC

Source: Fieldwork (2004); Kudumbashree (2004)

a group is 10, although training units have lesser numbers of members. In addition, employment is further provided to women and men from similar backgrounds, again demonstrating the welfare dimensions of the enterprises. There is also a genuine perception of the women themselves that they have been empowered; and as active agents in society, and as revealed from discussions, a number of women from the units having been participating in community level affairs (Arun et al., 2004).

Human capital is quite high among these entrepreneurs, despite belonging to economically poor families. On an average, women possess under-graduate and post-graduate degrees, with further technical qualifications in ICT such as electronics, Post-Gradate Diploma in Computer Applications (PGDCA) and Diploma in Computer Applications (DCA). On joining the enterprise, many have increased their capabilities in both computing as well as personal skills such as language, communication and business skills, and competencies in the ICT sector which is traditionally the preserve of men. In the Training Sector, some have gone on the gain teaching qualifications in computing, (i.e., Computer Teachers Training Course (CTTC)) which is seen as a very proud achievement as summarised by Preeta, the group leader from the Training Unit, "for women like me from both economically and socially deprived groups (in terms of caste), the status of a teacher is a dream come true."

The main source of finance of these enterprises comprises of institutional loans, state subsidy and group contribution, all of which is facilitated by the Kudumbashree institution. Some units, like the Data entry Unit, started in 1999 has repaid all its loans and surplus funds are being diverted for reinvestment in the business, venturing up the value chain in specialised services such as Web designing. The main customer/client base for these enterprises is largely the state sector, enforced through supportive policies which include digitisation of all state records, provision of ICT infrastructure and equipment, increasing human capital of school students in ICT training, all which are to be provided through Kudumbashree units. It must be recognised that development of the Kudumbashree IT units has required a significant degree of institutional support from government departments, banks, other financial intermediaries, and other local organisations. In order that these enterprises run effectively and sustainable in the long run, and without much support, it is further crucial to formulate policies that support such small enterprises, in terms of both generic and specific needs of ICT-based enterprises for women.

CONCLUSION AND FUTURE TRENDS

Often the compelling nature of technological change is crucial in the economic shaping of technology, guided by economic factors and market competition ignored roles of social actors and interactions at both macro and micro levels. Social relations affect and shape technology, with a number of forces responsible to the economic and social shaping of technology, including the nation-state, markets, and other institutions. This article has presented case studies from South India that provides the basis for a qualitative and exploratory investigation of the ways in which interventionist ICT initiative, Kudumbashree, illustrate the capacity of gender-focused, locally-owned and participative–and for rather different outcomes.

Intervention certainly has its disadvantages and—despite the existence of these ICT-based enterprises for five years—questions are still raised about sustainability. The range of competencies developed is considerably greater because women have been allowed to break out of the traditional stereotypes of inequalities of power and responsibility and thus been able to make at least a start on breaking down some of the social, political, and even institutional bases of gender inequality as seen from a GEM perspective. The interventionist approach to using ICTs for women's development—as represented here by Kudumbashree—is by no means a panacea. It arises from a particular set of institutional arrangements and political priorities that cannot be wholly transplanted to other contexts, but needs to be adapted according to location specific characteristics. Some generic lessons are valuable from the Kudumbashree example, such as enabling policies and institutions and locally participatory nature of initiatives. It also pinpoints to the fact that choices of technological change (such as ICTs) and strategies are continuously shaped and renegotiated in local contexts as technological change is to be actively shaped rather than passively responded to (Mackenzie & Wajcman, 1999). In adopting a gendered perspective, the non-gender-neutral nature of technology, ICT engagement has to go beyond mere participation, with gender relations that are embedded in their environmental context.

REFERENCES

APC-WNSP (Association for Progressive Communications—Women's Networking Support Program).

(2002). *Gender and ICT: Towards an analytical framework*. Retrieved from http://www.apc women.org/work/research/analytical-frame work.htm

Arun, S., & Arun, T. (2001). Gender at work within the software industry: An Indian perspective. *Journal of Women and Minorities in Science and Engineering*, *7*(3), 217-232.

Arun, S., Heeks, R., & Morgan. S. (2004). *ICT Initiatives, women, and work in developing countries: Reinforcing or changing gender inequalities in South India*. Development Informatics Working Paper Series No 20. Institute for Development Policy and Management. Manchester: University of Manchester.

Government of Kerala. (2003). *Economic review*. State Planning Board. Kerala, India: Trivandrum.

Hafkin, N., & Taggart, N. (2001). *Gender, information technology, and developing countries: An analytic study*. Washington, DC: WID Office U.S.AID.

Harding, S. (1987). Introduction: Is there a feminist method? In S. Harding (Ed.), *Feminism and methodology*. Milton Keynes, UK: Open University Press.

Kline, R., & Pinch, P. (1999). The social construction of technology. In D. Mackenzie & J. Wajcman (Eds.), *The social shaping of technology*. Philadelphia: Open University Press.

Kudumbashree. (2004). *Concept, organisation, and activities*. Kerala: State Poverty Eradication Mission (SPEM).

Mackenzie, D., & Wajcman, J. (1999). *The social shaping of technology*. Philadelphia: Open University Press.

Parayil, G. (2000). Introduction: Is Kerala's development experience a "model"? In G. Parayil (Ed.), *The development experience: Reflections on sustainability and replicability*. London; New York: Zed Books.

Strum, S., & Latour, B. (1999). Redefining the social link: From baboons to humans. In D. Mackenzie & J. Wajcman (Eds.), *The social shaping of technology*. Philadelphia: Open University Press.

TechnoWorld Digital Technologies. (2003). Unpublished Annual Report. TechnoWorld Digital Technologies. Kerala: Trivandrum.

Tornquist, O. (2000). The new popular politics of development: Kerala's experience in the development experience. In G. Parayil (Ed.), *The development experience: Reflections on sustainability and replicability*. London: Zed Books.

Trauth, E. M. (2002). Odd girl out: An individual differences perspective on women in the IT Profession. *Information Technology and People, 15*(2), 98-118.

Wajcman, J. (2004). *Techno-Feminism*. Cambridge, UK: Polity Press.

Webster, J. (2005, February 1). *Women in IT professions: Corporate structures, masculine cultures*. Paper presented to 3rd European Symposium on Gender and ICT: Working for Change, Manchester, UK.

KEY TERMS

Ayalkootam: Self-help groups of around ten women from poor families.

Caste: Rigid social groups differentiated by descent in the Hindu Society.

Empowerment: Increased capability at various levels, (e.g., economic, political and social).

Kudumbashree: A poverty eradication programme aiming for the prosperity of Family through helping women.

Micro-Enterprises: Small enterprises, defined by size of investment, employment, and scale of business.

Panacea: Universal solution.

Vidyashree: Small enterprise units that undertake ICT training in Schools (*Vidya meaning education*).

ENDNOTE

[1] Kerala rates relatively highly—as least compared with other Indian states—on various social development indicators (Parayil, 2000). In many respects, Kerala is seen to be a very particular model of development based on specific institutions, interventions, and historical processes (Tornquist, 2000).

The Impact of Gender and Ethnicity on Participation in IT

John Paynter
University of Auckland, New Zealand

INTRODUCTION

The excitement of information technology is not only within the discipline itself. Advances in computers have led to leaps in almost every academic discipline and changed the very nature of our everyday lives. The knowledge revolution has resulted in rapid changes to the way we work and live. This includes the offering of an increasing range of career opportunities that did not exist before. Computing is one of the fastest growing industries, but since most jobs remain dominated by males, women remain a major latent source of talent for the technology field.

BACKGROUND

Participation in IT

Information technology (IT) serves broad needs of society. The technology workplace should reflect the interests of both men and women. Assuming entry to the IT career domain is restricted by tertiary qualifications, female's low enrolment rate in computer related subjects would lead to ineffectual future workforce planning. A survey of the top 150 public hi-tech companies in the Silicon Valley found that only four had female CEOs (Raghaven, 2001). In the United States, the number of female computer science graduates has fallen from a 1985 peak of 35.8% to 27.5% in 1994 (Raghaven, 2001). The United States National Science Foundation (1999) statistics show that the proportion of women receiving bachelor's degrees in computer science dropped from 37% to 27% between 1984 and 1997 (Anonymous, 2000). According to research conducted by Arthur Anderson, young men are five times more likely than young women to select computer science or computer related majors in schools (Cohen, 2001). Also, women are not entering university computing and information technology courses at the same rate

as men. If this situation persists, the gender imbalance in the computer technology industry could worsen. Women are, therefore, excluded from the exciting prospects promised by information technology.

There are issues of stereotyping and false perceptions permeating throughout society that women are less competent in technology compared to their male counterparts. These misconceptions could be one of the reasons for the low enrolment rate for women. Previous research suggests that girls receive little encouragement to explore computers early in their schooling (Henwood, 2000). Some suggest that there are few female role models in technical industries (Cockcroft & Cunningham, 1995). These problems may each contribute to part of the overall problem of declining female participation. Identifying reasons leading to the problem helps us to understand the gender imbalance. Surveys have been conducted to see if the genders and domains of the case studies influence student participation in IT (Wong & Paynter, 2001). However, no consensus could be drawn from the results. Seeking ways to increase women's participation remains an important but yet unsolved task.

Gender imbalance is not a new topic in the information systems (IS) research realm. Much effort has been spent in identifying remedies to combat the problem. Evidence shows that women of Asian ethnicity significantly outnumber other ethnic female students, both in Australian and New Zealand IT degree studies (Cockcroft & Cunningham, 1995; von Hellens & Nielsen, 2001). European studies, too, show an ethnicity bias with female participation decreasing in Western European countries but not in developing nations, including Eastern Europe (Schinzel, 1999, 2002). It seems that cultural differences may be more influential than gender alone. This finding forms the background of this research. Social and cultural factors have to be considered together with inherent gender differences. The gender imbalance issue in the IT industry is a social as well as cultural construct that can be improved if its causes are being identified and therefore appropriate measures are being applied.

Cognitive Style and Learning

Kirton's Adaptor Innovation (KAI) Inventory is a 33-item questionnaire. It is a pre-tested and validated instrument. It consists of 33 questions in the form that looks like the Likert scale often used in Information System (and other) surveys. Kirton claims KAI measures creativity in his adaptation-innovation theory, which was first introduced in 1976. The adaption innovation theory studies individuals' cognitive styles. The theory classifies people's cognitive styles as lying on a continuum with adaptive and innovative on two opposite poles. The associated tool for the theory is called Kirton's Adaption Innovation (KAI) Inventory. KAI measures people's cognition and it is especially effective for identifying a person's learning preferences as well as problem solving styles. Creators have a high KAI, while Adaptors have a low one. Kirton's study states that each person has a preferred problem-solving style that remains stable over a lifetime.

Individuals develop coping mechanisms to deal with circumstances which are at odds with their preferred problem-solving style, but as soon as the situation allows, return to the preferred style; positive or negative outcomes are not necessarily the result of differences between adaptors' and innovators' problem-solving approaches but of situational factors. (Kirton, 1976 as cited in Osborne, 1995).

To assist practitioners, Kirton encourages a distinction between level (what is done) and style (how it is done). It is said that level may be affected by intelligence, knowledge and experience while style develops early in life and persists over time, regardless of age (Osborne, 1995).

Methodology and Findings

This research started by examining the ethnicity and gender composition of the students taking information systems courses at the University of Auckland

Information Systems Business School. It then used a survey instrument (KAI) to examine different cognitive styles to see if this had an influence on their participation. Students were initially surveyed during their lectures and follow up surveys were sent to those who were not present on the day the surveys were initially given.

The students taking the courses surveyed represent a mixture of genders and nationalities and ethnicities and have a variety of educational backgrounds. The first year information systems course is typically a mixture of IS and non-IS majors, whereas the second and third year classes have a more technical orientation (e.g., computer science, engineering, information systems). For female students it was found that year I students had the highest KAI, while the mean KAI of stage III students was the lowest. The stage II mean KAI score was higher than those of stage III and lower than those of stage I so that there was no statistical significance between stage II and the other two groups. This finding was interesting and it led us to ask what type of female students the university is retaining. As Kirton (1978) suggested that people's problem solving styles are consistent over time, it is unlikely that female students become more "adaptive" as they proceed to higher levels. In other words, it is likely that the Information Systems major is attracting and retaining females of "adaptive" problem solving styles. Enrolment patterns show that different genders have different preferences in selecting their major. Learning styles are deter-

Figure 1. Impact of course and gender on KAI

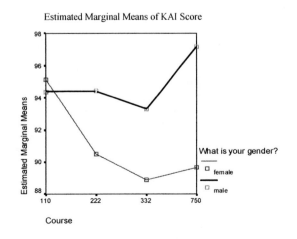

mined by cognitive styles, and different genders have different learning styles. Gender has an effect on influencing individuals' cognitive styles. This is perhaps the most important finding of the work to date. It implies that innovative females may not be attracted to Information Systems courses beyond the first year (Figure 1).

FUTURE TRENDS

The observations made about IT participation in the Australasian universities are not unique. In Europe (Schinzel, 2002) it appears that it is the children of recent immigrants who gravitate to the IT field. Indeed, 50% of those of European ethnicity taking the IS program at the University of Auckland were recent immigrants.

The low participation of Pakeha (European ethnicity) women in IT courses and then in the work force does nothing to overcome the fears of those who feel that Asians dominate the courses and the industry. Affirmative action is needed to promote the place of the Pakeha females in IT courses and the industry to overcome the stereotypes that seem to be developing (Cockcroft & Cunningham, 1995; Wong & Paynter, 2001). At this stage it is not clear what action can be taken in the courses themselves. One possibility is to reintroduce case studies that themselves can promote participation across cultures and gender. Female students may feel more at home in the social atmosphere of discussing cases (a role typical of analysts) rather than in lab-based exercises behind a computer (Wong & Paynter, 2001). Another approach is to use role models but there is little evidence to suggest that this works and most accounts are anecdotal (von Hellens & Nielsen, 2001).

The less than expected growth in electronic commerce has seen a downturn in the IT industry. IT appears less attractive to potential students. In Auckland we have become heavily dependent upon the foreign student market. This is exacerbated by the low participation by the New Zealand born students. It is clear that we must restructure and reposition our current courses. This is a challenge for educators world-wide.

Most of the Asian students return to their homeland after completing their degrees. Even those who came to New Zealand earlier in their lives for secondary schooling and obtained Citizenship or Permanent Residence (PR) often do not find satisfactory employment in New Zealand. We have started interviewing those who have entered the IT workforce in New Zealand (Chan, 2004) and Hong Kong (due to its accessibility) in order to reconcile their educational and industry aspirations.

CONCLUSION

This is an exploratory study that aims to understand the factors influencing the ethnic and gender imbalances in IT participation. Judgment sampling was employed to distribute the KAI survey. Future research should employ a probability sampling to ensure that the findings can be generalised. One approach would be to survey final year courses in other disciplines where the gender and ethnicity patterns differ. We will also continue to interview IT workers and students in Hong Kong and New Zealand, as well as conduct focus groups in New Zealand to look at the issues arising from this study.

REFERENCES

Anonymous. (2000). Technology's gender gap. *The New York Times*, p. 26. Retrieved from http://www.nsf.gov/statistics/nsf01325/

Chan, K. Y. C. (2004). *Participation in IT training: The perceived skill set required*. Unpublished Master of Commerce thesis, University of Auckland.

Cockcroft, S., & Cunningham, S. J. (1995). *Gender and other social issues* (p. 336-338). Dunedin: University of Otago.

Cohen, S. (2001). Welcome to the girls club. *Infoworld*, pp. 55-58.

Henwood, F. (2000). Exceptional women? Gender and technology in U.K. higher education. *IEEE Technology and Society Magazine*, pp. 21-27.

Kirton, M. (1978). Have adaptors and innovators equal levels of creativity? *Psychological Reports, 42*, 695-698.

Kirton, M. (1987). *KAI manual.* Hatfield, UK: Occupational Research Centre.

Osborne, R. L. (1995). Adaptors and innovators: Styles of creativity and problem solving. *Management Decision, 33*(8), 60.

Raghavan, B. S. (2001). India: IT for women. *Businessline*, p. 1.

Schinzel, B. (1999). The contingent construction of the relation between gender and computer science. In A. Brown, & D. Morton (Eds.), *Proceedings of the 1999 International Symposium on Technology and Society: Women and Technology: Historical, Societal, and Professional Perspectives* (pp. 299-312). New Brunswick, NJ: Rutgers University.

Schinzel, B. (2002). Cultural differences of female enrollment in tertiary education in computer science. In K. Brunnstein & J. Berleur (Eds.), *Human choice and computers,* Boston, Dordrecht, London, Kluwer Academic Publishers; *Proceedings of the 17th World Computer Congress—TC9 Stream; 6th International Conference on Human choice and Quality of Life in the Information Society, (HCC-6),* Montreal, Quebec, Canada (pp. 283-292).

von Hellens, L., & Nielsen, S. (2001). Australian women in IT. *Communications of the ACM, 44*(7), 46-52.

Wong, W., & Paynter, J. (2001, December 5-8). Gender effects on IT participation. In *Proceedings of the 3rd Australia New Zealand Management Academy Conference,* Auckland, New Zealand (Paper 34, available on CD).

KEY TERMS

Adaptor: Someone who is more suited to following instructions, doing things by rote. It is considered that much of IT is learned and done this way (e.g., modifying programs, writing documentation, doing database queries in SQL). In fact, with the emphasis on templates and reuse this will become even more the norm.

Gender Imbalance: It is perceived that the IT industry and enrolment in IT courses are dominated by males. This is exacerbated by the lifestyle (nerdish long hours behind a computer screen), role models, and even the gender of text book writers.

Innovator: Someone who prefers starting things from new. Hence, there is room for creative flair. Kirton (1995) proposes that everyone falls into a creativity continuum ranging from doing things better (adaptors) to doing things differently (innovators).

Kirton's Adaptor Innovation (KAI): Kirton's Adaptor Innovation (KAI) Inventory is a 33-item questionnaire. It is a pre-tested and validated instrument. It consists of 33 questions in the form that looks like the Likert scale often used in information system (and other) surveys (Kirton, 1987).

Pakeha: "White" people, those of European descent compared to the native people (Maori) or more recent Polynesian immigrants.

Indigenous Women in Scandinavia and a Potential Role for ICT

Avri Doria
Luleå University of Technology, Sweden

Maria Udén
Luleå University of Technology, Sweden

INTRODUCTION

From a distance, the Sámi Network Connectivity initiative (SNC) does not necessarily appear as anything but another technical research project with certain science-fiction (sci-fi) connotations. It is aimed to create Internet connectivity for communications-challenged terrestrial settings using a protocol currently being developed for communications in space. However, while being a highly technical project, SNC emerged from an unexpected setting: an Indigenous women's initiative to save their traditional livelihood from threats of social and economic drain and to create better opportunities for women and youth to remain within the traditional community.

The first step towards the formation of SNC was taken in June 2001 when a group of women reindeer herders in Sirges Sámi Village in Jokkmokk, Norrbotten County in northern Sweden decided to start a gender equality project, Kvinna i sameby (KIS).[1] To the Sámi, reindeer herding serves not only as an economic base but also as a foundation for reproduction of cultural values. Already in the KIS planning stage, Susanne Spik, the project leader, contacted the Division for Gender and Technology at Luleå University of Technology (LTU) to invite scientific assistance from the early stage of the project. LTU is the regional technical university for northern Sweden and is situated in the Norrbotten County capital of Luleå 200 km southeast of Jokkmokk. Promoting women's possibilities to remain in reindeer herding and the traditional Sámi community, especially social and technical conditions for work and business development, were the focus in the discussions. An associated but separately funded project was subsequently formed by

LTU researcher Maria Udén. A solution to the project requirements came from a guest researcher at the computer science department, Avri Doria, an Internet systems architect. In spring 2002, after initial discussions with members of the Interplanetary Networking Research Group (IPNRG) at the NASA Jet Propulsion Lab, she contributed the proposal that came to be referred to as Sámi Network Connectivity. With a decision to accept this project, the establishment of SNC as both a technical idea and a concrete gender-based project became a prime goal for the cooperation between the women in Sirges and the scholars at LTU, and continued after the KIS project ended in December 2003. The SNC objective is to provide connectivity where other sources are not available, while making the local population part of the development of the technical system. To develop the technical solution space of SNC, the Sámi Network Connectivity proposition gained research funding from the Swedish national agency for innovation systems, Vinnova, for the period 2004 to 2006. This funding is distributed through the Vinnova program "New communication networks."

BACKGROUND

Being a technical project, it is not obvious how SNC relates to the understandings of the sex/gender and gender equality concepts, as these are maintained in women's movements and feminist theory. SNC is a result of a women's movement among the Sámi and will be shown also linked to the current feminist movement in academia. More than a unified position of gender issues, the common motivating factor shared by all participants in the SNC is a shared appreciation of grass-roots participation in technology development. To feminist researchers in science and engineering, formulating critiques of their mother disciplines is not a sufficient goal. The vision and expectation is to be able to present theoretical and methodological alternatives (Keller, 1992; Mörtberg, 2003; Trojer, 2002).

This has strongly affected the research scope of gender studies at LTU, where the presence of engineers, mathematicians, and systems and computer scientists has been substantial from the start. The SNC project is one among other activities aimed at changing the relations between gender and technology initiated in this environment. Internationally, the LTU research scope is consistent with aims and considerations expressed by, among others, Evelyn Fox Keller. It is characteristic, however, that feminist researchers engaged in science and technology continue to acknowledge difficulties in taking the step from observation and critique to presenting functional alternatives to/within them. Keller (1992) put the question of feminist interventions in science and their possible success as follows:

In short, feminist theory has helped us to re-vision science as a discourse, but not as an agent of change. And it is this latter question that I want to press on now. Since it is demonstrably possible to envision different kinds of representations, we need now to ask what different possibilities of change might be entailed by these different kinds of representations? (p. 76)

Though more than a decade has passed since Keller expressed these concerns, feminist methods for effectively acting as agents of change in science and technology are still barely developing, even in the field of Information and Communication Technologies (ICT), which have indeed generated a large body of feminist studies during the late 20th and early 21st centuries. Reasons behind this lack of progress are thought to be located in various social, cultural and economic factors, all of which are affected by symbolic, as well as material connotations, of sex/gender (Bratteteig, 2002; Mörtberg, 2003; Trojer, 2002). In this respect, the significance of the networks between engineers/scientists and the individuals and organizations that request and make use of their products and results, the patrons, must not be overlooked. These networks tend to be male dominated not only on the experts' side but also on the patrons' (Cockburn, 1985; Keller, 1992; Trojer, 2002; Udén, 2002).

CHALLENGES AND POTENTIALS

Sámi Lifestyle Today, its Challenges to ICT and the SNC Solution Space

Even if Sweden is indeed one of the world's most "Internet-connected" nations, the districts of concern to the reindeer herders are not as well off in this respect. The level of service and ICT access is significantly lower than in Swedish society at large. In 2002, The Swedish National Rural Development Agency investigated the infrastructure available in the Swedish Sámi herding communities, especially the summer lands. Among other reasons, the summer lands were chosen for the investigation as they are especially valuable for keeping the children's link with Sámi culture, and for both cultural and social reproduction in other respects. It showed that the majority of residents' camps in the summer lands have very little or no access to infrastructure, including post delivery, telephone and roads (Glesbygdsverket, 2002). Given that Sirges and its neighbor, Sámi Villages, to a large extent, operate in a large, 9,400 km² connected area of natural preserves and other protected areas, it is understood that installment of fixed infrastructure, such as major masts for mobile communications, are not wanted. This area of wilderness is known as Laponia, and listed by UNESCO as World Heritage. To the reindeer herding Sámi in Sirges and surrounding villages, Laponia is not wilderness, but their cultural landscape.

Today, the Sámi are an indigenous minority population incorporated within the Scandinavian and Russian national states, and their traditional lifestyle is challenged by conflicting demands. Many of these conflicts stem from the fact that maintaining economic and social sustainability makes it necessary to be part of modern society, which puts demands on being, more or less, resident in a fixed location, while their traditional lifestyles—in particular, reindeer herding—continues to require a more nature-based lifestyle and semi-nomadicity. (Haetta, 1993;

Jernsletten & Klokov, 2002) One basic assumption held within SNC is that access to the Internet could, to a certain degree, enable resolution of these conflicts. In fact, a venue for innovation is opened as the notion arises, that ICT is not genuinely available on the premises of Sámi semi nomadism, as this notion challenges popular understandings of ICT as eliminating boundaries in time and space, making place and time irrelevant, being limitless.[2] Yet, this potential is not only a myth but materially inherent in ICT, something that all of us who check our e-mail from hotels we stay at for a day or two as readily as from our homes or offices can benefit from. Perhaps this potential is even more valuable to a nomadic population than to others. If participating in local politics (which is vital to a minority population), making use of the new options for distance education, consulting healthcare services, generally keeping up business contacts and specifically running e-based business concepts would be possible from the grazing areas and in points of time adjusted to herding requirements, much of the strain on the individuals and on the community could be avoided. As herding is based on organic time and constant moving with the herds and the seasons, while the majority of society is based on the mechanical clock and steady settlement, the buffer capacity of ICTs and their innate capability of changing the implications of time and place carry the potential of making semi nomadism a more feasible lifestyle tomorrow than it is today.

The connectivity mix in the regions where the Sirges herders operate is constrained both by availability and possibility. While there is a mix of data delivery opportunities—for example, wired, wireless and digital television, in particular throughout the rims of the herding region—there are vast areas where none of these delivery mechanisms are available. Furthermore, the fact that much of the terrain is protected means that neither antennas nor cabling can be installed. The SNC solution to this challenge does not offer real-time services. Instead, providing robust connectivity is prioritized. The idea is that mobile relays periodically travel human byways to locations where gateways to the Internet are available, carrying data bundles that can be exchanged. Thus, connectivity is coupled to *presence*, relying on the movement and encounters of the population rather than being based on an even availability over a huge and periodically unused area. Current reports of the

solution space include Lindgren and Belding-Royer (2005), Lindgren and Doria (2005) and Lindgren, Doria, and Schlen (2004). Doria, Udén, and Pandey (2002) gives an overview of both technical and social impetus of SNC.

Women in the Sámi Villages and the Potential Role of ICT

When we present the SNC project, in almost every audience someone will ask why an indigenous people, and especially its women, should want high-tech, ICT development. To understand the implications for the semi-nomadic reindeer herders of Scandinavia, it is necessary to first acknowledge that there are aspects of a needs-and-demands analysis that would turn out the same having any activity built on field work in focus; for example, tour guiding and wildlife monitoring. For these, any computerized or Web-based system must be available in the field to genuinely be of use. Additionally, there are aspects of access to ICT that are specific for the Sámi reindeer herders as indigenous people. These aspects tend to turn out differently between women and men.

We have already mentioned that the reindeer herding communities are subjects of stress caused by conflicting demands from modern society and their traditional lifestyle. A critical Sámi women's movement has reported how women are often those who have to take on major parts of this mediating labor, and how the very limited resources available to women to fulfill the expectations put them in a situation of strain. Reindeer herding has also become increasingly a masculine matter, in part as a result of this split of life of traditional life and modern society and lack of coherence between the culture and economic base on the one side, and modern structural organization and demands on the other. This masculinization is not acceptable from a gender equality point of view, as women suffer both socially and economically from it, and it also threatens the vitality of the culture (Kråik 2002).

The vision of SNC is to be an active and positive part in bringing the potentials of ICT into use in a contemporary re-establishment of traditionally based nomadism. One example of possible use of ICT to resolve the conflict between traditional and modern is the compulsory school system that, though it is valuable insofar that it provides necessary and valuable education to all citizens, still has the specific disadvantage to the reindeer herding communities in that it hinders children of the reindeer herding families from being present in the grazing lands as much as is needed to gain traditional knowledge. As a consequence of the need for children to stay near the school in town, mothers' possibilities to migrate with the herds are limited as well (e.g., Ulvevadet & Klokov, 2004.). If Internet-based distance education were available, this conflict, the gendered effects of which are based in traditional division of labor among the Sámi, could be reduced. Thus, Sámi women, while enhancing a traditional identity, could gain from high-tech development.

This, however, is not an expectation free of reservations. Development of ICT for use in the grazing areas could also lead to the opposite extreme; increased strain and further marginalization of women in the Sámi Villages. It is reasonable to assume that the chance for gender-sensitive results of deployment of new means for communication rests upon the way in which women and men take part in development and employment. In this respect, the SNC process contradicts expectations on gender roles in the Sámi community as well as in Swedish innovation systems in general.

Difficulties that women and minorities may experience in taking part in ICT development were among the issues particularly addressed by The Working Group on Internet Governance (WGIG, 2005, pp. 2-3):

One weakness of present systems is that people who are excluded today may be in that situation partly because their involvement is structurally hindered in more or less all "normal" partnerships for development within their country. This can often be the case for minorities and women. These groups will have additional problems when compared to other local groups, in developing Internet use and in benefiting from the ICT potentials to improve their quality of life. ... Even if other paths may be open, e.g. access to technical expertise, generating the resources needed for implementing change can be dependent on relations with the same authorities that in other instances do not acknowledge the disadvantaged group as legitimate partners.

FUTURE TRENDS

Apparently, being a high-tech project, SNC emerged from an unexpected setting: an alliance between high-tech professionals, gender studies scholars and a locally situated indigenous people's gender equality project. New and unexpected actors and alliances stepping forward are not unknown in ICT development. Rather, such events have been intrinsic and altogether part of the process. The early development of the Internet, including the steps that made it available to other than elite scholars and military, grew from a mix of well-established and unexpected actors (Castells, 1996). Nevertheless, the significance of gender, ethnic identity and location has not been less in ICT sectors than elsewhere. At this point, it is too early to establish that SNC is typical for a coming stage of technical and organizational innovations, where yet more unexpected things happen and gender barriers and ethnic patterns are challenged. What we can note is that SNC to a certain degree has been successful, even a remarkably successful endeavour. Yet, at the start we envisioned technology transfer, with pilot development and deployment in cooperation with the requesting community. As it turned out, it has been more difficult to find resources for these activities than for the university-based research. We did not expect this, as there are European Union structural funds and various national and regional funds allocated for the development of the remote northern regions of Sweden.

CONCLUSION

The SNC process represents a notable novelty in terms of technology development. Significant factors are the active role of women as patrons; these women's rural location, ethnic identity and explicit gender equality agenda; and the operational part played by feminist scholars. The novelty of the process opens a rich array of research opportunities, including but not confined to representations and change in science, and the advance of alternative feminist paths for science and engineering. Not least, working in and from an alliance between women technicians and women patrons exposes pre-conceptions, skills and knowledge gaps of our own as well as of other actors. Certainly, we also note how structures may or may not support an endeavour such as ours. But the resources available to the SNC group have this far been allocated to the system development itself and to building strategic alliances to reach our aims of user participation and technology transfer.

In consequence, the positive research results achieved to this point primarily belong to the technical solution space. From the standpoint of feminist and gender theory, we note how concepts of gender and technology, and specifically understandings of their relations, may be exposed and challenged in an action-oriented endeavor such as SNC. Also in the case of SNC, concepts of ethnicity, location and tradition are at play.

REFERENCES

Amft, A. (2000). *Sápmi i förändringens tid* (dissertation). Sweden: Umeå University.

Bratteteig, T. (2002). Bringing gender issues to technology design. In C. Floyd, G. Kelkar, C. Kramarae, C. Limpangog, & S. Klein-Franke (Eds.), *Feminist challenges in the Information Age*. Verlag Leske; Budrich.

Castells, M. (1996). *The information age: Economy, society and culture*. Malden; Oxford: Blackwell.

Cockburn, C. (1985). *Machinery of dominance: Women, men and technical know-how*. London: Pluto Press Ltd.

Doria, A., Udén, M., & Pandey, D. P. (2002). Providing Internet connectivity to the Sami Nomadic Community. *Proceedings of the 2nd International Conference on Open Collaborative Design for Sustainable Innovation*. Retrieved from http://www.thinkcycle.org

Glesbygdsverket. (2002). *Service och infrastruktur i samiska sommarvisten*. Östersund, Sweden: Glesbygdsverket.

Haetta, O. M. (1993). *The Sami: An indigenous people of the Arctic*. Karasjok, Norway: Davvi Girji o.s.

Jernsletten, J-L., & Klokov, K. (2002). *Sustainable reindeer husbandry*. Tromsö, Norway: Center for Sámi Studies.

Keller, E. F. (1992). *Secrets of life, secrets of death: Essays on language, gender and science*. New York; London: Routledge.

Kråik, M. (2002). Sámi women equal rights—yesterday and tomorrow. *Taking Wing Conference Report. Reports 2002:12eng*. Helsinki: Ministry of Social Affairs and Health.

Lindgren, A., Doria, A., & Schlen, O. (2004, August). *Probabilistic routing in intermittently connected networks*. The First International Workshop on Service Assurance with Partial and Intermittent Resources (SAPIR 2004), Fortaleza, Brazil.

Lindgren, A., & Belding-Royer, E. M. (2005, July). *Multi-path admission control for mobile ad hoc networks*. The Second Annual International Conference on Mobile and Ubiquitous Systems: Networking and Services (MobiQuitous 2005), San Diego, CA.

Lindgren, A., & Doria, A. (2005). *Probabilistic routing protocol for intermittently connected networks (Internet draft)*. Retrieved from draft-lindgren-dtnrg-prophet-00.txt

Mörtberg, C. (2003). Heterogeneous images of (mobile) technologies and services: A feminist contribution. *NORA, 11(3), 158-169*.

Trojer, L. (2002). *Genusforskning inom teknikvetenskapen: en drivbänk för forskningsförändring*. Stockholm: Högskoleverket.

Udén, M. (2002). The impact of women on engineering: A study of female engineering students' thesis topics. *International Journal of Engineering Education, 18*(4), 458-464.

Ulvevadet, B., & Klokov, K. (Eds.). (2004). *Family based reindeer herding and hunting economies, and the status of management of wild reindeer/ caribou populations*. Tromsø, Norway: Centre for Saami Studies.

Vehviläinen, M. (2002). Gendered agency in information society: On located politics of technology. In M. Consalvo & S. Paasonen (Eds.), *Women and everyday uses of the Internet: Agency and identity* (pp. 275-291). New York: Peter Lang Publishing.

Wikipedia. (2006). *Information technology*. Retrieved from http://en.wikipedia.org/wiki/Information_technology

Working Group on Internet Governance (WGIG). (2005). *Draft WGIG issue paper on social dimensions and inclusion.* Retrieved from www.wgig.org/docs/WP-SocialDimensions.pdf

World Summit on the Information Society (WSIS). (2006). *First phase of the WSIS: Genevea plan of action* (ITU document WSIS-03/GENEVA/DOC/5-E). Retrieved from http://www.itu.int/wsis/documents/doc_multi.asp?lang=en&id=1160|0

KEY TERMS

Innovation System: A network of organizations, people and rules within which innovative exploitation of technology and other knowledge take place.

Interplanetary Networking Research Group: A former working group within the Internet Research Task Force that developed the bundling architecture for Delay Tolerant Networking.

Information Technology (IT) or Information and Communications Technology (ICT): The technology required for information processing. "In particular, ... the use of electronic computers and computer software to convert, store, protect, process, transmit and retrieve information" (Wikipedia, 2006).

Sámi: The indigenous people of Scandinavia. According to official estimations, there are 70,000 to 80,000 Sámi.

Sámi Village: Grazing community whose members are entitled to let their reindeer graze within its area.

Working Group on Internet Governance (WGIG): Established by the United Nations Secretary-General in order to present a report "for consideration and appropriate action for the second phase of the World Summit on the Information Society WSIS in Tunis 2005" (World Summit on the Information Society [WSIS], 2006).

ENDNOTES

[1]   English: Woman in the Sámi Village.

[2]   This popular understanding has been frequently referred to in marketing, for instance, in promoting laptops as enabling working from "any" location—for example, from home—and also referred to in public policy documents. For instance, the Swedish Government's Proposition 1999/2000:86, "An information society for all" stated that: "IT represents a new base technology comparable to e.g. electricity. It is characterised by speed and interaction, and it is *limitless*" (our italics).

The Influences and Responses of Women in IT Education

Kathryn J. Maser
Booz Allen Hamilton, USA

INTRODUCTION

This article highlights findings from an empirical study that explores the nature of female underrepresentation in information technology. Specifically, this research focuses on (a) identifying key sociocultural factors that can facilitate the pursuit of IT at the undergraduate level, and (b) testing Trauth's (2002) Individual Differences Theory of Gender and IT through a comparison of female responses to the social construction of IT. To answer the author's research questions, interviews were conducted with 10 female seniors in an IT department at an American university in the mid-Atlantic region (MAU).[1]

Although experiences with social factors vary, comparing the stories of women who have successfully navigated their way into and through an IT undergraduate degree program reveals common influences and motivations. In addition, though some common factors may facilitate female entry into the field, the Individual Differences Theory of Gender and IT explains that women will react differently to the social constructions of gender and IT. By gaining a better understanding of the gender imbalance, applying appropriate theories to explain the problem, and uncovering the challenges that women of our society face in their entry to the field of IT, collegiate programs can more effectively implement strategies that will improve the recruitment and retention of female students.

BACKGROUND

IT-related undergraduate degree programs such as computer science (CS), management information systems (MIS), and information-science and -technology programs are important gateways to the IT industry, providing valuable exposure and experience to students interested in pursuing IT careers.

Research suggests that women are entering undergraduate IT programs in smaller numbers (e.g., Camp, 1997; Freeman & Aspray, 1999) and may be doing so with less formal and informal IT experience (e.g., Craig & Stein, 2000; Fisher, Margolis, & Miller, 1997; Margolis & Fisher, 2002; Teague, 1997). Thus, education at the undergraduate level is critical in the foundation of their skills, their interests in IT, and their pursuit of work in the field. Moreover, actively recruiting and retaining females in IT-related undergraduate degree programs can have a significant impact on the diversification of the IT workforce. As Margolis and Fisher (2002, p. 3) explain, "women must be part of the design teams who are reshaping the world, if the reshaped world is to fit women as well as men."

This study first focuses on identifying sociocultural factors influential in women's decisions to pursue IT at the undergraduate level. The social-construction perspective of gender and IT explains that, reflective of the social norm in America, cultural expectations and influences often convey the message that women are unsuitable for the IT world (e.g., Trauth, 2002; von Hellens, Nielsen, & Trauth, 2001). By the time young women reach college, there is evidence of the effects of these social norms and expectations. For example, in the years prior to college, certain studies have revealed that, in comparison with males, females exhibit lower levels of self-efficacy in computing, are less likely to explore computing independently through informal channels (e.g., within peer groups, computer camps, and clubs), and elect to take advanced computing courses less frequently; in addition, some women have misconceptions about the IT workforce and IT work (e.g., Beise & Myers, 2000; Craig & Stein, 2000; Fisher et al., 1997; Margolis & Fisher, 1997, 2002; Nielsen, von Hellens, & Wong, 2000; Symonds, 2000; Teague, 1997; von Hellens, Nielsen, Doyle, & Greenhill, 1999; Woodfield, 2000).

An examination of the factors that enable women to confront and circumvent these social barriers is an important part of understanding the gender imbalance; however, it should not be assumed that all women have the same reactions to these barriers. The Individual Differences Theory of Gender and IT embraces the notion that gender is a fluid continuum rather than a dichotomy. This theory focuses on women as individuals, having distinct personalities, experiencing a range of sociocultural influences, and therefore exhibiting a range of responses to the construction of the IT field (Trauth, 2002). Comparing and contrasting females' responses to the social construction of IT tests the individual-difference theory of gender and IT.

RESEARCH APPROACH

This research focuses on women at a critical point of IT entry: the undergraduate level of education. In examining the trends of female underrepresentation discussed in the literature and the theoretical perspectives used to explain the problem, the following research questions emerged.

1. What significant sociocultural factors in the lives of women are influential in their pursuit of IT at the college level?
2. How similar are female responses to the social construction of gender and IT?

To investigate these questions, in-depth interviews were conducted with a sample of 10 female seniors in MAU's IT department in the spring of 2003. The IT department at this university was chosen because of its proactive stance with respect to the recruitment and retention of women students. The department also has a diversity committee and a student organization, Women in Information Technology (WIT), that was established to provide support and mentoring for female students in the program. At the time of these interviews, the student enrollment in the department was 21% female. Interviews were open ended and lasted approximately 40 minutes in duration. The qualitative format was selected as it was most appropriate for capturing in detail the participants' broad range of influences and experiences. Interview questions were derived from the themes of family background, educational history, personal traits and interests, discovery and selection of the IT program at MAU, experiences in the program, and future plans.

FINDINGS

Comparing Sociocultural Influences

In comparing participant experiences, the women study reported modest levels of formal education and informal experimentation in IT; these experiences made little impact on their decisions to pursue the IT degree. Participants consistently described their education in high-school computer classes as basic. The two women who elected to exceed the minimum computing requirements and complete C^{++} classes felt they lacked a clear understanding of the extent to which the language could be applied in real-world scenarios. On the whole, these high-school computer classes served the purpose of familiarizing these women with computers, but did little more. Although a few of the participants were aware of certain IT careers, the majority did not have a clear and complete understanding of the IT field prior to college. In terms of computing exposure and use in the home, experiences were quite consistent and corresponded strongly with the literature (e.g., Margolis & Fisher, 2002). The primary functions of home computers were education and communication: word processing for homework, and e-mail and instant messaging for chatting with friends.

Family influence and encouragement was a key social factor identified as impacting the participants' decisions to pursue the IT program at MAU. Despite differences in family environments, common to each of the women's experiences was a high level of parental academic support, encouragement, and expectation. The participants had mothers, fathers, and siblings that were, to varying degrees, actively involved in their academic careers. Many of the participants were pushed for academic achievement, and many were also specifically encouraged to choose the IT program at MAU. Other participants reported less direct academic involvement, though expectations and encouragement remained strong. This encouragement, with little exception, helped the women create high personal expectations for their future careers.

In developing an understanding of gender roles, the participants grew up in homes where mothers and fathers assumed diverse roles and responsibilities. Regardless of their family environments (e.g., dual income, sole breadwinner, single parent), the children in these households, male and female, were treated equally. The participants were raised believing they were able to achieve whatever they wished, and that gender was not a factor that should steer them in one direction or deter them from another. This confidence facilitated their selection of MAU's IT program. This confidence was also revealed in the way the women dealt with male domination in the IT department; half of the participants reported being largely unaware of the gender imbalance, and the remaining half, though initially intimidated by the experience gap, learned quickly that they were equally capable of achieving success in the program.

The presence of role models was another significant factor that influenced participants' decisions to pursue IT. Lacking a significant amount of formal and informal experience in IT, the women's IT understanding was strongly correlated with the presence of a role model. Consistent with the literature (e.g., Beise & Myers, 2000; Craig & Stein, 2000; Symonds, 2000; Teague, 1997), IT role models affected how the women perceived and related to the field, exposed them to opportunities in the field, and helped them develop an interest in IT work. The majority of females with IT role models entered college with a general understanding of the field they wanted to pursue.

Specific characteristics of MAU's IT program were also influential in the participants' decisions to pursue IT. In particular, the program was perceived by the women as new, exciting, cutting-edge, and as offering a wide range of both technical and nontechnical business-related educational opportunities. A number of the participants also described positive experiences with IT program advisors. Finally, the fact that the IT program was formed with close ties to industry and emphasized postgraduation employment opportunities was a principal factor in the participants' decisions.

Exploring Individual Responses

Findings regarding personal future expectations and outlooks provide clear support for the Individual Differences Theory of Gender and IT. Individual differences were most clearly revealed through the participants' outlooks on female participation in IT and expectations for their own futures.

Although the participants arrived at the same source of IT education, differences in their interests, values, and priorities will cause some women to maintain their participation in the IT industry, and others to reevaluate and change careers. Half of the women in this study strongly believed IT was the field they wanted to pursue, did not envision themselves switching careers, and were committed to balancing their work and family lives. The remaining half of the participants expressed uncertainty over whether or not they would remain in the IT field. Some of these women were unsure they would enjoy the IT industry and indicated a desire to explore other fields, while others anticipated an incompatibility in balancing an IT career with expectations for family commitments in the long term.

Differences in formative experiences have also led these women to hold a variety of opinions about female participation in IT: differences in their explanations of the gender imbalance, and differences in their opinions on how and whether or not the issue should be addressed. In offering explanations for the underrepresentation of females in IT, opinions were split between those who believed women were simply less interested in the technical subject matter, and those who believed the imbalance persists due to a lack of female exposure to the field. Additionally, some of the women felt that the gender imbalance in IT was a significant issue and that, in certain situations, intervention was necessary to provide support for females. However, the perspective most frequently described by this group of participants was the belief that hypersensitivity about gender issues can be problematic and can place unnecessary emphasis on the division between the sexes.

CONCLUSION

The analysis of females in the IT program at MAU reveals high levels of encouragement from family, exposure to IT through role models, and balanced perceptions of gender roles and expectations to be the primary social factors facilitating participant decisions to pursue this IT-related undergraduate degree program. Specific characteristics of MAU's IT program were also influential in attracting the participants to the department. In particular, the program was perceived by the women as a cutting-edge program with a comprehensive curriculum, and as capable of providing access to great career opportunities. Finally, in investigating participant outlooks on female participation in IT and expectations for their own futures, this research found strong support for the Individual Differences Theory of Gender and IT.

The participants' stories also revealed that their high schools were an untapped source of potential influence, exposure, and encouragement. Findings suggest that improved IT education in high school could provide females with exposure to the field they might not receive elsewhere, which would greatly support the development of interest in IT. Findings also indicate that high-school IT courses should be developed with an emphasis on both technical knowledge and real-world business applications of the subject matter. Additionally, women should be actively encouraged to participate in these courses and explore computing beyond minimum requirements. Finally, high school would be an appropriate place to provide young women with exposure to female role models in the industry, a factor that is particularly important when role models are not present within the home.

Finally, in discussing the women's expectations for the future, implications for female retention in the IT industry emerged. As exemplified by this participant group, there are both unpredictable success stories and unexpected stumbling blocks. One participant, for example, entered MAU with great uncertainty and little career orientation. Having been strongly encouraged by her father to achieve academically and pursue the IT program, she secured a summer internship where she discovered a specific technical focus within IT she enjoyed and became intent on pursuing. In contrast, another woman described growing up as a tomboy and was the only participant to exemplify the "boy-wonder syndrome"[2] (e.g., Margolis & Fisher, 1997). Her conditions for IT selection were very favorable: a strong interest in technology, exposure to IT, strong role models, encouragement, and confidence. Yet, she felt her participation in the industry was temporary because of her desire to pursue other interests she viewed as more compatible with the kind of family environment she wants to create. To increase female participation in IT, we should first strive to make conditions for the selection of IT education most favorable so that with exposure and experience in the field, women may find their niche among the many lines of work that IT has to offer. Retention strategies at the undergraduate and industry levels are the next step; however, they are beyond the scope of this research.

REFERENCES

Beise, C., & Myers, M. (2000). A model for examination of underrepresented groups in the IT workforce. In *Proceedings of the 2002 ACM SIGCPR Computer Personnel Research Conference*, Kristiansand, Norway (pp. 106-110).

Camp, T. (1997). The incredible shrinking pipeline. *Communications of the ACM, 40*(10), 103-110.

Craig, A., & Stein, A. (2000). Where are they at with IT? In E. Balka & R. Smith (Eds.), *Women, work and computerization: Charting a course to the future* (pp. 86-93). Boston: Kluwer Academic Publishers.

Fisher, A., Margolis, J., & Miller, F. (1997). Undergraduate women in computer science: Experience, motivation and culture. In *Proceedings of the 1997 ACM SIGCSE Technical Symposium*, San Jose, CA (pp. 106-110).

Freeman, P., & Aspray, W. (1999). *The supply of information technology workers in the United States*. Washington, DC: Computing Research Association.

Margolis, J., & Fisher, A. (1997). Geek mythology and attracting undergraduate women to computer science. *Impacting Change through Collabora-*

tion: Proceedings of the Joint National Conference of the Women in Engineering Program Advocates Network and the National Association of Minority Engineering Program Administrators. Retrieved from http://www.cs.cmu.edu/~gendergap/working.html

Margolis, J., & Fisher, A. (2002). *Unlocking the clubhouse: Women in computing.* Cambridge, MA: The MIT Press.

Nielsen, S., von Hellens, L., & Wong, S. (2000). The game of social constructs: We're going to WinIT! In *Panel Presentation: Proceedings of the International Conference on Information Systems,* Brisbane, Australia.

Symonds, J. (2000). Why IT doesn't appeal to young women. In E. Balka & R. Smith (Eds.), *Women, work and computerization: Charting a course to the future* (pp. 70-77). Boston: Kluwer Academic Publishers.

Teague, J. (1997). A structured review of reasons for the underrepresentation of women in computing. In *Proceedings of the Second Australasian Conference on Computer Science Education,* Melbourne, Australia (pp. 91-98).

Trauth, E. M. (2002). Odd girl out: An individual differences perspective on women in the IT profession. *Information Technology & People, 15*(2), 98-118.

Trauth, E. M., Quesenberry, J. L., & Morgan, A. J. (2004). Understanding the under representation of women in IT: Toward a theory of individual differences. *Proceedings of the ACM SIGMIS Computer Personnel Research Conference,* Tucson, AZ (pp. 114-119).

Von Hellens, L., Nielsen, S., Doyle, R., & Greenhill, A. (1999). Bridging the IT skills gap: A strategy to improve the recruitment and success of IT students. In *Proceedings of the 10th Australasian Conference on Information Systems,* San Diego, CA (pp. 116-120).

Von Hellens, L., Nielsen, S., & Trauth, E. M. (2001). Breaking and entering the male domain: Women in the IT industry. *Proceedings of the ACM SIGCPR Computer Personnel Research Conference.*

Woodfield, R. (2000). *Women, work and computing.* Cambridge, MA: Cambridge University Press.

KEY TERMS

Boy-Wonder Theory: The belief that true scientific talent, interest, and achievement must be exhibited early in one's lifetime.

Individual Differences Theory of Gender and IT: The perspective of gender and IT that argues that "women as individuals experience a range of different socio-cultural influences which shape their inclinations to participate in the IT profession in a variety of individual ways" (Trauth, 2002, p. 103).

Social-Construction Perspective of Gender and IT: The perspective of gender and IT that attributes female underrepresentation to societies' incompatible constructions of femininity and the IT field.

Tomboy: A female considered boyish or masculine in behavior or manner.

ENDNOTE

[1] The name of this university has been changed.

[2] As Sheila Tobias explains, "one of the characteristics of the ideology of science is that ... both scientific talent and interest come early in life—'the boy wonder syndrome'. If you don't ask for a chemistry set and master it by the time you're five, you won't be a good scientist" (Margolis & Fisher, 1997).

Institutional Characteristics and Gender Choice in IT

Mary Malliaris
Loyola University Chicago, USA

Linda Salchenberger
Northwestern University, USA

INTRODUCTION

While the issue of attracting women to information technology professions has been studied extensively since the 1970s, the gender gap in IT continues to be a significant social and economic problem (Thom, 2001). Numerous research studies have been conducted to understand the reasons for the gender gap in IT (Gurer & Camp, 2002; Sheard, Lowe, Nicholson, & Ceddia, 2003; von Hellens, Nielsen, & Beekhuyzen, 2004). Universities and colleges have developed a variety of programmatic efforts to apply gender gap research results, implementing strategies that increase female undergraduate enrollment in computer science programs (Wardle & Burton, 2002). Yet, individual successes have not translated into any significant change in the overall percentages of women choosing IT. An analysis of current choices of women in their selection of four-year undergraduate institutions reveals yet another alarming trend—young women are not choosing to study IT at the traditional academic four year institutions that would best prepare them for the IT professional careers of the future.

To complicate matters, the information technology job market is changing rapidly. For example, some well-documented IT trends that are causing such shifts are outsourcing, the commoditization of IT, the effect of the dot com bust on the job market, and most importantly, the integration of IT into the fundamental economic, social and cultural fabric of our society. IT now permeates every aspect of professional work, even the traditional female-oriented occupations such as nursing and teaching. This integration of IT into the professions must guide the development of a new set of strategies to insure that women have equal opportunities and access to

the benefits of an education that prepares them for professional careers. It is in the best interest of the IT profession and our society in general to help young women make choices that include the pursuit of information technology.

BACKGROUND

The under representation of women in IT is a critical issue of equity and access for women due to the pervasiveness of computing in our society, the many economic opportunities afforded those who have technology skills and knowledge, and value of diversity for this profession (Cohoon, 2003). Although job opportunities in technology companies and technology-oriented industries have recently declined, the need for advanced technology skills in mainstream business careers and entrepreneurship remains critical (Thibodeau & Lemon, 2004). Nearly 75% of future jobs will require the use of technology, 8 of the 10 fastest growing occupations between 2000 and 2010 will be computer-related. The annual mean salary for computer and technology occupations remains significantly above average compared to all occupations (U.S. Department of Labor, 2004). Thus, the IT gender gap translates into salary and employment inequities.

Table 1 shows that in 1996, women were 41% of the IT workforce compared to 34.9% in 2002, yet they accounted for 46% and 46.6% of the overall workforce in 1996 and 2002, respectively. Note that, in 1996 and 2002, the higher percentage of females was due largely to greater numbers of women in Data Entry and Computer Operator positions, jobs that required less formal education and experience, and provide lower pay. In fact, in both years, women

Table 1. Women in the IT workforce vs. overall workforce (1996 and 2002)

2002 Total Employed (thousands)	2002 Total	2002 % Men	2002 % Women	1996 % Men	1996 % Women
Electrical and electronic engineers	677	89.7	10.3	92	8
Computer systems analysts and scientists	1,742	72.2	27.8	72	28
Operation and systems researchers and analysts	238	51.3	48.7	57	43
Computer programmers	605	74.4	25.6	69	31
Computer operators	301	53.2	46.8	40	60
Data entry keyers	595	18.3	81.8	15	85
Total IT occupations	4,158	65.1	34.9	59	41
All Occupations	136,485	53.4	46.6	54	46

Source: Bureau of Labor Statistics

Table 2. Computer/information science bachelor's degrees awarded

Year	Degrees awarded			
	Total	Men	Women	% Women
1986	42,195	27,069	15,126	35.8
1987	39,927	26,038	13,889	34.8
1988	34,896	23,543	11,353	32.5
1989	30,963	21,418	9,545	30.8
1990	27,695	19,321	8,374	30.2
1991	25,410	17,896	7,514	29.6
1992	24,958	17,748	7,210	28.9
1993	24,580	17,629	6,951	28.3
1994	24,553	17,533	7,020	28.6
1995	24,769	17,706	7,063	28.5
1996	24,545	17,773	6,772	27.6
1997	25,393	18,490	6,903	27.2
1998	27,674	20,235	7,439	26.9
2000	37,388	26,914	10,474	28.0
NOTE:	Data not available for 1999			

account for over 81% of the data entry positions. The current lack of women in the IT workforce is in part a consequence of women not choosing IT undergraduate degree programs or dropping out of these majors.

One traditional path into the IT profession is the completion of an undergraduate degree in Information Technology. However, the percentage of undergraduate degrees awarded to women in computer science and information technology as reported by the National Center for Education Statistics has declined since 1986 (See Table 2). It is well known that one approach to moving women into IT is through the educational pipeline, that is, motivating young women to explore these career paths early in life and to choose IT degree programs.

Despite the benefits of professional technology careers and the advancements of women in many other fields, little progress has been made in moving women through the educational pipeline in computer science (Camp, 1997). In fact, less than 33% of participants in computer courses and related activities in high schools are girls (AAUW, 2000).

The extensive literature on this topic (Beyer, Rynes, & Haller, 2004, Gurer & Camp, 2002; Klawe & Leveson, 1995) provides us with many reasons why IT is not attractive to young women. Potential causes include: unsupportive academic environment, the perception of computing as a male-oriented profession, gender differences in how students assess their own performance, lack of role models and insufficient critical mass of female students and faculty to build community.

Colleges and universities face additional challenges in recruiting women. Because of the pipeline issue, women are often less experienced in computing when they enter college, computer science department cultures and software are typically male-oriented and don't appeal to women, and there is a lack of visibility regarding the social value of computing that would appeal to women. Furthermore, while some institutions have been successful in recruiting females to undergraduate computer science programs (Fisher & Margolis, 2002; Roberts, Kassianidou, & Irani, 2002), the percentage of women in these disciplines for most institutions continues to decline (ITAA, 2002). Cohoon (2001) argues that, based on her investigation of the University of Virginia's CS department, the characteristics and practices of computer science departments affect female retention at the undergraduate level and inherent female characteristics are an insufficient explanation of women's under representation in computer science. In fact, women themselves tell us why they are not choosing IT, often indicating they find IT uninteresting or perceive that it is more difficult academically than other professions such as surgery and law (Weinberger, 2004). Individual characteristics and environmental influences are explored to provide a perspective on women in IT in Trauth (2002) in order to better understand women's lack of involvement in IT.

Numerous recommendations to assist educational institutions in attracting women to undergraduate degree programs in IT appear in the extensive body of research on gender and IT (Baum, 1990; Cohoon, 2003; Cuny & Aspry, 2000; Wardle & Burton, 2002). Colleges and universities and academic departments that have been successful in increasing the number of women in technology have shared their strategies for recruiting and retaining female undergraduate students (Margolis & Fisher, 2002). They encourage institutions to establish and fund university programs and policies to expand the recruitment pool, provide a supportive climate with appropriate student services, broaden (not weaken) admission requirements, offer bridge programs, educate parents and teachers on gender issues, expand undergraduate research opportunities, and build supportive communities of learning through role models and mentoring.

Nevertheless, these successes are not widespread and the question of why women are not choosing computing as a major is a question that may benefit from institutional research. Exploring the characteristics of institutions that have been successful in attracting women may help us to better understand the choices that women are making when they do choose to pursue an undergraduate degree in IT.

USING DATA MINING TO DISCOVER WOMEN'S EDUCATIONAL CHOICES

Data mining refers to a set of techniques used to search large amounts of data for patterns. Rather than specifying a hypothesis, selecting a sample, and performing a test of the hypothesis, data mining instead searches the data for patterns that occur within it. Thus, it is a set of data driven techniques. The knowledge contained within the data set gives shape to the model. Three of the most used data mining techniques are cluster analysis, association analysis, and decision trees. In this analysis, we only used descriptive analysis as a first step in analyzing our data. Descriptive analysis was used to search a

large data set for patterns and associations related to educational choices of women.

When a young woman leaves high school for college, she is making a career decision that influences the path she takes for the rest of her life. Not only the choice of major, but also the choice of the institution can have long-lasting effects on the possibilities available thereafter. The following preliminary research study compares institutions where women, in large numbers and in very small numbers, have chosen to concentrate in IT-related majors.

This study uses IPEDS (Institutional Post-Secondary Educational Statistics) data for 2000-2001. The purpose of this project is to discover factors associated with the type of institutions that have demonstrated success in attracting women to IT programs, using data mining techniques. The comprehensive IPEDS data set contains variables related to characteristics of the institution, including enrollment numbers by academic discipline using a variable called the CIPCODE (Classification of Instructional Program Code) reported by institutions of higher education. CIP codes indicating specializations in information systems, computer science and information technology were used in this study. The number of female students across these IS/IT CIP codes was calculated to give a total for each institution. This sum was divided by the total number of students at the institution, then multiplied by 100, to generate our target variable: Percent of women in IT/IS at the institution. This yielded a total of 985 institutions for our analysis. Institutions that had missing values on the variables considered, including both Not Reported and Not Applicable, were deleted.

The variables from the IPEDS data that we examined for this study, based on previous studies, included the following sixteen variables: accreditation of the institution, requirement of secondary school GPA for applicants, requirement of test scores for applicants, affiliation of institution, athletic association of the institution, provision of a meal plan, type of calendar system, highest degree offered, whether institution has a hospital, degree of urbanization, region code, on-campus housing, sector of the institution, availability of on-campus jobs, placement services, and variation of tuition. Four-

Table 3. Characteristics of the top and bottom institutions with respect to female representation in IT programs

Group		Bottom	Top
% Women in IT		< .0599	> .7817
Total in Group		119	121
Avg # Students		7947	4134
Category			% in Category
Affiliation	Private, for profit	0.0	30.6
	Private, NFP, not relig.	21.0	31.4
	Private, NFP, religious	29.4	24.0
	Public	49.6	14.0
Highest Degree	Bachelors	11.8	33.1
	Masters	49.6	37.2
	Doctorate	38.7	29.8
Accredited	No	11.8	22.3
	Yes	88.2	77.7
Placement	No	3.4	22.3
	Yes	96.6	77.7
Dorms	No	2.5	33.9
	Yes	97.5	66.1
GPA Required	Neither	5.9	25.6
	Recommend	18.5	14.9
	Required	75.6	59.5
Test Required	Neither	0.0	10.7
	Recommend	2.5	17.4
	Required	97.5	71.9
Ath. Assoc. Member	No	3.4	35.5
	Yes	96.6	64.5

year institutions that offered an undergraduate degree in information technology in 2000-01 composed the data set to be analyzed.

All institutions were sorted by the value of the target variable. The group with the greatest percentage of women majoring in IT/IS areas is hereafter called the top group. The bottom group includes institutions with the smallest number of women majoring in IT/IS areas. In the top group of schools, the proportion of females is 0.78% or higher and there are 121 schools in this category. The bottom group is the set of schools with the lowest female representation, that is, under .06% and it contains 119 colleges and universities. Though the sizes of these two groups are about the same, they have very different sets of values on the IPEDS category variables as shown in Table 3. In our analysis, either we will only focus on those which provided new insights or great contrast between the two groups.

What are the characteristics of institutions that have the greatest percentage of women IT majors in their overall undergraduate population? The vari-

ables which show the greatest differences between the groups are: size, affiliation, highest degree offered, accreditation, placement services offered, dorms, athletic association, and entrance requirements. From the data, we observe that 22.3% of the top group are non-accredited while only 11.8% of the bottom group are non-accredited. Of the schools with the lowest female representation, 49.6% are public institutions, 0% are for profit, 21% are private, non-profit, nonreligious and 29.4% are private, non-profit, religious institutions. For those with the most females in IT, these numbers are 30.6% public, 31.4% for profit, 24% private, nonprofit, nonreligious and 14% private, nonprofit, religious. Thus, for the group with the best female representation, they break out into fewer publics (14% vs. 49.6%) and more private nonreligious institutions (31.4% vs. 21%). For the bottom group, 97% of the schools belong to an athletic association, but for the top group, this drops to 64.5%. We also note that 97.5% of the bottom group offer dormitories, while only 66% of the top group are schools are residential. The data also shows that 22.3% of the schools with high representation of women do not offer placement services as compared to 3.4% for schools with lower numbers of females in IT. Finally, 33% of the schools that have higher concentrations of women in IT offer nothing higher than a bachelors' degree, while 88.2% in the group not chosen by women offer masters' and doctoral programs.

Looking at this data in total, we see that the top group institutions tend to be smaller in size, private, and do not offer doctoral degrees. They are less likely to have any athletic association, to require an admissions test or secondary school GPA, or to provide dormitories.

FUTURE TRENDS

As many of the previous studies examining the gender gap indicate, this is a complex problem and one that requires a comprehensive, yet focused, institutional, and departmental strategy in order to bring about significant change. One critical component that will impact the agenda to increase IT enrollments is the changing face of the IT profession over the next decade. The traditional approaches employed must be supplemented with new strategies, addressing future IT trends such as outsourcing, the commoditization of IT, the effect of the dot com bust on the job market, the integration of IT into the fundamental economic, social and cultural fabric of our society, and the cycles of interest in academic areas of the current and future generations of undergraduate students.

There are multiple challenges—understanding why girls and women make the educational choices they do, attempting to change the culture to help them make choices that will prepare them for technology careers, and predict how the IT profession will change in the short run and the long run. In the short run, we see that the global economy and new impetus on IT to deliver business value are changing jobs from those that require traditional programming and software development skills to jobs such as project management and application integration that require teamwork and organizational skills (Thibodeau, 2004). In the long run, we need to design new ways to educate girls and women on the opportunities of using IT in many non-IT professions and engage them in interesting, challenging and meaningful work requiring technology skills. For example, areas such as nursing, teaching, marketing, and human resource management that have a large female professional staff are becoming increasingly IT-oriented. Specific examples are medical informatics in healthcare, customer relationship management in marketing, and meeting the needs of K-12 tech-savvy students.

Table 4. Recommendations for attracting female students to undergraduate IT programs

- Offer pre-college experiences such as summer technology camps
- Educate local high school counselors and teachers about women and IT careers
- Align 4 year programs with community college programs
- Provide "gap" educational programs to prepare students coming from institutions with different academic standards
- Develop baccalaureate programs that provide the skills needed for today's IT workforce
- Offer an array of ways to major in technology—CS, MIS, IS
- Hire faculty and staff who can provide insights into the variety of IT-related careers

CONCLUSION

Based on the research presented here, there are several conclusions and consequent recommendations for reducing the gender gap in IT. First and foremost, we need to educate girls and young women while they are in elementary and high school to help them make better choices regarding their future and lives beyond college. The pipeline begins at a very young age. Stereotypes and barriers are established early in life. If they are getting their degrees at unaccredited, non-PhD granting schools, we will not see them at research universities in the future, adding to further decline in the pipeline. Specific recommendations are shown in Table 4, incorporating what we have learned about the gender issue.

First and foremost, we must provide interventions early in girls' lives through programs that reach them and their teachers, parents, and counselors. If young girls are choosing to attend nontraditional and two-year programs, then we must connect with them at the community college level and align ourselves with these institutions. Aggressive recruiting and then providing "gap education", i.e., classes, workshops or seminars to bridge the gap between their educational background and those of the students who follow a more traditional path is the next recommendation. Finally, the ultimate challenge will come for all IT students as we better align our undergraduate curricula with the skills and knowledge needed to succeed in the next generation of computing.

REFERENCES

American Association of University Women (AAUW). (2000). *Tech-saavy: Educating girls in the new computer age.* Washington, DC: AAUW Press.

Beyer, S., Rynes, K., & Haller, S. (2004). Deterrents to women taking computer science courses. *IEEE Society and Technology Magazine, 23*(1), 21-28.

Camp, T. (1997). The incredible shrinking pipeline. *Communications of the ACM, 40*(10), 103-110.

Cohoon, J. (2001). Toward improving female retention in the computer science major. *Communications of the ACM, 44*(5), 108-114.

Cohoon J. (2003, May 3-10). Must there be so few? Including women in CS. *ICSE '03: Proceedings of the 25ᵗʰ International Conference on Software Engineering* (pp. 668-674), Portland, OR.

Cuny, J., & Aspry, W. (2002) Recruitment and retention of women graduates in computer science and engineering. *SIGSCE Bulletin, 34*(2), 168-174.

Fisher, A., & Margolis, J. (2002). Unlocking the clubhouse: The Carnegie Mellon experience. *Communications of the ACM, 34*(2), 79-83.

Gurer, D., & Camp, T. (2002). An ACM-W literature review on women in computing. *SIGSCE Bulletin, 34*(2), 121-127.

Information Technology Association of America. (2003). *Report of the ITAA Blue Ribbon Panel on IT Diversity*. Presented at the National IT Workforce Convocation, May 5, 2003, Arlington, VA. Retrieved from http://www.itaa.org/workforce/docs/03divreport.pdf

Klawe, M., & Leveson, N. (1995). Women in computing: Where are we now? *Communications of the ACM, 38*(1), 29-35.

Margolis, J., & Fisher, A. L. (2002). *Unlocking the clubhouse: Women in computing*. Cambridge, MA: MIT Press.

National Center for Education Statistics. (2001). *Digest of Education Statistics*. Retrieved from http://nces.ed.gov/pubsearch/pubsinfo.asp?pubid=2002130

Roberts, E. S., Kassianidou, M., & Irani, L. (2002). Encouraging women in computer science. *SIGSCE Bulletin, 34*(2), 84-88.

Sheard, J., Lowe, G., Nicholson, A., & Ceddia, J. (2003). Tackling transition: Exposing secondary school students to tertiary it teaching and learning. *Journal of Information Technology Education, 2,* 165-180.

Thibodeau, P., & Lemon, S. (2004, March). R&D starts to move offshore: Outsourcing evolves beyond low-wage programming jobs. *Computerworld, 38*(9), 16.

Thom, M. (2001). *Balancing the equation: Where are the women & girls in science, engineering, and technology*. Washington, DC: National Council for Research on Women.

Trauth, E. (2002). Odd girl out: The individual differences perspective on women in the IT profession. *Information Technology and People, 15*(2), 98-117.

U.S. Department of Labor, Bureau of Labor Statistics. (2004). *Tomorrow's jobs*. Occupational Outlook Handbook 2004-2005 Edition. Retrieved from http://www.bls.gov/oco/oco2003.htm

U.S. Bureau of Labor Statistics, Employment, and Earnings. (2003). *Table # 619, Employment by Industry: 1980 to 2002*. U.S. Census Bureau, Statistical Abstract of the United States, 2003. Retrieved from http://www.census.gov/prod/2004pubs/03statab/labor.pdf

U.S. Bureau of Labor Statistics, Employment, and Earnings. (2003). *Table # 620, Employment Projections by Industry: 2000 to 2010*. U. S. Census Bureau, Statistical Abstract of the United States, 2003. Retrieved from http://www.census.gov/prod/2004pubs/03statab/labor.pdf

von Hellens, L., Nielsen, S., & Beekhuyzen, J. (2004). An exploration of dualisms in female perceptions of IT work. *Journal of Information Technology Education, 3,* 103-116.

Wardle, C., & Burton, L. (2002). Programmatic efforts encouraging women to enter the information technology workforce. *SIGSCE Bulletin, 34*(2), 27-31.

Weinberger, C. (2004, Spring). Just ask! Why surveyed women did not pursue IT courses or careers. *IEEE Society and Technology Magazine, 23*(2), 28-35.

KEY TERMS

Accredited Institution: Institution that is accredited by national institutional or specialized accrediting agency that establishes operating standards for educational or professional institutions and programs, determine the extent to which the standards are met, and publicly announce their findings.

Computer Science: Study of data, computation, and information processing, including methodologies, processes, hardware, software, and applications.

Data Mining: Search of large databases for patterns and trends using a variety of techniques implemented by computer software, sometimes referred to as KDD.

Gender Roles: Professional or social roles associated with males or females that are socially acceptable and considered to be the norm.

Gender Studies: Theoretical and empirical work that focuses on gender in society.

Information Technology: The technology associated with information processing, including computer hardware and software used to store, process and transmit data and information.

IPEDS: The Integrated Postsecondary Education Data System (IPEDS), established as the core postsecondary education data collection program for the National Center for Education Statistics, is a system of system of surveys designed to collect data from all primary providers of postsecondary education.

KDD: Knowledge discovery and data mining; finding applicable insights through the analysis of large amounts of data.

Pipeline: Channel that moves an object from start to finish; used metaphorically to represent the number of qualified individuals who move from one stage in the educational process to another.

Private, For-Profit Institution: A private institution in which the agency in control receives compensation other than wages, rent, or other expenses for the assumption of risk.

Private, Nonprofit Institution: A private institution in which the agency in control receives no compensation other than wages, rent, or other expenses for the assumption of risk. These include both independent and those affiliated with religious organizations.

Public Institution: An educational institution whose programs and activities are operated by publicly elected or appointed officials and supported primarily by public funds.

The Intersection of Gender, Information Technology, and Art

Linda Doyle
University of Dublin, Ireland

Maryann Valiulis
University of Dublin, Ireland

INTRODUCTION

The interdisciplinary field of art and technology is now well established in artistic and academic communities (Wilson, 2001). However, this article will focus on how the combination of technology and art can be used to facilitate the expression of thoughts, the experience of ideas and the explorations of concepts dealing with gender. A research project called the Art of Decision, which focuses on women in decision making, is used as a means of investigating the ways in which creative technologies can illuminate aspects of gender studies.

BACKROUND

Creative Technologies

In the context of the research presented here, information technology (IT) is defined very broadly as an entire array of mechanical and electronic devices that aid in the storage, retrieval, communication and management of information. It includes all computing technologies and mobile and fixed communication technologies, but it is not restricted to those areas. Smart materials that change attributes on the basis of input stimuli and that can be used to present and display information or react to information, holographic systems, sensors, audio technologies, image technologies, video technologies and many more are all of interest. In this article, the term "creative technology" is used to describe the combination of these types of technologies with artistic practices and methods or the use of these technologies in an artistic manner or in a mode that follows a particular artistic aesthetic. The use of technology for artistic expression is widespread, and while very many works of art can be of a political nature, the concept of using art and technology in the construction of purpose-build systems for exploring gender questions is novel.

Gender

Gender is a complex category of analysis that defies simple definition. It can be viewed as the result of socialization—the emphasis of 1970s/1980s second-wave feminist theorists (Nicholson, 1997) or more currently of performance, of the repetition of doing gender, "the repeated inculcation of a norm" (Salih, 1993, p. 139). This article endorses the view of gender as a result of the interaction between biology and the social environment—what Anne Fausto Sterling calls the "complex web" (Sterling, 2000). It endorses her repudiation of the sex-gender or nature-nurture divide that she claims fails to "appreciate the degree to which culture is a partner in producing body systems ..." (Sterling, 2005, p. 1516). This entry also reflects the view of Caroline Ramazanoglu, who takes "gender to include: sexuality and reproduction; sexual difference, embodiment, the social constitution of male, female, ... masculinity and femininity" (Ramazanoglu, 2002, p. 5). Finally, it appreciates the views of Alsop, Fitzsimons and Lennon, who hold a multifaceted view of gender that includes gender as a "feature of subjectivity," as "cultural understandings and representations of what it is like to be a man or a woman" and "as a social variable structuring the pathways of those so classified within society" (Alsop et al., 2002, p. 3). What must be emphasized in all these definitions is that gender intersects and interacts with other factors of identity, such as class, race or sexual orientation.

A definition of gender must include a theory of power. Gender is not a neutral concept, but rather, different degrees and kinds of power attach itself to genders in specific ways. Again, it is important to understand the power of gender as it intersects with all the other human differences. For example, in its simplest form, the traditional white male middle-class gender speaks of political power.

Power, however, is important in other ways. It is integral to the joining of creative technologies and gender, and in this context is defined as the power to produce, authorize and impart knowledge. Often, traditional science and technology have assumed the air of impartiality and objectivity, which gave them the veneer of having produced "authoritative knowledge." Joining gender to art and technology is to problematize questions of objectivity, authority and knowledge production.

Using new creative media in an exploration of gender and power opens up new possibilities for studying that relationship. This is of particular importance in an age that often considers it trendy to speak of postfeminism, of the "idea that feminism has had its day" (Davis, 2004, p. 140). It is in this context that this article argues that the joining of gender and art and technology through the use of feminist methodology can invigorate a discussion about gender and allow for the presentation of material on gender in new and exciting ways.

THE INTERSECTION OF GENDER, IT, AND ART

The use of creative technologies with its flexibility, crossing of boundaries, multidisciplinarity and interdisciplinarity lend themselves to feminist inquiry and provide a space to develop feminist research. At the most basic level, the tools available to us allow material on gender to be presented in a new and exciting way. While this, of course, applies to material of any nature, the use of these techniques in the gender sphere is particularly appropriate.

Gender studies are underpinned by feminist research methodologies. Feminist methodology is interdisciplinary and multidisciplinary, drawing insights from different fields and weaving them together through an understanding of feminist theory. For example, Ramazanoglu's definition of feminist methodology (Ramazanoglu, 2002) speaks about feminist methodology as being grounded in women's experience and seeks to analyze connections among ideas, experience and material reality. DeVault, on the other hand, discusses the need for "excavation ... that is to find what had been ignored, censored and suppressed, and to reveal both diversity of actual women's lives and the ideological mechanisms that have made so many of those lives invisible" (DeVault, 1999, p. 30). Reiharz includes in her definition an emphasis on multiplicity of methods and perspectives, of being transdisciplinary, of the effort to create social change, of being inclusive (Reiharz, 1992). Jenkins et al. see "the concept of power as central to feminist research" as well as noting the importance of "how the researcher and the researched have been gendered, sexualized, raced and classed" (Jenkins et al., 2003 p. 2, 4).

Irrespective of the exact definition used, feminist methodologies incorporate the desire to give women an opportunity to tell their stories, express their views and have their voices heard. In essence, it is women-centered. It acknowledges that researcher and researched are "gendered, sexualized, raced and classed," and both bring these characteristics into the research project (Jenkins, 2002). We consider the interaction of gender with other characteristics of identity as vital to understanding the complexities of research. To summarize very broadly, feminist research methodology: (1) places major emphasis on valuing a variety of viewpoints, (2) is highly concerned with remaining true to the voices of both those who research and are researched, (3) embraces complexity of argument, and (4) incorporates elements of social responsibility and a desire for social change. IT and, in particular in this case, the combination of the technology with artistic practices and methods, can play a major role in the first of these three essential criteria, as shown in Table 1.

As can be seen from that table, the negative potential of the technology is also listed. However, the existence of these very obvious negative factors also has a role to play. Often, traditional science and technology have assumed the air of impartiality and objectivity that gave them the veneer of having produced "authoritative knowledge." In the current

Table 1. Feminist methodology criteria and technology impacts

Feminist Methodology Criteria	Technologies of Interest	Positive	Negative
Capturing of Variety of Viewpoints	*Communication Technologies* Fixed networks, wireless networks (mobile communication systems, 2G, 3G, IEEE 802.11, Bluetooth, Zigbee, WiMax and so forth, ad hoc networks and so forth) *Communication Applications* Web, e-mail, sms, voice, instant messaging, blogs, moblogs and so forth *Communication Interfaces* Non-traditional input devices, Haptic interfaces and so forth	Wider access for broader ranges of people to participate in research, debate, conversation and so forth. More possibilities for viewpoints to be captured.	Digital divide that excludes large portions of the population from the online sphere.
Remaining True to the Voices	*Devices* for capturing voice, video and text and widely available *multimedia applications* for the manipulation and or creation of content.	Multimodel means of self-expression reducing the need for verbal or written expression only. Possibility for complete and unedited representation of the voice.	The ability to manipulate and distort and misrepresent data more easily than ever.
Embracing of Complexity	*Applications* that allow non-linear presentation of information (hypertext, etc.), applications for complex visualization of systems, immersive environments, virtual environments, gaming technologies and so forth.	The opportunity to deal with complex themes and topics in accessible and interesting ways.	Information overload.

world, this is very much not the case. Exploring gender with creative technologies opens the discussion on how knowledge is produced, authorized and imparted. It helps to problematize questions of objectivity, authority and knowledge production.

The Art of Decision

The ideas introduced in this article are illustrated through an example of a research project undertaken jointly by the Centre for Gender and Women's Studies and the Department of Electrical Engineering at Trinity College in Dublin, Ireland. This project, known as the Art of Decision, was funded by the Irish government through the Department of Justice, Equality and Law Reform and the European Union. The project aims to bring more women into political decision-making. To show how creative technologies and gender can combine in an effective way, a large interactive multimedia exhibition was specially designed and built. The exhibition took place in a Dublin city center warehouse during May 2005 and was open to the public. The exhibition comprised a series of rooms that immerse visitors in situations that invite

them to reconstruct their perceptions of political structures and political involvement as women. Fionnuala Conway (artist and technologist) and Jane Williams (gender studies researcher), both from Trinity College, played a major role in the design of the Art of Decision. Figure 1 shows a schematic of the exhibition. Visitors enter the reception area and wander around the space as they please.

Figure 1. Art of Decision exhibition plan

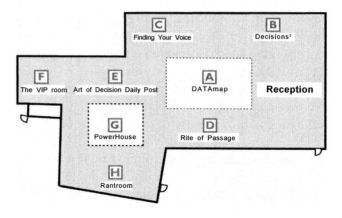

DATAmap

This is a large-scale (47x20 meter) interactive map of Ireland designed to present statistical data on the gender balance on Irish State bodies in more than 70 locations around the country. Users walk on the map and sensors embedded in the map trigger the associated area information on surrounding screens. The DATAmap presents information in a visually interesting and memorable way, depicting the statistics on women and men as pairs of symbols from everyday life – knives and forks, matches and flames, while suggesting a gender for each object. The presentation of the information in his manner highlights the arbitrariness of gender categories and opens discussion on gender categories.

Decisions, Decisions, Decisions

This is a short documentary film where nine people present their perspectives on decision-making. The film is screened in three parts, all three parts running concurrently and in the same location. As the user approaches the three screens, a jumble of voices emerges and it is only by standing in the correct location that sense can be made of the film. In many ways, this project reflects the conflicting nature of decision-making, the "'messiness" of the process and the fact that neither men nor women have a monopoly on decision-making procedures.

Finding your Voice

These are simple audio pieces that present unedited stories of two activist women. The stories are presented in two small, intimate spaces as audio installations with lighting design that responds as the stories unfold. The actual voices of the women are heard and their stories are told in their own, unedited words.

Rite of Passage

Images of the visitors to the Rite of Passage are digitally captured and their faces are superimposed on a large digital mural of figures from political life in Ireland, using image signal processing techniques. The power that attaches itself to gender, in this case the masculine gender, is evidence from the dominance of white middle-class men. As women's faces are superimposed, the balance shifts and we are able to see a more gender-balanced picture.

The Art of Decision Daily Post

This is a giant interactive newspaper projected on a large wall of the room. A headline is automatically pulled at random from an online daily newspaper and displayed as the Art of Decision Daily Post headline and is based on technology by Doyle, Conway and Greene (Doyle et al., 2003). Visitors to the room can text their reactions from their mobile phones to the headline and the reactions appear as text of the newspaper. This is particularly relevant for gender concerns, as it can trigger a spontaneous discussion on current issues important to gender studies.

The VIP Room

The VIP room contains interactive pieces that allow visitors to explore power relations through the manipulation of graphical representations of people's understanding of power. Drawings from participants are converted to digital format and displayed on the walls of the room. The images can be rotated and explored through control with joysticks. This exhibition illuminates the way in which women see the flow of power and the ways in which the power attaches itself to gendered institutions.

PowerHouse

The PowerHouse is a photographic exhibition presented in a set-designed caricature of a home, its garden and street. Seventy anonymous participants were given disposable cameras and asked to take photos that represent their ideas of power and include comments on the photographs. The visitor is invited to find these photos and comments in the PowerHouse. The research participants' evaluations of how their attitudes to power changed over the course of the project are also presented over speakers in the PowerHouse. The images can be viewed as on online exhibition at www.imagesofpower.net, where viewers are encouraged to contribute their own comments on the photographs. Using photographs to capture ideas of power provides a different way to capture what power means. Having a camera for a period of time to capture the images creates a persistent alertness to the notion of power. The "voices" of women are heard through the images they chose to take. This type of approach allowed a wide range of women to participate without the need to be able to write skillfully or express themselves eloquently. It also meant that the unedited view of the participants, namely, their photographs, could be presented without interpretation by the researchers.

Rant Room

This is the last exhibit in the journey and is intended as a resting space. Visitors can relax here and text or mail their opinion and comment on the exhibition. The comments board is continuously updated in the space and online during the exhibition so that visitors have the opportunity to send their comment from any remote location via text and e-mail and see it on screen in the space. This exhibition simulates discussion on gender issues. It is an excellent mechanism for allowing voices to be heard in an uncensored manner and for collecting data on views about gender.

FUTURE TRENDS

The ability to stimulate thinking and discussion about gender can increase with the growth in creative technologies. This is particularly important among the younger segment of the population. What was quite striking in the Art of Decision project was the appeal it had to second-level students (students in the 12-18-year age group). They responded positively to the presentation of ideas about gender, gender imbalance and the arbitrariness of gender categories. Thus, joining creative technologies and gender in an interactive manner has the potential to revitalize interest in and discussion about gender among younger cohorts of men and women.

CONCLUSION

Using creative technologies to explore gender enables the researchers to design research projects that cross traditional boundaries and create spaces for women to participate in an active and engaged manner. Using the principles of feminist methodologies focuses the research projects on being women-centered and about doing research in the interests of women. They involve the researched in an active and engaged manner in the evolution of the project. Some allow the participants to change the direction of the research by their input. All stress agency. Moreover, these projects are about raising awareness and consciousness, about presenting information in a new and engaging manner. They tackle issues of power imbalance and work towards bringing new faces into a revitalized feminist debate. When feminist methodology is linked to creative technology, the results are powerful and our understanding of the operation of gender is magnified.

On a theoretical level, the joining of gender, feminist methodologies and creative technologies is most exciting. Creative technologies disrupt the traditional notions of authority and the authorization of knowledge. Because it is technology and technology is allied to science, there is an expectation of authority, validity and objectivity. But creative technology problematizes these expectations and instead focuses on issues of subjectivity, of nonlinear thinking, of multiplicities and imaginings. The questioning of traditional authority characteristic of feminist methodology is given added weight when allied to creative technologies. It reinforces one of the beliefs of feminist methodologies that the authorization of knowledge is not an objective process based on a detached analysis. Rather, it stresses the fact that knowledge and the authorization of knowledge is political and provides the mechanism to reveal that this authorization is raced, classed and sexed.

Moreover, linking creative technology to gender provides opportunities for many, especially those not necessarily comfortable with technology, to not only understand the technological dimension to their project, but to take ownership of it. It calls into question the divide that places technology as a "masculine" tool and allows men and women to participate in this field.

By harnessing the power and potential of creative technologies to feminist methodologies, a richer, stronger, more dynamic understanding of gender may emerge.

REFERENCES

Alsop, R., Fitzsimmons, A, & Lennon, K. (Eds.). (2002). *Theorizing gender.* Polity Press.

Davis, K. (2004). Editorial. *The European Journal of Women's Studies, 11*(2), 139-142.

DeVault, M. (1999). *Liberating method.* Philadelphia: Temple University Press.

Doyle, L., Conway, F., & Greene, K. (2003, July/August). Mobile graffiti, a new means of political engagement. In *Proceedings of the International Conference on Politics and Information Systems Technologies and Applications (PISTA 2003),* Orlando, FL.

Fausto Sterling, A. (2000). *Sexing the body: Gender politics and the construction of sexuality.* New York: Basic Books.

Fausto Sterling, A. (2005). The bare bones of sex: Part I—sex and gender. *Signs, 30*(2), 1491-1529.

Jenkins, S., Jones, V., & Duon, D. (2003). Thinking/doing the "F" word: On power in feminist methodologies. *ACME: An International E-Journal for Critical Geographies, 2*(1), 58-63.

Nicholson, L. (Ed.). (1997). *The second wave: A reader in feminist theory.* London; New York: Routledge.

Ramazanoglu, C. with Holland, J. (2002). *Feminist methodology: Challenges and choices.* London: Sage Publications.

Reinharz, S. (1992). *Feminist methods in social research.* Oxford: Oxford University Press.

Salih, S. (Ed.). with Butler, J. (2004). *The Judith Butler reader.* Oxford, UK: Blackwell.

Wilson, S. (2001). *Information arts.* Cambridge: MIT Press.

KEY TERMS

2G, 3G: Second-generation and third-generation mobile phone networks (cellular networks).

Ad Hoc Network: A collection of nodes that form a network on an as-needed basis without the need for any preexisting infrastructure.

Blog: This is the short form of Weblog. A Weblog is a personal journal published on the Web. These journals typically contain informal thoughts of the author in the form of posts or short dated entries in reverse chronological order.

Bluetooth: A technology specification for small form-factor, low-cost, short-range radio links between mobile PCs, mobile phones and other portable devices.

Gender: Gender is primarily defined as the interplay between biology and culture in which definitions of femininities and masculinities are developed and dispersed in accordance with their norms. These definitions are neither unity nor static, so that one can speak of hegemonic masculinities or emphasized femininities.

IEEE 802.11: A wireless local area network standard.

Instant Messaging (IM): A text-based computer conference over the Internet between two or more people who must be online at the same time. When you send an IM, the receiver is instantly notified that he or she has a message.

Moblog: This is similar to a blog except in the case of a mobile Weblog (or moblog)—the content is posted to the Internet from a mobile or portable device, such as a cellular phone.

SMS: Short Message Service, or also known as text. SMS facilitates the sending of text messages on mobile phone systems.

WiMAX: A standard for delivering point-to-multipoint broadband wireless access. Specifically, WiMAX is an acronym that stands for Worldwide Interoperability for Microwave Access.

Zigbee: This is a proprietary set of high-level communication protocols designed to use small, low-power digital radios based on the IEEE 802.15.4 standard for wireless personal area networking.

Introducing Young Females to Information Technology

Michaele D. Laws
ETSU GIST, USA

Kellie Price
ETSU GIST, USA

INTRODUCTION

The difficulties in recruiting females into information technology and computer science (CS) have been well documented. Engineering disciplines have faced the same problem for many years. Some of the main underlying issues include unsupportive classroom environments (Hall & Sandler, 1982), gender-related perceptions of performance, a lack of role models, and inadequate peer communities (Zappert & Stansbury, 1984). Other contributing factors are the amount of positive computing experience gained prior to enrollment at the university level (Robers, Kassianidou, & Irani, 2002) and self-confidence. Research provides significant evidence to indicate that, even though females perform at the same levels as their male counterparts, they have less confidence in their abilities (Arnold, 1993; Fisher, Margolis, & Miller, 1997; Sax, 1994; Strenta, Elliot, Matier, Scott, & Adair, 1994). This lack of confidence keeps many females out of the technical classes. Finally, those females that do enter IT or CS courses may come to the discipline with multiple interests and, consequently, feel out of place at times among their more single-minded male counterparts (Widnell, 1988).

While it is predicted that 8 of the 10 fastest growing occupations from 2000 to 2010 will be in the IT or CS fields, it is expected that women will not be equally represented within these occupations (http://www.bls.gov/oco/ocos267.htm; Camp, 1997). The underrepresentation of women in computer science was given priority in the June 2002 special issue of *SIGCSE Bulletin* dedicated to women and computing, bringing focus to previous and current research regarding this dilemma. One particular factor highlighted in this bulletin is that changing this male-dominant field requires the crucial step of targeting young females in an effort to dispel stereotypical ideations and gender bias associated with computer science, thus attracting more women to the profession (American Association of University Women Educational Foundation, 1999).

Girls in Science and Technology (GIST) is a free science and technology camp at East Tennessee State University (ETSU) making efforts to change these trends. The primary goal of the girls-only GIST camp is to introduce females to the fields of information technology, computer science, and math by providing discipline-related activities, enhancing teamwork competency, connecting females with women mentors working in the field, and creating a challenging yet fun atmosphere free from male competition. The hope is that this exposure will instill technical confidence and aptitude in the young females that will last through their college careers, giving them a positive outlook on information technology.

BACKGROUND

The enrollment statistics at ETSU for undergraduate degree-seeking students indicate that women are underrepresented within the IT and CS majors. In fact, these statistics show that women lost ground over the last 5 years. Note the downward trend in the percentages of females enrolled in IT- or CS-related majors at ETSU shown in Figure 1.

An analysis of data from the Office of Institutional Effectiveness and Planning at ETSU shows female enrollment at 21.4% vs. male enrollment at 78.6% in the fall of 2000. However, in 2004, female enrollment dropped by 6.3% compared to an in-

Figure 1.

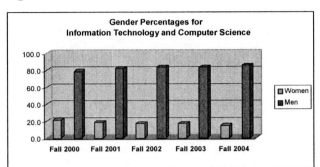

crease of 6.7% in male enrollment (Baxley, 2004). The underrepresentation of women in IT and CS at ETSU parallels nationwide research suggesting that women are not equally represented in computer-related fields (Camp, 1997). It is our goal to provide exposure and opportunities to females, especially rural, underprivileged females in East Tennessee, and help reverse this trend.

The GIST camp offers a different perspective in introducing young females to IT. Previous research targeting females, primarily between the ages of 12 and 16, has shown the effectiveness of science and technology camps in providing a climate free of male competition and gender bias (Countryman, Feldman, Kekelis, & Spertus, 2004). Additionally, other research notes the importance of implementing science and technology programs within elementary schools so young females will not lose interest or feel less competent in IT and other predominately male areas when reaching middle and high school (Entwistle, 2002). Based upon research and prior experience, leaders of the GIST camp are now targeting females between the ages of 10 and 13 by providing a free science and information-technology camp for middle-school females.

MAIN THRUST OF THE ARTICLE

GIST Camp Activities

Participants in the GIST camp are divided according to age to accommodate the developmental stages of the different age groups. Each 5-day camp session is held Monday through Friday. Camp activities for each week are structured in similar fashion. Each day consists of computer lab time, team-building

exercises, and a science experiment or science tour, with female professional speakers visiting several times per week. The order of activities varies slightly during the week to accommodate the schedules of speakers and departments hosting science tours. However, there is a deliberate alternation between time spent in front of the computer and time spent physically active. Figure 2 contains a schedule from a sample day in the camp. The camp activities lead to our desired outcome of participants gaining exposure to science and information technology in an atmosphere encouraging uninhibited exploration and experimentation related to these fields.

Evolution of the Camp

A pilot summer-camp program was initiated by two faculty members from the Department of Computer and Information Sciences at ETSU in 2000. The camp consisted of two week-long sessions and included a broad age range of females, from 9 to 16 years. Beginning in June of 2002, the camp was started on an annual basis, targeting females ages 10 to 15.

After the pilot camp was completed, the researchers formulated two important questions: (a) What is the appropriate age range for this kind of program, and (b) what are the appropriate kinds of activities for the selected age ranges? After holding the camp for 4 years, the researchers feel they have a working answer for both of these questions.

The researchers decided to approach the question of an ideal age range for a summer IT camp in multiple stages. The first stage would be to target the age group of 12 to 13 years. If this proved to be successful, the program would then be expanded incrementally to include younger and older females. As long as each new session was successful, then new groups would be added to expand the range of girls included in the program.

Figure 2.

Sample Daily Schedule	
Lab Time	9:00 a.m. – 9:50 a.m.
Team-Building Exercise	10:00 a.m. – 10:50 a.m.
Lab Time	11:00 a.m. – Noon
Lunch	Noon – 12:30 p.m.
Speaker	12:30 p.m. – 1:15 p.m.
Science Tour	1:30 p.m. – 2:15 p.m.

To determine age-appropriate computer activities, researchers decided to choose projects, tasks, and programs that would demonstrate the wide range of topics in IT while staying true to the discipline. The scope of these tasks and projects would evolve in a trial-and-error form after each day of the camp and from year to year.

In 2002, one week-long session was offered for ages 12 to 13. No formal data were collected from this year, but the camp hosted approximately 20 females. The researchers used this first year to refine group and lab activities. The campers were easily able to sit and complete specific project-oriented tasks in front of the computer and were not inhibited when participating in group events.

Due to the success of the 2002 camp, a new week-long session was added in 2003 for ages 10 to 11. Additionally, the size of the sessions was expanded from 20 campers per session to 30 campers per session, and the researchers began to collect formal data about the camp sessions. Recruitment efforts, which included sending information to school counselors and advertising in the local paper, provided sufficient exposure to fill both sessions. While the 12- to 13-year-old females worked well with limited supervision or help and were able to problem solve effectively, the 10- to 11-year-old females required more individual attention and were easily frustrated upon encountering problems difficult for them to solve, whether hardware or software related. The difference in the responses of the two groups was attributed to the amount of prior computer exposure and general age-related maturity.

Based on experience from the 2003 camps, computer activities for the 10- to 11-year-old age group were altered to be more creative or game related, and simplified significantly for future camps. For example, an activity involving the development of a multipage Web site in Microsoft Front Page© was modified so that campers were developing a single Web page using Microsoft Word© instead. The goal for the youngest camp was modified from being project oriented to providing general exposure and fun with computers.

Based on the initial planned approach, the camp was expanded in 2004 to include 1 week for campers ages 14 to 15, and the number of campers per session was modified to 20 for the 10- to 11-year-olds and 25 for the 12- to 13-year-olds. Although the same re-cruitment tools were used as in previous years, recruitment for the 14- to 15-year-old session was noticeably more difficult. While there were waiting lists for the 2 weeks involving 10- to 13-year-olds, only 12 females signed up to attend the 14- to 15-year-old session. As an additional recruitment effort, personal visits were made to the schools targeting 14- to 15-year-olds. Researchers found during these visits to local school campuses that the 14- and 15-year-old females were very disinterested in the camp, and they heard many comments such as, "I hate math," or "Computers are boring." However, feedback showed that the females who attended the 14 to 15 session had a positive experience. In the comments section of the postcamp survey, one camper said, "I have really enjoyed this camp. I wish I could just keep coming to it year after year! It's been great," and another noted, "I hope this camp is offered next year. If it is you can expect to see me there!!!" In fact, 100% of the campers agreed or agreed strongly that they knew more about careers in computers, science, and technology and were more interested in these types of careers as a result of attending the camp (Laws, Loyd, & Price, 2004).

Following the outlined strategy for determining the appropriate age range for the camp, the researchers decided to eliminate the 14- to 15-year-old session for 2005 and concentrate on the younger ages. This decision was based mainly on the negative feedback experienced during recruiting activities. It seems that by ages 14 to 15, many females have already formed negative impressions of technology as a career or even as a fun hobby. Researchers decided to concentrate on the younger ages to help prevent the formation of these kinds of negative opinions.

The overwhelming positive response, excitement, and attendance of the 10- to 13-year-old females seem to indicate that reaching the females at the critical age range of 10 to 13 is ideal for this kind of camp. They are old enough to perform given technology tasks, but not so old that they have already formed negative opinions about computers and technology. Figure 3 shows that the overwhelming majority of campers aged 10 to 13 surveyed in 2003 and 2004 agreed or strongly agreed when asked if "science, computer and technology activities are more interesting to them because of the GIST camp" (Laws et al., 2004).

Figure 3.

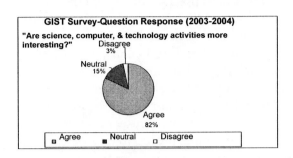

The 2005 summer program consisted of three 5-day sessions with 20 to 25 females per session, serving at least 65 females within the age group of 10 to 13. Due to difficulty in maintaining age boundaries, enrollment criteria were changed from camper age to school grade. The three sessions were offered for rising fifth graders, rising sixth graders, and rising seventh graders. New activities with Lego Mindstorm© robots and Alice (http://www.alice.org) were introduced with much success.

Factors for Success

According to published research as well as data that we have gathered from the GIST camps, there are several factors that can affect female success in technical areas. These include exposure to role models, comfort level, confidence, and one-on-one instruction.

According to a survey of ninth- to twelfth-grade girls conducted by the Garnett Foundation, girls are less likely to pursue computer careers because there are not enough role models (Jepson & Peri, 2004). Therefore, the GIST camp sessions offer exposure to female role models working in the field of science and technology, and other material related to career options.

The case study by Shrock and Wilson (2001) showed comfort level as the number-one factor in contribution to success in an introductory programming course. The camp strives to provide a supportive atmosphere to discuss issues hindering females from reaching their potential in school or pursuing possible interests in science and technology. The all-female atmosphere is critical because it allows young females to be assertive in the education process and helps them gain confidence in their

technical skills. Studies show that females are not normally assertive in technical arenas if males are present (Cuny & William, 2002).

Campers learn various IT skills during the camp, where they are challenged with age-appropriate activities. The session sizes are limited to a maximum of 25 campers, which allows for one-on-one interaction between the staff and the campers. This facilitates better material comprehension and a more positive experience for the campers. One-on-one tutoring is shown to be much more effective than classroom instruction and other computer-aided tutorials (Heffernan, 1988).

Measuring Success

A 16-question survey is used to measure the attitudes of GIST participants at the close of each camp session. Analyses of the survey results from camp participants help us determine whether we accomplish the following:

- Was the camp a success? Did the camp remove the "mystery" of computers, science, and technology; help the campers form bonds of friendship; and provide a fun yet challenging atmosphere for learning?
- Did the participants gain self-confidence regarding science and technology because of the camp?
- Was there a change in the participants' perception of computers, science, and technology because of the camp?

Throughout the camp sessions, answers to all 16 questions were very positive. The question that sums up the overall attitude and success of the camp is, "Should this camp be offered again next year?" Figure 4 shows the results of this survey question for all sessions in 2003 and 2004. The feedback shows that the camps are very successful.

The participant survey as well as a survey for parents and guardians is completed online through a Web site dedicated to the camp. Completing the surveys in this format contributes to the goals of encouraging females to participate in and utilize current technology.

The dissemination of results of the camp is accomplished through press releases, brochures, a

Figure 4.

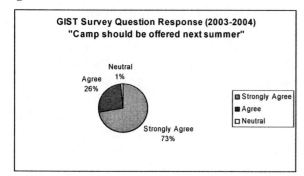

GIST Survey Question Response (2003-2004)
"Camp should be offered next summer"

Web site, and a promotional video. These are used to provide outcome effectiveness to the surrounding community, thereby increasing future participation in the camp. Results of these camps are presented at conferences nationwide in the form of papers, panels, and tutorials to encourage professionals at other institutions to provide similar programs for young females in their areas.

The current Web site dedicated to the ETSU GIST camp is located at http://cscidbw.etsu.edu/gist.

FUTURE TRENDS

GIST is intended to be an annual event for at least the next 5 years. Participation in GIST should help the participants become part of a supportive, vertically integrated network consisting of themselves, female graduate and undergraduate ETSU students, ETSU mathematics and computer-science faculty, and distinguished female professionals external to ETSU.

By sharing information about GIST, we intend to reach beyond the East Tennessee area, creating partnerships with organizations and schools. We hope to eventually be able to collaborate with others to offer various activities that will continue to nourish young females throughout the year.

CONCLUSION

Learning from the success of other programs throughout the country, a plan was developed and implemented at ETSU to introduce young females to technical fields in an atmosphere full of fun and free of gender bias. The data collected during the camp are evidence that the participants gained an in-

creased awareness of science and technology, offering a potential positive impact on their future.

As a result of this study, it has been determined that science and technology camps can be more effective when targeting females ages 10 to 13.

According to Cohoon (1999), there must be females already in the discipline in order to attract more females and retain them in the discipline. Therefore, collaborative research efforts among faculty, students, and professionals will persist each year as the summer camp continues in hopes of identifying a positive change in the recruitment and retention of females in science and technology at ETSU.

REFERENCES

American Association of University Women Educational Foundation. (1999). *Gender gaps: Where schools still fail our children*. New York: Marlow & Company.

Arnold, K. (1993). Academically talented women in the 1980s: The Illinois valedictorian project. In K. Hulbert & E. Schuster (Eds.), *Women's lives through time*. San Francisco: Jossey-Bass.

Baxley, R. (2004). *Gender representation by major*. East Tennessee State University Institutional Effectiveness and Planning Office.

Camp, T. (1997). The incredible shrinking pipeline. *Communications of the ACM, 40*, 103-110.

Cohoon, J. M. (1999). Toward improving female retention in the computer science major. *Communications of the ACM, 44*, 108-114.

Countryman, J., Feldman, A., Kekelis, L., & Spertus, E. (2004). Developing a hardware and programming curriculum for middle school girls. *SIGCSE Bulletin, 34*, 46.

Cuny, J., & William, A. (2002). Recruitment and retention of women graduate students in computer science and engineering: Results of a workshop organized by the Computing Research Association. *SIGCSE Bulletin, 34*, 168-174.

Entwistle, M. (2002). Augustana women in computer science: A program to encourage women in the pursuit of technical education and careers. *The Journal of Computing in Small Colleges, 17*, 206-215.

Fisher, A., Margolis, J., & Miller, F. (1997). Undergraduate women in computer science: Experience, motivation, and culture. *ACM SIGCSE Technical Symposium*.

Graham, S., & Latulipe, C. (2003). CS girls rock: Sparking interest in computer science and debunking the stereotypes. *Proceedings of the 34th SIGCSE Technical Symposium on Computer Science Education*, 322-326.

Hall, R., & Sandler, B. (1982). *The classroom climate: A chilly one for women?* Washington, DC: Association of American Colleges, Project on the Status and Education of Women.

Heffernan, N. (1998). Intelligent tutoring systems have forgotten the tutor: Adding a cognitive model of human tutors. *Conference of Human Factors and Computing Systems: Conference Summary on Human Factors in Computing Systems*, 50-51.

Jepson, A., & Peri, T. (2004). Priming the pipeline. *SIGCSE Bulletin, 34*, 36.

Laws, M., Loyd, R., & Price, K. (2003-2005). *Girls in science and technology summer camp.* East Tennessee State University, Department of Computer and Information Sciences.

Magoun, D., Eaton, V., & Owens, C. (2002). IT and the attitudes of middle school girls: A follow-up study. *Educational Computing Conference Proceedings*. Retrieved February 12, 2005, from http://ccenter.uoregon.edu/conferences/necc2002/program/research_papers.php

Robers, E., Kassianidou, M., & Irani, L. (2002). Encouraging women in computer science. *SIGCSE Bulletin, 34*, 84-88.

Sax, L. J. (1994). Predicting gender and major-field differences in mathematical self-concept during college. *Journal of Women and Minorities in Science and Engineering, 4*, 291-307.

Shrock, S., & Wilson, B. (2001). Contributing to success in an introductory computer science course: A study of twelve factors. *Proceedings of the 32nd SIGCSE Technical Symposium on Computer Science Education*, 184-188.

Sivilotti, P., & Demirbas, M. (2003). Introducing middle school girls to fault tolerant computing. *Proceedings of the 34th SIGCSE Technical Symposium on Computer Science Education*, 327-331.

Strenta, C. R., Elliot, M., Matier, M., Scott, J., & Adair, R. (1994). Choosing and leaving science in highly selective institutions. *Research in Higher Education, 35*(5), 513-547.

Widnall, S. E. (1988). AAAS presidential lecture: Voices from the pipeline. *Science, 241*, 1740-1745.

Zappert, L., & Stansbury, K. (1984). *In the pipeline: A comparative analysis of men and women in graduate programs in science, engineering, and medicine at Stanford University.* Stanford University, Institute for Research on Women and Gender.

KEY TERMS

Comfort Level: The level of feeling at ease, without inhibition, or the freedom to be oneself.

Dissemination: The process of distributing something, such as information. There are various methods of distribution that can be used including print, Web, and video.

Gender Bias: An unfair act or policy stemming from prejudice based on a person's gender.

GIST: Girls in Science and Technology.

Modified Likert Scale: A 5- or 7-grade rating scale with the *undecided* option removed. The scale measures the strength of agreement with a clear statement. It is often administered in the form of a questionnaire and used to gauge attitudes or reactions.

SIGCSE: ACM Special Interest Group on Computer Science Education.

Stereotypical Ideations: Forming ideas that conform to a presupposed type, idea, or convention.

Vertically Integrated Network: Relationships formed between people of various age groups and professional or educational classifications who share a common interest or goal.

Issues Raised by the Women in IT (WINIT) Project in England

Marie Griffiths
University of Salford, Greater Manchester, UK

Karenza Moore
University of Salford, Greater Manchester, UK

THE UK IT SECTOR: THE CONTEXT OF THE WINIT PROJECT

This article explores several issues raised by the European Social Fund (ESF) Women in IT (WINIT) project (February 2004 to February 2006) which focuses on women in the IT industry in England. The project consists of an online questionnaire aimed at women currently in the IT sector in England and those wishing to return to IT following a career or "carer" break (a break to care for children, or sick or elderly relatives). The WINIT team aims to target 750 respondents in order to collect and analyse data from a demographically diverse group on a range of issues including perceptions of fairness of pay, promotion prospects and future career aspirations. In addition the WINIT team are currently conducting a series of in-depth interviews with women in the IT industry in order to gain a rich understanding of these women's perceptions of, and experiences in, IT in England.

In order to explore the issues raised by the WINIT project it is important to consider the wider historical and contemporary socio-economic backdrop of individual women's experiences. The IT industry in Britain has experienced considerable expansion over the past twenty years. In November 2004 it was estimated that the IT workforce consisted of 1.2 million people (580,000 in the IT industry, with an additional 590,000 IT professionals in other sectors). There are also an estimated 20 million people in Britain using IT in their everyday work. All the above figures are predicted to grow between 1.5% to 2.2% per annum over the next decade (e-skills UK/Gartner, 2004). In terms of gender, in spring 2003 it was estimated that 151,000 women were working in IT occupations compared with 834,000 men, whilst in the childcare sector, there were less than 10,000 men working in these occupations, compared with 297,000 women (Miller, Neathey, Pollard, & Hill, 2004). To clarify, it is estimated that only 1 in 5 of the IT workforce in Britain is female (e-skills UK/Gartner, 2004). Such statistics indicate a classic case of *horizontal* occupational segregation. However, it must be noted that all statistics regarding the IT industry should be treated with caution given the problems of defining the sector (von Hellens, Nielsen, & Beekhuyzen, 2004).

In the UK, figures from the Office of National Statistics (ONS) indicate that women accounted for 30% of IT operations technicians, but a mere 15% of ICT Managers and only 11% of IT strategy and planning professionals (Miller, Neathey, Pollard, & Hill, 2004). Although women are making inroads into technical and senior professions there remains a "feminisation" of lower level jobs, with a female majority in operator and clerical roles and a female minority in technical and managerial roles (APC, 2004). Again this is a classic case of *vertical* gender segregation with women more strongly represented in lower level IT occupations than in higher status and higher paid ones (Miller, Neathey, Pollard, & Hill, 2004, p. 69). There is a relatively narrow gender pay-gap in the IT sector in comparison with all occupations. According to the ONS (2003), the gender pay-gap amongst ICT professionals in terms of hourly earnings stands at 7.5%, which is slightly narrower than the figure for all professional occupations.

WINIT'S THEORETICAL FRAMEWORK

The under-representation of women in the sector has been the focus of various initiatives in the UK over the last 30 years. These initiatives predominately draw on liberal feminist approaches to the women in computing "problem". Perhaps the most notable aspect of the 'liberalist' agenda is the recommendations for action advocated. The liberal feminist approach to the "problem" of women in computing, typified by Women in Science and Engineering (WISE) and Science, Engineering, and Technology (SET) discourses (Henwood, 1996) highlights the need to improve access to ICT, the need to encourage more women onto computing courses, and the need for better Equal Opportunities and Managing Diversity legislation. It is suggested that better gender equity will bring economic benefits for specific employers and for the UK economy, with the IT "skills gap" being narrowed through the greater participation of women in the IT industry (e-skills UK/Gartner, 2004).

There have been many criticisms of the liberal feminist approach in general terms and in terms of the actions advocated to address gender imbalances in IT settings (Cockburn, 1986). These include its tendency towards technological determinism, given that it leaves "technology" largely untroubled and views technology as "neutral" (Faulkner, 2000). The "individualism" of the liberal feminist approach to the "problem" of women and technology has also been highlighted as problematic, situating the "problem" as it does with the "failure" of women to realise the (liberating) potential of technologies (such as the Internet), their "failure" to properly engage with these technologies in home and workplace settings and their "lack" of awareness of the myriad of career options made available through technological engagement. We suggest, with others (Clegg & Trayhurn, 1999), that there is more to the women and computing "problem" than getting more women into the IT industry and into particular (high-paid, more prestigious) posts, although this is of course important.

With the contextualisation of the woman and computing "problem" comes the highlighting of the unsuitability of the IT workplace for many women; the long hours and presenteeism (Simpson, 1998) culture that exists within IT, negative perceptions of part-time workers in the IT sector (DTI, 2004) and of part-time work more generally (Epstein, Seron, Oglensky, & Saute, 1999), the instability of the IT market, and the deeply ingrained "masculine culture" of IT—these aspects need to change before (some) women can comfortably find a place within the IT industry. Rather than women, and for example older workers, being forced to "adapt" to the current IT culture, it is suggested that the IT industry needs to broaden its appeal to a more diverse pool of talent (Platman & Taylor, 2004; Women & Equality Unit, 2004). Having explored some of the work on gender and technology which has informed WINIT research, we now move on to some of our main findings at this initial stage.

THE WINIT SURVEY: INITIAL FINDINGS FROM 111 FEMALE ICT PROFESSIONALS

The WINIT team used contemporary literature, and expertise from academic and industry practitioners, to generate pertinent survey themes and questions. The online WINIT survey is securely hosted at the University of Salford. It went live in autumn 2004, and will remain so until autumn 2005. The WINIT team promoted the survey URL to a wide variety of women's forums, networking groups, special interest groups (i.e., BCS [British Computer Society] Women), IT recruitment agencies and female academics. This means that female ICT professionals who completed the survey self-elected to do so. Given that this is an online survey there was no pre-defined sample.

The majority of the initial 111 respondents were aged between 30-34 years of age (20%) while the second largest age group (16%) were aged between 25-29 years of age. This reflects the predominance of relatively young people within the industry (Platman & Taylor, 2004). In terms of living arrangements, 59% were living in a couple which incorporated being married, remarried, and co-habiting. Geographically initial respondents were predominately located in London and South-East England (40%), with North-West England (12%) having the second highest proportion of respondents, and Yorkshire a close third (11%). Fifty-nine percent of women in our initial cohort indicated that they had no children; while 40% had one or more. Seventy percent of respondents had no children living with them. Of the women who indicated that they had adapted their working practices as a result of having children, a shift to part-time work was the most common change. Behind such changes in working practices lie nuanced gendered experiences and gendered patterns of work and care. "Flexible" (but ultimately "feminised") part-time work remains of low-status within the IT industry (DTI, 2004). The low status of part-time work may curtail female part-timers IT career progression (Kodz, Harper, & Dench, 2002).

This said, 84% of women in our cohort had full-time or full-time flexi-time roles. This is significant given that 59% of the respondents had no children. Is this an indication of the possibility that for women in full-time IT positions raising a family is a difficult challenge to meet? A recent study (Gatrell, 2004) examining changes in family and working practices identified that highly qualified working women with children are suffering hidden discrimination from their employers despite current UK government work-life balance initiatives. Such "career vs. carer" difficulties were recognised by WINIT respondents, with one woman (Respondent No. 7) explaining, "It's all very well a company having a work-life policy or suggesting that they will try to support part-time/ flexible working. What is needed is for them to actually act on this and prove they support it. I cannot see much evidence of this at present. I would hate to leave IT but in due course I hope to start a family and this will definitely take the highest priority". This respondent's concerns, about the un-family friendly policies and practices of many IT companies, are unfortunately supported by the aforementioned report (Gatrell, 2004), which found that organisations in many different sectors are still reluctant to employ working mothers. This problematic of combining home/caring and work responsibilities is exacerbated by the need for IT professionals to keep up with the rapid rate of change in the industry, making even relatively short career breaks risky.

Our initial cohort consists of women in a diverse selection of occupations in IT at a variety of different levels of seniority. A selection of our respondents included a Senior Software Engineer, a Senior Database Analyst, a Head of ICT, a Managing Director and a Professor of Software Engineering. These senior roles indicate that (some) women are progressing in their chosen profession. 59% of WINIT respondents believed that they have the same chance as promotion as their male colleagues, with only 36% disagreeing and 5% choosing "Do Not Know".

WOMEN'S PERCEPTIONS OF THEIR WORKING ENVIRONMENT

The WINIT survey includes questions exploring women's perceptions of their working environment including the support they received from their colleagues and line managers, the pay they received, how comfortable they felt talking about personal issues in the workplace and so on. Respondents were asked how they felt regarding the nature of their current IT employment in relatively general terms. Many respondents commented that they perceived their situation to be "fine", "very happy", "comfortable", "I love the work and the people", "convenient", "I love it" and "good conditions". However there was a contrary trend of disappointment and dissatisfaction; "I feel under-used and stingily metered out", "do not enjoy the post I currently occupy", "the extra level of hierarchy has restricted my growth in the company", "I would like to work part-time but that option is not available to me", "overburdening", "'potentially a great job but the workload is way beyond what should be expected", "overworked and underpaid". WINIT respondent No. 47, who had recently left the software industry to do a PhD, offered a snapshot of her former workplace saying, "Too many decisions and discussions took place down the pub. I have worked in IT over the last 20 years and still had to put up with people commenting on how unusual it was to see women writing software."

Our group of women are generally satisfied with their working environment. Sixty percent of WINIT respondents believed that their pay package reflects their workload, 61% believe that their pay packet reflects their current skill set, with 50% agreeing that their salary mirrors their highest qualification. Overall, 49% perceived that their individual position in the IT industry is reflected in their pay package, and 55% believe it is reflected in their position within their organisation. As a means to assess our respondents' perceptions of their salary, we asked whether there are any differentiating factors between their pay with that of male colleagues of a similar level. Sixty-two percent said it was comparable, and 77% said their female colleagues' salary was comparable, indicating a small discrepancy between (perceptions of) male and female pay amongst our respondents. We have thus far found few adverse trends in relation to female perceptions of their working environment. Seventy-two percent of our initial respondents believed that they are valued at work, while 77% of agreed that they are valued as part of a team. The data from this initial group of respondents conflicts somewhat with the "haemorrhaging" of women from the IT industry (IBM/George, 2003) but may reflect the fact that our respondents do, in the majority, work full-time and in the majority have no children. The Women in IT Forum has identified the retention of older and more experienced women as vital to the IT industry and suggests that flexible working initiatives would allow more of such women to remain in the sector (DTI, 2004).

A further theme was whether long-hour cultures and flexible working practices co-exist in the UK IT industry as reported in the DTI flexible working report? (DTI, 2004) There are comparable findings (with the above report) in the WINIT survey regarding a conflict between current long-hours culture and support for work-life practices. Sixty-seven percent of WINIT respondents state that there are flexible working initiatives in place within their organisation. This said, 64% report a long-hours culture. The IT Industry DTI report (2004) claims that 51% of IT professionals adopt a flexible working schedule but inconsistently 65% are working over 10 hours a day. Respondents felt that working flexibly would lower their pay, their status and diminish their promotional opportunities (DTI, 2004). This phenomenon is identified as the "take-up gap" by research conducted by the Institute of Employment Studies (Kodz, Harper, & Dench, 2002). Work-life balance initiatives such as part-time work, career breaks and job shares are in place in the IT industry, but heavy workloads and mangers' nega-

tive reactions formed barriers to the "take-up" of these options (Kodz, Harper, & Dench, 2002).

FUTURE ACTIONS, ASPIRATIONS, AND PERCEPTIONS

Thus far we have discussed initial WINIT findings in terms of the demographics of our initial respondents, their management of carer and career "clashes", the possibility of a long-hours culture in IT which may hinder attempts to adopt more "flexible" and/or "family-friendly" working practices, our respondents' perceptions of their working environment including promotion prospects and pay equity, and their overall satisfaction and/or dissatisfaction with their IT careers. But what of their *future* actions, aspirations and perceptions? (von Hellens, Nielsen, & Beekhuyzen, 2004).

In terms of personal career trajectories, responding to the statement "I can imagine myself working in the IT industry in the future", the majority of women (38%) said they strongly agreed, 17% moderately agreed, 12% slightly agreed and 17% agreed. This is in contrast to the 5% who slightly disagreed with the statement, the 5% who moderately disagreed, the 3% who strongly disagreed and the 3% who said "don't know". However these women's personal optimism did not match their overall pessimist view that the IT industry's image was unlikely to become more "female friendly" in the future. In response to the statement "The IT industry [will have] a female friendly image in the future", a mere 2% strongly agreed, 10% moderately agreed, 11% slightly agreed and 11% agreed.

Improving the IT industry's image to make it more "female-friendly", while admirable, may be somewhat problematic. To improve the image of the IT industry in England should not involve solely concentrating on *appearances* and negative perceptions. The image of the industry will only change if more is done to support women (currently) working in the industry who wish to combine home and family life. It is in this sense that we require initiatives which tackle the lack of affordable child-care facilities available in IT workplaces (and of course workplaces more generally), the lack of work-life balance initiatives (IBM/George, 2003) and the dearth of (desirable and respected) part-time IT positions, which may suit older workers and working parents in particular (Platman & Taylor 2004). It is in this sense that the under-representation of women in IT should not be configured solely as a "women's problem" but as problem of the industry itself and a problem related to the (unequal) gendering of domestic and care (i.e., parental) work. A socio-cultural contextualisation of the problem then shifts the focus from an essentialist notion of "woman" to the constraints of wider gendered society and gendered organisations in relational, interactional and institutional terms.

CONCLUSION

The women in the initial WINIT cohort are a heterogeneous group from a variety of educational backgrounds in a broad range of positions at all levels within the IT industry in England. Attention to this heterogeneity provides us with a nuanced view of women's experiences in the IT industry and offers a solid base on which to build further WINIT research. The under-representation of women in the IT industry as highlighted by the liberal feminist position clearly needs to be tackled. However, we have demonstrated in this paper that we also need to use other feminist approaches to gender and technology (Faulkner, 2000) in order to trouble these two inter-related terms. In so doing we can tackle the issue of under-representation without assuming that simply encouraging more women into the industry, say by improving its image, will be sufficient to tackle the continued complicated socio-cultural construction of IT as a "masculinised domain".

REFERENCES

Association for Progressive Communications (APC). (2004). *Gender and information and communication technology: Towards an analytical framework.* Retrieved January 2004, from http://www.apcwomen.org/work/research/analytical-framework.html

Clegg, S., & Trayhurn, D. (1999). Gender and computing: Not the same old problem. *British Educational Research Journal, 26*(1), 75-89.

Cockburn, C. (1986). Women and technology: Opportunity is not enough. In K. Purcell, S. Woods, A. Wharton, & S. Allen (Eds.), *The changing experience of employment: Restructuring and recession.* London: Macmillan.

Department of Trade and Industry (DTI). (2004). *Flexible working in the IT industry: Long hour cultures and work-life balance at the margins?* Retrieved November 2004, from http://www.dti.gov.uk/industries/electronics/flexwork-it04.pdf

e-skills UK/Gartner. (2004, November) *IT insights: Trends and UK skills implications.* London: Researved, e-skills UK/Gartner Inc. Retrieved from http://www.eskills.com/Research/itinsights/1055#Trends

Epstein, C. F., Seron, C., Oglensky, B., & Saute, R. (1999) *The part-time paradox: Time norms. professional life, family. and gender.* New York: Routledge.

Faulkner. F. (2000). The technology question in feminism: A view from feminist technology studies. *Women's Studies International Forum, 24*(1), 79-95.

Gatrell, C. (2004). *Hard labour: The sociology of parenthood and career* (1st ed.). Milton Keyes: Open University Press.

Golding, P. (2000). Forthcoming features: Information and communication technologies and the sociology of the future. *Sociology, 34*(1), 165-184.

Henwood, F. (1996). WISE choices? Understanding occupational decision-making in a climate of equal opportunities for women in science and engineering. *Gender and Education, 8*(2), 199-214.

IBM/Women in IT Champions/George, R. (2003). *Achieving workforce diversity in the e-business on demand era.* Retrieved August 2004, from http://www.intellectuk.org/sectors/it/women_it/2003/Achievingworkforcediversity.pdf

Kodz, J., Harper. H., & Dench, S. (2002). *Work-life balance: Beyond the rhetoric.* London: Institute for Employment Studies.

Miller, L., Neathey, F., Pollard, E., & Hill, D. (2004). *Occupational segregation, gender gaps, and skills gaps*. Working Paper Series, No. 15. Manchester, UK: Equal Opportunities Commission.

Office of National Statistics (ONS). (2003). *New earnings survey*. Labour Market Trends. Retrieved August 2004, from http://www.statistics.gov.uk/

Platman, K., & Taylor, P. (2004) *Workforce ageing in the new economy: A comparative study of information technology employment*. Cambridge: University of Cambridge. Retrieved from http://www.wane.ca/PDF/Platman&TaylorSummary Report2004.pdf

Simpson, R. (1998). Presenteeism, power and organisational change: Long hours as a career barrier and the impact on the working lives of woman managers. *British Journal of Management, 9*(Special Issue), 37-50.

Von Hellens, L., Nielsen, S. H., & Beekhuyzen, J. (2004). An exploration of dualisms in female perceptions of work. *Journal of Information Technology Education, 3*, 103-116. Retrieved from http://jite.org/documents/Vol3/JiteContentsVol3.pdf

Women and Equality Unit. (2004). *Encouraging diversity in the boardroom*. Retrieved November 2004, from www.womenandequalityunit.gov.uk/boardroom_diversity

KEY TERMS

Carer vs. Career: The difficulty some women experience in trying to manage domestic/caring responsibilities alongside their career. This difficulty should be viewed in the context of the continuation of the expectation that women are primarily responsible for the home and for children.

Flexible Working: Flexible working includes a variety of differing options with a reported aim of enabling people to better balance home life, family responsibilities, and working practices. Flexible working options are thought to include; part-time work, job-shares, flexi-time, time of in lieu, term time working, home, remote and teleworking, compressed hours and annualised hours.

Masculinised Domain: Suggests that within a given sphere of social life (i.e., the IT workplace) men tend to dominate proportionately (i.e., the under-representation of women in IT) and symbolically (i.e., that technology and masculinity are co-produced and that cultural images of technology are associated with hegemonic masculinity).

Presenteeism: The social/peer pressure to be seen to be at work beyond the call of duty and beyond contract stipulations, possibly to improve promotion prospects.

Take-Up Gap: The gap between the availability of flexible working practices (particularly part-time contracts) and the number of employees who opt for these 'work-life balance' initiatives.

Troubling "Technology": From work in social and feminist studies of technology, challenging the assumptions that technology is "neutral" and suggesting that technologies are socially constructed.

IT for Emancipation of Women in India

Anil Shaligram
One Village One Computer Project Trust, India

INTRODUCTION

At "One Village One Computer Campaign" (1V1C) in India we are resolved to tackle the gender question using information technology. The strategic slogan is "Age old problems, Youthful movement". Gender equality is sought in the context of the fight against a digital divide that is expressed through the problems of underdevelopment and exclusion. The approach is based on introduction of organizational innovations to raise human capital and social capital in the rural communities and connect them with each other and the world over through a knowledge network. In the hands of women, this becomes a weapon to fight against gender inequality and discrimination.

Through the use of information technology, a community centric approach can help rural India to combat social problems. In contemporary times where information, knowledge is the key to development and progress, IT can be used to combat the development concerns of rural India, while keeping local communities and their involvement and empowerment at the forefront of the process. As a technology IT is best suited for the "gendered" sex to empower themselves with education, information, knowledge, skills and so forth, and connect themselves with other rural communities and overcome physical isolation through IT network.

For resolution of gender problem, individualized IT empowerment has extremely marginal relevance, whereas tele-center like models based on private proprietorship has also very little success. IT Enabled Women's Social Network can be a solution in bridging the digital divide and gender problem. 1V1C campaign shows that it is possible to build such networks in remote villages and reach the most downtrodden and even illiterate women.

BACKGROUND

One Village One Computer started its work in village Mod, District Nandurbar of the state of Maharashtra in India, in the year 2000. A database of 3,000 landless laborers was created. The problems confronting destitute senior citizens, women, and patients were identified during the collection of this data and its processing. This led the laborers to organize agitation and make structured presentation of their specific health problems related to women, old people's pension entitlements, to the local health and development authorities. Thus, the problems were resolved immediately.

In the same year, this method was used in case of tribal women from Thane and Pune districts of the state of Maharashtra. Extensive data regarding ration cards for public distribution system (PDS), availability of food grains, functioning of ration shops under PDS, distribution of kerosene, black marketing of rations goods etc was collected. This concrete information could ensure that all deserving families obtained ration cards. This also helped in restoration of over 2500 ration cards, which had been arbitrarily cancelled by the authority.

1V1C project is in operation in 18 districts in the state of Maharashtra. Plans are drawn to ensure the spread of 1V1C in all the districts of the state of Maharashtra. 1V1C is collaborating with active people's organizations formed by peasants, landless labors, women, students, and youth.

1V1C is supported by the USA based organizations of people of Indian origin, such as Maharashtra Foundation and Asha for education. Ashoka Innovators for the public is supporting the project through the social entrepreneur fellowship program. People from IT, management, social research, media background from India and abroad contribute resources, knowledge inputs, and voluntary efforts.

A STRUGGLE FOR BRIDGING DIGITAL AND GENDER DIVIDES

An Overview of 1V1C

In this increasingly unequal world, one need not talk of the spread of hi-tech technology and expensive investments, like most IT providers do. Instead we can think of taking to rural India the very basic core IT applications, which urban educated people often take for granted, thereby making a positive impact on the many lives of rural India through the use of very basic techniques and inputs of the IT that are enough to assist rural India to develop.

1V1C uses readily available applications such as graphic designing, word processor, spreadsheet, presentation, e-mail, Web pages. 1V1C has developed Indian languages solutions through its technology partner Akruti Software. In most of the Indian villages connectivity is not available, so 1V1C depends on inexpensive stand-alone computers instead of heavily investing on connectivity technologies as is done by most of the IT for development projects. Hence 1V1C model is adopted to the existing level of technology, as well as it is adaptable to variety of social and geographic situations. Reliance on existing technologies and alliance with ongoing social movements makes 1V1C village centers cost effective, community supported and immediately beneficial to the villagers.

Philosophy

1V1C is a development strategy for introducing IT in rural areas as a tool for finding solutions for simple problems faced by rural communities. 1V1C has developed new methods of mass IT education and training for village youth and formation of community owned IT Centers. 1V1C strives to facilitate the vertical knowledge flows. It is the flow of subject knowledge held by experts (located at a distance) to the contextual knowledge held by people and vice versa, that can lead to development. 1V1C networks with existing social movements, non-government organizations (NGO) and community-based organizations (CBO). The activities of 1V1C lead to the development of sufficient social capital assets in a given locality making possible the formation of IT Center through community resources. This opens the window of the world to the local community while keeping their basic characteristics intact. It offers them access to the world level cutting age knowledge, which they can use, after appropriate contextualization, for their development. This also facilitates transmission of their traditional as well as and newly developed knowledge to the world community. 1V1C enables the social processes through creation of developmental software. This software is made available in the public domain through as free software.

Methodology

1V1C's efforts have been towards teaching and organizing of village communities to collect information and data relevant to their issues and concerns. The communities where 1V1C works have effectively tackled problems such as rural unemployment through organized and systematic data collection. The information regarding extent and prevalence of unemployment in the villages is used to demand more work under the government's employment guarantee scheme. Local communities have also been taught the value of the principles and techniques of IT such as systematic and reliable data collection and fact finding on their core issues like the below poverty line (BPL) numbers in villages, and demographic information which in turn when used effectively have accelerated people's struggles on issues such as rationing, accessing housing schemes, and old farmer pension schemes.

Another issue was addressed in one of the 1V1C operating villages, where the health of the women and children was badly affected. Here too, the team of people trained by 1V1C engaged in systematic data collection on the occurrence of the problem, which when analyzed and presented to the local health authorities was evidence enough to convince the state health department to organize health camps and check ups and ensure the right to health of community. Such demonstrated attempts have proved the value of reliable data collection and the science of information technology to the local community.

1V1C's work comprises organizing training camps where basic computer skills are taught and it is also used to generate an interest among the local communities on the various uses of IT to benefit their own lives. Interested villagers form CSCs (computer support committees) and learn to use the IT methodology and techniques to try to solve their social concerns. Once successful, the village communities attempt to use the IT to solve many more social issues and gradually move towards forming an IT Center.

The 1V1C process gives village leaders and communities at large a new boost to tackle their issues and since by conception the process is rooted in the involvement of rural communities, it also contributes to leadership development, capacity building, and human capital, resulting in overall social capital building in society.

Information Cooperatives

The whole system operates on a cooperative basis. Any individual or a set of individuals does not own the IT Seva Kendra. The local people are asked to form a cooperative-like structure or are part of an already existing people's organization. Further, a core committee is elected whose membership is by rotation and it is the responsibility of this committee to manage the center. They engage volunteers to manage these centers, and the center usually develops around an existing people's organization already working in that area. The elected CSC, particularly youth and women members are responsible for running and managing the center. 1V1C team plays a crucial role in developing the capacity and ability of these elected people to learn and use information technology to solve community issues, who in turn train and assist the village people to do the same. No one person or groups of persons owns the center or its resources, including the products and learning that are used and developed. The information generated belongs to the community as a whole and hence they have full ownership and access to the same.

Networking

While the democratic way of functioning within villages and districts works through the above mechanism, the idea of 1V1C is to help villages connect beyond themselves to other villages, districts, and maybe even beyond in the near future as and when technology advancement and accessibility grows. The idea is to form a widespread and deep 1V1C network (a social network) wherein knowledge, experiences, and models can be shared across villages and districts and even states, leading to actual self-empowerment and change, by the people and for the people of rural India. The conception of IT Center requires the local population to be convinced of the value of information technology for their villages and districts and hence their willingness to invest in the same. This is to ensure that the local village communities gain greatest stakes in this investment, which works for them.

Women and 1V1C

Information technology can be a decisive weapon for women in their fight for emancipation. In India women are doubly oppressed and in modern times are mostly engaged in some productive occupation. They may be working as farm laborers, household domestic workers, servants in government departments and offices, lunch and eatable suppliers, domestic cigarette rollers, and so forth. Female literacy is lagging substantially behind male literacy. Almost half of the women are illiterate; hence when we speak about illiteracy in India, it is mostly about women. Women are engaged mostly in unskilled work in agriculture as well as in industry that is valued less. They have to face the main brunt of unemployment and poverty. They are deprived of mainstream knowledge, science, and technology. It is a well-known fact that women are paid less than men for similar work throughout the world. Socially women are given secondary status to men.

The Training Curriculum for Women

The training curriculum covers not only the necessary IT tools in the local language, but presentations and discussions on social issues, gender issues, and capacity building exercises to build leadership skills of women volunteers. Social issues are integrated in the skill building exercises. For example, in one of the training programs, participants were asked to provide information on the educational status of girls in their families using a word processor, and they were then taught to make presentations using the same information. This highlighted the problems faced by the girls in completing education, and prompted the male participants to reflect on issues of gender inequality in the family.

Thus, we could experiment and develop the concept of mass community IT training in primary computer applications and computer usage, and integrate this training organically with social thinking and social issues. This process of training saw the village youth gaining confidence and self-esteem. They started thinking about their community's problems and exploring innovative ways of using IT in solving them.

The objective of the training curriculum is to create *IT Enabled Women's Social Networks*. The starting point is gradually training a large number of young women from villages. They are equipped with the necessary IT tools to build leadership skills and social entrepreneurship skills. 1V1C has developed and implemented the concept of *training of trainers* to train large number of village women and community volunteers. Thus, a substantial amount of human knowledge capital is created at the local level. This human capital forms the basis for building of *IT Seva Kendras* (IT Service Centers).

IT Seva Kendra

The IT Seva Kendra's primary role is to link the village community to the world. The Kendra have a symbiotic relationship with the community, with both elements inspiring, nurturing, and drawing from each other. Thus, the Kendra is part of the community and contributes to its growth, while at the same time the community supports the Kendra to grow and be sustainable. IT Seva Kendra becomes a tool, a nucleus for the community to progress and fight against its backwardness and isolation. It is this interactive and incremental process of knowledge transfer that makes the IT Seva Kendras stand apart from the much publicized tele-center models across the world. It is possible to build Women's IT enabled knowledge based rural social networks in backward areas. If IT were successfully and extensively diffused at the village level, it would enable democratic participation of people in a variety of issues. These social nets would form a horizontal structure.

The "Women Only" Program

Fifteen training programs were conducted up to the end of 2003. The first five days training program was held in March 2002 and had 62 participants. These participants represented peasants, landless laborers, women, students, and youth organizations working in 16 districts of Maharashtra. Four training programs were subsequently conducted for the New Bombay youth. Additionally, one training program was organized exclusively for girls and women in the village Murbi, near New Bombay. Altogether 72 boys and girls were trained in these camps. The youth trained at these programs started to assist the trainers in the ongoing training camps. This led to launching of IT Seva Kendras in several districts of Maharashtra state.

In all these training camps, there were women participants. However, we found it necessary to experiment with a "women only" training camp. We provided a "safe", women only, space where young girls/women could come together, express themselves, share, and learn without any inhibitions.

A "women only" five days training program was held at district Wardha in the state of Maharashtra in June 2004. Sixty young women volunteers active in a social movement against illicit liquor brewers and drunkards attended it. They resolved to use IT in the struggle against this problem, which is spreading like an epidemic in rural areas and causing severe social problems. In this camp illiterate women learnt computer applications.

Women March to Seize IT

At a training camp held for Karad, district Satara of the state of Maharashtra, two illiterate village women worked on computer. When Shantabai and Shankuntalabai—two illiterate women from village Goleshwar, district Karad, decided to attend 1V1C program, villagers laughed at them. They wondered how these women could learn computers when they didn't know how to read or write. However, the women vowed to return to the village armed with computer skills. They spent 20 hours a day during the 5-day camp to learn about computers. After the camp, they returned to the village and started teaching other women. Soon the villagers, who were skeptical earlier, approached for help. They were interested in seeking solutions to the sanitation problem in the village. The two women logged on to various Web sites, which offered solutions, like how to construct low cost toilets, etc. Says Shantabai, "Now people don't dare to laugh at us. On the contrary they take us seriously."

At Manvat in Parbhani district, as soon as a computer was installed in the village, a group of curious women, who worked as sweepers, came to see how it worked. It was explained to them how they could use the computer to improve their lives. The women got interested and 30 of them, mostly illiterate, attended a 5-day computer-training camp. 1V1C members advised them to collect all the relevant data about their working conditions and with the help of their school-going children, feed them into the computer. In 2003, on the basis of this information, the women filed cases before the Right to Health Commission in Mumbai for accident and injury claims. Emboldened by the commission's positive response, about 700 women from Manvat town gathered under the leadership of these sweeper women and participated in a protest march to highlight their grievances and demands—lack of medical facilities and housing, unemployment, and pensions for old women workers.

1V1C held three computer-training camps for rural people at Agroli in New Bombay in May 2004. The first camp included men and women who were trained as trainers by the seven women of Murbi village. This batch further trained 138 women from different parts of Maharashtra. The batch learnt to use the keyboard in the local language. Besides, they also learned the art of public speaking, which was found essential for voicing their grievances.

Using IT for Women's Issues

Extensive surveys have also been conducted on the impact of the dowry system resulting in increasing violence against women leading to bride burning, desertions, and bigamy. In Maharashtra there are over 600,000 deserted women, most of them living below the poverty line.

A database of more than 30000 domestic women workers is created and 1V1C has developed a Domestic Women Workers' Software (KamwaliBai

software) to analyze the data. This software is useful in formulating and resolving complex problems faced by these women.

CHALLENGES AND FUTURE TRENDS

1V1C proposes to take up challenging job of development of full-fledged training module for illiterate women to teach IT literacy, which could lead to primary literacy. The second challenge is to link women's social network with that of the self-help groups (SHG) (as part of micro credit movement). Third task is to link with elected women representatives at the grass root level. 1V1C also expects to deal with issues faced by slum dwelling women from cities and towns. Domestic Women Workers' Software is a step in that direction.

CONCLUSION

From beginning 1V1C has always kept the gender question on the forefront. Learning of the 1V1C campaign shows that gender issues and women's participation should be taken up from the very inception and special attention should be given to nurture the leadership of women activists.

Keeping computers at the center stage of the entire process is very much successful so far and has shown definite results. Training in skills, leadership and social entrepreneurship has definitely helped. The spontaneity of the whole process is encouraging. Young women activists are coming up to take responsibility for running of the IT Centers on voluntary basis. A group of women trainers have emerged who are conducting training camps not only in their own villages but elsewhere as well. Women's leadership is proving to be an asset for the entire communities.

Armed with databases, facts and figures, and quantitative analytical tools in their hands women feel more confident to press and lobby for their demands and achieve greater successes. They are able to analyze their problems at micro as well as macro levels, which help them formulate their demands in a better and effective way. It also helps them to acquire various types of skills, and develop a rational, scientific outlook that is necessary for their advancement and leadership development.

KEY TERMS

1V1C: One Village One Computer Campaign in India.

Akruti: A brand name of Indian language software. (Literal meaning of Akruti is graphic figure.)

CBO: Community based organizations.

CSC: Computer Support Committees formed around 'IT Seva Kendra' to work on specific community issues and consisting of village volunteers.

IT Seva Kendra: It is a community owned cooperative entity that nurtures learning and innovation in the community. IT Seva Kendra is a window to the world for the village community through whom they can access cutting edge knowledge, modify and use that for their own benefit, publish their own implicit and local knowledge and share their aspirations to the world.

KamwaliBai Software: Proper name of the domestic women workers software developed at 1V1C (Literal meaning, Kamwali=domestic worker, Bai=woman).

IT Work in European Organisations

Juliet Webster
Work & Equality Research, UK

INTRODUCTION

Employment in IT professions has increased greatly in recent years. Aside from the crisis of the dot.com crash in 2001, there has been significant growth in hardware manufacturing and particularly in software and IT services. In the European Union, employment in computer services doubled between 1997 and 2001, and grew by 10% in 1998 alone.

This pattern has not been matched by a parallel increase in women's participation in IT work. Women's employment in IT has remained resolutely around an average of 28% across the EU; in the professional areas of IT work (as opposed to clerical and other non-professional occupations), women made up only 17% in 2001 and their representation is in fact declining (Millar, 2001; Millar & Jagger, 2001; Webster & Valenduc, 2003).

It is an issue of some concern to policy makers, employers, and indeed gender equality practitioners that, despite more than 20 years of attempts to attract women into this comparatively well paid and privileged area of the labour market, women remain such a small and, worse, apparently declining, proportion of IT professionals. Why are women still so poorly represented in IT professions in the EU? What is the nature of working life in IT and what are the working conditions like? Why have more than 20 years of initiatives to get more women into technology professions had so little apparent impact?

BACKGROUND

This article summarises the results of a European research project which attempted to answer these questions, focussing on the situation in seven EU countries: Austria, Belgium, France, Ireland, Italy, Portugal, and the UK. Entitled "Widening Women's Work in Information and Communication Technologies" (WWW-ICT), the project combined biographi-cal interviews with female and male IT professionals with case studies of employing organisations in the IT services sector, and was conducted between 2002 and 2004.

As Table 1 shows, women still made up less than one-fifth of IT professionals in these countries in 2001, with the exception of Ireland. Indeed, IT professionals in Europe are typically male, young (in their mid twenties), and without domestic responsibilities. The majority of women working in the sector are also young and childless. These employees are among the most favoured in the labour market. Wages are relatively high, and many IT workers are paid in a combination of cash and share options. Moreover, employment contracts involving individually agreed pay, terms and conditions replace the fixed pay grades traditionally found elsewhere. Performance-related pay or bonus schemes are common (Valenduc et al., 2004).

Employment is predominantly on full-time permanent contracts. Part-time employment and flexible working arrangements are very unusual, though

Table 1. Employment in IT professions (ISCO213) in the WWW-ICT countries, 2001 (thousands of employees)

	Female	Male	% Female
EU15	265.4	1264.8	17%
Belgium	8.8	49.7	15%
France	50.3	250.3	17%
Ireland	6.0	14.6	29%
Italy	(2.0)	9.0	(18%)
Austria	(1.4)	8.8	(14%)
United Kingdom	63.5	351.1	15%

Note: Data in brackets and on Portugal are considered unreliable by Eurostat
Source: Eurostat, data from the Labour Force Survey, quoted in Valenduc et al. (2004)

they are more common among female employees. Full-time working often means long working hours. Project work can be unpredictable, involving tight deadlines, so evening and weekend working is common. Working hours often exceed those laid down in employment contracts, though overtime is rarely paid for. Employees can arrive at and leave work according to their own preferences, but this tends to translate into long hours, which are often self-imposed (Mermet & Lehndorff, 2001). Consequently, in France, for example, the implementation of the 35-hour working week has been very problematic in this sector; even the imposition of the legal limit of 39 hours was fraught with difficulties. Given these kinds of working patterns, it is unsurprising that the sector employs predominantly young men able (and apparently willing) to provide the total availability needed by their employers.

In employment and industrial relations, the IT sector is a world away from traditional companies. Trade union membership and collective bargaining are weak, and there is corporate antipathy or hostility to unions. There are particularly low levels of unionisation on U.S. owned green-field sites. The fact that computer services employees are young, highly skilled, and up until recently, operating in a favourable labour market, also militates against trade unionism. Even in countries with strong collective bargaining frameworks (for example, Belgium, France), union membership is low and employment relations are highly individualised. Pay and conditions are agreed bilaterally, and often kept confidential from other employees. Pay is based partly on performance assessed through individual appraisals carried out periodically by line managers. Performance systems, bonus systems, and stock options have been relatively lucrative for IT professionals, but since the 2001 downturn, they have been more vulnerable to the vicissitudes of the stock and labour markets. Communication—not consultation—is carried out on a one-to-one basis between employers and employees (Valenduc et al., 2004). This is the context within which we attempt to understand the under-representation of women in the professional areas of IT.

MAIN THRUST OF THE ARTICLE

The Organisation of Work

IT companies tend to be flat structures with few hierarchical layers. It is common for IT professionals to be organised into project teams, led by a project manager. These teams may be temporary, operating only for the duration of the project, or semi-permanent. They may consist of interdependent workers with complementary skills, or individuals with the same skills working independently of one another within the team. Women are often undervalued in interdependent teams, where their technical skills are taken for granted relative to the interpersonal or team-working skills of their male counterparts (Woodfield, 2000).

Working Time and Work-Life Balance

IT work is predominantly full-time work. Much of it is deadline-driven, particularly where it is governed by project timetables or client demands. Long working hours are the norm, as is availability to the company and to clients. Hot-desking and client-based working are common among IT professionals, as is home-based working, with systems provided by employers. This can extend working hours; it is common for IT professionals to work at "unsocial" working hours—late at night after children are in bed, or very early at weekends.

Working hours? They are exaggerated because no one can say a simple "no" to the client. This is the company's policy. You have to give all your availability and energy to the firm: working overtime and sometimes also at home after work. (Marta, Italian IT company, quoted in Webster [2004] Case Studies of Work Organisation [WWW-ICT Deliverable No. 7], www.ftu-namur.org/www-ict)

Part-time working is very unusual in this sector, and is principally done by women returning from maternity leave. It has been found to severely limit progression prospects, with companies demoting and marginalising part-timers. Other family-friendly working arrangements are rare in European IT companies, some of which regard families as problems that divert employees from their work. Work-life balance policies may be used in a tight labour market to attract a wider pool of job applicants. In recessionary conditions, there is no need for such arrangements.

Informal flexible working arrangements do exist; employees are often allowed by their companies to take time off when they need to, as long as their work is done. In practice, this usually means more time spent at work, rather than less. In general, reconciliation between professional and private life is difficult for employees (of both sexes) in IT professions (Webster, 2004). It is not clear whether the industry attracts young, single people because they are the only employees who can manage these types of working time demands, or whether the working time arrangements have evolved in response to the type of employees who predominate in the sector.

Employee Development and Women's Progression

Most IT professionals in Europe have first degrees, and many have higher degrees. On-the-job learning and skills maintenance are considered critical in the IT professions as the means by which professionals build their knowledge of the most recent technical developments. Business and management skills take on increasing importance in the career development of IT professionals, and technical skills become less prominent.

In large companies, there are employee development opportunities, particularly in comparison with smaller organisations which have fewer resources for this. However, increasingly there is a tendency across the sector for training and development to be individualised—for training to be managed and conducted by the individual employee using computer-based learning, the Internet, and interaction with peers, with low levels of intervention by the employing organisation. The individualisation of training departs from formal, supply-driven systems, focuses more on the individual learning requirements of employees, and places much more autonomy in their hands. However, it can be difficult for employees with domestic commitments to find time for learning outside of normal working.

When I see IT professionals, programmers in fact, they have to constantly continue training. I think that that side of the job is difficult to balance with family life. In fact, those who do so aren't married, and don't have kids. (Computer graphic artist, Belgium, quoted in Webster [2004] Case Studies of Work Organisation [WWW-ICT Deliverable No. 7], www.ftu-namur.org/www-ict)

The IT sector is a relatively privileged place for women to work. Pay and autonomy are high, and there are considerable opportunities for progression, along two basic career trajectories: a technical career path and a management career path. The latter is the most common career pattern for IT professionals, and includes possibilities to move into project management, team management or business management. Employers often assume that women are more comfortable in management than in technical roles. In fact, women are very much attracted to technical work and enjoy doing it, because it is "creative." Creative work can mean coding and programming, designing and developing a Web site or service, or developing an overview of a project through project management work. Solving problems is one of the most satisfying aspects of the work. Yet women are sometimes directed away from technical work and towards project or business management, on the assumption that this sort of work is particularly closely compatible with their assumed interpersonal and organisational skills.

Nevertheless, women remain significantly under-represented in managerial and particularly executive positions in IT (see also Panteli, Stack, & Ramsey, 2001; Tijdens, 1997). First, informal and opaque progression arrangements persist in the IT professions. These include "promotion through visibility," in which participation in informal social activities (football clubs and pub evenings, for example) raises visibility and so confers advantages on certain employees, usually men (see also Tierney, 1993). The WWW-ICT study also found direct discrimination against women by male managers, on the basis of assumptions about their availability for, and commitment to, their work, particularly on and after maternity. Panteli, Stack, and Ramsey (2001) similarly report employers giving women less responsibility and allowing their marginalisation in organisational cultures.

Women also commonly understate their own skills and knowledge, and deselect themselves from eligibility for promotion opportunities. Self-advocacy, although in principle empowering, can disadvantage those without strong self-confidence, women in particular. It can also be problematic if their self-confidence, rather than their other skills and qualifications, are the basis on which employees are assessed and promoted.

Good employers understand the need to implement consistent policies for recruitment, training, appraisal and development, in order to improve women's recruitment and, crucially, their retention in IT. Such coherent policies communicate clear messages to women about potential career routes, and provide the infrastructural channels through which they may move. Some companies also run "fast track" progression systems in conjunction with specific schemes for developing women, through mentoring and other confidence-building initiatives. In general, organisations with awareness of how gender operates within and beyond their own spheres are most likely to recruit and promote women into senior positions. However, even the most equality-conscious companies have internal conflicts between their equality agendas and their other organisational practices, particularly during periods of restructuring. Women IT professionals seem to leave the profession in disproportionate numbers at maternity, and then again in mid-life; either they are disproportionately targeted by organisational redundancy programmes or they voluntarily leave their jobs in search of other working arrangements, just at the point when they might be entering senior management and executive positions (George, 2003).

A major obstacle to women's representation in IT professions—one which cuts across very many well-intentioned corporate equality programmes—lies in the working time arrangements and culture of the profession. Long working hours particularly affect people in technical roles, who have to be available to their employers and their clients, and those in senior management. Moreover, they are part of the IT working culture even in countries which do not otherwise have a "long hours culture", and in which the European Working Time Directive has been adopted without quibble. Even in companies with strong gender equality programmes, promotion into senior positions appears to depend upon the ability and willingness to work long hours. This transmits implicit messages from senior executives to more junior staff that such working patterns are necessary for career advancement—messages which fundamentally contradict those that they wish to convey through their other equality initiatives. This may discourage people—of both sexes—who are unable to engage in it from pursuing promotion possibilities in their organisations. In general, of course, it is women who are primarily disadvantaged by long working hours.

FUTURE TRENDS

Since the early 1980s, when computing first emerged as a significant new area of work, there have been widespread attempts by public authorities, voluntary organisations and private sector employers, to attract and retain women into computing professions. Most of these initiatives were informed by the idea that "adding women in" to technological jobs would address the exclusion of women from technology (Henwood, 1993). The context within which these initiatives were pursued—wider corporate strategies and practices concerning organisational and technological changes—were often, however, overlooked and, consequently, many initiatives were ineffectual.

Developments in the IT sector at the beginning of the 21st century have created a difficult environment for improving women's representation. The sector has undergone an almost unprecedented downturn. Over 100,000 employees and contractors have been made redundant in the UK since the middle of 2001 (E-Skills Bulletin, 2004), organisations have been restructured, while programming functions are now routinely outsourced to third countries (India, Israel, Romania, for example). Many of these events are extremely detrimental to both women's numerical representation and the quality of their working lives in IT.

Hacker (1989) noted that the process of organisational and technological change in AT&T in the 1980s ultimately undermined her attempts to pursue equality initiatives. Similarly, restructuring programmes in European IT companies are proving extremely hostile to more localised, decentralised equal opportunities programmes. Recessionary conditions (such as those following the dot.com crash of 2001) seem to prompt, in the large corporations at least, a reassertion of highly centralised decision-taking, authoritarian and bullying management styles, an abandonment of corporate commitments to equality and a return to conventional fiscal performance measures which allow no leeway for longer-term projects.

In this context, training and development budgets are commonly cut back, with particularly negative consequences for women, who find generally it difficult to pursue these activities in their own time. In corporate redundancy programmes, middle-aged women may be more vulnerable than their male counterparts. Competitive pressures wrought by an economic downturn cut across well-intentioned and well-structured equal opportunities strategies and are ultimately more influential on corporate behaviour. The retrenchment by corporations also reduces the pressure on them (at least during periods of skills shortage) to draw from a wide a portion of the labour market as possible.

Nor are these merely temporary responses to contemporary competitive conditions. A profound change in the conduct of IT organisations is taking place, and this is gaining ground across the European IT sector. Despite rhetorical emphasis on teams and team working, the organisation of both employment and work processes are becoming increasingly individualised. In HR management, there is increasing emphasis on employees' personal qualities, including self-direction, self-management and self-advocacy, and on placing responsibility for employee development with the individual rather than the organisation. Collective bargaining is being displaced by the setting of pay and employment terms and conditions on a unilateral basis between management and employee. Trade unionism is discouraged by employers and seen as irrelevant by many employees.

CONCLUSION

Employment conditions in IT professions are not woman-friendly. Long working hours, lack of structured training and development, promotion systems based on availability and visibility, and persistent chauvinistic assumptions about women's commit-

ment to the work, which underpin management practices, are all factors that have combined to prevent women from progressing in the IT professions. Equality and diversity programmes have admittedly attempted to address these issues, often through adjusting recruitment, training and development systems, improving progression systems and modifying working time demands. This was feasible when organisations were motivated to improve their record on women's participation in the IT professions, through skills shortages or tight labour markets. However, when this is less of an incentive and as the sector becomes increasingly governed by cost cutting through new forms of efficiency management, the project of improving the conditions of women's participation in IT is sacrificed to the cause of improving shareholder value through fiscal performance improvements. This exemplifies the profound tension between the objectives of equality and social cohesion with the imperatives of the market, a tension which may explain the apparently negligible impact of initiatives to get more women into technology. With IT services becoming increasingly competitive, globalised, and rationalised, what is the longer-term scenario for women in IT? This is a question for future cross-national comparative research.

REFERENCES

E-Skills Bulletin. (2004). *Quarterly review of the ICT labour market*. London: E-Skills UK.

George, R. (2003). *Achieving workforce diversity in the e-business on demand era*. Portsmouth, UK: IBM.

Hacker, S. (1989). *Pleasure, power, and technology*. London: Unwin Hyman.

Henwood, F. (1993). Establishing gender perspectives on information technology: Problems, issues, and opportunities. In E. Green, J. Owen, & D. Pain (Eds.), *Gendered by design: Information technology and office systems*. London: Taylor and Francis.

Mermet, E., & Lehndorff, S. (2001). *New forms of employment and working time in the service economy (NESY)*, Country Case Studies Conducted in Five Service Sectors (ETUI Report 69). Brussels, Belgium: European Trade Union Institute.

Millar, J. (2001). *ITEC skills and employment—assessing the supply and demand: An empirical analysis* (Issue Report No 11). Socio-Economic Trends Assessment for the Digital Revolution (STAR) project report. Brighton, UK: Science Policy Research Unit.

Millar, J., & Jagger, N. (2001). *Women in ITEC courses and careers*. Brighton, UK: Science Policy Research Unit and Institute for Employment Studies.

Panteli, N., Stack, J., & Ramsey, H. (2001). Gendered patterns of computing work in the late 1990s. *New Technology, Work, and Employment, 16*(1), 3-17.

Roche, W. K., & Gunnigle, P. (1995). Competition and the new industrial relations agenda. In B. Leavy & J. S. Walshe (Eds.), *Strategy and general management: An Irish reader*. Dublin: Oaktree Press.

Tierney, M. (1993). Negotiating a software career: Informal work practices and "the lads" in a software installation. In K. Grint & R. Gill (Eds.), *The gender-technology relation: Contemporary theory and research*. London: Taylor and Francis.

Tijdens, K. (1997). Gender segregation in IT occupations. In A. F. Grundy, D. Kohler, V. Oechtering, & U. Petersen (Eds.), *Women, work, and computerization—Spinning a Web from past to future*. Berlin: Springer.

Valenduc, V., Vendramin, P., Guffens, C., Ponzellini, A., Lebano, A., D'Ouville, L., et al. (2004). *Widening women's work in information and communication technology*. Namur, Belgium: Work and Technology Research Centre.

Webster, J. (2004). *Case studies of work organisation* (WWW-ICT Deliverable No. 7). Retrieved from http://www.ftu-namur.org/www-ict

Webster, J., & Valenduc, G. (2003). Mapping gender gaps in employment and occupations. In P. Vendramin, G. Valenduc, C. Guffens, J. Webster, I.

Wagner, & A. Birbaumer (Eds.), *WWW-ICT Conceptual Framework and State of the Art* (WWW-ICT Deliverable No 1). Retrieved from http://www.ftu-namur.org/www-ict

Woodfield, R. (2000). *Women, work, and computing*. Cambridge, UK: Cambridge University Press.

KEY TERMS

35-Hour Working Week: Law enacted in France in 1998 which provided for the introduction of a statutory 35-hour week from January 2000 (2002 for smaller companies). The legislation was relaxed in March 2005.

Business Management: In matrix organisations, management of separate businesses within the organisation is devolved, and involves overall responsibility for the financial, marketing, planning, human resource management and project implementation tasks involved in the business.

Biographical Interview: Interview which aims to understand a person's biography or developmental trajectory, covering significant episodes, important events and the role of relevant others.

Development Work: Development work involves the programming of software, at different levels of complexity.

Dot-Com Crash: Sudden crash in the share prices of internet companies in 2001 after return on investments were not met by many companies, with repercussions throughout the IT sector.

European Working Time Directive: The European Working Time Directive was first adopted in 1993, and aims to ensure that workers are protected from working excessively long hours, having inadequate rest or disrupted work patterns. It includes provision for a maximum 48 hour working week.

Project Management: General term covering a group of jobs involving leading and managing a team charged with a specific set of tasks.

IT Workforce Composition and Characteristics

Joshua L. Rosenbloom
University of Kansas & National Bureau of Economic Research, USA

Ronald A. Ash
University of Kansas, USA

LeAnne Coder
University of Kansas

Brandon Dupont
Wellesley College, USA

INTRODUCTION

Women are under represented in the information technology (IT) workforce. In the United States, although women make up about 45% of the overall labor force they make up only about 35% of the IT workforce. (Information Technology Association of America, 2003, p. 11). Within IT, women's representation declines as one moves up to higher-level occupations. While women are relatively more numerous among data entry keyers and computer operators, they are relatively less likely to be found in high-level occupations like systems analysts and computer programmers.

The relatively low representation of women in IT fields parallels a broader pattern of gender differentials in other scientific and technical fields. In all science, technology, engineering, and mathematics fields combined, women held 25.9% of jobs in 2003. Women's representation varies widely by sub-fields, however; 65.8% of psychologists and 54.6% of social scientists are women, but only 10.4% of engineers, and 37.4% of natural scientists (Commission on Professionals in Science and Technology, 2004, p. 2).

Over the course of the past 100 years, there has been a dramatic change in women's economic role. In 1900, only one in five adult women worked outside the home, and most of these were young and unmarried (Goldin, 1990). Since then, male and female labor force participation rates have tended to converge. Between 1900 and 1950 there was a gradual expansion of women's labor force participation.

After World War II the pace of change accelerated sharply as more married women entered the labor force. During the 1960s and early 1970s a series of legal changes significantly broadened protection of women's rights ending essentially all forms of overt discrimination (Fuchs, 1988; Long, 2001, p. 9-10). The removal of these barriers in combination with the availability of cheap and reliable birth control technology greatly facilitated the entry of women into higher education, and technical and professional positions (Goldin & Katz, 2002).

Nevertheless, as the figures cited at the outset reveal, women's participation in IT and other technical fields has not increased as rapidly as it has in less technical fields. And in striking contrast to the general trend toward increasing female participation in most areas of the workforce, women's share of the IT workforce in the United States has actually declined over the past two decades. Any effort to explain gender differences in IT must begin with an understanding of how the number, characteristics, and pay of women in IT have evolved over time, and across different sub-fields within IT. This chapter provides a foundation for this analysis by documenting recent changes in the number of women employed in IT, their demographic characteristics, and relative pay.

BACKGROUND

A discussion of the gender composition and characteristics of the IT workforce must begin by clarifying what is meant by IT. This is difficult because IT encompasses a broad array of products and activities related to computing and communications in the modern economy (Freeman & Aspray, 1999, p. 29-31). Although many workers make use of IT in their jobs, most studies agree that only those workers who are responsible for creating IT hardware and software should be included in the IT workforce, while those who are primarily users of these products should be excluded (In addition to Freeman & Aspray, see Ellis & Lowell, 1999, p. 1; National Research Council 2001, p. 44-54).

Whatever conceptual definition one adopts, however, its application is limited by the classification schemes used by agencies engaged in collecting data on different elements of the workforce. In what follows we will focus on those IT occupations that are enumerated in the Bureau of Labor Statistics' Current Population Survey (CPS). The CPS data cover computer systems analysts, computer programmers, operations and systems researchers, computer operators, and computer operators supervisors. These occupations constitute more or less what the National Research Council (2001, p. 48) has termed "Category 1" IT occupations: those involved with the creation of new products, services and applications. CPS data do not permit us to measure or describe the characteristics of the National Research Council's "Category 2" occupations: those involved in the application, adaptation, configuration, support or implementation of IT products or services (National Research Council 2001, p. 49). Because occupational titles do not adequately capture the IT content of the support activities of many of the technicians and other occupations included in this group it is more difficult to adequately measure its size or demographic characteristics.

THE SIZE, COMPOSITION, AND CHARACTERISTICS OF THE IT WORKFORCE

An Overview of IT Labor Market Conditions

The rapid and sustained decline in the cost of computers over the past two decades has been a prominent factor in the reorganization of work in the United States. Between 1984, near the beginning of the personal computer era, and 2001 the quality-adjusted price of computers fell at an average annual rate of 16%, resulting in an 18-fold drop in price (U.S. Department of Commerce; cited in Weil, 2005, p. 263). As personal computers diffused into widespread use, mini-computers vanished from the market, and sales of large corporate mainframes languished. Shifting markets and the changing needs of users resulted in significant shifts in the software industry. Growing consumer markets fostered growth of the packaged software industry, and created whole new categories of software. Since the early 1990s, the spread of the internet and the increasing importance of networked computing have initiated a new round of changes in the IT industry (Mowery & Rosenberg, 1998). Adding to demand pressures during the late 1990s was global concerns about the Y2K problem.

Strong demand for IT professionals contributed to a rapid expansion of the IT workforce and rising relative pay. From 1983 to the peak of the technology boom in 2000, the IT workforce more than doubled in size, increasing from 1.47 million to 3.13 million persons. To put this in perspective, during this same period the total U.S. labor force increased by just 34%, from 99.5 million persons to 132.2 million persons (these figures and all the subsequent statistics are derived from the authors' computations based on data from the Current Population Survey's merged outgoing rotation groups). Despite the loss of more than 200 thousand IT jobs in the next two years, the IT labor force in 2002 was still 96% larger than it had been in 1983. To draw more workers into IT jobs relative pay had to rise substantially. In 1983 the median hourly wage of full-time IT professionals was about 20% above that for all non-IT occupations. By the late 1990s the wage gap had more than tripled, so that IT professionals earned more than 60% more than did workers outside of IT.

The growth of IT employment coincided with important changes in the type of jobs performed by IT professionals. Most obviously, as the importance of mainframe computers diminished, the number of computer operators fell substantially. From a peak of 962 thousand computer operators in 1986, the number of computer operators had fallen to just over 300 thousand by 2002. From being close to half of all

IT professionals in the mid-1980s this category of workers fell to under 11% of the IT workforce by 2002. Offsetting this decline was the extremely rapid growth in the number of computer systems analysts and scientists. This segment of the IT labor force grew from 273 thousand in 1983 to more than 1.7 million in 2002. By the latter year, this category of workers constituted over 60% of all IT professionals, up from less than 20% in the early 1980s.

Gender Differences in Employment, Earnings, and Hours

Contrary to the trends in most of the U.S. labor force, the share of women in the IT workforce has declined substantially over the past two decades. In 1983 women made up slightly more of the full-time IT workforce (43%), than they did of all full-time non-IT workers (40%). By 2002, however, the share of women in IT had fallen sharply, dropping to 30%, while the share in the non-IT workforce had risen to over 49%.

The decline of female representation in IT is troubling, but much of this decline can be accounted for by the declining number of computer operators. Removing this group, the share of women in other IT occupations has remained quite stable at around 28 to 29% of the workforce. Thus the falling share of women reflects the growing importance within IT of occupations that have traditionally been dominated by men (and, implicitly, the failure of more women to enter these traditionally male-dominated fields).

As is true more generally, women in IT earn less than men do. Indeed the gender wage gap in IT is quite similar to that in the rest of the labor force. In 2002, women in IT earned 82.5% as much per hour

as men, while in the rest of the labor force they earned 82.8% of what men did. Average pay for computer operators is considerably lower than for other IT occupations, so the concentration of women in this field tends to magnify the gender pay gap. Excluding computer operators, women earned about 86% of what men did in the remaining IT occupations. This pay ratio has been approximately constant over the past two decades, increasing only from 83% in the early 1980s.

IT occupations are often characterized as involving long hours and requiring a significant time commitment. One reflection of this is the higher proportion of both men and women in IT who work full time. In 2002, 95% of men and 91% of women in IT worked full-time. In non-IT jobs 87% of men and just 73% of women worked full-time. As a result the average woman in IT worked more than three additional hours per week than did the average woman in a non-IT job (39.5 hours compared to 36.2 hours). The longer hours in IT may be one factor that discourages women—especially those with young children—from going into or staying in the field.

Gender Differences in Demographic Characteristics

Table 1 summarizes a variety of demographic characteristics for IT and non-IT occupations broken down by gender. As the table reveals, IT workers tend to be somewhat younger than the rest of the labor force. This is especially true for male IT workers, who are on average more than three years younger than their non-IT counterparts, but female IT workers are also younger than women in non-IT

Table 1. Selected demographic characteristics of information technology and non-information technology workers, 2002

| | Information Technology | | Non-Information Technology | |
	Male	Female	Male	Female
Average age	37.9	39.9	41.0	41.0
Percent with Bachelors Degree	50.0	39.4	19.4	21.0
Percent with more than Bachelor's degree	19.0	13.6	10.5	10.1
Percent married, spouse present	64.9	53.4	64.7	54.3
Percent never married	26.2	27.4	22.6	22.7
Percent living with one or more of their own children	58.2	54.2	56.3	53.8

Source: Authors' calculations from Current Population Survey merged outgoing rotation group data

occupations. Reflecting the high levels of training needed to enter IT professions, many more workers in IT jobs have bachelors degrees or higher. Fully 2/3 of men and more than half of women in IT occupations have at least a Bachelors degree, compared to 30% of men and 31% of women in non-IT occupations.

In contrast to the differences in age and education levels, the percent of workers who are married with spouse present is relatively similar between IT and non-IT occupations. It is true, however, that IT workers are somewhat more likely to have never been married than is true for those in non-IT occupations, but it seems likely that this is due to the fact that IT professionals are younger than the non-IT workforce. Reflecting the fact that married women are still more likely to exit the labor force than are married men, within both groups working women are less likely to be married with their spouse present than is true for men. On the other hand, the proportions of workers with one or more of their own children present in the household is quite similar between IT and non-IT occupations, suggesting that this pattern is similar for both IT and non-IT workers.

FUTURE TRENDS

After nearly two decades of explosive growth and transformation, the expansion of the IT workforce came to an abrupt halt with the collapse of the technology bubble in 2001. For the past several years the number of IT workers has been declining. This decline is generally expected to be temporary, and most forecasts anticipate that employment in IT occupations will continue to grow more quickly than in the labor force generally, though the differential is unlikely to be as large as it was in the past (U.S. Bureau of Labor Statistics 2004).

In the past few years there has been increasing concern about the role of off shoring in IT job losses. There have been numerous reports of companies exporting technical support and programming jobs to suppliers in India, China, and other low-wage countries with well-educated labor forces. Given the large international differences in wages, shifting some tasks to Asian countries is an attractive option for U.S. companies seeking to cut labor costs. But it

is important not to overstate the potential impact of this trend. Off shoring is most effective when the tasks to be performed have been routinized. These, in turn are the sorts of jobs that are most in-danger of being automated in any event. Jobs requiring specialized knowledge of business practices and discretionary decisions are likely to continue to be performed in proximity to customers, thus ensuring that the vast majority of higher level IT jobs, such as those performed by systems analysts, will remain in the United States (Edwards, 2004).

While this suggests that IT job losses in the United States due to off shoring may be small, it also suggests that the composition of IT jobs will remain biased towards those high skilled jobs that contain relatively few women. Thus prospects for increasing the representation of women in IT appear relatively bleak. If relatively few women have been drawn into the rapidly growing field of computer systems analysts and scientists during the period of rapid expansion in employment, opportunities for women are likely to remain limited in the future as aggregate growth slows. More research is needed to understand why women have tended to avoid these higher-level IT jobs, and to identify those dimensions of education, hiring, and retention that have produced such large gender gaps in representation.

CONCLUSION

During the past half-century gender differences in the labor market have closed substantially. Overall, women's labor force participation behavior has come increasingly to resemble that of men, so that today women constitute approximately half of the U.S. labor force. Although a gender earnings gap remains today, the size of this gap has been reduced considerably, and after accounting for differences in education, experience, and other characteristics it is smaller than indicated by unadjusted comparisons.

Set against the background of these broad labor market changes, gender differences in Information Technology are striking. While total employment in IT has grown rapidly, women's share of employment across all IT occupations has fallen substantially over the past two decades. The absence of women does not reflect an absence of financial incentives. Gender pay gaps in IT have paralleled those in the workforce generally. Since pay in IT occupations has grown quite quickly women could realize significant financial rewards from moving into IT occupations.

Although the growing gender gap in IT employment is largely due to changes in the mix of IT occupations that has increased the numbers of computer systems analysts and scientists, the fact remains that women hold less than 1/3 of such jobs today, about the same proportion as they held 20 years earlier. The persistent under representation of women in these higher-level IT occupations is an as yet unexplained phenomenon that requires further study.

REFERENCES

Commission on Professionals in Science and Technology. (2004). Supply and demand. *CPST Comments, 41*(7).

Edwards, B. (2004, November). A world of work: A survey of outsourcing. *Economist, 13*.

Ellis, R., & Lowell, B. L. (1999). *Core occupations of the U.S. information technology workforce*. IT Workforce Data Project: Report I. Commission on Professionals in Science and Technology. Retrieved December 14, 2004, from http://www.cpst.org/IT-1.pdf

Freeman, P., & Aspray, W. (1999). *The supply of information technology workers in the United States*. Computing Research Association. Washington, DC: CRA. Retrieved December 14, 2004, from http://www.cra.org/reports/wits/it_worker_shortage_book.pdf

Fuchs, V. R. (1988). *Women's quest for economic equality*. Cambridge, MA; London: Harvard University Press.

Goldin, C. (1990). *Understanding the gender gap: An economic history of American women*. NBER Series on Long-Term Factors in Economic Development. New York and Oxford: Oxford University Press.

Goldin, C., & Katz, L. F. (2002). The power of the pill: Oral contraceptives and women's career and marriage decisions. *Journal of Political Economy, 110*(4), 730-770.

Information Technology Association of America. (2003). *Report of the ITAA Blue Ribbon Panel on IT Diversity*. Retrieved December 14, 2004, from http://www.itaa.org/workforce/studies/diversity report.pdf

Long, J. S. (2001). *From scarcity to visibility: Gender differences in the careers of doctoral scientists and engineers*. National Research Council, Committee on Women in Science and Engineering. Washington, DC: National Academy Press.

National Research Council, Committee on Workforce Needs in Information Technology. (2001). *Building a workforce for the information economy*. Washington, DC: National Academy Press. Retrieved December 14, 2004, from http://books.nap.edu/html/building_workforce/

Mowery, D. C., & Rosenberg, N. (1998). *Paths of innovation: Technological change in 20th Century America*. Cambridge: Cambridge University Press.

U.S. Bureau of Labor Statistics. (2004). *Occupational outlook handbook 2004-05 Edition*. Retrieved December 3, 2004, from http://www.bls.gov/oco/home.htm

Weil, D. N. (2005). *Economic growth*. Boston: Pearson-Addison Wesley.

KEY TERMS

Current Population Survey: A monthly survey of approximately 50,000 households administered jointly by the U.S. Census Bureau and Bureau of Labor Statistics to gather information about employment status and demographic characteristics of individuals.

Full-Time Worker: An individual who works an average of 35 or more hours per week.

Labor Force Participation: In the United States labor force participation is assessed based on the response to questions asked as part of the Current Population Survey. An individual is said to participate in the labor force if he or she is over 16 and either performed paid work, or engaged in a variety of job-seeking activities in the week prior to the survey.

Labor Force Participation Ratio: The ratio of the number of workers participating in the labor force to the total population, aged 16 or over.

Off-Shoring: The practice of relocating jobs previously performed in the United States to other, countries; typically low-wage Asian countries like China and India.

Technology Bubble: The period of time during the mid- to late-1990s when investment in internet based companies boomed. The precise dating of the beginning of the bubble is difficult, but it is generally agreed that the bubble came to an end when U.S. stock markets reached a peak in early 2001.

Wage Gap: The difference in pay between two groups of workers, such as Blacks and Whites, or Men and Women. The wage gap can be measured either in absolute terms or as a ratio or percentage difference.

IT Workplace Climate for Opportunity and Inclusion

Debra A. Major
Old Dominion University, USA

Donald D. Davis
Old Dominion University, USA

Janis V. Sanchez-Hucles
Old Dominion University, USA

Lisa M. Germano
Old Dominion University, USA

Joan Mann
Old Dominion University, USA

INTRODUCTION[1]

Our program of research is rooted in organizational psychology and employs a climate perspective to understand women's experiences in the information technology (IT) workplace. An appropriate climate can help a workplace effectively attract and retain a diverse employee base (Miller, 1998). Climate consists of employees' perceptions of workplace events, practices and procedures, including which behaviors are expected, supported and rewarded (Schneider, Wheeler, & Cox, 1992). Climate requires a referent to have meaning, and there is not a single climate within an organization. For example, there are climates for safety, innovation and customer service. Our focus is climate for opportunity and inclusion (Hayes, Bartle, & Major, 2002).

Climate for opportunity is defined as an individual's overall perception of the fairness and inclusiveness of the workplace in terms of the processes used to allocate opportunities and the resulting distribution of opportunities. Opportunities include hiring, assignments, promotions, pay, power, authority, awards and training. By creating an *inclusive* work environment, employers can capitalize on the benefits of diversity. Although definitions vary in this emerging literature (cf. Miller, 1998; Mor-Barak

& Cherin, 1998; Pelled, Ledford, & Mohrman, 1999), most agree that inclusion means ensuring that everyone in an organization's diverse workforce feels a sense of belonging, is invited to participate in important decisions and feels that his or her input matters. Exclusion leads to turnover, reduced organizational commitment and decreased job satisfaction (Greenhaus, Parasuraman, & Wormley, 1990). Moreover, prospective employees are more likely to be attracted to inclusive organizations (Powell & Graves, 2003).

Our research suggests that three key factors predict inclusive climate: (1) good working relationships between supervisors and IT employees, (2) supportive coworkers, and (3) an organizational culture that supports balance between one's work life and personal life (Major, Davis, Sanchez-Hucles, & Mann, 2003). In turn, IT employees respond to an inclusive climate with better performance, greater job satisfaction, heightened commitment and increased likelihood of remaining with the current employer and staying in the IT field (Major et al., 2003). In this article, we focus on the gender differences and similarities regarding IT employees' perceptions of: (a) inclusion and climate for opportunity, (b) workplace relationships, and (c) satisfaction and commitment.

BACKGROUND

Research Methodology

Our Web-based survey was completed by 916 IT employees from 11 companies. (See Major & Germano, 2006, for a detailed description of participants and measures.) Due to missing data, the sample size for most analyses reported here is 872; exceptions are noted.

Key Research Findings: Gender Similarities and Differences

In comparing men's and women's experiences in the IT workplace, we found a number of similarities and differences. Means and standard deviations, along with the results of independent samples t-tests and Cohen's effect size (d), are presented in Table 1. We used the results of t-tests to estimate mean differences between the responses of men and women. Because our sample is large and the significance of the t statistic is influenced by sample size, we also calculated d, which is uninfluenced by sample size, to estimate the magnitude of the gender effect. These two statistics together tell us whether there is a significant difference between men and women (t statistic) and how meaningful the magnitude of this difference is (d statistic). The sign of the effect size (positive or negative) merely reflects the direction of the gender difference (positive effect size when the mean score for men is greater than the mean score for women).

Inclusion and Climate for Opportunity

Men and women did not differ on the belonging and participation dimensions of inclusion, suggesting that both feel equally welcome in the IT work environment and both are equally likely to be a part of decision making. However, when it comes to having an influence in the environment (i.e., feeling that one's contributions actually have an impact), men were significantly higher than women. Men were also more likely to perceive a positive climate for opportunity than women. That is, women were less likely than men to feel that opportunities are provided without regard to gender and ethnicity.

Workplace Relationships

Effective interpersonal relationships allow individuals to feel adjusted and anchored in work contexts that might otherwise be overwhelming and unwelcoming (Kahn, 1996). Employees' relationships with their mentors, coworkers and immediate supervisors are particularly important.

Mentors

Mentors are senior individuals with advanced expertise and knowledge who assist in providing upward support and mobility to their protégés' careers (e.g., Wanberg, Welsh, & Hezlett, 2003; Ragins & Cotton, 1999). Mentors typically offer both career development and psychosocial support. Having a mentor has a positive influence on numerous career outcomes for protégés. Compared to their nonmentored counterparts, those with mentors have higher job performance ratings, are promoted more frequently, have higher satisfaction with their jobs and have higher incomes (e.g., Allen, Eby, Poteet, Lentz, & Lima, 2004; Ragins & Cotton, 1999).

In general, research shows that men and women have equal access to mentors (O'Neill, 2002). In our study, women were actually more likely than men to report having at least one mentor. However, this finding may be due to gender differences in conceptualizing mentoring. Follow-up focus group discussions with men and women in our sample (see Major & Germano for a description) suggested that, despite being provided with the same definition of mentoring on the survey, women may have had a broader interpretation of mentoring (i.e., any helping behavior) than men. Among men and women who reported having at least one mentor, there was not a statistically significant gender difference in level of satisfaction with mentoring received, although the effect size suggests that men may be less satisfied with mentoring than women.

Coworkers

Research shows that supportive coworkers are beneficial in a variety of ways. Coworker support is associated with reduced stress, greater organizational commitment, higher job satisfaction and re-

Table 1. Gender comparisons: Means, SDs, T-statistics & Cohen's effect size

Dependent Variable	Men			Women			Independent samples t test			
	N	Mean	SD	N	Mean	SD	t	df	p	d
Inclusion:										
• Belonging	530	4.17	0.72	344	4.14	0.77	0.57	872	0.57	0.04
• Participation	530	4.14	0.73	344	4.05	0.82	1.79	872	0.07	0.12
• Influence	530	3.83	0.87	344	3.67	0.93	2.72	872	0.01	0.19
Climate for Opportunity	530	4.51	1.09	344	4.32	1.10	2.39	872	0.02	0.17
Workplace Relationships:										
• Affective Coworker Support	530	3.84	0.73	344	3.91	0.75	-1.40	872	0.16	-0.10
• Instrumental Coworker Support	530	3.99	0.75	344	4.09	0.69	-1.98	872	0.05	-0.14
• Leader-Member Exchange	530	3.48	0.92	344	3.57	0.93	-1.46	872	0.14	-0.10
• Currently Being Mentored	519	0.47	0.50	333	0.61	0.49	-4.16	850	0.00	-0.29
• Satisfaction with Mentoring	187	4.03	0.77	145	4.16	0.63	-1.65	330	0.10	-0.18
Satisfaction:										
• Overall Job Satisfaction	530	5.07	0.98	344	5.18	0.99	-1.73	872	0.08	-0.12
• Satisfaction with Supervision	530	5.11	1.52	344	5.27	1.48	-1.58	872	0.11	-0.11
• Satisfaction with Job Security	530	4.62	1.66	344	4.85	1.45	-2.10	872	0.04	-0.15
• Satisfaction with Pay	530	4.79	1.29	344	4.94	1.27	-1.74	872	0.08	-0.12
• Satisfaction with Social Environment	530	5.60	0.92	344	5.62	0.96	-0.30	872	0.77	-0.02
• Satisfaction with Growth Opportunities	530	5.07	1.18	344	5.15	1.20	-0.99	872	0.32	-0.07
Organizational Commitment	530	5.23	1.15	344	5.24	1.12	-0.01	872	0.99	0.00
Career Commitment	530	3.56	0.74	344	3.31	0.73	4.78	872	0.00	0.33

Note. d is Cohen's effect size. Most constructs were measured using a 5-point scale with a few exceptions. Climate for Opportunity was measured on a 6-point scale. Organizational Commitment and all of the Satisfaction constructs were measured on 7-point scales. Currently Being Mentored represents the percentage of the sample that indicated having a mentor at the time of the survey.

duced intentions to quit (Baruch-Feldman, Brondolo, Ben-Dayan, & Schwartz, 2002; Ducharme & Martin, 2000; Lee, 2004). There are two main types of coworker support: affective and instrumental. Affective support is a form of social support that coworkers offer by being sympathetic, listening to problems, and expressing care and concern. Instrumental support is more tangible helping behavior demonstrated by assisting with work responsibilities, switching schedules and other similar behaviors. Our results showed no differences in the amount of affective coworker support that men and women working in IT reported receiving. However, women reported receiving significantly more instrumental support from coworkers than men did.

Supervisors

Leader-Member Exchange (LMX) has been widely used to characterize the quality of supervisor-subordinate relationships on the basis of mutual respect, trust and loyalty (see Graen & Uhl-Bien, 1995 for a review). Although it can be assessed from both the supervisor and subordinate perspectives, research most commonly uses subordinates' reports of LMX (Gerstner & Day, 1997), as we do in this research. Research over the past 25 years has shown that LMX is positively related to job satisfaction, organizational commitment and performance (e.g., Gerstner & Day, 1997; Major, Kozlowski, Chao, & Gardner, 1995).

Table 2. Impact of supervisor and subordinate gender on supervisor-employee relationship

Employee Gender	Supervisor Gender	
	Men	Women
Men	3.39 (.92) n = 374	3.70 (.87) n = 154
Women	3.56 (.97) n = 175	3.58 (.89) n = 168

Note. Cells indicate means (SD) for LMX.

As shown in Table 1, there were no mean differences in the levels of LMX that men and women in IT reported. To explore the potential effects of gender dynamics on LMX, we examined the quality of supervisory relationships as a function of supervisor and subordinate gender. Results of a 2x2 ANOVA show gender of supervisor matters little for female IT employees, but matters considerably for male IT employees—the interaction yielded $F_{1,867} = 5.97$, $p = .02$, $\eta^2 = .01$. The quality of the supervisor-subordinate relationship is best for male IT employees when they have female supervisors. Mean differences are shown in Table 2.

Satisfaction and Commitment

Results show that overall, men and women working in IT are equally satisfied with their jobs. They are also equally satisfied with more specific aspects of their jobs, including supervision, pay, the social environment and opportunities for growth. Women, however, report statistically significant greater satisfaction with their job security than men.

Our findings indicate that men and women in IT are equally committed to their employing organizations. However, women reported significantly less commitment to a career in IT than men. This is not to say that women are less committed to working. Instead, they are less likely to view IT as their "life's work" than men. This gender difference has the largest effect size ($d = .33$) of any in our sample.

To further explore the gender difference, we conducted a hierarchical multiple regression analysis to ascertain whether or not other observed differences between men and women in our sample accounted for the career commitment gender difference. In step 1 of the regression, we included the demographic characteristics on which men and women differed (see Major & Germano, 2006), including relationship status, number of children, IT relatedness of educational degree, salary, organizational tenure, years worked in IT and hours worked per week. In step 2, we entered influence and climate opportunity since women reported less of both than men. In the final step of the hierarchical regression equation, we entered gender. Results shown in Table 3 indicate that IT-related degree, salary, organizational tenure, hours worked per week, influence and climate for opportunity are all significant predictors of career commitment. However, even after controlling for these variables, women's commitment to a career in IT is lower than men's.

FUTURE TRENDS

Overall, our results show more similarities than differences between men and women in IT. Women and men enjoy equally supportive workplace relationships (e.g., similar levels of LMX and affective coworker support), and in some instances, women report modestly greater support than men (e.g., instrumental coworker support). In supervisory roles, women are effective and are better able than men to establish high-quality relationships with male subordinates. Notably, female subordinates report equally effective relationships with both male and female supervisors. Nonetheless, women still face barriers to inclusion and opportunity in IT. In particular, women are less likely than men to believe that opportunities are fairly distributed without regard to race and gender. Although men and women report similar levels of belonging and participation, women feel that they have less workplace influence than men. In other words, women feel invited to participate but perceive that their input has less of an impact than men's input.

One of our most meaningful findings is the gender difference in commitment to an IT career; women report lower commitment than men. This difference in commitment is not due exclusively to less inclusion, less opportunity or other demographic differences. Interestingly, our results show that organizational tenure is inversely related to commitment to an IT career (see Table 3). Perhaps there

Table 3. Effects of gender on career commitment after controlling for demographic variables, influence, and climate for opportunity

Variables	β	T	R^2	$?R^2$
Step 1:			.09*	
Relationship Status	.05	1.327		
Number of Children	.05	1.323		
IT Degree	-.19	-5.364*		
Salary	-.11	-2.521*		
Organizational Tenure	-.11	-2.892*		
Years Worked in IT	.04	1.037		
Hours Worked per Week	.19	5.110*		
Step 2:			.18*	.09*
Influence	.15	3.849*		
Climate for Opportunity	.21	5.383*		
Step 3:				
Gender	-.09	-2.665*	.19*	.01*

Note. N=762. Relationship Status coded: single (including divorced, separated and widowed) = 0 and married or living with partner = 1. IT Degree coded: holds IT related degree = 0 and holds non-IT related = 1. Gender coded: men = 0 and women = 1. Standardized regression weights are reported for the step in which variables were entered. *p < .05.

are trade-offs between commitment to a particular organization and commitment to an IT career, especially for women. It may be that the longer a woman works for her employer, the more invested she becomes in that organization, regardless of whether her work continues to be IT related. Because reduced commitment to an IT career is likely a precursor to leaving the IT workforce, developing a more comprehensive understanding of the factors that predict career commitment is essential.

CONCLUSION

The underrepresentation of women in IT has been attributed to several factors, including long hours, little work-life balance and a workplace that is inhospitable to women (Lambeth, 1996; Panteli, Stack, & Ramsay, 1999), few opportunities for social interaction (Misic & Graf, 1999) and reluctance of male supervisors to coach and mentor women subordinates (Ragins, 2002). Our research findings suggest that these factors may exert a weaker influence on retention than once thought. IT workers in our sample did not report working exceptionally long hours (Major, Cardenas, Davis, Germano, & Mickey, 2004). Moreover, women reported support from coworkers and supervisors. In terms of inclusion, women in our sample reported belonging and participating, but they do not believe that their

inclusion leads to influence. More work is needed to increase the influence of women in IT departments. See Major and Germano (2006) for a description of our intervention methodology.

REFERENCES

Allen, T. D., Eby, L. T., Poteet, M. L., Lentz, E., & Lima, L. (2004). Career benefits associated with mentoring for protégés: A meta-analytic review. *Journal of Applied Psychology, 89,* 127-136.

Baruch-Feldman, C., Brondolo, E., Ben-Dayan, D., & Schwartz, J. (2002). Sources of social support and burnout, job satisfaction, and productivity. *Journal of Occupational Health Psychology, 7,* 84-93.

Ducharme, L. J., & Martin, J. K. (2000). Unrewarding work, coworker support, and job satisfaction: A test of the buffering hypothesis. *Work Occupations, 27,* 223-243.

Gerstner, C. R., & Day, D. V. (1997). Meta-analytic review of leader-member exchange theory: Correlates and construct issues. *Journal of Applied Psychology, 82,* 827-844.

Graen, G. B., & Uhl-Bien, M. (1995). Relationship-based approach to leadership: Development of leader-member exchange (LMX) theory of leadership over 25 years: Applying a multi-level multi-domain perspective. *Leadership Quarterly, 6,* 219-247.

Greenhaus, J. H., Parasuraman, S., & Wormley, W. M. (1990). Effects of race on organizational experiences, job performance evaluations, and career outcomes. *Academy of Management Journal, 33,* 64-86.

Hayes, B. C., Bartle, S. A., & Major, D. A. (2002). Climate for opportunity: A conceptual model. *Human Resource Management Review, 12,* 445-468.

Kahn, W. A. (1996). Secure base relationships at work. In D. T. Hall (Ed.), *The career is dead—Long live the career* (pp. 158-179). San Francisco: Jossey-Bass.

Lambeth, J. (1996, May 30). Time to come to terms with flexible working? (IT professionals face fixed-term contracts). *Computer Weekly,* 18.

Lee, P. C. B. (2004). Social support and leaving intention among computer professionals. *Information & Management, 41,* 323-334.

Major, D. A., Cardenas, R. A., Davis, D. D., Germano, L. M., & Mickey, S. K. (2004, August). Managing work-family conflict in the IT workplace. In J. Cleveland (Chair), *Work & family: Constructing a view using multiple methods, occupations, cultures.* Symposium presented at the 112th Convention of the American Psychological Association, Honolulu, HI.

Major, D. A., Davis, D. D., Sanchez-Hucles, J., & Mann, J. (2003, October). Climate for opportunity and inclusion: Improving the recruitment, retention, and advancement of women and minorities in IT. *Proceedings of the National Science Foundation's ITWF & ITR/EWF Principal Investigator Conference* (pp. 167-171). Albuquerque: The University of New Mexico.

Major, D. A., & Germano, L. M. (2006). Survey feedback interventions in IT workplaces. In E. Trauth (Ed.), *Encyclopedia of gender and information technology.* Hershey, PA: Idea Group Reference.

Major, D. A., Kozlowski, S. W. J., Chao, G. T., & Gardner, P. (1995). A longitudinal investigation of newcomer expectations, early socialization outcomes and the moderating effects of role development factors. *Journal of Applied Psychology, 80,* 418-431.

Miller, F. A. (1998). Strategic culture change: The door to achieving high performance and inclusion. *Public Personnel Management, 27,* 151-160.

Misic, M., & Graf, D. (1999). The interpersonal environments of the systems analyst. *Journal of Systems Management, 44,* 12-16.

Mor-Barak, M. E., & Cherin, D. (1998). A tool to expand organizational understanding of workforce diversity: Exploring a measure of inclusion-exclusion. *Administration in Social Work, 22,* 47-64.

O'Neil, R. M. (2002). Gender and race in mentoring relationships: A review of the literature. In D. Clutterbuck & B. R. Ragins (Eds.) *Mentoring and diversity: An international perspective* (pp. 1-22). Oxford: Butterworth-Heinemann.

Panteli, A., Stack, J., & Ramsay, H. (1999). Gender and professional ethics in the IT industry. *Journal of Business Ethics, 22,* 51-61.

Pelled, L. H., Ledford Jr., G. E., & Mohrman, S. A. (1999). Demographic dissimilarity and workplace inclusion. *The Journal of Management Studies, 36,* 1013-1031.

Powell, G. N., & Graves, L.M. (2003). *Women and men in management* (3rd ed.). Thousand Oaks: Sage.

Ragins, B. R. (2002). Understanding diversified mentoring relationships: Definitions challenges and strategies. In D. Clutterbuck & B. R. Ragins (Eds.), *Mentoring and diversity: An international perspective* (pp. 23-53). Oxford: Butterworth-Heinemann.

Ragins, B. R., & Cotton, J. L. (1999). Mentor functions and outcomes: A comparison of men and women in formal and informal mentoring relationships. *Journal of Applied Psychology, 84,* 529-550.

Schneider, B., Wheeler, J. K., & Cox, J. F. (1992). A passion for service: Using content analysis to explicate service climate themes. *Journal of Applied Psychology, 77,* 705-716.

Wanberg, C. R., Welsh, E. T., & Hezlett, S. A. (2003). Mentoring research: A review and dynamic process model. In G. R. Ferris & J .J. Martocchio (Eds.), *Research in personnel and human resources management* (Vol. 22, pp. 39-124). Greenwich: Elsevier Science/JAI Press.

KEY TERMS

Affective Coworker Support: A form of social support that coworkers offer by being sympathetic, listening to problems, and expressing care and concern

Career Commitment: Describes one's attachment to one's profession or vocation

Climate for Opportunity: An individual's overall perception of the fairness of the organization in terms of the management processes used to allocate opportunities, including interpersonal treatment, and the distribution of opportunities in the organizational context (Hayes, Bartle & Major, 2002)

Inclusion: The extent to which employees feel that they are part of important work activities. Inclusion is comprised of three components: belongingness, participation and influence. Belongingness is the degree to which employees feel that their workgroup members accept them. Participation is the degree to which employees feel that the workgroup invites them to take part in group discussions and decisions. Influence is the extent to which employees feel their participation actually has an impact on decisions.

Instrumental Coworker Support: Tangible helping behavior offered by coworkers in response to specific needs; for example, assistance with work responsibilities and switching schedules.

Job Satisfaction: Represents the employee's emotional reaction to his or her job overall and to specific aspects of the job, including supervision, job security, pay, social environment and growth opportunities. It is possible to have different emotional responses to each aspect of one's job.

Leader-Member Exchange: Describes the perceived quality of the relationship between employees and their immediate supervisors in terms of mutual respect, trust and confidence. May be assessed as either a supervisor or a subordinate perception; the latter being more common. Relationships are typically described as either high- or low-quality exchanges.

Organizational Commitment: Describes an employee's loyalty and attachment to one's employing organization.

ENDNOTE

[1] This material is based upon work supported by the National Science Foundation under Grant No. 0204430. The authors would like to acknowledge Thomas D. Fletcher for his assistance with data management and analyses.

Making Executive Mentoring Work in IT

Shari Lawrence Pfleeger
RAND Corporation, USA[1]

Norma T. Mertz
University of Tennessee, USA

M

INTRODUCTION

Although it is a relatively young discipline, information technology has a lack of gender diversity that is similar to many older sciences. For example, the 33rd annual Taulbee survey of computer science graduates indicates only 20% of those enrolling in computer science doctoral programs are women, and only 16.8% of those receive PhDs; these rates have been the same for the last few years (Zweben & Aspray, 2004). Moreover, once women move into computing careers, they can have a difficult time moving up the career ladder. For example, women's advancement in academia has been disappointing: 19% of the computer science faculty in the United States are female, but only 8.6% of full professors and 12.3% of associate professors are women (Zweben & Aspray). Similar figures are reported for women in industry as they hit the glass ceiling (Morrison, White, & van Velsor, 1987), but women in some countries may be catching up. For example, "pay and prospects for women in IT are the best they have ever been" in the United Kingdom: They achieved higher pay increases than men across all sectors for the 8th year running, but are still behind (Mortleman, 2004).

Thus, there is still room for women at the top. According to Corporate Women Directors International (2004), "The glass ceiling in corporate directorships is solidly in place." Indeed, only 7.5% of Fortune Global 200 boards have three or more women serving on them. Similarly, a recent survey sponsored by the UK Department of Trade and Industry and Shell revealed that a third of the boards of British companies still have no females (Cranfield School of Management, 2004).

BACKGROUND

A 2000 study sponsored by the American Association of University Women (2000) notes that women comprise only 17% of the high-school students who take advanced-placement exams in computer science and only 28% of those with undergraduate degrees in computer science. Indeed, fewer women are expressing interest over time (see Figure 1). A major problem in attracting and keeping women and minorities in computer science (and other disciplines) is the lack of role models at all levels, and in particular at senior levels. In the early 1990s, we investigated what makes a good role model or mentor.[2]

MENTORING

Mentoring has long been associated with career advancement in business. Indeed, not only does "everyone who makes it 'have' a mentor" (Collins & Scott, 1978), but everyone needs a mentor. Recently, mentoring has become associated with efforts to increase the representation of underrepresented groups, such as women and minorities, in fields such as IT in which their presence at higher levels of such organizations has been notably absent. Professional and institutional calls for addressing the situation, whether responding to law or pressure, have led to creating projects and processes for changing the profile of top leaders and for enhancing the likelihood that women and minorities will advance. For example, the Computing Research Association's Committee on the Status of Women in Research has for 10 years conducted a distributed mentoring project:

Figure 1. Computer science listed as probable major among incoming freshmen

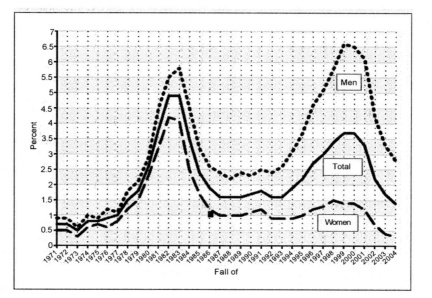

Source: HERI at UCLA

... to increase the number of women entering graduate studies in the fields of computer science and engineering. This highly selective program matches promising undergraduate women with a faculty mentor for a summer research experience at the faculty member's home institution. Students are directly involved in a research project and interact with graduate students and professors on a daily basis. This experience is invaluable for students...increasing their competitiveness as an applicant for graduate admissions and fellowships. (Committee on the Status of Women in Computing Research [CRA-W], 2006)

As organizations consider the underrepresentation problem, they almost invariably institute mentoring projects, pairing entry- and junior-level women and minorities with more senior-level members of the organization, most often the majority of whom are white and male. Despite the programs reported in the literature, and mentoring's ancient lineage, we know relatively little about the nature of such relationships and factors that contribute to its success. This situation is as true of traditional mentoring pairs (naturally occurring between junior and senior white males) as it is of relatively newly developed, ar-

ranged cross-gender or cross-race mentoring pairs (Mertz, Welch, & Henderson, 1988, 1990).

As part of a project sponsored by the National Science Foundation and the Association for Computing Machinery to institute and study mentoring for the career advancement of women and minorities in computer science in academia and industry, 15 pairs of mentors and protégés were studied over 18 months. All of the mentors and protégés agreed to participate, and their organizations were committed to the project and to the goal of advancing women and minorities in their organization.

Research Method

Approval was sought from highly placed members of the organizations' administrative or management chains. This buy-in was intended to maximize the likelihood of commitment to the project and to the mentoring process. Three industry organizations and two academic institutions agreed to participate in the project, to commit to mentoring women and minorities in their organizations, and to arrange for mentoring pairs of junior-level, promising women and minorities and senior-level persons. After an intensive workshop that examined the nature and intent of mentoring, the need for mentoring women and minorities, and the nature of the project, each pair set goals and objectives for the process and worked through anticipated problems and concerns together and with the workshop facilitators. Attitudinal surveys (toward mentoring and the need for mentoring women and minorities) and a preference instrument (Myers-Briggs Type Indicator) were administered to all participants before the start of the workshop. In addition, participants were interviewed individually, in depth, about their backgrounds, career aspirations, expectations, and concerns. During the project study period, each site was visited and each participant interviewed individually at least twice, the last interview occurring at the project's conclusion. These interviews explored events during the mentoring process, satisfaction with the relationship, problems or concerns, and generally how the relationship was faring. At the end, participants were asked to reflect on the entire process to assess its success (e.g., in terms of effectiveness) and their feelings. They also completed an attitudinal survey to complement the one completed before the workshop.

Findings

Of the 15 mentoring pairs, only 3, all in industry, were found to have been successful in terms of the criteria established for measuring the success of the process:

- Both members of the pair perceived the experience to be successful and reported that the experience had value.
- The experience involved actions and attitudes, particularly on the part of the mentor, specifically designed to help the protégé advance.
- The experience involved a professional relationship between mentor and protégé beyond simple advising or supervising, and beyond that which was required by the project. The mentor and protégé were involved in the enterprise together willingly, trusted one another, and felt their relationship would likely continue.

Two other pairs were partially successful (met two of the three criteria), two pairs were unsuccessful (met none of the criteria and indeed were dismal failures), and the remaining pairs were neither successful nor unsuccessful. They met at best one criterion, but were universally characterized by dissatisfaction or disappointment. For detailed information about the project, study, and its findings, see Pfleeger and Mertz (1994, 1995).

This article addresses what was learned about mentoring for career advancement from the relationships studied: what characterized successful mentoring relationships, what differentiated them from less successful and unsuccessful relationships, and what these findings suggest for implementing similar programs. The authors recognize that the small number of pairs studied limits the generalizability of the findings and lessons learned. Nevertheless, the depth of study of each pair warrants their acknowledgement and potential value.

What Characterized Successful Pairs?

First, they met regularly. There was no particular schedule; they met without having been prompted to meet. Beyond meeting, each protégé felt comfortable calling the mentor to arrange a meeting; this access represented a relationship another similarly placed person might not have had. Indeed, all protégés in successful pairings felt they could (but did not) call the mentor at home if necessary.

In two of the three successful pairings, the mentoring meetings were facilitated because the mentor and protégé worked in the same organizational division or line, or had work-related reasons to get together. However, this condition was not a necessary one, as one pair had members who neither worked in the same organizational division or on related projects, nor were even interested in the same kind of work, except in the most general terms. In this case, the mentor's commitment to the project, to its broad purposes, and to doing well in whatever was undertaken may have been the driving force underlying the success of the pair. That is, the mentor's attitude mandated success, so the pairing was successful. In the other two cases, it was a lot easier for the pairs to interact without making extraordinary efforts.

Second, in each of the three successful cases, the relationship between the pairs was characterized by respect for one another. Each saw the other as competent and seriously committed to work and to excellence, to being successful in moving up in the organization, and to doing so because of competence and past successes. There was compatibility in the values held as well as in career goals, and shared values made the participants comfortable in talking with one another.

Third, the relationship was focused on the protégé and what was needed to be successful. Where the mentor and protégé performed similar work, the mentor often provided opportunities for the protégé to have additional responsibility, to gain visibility for work done, to be recommended for challenging projects, and to participate in activities or projects that would stretch, challenge, or grow the protégé. In one case, the mentor shared with the protégé a network of contacts garnered from years of working in the field and, where appropriate, took the protégé to meetings with those contacts.

In the pair whose participants did not work in the same divisional line, the mentor accessed information, shared knowledge and experience about the company and how things worked within it, spoke to the protégé's superiors about the protégé's progress, and opened doors for the protégé by virtue of the mentor's high position.

Fourth, the mentor and protégé became associated with one another in the minds of others; their relationship was known and noted within the organization. None of the three mentors had any problem with this perception. Indeed, they saw some benefit in this association. The protégés came to hold the status of fast-trackers: people who have been identified as having potential and who are being considered for future promotion. Protégés seen as fast-trackers were legitimized on the basis of their perceived potential, not (as in the case of others involved in the project) because they were women or minorities.

What Characterized Unsuccessful Pairs?

In pairs that were unsuccessful, the mentors and protégés did not share a common perspective about mentoring or about what should go on in the name of mentoring. While the protégés were concerned about getting ahead, one mentor had severe reservations about the need to mentor and especially to mentor women or minorities. Furthermore, this mentor saw the role as listening to problems the protégé might bring and telling the protégé what to do. Another,

while keenly aware of the value of mentoring, saw the role as helping the junior person to understand how to fit in (fitting in being the path to eventual success), not necessarily how to get ahead. Moreover, all but the successful pairs, particularly the unsuccessful pairs, did not seem to share the same values or views of the world, which contributed to problems in the relationships.

The unsuccessful pairs had difficulties in communicating with one another and found, over time, that these difficulties increased. In one case, the mentor felt that the protégé did not approach the mentor with problems, resented it, and concluded that no further interaction was required. In part, this mentor felt that the protégé had not fulfilled the responsibilities of a protégé. Interestingly enough, the protégé did not trust the mentor and did not believe one should go to a mentor with problems.

These two unsuccessful pairings were quite different from one another but were equally disastrous. In one case, the mentor did not respect the protégé; in the other, the protégé did not respect the mentor. Nothing about their interactions changed these perceptions. Indeed, although neither party of either pairing said so directly, they appeared to have less respect for one another after the experience than before, and the mentors' attitudes toward the project and toward mentoring were more negative after the experience. Their interactions disintegrated quickly, resulting eventually in a complete lack of communication between mentor and protégé. Neither pair engaged in any conversation about this sad state of interaction; the pairs just let the relationship die.

To greater or lesser degrees, the pairs that were neither successful nor unsuccessful shared some of the same characteristics as unsuccessful pairs. There was little clarity or consensus between mentor and protégé about the role of the mentor or the goals of the mentoring, despite identifying these clearly at the inception of the project. The participants in the pairing did not develop a comfortable rapport or find a common basis for interacting. This dissonance was exacerbated by situations in which there was no basis for their working together, or where the mentor did not necessarily see a need to increase the number of women and minorities in the field or company. Some of these mentors did not see why the protégé had been chosen to participate, had reservations about the competence of the protégé, or were uncertain about why the protégé had been assigned to them. In other words, the mentors had no initial commitment to the protégés and saw no reason to become committed to them. Several of these mentors explained that they were very busy with their own work and had no organizational accounting code to which to charge the time spent in mentoring. In general, the unsuccessful mentors perceived no organizational or personal benefit to be gained from mentoring.

TRENDS

While advancement was not a criterion for assessing the success of the mentoring experience, given that it was the focus of the project, one might reasonably ask what happened to the protégés? Was there any relationship between involvement in the project or in a successful mentoring pair and promotion or recognition within the organization?

Of the 10 industry protégés, 5 received promotions after the mentoring project's inception, and 2 received salary increases (one in conjunction with a promotion). One protégé was a finalist for a coveted department-head position. Although the protégé was not ultimately chosen, having been a finalist was considered a great coup. Two protégés were accepted in a highly competitive, company-sponsored program that provided paid time off to pursue an advanced degree (one in conjunction with a promotion). In the academic setting, where achieving tenure and promotion are the marks of advancement, the study did not proceed long enough to learn the fate of the protégés.

It is not clear to what extent any changes in the status or position of the protégés can be attributed to the mentoring project or anything done in its name. The changes might have occurred if the project had never existed. However, it is interesting to note that of the three pairs that were successful, two protégés were promoted and the third got a level increase and was a finalist for promotion.

CONCLUSION

Although the limited number of participants lends tentativeness to our conclusions, several clear ob-

servations can be made. First, the success of mentoring programs depends on the degree of commitment of the participants and of their organizations. It is not enough for an organization to profess interest in and to value gender diversity by establishing a mentoring program. Mentors must be convinced of the protégé's excellence and be willing to invest time and honesty in the mentoring relationship. Likewise, the protégé must be able to trust the mentor to a higher degree than in an ordinary professional relationship.

At the same time, the organization must demonstrate formally that it values the protégés, the mentors, and the mentoring program itself. Mentoring cannot work as an afterthought. The most successful pairs brought energy to their relationship and made time for it, sanctioned by their home organizations. Even when mentor and protégé did not work in the same division of the organization, organizational and personal commitment overcame what otherwise would have been insurmountable obstacles.

Third, the careful selection of pairs is essential. It is not enough to identify good mentors and good protégés, and it is not necessary to match gender or racial characteristics. In the right pairing, each member must share an appreciation of what is important to their organization and to their profession; they must have a similar worldview. For this reason, preparation and training are needed to help the participants get to know each other, to assist them in understanding the nature and degree of the commitments they are about to make, and to explain their roles and responsibilities.

In particular, the participants must have a strong commitment to advancing women and minorities and an understanding that advancement does not just happen. The mentors should believe that senior people must invest in junior people as part of their professional responsibilities, regardless of issues of underrepresentation. That is, the mentors and their organizations must believe that the best people are encouraged and developed by investing in and nurturing them.

Our mentoring program was designed to enable different pairs to communicate if problems arose. However, once the mentoring pairs were established, there seemed to be little interpair communication. Indeed, the presence of another pair in the same organization did not make a difference; if a pair did not form a successful relationship, the other pair was not consulted for help or advice. Thus, a central mentoring advice service may be useful in helping organizations get started and in assisting pairs to overcome problems that arise.

Finally, mentoring is very difficult. More pairs failed than succeeded. Different organizations used different techniques for choosing participants and assigning pairs. Where the basis for selection and pairing was not achievement and respect, the mentoring did not work. Thus, those interested in establishing mentoring programs must consider two key questions before starting any mentoring: What does excellence mean in an organization, and what engenders respect in the organization and in computing at large? The answers to these questions depend on the organizational context and values. Therefore, different organizations are likely to have different mentoring programs. The common thread is an appreciation of gender and racial diversity as well as a desire for excellence. Just because mentoring is hard does not mean that it is not valuable. In fact, it means the reverse: that organizations with successful mentoring programs should be congratulated for their commitment and viewed as role models for the rest of the profession.

REFERENCES

American Association of University Women. (2000). *Tech-savvy: Educating girls in the new computer age.* Retrieved from http://www.aauw.org/

Collins, E. G. C., & Scott, P. (1978). Everyone who makes it has a mentor. *Harvard Business Review, 56*(4), 89-101.

Committee on the Status of Women in Computing Research (CRA-W). (2006). *CRA-W Distributed Mentor Project (DMP) objective.* Retrieved from http://www.cra.org/Activities/craw/dmp/index.php Corporate Women Directors International. (2004). *Women board of directors of Fortune Global 200 companies.* Retrieved from http://www.globe women.com/cwdi/order_form.htm

Cranfield School of Management. (2004). *The female FTSE report 2004.* Center for Developing Women Business Leaders. Retrieved from http://

www.som.cranfield.ac.uk/som/research/centres/cdwbl/downloads/FT2004FinalReport.pdf

Higher Education Research Institute. (2005). *The American freshman: National norms for 2004.* University of California at Los Angeles.

Mertz, N. T., Welch, O. M., & Henderson, J. (1990). *Executive mentoring: Myths, issues, strategies.* Newton, MA: WEEA Publishing Center.

Morrison, A. M., White, R. P., & van Velsor, E. (1987). *Breaking the glass ceiling.* Reading, MA: Addison-Wesley.

Mortleman, J. (2004, September 17). IT women smash glass ceiling. *Computing On-Line.* Retrieved from http://www.computing.co.uk/news/1158163

Pfleeger, S. L., & Mertz, N. (1994). *Final report on the ACM/NSF executive mentoring project.* New York: ACM.

Pfleeger, S. L., & Mertz, N. (1995). Executive mentoring: What makes it work? *Communications of the ACM, 38*(1), 63-73.

Zweben, S., & Aspray, W. (2004). 2002-2003 Taulbee survey: Undergraduate enrollments drop; department growth expectations moderate. *Computing Research News, 16*(3), 5-19.

KEY TERMS

Career Ladder: The hierarchy of job categories possible for an employee.

Fast-Tracker: Person identified as having potential and considered for rapid promotion.

Glass Ceiling: Obstacles that prevent an employee from moving up the career ladder beyond a certain point.

Mentor: Senior person who provides advice and counsel to a less experienced junior person.

Pair: A designated relationship between a mentor and a protégé.

Protégé: Junior person who is identified by a mentor as deserving of advice and counsel.

Role Model: Person whose actions and accomplishments provide goals for others.

Underrepresented Group: Group disproportionately underrepresented in a particular profession or activity compared with its availability.

ENDNOTE

[1]   This work was performed under a grant from the National Science Foundation to the Association for Computing Machinery. It was not a project of the RAND Corporation.

[2]   Additional information about mentoring can be found in the following resources:

Boards still lack female faces. (2004, December 7). *Accountancy Age*. Retrieved from http://www.accountancyage.com/news/1138886

Mertz, N. T., Welch, O. M., & Henderson, J. (1988). Mentoring for top management: How sex differences affect the selection process. *International Journal of Mentoring, 2*(1), 34-39.

M

Making of a Homogeneous IT Work Environment

Andrea H. Tapia
The Pennsylvania State University, USA

INTRODUCTION

What is the responsibility of the information technology (IT) industry in addressing gender issues? Exploring recruitment and retention issues that exist for women are crucial for increasing the capacity and diversity of the IT profession.

An understanding of the underlying causes of gender under representation in the IT profession is needed to develop effective workplace human resource strategies to attract and retain more of this underrepresented group. Unfortunately, while there is a documented need for a deeper understanding of the imbalance in this field, there is a lack of adequate data, methods and theory to provide a basis for explanation and prediction. Despite numerous efforts to recruit and retain women into both educational programs in IT and the IT workforce, these efforts have largely proved unsuccessful.

Women remain acutely underrepresented at the higher-paying professional and managerial levels (National Science Foundation, 2000; National Action Council for Minorities in Engineering, 2001-2002; Annenberg Public Policy Center, 2001; ITAA, 2003; Geewax, 2000; Spender, 1997). While women now represent a significant proportion of the labor force, they continue to be underrepresented in the IT workforce. Women have made few gains in employment numbers in the sector between 1996 and 2002. The Information Technology Association of America (ITAA) (2003) reported that the percentage of women in the overall IT workforce actually dropped from 41% to 34.9%. The underrepresentation of women in the IT workforce can be attributed to a "pipeline" issue. Women earn significantly fewer undergraduate degrees in computer science and engineering than their representation in the United States (U.S.) population. (Camp, 1997; Freeman & Aspray, 1999; U.S. Department of Education, National Center for Education Statistics, 2002).

BACKGROUND

Theoretical Perspective

The theoretical standpoint from which IT recruitment has been viewed in this article is one encompassing the social construction of gender in the IT-enabled workplace. Many of the processes that take place within organizations on a daily basis are imbued with gender attitudes and behaviors, and have strong implications for power, exploitation and control in the workplace. Gender can be seen as a set of patterned, socially produced differences between male and female, which usually involve the subordination of women, concretely or symbolically (Acker, 1992, 1998, 1999). The social construction of gender in the workplace perspective posits the development of and maintenance of a masculinized IT culture that systematically excludes women from IT work and all educational and professional steps leading up to IT work (Trauth, 2002; von Hellens, 2001; Tapia, 2002). This view attributes the problem to the construction of IT as a "man's world." This perspective, although recognizing that there are no universally male or female cultural traits, emphasizes that within the IT workplace certain social characteristics are gathered together in a unit that has come to be seen as "male" and the excluded traits as "female." Female IT workers are faced with two choices: to masculinize themselves and "fit in" or to challenge the cultural system and attempt to feminize the workplace (Cockburn, 1983, 1988; Cockburn & Ormrod, 1993; Wajcman, 1991; Adam, Emms, Green & Owen, 1994; Balka & Smith, 2000; Eriksson, Kitchenham & Tijdens, 1991; Hacker, 1981, 1989, 1990; Hovenden, Robinson, & Davis, 1995; Murray, 1993; Glastonbury, 1992; Lovegrove & Segal, 1991; Spender, 1995; Star, 1995; Webster, 1996; Woodfield, 2000).

Three Case Studies

Three IT companies who fit the description of a dot-com were examined at various points during their life cycle: *Headsup.com*, *Contentman.com* and *Ebiz.com*. To read more about these studies please see Tapia (2003, 2004). Briefly, the three companies were overwhelmingly male (82%, aggregate data). For example, after 1 year of existence, Headsup.com had 80 employees, 19 of whom were female. After 18 months, 30 employees left, including 15 out of the 19 women. After 23 months, the company had 20 employees, none of which were female. After 24 months, the company officially folded.

Gendered Recruitment

During the era known as the dot-com bubble, how successful was the IT industry in terms of recruiting and retaining women? What lasting effects can we see from the dot-com bubble's social changes on women's role in the future of the IT industry (see Tapia, 2003, 2004)? The dot-com era can be characterized as a time in which the IT industry was facing an acute shortage of employees, and yet, as the analysis shows, chose to create a culture that made it very difficult for female employees to be hired, trained and retained. I argue that in a few small start-up firms, the organizational culture created during this era made it nearly impossible for female employees to be recruited and retained. These cultures may have satisfied the immediate needs of the small start-up IT firm but were disastrous for traditional organizational measures that protected, recruited and retained women in the workforce, and that make the IT workplace hospitable to a variety of people, including women.

The dot-com era and its accompanying get-rich-quick mentality led to unconventional hiring practices, which led to the hiring of homogeneous populations, excluding women, people of color and older professionals. The essential problem with this is that IT has become associated with many material and immaterial benefits in society. This method of hiring systematically selected individuals of one race, gender, age, background, culture and class granted those benefits to a select few.

During this time, technical professionals believed they were capable of starting and managing their own business with no training, no business or managerial skills and no professional human resources help. This led to a lack of protective organizational norms and values. This opened the door to the creation of a hostile work environment. Protective organizations, such as a professional human resources staff, were seen as unnecessary fat in a lean, agile, fast-moving organization. The prevailing mentality fostered a short-timer culture in which anything was acceptable, since most employees would be gone in a matter of months. The rules and policies that would have protected women and minorities from the creation of a hostile work environment were seen as slowing down the process.

According to case data, all three companies used three recruitment strategies: online employment listings (such as Monster.com), personal references and recruitment parties. The most successful method of hiring the most employees was by far recruitment parties. However, the employees who were hired through personal contacts (i.e., friends of the owners) were the most long-lived at each company. The initial growth spurt for each company was accomplished via personal references. For example, during this growth phase, Headsup.com grew from 5 to 20 employees, all male. When asked, the additional 15 employees stated they were friends of the original five. They had known them at college, or in a past job or through a friend of a friend. For Contentman.com, the original 20 employees were very homogeneous. They were all computer programmers, all male, all in their mid to late 20s, and they devoted almost all of their time to work activities. Their time had no competition from external forces, such as families, pets, girlfriends or hobbies.[1]

Anticipating that they needed to grow fast, each company developed the second method of hiring: the college tour recruitment party. I argue that the development of this recruitment party was an extension of the first effort to hire more individuals who were just like themselves, but on a larger scale without the personal ties. The first phase of the recruitment party was to send invitations to universities' management information systems (MIS), computer science (CS) and information systems (IS) departments and computer labs on campus. This is significant, because the arenas in which these companies were seeking new employees (MIS, CS, IS) were already strongly male dominated.

At these parties, the owners attempted to convey financial success and a fun, youthful, relaxed working environment. They provided very expensive, sophisticated food and drinks. At the same time, they also provided video gaming systems and wall-projection units for the candidates' entertainment. The owners/managers would circulate among the invited candidates, chatting and what they call "geeking," discussing technical issues in a fun, lively banter, with a one-up-man-ship style, seeking which person knew the most obscure technical facts. They ate, they chatted, they drank and they played one-on-one fighting games like "Soul Caliber" against one another for hours. No formal interviews were ever held, no formal questions about the applicants' technical expertise were ever asked, no credentials or references were ever examined or checked; yet the majority of new employees were hired through these recruitment parties. It became apparent that these organizations used these recruitment parties as a cultural sifter, sifting out those that were the closest match culturally to the owners/managers and original formative employees. To be hired by one of these companies, the candidate must first pass a series of "tests" that, while intentional or not, weeded out almost all women and people of color. The result was a very homogenous workforce.

Eventually, each company hired a few female computer programmers through recruitment parties, and several more women were hired locally through personal contacts. The facts are that the owners/managers of these three companies did not hire a significant number of female employees, the female employees that were hired left (or were laid off) before the business closed, and these departures may point to the existence of a hostile work environment that drove them out (see Tapia 2003, 2004 for more detains on this hostile work environment). Another possible explanation is that the women who left saw reality a bit more clearly than the men that stayed. They saw the layoffs and the beginnings of organizational failure and got out before the doors were closed for good. In the case of Contentman.com, the company's first closure was dramatic in that the

doors were closed to all remaining employees before the start of work. The employees were left at the door without paychecks owed them, their personal items, their retirement investments, their health benefits and any sort of notice or explanation.

CONCLUSION

The administration of these dot-com era companies intended to enter the market with a single product, make as much money as possible—as quickly as possible—and get out of the market just as quickly with their millions. They had no intention of "changing the system," nor any concerns about quality or diversity. These owners expressed a concern about external pressure that they felt was exerted upon them by a high-speed, high-change industry with millions of dollars to win or lose.

This need for speed, agility, incredible outputs of time and effort led owners/managers to hire employees who were a cultural match to themselves. They hired a homogenous workforce who were highly technically competent, likely to put in long hours, required little training, socially unfettered and in all ways flexible. Their most successful hiring efforts were aimed at university centers filled with young men just like themselves. Their office cultures rewarded social behaviors that fostered competition, extreme efforts and the degradation of the weak (and feminine).

These categories, however, were not exclusively tied to biological sex; there were indeed "masculinized" women in the form of a few female programmers and "feminized" males in the form of the servile assistant to the system administrator. These categories were powerful enough to associate a higher status with and create an opportunity structure for those employees who exhibited masculine behaviors (as defined by these organizations). This presented a clear dilemma for women who may have sought entrance into the IT workforce of the dot-com era, what had come to be known as a highly lucrative industry. Given that much of the hiring was done through recruitment parties in principally male environments and through male-to-male word of mouth, women were excluded from the hiring process. If they were hired, against all odds, they were forced to adopt "masculinized" cultural behaviors to fit in. Those that didn't were seen as outside the system (quite literally in many cases) and were treated as outsiders.

The homogeneity of IT departments increases the likelihood of the development of unacceptable anti-social behavior. Diversity of skill sets is important for the economic success of businesses in the industry. In addition, homogeneous IT worker populations lead to a lack of creativity, stagnation and potential business failure (Florida, 2002). Real intellectual and social diversity should foster constructive dissent. Homogeneous IT departments can also breed groupthink (Janus, 1972), in which the group can make bad or irrational decisions as each member attempts to conform his or her opinions to what they believe to be the consensus of the group.

FUTURE TRENDS

The future of recruiting women into IT work is unclear. Times are changing for the IT employee. Since the bursting of the dot-com bubble, the market has been slowly moving from a *sellers' market* to a *buyers' market,* in which the hiring organization has the upper hand in the employment relationship. Competition has increased for IT applicants for the first time in several years.

Prior to the bursting of the dot-com bubble, employers hired IT staff based on the potential of the new hires, rather than the degree, experience or credentials the hire already possessed. According to the ITAA (2004), at the turn of the century, hiring IT managers tended to play down the importance of a college degree, letting practical experience replace more formal education and training credentials.

The crisis in the supply of IT workers has not just happened as predicted. Although supply has remained low for several reasons, such as the massive retirement of technical baby boomers and the steady drop-off of students graduating with a degree in a computer-related field, demand has also dropped. In addition, since 2000, approximately 100,000 computer software and services jobs have moved offshore, including both domestic jobs eliminated and new jobs created overseas (ITAA, 2004).

Due to increased competition, the threshold to enter a career in IT has been low in the past, but is

rising. Like other occupational groupings, employees in the IT sector are becoming increasingly professionalized. This professionalization raises the bar for entry into the field by increasing the value of formally obtained education, training and certifications.

This is a double-edged sword for women seeking to enter the IT field. On the positive side, a woman in possession of a degree or certification in IT is more likely to be on equal footing with all other candidates possessing similar skill markers. Culture matters less in a tight market, so highly skilled women are more likely to be recruited. Employers are also seeking to hire employees for longer tenures, therefore investing more in terms of training into the employee and in work-family friendly policies, increasing the likelihood of retention. On the negative side, since competition is growing among IT employees and potential recruits, those without formal IT training, degrees and certifications will have a much harder time entering the field. If the trend in the pipeline continues and women enter IT education programs in fewer and fewer numbers, this plus increased professionalization and competition will effectively weed out most women from the IT workforce.

REFERENCES

Acker, J. (1992). Gendering organization theory. In Mills & Tancred (Eds.), *Gendering Organizational Analysis.*

Acker, J. (1998). The future of gender and organizations. *Gender, Work, and Organizations, 5*(4), 195-206.

Acker, J. (1999). Gender and organizations. In J. Saltzman Chafetz, (Ed.), *The handbook on gender sociology.* New York: Plenum.

Adam, A., Emms, J., Green, E., & Owen, J. (Eds.). (1994). *Women, work and computerization: Breaking old boundarie—building new forms.* Amsterdam: North-Holland.

Adam, A., Howcroft, D., & Richardson, H. (2001). Absent friends? The gender dimension in IS research. In N. L., Russo, B. Fitzgerald, & J. L.

DeGross (Eds.), *Realigning research and practice in information systems development: The social and organizational perspective* (pp. 333-352). Boston: Kluwer Academic Publishers.

Annenberg Public Policy Center. (2001). *Progress or no room at the pop? The role of women in telecommunications, media, and e-companies.* Retrieved March 14, 2001, from www.annenberg publicpolicycenter.org/04_info_society/women_ leadership/telecom/2001_progress-report.pdf

Balka, E., & Smith, R. (Eds.). (2000). *Women, work and computerization: Charting a course to the future.* Boston: Kluwer Academic Publishers.

Cockburn, C. (1983). *Brothers: Male dominance and technological change.* London: Pluto Press.

Cockburn, C. (1988). *Machinery of dominance: Women, men, and technical know-how.* Boston: Northeastern University Press.

Cockburn, C., & Ormrod, S. (1993). *Gender and technology in the making.* London: Sage.

Eriksson, I. V., Kitchenham, B. A., & Tijdens, K. G. (Eds.). (1991). *Women, work and computerization: Understanding and overcoming bias in work and education.* Amsterdam: North-Holland.

Florida, R. (2002). *The rise of the creative class.* New York: Basic Books.

Geewax, M. (2000). If women ruled ... Female techies imagine a world. *The Atlanta Journal,* August 27, D1.

Glastonbury, B. (1992). *The integrity of intelligence: A bill of rights for the information age.* Basingstoke.

Hacker, S. (1981). The culture of engineering: Woman, workplace and machine. *Women's Studies International Quarterly, 4*(2) 341-353.

Hacker, S. (1989). *Pleasure, power, and technology.* Boston: Unwin and Hyman.

Hacker, S. (1990). *Doing it the hard way: Investigations of gender and technology.* Boston: Unwin and Hyman.

Hovenden, F. Robinson, H., & Davis H. (1995). The software maverick: Identity and manifest destiny. In

The subjects of technology: Feminism, constructivism and identity. Brunel: Center For Research Into Innovation, Culture and Technology.

Information Technology Association of America (ITAA). (2003, May 5). Report of the ITAA Blue Ribbon Panel on IT diversity. *Proceedings of the National IT Workforce Convocation,* Arlington, VA.

Kunda, G. (1992). *Engineering culture.* Philadelphia: Temple University Press.

Lovegrove, G., & Segal, B. (Eds). (1991). *Women into computing: Selected papers 1988-1990.* London: Springer-Verlag.

Murray, F. (1993). A separate reality: Science, technology and masculinity. In E. Green, J. Owen, & D. Pain (Eds.), *Gendered by design: Information technology and office systems* (pp. 64-81). London.

National Action Council for Minorities in Engineering. (2001-2002). *The state of minorities in engineering and technology,* 40-44

National Science Foundation. (2000). *Women, minorities and people with disabilities in science and engineering 2000,* 53.

Perlow, L. (1998). Boundary control: The social ordering of work and family time in a high tech corporation. *Administrative Science Quarterly, 43,* 328-357.

Perlow, L. (1999). Time famine: Toward a sociology of work time. *Administrative Science Quarterly, 44,* 57-81.

Spender, D. (1997). The position of women in information technology, or who got there first and with what consequences? *Current Sociology, 45*(2), 135-147.

Star, S. L. (Ed.). (1995). *The cultures of computing.* Oxford: Blackwell Publishers.

Tapia, A. (2003, April 10-12). Hostile_work_ environment.com, computer personnel research. *Proceedings of the ACM SIGCPR/MIS Conference,* Philadelphia, PA.

Tapia, A. (2004, Summer). The power of myth in the IT workplace: Creating a 24-hour workday during the dot-com bubble. *Information Technology and People*.

Tapia, A., & Kvasny, L. (2004, April 22-24). Recruitment is never enough: Retention of women and minorities in the IT workplace. In *Proceedings of the ACM SIGCPR/MIS Conference*, Tucson, AZ.

Tapia, A., Kvasny, L., & Trauth, E. (2003). Is there a retention gap for women and minorities?: The case for-moving in versus moving up. In C. Shayo (Ed.), *Strategies for managing IS/IT personnel*.

Trauth, E. M. (2002). Odd girl out: An individual differences perspective on women in the IT profession. *Information Technology & People, 15*(2), 98-118.

von Hellens, L., Nielsen, S., & Trauth, E. M. (2001). Breaking and entering the male domain: Women in the IT industry. In *Proceedings of the 2000 ACM SIGCPS Computer Personnel Research Conference*.

Wajcman, J. (1991). *Feminism confronts technology*. University Park: The Pennsylvania University Press.

Wajcman, J. (2000). Reflections on gender and technology studies: In what state is the art? *Social Studies of Science, 30*(3), 447-464.

Webster, J. (1996). *Shaping women's work: Gender, employment and information technology*. London: Longman.

Woodfield, R. (2000). *Women work and computing*. Cambridge: Cambridge University Press.

KEY TERMS

Dot-Com Bubble: The disproportionate market hysteria created under the premise of the endless investment possibilities of the Internet and services related with it.

Geeking: It is to sit online and read mail, news, chat and consult other sources of information primarily related with computers and technology.

Groupthink: Groupthink is a term coined by psychologist Irving Janis in 1972 to describe a process by which a group can make bad or irrational decisions. In a groupthink situation, each member of the group attempts to conform his or her opinions to what they believe to be the consensus of the group.

Hostile Work Environment: A hostile work environment exists in a work place when an employee experiences harassment and fears going to work because of the offensive, intimidating or oppressive atmosphere generated by coworker(s).

Human Resources: It is the part of management that deals with the attraction, retention and development of employees.

Sexual Harassment: It is unwelcome sexual advances, requests for sexual favors and other verbal or physical conduct of a sexual nature that take place in a work or work-related environment.

Short Timer Culture: It is a expression used to describe the small interval of time that nowdays the customs and traditions take to appear and fade away.

Start-Up Firm: It is a company that fits in one of the following criteria: Time in business does not exceed 4 years, no public offering, annual sales do not exceed $1,000,000, or a minimum of one full-time employee and no more than 12.

Work Life Gap: Differences among amount/quality of time that a person dedicated to activities related with the work environment and his or her personal life.

Zero Drag Employee: It is a term for the employee who is available at a moment's notice. The ideal zero-drag employee is young, unmarried, and childless with no responsibilities and an eagerness to do well.

ENDNOTE

[1]   This is known as zero drag (Perlow, 1998; Kunda, 1999).

Managerial Careers, Gender, and Information Technology Field

Iiris Aaltio
Lappeenranta University of Technology, Finland

INTRODUCTION

Careers are organizational and institutional, and they have know-how-based contexts. Managerial careers from a gender perspective, gendered "blind spots" in organizations and the invisibility of women in management have been an object of study since the 1970s. Gender is a part of socially constructed individual identity. Gendered identities in organizations are defined and redefined in relationships as people become socially constructed through work groups, teams and interactions. Because of this social construction, femininity and masculinity grow into human behavior and outlook. Understanding gender as an activity and a term in the making (Calás & Smircich, 1996), it is a constitution of an activity, even when institutions appear to see woman and man as a stable distinction (Korvajärvi, 1998). Beyond work-life and organizations, there are multiple institutional and gendered structures. The information technology (IT) industry and companies are also an institutional construction with gendered dimensions, and they also participate on the creation of femininity and masculinity.

Career can be seen as a conceptual artefact that reflects a culture and rhetorical context in its use. It is a kind of window to a network of values, institutions and functions, where actual careers are made. Usually, the formal organization is based on neutrality and equality, but a closer look reveals the deeper social structures that make it different to women and men. There is a concept of an abstract and neutral worker, and this worker is supposed to be highly competent, work-oriented and available, committed to work-life without any knit to private life. These characteristics support a good career climb in an organizational hierarchy, and many of these characteristics better suit men than women (Metcalfe & Altman, 2001). For instance, home responsibilities make often working hours less flexible for women than men. The notion of an essential person with no gender characteristics does not recognize these issues, whereas taking gender as a research topic shows that work-life as a context differs between women and men.

BACKGROUND

Managerial Positions within the IT Industry

Organizational structures, including managerial positions, are gendered by nature. Overall, there is a high degree of vertical segregation, which means that there are few women in managerial positions compared to men (Acker, 1992). According to the United Nations' *World's Women 2000* report, women's share of the administrative and managerial labor force is less than 30% in all regions of the world. This is true also in Nordic countries, where the participation of women in work life is almost 70% and has a long tradition. Women also hold only 1% to 5% of all the top executive positions (Wirth, 2001), and the numbers seem to change very slowly. In the European Union countries, women's share has barely changed since the early 1990s, and has remained at a less than 5% level (Davidson & Burke, 2000). This division of managerial top positions is called glass-ceiling phenomena, and it exists world wide (Powell, 1999): "The higher the managerial position the fewer the women." As a result of this, women and the highest economic power become separated.

Taking a closer look at the numbers, the least amount of women in top positions are found in male dominated areas, such as heavy industry and construction business, where the amount of female leaders is less than 10%. IT is also a male-dominated field. There are few female directors in an organization that employs mostly men (Kauppinen & Aaltio-

Marjosola, 2003). The number of female managers has increased slowly. In many fields, like IT, it is still low (Ahuja, 2002). Women's and men's work in organizations also differ from each other by nature; that means women and men end up doing different kinds of work horizontally.

Current statistics indicate that women account for about 25% of technology workers in the European workforce and about 20% of those in the United States' (U.S.), and that there looks to be a polarization in the type of work women and men do. The majority of women are employed in routine and specialist work, like clerks, while men are engaged in analytical and managerial activities. In the studies, overall 10% of males and only 3% of females within IT had achieved senior managerial positions (Ahuja, 2002). Salary gaps for women and glass-ceiling perceptions are reported as well in this research. Despite that the IT profession has grown in recent years, there remains a gender imbalance and, in some cases, even evidence of a decline in female workforce numbers (Ahuja, 2002).

Managers and leaders have identities that become constructed within special circumstances, and IT constitutes a particular background for identities to grow in. As stated by Davis (1995), organizations and their activities are cultural constructs arising from the masculine vision of the world, and IT's close connections to the male-dominated technology field and its high numbers in male participation makes its connection to masculinity evident. The glass ceiling in the IT field might even be stronger that in others, because there is evidence that women there tend to be stereotyped as staff, the ones who don't take risks, rather than "line" people; whereas men are the innovators and designers (D'Agnostico, 2003; Russell, 2004). This results on men's career outcomes including higher managerial positions.

The segregation of work is based on the classical stereotypes of women's and men's behavior and orientations. Men are oriented towards technical and industrial work, whereas women are engaged in occupations where one needs caring ability and social integration, such as teachers and nurses. Ideals for men's and women's work differ from each other and carry stereotypes (Aaltio, 2002). Women and men are easily valued differently because of their gender. In society, there are different places for women and men, and this holds both in families and in work organizations. Men historically relate to the public and women to the private spheres of life (Acker, 1992). By extending their roles and breaking into public institutions, women challenge the prevalent male ideals and bring private issues into public and institutional spheres. Career opportunities mean different things for men and women and, therefore, challenge them differently.

It is also notable that the relationship between gender identity and participation in the IT profession is not the same for all women, nor is it based on monolithic values that are the same for all women. Each individual woman experiences societal influences differently and brings her personality and characteristics into the field (Trauth, 2002), even if women's institutional backgrounds are similar.

WOMEN'S CAREERS WITHIN THE IT FIELD

Discourses of Career: Interviews from IT Managers

In the study, five managers were interviewed, two of them women and three men. They all work in a large, successful IT industry company in Finland, are young or early middle-aged, and are labeled as high achievers with a promising future in the company. In the interviews, it was asked how they see their managerial career development up until now and in the future. The data gathered was narrative by nature, because a lot of space was given to open talk around the topics raised at the interview (Aaltio-Marjosola, 2002).

As one of the managers describes: "Professional development is my area, and I am globally responsible [for] this part of the business. The culture here in the company is, however, where I start from. In my close network, we have a team of 10 people, but globally, it is a big circle, of course." As seen in the citation, instead of manager-subordinate relationships, she emphasizes the close network and teamwork orientation. Another manager illustrates his job with no clear subordinate relationships: "I came here to coach a team with a few people and to coordinate things. This is more or less process management; I have to think first and then make things simple and concrete, communicate them to the others. I work closely with the head of the company in some projects, in addition. In fact, I do not see that there is a clear organization where I am a leader."

Further, one of the IT managers describes her job: "I am a personnel consult, even if I also take care of the unit's personnel management, and my work is multiple of things. For instance, I also do training and lecturing in our inner training seminars." In spite of the high organizational position, she sees her job to be consultative, including support of teams.

It is also typical to describe career development that includes multiple paths, work in many organizations and rapid movements during the career. One of the managers described his work: "I find that my career is more or less a bumpy road, not going upward all the time. I have started two times from a beginning in my job and found that you have to take your space in every new job no matter what is your position when you start. You have to find your fit to the values of the enterprise, feel they are near to yours, to get the right start." Still another manager describes his career development: "My job is changing. I will take care of the process development here, coordinate and harmonize things. I have to move as rapidly as the whole company does, this is the way it comes here. The number of my subordinates will be like 50 or 100 altogether, I am not quite sure how many they are, in fact."

Career, in the descriptions of these managers, is advancement of abilities, seeing oneself as an important and integral person in a certain close network, and giving good support for subordinates. None of the interviewed managers started their description of their work from the organizational ladder or other kind of embodiment of hierarchy. This shows the IT industry's development, being at the top of companies that themselves develop work practices and communication styles that break traditional ones with hierarchy and high formal authority. Much of the work is based on projects and close teamwork. Both men and women adapt into this if they seek to be successful in the company and get advancement.

Managerial Careers of Women within IT

Today's careers are based on a variety of choices; careers are "boundaryless" due to multiple routes and individual choices. To predict career development is more difficult than it used to be. Individuals'

knowledge and skills give them good possibilities to move from one job to another, and organizations seldom serve for a life-time basis in a person's career development. Flexibility and adaptability of careers have increased, and diverse assignments are common instead of a clear and one-path kind of career development (Arnold, 2001; Storey, 2000). Work will become rich from difference and requires abilities and motivation typical for women, who have always combined many roles, private and public, and thus learned flexibility.

For instance, Ahuja (2002) suggests that there are barriers to female managers' careers within IT, and that their minor positioning in the IT industry is due to an "old boys" network. There is a large pool of qualified and experienced male professionals, whereas the pool of women is still few. This counts for a lack of female role models and mentors for the younger generations, and discriminatory practices remain. There is also evidence that women do not see IT as an attractive option, and even if they do, they are technically less equipped to come over. There are few managerial role models for women in IT industry companies, and women would benefit from career consulting.

Himanen (2001) argues that computer hackers will become heroes of the information society, and the heroes of IT are and will in the future be men, as seen in, for example, Silicon Valley, where there are very few women in higher-level positions. However, men also represent a variety rather than a unity. Taking the fluid notion of gender, we can see that masculinities are carried by organization cultures and are not unitary. In the changing work life, advancement also requires new skills and attitudes from men, not only for women as a unitary category.

Also, family-work contrast barriers were reported. A survey (Prencipe, 2001) outlined attitudes of IT women leaders, half of them in senior positions. The study showed a positive personal valuation of one's work, but also that balancing personal and professional life was still complex for most of them. Reported stress was due to working late, constant change and sometimes reported discrimination in a social environment consisting of mainly young, often male colleagues (Prencipe, 2001). When managerial career advancement starts from lower professional levels, factors like role stress tend to decrease job satisfaction, resulting in lower career outcomes of women (Igbaria & Chidambaram, 1997). In addition, at the institutional level, where changes are very slow, women and technology are still a combination that arouses suspicion.

FUTURE TRENDS AND NEEDS OF RESEARCH

Challenges of IT women managers' career development are in line with the results of earlier gender studies. In addition, the attractiveness of the company cultures for women mangers is also questionable, because of the cultural codes that might be gender-biased, repulsive and more favorable for men to follow. These cultural codes need to be studied and better understood.

As a part of high technology, IT has taken the world by storm and is changing the way businesses learn, as well as the nature and characteristics of work. The implementation of IT within and across organizations is reducing the importance of hierarchy and command-and-control authority systems that structure power within them. High technology changes the traditional managerial and communicational style from vertical to horizontal (Zeleny, 1990). In high technology-oriented industries, power is connected with expertise, which may break down the traditional hierarchy in an organization (e.g., Gunz, Evans, & Jalland, 2002). Research is needed on the special nature of work in IT companies, in order to understand which kind of management they really need, in terms of good results and higher well-being of the work force.

IT enterprises are themselves examples of "new" ways of management and doing work. It has been stated that the IT industry is in transition between the old sense of identity and the new one (Colwill & Townsend, 1999). As argued, the old culture was directed towards providing the answers, not at meeting the needs of the users. Working in intense networks is increasing and good communication skills are highly needed in the future. Gender-biased cultural expectations and gender stereotypes still work as a barrier for female managers' careers in IT. However, the changing nature of IT work and its prevalent practices make it open to diversity of expertise and values. Both women and men can

benefit from these developments if the IT field itself learns gender-equality and tolerance.

The future's industries also compete to commit good expertise, and women are a resource. Multiple skills and their good management are needed, and new solutions to old problems will be searched. This also will advance women's managerial careers.

CONCLUSION

IT-field companies appreciate technology development. They have to learn understanding of social and cultural aspects that, however, are the true background for their development in the future. The field should attract both capable women and men. While there are a few managerial role models for women in the IT industry, women involved would benefit from career consulting and mentoring programs (Ahuja, 2002). The use of tele-work and e-leadership might also positively affect women in the industry, because they support combining family and work life (Avolio, Kahai, & Dodge, 2000; Beasley, Ewuuk, & Seubert, 2001). This kind of supportive activities may attract more women to the field and advance women's managerial careers. Organizational cultures that grow towards multiplicity of values and a variety of femininities and masculinities may be better work environments, especially for women but also for men in the future.

REFERENCES

Aaltio-Marjosola, I. (2002). Interviewing female managers—Presentations of the gendered selves in contexts. In I. Aaltio-Marjosola & A. Mills (Eds.), *Gender, identity and the culture of organizations* (pp. 201-219). London: Routledge.

Acker, J. (1992). Gendering organizational theory. In A. J. Mills & P. Tancred (Eds.), *Gendering organizational analysis* (pp. 248-260). Neewpury Park: Sage Publications.

Ahuja, M. K. (2002). Women in the information technology profession: A literature review, syntesis and research agenda. *European Journal of Information Systems, 11*(1), 20.

Arnold, J. (2001). Careers and career management. In N. Anderson, D.S. Ones, H. K. Sinangil, & C. Viswesvaran (Eds.), *Handbook of industrial, work and organizaitonal psychology* (p. 2). London: Sage Publications.

Avolio, B. J., Kahai, S., & Dodge, G. E. (2000). E-leadership: Implications for theory, research, and practice. *Leadership Quarterly, 11*(4), 615.

Beasley, R. E., Ewuuk, L.-D., & Seubert, V. R. (2001). Telework and gender: Implications for the management of information technology professionals. *Industrial Management & Data Systems, 101*(8/9), 477-482.

Calás M., & Smircich, L. (1996). From The woman's point of view: Feminist approaches to organization studies. In I. Clegg, R. Stewart, C. Hardy, & W. R. Nord (Eds.), *Handbook of organization studies* (pp. 218-258). London: Sage Publications.

Colwill, J., & Townsend, J. (1999). Women, leadership and information technology. The impact of women leaders in organizations and their role in integrating information technology with corporate strategy. *The Journal of Management Development, 18*(3), 207.

D'Agostino, D. (2003, October). Where are all the women IT leaders? *EWeek.* Retrieved from https://shop.eweek.org/eweek/

Davidson, M. J., & Burke, R. J. (2000). *Women in management: current research issues* (p. II). London: Sage.

Davis, C. (1995). *Gender and professional predicament in nursing.* Buckingham; Philadelphia: Open University Press.

Gunz, H. P., Evans, M. G., & Jalland, R.M. (2000). Chalk lines, open borders, glass walls, and frontiers: Careers and creativity. In M. Peiperl, M. Arthur, & N. Anand (Eds.), *Career creativity. Explorations in the remaking of work* (pp. 58-76).

Himanen, P. (2001). *The hacker ethic and the spirit of the information age.* New York: Random House.

Igbaria M., & Chidambaram, L. (1997). The impact of gender on career success of information systems professionals. A human-capital perspective. *Information Technology & People, 10*(1), 63.

Kauppinen, K., & Aaltio-Marjosola, I. (2003). Gender, power and leadership. In C. & M. Ember (Ed.), *Encyclopedia of sex and gender.* New York: Kluwer Academic Publications.

Korvajärvi, P. (1998). *Gendering dynamics in while-collar work organizations.* University of Tampere. Acta Universitatis Tamperensis 600. Vammalan Kirjapaino Oy.

Mettcalfe, B., & Altman, Y. (2001). Leadership. In E Wilson (Ed.), *Organization behaviour reassessed. The impact of gender.* London: Sage.

Powell, G. N. (1999). Reflections on the glass ceiling. Recent trends and future prospects. In G. N. Powell (Ed.), *Handbook of gender and work* (pp. 325-345). London: Sage.

Prencipe, L. W. (2001). Despite some successes, women still face obstacles in reaching the IT heights. *Info World, 23*(17), 62.

Russell, J. E. A. (2004). Unlocking the clubhouse: Women in computing. *Personnel Psychology, 57*(2), 498.

Shoba, A., & Arun, T. (2002). ICTs, gender and development: Women in software production in Kerala. *Journal of Internationl Development, 14*(1), 39.

Storey, J. A. (2000). Fracture lines. In A. Collin & R. A. Young (Eds.), *The future of career.* Cambridge, UK: Cambridge University Press.

Trauth, E.M. (2002). Odd girls out: An individual differences perspective on women in the IT profession. *Information Technology & People, 15*(2), 98.

Wirth, L. (2001). Shaping women's work: Gender, employment and information technology. *International Labour Review, 140*(2), 216.

Zeleny, M. (1990). High technology management. In H. Noori & R. E. Radford (Eds.), *Readings and cases in the management of new technology: An operations perspective* (pp. 14-22). Englewood Cliffs: Prentice-Hall.

KEY TERMS

Career: Consists of the sequential choices made by a person. It is a developmental process of professional identity and personality.

Culture: A human process of constructing shared meaning that goes on all the time and is based on human unique capacity for self- and other-consciousness.

Doing Gender: Concerned with the constitution of gender through interaction, and makes gender a consequence of an ongoing activity.

Gender: An integral part of socially constructed individual identity that constitutes and restructures a multitude of cultural and social phenomena.

Gender Bias: To generalize essential individual behavior without gender dimensions.

Gender Stereotypes: Refer to the traits and behaviors believed to occur with differential frequency in the two gender groups of women and men.

Glass Ceiling: A metaphor that describes the tendency of women get excluded from top management in organizations in spite of their occupation of middle-management positions.

Horizontal Gender Segregation: Refers to the structure of the workforce that tends to become separated into women's and men's work.

Organizational Culture: A collective construction of practices, meanings and expressions that can be seen developed in interaction within organizational social spheres.

Vertical Gender Segregation: Refers to the structure of organizations that position women on lower managerial levels compared to their male counterparts.

Matrix

Irina Aristarkhova
*National University of Singapore, Singapore, and
The Pennsylvania State University, USA*

INTRODUCTION

1. Matrix = Womb.
2. *The Matrix is everywhere, it's all around us, here, even in this room. You can see it out your window, or on your television. You feel it when you go to work, or go to church or pay your taxes. It is the world that has been pulled over your eyes to blind you from the truth … that you, like everyone else, was born into bondage … kept inside a prison that you cannot smell, taste or touch. A prison for your mind. A Matrix.* (Wachowski & Wachowski, 1999)
3. *What is Matrix? Simply … the "big Other," the virtual symbolic order, the network that structures reality for us.* (S. Zizek, 1999)

What is Matrix? In the past years, the notion of the Matrix has become dominant in figurations of cyberspace. It seems as if it is the most desirable, the most contemporary and fitting equation; however, its gendered etymology is rarely obvious. On the opposite, the gender of the matrix as a notion and term has been systematically negated in such disciplines as mathematics, engineering, film studies or psychoanalysis. It is necessary thus to explore and critique the Matrix as a most "fitting" metaphor in/ for cyberspace that has conceived it (cyberspace) as a free and seamless space very much like the maternal body (Aristarkhova, 2002). The challenge today, therefore, is to reintroduce the maternal as one of embodied encounters with difference, to recover the sexual difference and gender in the notion of matrix with reference to cyberspace and information technologies that support it.

BACKGROUND

There is nothing new in this equation of matrix and cyberspace. This equation points out to a long history

of use of maternal body as a source of "making sense" space as a foundational category (in addition to time). "Space" enables introduction of other notions, such as extension, arrangement, geography and body, among others. However, the origin of "space" itself is usually found in the maternal body, such as the case with "matrix" or its related notion: "chora." Once again, place for cyberspace has been found in a woman's body that has been misplaced, in this first and unique place (Irigaray, 1985), a house/ home (Levinas, cited in Derrida, 1999; Derrida, 1997,1999) or container (Aristotle, cited in Irigaray, 1985). While some might celebrate this fact as effecting a "feminization of the cyberspace imaginary" and thus potentially empowering women, others caution us that it follows the Western tradition of depriving woman of her own place, treating cyberspace, in fact, like her (body): an instrument, as a dismembered tool waiting to provide a place *for* man; his cultural, technological and political aspirations (Irigaray, 1985; Plant, 1997). Whichever way one decides on how such imaginaries empower women, it is particularly noteworthy that this gendered nature of the notion of the matrix has been historically and discursively neutralized by constant references to its infinite openness and indifference to difference, sexual or otherwise in these new technologies (e.g., in films like *The Matrix*, and their postmodern formulations). Despite the occasional and even foundational references to the gendered nature of the matrix, little has been done to theoretically recover its positive attributes for rethinking cyberspace as such.

We can name at least three associations that currently operate between notions of cyberspace and the matrix, making it so appropriate for representations of cyberspace:

1. Both are seen as infinite and ever expanding, where expansion is itself their function (as in mathematics, where the initial matrix forms the

basis for serial and cumulative development; or in contemporary cybertheory and cyberpunk literature, where cyberspace is often assumed to be limitless and fully imaginary, to be filled with any desirable content).

2. They are supposed (and wanted?) as empty spaces, passively waiting to be filled and occupied—a fact that also lands to its being conceptualized as *virtual* vis-à-vis real. It is simply "out there," without having its own place, though providing a place for everything. As Doug Mann and Heidi Hochenedel define it, after Baudrillard (1994), "it is a desert of the real in which hyper real simulacra saturate and dominate human consciousness," it is "a map without territory" (Mann & Hochenedel, 2002). Being appropriated by phallocentric imaginary, matrix has become an empty space to be filled with any content, psychological, scientific, artistic or philosophical theorizations. It does not anymore belong to a body marked by sexual difference and gender.

3. Ultimately, both have been disembodied. Cyberspace has been invented as being nowhere and everywhere, which has no corporeal reference or geographical location. It is a place of ultimate escape, where we can explore our desires, anxieties and fears to become more stable, normal and healthier (in earlier social science literature, some assumed that exploring identity swapping in cyberspace would allow teenagers to overcome their fears of sexuality and "opposite sex").

These characteristics imply that the "matrixial," therefore, is indifferent to difference, that its infinite openness does not impose barriers on/to entry and participation. And also, participation is understood to be free and on equal terms. The matrix provides a sense of limits and spherical closure to limitless, borderless imaginary of cyberspace.

Thus, I argue there is a tension between the generative (as abstract) vs. maternal (as embodied) in definitions and representations of matrix as cyberspace. The appropriation of corporeal matrix and its relation to woman's body and subjectivity through scientific, philosophical and aesthetic reductions and abstractions in Western culture has been instrumental in producing cyberspace, fantasizing it

as "self-reproducing," matrix-perfect mega-computer. In fact, these domestications of the notion of the matrix serve to disarticulate it from its relationship to embodied sexual difference, and are the matrixial as matricidal economies of cyberspace.

CONCLUSION

Thus, the issue at stake here is not so much a celebration of matrix as something that derives and undertsands woman's power as man's dependency on the maternal and the feminine, but rather, how the notion of the matrix serves as this mimicry of the maternal in cyberspace, as something that can be easily detached and performed without any references to sexual difference and gender. Therefore, a cyberfeminist critique of the certain recent appropriations of the notion of matrix is necessary in order to find alternative (to matricidal) formulations and images of spaces generated with the advent of information technologies.

REFERENCES

Aristarkhova, I. (2002). Hospitality—chora—matrix—cyberspace. *Filozofski Vestnik, XXIII*(2), 27-42. Special Issue "The Body."

Baudrillard, J. (1994). *Simulacra and simulation* (translated by S.-F. Glaser). Ann Arbor: University of Michigan Press.

Derrida, J. (1997). *Choral works* (translated by I. McCloud). New York: The Monacelli Press.

Derrida, J. (1999). *Adieu to Emmanuel Levinas* (translated by P. -A. Brault & M. Naas). Stanford: Stanford University Press.

Irigaray, L. (1985). *Speculum, of the other woman* (translated by G. C. Gill). Ithaca: Cornell University Press.

Mann, D., & Hochenedel, H. N. (2002). *Evil demons, saviors, and simulacra in The Matrix*. Retrieved September 8, 2005, from http://publish.uwo.ca/~dmann/Matrix_essay.htm

Plant, S. (1997). *Zero and ones: The matrix of women and machines*. London: HarperCollins.

Wachowski, A. (Writer/Director), & Wachowski, L. (Writer/Director). (1999). *The Matrix* [Motion picture]. United States: Warner Brothers.

Zizek, S. (1999) *The Matrix, or, the two sides of perversion*. Retrieved September 9, 2005, from http://container.zkm.de/netcondition/matrix/zizek.html. Published as *Enjoy your symptom: Jacques Lacan in Hollywood and out* (2nd ed., 2001). New York: Routledge.

KEY TERMS

Chora: A Greek philosophical term referring to that which gives place and enables spatial dimension. Chora might take a variety of meanings, including a container, an interval (like a zero), a variable (like an X) and a receptacle. It could also mean a nurse, a maternal space. Today, it is mostly used in philosophy and architectural theory to discuss the concept of "space" and how space comes about.

Cyberfeminism: A recent movement in art, literature and academia, cyberfeminism deals with the relation between gender and technology, or gender and machines. It covers a wide range of topics and practices, such as gaming, reproductive and biotechnologies, telecommunications, net communities and cyborg studies, among others.

Imaginaries: Introduced first in psychoanalytic theory, today the concept of imaginary is mostly used in sociology and other social sciences to refer to a system of values, laws and institutions "imagined" collectively within a certain social and cultural context in relation to a topic or issue. Here, it is used in plural to emphasize that there are varieties of imaginaries within a given context.

Matricidal: That which leads to the annihilation of the mother or of the maternal in cultural, social, economic, physical or political sense. Here, it refers to the substitution (as annihilation) of the maternal within studies of techno- and cyber-spaces by the concept of the matrix.

Mimicry: A biological concept, meaning a behavior of camouflage and self-concealment for the purpose of survival, has been adopted within psychoanalytic and feminist theory to refer to a desire for others through a response of fascination and an impulse to mimic.

Maturity Rather than Gender is Important for Study Success

Theresa McLennan
Lincoln University, New Zealand

INTRODUCTION

For a number of years there has been concern, particularly in the western world, about the low female participation and retention rates in computer science (Bernstein, 1994; Clarke & Chambers, 1989; Clarke & Teague, 1994; Durndell, 1991; Sturm & Moroh, 1994). Studies in New Zealand found similar trends (Brown, Andreae, Biddle, & Tempero, 1996; Ryba & Selby, 1995; Toynbee, 1993). More than ten years after these concerns were first raised, these problems largely still exist (Weston & Barker, 2004) although successful strategies are being reported (Cohoon, 2002; Fisher & Margolis, 2002; and other authors in *Women and Computing*).

To some extent at Lincoln University, New Zealand, the situation has always been different. Computing classes at all levels usually have a reasonable proportion of women (typically 25-40%). Furthermore, success in the first year programming class has been modeled for five cohorts and has been found to have no direct relationship to gender. The most consistent finding in these models is that older students are more likely to be successful than younger students. As well as summarizing the longitudinal study, key findings of interviews with some recent mature aged, female computing graduates are also included.

BACKGROUND

Lincoln University, with only 4,500 students, is by far the smallest of New Zealand's eight universities. It has a reputation for being "small and friendly". In addition to receiving compulsory course advice there is also considerable informal mentoring of undergraduate students. There are approximately equal numbers of male and female students and about 40% of students on campus are mature aged.

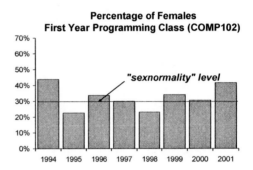

Figure 1. Female participation rate in COMP102 from 1994 to 2001

The computing degrees and diplomas offered are in applied computing, rather than computer science, and applied disciplines are known to appeal to women (Kossuth & Leger-Hornby, 2004). The introductory programming class (COMP102) has typically had about 120 students each year. This is a medium size class by New Zealand standards. It is usually at least 30% women (Figure 1) and can therefore be considered to be "sexnormal" (Byrne, 1993). Not all students studying COMP102 intend to major in applied computing and approximately 60% continue on to advanced level programming classes.

MODELING SUCCESS IN COMP102

For the five cohorts studied, the 120 or so students in COMP102, were surveyed to find their ages, genders, likely majors, expectations from this subject, and computing and educational backgrounds. Success in the subject as measured by final marks and grades, has been modeled using linear, logistic, and ordinal regression techniques (Agresti, 1990) for the 1994 (McLennan, Young, Johnson, & Clemes, 1999) and 1998 students (McLennan, Clemes, Young, & Kamikubo-Gould, 2000). A longitudinal study of the 1994 and 1998-2001 students, using artificial

Figure 2. Ordinal regression model for predicting the probability of passing COMP102 for 1994 students (adapted from McLennan et al., 1999)

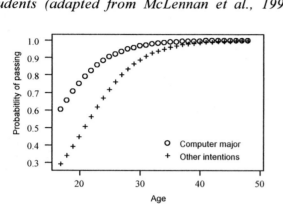

Figure 3. Box plots showing different age distributions for male and female students in 1998 (adapted from McLennan et al., 2000)

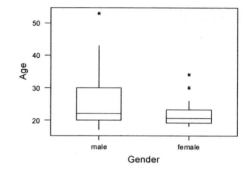

neural networks to model success, has also been completed (Li, Samarasinghe, & McLennan, 2002).

In the regression models the most significant factor contributing to success, was age, with older students more likely to pass than younger students. Not surprisingly, students with higher expectations and those intending to major in computing were also likely to do better. Figure 2 shows the probability of passing COMP102 for the 1994 students.

Given the prevailing view of the time, that computing was the domain of young males, it was surprising to also find in the 1994 study that gender had no direct bearing on likely success. This pleasing finding has been confirmed in all subsequent models. In the 1998 study there appeared to be a gender effect, because the average mark was 74% for men and 67% for females. This can largely be explained by the different age distributions for the sexes (Figure 3). A further factor was evidence that women had lower expectations and lower expectations were associated with lower marks.

Multi-layer perceptron artificial neural network models (Jain, Mao, & Mohiuddin, 1996) were used to model the final marks of students from 1994 and 1998-2001. Each year was modeled separately and for every year gender was not significant in predicting likely success. There were marked variations from year to year and age is the only consistently strong contributor for each model with older students always doing better. Viewed overall, a student's expectation of getting a good grade and having a

strong computing background were also important. These results were not entirely consistent with the regression models where computing background was not a significant factor.

About one third of the students in the longitudinal study were of mature age. In the later years roughly 40% of the mature students were university graduates from other disciplines who were now studying for Graduate Diplomas in Applied Computing. It is not surprising that these highly motivated, older students, with a history of studying success did so well in general. However, this qualification did not exist prior to 1999, so it doesn't explain the success of older students in 1994 and 1998.

INTERVIEWS

In 2002, 17 mature-aged women who had recently graduated with majors in applied computing were interviewed to provide further insights on studying in a non-traditional area (McLennan, 2003). They ranged in age from 26 to 48 and had studied COMP102 between 1999 and 2001. Their median age when studying COMP102 was 36 and their average mark was 81% (compared with 22 and 61%, respectively, for the entire classes).

Questions were asked about their reasons for returning to studying and their study experiences. Almost all had returned to study for employment related reasons. Some wanted to change career direction and others, who wanted to re-enter the workforce, saw computing as providing challenge and financial security. In this sample, only five had completed degrees and of these women four had originally intended to major in other subjects. All of the others were already university graduates who were studying for graduate diplomas rather than degrees. Some of this latter group intended to pursue careers applying computing in their original discipline area. Invariably, they felt that studying applied computing was a more appropriate choice than computer science.

Getting started, both in COMP102 and higher level subjects, was sometimes an issue. For some, including a few university graduates, it was a completely new way of thinking. However, almost all of the women achieved higher grades in their computing subjects than they had expected. Some described the pleasure of studying with other mature students whereas others mentioned supportive staff and female role models. Group work was considered both a positive and a negative by these ex-students. Five women said without prompting that family or personal health problems had interfered with their studies.

Almost all had gone on to suitable employment although not all were in computing jobs. Many thought that their maturity was a positive factor when applying for jobs. None reported having current gender issues in their workplaces. The youngest said that she had previously encountered a male with a "pride factor". Pleasingly, several who worked as software engineers in strongly male dominated firms, reported that these were very "family friendly" workplaces.

CURRENT AND FUTURE TRENDS

The longitudinal study described above was discontinued in 2002. No attempts have been made to statistically model success in more recent classes largely because there has been a marked change in the educational and ethnic backgrounds with about 80% of COMP102 students being international students. Prior to 2002 only about 1/3 of the students in COMP102 were of international origin. One of the variables in the longitudinal study was whether or not the language of instruction at high school was English. It was not a significant factor for predicting success in any of the models.

Since 2002 the international students have typically been younger and less qualified than before. The majority have been of Chinese nationality and only a small proportion have studied at senior high school level in English speaking countries. Many have had difficulty with the level of English comprehension required to study COMP102. This has been reflected in lower average marks for the 2002 to 2004 classes compared with earlier years. However the average marks for males and females, while lower than previously, were still not significantly different from each other. It is also apparent that older students have continued to do well in COMP102.

The 2004 class (about 100 students) was the smallest for some years and it was only 24% female. The roll for 2005 has not been finalized, but the size and proportion of women will be similar to 2004. Some of the drop in class sizes for 2004 and 2005 can be explained by a decrease in Graduate Diploma in Applied Computing enrolments. It is hoped that a targeted advertising campaign will attract more women and graduate diploma students into COMP102 in the future.

CONCLUSION

Lincoln University, New Zealand, offers qualifications in applied computing rather than computer science. Both the introductory computer programming class and higher level computing classes required for this major usually have a reasonable proportion (25-40%) of female students. Success in the introductory class is not directly related to gender. Over a number of years it has been consistently found that older students do better than younger students. Not surprisingly, highly motivated students usually do better than less motivated students. This is variously reflected in the different models as either an intention to major in computing or by an expectation of obtaining a high grade. There are inconsistent results as to whether a good general knowledge of computing, prior to studying programming, is important.

A small group of mature, female, applied computing graduates have been interviewed. They had almost all chosen to study computing for career reasons. Most had achieved higher grades than they expected although many had trouble getting started and were grateful to be studying in a supportive environment. In general these women had found suitable positions on re-entering the workforce.

REFERENCES

Agresti, A. (1990). *Categorical data analysis.* New York: Wiley.

Bernstein, D. (1994). Effects of computer ownership and gender on computing skills and comfort: Two countries' experiences. *New Zealand Journal of Computing, 5*(1), 77-84.

Brown, J., Andreae, P., Biddle, R., & Tempero, E. (1996). Women in introductory computer science: Experience at Victoria University of Wellington. *Technical Report CS-TR-96/1.* Wellington, New Zealand: Department of Computer Science, Victoria University.

Byrne, E. (1993, September 2-4). Women, science, and the snark syndrome: Myths out, policy strategies in. *Proceedings of the Women's Suffrage Centennial Science Conference* (pp. 18-28), Victoria University, Wellington, New Zealand. Wellington: The NZ Association for Women in the Sciences.

Clarke, V., & Chambers, S. (1989). Gender-based factors in computing enrolments and achievement: Evidence from a study of tertiary students. *Journal of Educational Computing Research, 5*(4), 409-429.

Clarke, V., & Teague, J. (1994). Encouraging girls to study computer science—Should we even try? *Australian Educational Computing, 9*(1), 17-22.

Cohoon, J. (2002). Recruiting and retaining women in undergraduate computing majors. *Inroads, SIGCSE Bulletin, 34*(2), 48-52.

Durndell, A. (1991). The persistence of the gender gap in computing. *Computers and Education, 16*(4), 283-287.

Fisher, A., & Margolis, J. (2002). Unlocking the clubhouse: The Carnegie Mellon experience. *Inroads, SIGCSE Bulletin, 34*(2), 79-83.

Jain K., Mao J., & Mohiuddin K. (1996). Artifical neural networks: A tutorial. *Computer, 29*(3), 31-44.

Kossuth, J., & Leger-Hornby, T. (2004). Attracting women to the technical professions. *Educause Quarterly, 27*(3), 45-49.

Li, L., Samarasinghe, S., & McLennan, T. (2002, January 30). Using neural networks to model characteristics of successful students in introductory computer programming at Lincoln University. *Science 2002@Lincoln*, Lincoln, New Zealand. Lincoln: Lincoln University.

McLennan, T., Young, J., Johnson, P., & Clemes, S. (1999). Success in an introductory programming class: Age and agenda intentions are more important than gender. *Gates, 5*(1), 20-29.

McLennan, T., Clemes, S., Young, J., & Kamikubo-Gould E. (2000, July 21-22). Age and expectations? Attributes of successful students in an introductory programming class revisited. *Living I.T., Proceedings of WIC2000, The 6th Australasian Women in Computing Workshop* (pp. 35-43), Brisbane, Australia. Brisbane: Griffiths University.

McLennan, T. (2003, February 19-20). Women with new careers: A follow-up study of mature age, recent computing graduates. *Participation, Progress and Potential, Proceedings of the 2003 Australian Women in IT Conference* (pp. 47-52), Hobart, Australia. Sandy Bay: University of Tasmania.

Ryba, K., & Selby, L. (1995). *A study of tertiary level information technology courses: How gender inclusive is the curriculum?* Wellington, New Zealand: Ministry of Education.

Sturm, D., & Moroh, M. (1994). Encouraging the enrollment and retention of women in computer science classes. *Proceedings of the 15th National Educational Computing Conference* (pp. 267-271), June 1994, Boston, USA.

Toynbee, C. (1993). On the outer: Women in computer science courses. *ACM SIGACT News, 24*(2), 18-21.

Weston, T., & Barker, L. (2004). *Institutional factors and representation of women in computer science program.* Retrieved December12, 2004, from http://www.colorado.edu/ ATLAS/evaluation/ite/arw.html.

Women and Computing. (2002). *Special Issue, Inroads, SIGCSE Bulletin, 34*(2).

KEY TERMS

Applied Computing: Study programmes in applied computing focus on evaluating and applying existing information technology techniques to solve real world problems. Subjects studied include systems analysis and design, programming, databases, hardware, operating systems, networking, end user computing and simulation as well as domain knowledge from areas of application.

Bachelor of Applied Computing: A three year, post high school, degree where students as well as completing a major in applied computing must also complete a minor in another discipline area.

Female Participation Rate: The percentage of students in a class who are female.

Female Retention Rate: The percentage of females in an introductory class who continue on to study at a higher level.

Graduate Diploma in Applied Computing: A one year course enabling graduates from other disciplines to study undergraduate subjects and complete a major in applied computing.

Mature Student: A mature student is defined as a person twenty five or older studying at tertiary level.

Sexnormal: There is a critical threshold in the proportion of female enrolments in a class or institution. Above this threshold, considered to be about 33%, female enrolments are considered to be sexnormal. Below this threshold female enrolments are considered to be untypical, abnormal or exceptional.

Mentoring Australian Girls in ICTs

Jenine Beekhuyzen
Griffith University, Australia

Kaylene Clayton
Griffith University, Australia

Liisa von Hellens
Griffith University, Australia

INTRODUCTION

In *Australia*, the participation rate for females in information and communication technology (ICT) courses in secondary, vocational. and higher *education* is significantly lower than that of males, and is decreasing (Thorp, 2004). In Queensland, Australia, only 20% (at most) of ICT *students* and employees are female, with the IT first preferences for tertiary admission down 22% for 2004 enrolments (Thorp, 2003). This downturn is in line with the trend in other Western countries and reflects the general lack of interest in ICT education amongst *adolescents*.

Recent Australian research confirms the importance of *role models* and *mentors* when adolescents are considering career options (Clayton, 2005). The importance of implementing sustainable strategies, such as *mentoring programs*, to rectify this imbalance cannot be understated. Jepson and Peri (2002) believe that mentoring programs should commence at middle and *high school*. Early mentoring programs are valuable as girls have fewer ICT role models and mentors in the classroom, industry and computer games (Carey, 2001). Mentors in these programs need to provide an accurate portrayal of the broad range of careers available in the ICT field (Klawe, 2002).

To date, a number of mentoring programs and intervention activities have been and continue to be undertaken in Queensland. This article presents three different mentoring programs the authors have been involved in and discusses the challenges involved in implementing these strategies. The first two programs discussed are for high school students and the third is for *university* students in ICT degree programs (von Hellens, Beekhuyzen, & Nielsen, 2005).

Adding to the complexity of this problem, funding to implement programs aimed at increasing female participation in ICTs may be difficult to justify due to the problems of measuring the effectiveness in achieving this goal. Australian researchers recognize this problem and are concerned about the absence of ongoing evaluation of programs to encourage girls into ICTs (Lang, 2003). While this chapter makes recommendations for implementing new strategies based on the experiences discussed, more work needs to be done on how to evaluate the efficacy of ongoing and future strategies.

BACKGROUND

Existing research on mentoring has focused mainly on mentoring programs within the workplace. According to this research mentoring schemes can help companies improve the gender balance of their staff and to develop a stable corporate culture (Limerick, Heywood, & Daws, 1994). According to mentoring expert Ann Rolfe-Flett (2002), a genuine outcome of mentoring programs in Australia is increased student retention. However, there is no clear Australian evidence connecting mentoring to improvements in recruitment of students to specific areas of study.

Mentoring programs in schools are aimed at achieving the fullest potential of each student in emotional health, academic achievement, interpersonal relationships and vocational knowledge through a positive relationship with at least one adult (MacCallum & Beltman, 1999). Although there is a diverse range of models of mentoring programs operating in Australian schools, there are no examples of students being mentored specifically on ICT career choices. The ICT mentoring programs discussed in this chapter are the first to address the lack of research in this area.

HIGH SCHOOL MENTORING

Mentoring Program 1

The first example of high school mentoring is the program for Year 11 students established in 2001; a collaboration between information and processing technology (IPT) teachers in a Brisbane high school and researchers at Griffith University (GU). The focus was specifically on female students although the program also included male students enrolled in the IPT class.

When selecting role models and mentors, care needs to be exercised in that females with a specific role in the industry are included and they can demonstrate attainable levels of achievement (Standley & Stroombergen, 2001). In keeping with this recommendation, 28 mentors were recruited, comprising ICT professionals, academics and recent graduates. Groups of 4 to 5 students were allocated a mentor who would assist the students with analysis and design of a programming assignment and provide general advice about the nature of ICT work and the skills needed to succeed as an ICT professional. Presentations of these assignments and past student work occurred during a concluding breakfast function. A research officer acted as a liaison between mentors and students and maintained contact with the schoolteachers.

Focus group interviews were conducted with the participating students to get in-depth comments about their mentoring experiences and changes in perceptions about ICT education and work. Feedback was also sought from teachers and mentors. At the end of the mentoring program, an online survey was used to collect demographic data and perceptions of ICT education and work from the Year 11 and 12 students. This data provided information on gender balance in the ICT course and the students' perceptions of ICT education.

Most mentors believed it was a worthwhile initiative and would be willing to support the program in the future. Students were eager to meet their mentors face-to-face and used the time to find out more about ICT study and where it can lead. This helped fit the IPT subject into the bigger picture and give it a real-life context. The mentors helped students with problem solving, teamwork and with their class assignment by sharing their expert knowledge on ICT project principles, risk management, and business systems.

The program provided more accessible role models for female students and provided a strong positive image to female students and corrected the widely held view that the ICT industry is intrinsically a male domain. Mentors became role models for female students, offering general information and advice about the industry and providing examples of the diversity of career paths within the industry and the variety of training options and skills that are considered valuable, all of which will positively promote careers in the ICT industry.

The main problem of the program was the lack of opportunity for students and mentors to meet before the program started. This would have provided a means of breaking the ice and perhaps facilitated communication and interaction during the program. Some groups made little or no effort to contact their mentors, while on the other hand, some mentors did not respond to student emails. These incidents indicated somewhat unexpected project management problems that could be corrected in future mentoring programs by more careful monitoring of communication between students and their mentors.

This program demonstrated that such a program could be a viable way to challenge female students' perceptions of ICT education, and to make ICT a more attractive career option. However, to confirm this finding, an IT mentoring program should also involve longitudinal studies on female students' perceptions of IT work and education. Such a study would help gain a better understanding of the way women help configure the institutional realm of ICT work and how this understanding can be passed on to school students.

The positive experience from this program helped Queensland Government's Office for Women and GU establish an ICT mentoring program for several high schools in 2004; the Get SET (Science Engineering and Technology) project discussed next in this article. Our experience suggests that mentoring, interactions with professional ICT organizations, and professional ICT women talking to females during their ICT education can improve the perceptions of ICT.

Mentoring Program 2

In 2002, funding was sought and obtained from the Office for Women and Education Queensland to provide a schools-based mentoring program to encourage female students to consider Science and ICT as career options. Fourteen mentors were recruited with Science, IT or Science and ICT degree experience.

Two very different schools were chosen to participate in this project. Students from one school (S1) come from a wide range of socio-economic, cultural, and ethnic backgrounds. S1 is a large school with a reputation for academic, performing arts and sporting success. One Year, a year 10 Science class of 26 students (23 female) was chosen to participate. Approximately 25 female year 10 ICT students were to participate, but logistical difficulties arose with this arrangement and year 11 and 12 students were substituted.

The second, smaller school (S2) is based in a low socio-economic area and is working hard initiating successful programs to assist with various social issues. Eleven high achieving female students were chosen to participate during interschool sport time.

Throughout the year, students from both schools were invited to a number of GU campuses to experience the university environment and facilities. Academics gave presentations about their research areas and the mentors demonstrated projects that they were working on. Industry visits for the students were organized with local companies. Teachers at S1 refused the industry visits citing that it was too much work and too difficult to release the students from normal classroom activities or that there was a lack of interest from the students. The teacher, students, and their mentor at S2 attended industry visits during school hours where they were able to talk with scientists and ICT professionals at work.

Students at both schools were involved in projects with the assistance of the mentors. The S1 Science class was divided into teams of four and worked on various research projects, including the creation and experimentation with lasers made using electronic kits. The students at S2 investigated the water quality of the Logan River, which abuts the school. The students conducted field experiments as well as more sophisticated laboratory experiments under the supervision of the laboratory staff, mentors and teacher. A Web page was also created to describe the project, the field trips and the experiment results. Mentors supported the students throughout the design and implementation of the Web site including interface design techniques and programming. In conjunction with classroom contact time, mentors and students at both schools were encouraged to communicate with email and instant messaging (IM) technologies. At the presentation night, students from both schools displayed their work to mentors, peers, parents, and other guests.

Throughout the year, all students working on the practical projects with the mentors responded very positively about Science and ICT and these successes highlight the need to integrate mentoring within the curriculum. It was also pleasing that many of the students who participated at S2 were eager to mentor other students in future years. This was an significant outcome as Margolis and Fisher (2002, p. 115) regonize that "… some of the best recruiters of girls are other girls."

A final critical issue is that even though the structure of the program may be correct, red tape may impede the project. To overcome this, there must be a project champion with status and power at the school who is willing to help surmount these difficulties.

UNIVERSITY MENTORING

The ICT chapter of the GU Alumni Association is now in its fourth year of running a successful mentoring program intended to guide and inform first year ICT students in their studies and their future careers. The mentoring program calls upon

M

Alumni members (ICT graduates) working in the ICT industry who are matched with first year ICT students. It encourages an environment for mentors to share their knowledge and experience in the industry by helping students with difficulties they experience with transition to tertiary education, planning their degree program, assessing career options, understanding the ICT industry, making contacts, and entering the workforce.

The pilot program was launched in 2002 with a $12,000 grant from the Queensland Government's Information Industries Bureau (IIB) and is supported by the Australian Computer Society. Feedback is used to continually improve the program which continued successfully in 2003, 2004, and 2005. Due to the expensive nature of running such programs, this program has been limited to 15 participants each year. Since 2003, the program has run with minimum monetary input and mostly on the goodwill of its participants and volunteer committee members. Each year a mentoring coordinator is appointed by the committee which has proven to be an integral role for the success of the program.

In addition to one-on-one mentoring, the ICT program involved mentor training, networking workshops, and social functions for the mentors and students, and was free of charge to all participants. The events held as part of the program do contribute somewhat to ongoing funding for the program. Events held during the program allowed students to come together early throughout the semester in an environment that encouraged academic and career achievement and allowed students to meet like-minded students as well as benefiting from interaction with the mentors.

The Opening Event for each year is a combination of a short formal session for training and briefing, and a relaxed session where mentees and mentors meet over food and drinks, enjoy the challenge of a trivia quiz, and spend time getting to know each other. A credible ICT industry professional presents a practical guide for people to become better at career networking in the aptly named "Art of Networking" mid-year event. It is believed that approximately 80% of ICT jobs in Australia are gained through word of mouth, thus strengthening the need for these essential skills. The Closing Events provide a final opportunity for participants to meet, reflect on the program, complete their evaluations and provide feedback.

Exploring Mobile Communication Technology in Mentoring

The continuing Alumni program offers participants the chance to integrate mobile communication technology into their mentoring. According to Gürer and Camp (2002), telementoring, can be used to enhance the mentoring process by supporting students while breaking down the barriers of distance and time. Since 2004, students have been part of telementoring experiments using mobile phones with Multimedia Messaging (MMS) features. Phones are given to mentor and mentee for a 10-day period. The exchange of MMS messages is emphasized, as visual information sharing enhances the ability to report on the mentor's and mentee's everyday working environment.

The results of preliminary studies in 2003/2004 are encouraging and suggest that mobile communication technology has possible value for mentoring programs. In increasing the frequency and density of communication between the participants, communication between mentor and mentee included direct information of work issues, but also descriptions of atmosphere and opinions related to mentors' duties. Both mentors and students reported increased frequency and flexibility in communication. This was highly valued because of the time constrains of both parties. Mentors also emphasized the easy access and speed of use of the technology, as sending a message with a mobile phone was regarded as more easy and flexible than e-mail, which took more time and was limited to the work situations with a computer. Also, participants felt that more frequent and often more informal interactions resulted in a close and relaxed relationship (Hakkila & Beekhuyzen, 2005).

FUTURE TRENDS

From the experience that the authors have gained in running school and university based mentoring programs, a number of recommendations can be made to assist new entrants establish successful mentoring programs for girls and women in ICT. Firstly, funding is an important issue. Mentoring programs can be expensive to set up and run, often with no economies of scale possible, as most programs require tailoring to suit individual need. Furthermore, the number of females in ICT careers and study continues to decline and pressure is often put on existing ICT students and professionals to participate in these programs. Adequate remuneration is also required for these often high-achieving mentors who may need to attend during working hours or take time from their busy study schedule. Secondly, it is also critical that the mentoring coordinator is successful in building and maintaining of relationships and networks with and between stakeholders to ensure continuing success. Moreover, evolving communication technologies such as email mobile phones and IM should be fully exploited in encouraging communication between students and mentors. Thirdly, school based mentoring programs also need to be integrated into the curriculum to ensure maximum reach and that programs have more impact and acceptance from the general school community. Finally, even though the structure of the program may be correct, red tape may impede the project. To overcome this, there must be an internal project champion with status and power who is willing to help surmount these difficulties.

CONCLUSION

This article discussed three Australian ICT mentoring projects and offered recommendations for future mentoring projects based on the authors' experiences. Recommendations include the need for: adequate and realistic funding to run these projects; a coordinator to build and maintain good relationships and networks; increasing use of evolving technologies for mentor/student communication; integration of mentoring projects into the school curriculum; and a project champion with power and status to overcome bureaucratic difficulties.

The programs outlined in this paper generally provided positive results in encouraging females to excel and be leaders in their ICT careers. However, long-term evaluation of all ICT mentoring projects is a neglected area of research. Considering the cost

of these programs, it is essential to prove the benefit of these programs to justify and support the need for continuing funding of programs.

REFERENCES

Carey, P. (2001). *Girls and technology in the secondary school.* Perth: Catholic Education Office of Western Australia.

Clayton, K. (2005, June 23-24). *Engaging our future ICT professionals: What is the missing piece of the puzzle?* Paper presented at the Women, Work & IT Forum, Brisbane.

Gürer, D., & Camp, T. (2002). *Investigating the incredible shrinking pipeline for women in computer science—Final Report (version 4).* Retrieved February 16, 2003, from www.acm-w.org/documents/finalreport.pdf

Hakkila, J., & Beekhuyzen, J. (2005). Using mobile communication technologies in student mentoring: A case study. In C. Ghaoui (Ed.), *Encyclopedia of human and computer interation.* Hershey, PA: Idea Group Reference.

Jepson, A., & Peri, T. (2002). Priming the pipeline. *ACM SIGCSE Bulletin, 34*(2), 36-39.

Klawe, M. (2002). Girls, boys, and computers. *ACM SIGCSE Bulletin, 34*(2), 16-17.

Lang, C. (2003, February 19-20). *Flaws and gaps in the women in computing literature.* Paper presented at the 2003 Australian women in IT conference, Hobart, Tasmania.

Limerick, B., Heywood, E., & Daws, L. (1994). *Mentoring: Beyond the status quo? Mentoring, networking and women in management in Queensland. A women in business and industry project*: Brisbane, Queensland, Australia: Queensland University of Technology.

MacCallum, J., & Beltman, S. (1999). *International Year of the Older Persons Mentoring Research Project Report.* Perth: Murdoch University.

Margolis, J., & Fisher, A. (2002). *Unlocking the clubhouse: Women in computing.* Cambridge, MA: The MIT Press.

Rolfe-Flett, A. (2002). *Mentoring in Australia: A practical guide*. Australia: Pearson Education.

Standley, J., & Stroombergen, L. (2001). *21st Century Women and Understanding the Future Research Report*. University of Nottingham Research Report.

Thorp, D. (2003, November 17). IT still failing to attract students. *The Australian*, 135.

Thorp, D. (2004, March 23). Geeks need a makeover to attract girls. *The Australian*, 35.

von Hellens, L., Beekhuyzen, J., & Nielsen, S. (2005, June 23-24). *Thought and action: The WinIT perspective—Strategies for increasing female participation in IT*. Paper presented at the Women, Work & IT Forum, Brisbane.

KEY TERMS

Information Processing and Technology (IPT): A Queensland Studies Authority elective subject offered over four semesters as part senior secondary schooling. It aims to develop: student awareness and knowledge of IT; problem solving skills; communication skills, and critical thinking and analysis skills. It incorporates many areas including: social and ethical issues in IT; human-computer interaction; information systems; intelligent systems; and software and systems engineering.

Information Technology Field: The technical and people oriented area of information technology utilization in organizations and in society. Information technology education provides core technical skills for the analysis, design, and development of computer-based information and their subsequent implementation in organizations and the evaluation of such systems.

Instant Messaging (IM): A synchronous real-time method of communication over the Internet between two or more people. It is similar to a phone or conference call, except messages are text based. Contact details are swapped and added to individual contact lists.

Mentee: A person studying in high school or First-year University and interested in the area of IT. The mentee is involved in a mentoring program in which their role is to interact (with a view to getting help from) their mentor.

Mentor: A person working in IT and knows about the IT industry who is able to advise younger people about study options work opportunities in the IT industry. The mentor is involved in a mentoring program in which their role is to interact (with a view to helping) their mentee.

Mentoring: A process that allows a relationship to be created between a mentee and a mentor with a view to allowing the free flow of information leading to guidance, advice and encouragement.

Mentoring Program: A formalised and focused effort to bring mentors and mentees together in an encouraging, learning environment.

Multimedia Messaging Service: A mobile phone technology that allows people to communicate with any combination of images, videoclips, text, and audio through their mobile phone.

Secondary School: A stage of a youth's general schooling including Years 8 through 12 where students move between classrooms and are taught by a different teachers in various key learning areas. In junior secondary school, Years 8 to 10, most students follow similar syllabus although some choice is available in Year 9 and 10. Students may choose to continue into senior secondary school, Years 11 and 12, and choose six subjects offered by their school. Some secondary schools also now offer a range of accredited vocational subjects, traineeships, and apprenticeships.

Migration of IT Specialists and Gender

Elena Gapova
European Humanities University, Belarus

INTRODUCTION

The purpose of this article is to analyze "after the shift," which occurred in the second half of the 20th century, from a goods-producing society to an information or knowledge society, as information technology (IT) began to be seen as a most important asset of contemporary nations. Bell argued in 1973 that in the new social order, knowledge and information would replace industrial production, and would become the "axial principle" of social organization (Bell, 1973). By the end of the 20th century, IT has also become a truly global phenomenon, involved with the reconfiguration of the labor market and human and material resources from all over the world. Gary Becker, the 1992 Nobel laureate in economics, pointed out that the United States' (U.S.) Silicon Valley currently employs 1 million people, of whom 40% have at least a bachelor's degree and more than one-third are foreign-born.

In the new information economy, special importance is assigned to IT researchers and developers, who belong to the global group of "knowledge workers." In the post-industrial era, IT workers have skills that allow them to compete in the global labor market, as IT jobs, by their very nature, are not tied to any particular culture and "can work" anywhere. At the same time, IT production is labor-intensive, and many first-world nations (Britain, Germany, France, Ireland, the U.S.), which have undergone a reduction in birthrates, feel that their own human resources are not sufficient for its development. In 2000, the American Institute for Electric and Electronic Engineers (IEEE) recognized that "With declining numbers from national engineering graduate programs, the U.S. has no option but to satisfy the growing need for the engineering professionals from abroad" (Institute, 1999). To bring professionals into the country, the U.S., the biggest IT developer, introduced an employer-based H1-B visa program for specialty occupations (e.g., computer professionals, programmers or engineers).

BACKGROUND

In the U.S., visa petitions by IT specialists are approved for up to 3 years and may be extended to 6 years. During this period, the employee cannot change the employer, but (potentially) may get a permanent residence permit (i.e., a Green Card). The 1990 ceiling for admissions was set at 65,000 a year, and in 1997, "for the first time, the maximum limit was reached by the end of the year; in 1998 the ceiling was reached in May" (Immigration, 2003, p. 45), and employers complained of shortages. The 1999 limit of 115,000 was exceeded by 20,000, and in October 2000, the U.S. Congress passed the American Competitiveness in the 21 Century Act, increasing the annual limit to 195,000 for 2001, 2002 and 2003 (Immigration, 2003). Following that year, H1-B "cap" was set to return to 65,000 in fiscal year 2004, and U.S. Citizenship and Immigration Services received enough H1-B petitions, issued by U.S. employers, to meet the congressionally mandated number on February 17, 2004 (USCIS, 2004).

According to various sources, India provides 33% to 47% of U.S. high-tech employees with H1-B visas. The next-biggest supplier of IT developers is China, with about 9%, with Japan, Taiwan, Great Britain, Canada and South Korea providing 2% to 3% each. In recent years, specialists from Eastern Europe, mainly from Belarus, Russia and Ukraine, have also become a visible group. These nations are now becoming aware of the "brain drain" to the West (International, 2002; Ferro, 2004).

American society is experiencing profound effects from and is concerned with this type of migration. There is controversy over whether the system brings more benefits than losses (Saxenian, 2002) and how it may affect the most vulnerable, mainly older, U.S. IT workers, who may not be retrained but "substituted" by younger, educated foreign nationals (O'Lawrence, 2001). Responses to Senator Phil Gramm's introduction of a bill to raise the number of temporary high-tech guest workers were published in the IEEE newsletter, *The Institute,* in 1999 under the headline "Stop the Insanity of H-1B!" (Institute, 1999). The conflict in how to view the H1-B program is part of a much larger issue; in the era of mobile labor force, individual states stopped being basic units of capitalism, while the government can only protect their workers within the frameworks of national systems of social justice (Rorty, 1998).

CURRENT TRENDS

The employment-based relocation of IT specialists to the U.S. is a highly gendered phenomenon. Spouses (and children) are only allowed to follow relocating programmers as "dependents" on H4 visas, which do not include the right to work. Overtly gender-neutral, the system is based on the assumption that programmers are male, for their professional spatial mobility is more socially acceptable than women's: Men are not supposed to follow women as nonworking "dependents," and such cases are rare (Gapova, 2004). Thus, the H1-B system derives from the idea of a certain family pattern, reflecting and strengthening an underlying gendered division of labor. While IT workers (i.e., men) relocate as professionals, spouses (i.e., women) follow them as caretakers and providers of intimacy.

In the globalized world, the value of human intimacy and chains of care is high (Rotkirch, 2000; Parrenas, 2001). Sometimes the relocation prospect serves as a "catalyst" to move from partnership to legal marriage, which otherwise might not have taken place. Men, unhappy about being on their own in a strange country, are often doubtful about their value in the U.S. marriage market and how to find new partners there. When interviewed, most post-Soviet H1-B visa holders emphasize the value, in the foreign lands, of the intimacy and human bondage that women provide, and many stress the need of a loyal partner as an important precondition for their very successful professional functioning (Gapova, 2004).

Women's consent to follow as "dependents" may be conditioned by several considerations, the following two being most important: (1) their own professional status and career opportunities at home; and (2) the age of children, of whom they take care

more than men do. Wives with a (professional or advanced) degree and realistic career options view relocation as not bringing them personal professional gains, and such couples tend to reject the idea. Most women, though, being in their late 20s or early 30s, are too young to have developed a real career, so it looks like "there's nothing to sacrifice." Also, the money that the family can make under the new arrangement is a factor. As IT jobs are better paid than those done by women (whose occupation tend to be more bound to teaching, culture, healthcare, etc.) back home, it is women's jobs that are normally sacrificed "for family's sake." The leap form a dual career to a single earner family, conditioned by the H1-B system, is justified by a much bigger male wage (Gapova, 2004).

The individual social mobilities in such couples are "opposed" to each other. The man's social wealth derived from his work status is rather high and his class mobility tends to be upward: He is a professional in a prestigious field and the breadwinner. The woman's social mobility is contradictory, simultaneously being upward and downward. While the family's general financial situation improves, women on H4 visas depend on the male wage and have certain financial stability only as family members. Their occupational difference is converted into status inequality. Gapova (2000) writes about the vulnerability of post-Soviet H4 visa holders. Assisi (2004) claims human rights violations among H1-B visa holders' spouses, and Raj (2003, 2004) states that partners of South Asian women on dependent spousal visas may use immigration laws prohibiting them from working to limit their autonomy, or even resort to violence.

CONCLUSION

Global production of IT is involved with movement of skilled labor across space; namely, the physical migration of (mostly male) high-tech professionals to North America and Western Europe from post-socialist countries, India and China on specialty professional visas. A certain concept of gender roles underlies the seemingly neutral migration arrangement. The system is constructed to strengthen a certain family form, globally producing men as professionals and women as biological reproducers and primary caregivers, thus ensuring the (re)production of IT labor force and a certain gender hierarchy, which may be partially dealt with through policy solutions.

REFERENCES

American Institute of Electrical and Electronic Engineers. (1999, November). *The Institute.*

Assisi, F.C. (2004). *Human rights violations among H1-B visa holders' spouses.* Retrieved from http://www.indolink.com/displayArticleS.php?id=102104095811

Bell, D. (1973). *The coming of post-industrial society: A venture in social forecasting.* New York: Basic Books.

Ferro, A. (2004). Romanians email from abroad: A picture of the highly skilled labour migrations from Romania. *UNESCO-CEPES Quarterly, 23*(3).

Gapova, E. (2000, Fall). A glimpse into the lives of foreign high-tech wives in the United States. *The Journal of the International Institute, 8*(1), 13.

Gapova, E. (2004). Zheny russkih programmistov, ili zhenshchiny, kotorye edut vsled za muzhchinami. In S. Oushakin (Ed.), *Semeinye uzy: Modeli dlya sborki* (pp. 409-431). Moscow: NLO.

International Mobility of the Highly Skilled. (2002). *OECD.* Retrieved from http://www.oecd.org/migration

O'Lawrence, H., & Hickey, W. (2001). *An evaluation of H1-b tech visa's and their effect on the US workforce.* IVETA Annual Conference 2001: Improving VET Systems. Retrieved from www.iveta.itweb.org/Papers/OLawrence%20H1B%20NOW.pdf

Parrenas, R. S. (2001). *Servants of globalization: Women, migration and domestic work.* Stanford: Stanford University Press.

Raj, A., & Silverman, J. (2003). Immigrant South Asian women at greater risk for injury from intimate partner violence. *AJPH, 93* (3), 435-7.

Raj, A., Silverman, J., McLeary-Sills, J., & Liu, R. (2004). Immigration policies increase South Asian immigrant women's vulnerability to intimate partner violence. *Journal of the American Medical Women's Association, 60,* 26-32.

Rorty, R. (1998). *Achieving our country. Leftist thought in twentieth-century America.* Cambridge: Harvard University Press.

Rotkirch, A. (2001). The internationalization of intimacy: A study of the chains of care. In *Proceedings of the 5th Conference of the European Sociological Association Visions & Divisions.* Retrieved from http://www.valt.helsinki.fi/staff/rotkirch/ESA%20paper.htm

Rein, M. L. (2003). *Immigration and illegal aliens. Blessing or burden?* (Information Plus Reference Series). Information Plus.

Saxenian, A. (2002, Winter). Brain circulation. *Brookings Review, 20*(1), 28-31.

U.S. Citizenship and Immigration Services. (2004, February 17). Press release (revised February 19, 2004). Retrieved from http://uscis.gov/graphics/publicaffairs/newsrels/h1bcap_NRrev.pdf

KEY TERMS

Brain Drain: The loss of skilled intellectual and technical professionals through the movement to more favorable geographic, economic or professional environments. The term originated about 1960, when many British scientists emigrated to the U.S. for a better working climate. Also termed human capital flight.

Green Card: Permanent residence permit. Issued to different categories of people, including those coming to the U.S. by marriage or as family members, priority workers, skilled workers, members of the professions holding advanced degrees or persons of exceptional ability, refugees and others.

H1-B Visa Program: Intended to bring into the country workers in specialty occupations; for example, computer professionals, programmers or engineers. H1-B is a temporary work visa for skilled professionals.

H-4 Visa: May be granted to a spouse and any unmarried children younger than 21 years of age of an H1-B holder. A person with an H-4 visa may not work unless the person qualifies for an H1-B visa and is approved for a "Change of Status" by the "Immigration and Naturalization Service. H-4 holders are able to acquire a driver's license, open a bank account and go to college.

Knowledge Worker: Anyone who works for a living at the tasks of developing or using knowledge. A term first used by Peter Drucker in 1959, the knowledge worker includes those in the IT fields, such as programmers, systems analysts, technical writers, academic professionals, researchers and so forth. The term is also frequently used to include people outside of IT.

M

Motivating Women to Computer Science Education

Roli Varma
University of New Mexico, USA

Marcella LaFever
University of New Mexico, USA

INTRODUCTION

The problem of disproportional representation of women in the computer science (CS) field in post-secondary education has become a major concern (AAUW, 2000; Camp, 2002; Carver, 2000; Varma, 2003). Currently, universities are increasing their focus on retaining women into CS programs. However, the number of women in that field remains low in proportion to males, and many women who are recruited often drop out or switch majors before completing their degree in CS (National Science Board, 2004, pp. 2-6, 3-17). In order to promote retention, it is important to compare possible differences in learning motivation between males and females in CS, examine changes in motivations across the span of CS study, and assess whether recruitment messages and program structures are matched (or mismatched) to the motivations of females. This article investigates the motivations for women to enter into, remain in, and continue the study of CS at the post-secondary level.

BACKGROUND

In recent years, a number of researchers (Chory-Assad, 2002; Kerssen-Griep, Hess, & Trees, 2003; Noels, Clement, & Pelletier, 1999; Postlewaite & Haggerty, 2002; Volet, 2001) have specifically concentrated on motivations for learning in the classroom and the factors that match teaching techniques with student success and satisfaction. Motivation, in the context of learning, refers to stimulation that drives students to derive academic benefits from classroom activities. In a learning setting, motivation can also be described as either trait motivation, a general level of desire to learn across all learning situations, or state motivation, a general level of desire to learn in a particular class, task, or content area (Anderson & Martin, 2002). The present study probes state motivations rather than trait motivations because of the focus on motivations that are particular to choosing and continuing study in the CS field.

Several scholars have posited a variety of theoretical constructs centred on state motivation. One such construct is the achievement goal theory (Dweck & Leggett, 1988), which reasons that goals are either ego oriented, wanting to gain favourable judgments of competence through social comparison, or task oriented, wanting to be competent and master a skill through effort based on internalized standards. In this construct, the general attitude towards reaching the goal is important. Another construct is self-determination theory (Deci & Ryan, 1985), which includes categories of intrinsic and extrinsic motivations. An educationally based construct is that of Pintrich, Smith, Garcia, and McKeachie (1991) who developed the Motivated Strategies for Learning Questionnaire. The bases for these scales are internal and external goal orientations. This instrument is currently the measurement standard for motivation in education.

Demonstrating how the interaction between internal and external attitude orientations and rewards might create a broader range of motivational categories requires a more complete explication. Vallerand and Bissonette (1992) posit a matrix which puts forward three types of extrinsic motivation: (1) external regulation (influences from means outside of the individual such as reward and punishment), (2) introjected regulation (results from outside pressure that the individual has internalized such as guilt or desire to impress others), and (3) identified regulation (whereby the individual feels that something is personally worthwhile and relates to their value system). This matrix has been related to second language learning, a learning situation similar to CS because it involves a very specific content area where motivational factors may be highly determinate in the success or failure of learning.

Volet (2001) modified Pintrichs et al.'s (1991) "Self-Efficacy and Expectancy of Success" as a measure of motivation. Self-efficacy describes a student making a judgment about his or her own ability to be successful in a learning task. Self-efficacy is posited to be an important motivation for both entering into and continuing in a particular learning context. The Williams and Ivey (2001) case study of motivations in math education also concentrated on an internal motivation orientation that includes self-efficacy as a factor. They highlighted an internal perception of usefulness as an essential part of the motivational matrix. A perception of usefulness is whether the student perceives that the particular skill to be learned will have a current or future utility for them. As with math, a perception of usefulness may also be an important motivational factor in continuing in the study of CS.

Margolis and Fisher (2002) posit that males and females have different motivations for entering the study of CS. They developed a set of seven motivational factors for the study of CS: enjoyment, versatility, math/science related, employment, encouragement by others, exciting field, and the quality of CS department. While both males and females list enjoyment as their top motivation, the most important difference is that females list the versatility (utility and purpose) of computing as their secondary reason, while male's rate this motivation as sixth. In programming, males and females named self-efficacy as a motivation, but males cited this to a lesser degree.

Yet scholars have not investigated the role of motivation in both the recruitment and retention of women in CS program. As a synthesis of the various literatures on motivation, and with the specific motivation for CS, this study offers a motivation matrix that can be utilized to measure motivations across time because it encompasses a broad range of state motivational behaviors within a restricted number of concepts. This matrix includes three intrinsic and three extrinsic motivations, listed as intrinsic-self, intrinsic-social, intrinsic-economic, extrinsic-self, extrinsic-social, and extrinsic-economic.

To analyze motivation in both the recruitment and retention of women in CS, it is essential to investigate possible changes in motivations over time, given the interaction of other factors such as success in the classroom or desires for challenge and fun. Time parameters in the present study are before enrolment and during CS coursework.

METHOD

The present study hypothesizes the following relationships:

- **H1:** Females and males will differ significantly on intrinsic-self motivation in CS study.
- **H2:** Females and males will differ significantly on intrinsic-social motivation in CS study.
- **H3:** Females and males will differ significantly on intrinsic-economic motivation in CS study.
- **H4:** Females and males will differ significantly on extrinsic-self motivation in CS study.
- **H5:** Females and males will differ significantly on extrinsic-social motivation in CS study.
- **H6:** Females and males will differ significantly on extrinsic-economic motivation in CS study.
- **H7:** Motivations to study CS will differ across time based on gender.

The participants in the present study were students in CS at four institutions of higher education designated as minority-serving institutions because existing studies have focused mostly on non-minority institutions. The total sample size was 66, which included 35 female and 31 male participants. The sample was ethnically diverse with 22 White (11 female, 11 male), 15 African American (seven female, eight male), 10 Hispanic (five female, five male), 10 Native American (eight female, two male), and nine Asian American (four female, five male) participants.

The data for this study was gathered in 2002-2003 through in-depth interviews, as part of a larger project on women in information technology. Each student was asked the same 61 questions and 15 of those questions provided the specific data about motivations to study CS. Each interview was audio taped and transcribed verbatim. Random sampling was used to select subjects representing sufficient numbers of women and men. However, purposive sampling was used when the numbers of students majoring in CS was small (e.g., Native Americans).

A content analysis coding scheme was developed based on six motivation variables: (1) *Intrinsic-self*—"I love the challenge," "Computers are interesting;" (2) *Intrinsic-social*—"I want to be able to use it to help my community," "I'll do programming if it relates to human rights;" (3) *Intrinsic-economic*—"I've always been good at," "I made it work," "I played with it until I figured it out;" (4) *Extrinsic-self*—"I can use it no matter what work I do after this," "They teach you how to think so you can apply it to any situation," "It's something practical;" (5) *Extrinsic-social*—"I want to show that I am just as good as the guys," "I do it because I have to," "My dad really encouraged me;" and (6) *Extrinsic-economic*—"I can make a lot of money," "It will be easy to get a job," "I need a good grade."

One category was designated for each type of motivation. This created six categories. Any statements that could be coded in any of the six categories were coded only once in a single category, creating an exclusive coding system. Each respondent was designated with a numeric label (1-66) and each interview question was given an alphabet designation. Therefore, each coded statement was given an alphanumeric label. Designation of the two phases of study was accomplished by separating the types of interview questions into two categories—motivations related to the pre-study stage and motivations during CS study. Two trained coders coded the interviews to ensure coded data are consistent with each other. Intercoder reliability (Lombard, Snyder-Duch, & Bracken, 2002) for each category was assessed using Scott's P, and reliability was estab-

lished between coder one and coder two. Reliability for intrinsic-self was 0.94; for intrinsic-social was 0.87; for intrinsic-economic was 0.755; for extrinsic-self was 0.925; for extrinsic-social was 0.97; and for extrinsic-economic was 0.80. Overall, reliability was 0.88. All of these values are within the acceptable range for reliability. A total of 495 items were coded.

FINDINGS AND DISCUSSION

Demographic information was gathered in order to self-report socio-economic background of parents, age range for traditional or non-traditional student, prior exposure to computers, marital and family status, occupations of parents, year in school, educational major, student status (full or part time), and employment. A cross tab calculation was performed on all of the demographic variables in relation to gender to check for distribution across the sample. No significant relationships were found, eliminating these for consideration as intervening variables.

Hypotheses two, six, and seven were supported; hypothesis one was not supported but a near-significant difference was noted; hypotheses three, four, and five were not supported (Table 1). The second hypothesis predicted that females and males would differ significantly on the measure of intrinsic-social motivation. There was a significant difference between females and males in intrinsic-social motivation in the enrolment phase of CS (X^2=5.128, p<.05). Males were more likely to cite motivations for enrolment that indicated the importance of CS as personally worthwhile and relating to their own value system. There was no significant difference in this measurement during CS study

Table 1. Bivariate relationship for gender and motivation for pre-enrolment and during study in computer science

	Intrinsic-Self χ^2	Intrinsic-Social χ^2	Intrinsic-Economic χ^2	Extrinsic-Self χ^2	Extrinsic-Social χ^2	Extrinsic-Economic χ^2
Gender- Pre-Enrolment	0.020	5.128	0.002	0.004	0.712	0.088
Gender- During Study	3.41	0.374	0.649	0.122	0.560	4.71

Note: Significant relationships (p<.05) are shaded.

phase. Hypothesis six predicted that females and males would differ significantly on the measure of extrinsic-economic motivation. There was a significant difference between females and males in extrinsic-economic motivation during the CS study phase (X^2=4.71, p<.05). Males were more likely than females to cite the anticipation of a tangible positive result (a job) as a motivation for continuing the study of CS. The first hypothesis that predicted females and males would differ significantly on the measure of intrinsic-self was not supported, but showed a near-significant difference (X^2=3.411, p<.05).

Hypothesis seven posits that motivations differ depending on whether they are measured when contemplating enrolment in CS or during actual engagement in CS study and it was supported (Table 2). For females, there was a drop in their motivation based on intrinsic reward (In-Self) from 63% pre-enrolment to 48% during study. Statements regarding loving the challenge or thinking that computers were interesting dropped by 15%, while for males these statements increased by 3%. Motivation statements also decreased for females but went up for males judging tangible material rewards (extrinsic-economic). Females' statements dropped by 3%, while males' statements increased by 25%. This difference is consistent with the difference in the

respondents reporting work in the field of computers during study. Only 57% of females had related jobs while 87% of males did. Motivation based on a match between personal values and the study of CS (Intrinsic-social) increased 29% for females, while there was a slight drop (3%) for males. Motivation based on females' perceived fit between their own values and what CS could do for them increased dramatically once they were in the field of study. There was an increase for both males and females from pre-enrolment to course study in their personal judgments of their own ability (intrinsic-economic) in CS. Although both females and males were at about the same level prior to enrolment, male judgments of success increased by 22% while female judgments of success increased by only 15%. The perceived utility of CS skills (extrinsic-self) also increased for both males and females, but less so for females (5%). Male motivation statements regarding perceived utility increased by 22%. Extrinsic social influences as a motivation (i.e., impressing friends and family), went down for both males and females once they started their studies. However, it fell more for males (39%) than for females (20%). Outside social influences remained a higher motivation for females than for males.

Table 2. Changes in motivation to study CS across time by gender

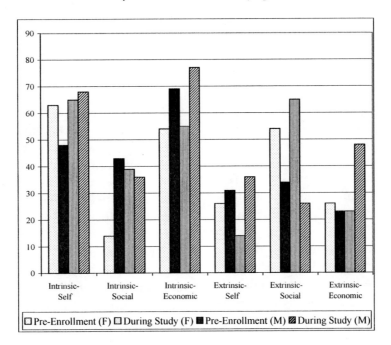

The purpose of the study was to investigate motivation factors among genders in the study of CS at two stages of post-secondary education and to investigate how motivations vary. First, this study proposed a new matrix of state motivations broad enough to measure consistently across time periods, yet confined to six categories of motivation. These six motivational constructs examined both intrinsic and extrinsic motivations and were measured across time. The results indicate that motivations change between contemplation of studying CS and the actual engagement in study. This suggests that varying strategies, aimed at different motivations, can be utilized to recruit and then to retain women in CS.

The findings in the current study correspond with the findings of Margolis and Fisher (2002) in which both males and females cited intrinsic-self (e.g., enjoyment) as their primary motive for enrolment. However, this did not hold true for motivation during study when ability (intrinsic-economic) became the top motivation for both. Additionally, Margolis and Fisher (2002) found that extrinsic-self (e.g., versatility) was the secondary motivation for females, but the sixth motivation for males. In contrast, the present study indicated no significant difference between males and females on this measure during enrolment or study. However, there was a significant difference between males and females in the enrolment motivation that Margolis and Fisher (2002) attribute to respondents wanting to enter a field that matches their values. Males were much higher on this measure in the enrolment phase and then levelled out with females during the study phase of CS. Females did not cite this as an initial motivation, yet this area had the highest increase for females once they began study in the field. Therefore, it may be that CS programs need to incorporate connections between CS skills and female social values, or demonstrate in practical terms how social values can be met through the field of CS.

Second, while there was no significant difference between males and females during enrolment based on judgments of intrinsic interest in computers and CS, there was near-significant difference once they began CS study. Female estimations of loving computers and finding the challenge rewarding dropped a great deal more than it did for males. To retain women in CS, this factor needs further assessment.

Finally, the appraisal of tangible material rewards during study showed a significant difference between males and females. Female judgments of tangible reward remained fairly stable while male estimations jumped upwards. As noted earlier, this coincides with information about males having more employment and internships in computer related fields than females while they are studying. There is some indication that work opportunities do not fit the lives and schedules of female students. Childcare and family responsibilities may be a factor in whether or not women get internships or available jobs. This could subsequently affect their motivation to remain in the CS field.

FUTURE TRENDS

In the past, efforts at recruiting and retaining women in CS have concentrated on providing early hands-on computer experiences and recruitment into programs. This approach supposes that a critical mass of women will provide a community of scholars that will support each other. Some of these efforts have been successful, but do not consider other possible factors, including the motivations that students have for studying CS and how these motivations are, or are not, matched to recruitment and retention strategies. Faculty, advisors, and administrators need to take a careful look at these factors when modifying programs that are not acting to retain women in CS field.

CONCLUSION

The study shows that it is not enough to look at a single time construct of motivation. As experiences and contexts change students modify their own estimations of the motivations that drive them. Recruitment techniques that concentrate on appealing to women's needs to enjoy and find a challenge in computers does not work to retain them once the reality of spending hours in front of a computer sets in. At that point, an increase in the activities that connect computing to both real world problems and real world employment need to be the focus of retention efforts.

ACKNOWLEDGMENT

This research was supported by a grant from the National Science Foundation (EIA-0120055).

REFERENCES

AAUW. (2000). *Tech-savvy: Educating girls in the new computer age.* Washington, DC: American Association of University Women.

Anderson, C. M., & Martin, M. (2002). Communication motives (state vs. trait) and task group outcomes. *Communication Research Reports, 19*(3), 269-282.

Camp, T. (2002). Special issue on women and computing. *SIGCSE Bulletin, 34,* 1-208.

Carver, D. L. (2000). *Research foundation for improving the representation of women in the information technology workforce: Virtual workshop report.* Arlington: National Science Foundation.

Chory-Assad, R. M. (2002). Classroom justice: Perceptions of fairness as a predictor of student motivation, learning, and aggression. *Communication Quarterly, 50*(1), 58-77.

Deci, E. L., & Ryan, R. M. (1985). *Intrinsic motivation and self-determination in human behaviour.* New York: Plenum.

Dweck, C. S., & Leggett, E. L. (1988). A social-cognitive approach to motivation and personality. *Psychological Review, 95,* 256-273.

Kerssen-Griep, J., Hess, J. A., & Trees, A. R. (2003). Sustaining the desire to learn: Dimensions of perceived instructional face work related to student involvement and motivation to learn. *Western Journal of Communication, 67*(4), 357-381.

Lombard, M., Snyder-Duch, J., & Bracken, C. (2002). Content analysis in mass communication: Assessment and reporting of intercoder reliability. *Human Communication Research, 28*(4), 587-604.

Margolis, J., & Fisher, A. (2002). *Unlocking the clubhouse: Women in computing.* Cambridge, MA: The MIT Press.

National Science Board. (2004). *Science and engineering indicators 2004.* Arlington, VA: National Science Foundation, NSB04-1.

Noels, K. A., Clément, R., & Pelletier, L. G. (1999). Perceptions of teachers' communicative style and students' intrinsic and extrinsic motivation. *The Modern Language Journal, 83*(1), 23-34.

Pintrich, P., Smith, D., Garcia, T., & McKeachie, W. (1991). *A Manual for the Use of the Motivated Strategies for Learning Questionnaire* (MSLQ). National Center for Research to Improve Postsecondary Teaching and Learning and the School of Education, University of Michigan, Ann Arbor.

Postlewaite, K., & Haggerty, L. (2002). Towards the improvement of learning in secondary school: Students' views, their links to theories of motivation and to issues of under- and over-achievement. *Research Papers in Education, 17*(2), 185-209.

Vallerand, R. J., & Bissonette, R. (1992). Intrinsic, extrinsic and amotivational styles as predictors of behaviour: A prospective study. *Journal of Personality, 60,* 599-620.

Varma, R. (2003). Special issue on women and minorities in information technology. *IEEE Technology and Society, 22*(3), 1-27.

Volet, S. (2001). Significance of cultural and motivation variables on students' attitudes towards group work. In S. Salili, Y. Hong, & C. Chiu (Eds.), *Student motivation: The culture and context of learning.* New York: Kluwer Academic.

Williams, S. R., & Ivey, K. (2001). Affective assessment and mathematics classroom engagement: A case study. *Educational Studies in Mathematics, 47,* 75-100.

KEY TERMS

Extrinsic-Economic: Refers to motivation that is determined through means outside of the individual such as a tangible positive result.

M

Extrinsic Motivation: Refers to as motivation that is determined through means outside of the individual; behaviours that are performed in order to arrive at some instrumental end.

Extrinsic-Self: Refers to motivation that is a result of projecting into the future as to whether a skill will have an utilitarian purpose.

Extrinsic-Social: Refers to motivation that is a result of an outside force that the individual has internalized such as guilt or desire to impress others.

Intrinsic-Economic: Refers to personal judgment of ability to do and be successful at a particular activity.

Intrinsic Motivation: Refers to the performance of an activity for the pleasure and satisfaction that accompany that action; fulfilling innate needs for competence and self-determination.

Intrinsic-Self: Refers to the performance of an activity for the pleasure and satisfaction that accompany that action.

Intrinsic-Social: Refers to motivation whereby the individual feels that something is personally worthwhile and relates to their value system.

Multi–Disciplinary, Scientific, Gender Research

M

Antonio M. Lopez, Jr.
Xavier University of Louisiana, USA

INTRODUCTION

Many phenomena of interest in education research are results of voluntary human action: whether a first-year college student elects to pursue a degree in information technology or not, whether the pursuit is in computer science vs. computer engineering, and whether the student will persist in a discipline throughout her or his college matriculation or change disciplines after a year or two. Although the human action is observable and can be tracked, the reasons an election is made and when it is made are not easily modeled. This article describes the design of a multidisciplinary, scientific study of gender-based differences, and ethnic and cultural models in the computing disciplines. The term computing disciplines is a collective one subsuming for ease of discussion the various disciplines that have evolved from the mid 20th century through the present 21st century, for example, computer engineering, computer science, computer information systems, information science, information technology, telecommunication systems management, and so forth. The researchers and study advisors formed a multidisciplinary team that is investigating in a scientific way the psychological, social, and educational rigidities that might exist between computing disciplines, and in so doing is developing different predictive models for women and ethnically underrepresented groups, in particular, African Americans. The article highlights recognized guiding principles for conducting scientific research in education and explains how the guiding principles have been implemented thus far in the study.

BACKGROUND

In 1959, as part of a prestigious lecture series at Cambridge University, Professor C. P. Snow, a recognized physicist and poet, gave a talk entitled "The Two Cultures" (Snow, 1964). Professor Snow expressed his concerns regarding the growing gap between the scientific and literary intellectuals of the time. He hypothesized that the reasons for the divide were a belief in educational specialization and a tendency to allow social forms to "crystallize." He suggested that once a cultural divide is established, all the social forces operate to make it more rigid and enduring.

Professor Snow's observations and hypotheses might well be applied generally to today's growing gap between women and men in the computing disciplines. However, the cultural divide may indeed go beyond gender differences, existing between specific computing disciplines (e.g., between computer engineering and information technology) or between types of institutions of higher education (e.g., between historically black colleges and universities [HBCUs] and predominantly white institutions [PWIs]), or it may exist in some other tier of the multilayered educational experience of the person (Lopez & Schulte, 2002). Researchers have documented the influence of culture: institutional culture (Barker & Garvin-Doxas, 2003), departmental culture (Meeden, Newhall, Blank, & Kumar, 2003), and computing discipline culture, that is, the human centric vs. math-science centric (Denning, 2001). The concept of culture seems to be associated with gender and ethnicity as well, playing a role in African American women attending HBCUs or PWIs (Constantine & Watt, 2002).

In a group of people, a culture develops from ongoing group discussions about values, meanings, expectations, and prevailing unwritten rules, thus affecting the perceptions, appraisals, and behavior of individual members of the group (Seel, 2000). Through learning and socialization, people internalize either consciously or unconsciously patterns of culture. The consequences of culture are that studies, say, involving the STEM disciplines (science, technology, engineering, and mathematics), may produce findings that may or may not translate into valid conclusions for all or some of the computing disciplines. Likewise, even studies in a specific computing discipline involving subjects from only PWIs may produce findings that may or may not transfer to HBCUs. The cultural divides in the computing disciplines exist because of the educational specializations that have been constructed. As Professor Snow suggested, the socialization process (i.e., the daily conversations and negotiations among the group members as well as between group members and nonmembers) will endeavor to make these divides more rigid. Consequently, extreme care must be taken in educational studies to delineate the conditions under which findings are applicable. Furthermore, large studies spanning a wide geographic area (e.g., the United States) over several years are needed to attempt to grasp the influence of culture in the computing disciplines.

MAIN THRUST OF THE ARTICLE

Forming a Multidisciplinary Team

Research in education must concern itself with the physical, social, and economic environments in which the research is conducted because contextual factors often influence research results in significant ways. For example, care must be exercised when using data collected from the computing disciplines before 2000 and 2001 (i.e., the dot-com bust and the 9/11 terrorist attack on the World Trade Center period) to draw inferences about the student population of the computing disciplines today. The economic markets that influence recruitment into the computing disciplines have changed significantly since that time. Thus, research on human action must take into account a number of contextual factors as well as the individual's understanding, intentions, and values. Consequently, studies must be multidisciplinary.

The remainder of this subsection describes the multidisciplinary team assembled under a National Science Foundation (NSF) grant to conduct a study on gender-based differences, and ethnic and cultural models in the computing disciplines. The principal investigators (PIs) are Antonio M. Lopez, Jr. (pro-

fessor and the Conrad N. Hilton endowed chair in computer science), Lisa J. Schulte (associate professor and chair of psychology), and Marguerite S. Giguette (professor and the BellSouth distinguished professor in computer science). All are from Xavier University of Louisiana.

The advisory board is a highly skilled, multidisciplinary group that can help address the physical, social, and economic environments.

Sylvia Beyer, associate professor of psychology, University of Wisconsin-Parkside. Her research focuses on gender differences in self-perceptions in male-dominated domains.

Doris Carver, professor of computer science and associate vice chancellor of research and graduate studies, Louisiana State University. She is an IEEE fellow and editor-in-chief of *Computer*.

Joanne Cohoon, research assistant professor in the Curry School of Education, University of Virginia. She is a sociologist who studies technology, gender, education, and their interaction.

Andrea Lawrence, associate professor and chairperson of computer science, Spelman College. She is the president of the Association of Departments of Computer and Information Sciences and Engineering at Minority Institutions.

Jane Margolis, research educationist, University of California Los Angeles, in the Graduate School of Education and Information Studies. She recently conducted an investigation of African American and Latino male and female high-school students' decisions to study (or not study) computer science in three public Los Angeles high schools.

Bradley Jensen, academic relationship manager, Microsoft, Inc.

Alfred Zenon, logistics engineer, Apogen Technologies, Inc.

Since Xavier University of Louisiana is not a Research I (Carnegie classification) institution, research partners were recruited to help the PIs develop the investigation methods to be used and to analyze the data collected. The research partners are as follows.

Madonna G. Constantine, professor and chair of the psychology department, Columbia University. She is a fellow of the American Psychological Association (APA) and codeveloper of the Cultural Congruity Scale. She is a consulting editor for the

Journal of Cultural Diversity and Ethnic Minority Psychology and an associate editor for the *Journal of Black Psychology*.

Robert W. Lent, professor and the director of counseling psychology for the Department of Counseling and Personnel Services, University of Maryland, College Park. He is an APA fellow and a codeveloper of the social cognitive career theory (SCCT). SCCT attempts to explain how people develop their academic and career interests, how they translate those interests into career choices, and what additional influences, such as cultural and environmental factors, contribute to their choices and achievements at school and work.

Frederick G. Lopez, professor and the director of training in counseling psychology for the Department of Educational Psychology, University of Houston. He is an APA fellow and codeveloper of the Academic Hardiness Scale. He is a former Fulbright senior scholar (Portugal), and a member of the editorial board of the *Journal of Counseling Psychology*.

Blueprint for Scientific Research

According to scholars at the National Research Council (Shavelson & Towne, 2002), the prevailing view with regard to findings from research studies in education is that the findings are of low quality and are endlessly contested. They state that one reason research in education is highly contested is the central role of values; people's hopes and expectations regarding education are tied to their hopes and expectations about the direction of society. Naturally, these values can be some overlapping set, superset, or subset of the values that are creating the various cultures surrounding education. One of the conclusions drawn is that in this complex world, there is an understandable attraction to the rationality and disciplined style of scientific research in education. The text gives six guiding principles that make a study of an educational process a scientific research study.

1. It poses significant questions that can be investigated empirically.
2. It links research to relevant theory.
3. It uses methods that permit direct investigation of the questions.

4. It provides a coherent and explicit chain of reasoning.
5. It replicates and generalizes across studies.
6. It discloses research to encourage professional scrutiny and critique.

Researchers studying the computing disciplines must take these guiding principles to heart.

Implementing Guiding Principle #1

"The Incredible Shrinking Pipeline" (Camp, 1997) was the title of a paper that drew attention to the decreasing population of college women majoring in computer science in the United States. Studies in the United States have continued to focus on the computer-science pipeline (Beyer, Rynes, Perrault, Hay, & Haller, 2003; Cohoon, 2001; Margolis, Fisher, & Miller, 2000). However, outside the United States, studies have taken a broader computing-disciplines approach (Adams, Bauer, & Baichoo, 2003; Cukier, Shortt, & Devine, 2001; Galpin, Sanders, Turner, & Venton, 2003). In fact, a Canadian study (Cukier et al.) found that a narrow definition of computing disciplines (e.g., just computer science) marginalizes women and their contribution, noting that the language of technology reflects and shapes the culture. Only recently have United States researchers in gender-based differences (Lopez, Schulte, & Giguette, 2005; Randall, Price, & Reichgelt, 2003) begun to differentiate between the various computing disciplines.

A list of variables that seem to have an impact on the computing disciplines' pipeline has emerged from these and other research publications (Beyer, De Kenster, Walter, Colar, & Holcomb, 2005; Cohoon, 2002; Kahle & Schmidt, 2004; Margolis & Fisher, 2002; Rowell, Perhac, Hawkins, Parker, Pettey, & Iriate-Gross, 2003). Some of these variables are collective self-esteem, computer self-efficacy, coping self-efficacy, gender roles, goals, interests, mathematics self-efficacy, outcome expectations, self-efficacy, social support and/or barriers, and stereotype threats. These variables need to be investigated in the specific context not only of gender but ethnicity and culture as well. Thus, some research questions are the following. Which of the previous variables, if any, predict the perseverance of African American women in computer engineering? Is it the same variable or set of variables that predicts the perseverance of African American women in information technology? How strong is the stereotype threat for African Americans in computer science vs. computer information systems?

Implementing Guiding Principle #2

The SCCT is at the heart of the present study. SCCT consists of three overlapping models aimed at explaining the processes through which people (a) develop basic academic and career interests, (b) make and revise their educational and vocational plans, and (c) achieve performances of varying quality in their chosen academic and career pursuits (Lent, Brown, & Hackett, 1994). Self-efficacy (in particular, coping self-efficacy), outcome expectations, interests, and goals play key roles within each of these three models, operating in concert with a variety of additional personal, contextual, and learning variables (e.g., gender, ethnicity, social support/barriers, etc.) to help shape people's career development (see Figure 1).

SCCT maintains that people develop interests in activities in which they believe they can perform effectively and for which they anticipate receiving positive outcomes. The theory focuses on several cognitive-person variables (e.g., self-efficacy, outcome expectations, etc.) and how these variables interact with other aspects of the person (e.g., gender, ethnicity, etc.). Contextual influences, in particular, proximal ones (e.g., role models, faculty encouragement, etc.), are important during decision making. Recent findings (Lent, Brown, Schmidt, Brenner, Lyons, & Treistman, 2003) indicate that

SCCT variables were strongly predictive of engineering students' persistence goals across genders and university types (i.e., HBCUs vs. PWIs).

Stereotype threat is another theory embedded in the study. Stereotype threat is defined as the danger of being viewed through the lens of a negative stereotype, or the fear of doing something that would inadvertently confirm that stereotype (Steele, 1997). Steele suggests that stereotype vulnerability might explain the decline of interest among some groups in specific fields, especially among women in male-dominated fields and among African Americans in academic settings in general.

There are other theories embedded in the present study, but space limits their exposure here. Suffice to say, the theories being supported by the collected data will be those propagated in future research.

Implementing Guiding Principle #3

The research methodology is a mixed design, primarily a three-year longitudinal study with a new sample of first-year undergraduates being introduced in the second and third years. The project seeks to involve a total of 50 institutions of higher education from across the United States. Twenty-five of these institutions are HBCUs and 25 are PWIs. Every year of the study, each institution will have approximately 70 of their computing-discipline undergraduates (both male and female) surveyed; these subjects will range from first-year students to seniors. For comparison each year, 30 first-year undergraduates from noncomputing disciplines (e.g., psychology, music, education, etc.) at each institution will also be surveyed.

Two colleagues from each of the 50 institutions help coordinate the study. These faculty members are critical to the success of this project because they serve as the vital link to the target population. At each institution, one colleague is from a computing discipline and the other from a noncomputing discipline. The computing-discipline faculty member will support the longitudinal segment of the project, while the noncomputing-discipline faculty member will support the successive independent samples of the noncomputing-discipline students.

The prepared surveys are on the World Wide Web, protected via a unique user log-in and a secure socket with certificate verification. Several versions

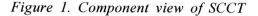

Figure 1. Component view of SCCT

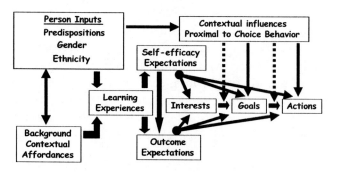

of the two basic surveys (one for computing-discipline participants and one for noncomputing-discipline participants) are used to balance order effects. Participants may complete the survey at their convenience. The survey takes less than 30 minutes to complete. On submitting the survey, an e-mail voucher for $10 is sent, whereby the participant can print it out and take it to the faculty member for authorization and collection of compensation.

FUTURE TRENDS

The data collection for the first year of the described study ended on January 31, 2005. The PIs and research partners are analyzing the data at the present time. Once the evidence is available, the team will endeavor to provide a *coherent and explicit chain of reasoning* between evidence and theory. Some theories may be strengthened, others modified, and yet others discarded. The longitudinal nature of the study will allow the possible *replication and generalization* of results over time. At the same time, others may elect to use the SCCT or stereotype threat in their own large studies, and this will allow further comparison of results across studies. Finally, papers written and published by members of the study's multidisciplinary team will disclose the research and encourage multidisciplinary *scrutiny and critique*.

The National Research Council guiding principles provide a strong framework on which to construct an understanding of the computing disciplines' pipeline. It is expected that the study described here will be one of many that advance such knowledge in a scientifically rigorous and self-correcting manner. Ultimately, the findings will affect policy makers who fund and/or direct future education in the computing disciplines.

CONCLUSION

This article describes the cultural lenses through which a multidisciplinary research team is studying the computing disciplines' pipeline. The study follows the National Research Council guidelines for scientific research in education. The article shows how the first three guiding principles were imple-

mented in the study. The future will see the implementation of the last three guiding principles.

ACKNOWLEDGMENT

This material is based upon work supported in part by the National Science Foundation under Grant No. HRD-0332780, Microsoft Inc. and Apogen Technologies Inc. Any opinions, findings, and conclusions or recommendations expressed herein are those of the author and do not necessarily reflect the views of the National Science Foundation, Microsoft, or Apogen Technologies.

REFERENCES

Adams, J., Bauer, V., & Baichoo, S. (2003). An expanding pipeline: Gender in Mauritius. *ACM SIGCSE Bulletin: Inroads, 35*(1), 59-63.

Barker, L., & Garvin-Doxas, K. (2003). Poster: The effects of institutional characteristics on participation of women in computer science bachelor's degree programs. *Proceedings of the ITiCSE, Thessaloniki, Greece* (p. 242).

Beyer, S., De Kenster, M., Walter, K., Colar, M., & Holcomb, C. (2005). Changes in CS students' attitudes toward CS overtime: An examination of gender differences. *ACM SIGCSE Bulletin: Inroads, 37*(1), 392-396.

Beyer, S., Rynes, K., Perrault, J., Hay, K., & Haller, S. (2003). Gender differences in computer science students. *ACM SIGCSE Bulletin: Inroads, 35*(1), 49-53.

Camp, T. (1997). The incredible shrinking pipeline. *Communications to the ACM, 40*(10), 103-110.

Cohoon, J. (2001). Toward improving female retention in the computer science major. *Communications of the ACM, 44*(5), 108-114.

Cohoon, J. (2002). Recruiting and retaining women in undergraduate computing majors. *ACM SIGCSE Bulletin: Inroads, 34*(2), 48-52.

Constantine, M., & Watt, S. (2002). Cultural congruity, womanist identity attitudes, and life satisfaction among African American college women attending historically black and predominantly white institutions. *Journal of College Student Development, 43*(2), 184-194.

Cukier, W., Shortt, D., & Devine, I. (2001). Gender and information technology: Implications of definitions. *Proceedings of ISECON* [CD]*, 18*, §06a.

Denning, P. (2001). The profession of IT: Who are we? *Communications of the ACM, 44*(2), 15-18.

Galpin, V., Sanders, I., Turner, H., & Venton, B. (2003). Computer self-efficacy, gender, and educational background in South Africa. *IEEE Technology and Society, 22*(3), 43-48.

Kahle, J., & Schmidt, G. (2004). Reasons women pursue a computer science career: Perspectives of women from a mid-sized institution. *The Journal of Computing Sciences in Colleges, 19*(4), 78-89.

Lent, R., Brown, S., & Hackett, G. (1994). Toward a unified social cognitive theory of career/academic interest, choice and performance. *Journal of Vocational Behavior, 45*, 79-122.

Lent, R., Brown, S., Schmidt, J., Brenner, B., Lyons, H., & Treistman, D. (2003, August 7-10). *Social cognitive predictors of engineering students' academic goals: Do sex or university type moderate predictor-criterion relations?* Paper presented at the meeting of the American Psychological Association, Toronto, Ontario, Canada.

Lopez, A., & Schulte, L. (2002). African American women in the computing sciences: A group to be studied. *ACM SIGCSE Bulletin: Inroads, 34*(1), 87-90.

Lopez, A., Schulte, L., & Giguette, M. (2005). Climbing onto the shoulders of giants. *ACM SIGCSE Bulletin: Inroads, 37*(1), 401-405.

Margolis, J., & Fisher, A. (2002). *Unlocking the clubhouse: Women in computing.* Cambridge, MA: MIT Press.

Margolis, J., Fisher, A., & Miller, F. (2000). The anatomy of interest: Women in undergraduate computer science. *Women's Study Quarterly, 28*, 104-127.

Meeden, L., Newhall, T., Blank, D., & Kumar, D. (2003). Using departmental surveys to assess computing culture: Quantifying gender differences in the classroom. In *Proceedings of the ITiCSE* (pp. 188-192), Thessaloniki, Greece.

Randall, C., Price, B., & Reichgelt, H. (2003). Women in computing programs: Does the incredible shrinking pipeline apply to all computing programs? *ACM SIGCSE Bulletin: Inroads, 35*(4), 55-59.

Rowell, G., Perhac, D., Hawkins, J., Parker, B., Pettey, C., & Iriate-Gross, J. (2003). Computer-related gender differences. *ACM SIGCSE Bulletin: Inroads, 35*(1), 54-58.

Seel, R. (2000). Culture and complexity: New insights on organizational change. *Organizations and People, 7*(2), 2-9.

Shavelson, R., & Towne, L. (Eds.). (2002). *Scientific research in education.* Washington, DC: National Academy Press.

Snow, C. (1964). *The two cultures: And a second look.* London: Cambridge University Press.

Steele, C. (1997). A threat in the air: How stereotypes shape intellectual identity and performance. *American Psychologist, 52*(6), 613-629.

KEY TERMS

Collective Self-Esteem: The value that individuals place on their own cultural groups.

Culture: A group-held set of values, meanings, expectations, and unwritten rules that affect the perceptions, appraisals, and behaviors of individual group members.

Gender Roles: A set of social, behavioral norms associated with a given gender.

Outcome Expectation: The extent to which one believes that one's action will bring about a certain result.

Self-Efficacy: An individual's judgment of her or his ability to produce a desired effect. Domains particularize the definition (e.g., coping self-efficacy, computer self-efficacy, and mathematics self-efficacy).

Social Barrier: Individuals or groups discouraging and/or working against one's effort to attain a goal.

Social Support: Individuals or groups encouraging and/or helping in one's effort to attain a goal.

Stereotype Threat: One's fear of doing something that would inadvertently confirm a socially held negative mental image of a group with which one shares certain characteristic qualities.

M

Native American Women in Computing

Roli Varma
University of New Mexico, USA

Vanessa Galindo-Sanchez
University of New Mexico, USA

INTRODUCTION

In the 1990s, a number of efforts had been made to increase the representation of women in computer science (CS) and computer engineering (CE) education, mostly to compensate for the expected shortfall of candidates from the traditional source: 18-year-old non-Hispanic white males. Yet, women remain underrepresented in the CS and CE disciplines. The underrepresentation of minority women is especially conspicuous and is absolutely glaring among Native American women. Though there are studies on the underrepresentation of women in CS and CE education, there are very few studies on minority women, and there is very little scholarly work on Native American women. Because Native Americans—officially classified as American Indians and/or Alaska Natives—are relatively small in number (1.5% of the U.S. population), they are seldom represented in assessments of gender and/or racial disparities in CS and CE education.

The educational attainment levels of Native American women have improved significantly over the last two decades. Despite these advances, the education level of Native American women remains considerably below the levels of the total population. They are less likely than the total population to graduate from high school, to enroll in college, and to graduate from college (Madrid, 1997). Native American women who do enroll in and graduate from college are less likely to be in science or engineering disciplines. Native American women who do graduate in science or engineering disciplines are less likely to be in CS or CE. For instance, in 2001, Native Americans earned only 271 bachelor's degrees in CS. Of these, Native American men earned 193 and women earned 78. Of incoming freshmen in 2002, only 4% of Native American men and 0.5% of Native American women intended to major in CS (National Science Foundation [NSF], 2004). This article discusses why so few Native American women pursue education in CS or CE disciplines after high school.

BACKGROUND

Most scholarly work on the underrepresentation of women in CS and CE education has been about the gender gap in science and engineering. It is generally assumed that many of the reasons that discourage women from science and engineering education also apply to CS and CE. When scholars have studied women in CS and CE disciplines, they have concentrated mostly on white women. If scholars have considered minority women, the focus has been on blacks and/or Hispanic women (e.g., American Association of University Women [AAUW], 2000; Howell, 1993; Margolis & Fisher, 2002; Martin & Murchie-Beyma, 1992; Moses, 1993; Spertus, 1991; Varma, 2002). These studies reveal gender bias in early socialization at home and in school, feelings of being deficient in mathematics and science, a lack of exposure to computers, the use of computers mostly for word processing, the masculine image of computers, and the absence of female role models—all of which contribute to the underrepresentation of women, including minority women, in CS and CE education. Though most of these barriers are likely to apply to Native American women, there may be additional historical and cultural factors that may play an essential role in their relative interest in CS and CE education.

Native Americans tend to maintain tribal traditions and connections to their tribal community. They are likely to live in what has been called "two worlds": the world of Native American ethos, which holds that sharing, generosity, and thinking as a group contribute to tribal community survival, and the world of American ethos, which values independence, individualism, and competition to enhance individual success (Benhaim & Stein, 2003). The Native American worldview emphasizes the importance of grasping the big picture before studying particular subjects (Megginson, 1990). Native Americans prefer harmony and group-oriented learning environments to environments that promote individual success (Anderson & Stein, 1992). As a result, Native American students may face more challenges in pursuing a major in a CS or CE discipline than whites, blacks, and Hispanics. Because of patriarchy, cultural values, and social norms, Native American women may have more problems studying CS or CE than Native American men.

This research explores different spectrums to explain the low representation of Native American women in CS and CE education at the undergraduate level. It is based on 50 in-depth interviews of Native American undergraduate students (25 females and 25 males) enrolled in a CS or CE program at six nontribal and tribal universities.

MAIN THRUST OF THE ARTICLE

Many interview students noticed that there are very few women studying CS or CE at their universities, and among the few women, very few are Native Americans. The majority of those who recognized a low number of women in CS and CE disciplines were female. Generally, students from nontribal sites were more likely to mention few women as being an issue than students from tribal sites. One female student said,

We are pretty rare in the computer program. Most of my classes, the ratio is like 1 to 10, 1 female per 10 male students ... These women are either white or Hispanic. I am the only Native woman in the class.

A male student observed, "I do not think there are any."

There are multiple reasons why there are few Native American women pursuing degrees in CS or CE programs. Because of the patriarchal way of life that dominates children's social and educational worlds, Native American women are historically seen as physically and intellectually less capable than men. Cultural and social notions about Native American women affect the way men view them in CS and CE programs. Although the majority of respondents (76%) indicated that they did not experience incidents related to gender in the CS and CE programs, gender bias and male preoccupations are prevalent among students. A closer look at the data shows that men's expectations and preconceived stereotypes about Native American women are more common than what the numbers might suggest.

For example, the majority of male students mentioned the gender bias was in favor of women. As one male student mentioned,

I often feel that [Native American women] have an advantage because of low male to female ratio ... In some sense they are more successful because they have all the resources from the smart guys who are always ready to help them out.

Another said, "I think [Native American women] get a lot more offers upon graduation than us because companies are trying to make their workforce diverse and it looks good to have Native American women." Another male student believed, "[Native American girls] receive favorable grades just because they are girls." These quotes show that what might appear as bias in favor of Native American women at first might not be the case at the end. For instance, the first student quoted mentions that girls are successful because they work with the "smart guys." In other words, Native American women succeed not because of their intelligence or hard work, but because of the help of somebody else, a smarter male specifically, and that special aid gives them an advantage over men.

Bias in favor of men differs greatly from bias in favor of women. While bias in favor of women relies on the help of others because of the inability of women to perform tasks by themselves, bias in favor of men is based on the simple fact that the student is male. As one female student said, "There is still a perception that males are bosses or think they have a better chance of getting further in their career." A male student believed, "Low representation of [Native American] women has to do with the scientific worldview. Because they don't have it, they encounter problems. Since most of us do have it, we don't encounter problems." So, a man's success in CS or CE depends solely on himself, whereas a Native American woman's success depends on acquiring the male scientific worldview. Such perceptions create an intimidating environment for Native American women. For the most part, more women from nontribal sites identified intimidation to be an issue than those from tribal sites. One female student said, "I have to always assure myself that I can do this, I am capable of doing this. I am doing the right thing by being here." Another echoed, "Sometimes, I am scared to speak, or ask questions, because what guys might think if I am wrong."

Yet, when asked if there is a difference in being a woman in a CS or CE program, almost one third of respondents said that there are no gender differences. Of those who believed that there are no gender differences, a large majority was women (58%). Nevertheless, a closer look at the responses of the female respondents shows that there is still a gender bias against women in CS and CE programs. For instance, one female respondent stated,

Well, I don't think that there is a difference in being a woman. I never thought about it. They treat me the same. I always get treated like one of the guys. I even forget that I am a girl.

This student mentions that there is no difference because she gets treated like "one of the guys" and she forgets she is a girl. Such statements show that even when women think that there are no gender differences, these differences are actually internalized. Women look up to men as the ultimate way to succeed in CS and CE programs. If there are no gender differences, why would a female student be pleased to be treated like a male? Or why would a female student forget that she is a female? By undermining their gender, female students have been able to transcend gender stereotypes. Some female students said, "[M]en look down at us just because we are females studying computers. Ironically, they have the power to look down at us."

Nevertheless, gender bias is not the only reason why Native American women are underrepresented in CS and CE programs. Cultural norms and early child socialization are important concepts to consider. There are strong patriarchal attitudes toward Native American women. As one female student said,

There is a disadvantage when you are a girl and you grow up in an unspoken way of life that doesn't allow girls to explore as much as boys would. So when girls get to college they might not have the skills to disassemble logic.

The underrepresentation of Native American women in CS and CE programs is in part a consequence of a historical Native American culture that favors men over women. Cultural patterns such as allowing men to "explore" more than women are closely linked to the different opportunities available to women and men. One male student acknowledged,

In general, computer science is not very attractive to women because it relates back to our culture. I am from Zuni and the women do not have a chance to learn the value of knowing how to program because of cultural aspects.

Another male student believed that "to do computer science, you need a scientific worldview. This is not what we teach our women." One female student regretfully said, "I just think a lot of parents don't encourage us to go in higher education or to study computers…They don't think it is really a woman thing to do."

Besides cultural aspects that do not encourage them to study CS or CE, Native American women face family and community challenges that other students may not. Family has been of paramount importance in Native American communities. The family structure tends to include extended family and several generations living in close proximity to each other. Women, especially grandmothers, play a key role in family affairs (Deloria, 1991). Native American women hold family together by taking care of elderly family members and/or children. They are responsible for exposing children to their traditions and ceremonies, and teaching Native American languages. Many females, mostly from nontribal sites, identified family affairs to be a barrier to success in CS or CE education. One female student stated,

Because it takes a lot of time, it takes a lot of effort on your part to study computer science; you have to figure out well in advance how you will handle your family matters and community events. Your study schedule has to have time for family and for community.

One male student described the situation like this:

Motherhood comes early for a lot of Native American females. So the success rate of women to get to the point of computer science is very low...Then family is always calling them back home for ceremonies or other matters so they just don't get through far enough into school to get to the point of actually doing computer science.

Cultural norms, early socialization, gender bias, and family matters are not the only factors creating obstacles for Native American women pursuing education in CS or CE; economics play a key role. The U.S. Bureau of the Census (2000) shows less education, lower earnings, more poverty, and poor health status among Native American women than the majority of the population. These economic difficulties experienced by Native American women are linked to the resources they are exposed to and the opportunities they have. It is well known that minority-serving schools, especially tribal schools, face the "digital divide" (Guice & McCoy, 2001). Still, because of socialization and cultural norms, Native American men get exposed to computers earlier than Native American women, and early exposure to computers leads to interest in a career in CS or CE. Female respondents cited a lack of exposure to computer technology more frequently than male respondents. Women from tribal sites cited a lack of resources more than those from nontribal sites. One male student acknowledged, "In order to be interested in computer science, women should be exposed to computers for a while. I mean, they are not cheap, and they are not everywhere. They are mostly in upper- to middle-class white households." One female student narrated, "Where I came from, it wasn't something that people thought

about. I never really had heard of computers until I went to high school."

When Native American women do decide to study CS or CE, they are not appreciated within their family and community, mostly because "the image of a computer scientist is of white male." Native American women are often stereotyped as operating outside traditional norms if they pursue degrees in CS or CE. Women who do enter CS or CE programs are seen as social outcasts, plain, and unfeminine. As one female student said, "They think that I am a nerd. I am a geek. I am intimidating." Another said, "My folks see me as nerdy. They complain that I no longer have time for them, for the community. It has been hard for me and for them." Many Native American women also do not enter CS or CE programs because it is a white-male-dominated field, and so Native American women wishing to pursue CS or CE are caught between filial obligations and pervasive stereotypes.

FUTURE TRENDS

Changing the underrepresentation of Native American women in CS and CE will be difficult to accomplish. Some factors can be altered, but others are more challenging to modify because they are historically and culturally based. Gender bias will take some time to change, but ingrained cultural traditions among Native Americans are unlikely to transform. Therefore, one must focus on goals that are more likely to be achieved, such as providing access to resources and early exposure to computers.

When students were asked what could be done to attract more Native American women to the program, none of the suggestions were related to those cultural and historical issues that make Native American women choose not to pursue CS or CE degrees. Nevertheless, the majority of students mentioned program changes such as more female faculty instructors, classes for females only, and an alternative approach to learning. As one male student said, "It might encourage them to see more female professors in computer science because as far as I know, a majority of the CS department is males, mostly males." Another female student said, "More female instructors. I haven't had a female instructor yet." Students also suggested a number of support services such as scholarships, tutoring, and more computers.

Although these are good techniques to possibly change the underrepresentation of Native American women in CS and CE programs, one should not forget that historically, Native Americans have not had equal opportunities in education. The federal government and the education department must address issues of poverty and income inequality in order to indirectly improve Native Americans' access to technological resources, which will consequently bolster their interest in CS and CE programs.

CONCLUSION

Native American women face several issues that are different from Native American men who pursue degrees in CS or CE. Because they face factors pertaining to both race and gender, Native American women have a long way to go to achieve equality in higher education in CS or CE programs.

ACKNOWLEDGMENTS

This research was supported by a grant from the Alfred P. Sloan Foundation (B2002-68). I would like to thank Julia Gilroy for her help in data analysis. I would also like to thank all the students who gave their valuable time.

REFERENCES

American Association of University Women (AAUW). (2000). *Tech-savvy: Educating girls in the new computer age.* Washington, DC: Author.

Anderson, L., & Stein, W. (1992). Making math relevant. *Tribal College, 3*(3), 18-19.

Benhaim, M. K. P., & Stein, W. J. (2003). *The renaissance of American Indian higher education: Capturing the dream.* Mahwah, NJ: Lawrence Erlbaum Associates.

Deloria, V. (1991). *Indian education in America.* Boulder, CO: American Indian Science and Engineering Society.

Guice, A. A., & McCoy, L. P. (2001, April). *The digital divide in Native American tribal schools: Two case studies.* Paper presented at the American Educational Research Association Annual Meeting, Seattle, WA.

Howell, K. (1993). The experience of women in undergraduate computer science: What does the research say? *SIGCSE Bulletin, 25*(2), 1-8.

Madrid, G. J. (1997). *Toward understanding the student integration: Experience of nine Native American and nine Hispanic women in a community college.* Albuquerque: University of New Mexico.

Margolis, J., & Fisher, A. (2002). *Unlocking the clubhouse: Women in computing.* Cambridge, MA: MIT Press.

Martin, C. D., & Murchie-Beyma, E. (Eds.). (1992). *In search of gender-free paradigms for computer science education.* Eugene, OR: International Society for Technology in Education.

Megginson, R. (1990). *Mathematics and Native Americans.* Washington, DC: Mathematical Association of America.

Moses, L. E. (1993). Our computer science classrooms: Are they friendly to female students? *SIGCSE Bulletin, 25*(3), 3-12.

National Science Foundation (NSF). (2004). *Women, minorities, and persons with disabilities in science and engineering 2004* (NSF 04-317). Arlington, VA: Author.

Spertus, E. (1991). Why are there so few female computer scientists? *The MIT artificial intelligence laboratory technical report 1315.* Retrieved October 12, 2005, from http://www.mills.edu/ACAD_INFO/MCS/SPERTUS/Gender/why.html

U.S. Bureau of the Census. (2000). *Census 2000 Summary File 4 (SF4)—sample data.* Retrieved March 12, 2004, from http://factfinder.census.gov/servlet/DTTable>

Varma, R. (2002). Women in information technology: A case study of undergraduate students in a minority-serving institution. *Bulletin of Science, Technology & Society, 22*(4), 274-282.

Wildcat, D., & Necefer, E. (1993). A Native American model. In E. B. Jones (Ed.), *Lessons for the future: Minorities in math, science and engineering at community colleges* (pp. 37-45). Washington, DC: American Association of Community Colleges.

KEY TERMS

Digital Divide: Refers to the socioeconomic gap between communities that have access to computers, the Internet, and computer-related technologies and those who do not.

Native American Cultural Traditions: Practices that include listening to elders; maintaining languages, ceremonies, and powwows; and having a self-perceived identity based on cultural traditions.

Native Americans: People officially classified as American Indians or Alaska Natives. They have origins in any of the original peoples of North and South America and maintain tribal affiliation or community attachment.

Two-Worlds Metaphor: The concept that Native Americans live in two worlds where there is cooperation vs. competition, group emphasis vs. individual emphasis, modesty vs. self-attention, the nonmaterialistic vs. the materialistic, harmony with nature vs. conquest over nature, the spiritual vs. the skeptical, and the aggressive vs. the passive.

N

Negotiating a Hegemonic Discourse of Computing

Hilde Corneliussen
University of Bergen, Norway

INTRODUCTION

The number of women within computer sciences is low in Norway, as in other Western countries (Camp & Gürer, 2002). Research projects have documented that girls and women use the computer less and in other ways than boys and men (Håpnes & Rasmussen, 2003). Even though variations between women and between men also have been documented through research, a dualistic image of gender and ICT has dominated throughout the 1990s (Corneliussen, 2003b). Worries about the "gender gap" related to computers have resulted in a number of initiatives to include girls and women in the "information society," but in order to do this in a successful manner we need knowledge about what it means to be a man or a woman with a relation to computers. How do men and women construct their own relations to computing?

BACKGROUND

This article presents a study of how male and female computer students perceive gender as meaningful in relation to computing, and how they create their own relations to computing (Corneliussen, 2003a).[1]

Empirical Material

The empirical material of the project is based on a study of seven men and 21 women who were students in a programming course[2] at the Department of Humanistic Informatics at the University of Bergen. During a period of three months, they were observed while working in the computer lab, they answered weekly questions on e-mail, including questions about their relation to computing, and most of them were interviewed in groups.

Gender

Gender is a social construction that gives norms, rules, and guidelines for men and women. Gender is experienced and performed by men and women. Simone de Beauvoir's description of gender as "what we do about what the world does to us" (Moi, 1999, p. 72) illustrates how gender is both a structure that we meet in the world, as well as what we do about it. Thus, in the main section, we will first look at how men and women *perceive* that gender has a meaning related to computers, and second, how they find their own positions as computer users.

Discourse Theory

The most important theoretical perspective in the project is poststructuralist feminist theory, mainly inspired by the historian Joan W. Scott's insistence on studying gender as a discursive structure (Scott, 1988). The analytical tool applied in this project has been elaborated through this theoretical perspective, with a special focus on cultural production of meaning, inspired by Ernesto Laclau and Chantal Mouffe's theory of discourse (1985). Two important concepts in the following presentation are "discourse" and "subject position." The concept of discourse refers to a limited and temporarily fixed meaning within one particular area—like the discourse of computing. Subject position refers to a discursive point of identification within a discourse. While a discourse gives the guidelines for how to understand a phenomenon, the subject position gives guidelines for the individual, about expected or accepted behaviour. The individual can either associate with, negotiate or reject a subject position.

A HEGEMONIC DISCOURSE

Research has documented that there are differences in how men and women's relations to computing are perceived (Lagesen Berg, Gansmo, Hestflått, Lie, Nordli, & Sørensen, 2002). This is also evident in this project; all the informants shared a series of conceptions of how gender and computing were related. Together, these conceptions comprise a "hegemonic discourse" which seems to suppress other and alternative conceptions about gender and computing.

A central part of this hegemonic discourse is the different expectations towards men and women's relations to the computer. Men are expected to have interest, experience and knowledge about computers, while women are expected *not* to have the same interest, experience, and knowledge. Men and women are also expected to engage in different activities; men in activities associated with the technical machine and playing with the computer, while women are expected to use the computer for a specific purpose, and for a limited number of tasks.

The hegemonic discourse thus creates two distinct subject positions; one associated with the computer skilled man, and one associated with the less computer skilled woman. It is important to emphasize that a subject position is not a description of "real" men and women, but rather a description of the *expectations* towards men and women. People use these expectations towards themselves and towards others. This does not mean that the individual always is in harmony with the discourse, a point further demonstrated in the following chapter.

NEGOTIATING THE HEGEMONIC DISCOURSE

All the informants articulate their own individual ways to describe or position themselves in relation to computing. However, by looking at how these positions are articulated it is possible to point to a pattern of seven different *positioning strategies*. A starting point for all the different strategies is the hegemonic discourse, but they differ from each other with regard to the position they aim at, and thereby also with regard to the degree of harmony with the hegemonic discourse. We will first look at the three positioning strategies among the men, before we turn to the women's strategies.

MALE POSITIONING STRATEGIES

"Rooted in a Room for Men"

In the first positioning strategy, the men have "roots in a room for men." They can display harmony with the masculine subject position in the hegemonic discourse. They have experience with computers since childhood, and they have a lot of knowledge about computers. They conform to many of the expectations towards men in the hegemonic discourse; they have acquired their own experience "together with the boys" and "as one of the boys." A close relation between boys and computers is described as "natural," and one of them even thinks that "people almost expect that a boy studies computing." This group of men use the hegemonic discourse as a positive reference to their own relation to the computer.

"Aiming at a Room for Men"

The next group of men is also aiming at "a room for men," but they can not display the same harmony with the masculine subject position. That is, except for being men. They do not have very much experience or knowledge about computers prior to the computer course, but to acquire more knowledge seems to be something that they have wanted, and it gives them a more "proper" relation to the computer. Expectations about men's close relationship with the computer becomes a positive force in their own relation to the computer. One of them thinks that he learns tasks on the computer faster because of "the 'taken for granted' assumption that computers are something I can-handle, because I am a boy." Another one illustrates how he might be associated with the masculine subject position without really being qualified, and he is able to "hide" in this position. This positioning strategy clearly demonstrates how men have the *possibility* of being associated with computer competence and a positive relation to the computer purely based on gender.

"Outside a Room for Men"

In the last positioning strategy among the men, it is rather a *distance* from the masculine position, which is the goal. In order to achieve this distance it is the *lack* of computer competence that is being emphasized: "I have a PC with a sound card that does not work, that probably tells you how much I have acquired in that area." This man also "hid" his previous experience with computers, and even though the informants several times were confronted with questions about their experience, it was not until the end of the term he revealed that he had been studying computing before! In other words, men seem to need an active strategy in order to distance themselves from the expectations towards men, as if the masculine subject position too easily is activated by gender alone.

We have seen that the men have various ways of positioning themselves in relation to computers. However, the differences has less to do with introducing new elements to the masculine subject position than how well they conform to—or want to conform to—the masculine position. They illustrate that they have the possibility to use this position as a positive description of themselves, as a goal to reach for, or as a position to "hide in," as if gender alone is sufficient to activate the association to computer competence.

FEMALE POSITIONING STRATEGIES

"A Limited Room for Women"

The subject position associated with women in the hegemonic discourse is in various ways *limited* compared to the masculine position. The first strategy among the women aims at "a limited room for women," closely associated with the feminine subject position in the hegemonic discourse. The computer is associated with boys, and is described as "boring, masculine, and a bit nerdy." It is "obvious that the boys have the best understanding of the technical stuff." Female lecturers in computing are not seen as positive role models, but rather as "dissenters'" who have crossed a gendered line. These women use the expectations towards women's limited computer competence when they position themselves: "I don't

understand any of that [computer programming], because I'm a woman!" A reference to gender is sufficient to activate the association to women's "limited" computer knowledge. These women do not challenge the hegemonic discourse, but rather use gender as an explanation of their own lack of computer knowledge.

"A More Open Room for Women"

The second positioning strategy among the women also emphasizes the difference between men and women's relations to computers. However, it differs from the previous strategy by aiming at "a more open room for women." These women describe themselves with a positive relation to computers, but still within certain limits compared to men. Many of them were sceptical in the start and attended the course with an idea about computers as boring and difficult to understand for women. However, as they learned more about computers they also "discovered" that they actually *could* learn, and that they actually *did* find computing enjoyable. Many of them also describe themselves as "addicted" after a short period at the computer course: "It feels like a new world has opened up to me ... and every day I think "How on earth is it possible to cope without knowing what I know today!??" It has to be a feeling close to something like going from being illiterate to being able to read ... I think that I have become addicted to the computer!!!" Many of the women in this group have become fascinated computer users, and they expressed pleasure about computer knowledge in general, as well as programming and other more technical related topics associated with men in the hegemonic discourse. An important part of their joy seems to be related to learning something within a field associated with men: "Maybe that is why I want to work with programming, because it is so masculine ... I feel sort of as if I were in a world that's a little bit forbidden." The former scepticism has turned to fascination. They are fascinated by being able to enter a field where women do not have a natural position, and they are about to create a more open and a more positive room for women within the discourse of computing.

"A Shared Room for Men and Women"

The third positioning strategy among the women aims at "a shared room" for both men and women. Gender is irrelevant for a person's abilities or possibilities within computing, these women claim. When they position themselves in a shared room, it is neither as identical to men nor as complementary, but rather as equals, with their own characteristics: "We are not men. We don't think as men. But we have values that are just as good as men's values." They have long experience with and knowledge about computers. However, they have also experienced being treated in accordance with the "limited expectations" towards women, and they need an active strategy in order to reject the hegemonic discourse: "No boy is allowed to tell me that I am not worth as much as he is, because then I'll tell him what I really think about that." This strategy is not about abolishing gender, but about establishing men and women as equals within computing. In addition, because the hegemonic discourse has a strong position, they need an active strategy to negotiate it; they need to "nag and make a fuss" to be heard.

"Woman in a Room for Men"

The fourth and last positioning strategy among the women aims at neither a room for women nor a shared room, but rather at "a room for men." These women use their experience with tasks and technical artefacts associated with men in order to position themselves in relation to computers: "Since I did not have a brother, my sister and I had to fill that 'gap' by learning practical tasks that traditionally often are performed by men." They describe the experience they have with masculine tasks as their advantage when they work with computers. By attending a computer course, they enter a masculine domain—like they have entered other masculine domains before.

CONCLUSION

We have seen how both men and women construct their own positions by negotiating the hegemonic discourse of computing. The positions constructed by the students are more varied than the gendered subject positions of the hegemonic discourse, and they illustrate the "individual's freedom" to negotiate the discourse. However, they also illustrate "the power of discourse." The hegemonic discourse creates certain limitations, and it establishes a norm which makes it easier to associate computer knowledge with men than with women.

The analysis illustrates what the informants "do about what the world does to them." The hegemonic discourse affects the different "spaces" that are available to men and women in relation to computing. Men can easily be associated with computers and computer knowledge purely based on gender. Women on the other hand meet a cultural perception of women as not competent in relation to computers. Thus, it is clear that men and women have different challenges to deal with in positioning themselves in relation to computing, and this requires more "hard work" for women than it does for men.

The tendency of women attending computer courses with a certain scepticism before they "realize" that they enjoy computing has also been found in other studies of computer students in Norway. Langsether claims that women who "dropped in" by accident or only chose one computer class to complete a grade at the University of Oslo found the study so interesting that they would have chosen a grade in computer science if they had not been "stuck" in their present path of study (Langsether, 2001). This indicates an unused potential for recruiting women to computer studies. One reason for women's hesitance seems to be the lack of cultural stories about women's pleasure in computing, which does not make it obvious for women that they would enjoy working with computers (Corneliussen, 2005). Another reason might be male enthusiasm or fascination with the technology, which has been documented as an important factor which "turns women off" and makes them distance themselves from the computer (Aune, 1996; Turkle, 1988). "Male enthusiast" has often acted as the norm for "proper" computer users, and this image undermines both differences among men as well as similarities between men and women.

FUTURE TRENDS

While stories about pleasure in computing seems to "stick" too well to men, the same stories do not seem to "stick to" women at all. It has often been claimed that we need positive role models for women within computing (Stuedahl & Braa, 1997). However, as we have seen here, a role model is not necessarily perceived as a *positive* role model if it does not have a "valid" cultural story to apply with. One of the big challenges for the future is to make available other subject positions than the hegemonic discourse's male and female positions, in particular to make visible cultural stories about women's pleasure in computing (Corneliussen, 2005; Nordli, 2003).

We often refer to "myths" about gender and computing, often seen as "unsettled questions" or "myths we have to reject." That might be true, but it is also true that we need to take these myths seriously—not as myths meaning something which is not true, but as cultural stories which contribute to the construction of gender in computing. The fact that these stories are not always in line with the "reality," does not make them less "real"—they still exist, and these cultural stories have real effects when they are perceived as a valid frame of reference for men and women trying to find their own position related to computers and to make themselves culturally understandable as computer users.

REFERENCES

Aune, M. (1996). The computer in everyday life. Patterns of domestication of a new technology. In M. Lie & K. H. Sørensen (Eds.), *Making technology our own?* (pp. 91-120). Oslo: Scandinavian University Press.

Camp, T., & Gürer, D. (2002). *Investigating the incredible shrinking pipeline for women in computer science*. Final Report—NSF Project 9812016.

Corneliussen, H. (2002). The multi-dimensional stories of the gendered users of ICT. In A. Morrison (Ed.), *Researching ICTs in context* (Vol. 3, pp. 161-184). Oslo: InterMedia Report.

Corneliussen, H. (2003a). *Diskursens makt— Individets frihet: Kjønnede posisjoner i diskursen om data (The power of discourse—The freedom of individuals: Gendered positions in the discourse of computing)*. Dr. art. thesis, Department of Humanistic Informatics, University of Bergen.

Corneliussen, H. (2003b). Konstruksjoner av kjønn ved høyere IKT—utdanning i Norge. *Kvinneforskning, 27*(3), 31-50.

Corneliussen, H. (2003c). Male positioning strategies in relation to computing. In M. Lie (Ed.), *He, she, and IT revisited. New perspectives on gender in the information society* (pp. 237-249). Oslo: Gyldendal Akademisk.

Corneliussen, H. (2005). Women's pleasure in computing. In J. Archibald, J. Emms, F. Grundy, J. Payne, & E. Turner (Eds.), *The gender politics of ICT*. London: Middlesex University Press.

Håpnes, T., & Rasmussen, B. (2003). Gendering technology. Young girls negotiating ICT and gender. In M. Lie (Ed.), *He, she, and IT Revisited. New perspectives on gender in the information society* (pp. 173-197). Oslo: Gyldendal Akademisk.

Laclau, E., & Mouffe, C. (1985). *Hegemony & socialist strategy. Towards a radical democratic politics*. London: Verso.

Lagesen Berg, V. A., Gansmo, H. J., Hestflått, K., Lie, M., Nordli, H., & Sørensen, K. H. (2002). *Gender and ICT in Norway. An overview of Norwegian research and some relevant statistical information*. Report for the EU project: Strategies of Inclusion: Gender in the Information Society (SIGIS).

Langsether, H. (2001). *Behov og barrierer for jenter på informatikkstudiet* (Hovedfagsoppgave). Trondheim: Senter for kvinne—og kjønnsforskning, NTNU.

Moi, T. (1999). *What is a woman? And other essays*. Oxford: Oxford University Press.

Nordli, H. (2003). *The net is not enough. Searching for the female hacker*. Trondheim: dr.polit.-avhandling, Institutt for sosiologi og statsvitenskap, NTNU.

Scott, J. W. (1988). *Gender and the politics of history*. New York: Columbia University Press.

Stuedahl, D., & Braa, K. (1997). *Where have all the women gone—From computer science?* Oslo: University of Oslo.

Turkle, S. (1988). Computational reticence. Why women fear the intimate machine. In C. Kramarae (Ed.), *Technology and women's voices. Keeping in touch* (pp. 41-61). New York: Routledge & Kegan Paul.

KEY TERMS

Articulation: Every social practise which contributes to putting meaningful elements in new relations to each other in such a way that their meaning is altered can be seen as an articulation (Laclau & Mouffe, 1985).

Discourse: Within discourse theory, a discourse is seen as a limited and temporarily fixed meaning within one particular area. A discourse is perceived as a fixed objectivity through "homogenisation of an interior" and "exclusion of an exterior."

Discourse Theory: Discourse theory as it originally is articulated by the political philosophers Ernesto Laclau and Chantal Mouffe (Laclau & Mouffe, 1985) is one among a wide variety of discourse analytical propositions. Its focus is on politics and the construction of meaning.

Gender: In a poststructuralist perspective, gender is seen as a discursive category based on "perceived differences between the sexes" (Scott, 1988, p. 42). Gender is a historical and social construction that both explains and provides norms, rules, and guidelines for men and women.

Positioning Strategy: The strategy employed to create an individual position in relation to the hegemonic discourse of computing.

Subject Position: In a discourse theoretical perspective, a subject position is a discursive point of identification within a discourse. The subject position gives guidelines for the individual.

ENDNOTES

[1] See also Corneliussen, 2002 for discussions about theory and method, and Corneliussen, 2003c for a presentation of the male informants.

[2] I was also teaching in this group, and my "double" position is discussed in Corneliussen, 2002.

Online Life and Gender Dynamics

Jonathan Marshall
University of Technology Sydney, Australia

INTRODUCTION

The study of gendered interaction online grows out of studies of gendered interaction off-line and will probably be found to be a cultural variable changing with off-line gender behaviour in different social groupings. However, this does not dispose of the issues of gender's influences on online behaviour, or of whether gender behaviour online is transformed in relation to behaviour off-line.

The relevance of gender may vary in different contexts: with class, religion, place, proportions, type of online forum, topic of discussion, and so on. These contexts could overwhelm gender identities existing outside them and their effects need to be investigated. Power ratios between people of various genders may also vary within different contexts and cannot be assumed in advance. Gender both enables and restricts behaviour; it is neither merely positive nor merely negative.

In the West (at least), gender seems to be constantly in flux and interrogation, and it is not surprising if such interrogations and uncertainties occur online. Despite this interrogation, gender in off-line life seems to be treated as an essential part of a person's being or identity, and it guides reaction to others.

BACKGROUND

History and Attitudes to Computers and the Internet

Gender can influence the ways people interact with computers even before they go online. Studies of education consistently show boys being given preference accessing computers by parents and teachers (Rajagopal & Bojin, 2003). Female students may feel as competent as males in using computers while being more negative about their own involvement with computers (Herring, 1993; Sophia, 1993).

Turkle (1984) suggests that different social groups bring differing modes of interaction to computers and, as a result, find them more or less satisfying. Thus, women might attempt to negotiate with computers, while men might try and control them . Furthermore, computers are usually posited as non-emotional (Turkle), and this supposed lack might induce some people (particularly women) to find interacting with computers less satisfactory than it is for people whose aims are primarily "results oriented" (Sophia, 1993, pp. 16-17). Women, as the default carriers of emotions and performers of emotional labour in the West (Cheal, 1988; Erickson, 2005), could define themselves in opposition to computers, particularly when their usage of them was relatively uncommon. Similarly, rather than deal with the complexities of real humans, the definitively masculine or nonemotional male might flee to computers, particularly if his masculine definition is not adequate in other areas (Turkle, 1984). However, relevant gender conventions may change; at one time, programming was considered a branch of secretarial work and handled by women (Wacjman, 1991).

As the Internet was initially largely constructed and used by males, the social customs developed may well have made it harder for women to use or approach it. Such effects might have altered over time, especially given the increase of women using computers and the Internet. However, detailed historical research of such changes is rare.

Type of Forum

Different types of online forums (i.e., mailing list, MOO [MUD (multiuser domain) object oriented], newsgroup, chat room, IRC [Internet relay chat] channel, blog) differ in the ways they structure communication and allow response or the use of power, and enable the ways in which gender can be an identifier. Some forums encourage people to play characters, or avatars, and some encourage people to use their own names. Large MOOs tend to be governed by committee, whereas mailing lists tend to be governed by debate (both on- and off-list) and the decision of the moderator. Newsgroups tend to be governed by argument, confrontation, and people withdrawing when they have had enough. As a result of these structures and organisations, different forums produce different types of experiences and behaviour (Marshall, 2004). Research in one kind of forum may produce different kinds of results than research in other types of forums. For example, anonymity, gender ambiguity, or cross-gender impersonation seem much less pronounced on mailing lists than on MOOs or IRC.

Presence is ambiguous online with people only appearing present when they type, therefore it may be more common for males on MOOs to engage women in conversation and try and gain their opinions to make sure they are there than it is off-line where the presence of a listener is so much more marked.

Communication Patterns

Susan Herring has carried out the most extensive studies of online interaction showing the replication of off-line communication patterns that tend to silence women or render women's talk marginal. In her early study of the discussion lists LINGUIST and Megabyte University, Herring (1993) found that women participated "at a rate that is significantly lower than that corresponding to their numerical representation." According to her research, "[w]omen constitute 36% of LINGUIST and 42% of MBU subscribers." Yet in a discussion on sexism, women constituted only 30% of posters. In more neutral discussions, only 16% of posters were women.

On three occasions, Herring (1993) found that "women's rate of posting increased gradually to where it equalled 50% of the contributions for a period of one or two days." Not enough information is given to discover whether this resulted from an increase in the number of women posting, or whether a few women became more active, or whether the number of men posting had declined. However, the reaction "was virtually identical in all three cases: a handful of men wrote in to decry the discussion, and several threatened to cancel their subscription to the

list." This certainly implies that some men could not cope with such a visible, or argumentative, female presence and did their best to stop it within the structural possibilities of mailing-list life.

Herring (1993) also claims there are distinctive gendered styles of interaction that reflect expectations in the off-line world. In order of decreasing magnitude, she found that men discussed issues, provided information, made queries, and wrote about personal things, while women wrote about personal things, made queries, then discussed issues, and least of all provided information. Herring also found that "the messages contributed by women are shorter ... a very long message invariably indicates that the sender is male," and that "messages posted by women consistently received fewer average responses than those posted by men ... [and] topics initiated by women are less often taken up as topics of discussion by the group as a whole."

This research suggests that women, and women's interests, were marginalised in the public activity of these groups, and that male dominance was replicated even without use of physical force and in an environment in which gender is claimed not to matter. However, when Herring (1994) looked at lists focused on "traditionally 'feminized' disciplines ... [she] found women holding forth in an amount consistent with their numerical presence on the list."

Herring did not investigate, except briefly, the ongoing interaction of male and female subjects; each utterance seems to be taken in statistical isolation. There is no investigation of overall trends or variation, or even of the ways in which people interact to coproduce the ambience of the list or recognise and reinforce gender.

Although many of Herring's results have been replicated (Herring, 2000), it is common for some parts not to be (Hatt, 1998; Savicki, Kelley, & Oesterreich, 1999; Savicki, Lingenfalter, & Kelley, 1996; Vaughn Trías, 1999), suggesting there may be other variables involved. It is only recently that Herring (2000) has suggested that different modes of Internet communication might have an effect. Engagement with such issues requires intensive fieldwork, rather than brief visits or abstract samples, because fieldwork better enables the researcher to know individuals and place them amongst local and wider social dynamics.

Flaming and Harassment

It is frequently alleged that women are less inclined to flame and argue than men (c.f. Baym, 1995), and are thus often driven out of online groups. Herring (1994) writes, "the simple fact of the matter is that it is virtually only men who flame" because male communication ethics "can be evoked to justify flaming." However, in her research, both men and women state that they do not particularly like flaming, and there is no indication of how she came to rate an e-mail as a flame. In practice, flaming tends to be seen as something that other people do, with people often regarding something they, or their group, have posted as not a flame even when others strongly identify it as one.

Therefore, this proposition about women and flames is hard to check, even though it is widely believed and would be widely expected due to off-line behaviours and beliefs. However, some researchers have suggested it is not the case. Witmer and Katzman (1998) concluded their statistical research on "newsgroups and special interest groups on the Internet and CompuServe" by writing, "The data do not support the ... hypothesis that men use more challenging language and flame more often than women" (p. 7). My own research suggested that women can get as involved in flame war as men (Marshall, 2004), and research at a girls' school in Melbourne was reported to show that girls had taken to the Internet to bully, exclude, and intimidate each other (Jones, 1998). Nevertheless, it is probably true that women, generally, do not engage in flame wars with men, and tend to avoid places where such events are common.

There are also well-documented stories about continual harassment of women by men demanding or requesting Netsex (Brail, 1996; Gilbert, 1996; Hall, 1996; Smith, McLaughlin, & Osborne, 1998; Spender, 1995). Campbell (1994) describes his experiences when he used a woman's account to get free access to a BBS. Immediately on logging on, he received 31 requests from one male whose message descriptions repeatedly included *sex* and with whom he had had no prior contact. Another man sent him his phone number. People who were initially helpful ended up wanting Netsex as recompense and so on. Another less direct form of harassment is that people presenting themselves as female on MOOs are often chal-

lenged to prove their gender (La Pin & Bharadwaj, 1998). This could lead to women downplaying, or neutralising, their gender and thus to the suppression of specifically gendered interests, irrespective of whether harassment is an attempt by some males to maintain their part of the Net as a male domain.

Harassment is also reputed to occur through the large number of sex jokes or obscenities being used. However, in at least one case in my research, sexual one-liners were used by women in an attempt to restore commonality after a group was riven by political disagreement during the lead up to the Iraq war. Because of its effects, this is an issue that needs more direct research and clearly will affect people's ease online.

Lowering flame and harassment while providing a safe place to speak has often been one of the reasons given for the founding of women-only forums. However, it is not always clear if this result arises because of the involvement of women alone, or because of strict control by the moderator (Herring, 2000). In some cases, it seems that women may be thrown off these lists for not behaving in an appropriately female way, as recorded in Hall's (1996) account of the SAPPHO list. The ideology of femininity becomes self-reinforcing by excluding those who do conform. Violence and discrimination can manifest in many ways other than through overt aggression.

Gender Anonymity

Supposed gender anonymity can be assumed to give people more freedom, but it can also be used to impose restrictions. I have observed cases in which people (usually male) react to challenges that they are behaving in an exclusionary male way by declaring that the challengers do not know what gender they really are even if they had been signing their names as male and had been recognised as male. Such tactics seem to attempt to degender male behaviour and imply that objections to it are sexist, further disempowering women.

Despite the alleged commonness of gender anonymity, the categorisation of people is one of the ways in which participants decide what kind of messages others are emitting, and whether they are likely to have much in common. As the importance of being able to categorise others often depends on what we are trying to achieve, researchers can get some idea of the importance of particular categories in different situations by investigating how often these categories are requested. People usually agree that the most popular category questions for Westerners online have to do with gender, age, and location. The question of "What do you do?" that locates class, prosperity, and probable education is far less common online than off (Ten Have, 2000). Locating gender seems especially important when people are seeking some level of private intimacy or emotional support.

Public and Private

There is often a level of ambiguity and uncertainty about what constitutes public and private domains online, and these may be associated with gender roles.

The ideas of public and private are ambiguous in much non-Internet space as well. The categories themselves may be contested, fluid, and ambiguous to start with. Saying that one part of the Internet is really public and another is really private may be possible on occasions, but most situations are not clearly marked and cannot be marked. Deciding what is what, or how the categories are deployed, is an ethnographic question, not one that can be decided in advance.

However, on mailing lists there are basically two forms of sociality: on-list and off-list. Off-list exchanges tend to be dyadic, approximating and suggesting the ideal Western English-speaking world the private intimacy of the couple. If this distinction between private (or intimate) and public (or communal) comes easily to Westerners, then it is worth investigating whether the traditional association of women with the intimate or off-list sphere and men with the public and on-list sphere has any effect upon behaviour and modes of exchange.

It has frequently been remarked that intimacy can appear to be established relatively quickly between relative strangers online who risk little in letting a distant person in on their secrets. If, as Cheal (1988) claims, gender is the prime way Western people establish intimacy, and this almost always involves women, we could expect that relationships directed at the private or off-list dyadic sphere will be influenced by, or even depend upon, gender

conventions. Establishing that at least one of the dyad is a woman allows closeness to manifest more easily. If it is common that maleness is identified with aggression and flaming, and that, in contrast, women are identified with more harmonious interaction, then such identifications probably increase the importance of gender in off-list exchange.

Non-Western Gender Behaviour

There are still relatively few accounts of online gendered behaviour from non-Western cultures. Miller and Slater's (2000) study of Internet use amongst people from Trinidad shows that people use the Internet for making friends (particularly of the other sex), to reestablish patterns of mother-daughter support when children are overseas, and to engage in the semisexual public communication style known as "liming," which the authors write might be considered sexist or racist in the countries of residence. People from Trinidad also remark on the speed in which intimacy can appear to be achieved online. This lack of studies can be expected to be filled very shortly.

FUTURE TRENDS AND CONCLUSION

Currently, the categorisation of either the self or others is unlikely to derive from the membership of Internet-based groups alone as there are few environments in which different online groups, or groups formed elsewhere online, interact with each other. Therefore, most such categorisation is going to involve off-line categories. It is possible that the more marked these off-line categories are in their being embedded in society, the more likely they will be exaggerated online, and the more these categories may overpower those senses of group membership that have no reinforcement other than in participation in the online group. This appears to be the case with both gender and political divisions, and it seems unlikely to change in the foreseeable future. There is not much evidence for the common Western propositions that the online world is separate from the off-line world, or that gender is unimportant online.

Not only have the numbers of people online been increasing rapidly, but the proportion of women and men online has been approaching equality in the Western English-speaking world. That it has not in other parts of the world indicates that off-line cultural factors are still important here. Twenty years ago, middle-class women could exist without much interaction with computers; today, this is much more difficult. It means that much research on gender has the problem of becoming history before it gets published.

REFERENCES

Baym, N. (1995). The emergence of community in computer-mediated communication. In S. Jones (Ed.), *Cybersociety* (pp. 138-163). Thousand Oaks, CA: Sage.

Brail, S. (1996). The price of admission: Harassment and free speech in the wild, wild west. In L. Cherny & E. R. Weise (Eds.), *Wired women.* Seattle, WA: Seal Press.

Campbell, K. K. (1994). Attack of the cyber-weenies. *Wasatch Area Voices Express.* Retrieved February 20, 2006, from http://kumo.swcp.com/synth/text/cyberweenies.html

Cheal, D. (1988). *The gift economy.* London: Routledge.

Erikson, R. J. (2005). Why emotion work matters: Sex, gender, and the division of household labor. *Journal of Marriage and Family, 67,* 337-351.

Gilbert, P. (1996). On space, sex and being stalked. *Women & Performance, 9*(1), 125-149.

Hall, K. (1996). Cyberfeminism. In S. Herring (Ed.), *Computer mediated communication: Linguistic, social and cross-cultural perspectives.* Amsterdam: John Benjamins.

Hatt, D. F. (1998). *Male/female language use in computer dyadic interactions.* Master's thesis, Laurentian University, Ontario, Canada. Retrieved February 20, 2006, from http://www.fortunecity.com/boozers/princ essdi/302/Lang.html

Herring, S. (1993). Gender and democracy in computer-mediated communication. *Electronic Journal of Communication, 3*(2). Retrieved February 20, 2006, from http://www.internetstudies.pe.kr/txt/Herring.txt

Herring, S. (1994). *Gender differences in computer-mediated communication: Bringing familiar baggage to the new frontier.* Keynote talk at the American Library Association annual convention, Miami, FL. Retrieved February 20, 2006, from http://cpsr.org/cpsr/gender/herring.txt

Herring, S. (2000). Gender differences in CMC: Findings and implications. *CPSR Newsletter, 18*(1). Retrieved February 20, 2006, from http://www.cpsr.org/publications/newsletters/issues/2000/Winter2000/herring.html

Jones, C. (1998, October 6). Bullies move into cyberspace. *Sydney Morning Herald,* p. 5.

LaPin, G., & Bharadwaj, L. (1998). *Pick a gender and get back to us: How cyberspace affects who we are.* Retrieved February 20, 2006, from http://www.fragment.nl/mirror/various/LaPin_G.1998 Pick_a_gender_and_get_back_to_us.htm

Marshall, J. (2004, April 14-15). Governance, structure and existence: Authenticity, rhetoric, race and gender on an Internet mailing list. In *Proceedings of the Australian Electronic Governance Conference 2004,* Centre for Public Policy, University of Melbourne, Victoria. Retrieved February 20, 2006, from http://www.public-policy.unimelb.edu.au/egovernance/papers/21_Marshall.pdf

Miller, D., & Slater, D. (2000). *The Internet: An ethnographic approach.* Oxford: Berg.

Rajagopal, I., & Bojin, N. (2003). A gendered world: Students and instructional technologies. *First Monday, 8*(1). Retrieved February 20, 2006, from http://firstmonday.org/issues/issue8_1/rajagopal/

Savicki, V., Kelley, M., & Oesterreich, E. (1999). Judgments of gender in computer-mediated communication. *Computers in Human Behavior, 15,* 185-194.

Savicki, V., Lingenfalter, D., & Kelley, M. (1996). Gender language style and group composition in Internet discussion groups. *Journal of Computer Mediated Communication, 2*(3). Retrieved February 20, 2006, from http://jcmc.indiana.edu/vol2/issue3/savicki.html

Smith, C. B., McLaughlin, M. L., & Osborne, K. K. (1998). From terminal ineptitude to virtual sociopathy. In F. Sudweeks, M. McLaughlin, & S. Rafeli (Eds.), *Network and Netplay: Virtual groups on the Internet*. Menlo Park: AAAI Press.

Sophia, Z. (1993). *Whose second self? Gender and (ir)rationality in computer culture*. Deakin University Press.

Spender, D. (1995). *Nattering on the Net: Women, power and cyberspace*. Melbourne: Spinifex.

Ten Have, P. (2000). *"Hi a/s/l please?" Identification/categorization in computer-mediated communication*. Paper presented at the Sociaal-Wetenschappelijke Studiedagen 2000, Session ICT & Huiselijk Leven. Retrieved February 20, 2006, from http://www.fragment.nl/mirror/Have2000/ASL-home.htm

Turkle, S. (1984). *The second self: Computers and the human spirit*. New York: Simon and Schuster.

Vaughn Trías, J. (1997). *Democracy or difference? Gender differences in the amount of discourse on an Internet relay chat channel*. Paper presented at the Popular Culture Association annual meeting, San Diego, CA. Retrieved November 3, 2001, from http://nimbus.temple.edu/~jvaughn/papers/pca99.html

Wajcman, J. (1991). *Feminism confronts technology*. Sydney, Australia: Allen and Unwin.

Witmer, D., & Katzman. (1998). Smile when you say that: Graphic accents as gender markers in computer mediated communication. In F. Sudweeks, M. McLaughlin, & S. Rafeli (Eds.), *Network and Netplay: Virtual groups on the Internet*. Menlo Park: AAAI Press.

KEY TERMS

Avatar: A graphic representation of a person's character in a multiplayer computer game or environment. By extension, it is often used to refer to the character itself.

Blog: A blog is basically a diary on the World Wide Web. Blogs generally link to other blogs, perhaps to indicate common interests or that the authors read each other's on occasions. Some blogs allow readers to post comments on the entries.

Chat Room: An Internet site that "contains" the participants in a real-time interaction. Those within the chat room can read what each other is typing. People enter the room to participate and can be banned from it. Some suppliers may allow people to temporarily, or permanently, set up their own rooms.

IRC (Internet Relay Chat): A way of communicating with others on the Internet. It requires that users have a software client allowing them to connect. An IRC channel is something like a chat room, except that it is not necessarily based on a particular computer. The person in charge of the channel has the ability to throw participants off, and to hand this right to others.

Mailing List: A way of communicating over the Internet as a group. The list is set up on a particular computer, and accepted mail written to that list is delivered to all the people registered with that list. Usually a person will be the list owner and is responsible for running the list. He or she will often be the moderator, be a person who participates in the list, and have the power to state the list rules and to remove people from the list.

MOO (MUD Object Oriented): A way of communicating with others over the Internet. It could be thought of as an interconnected set of spaces set up on a single computer. The term is here being used to cover the whole family of such setups descended from the original MUDs, a MUD being a multiuser dungeon or multiuser domain. A person using one of these Internet forums logs in and plays a specific named character, usually with a text description, and can write or accumulate programmed objects, rooms, or devices that add to the environment. This means that he or she has some prop support for playing the character.

Netsex: Sexual activity carried out over an Internet communication channel, usually by exchanging written descriptions of sexual behaviour.

Newsgroup: A way of communicating over the Internet. A newsgroup is generally a distributed form of communication resembling a message board but not sited on a particular computer. As such, it is notoriously difficult to exclude mails from newsgroups and many were rendered unreadable by the number of advertising posts received. The software used to read newsgroups varies, but they commonly used to organise groups by threads.

Online Life and Gender Vagueness and Impersonation

Jonathan Marshall
University of Technology Sydney, Australia

INTRODUCTION

Impersonation and Gender Categories

Gender is not always immediately obvious online and this has excited interest from early on (e.g., Bruckman, 1993; Curtis, 1997). Sometimes, people have drawn extreme conclusions from this vagueness. For example, Mark Poster (1997) suggests that "one may experience directly the opposite gender by assuming it and enacting it in conversations" (p. 223). McRae (1996) writes, "mind and body, female and male, gay and straight, don't seem to be such natural oppositions anymore ... The reason for this is simple: in virtual reality, you are whoever you say you are" (p. 245). Such statements imply that gender is simply a voluntary and unconstrained conscious performance. Other writers have concluded that such identity vagueness allows, or enhances, the formation of postmodern decentred or multiple selves (Kolko & Reid, 1998; Turkle, 1995).

These arguments suggest that, when online, people are free of off-line conventions, restrictions, and power dynamics, and can experience hidden aspects of themselves, or create themselves, through an act of will and performance. Frequently, these positions are surrounded by a conflicting moral discourse, either suggesting that the Internet promotes freedom and true self-expression, or that it promotes bad faith and betrayal.

However, easy voluntarism may not be common in practise. Although it is possible that people may present new identities, the categories they use and present within can remain unchallenged and may even intensify. After her praise of voluntarism, McRae (1996) points out that if someone plays a woman and wants to "attract partners as 'female' [they] must craft a description within the realm of what is considered attractive" (p. 250). Schaap (1999) likewise remarks on the relatively "strict rules on what constitutes a convincing female character and what a convincing male character." So, although the gender of the person online may not match their gender off-line, the gender they choose usually exaggerates the conventions of attractive or good gender construction. As Kendall (1996) writes, "choosing one gender or another does nothing to change the expectations attached to particular gender identifications" (p. 217). Even if gender is simply a matter of performance, people will not experience life as the other gender or class does because they have to indicate which category they are impersonating via conventions, and thus tend to experience cliché, and reaction to cliché, rather than normal complexity.

On MOOs (MUD [multiuser domain] object oriented), where Netsex can be important in reducing the ambiguities of presence and sustaining relationships, most women and men are adorned with an excess of the symbolism and roles of the gender and sexual discourse they participate within, and this may reinforce ideals of gender difference (Marshall, 2003). This seems to be the case even when people portray themselves as nonhuman. As an example of this supposed variance, McRae (1996) quotes a player on a kind of MOO in which people present themselves as anthropomorphic animals, saying there is a form of sex in which "the submissive partner is eaten at climax ... [B]ears and wolves are usually dominant. Foxes are sorta generally lecherous. Elves are sexless and annoyingly clever. Small animals are often very submissive" (p. 248). Even here, the relationship of size, bulk, aggression, and strength to dominance is not far from conventional constructions of male and female.

This requirement to indicate gender by conventional referents may also lead people to portray their off-line gender in conventional terms as well. Clark (1998) notes this clichéd gender emphasis in her study of online teenage dating, while Herring (2000) writes that she "found that nearly 90% of all gendered behavior in six IRC [Internet relay chat] channels indexed maleness and femaleness in traditional, even stereotyped ways; instances of gender switching constituted less than half of the remaining 10%." Conventions can also provide debate on women-only groups, where people can only be identified as female by their feminine behaviour unless they are checked by known links off-line.

As it is possible to ignore the gender of those who contradict our expectations of gender, those expectations may grow stronger for not being challenged.

BACKGROUND

Betrayal

In tandem with ideas of identity flexibility is the narrative that online life is full of cross-gender impersonation, and that as a result, interactions are potentially hedged with betrayal and disillusion (cf. Kolko & Reid, 1998).

These common narratives apparently contradict each other. If gender is so unimportant to online life, then why is impersonation such a source of anxiety and distress? The problem arises partly because life online is not separated from life off-line, and people commonly act as if the online needs verification by the off-line, where gender is important to the ways that people relate and the expectations people have of each other. These verification patterns are usually asymmetric; the off-line, which is more private or hidden, will usually be assumed to be true if it contradicts the online or more public sphere.

In general, it is the impersonation of women by men that causes anxiety, not the other way round, irrespective of whether the impersonation occurs amongst males or females, and this needs to be explained. Innocence of intention is rarely presumed. In a famous case, Van Gelder (1996) describes a male who was once mistaken for a woman by another woman who was "open in a way that stunned him." He then deliberately embraced a female persona, became intimate with many women, and helped them with their problems. When this was discovered, some of the women involved considered it violative. Van Gelder asks "why a man has to put on electronic drag to experience intimacy, trust and

sharing" (p. 546). However, this performer had constructed an elaborate fictional biography and engaged in Netsex with others before marrying an equally fictional off-line husband. Eventually the man, as himself, tried to make friends with his character's friends but failed. After the truth came out, only a few friendships carried over, with at least some of those who remained friends trying to see the similarities between the fiction and the persona (Stone, 1995). The case indicates that people may not wish to engage in intimate contact with a person who is not as they present themselves, contrary to the voluntarist or postmodernist argument.

Conventions of Identification

People online seem largely confident that they can identify a person's real off-line gender by that person's habits or styles of conversation. These identifiers are derived from customs based on off-line gender expectations. One such expectation is that women tend to use lots of emoticons to indicate or express emotional states as Western English-speaking female discourse is supposed to be more emotional, or deferential, than that of males, and some research has found greater emoticon use to occur (Witmer & Katzman, 1998). Aggressive or argumentative behaviour is usually considered a mark of masculinity.

Other methods of divining gender include discussing the kinds of things an off-line person of that gender could be expected to know. This often translates into some kind of product knowledge, such as panty-hose or ring sizes, for example (Suler, 1999), and does not always translate well across different cultures. Some reports indicate that people will read books on gender differences in speech use, either to improve their ability to identify a person's gender or to impersonate the other gender better (Wright, 2000).

When people apply gender-neutral pronouns to themselves, for example, in spivak, other people will not generally assume that they really are gender neutral or that their gender is a matter for privacy, and they will often try and find out what their real gender is (Kendall, 1996). Some researchers have reported that people who maintain vagueness about their real gender are "generally 'dropped' from the interaction" (O'Brien, 1999, p. 90). So, there can be punishment for being vague, as well as for being found out.

In general, people work hard at detecting the gender of others, and it is worthwhile to ask what kind of circumstances lead people to become concerned.

Frequency and Place

People generally assume that cross-gender impersonation is common. Elias (1997) reports that unnamed "experts estimate two out of three 'women' in many chat rooms, particularly the sex-oriented ones, are men" (Curtis, 1996).

It seems that impersonation occurs more frequently in MOOs, IRC, and chat rooms where people are identified by a character or avatar name than it does on mailing lists or in newsgroups where they are identified by e-mail addresses. Even here, it probably varies with the situation. It is perhaps most common in gaming, when differently gendered characters can have different kinds of, usually clichéd, advantages that are either built into the game, or arise from social factors, such as female characters receiving more help from male characters at lower levels, something that may not always be maintained at higher levels.

Herring (2000) writes that "claims of widespread gender anonymity have not been supported by research on online interaction." A major survey of online sex behaviour reported that "[o]nly 5% of the sample indicated pretending to be a different gender, and most of them (4%) said that they do so only occasionally" (Cooper, Scherer, Boies, & Gordon, 1999). Even on MOOs where impersonation might be expected to be common, Deuel (1996) writes, "Evidence suggests that most MOO participants represent themselves as their true gender or as neuter, with only a small percentage of players actually attempting to conceal or intentionally misrepresent their gender" (p. 133).

Survey data and conversations gathered during the author's own fieldwork suggested that gender impersonation did not have a widespread, long-term appeal. This is not to deny that some people may do it frequently, but to suggest that most people do not do it deliberately for prolonged periods of time. Sometimes, it does arise that the gender of a person can cause surprise. In my fieldwork site, the gender of one person was frequently mistaken, probably because of his gentleness, his responsiveness, and his

habit of typing emotions (e.g., *smile*), but he never masqueraded as female or as genderless, and indeed was commonly explicit about his gender. This implies that, for some people, reading the conventions of gender overrode his remarks on the subject. During my fieldwork, no one ever expressed surprise at discovering someone they thought was male was female.

There are some situations in which gender may be played down by people. For example, on a specially set-up list for 75 college students, "women tended to mask their gender with their pseudonym choice while males did not" (Jaffe, Lee, Huang, & Oshagan, 1995). This may be because it saved them from inequitable patterns of gendered interaction and harassment. However, it also destroys the capacity for a politics of gender.

Authenticity

If it is so rare, and people are so confident they can detect differences, why does gender impersonation cause so much anxiety?

It suggests that the true off-line gender of the other person is important for framing communication. It adds background that enables text to be read, suggests the possible styles of communication that can be engaged in, and enables people to know what is appropriate to say. Contexts that increase this importance should be investigated.

From the stories told, gender importance increases when the interaction has been dyadic, when intimacy has been invoked, and when the supposedly female person is found to be male. This arises as in the off-line world, most (nonhomosexual) intimate, emotional, or open relationships involve at least one woman (Cheal, 1988). Male bonding has become almost suspect, particularly if it involves intimacy. Finding out that a person you were intimate with was not female when you thought so almost automatically changes the relationship from the realm of intimate and private into a kind of public betrayal. Our private role and its vulnerabilities have broken into the public male domain.

Thus, people try to find out the truth about others with whom they interact intimately online. One of the few ways in which authenticity and truth can be shown is by reference to the body and its inchoate, underlying feeling nature. Thus, it is often the case that aggression is taken as more real, or as revealing more, than politeness, which is supposedly more distant from this primal nature. Not everyone has to follow the full problematics of authenticity, but it seems common, and people strive to find out the reality of others by reference to off-line or bodily factors, of which gender is one of the most important (Marshall, 2004).

Although people often claim that gender constructions should not detract from expressions of the authentic self, in practise, gender is a prime way of categorising others, determining what to expect from them, and allowing the interpretation of authenticity, so it can rarely be discarded, particularly when discourse shifts into the private or intimate realm.

The use of authenticity also emerges as it is common for people to see the use of fake gender as expressing some repressed, and therefore true, homosexuality. This is also a convention; homosexuality may have nothing to do with gender ambiguity. Curtis (1996) writes, "Some MUD players have suggested to me that such transvestite flirts are perhaps acting out their own (latent or otherwise) homosexual urges or fantasies" (p. 128). In an interview, Cooper (1996) states,

I'd rather see [impersonators] struggle with their sexual identity issues directly and get some clarification ... I'd rather they decide that they are gay or transsexual or transgendered and resolve their feelings one way or another so that they could become comfortable sharing their sexuality with a live partner.

Such views imply that gender impersonation is a psychological problem rather than a sociological phenomenon, and that the Internet allows people the delusion of escaping from their problems rather than facing them.

Performance

As should be clear, there is a large stream of either positive or negative constructivism popularly associated with Internet behaviour. Perhaps this is similar to U.S. tradition that holds that positive thinking can change your being, or

[t]hat freedom of choice includes everything: profession, family, religion, sexual preference, and above all the ability to change any of the options (in effect to rewrite one's life story) at almost any time. Admittedly, for many Americans this ultimate freedom is not available. But the ideal remains, and it is the ideal of a network culture. (Bolter, 1991, p. 233)

However, these ideas are more usually anchored in mentions of Foucault or Judith Butler than in references to Norman Vincent Peale. From the evidence, it is possible to be dubious as to the amount of reconstruction that occurs, and it seems probable that rather than expressing multiple selves, people tend to use authenticity and the idea of a hidden true self as their moral and interpretive guide to self and others. Finding out about the true self of the other becomes a problem that must be solved rather than a vagueness to be celebrated (Slater, 1998).

Following Butler (1990), gender may be considered as a performance and thus unstable. However, a performance can also be a manifestation of something held to be authentic and essentially true; it should not automatically imply fictiveness or liberated play. There is no reason why a performance should necessarily challenge convention. An actor or a musician is not entirely free to do as he or she chooses; even if there is no script or score, he or she must follow some kind of convention for the performance to achieve results. The same is true, as argued earlier, for gender.

Ideas around performance also seem confused in that a performative statement that creates what it announces (i.e., "You are now man and wife") is officially not a performance, and this is precisely because a known performance is taken as not real. The fact that movies may feature weddings does not undermine the idea of weddings themselves.

FUTURE TRENDS AND CONCLUSION

It is hard to predict the future, but it is probable that gender will continue to be of importance in online identification, that people will persist in trying to discover the authentic gender identities of those with whom they are becoming close, and that gender impersonation will still only have limited fields and places of application online, just as it does off-line.

REFERENCES

Bolter, J. D. (1991). *Writing space: The computer, hypertext, and the history of writing.* Hillsdale, NJ: Erlbaum.

Bruckman, A. (1993). Gender swapping on the Internet. In *Proceedings of INET'93.* Retrieved February 20, 2006, from http://www.cc.gatech.edu/elc/papers/bruckman/gender-swapping-bruckman.pdf

Butler, J. (1990). *Gender trouble: Feminism and the subversion of identity.* New York: Routledge.

Cheal, D. (1988). *The gift economy.* London: Routledge.

Clark, L. (1998). Dating on the Net: Teens and the rise of "pure" relationships. In S. Jones (Ed.), *Cybersociety: Revisiting computer-mediated communication and community* (pp. 159-183). Thousand Oaks: Sage.

Cooper, A. (1996). Intimacy and the Internet. *Contemporary Sexuality, 30*(9). Retrieved February 20, 2006, from http://web.archive.org/web/20010405172616/http://sex-centre.com/Internetsex_Folder/Intro_Internt_sex

Cooper, A., Scherer, C. R., Boies, S. C., & Gordon, B. L. (1999). Sexuality on the Internet: From sexual exploration to pathological expression. *Professional Psychology Research and Practice, 30*(2).

Curtis, P. (1997). MUDding: Social phenomena in text-based virtual realities. In S. Kiesler (Ed.), *Culture of the Internet.* Mahwah, NJ: Erlbaum.

Danet. (1998). Text as mask: Gender, play and performance. In S. Jones (Ed.), *Cybersociety 2.0* (pp. 129-158). Thousand Oaks, CA: Sage.

Deuel, N. (1996). Our passionate response to virtual reality. In S. Herring (Ed.), *Computer mediated communication: Linguistic and cross-cultural perspectives.* Amsterdam: John Benjamin.

Elias, M. (1997). Some find love online, but cyberdating has its risks. *USA TODAY.* Retrieved August 15, 2001, from http://www.usatoday.com/life/cyber/tech/ctb077.htm

Herring, S. (2000). Gender differences in CMC: Findings and implications. *CPSR Newsletter, 18*(1). Retrieved February 20, 2006, from http://www.cpsr.org/publications/newsletters/issues/2000/Winter2000/herring.html

Jaffe, J. M., Lee, Y.-E., Huang, L.-N., & Oshagan, H. (1995). *Gender, pseudonyms, and CMC: Masking identities and baring souls.* Retrieved from http://research.haifa.ac.il/~jmjaffe/genderpseudocmc/

Kendall, L. (1996). MUDder? I hardly know 'er! Adventures of a feminist MUDder. In L. Cherny & E. R. Weise (Eds.), *Wired women.* Seattle, WA: Seal Press.

Kolko, B., & Reid, E. (1998). Dissolution and fragmentation: Problems in online communities. In S. Jones (Ed.), *Cybersociety 2.0.* Thousand Oaks, CA: Sage.

Marshall, J. (2004a). The online body breaks out? Asence, ghosts, cyborgs, gender, polarity and politics. *Fibreculture Journal, 3.* Retrieved February 20, 2006, from http://www.journal.fibreculture.org/issue3/issue3_marshall.html

Marshall, J. (2004b). The sexual lives of cyber-savants. *The Australian Journal of Anthropology, 14*(3), 229-248.

McRae, S. (1996). Coming apart at the seams: Sex, text and the virtual body. In L. Cherney & E. R. Wiese (Eds.), *Wired women* (pp. 242-263). Seattle, WA: Seal Press.

O'Brien, J. (1999, April 19). Writing in the body: Gender (re)production in online interaction. In M. A. Smith & P. Kollack (Eds.), *Communities in Cyberspace* (pp. 76-104). London: Routledge.

Poster, M. (1997). Cyberdemocracy. In D. Holmes (Ed.), *Virtual politics: Identity and community in cyberspace.* Thousand Oaks, CA: Sage.

Schaap, F. (1999). *Males say "blue," females say "aqua," "sapphire," and "dark navy." The importance of gender in computer-mediated com-*

munication. Retrieved January 2002, from http://www.fragment.nl/texts/males_say_blue.txt

Slater, D. (1998). Trading sexpics on IRC: Embodiment and authenticity on the Internet. *Body & Society, 4*(4), 91-117.

Stone, A. R. (1995). *The war of desire and technology at the close of the mechanical age*. Cambridge, MA: MIT Press.

Suler, J. (1999). *Do boys (and girls) just wanna have fun? Gender-switching in cyberspace (v2.5)*. Retrieved February 20, 2006, from http://www.rider.edu/~suler/psycyber/genderswap.html

Taylor, C. (1991). *The ethics of authenticity*. Cambridge, MA: Harvard.

Trilling, L. (1974). *Sincerity and authenticity*. Oxford: OUP.

Turkle, S. (1995). *Life on screen: Identity in the age of the Internet*. New York: Simon and Schuster.

Van Gelder, L. (1996). The strange case of the electronic lover. In R. Kling (Ed.), *Computerization and controversy: Value conflicts and social choices* (2nd ed., pp. 533-546). San Diego, CA: Academic Press.

Witmer, D., & Katzman. (1998). Smile when you say that: Graphic accents as gender markers in computer mediated communication. In F. Sudweeks, M. McLaughlin, & S. Rafeli (Eds.), *Network and Netplay: Virtual groups on the Internet* (pp. 3-11). Menlo Park: AAAI Press.

Wright, K. (2000). *Gender bending in games*. Retrieved December, 2001, from http://www.womengamers.com/articles/gender.html

KEY TERMS

Authenticity: The convention whereby it is assumed that there is an underlying truth to people's actions, that they should, ideally, display this truth, and that this truth is uncoverable (see Taylor, 1991; Trilling, 1974).

Avatar: A graphic representation of a person's character in a multiplayer computer game or environment. The term is often used to refer to the character itself.

Emoticon: An icon depicting emotions. The term is usually used to refer to what are called smileys, which are glyphs that are read sideways. Thus, :-) represents the eyes, nose, and mouth of a face on its side. The form most usually deployed loses the nose, becoming :). Often, the term is extended to cover other textual expressions of emotion or bodily states, such as <OWWWW!!!>.

IRC (Internet Relay Chat): A way of communicating with others on the Internet. It requires that users have a software client that allows them to connect. An IRC channel is something like a chat room except that it is not necessarily based on a particular computer and is not controlled by people who are not part of the group.

Mailing List: A way of communicating over the Internet as a group. The list is set up on a particular computer, and accepted mail written to that list is delivered to all the people registered with that list. Communication tends to be asynchronous, and people will typically receive communications amidst the rest of their e-mail.

MOO (MUD Object Oriented): A way of communicating with others over the Internet. It could be thought of as an interconnected set of spaces set up on a single computer. Communication tends to be synchronous. The term is here being used to cover the whole family of such setups descended from the original MUDs, where a MUD is a multiuser dungeon or multiuser domain. A person using one of these Internet forums logs in and plays a specific named character, usually with a text description, and can write or accumulate programmed objects, rooms, or devices, which add to the environment. This means that he or she has some prop support for playing the character.

Netsex: Sexual activity carried out over an Internet communication channel, usually by exchanging written descriptions of sexual behaviour.

Newsgroup: A way of communicating over the Internet. A newsgroup is generally a distributed form of communication resembling a message board but not sited on a particular computer. As such, it is generally difficult to exclude mails from newsgroups, and many were rendered unreadable by the number of advertising posts received. The software used to read newsgroups varies, but they commonly used to organise the group by threads.

Spivak: Speaking spivak means a person uses the pronouns *e*, *em*, *eir*, *eirs*, and *eirself* instead of *I*, *he* or *she*, *hers* or *his*, and so forth. In some forums, this can be set into the program so that the conver-sion occurs automatically. For the origins of spivak on LambdaMOO, see Danet (1998).

Synchronous/Asynchronous: Terms used to describe the time factor of messages. A conversation is synchronous when it is conducted with the text and response being relatively close together. A conversation that is asynchronous is more likely to resemble an exchange of letters as the text and its response are separated in time.

Online Life and Netsex or Cybersex

Jonathan Marshall
University of Technology Sydney, Australia

INTRODUCTION

Netsex, or cybersex, may be thought of as the mutual textual simulation, or narration, of sexual activity between people online. Branwyn (1993, p. 786) divides Netsex into three different types. First is that in which people "describe and embellish real-world circumstances" such as touching themselves, taking their clothes off, and so on. They may or may not be performing these actions, but probably not if they are typing reasonably steadily. The second type involves "a pure fantasy scenario" in which people jointly create a story with relatively coherent expectations. This can be performed before an audience. The third type involves one party giving instructions to another who supposedly performs them. These techniques involve textual references to sexually charged notions of gender (anatomy, actions, clothing, and so on), which are frequently exaggerated to fit the story.

It is now possible to transmit real-time video pictures from a camera attached to a person's computer, and this may also be used for Netsex. However, people often express ambivalence about this, perhaps because it emphasises the distance between people, is not as mutually intense, or because it increases possible disjunctions. The disruption of expectations of narrative in Netsex is often a source of online humour. A final, often mentioned, but currently fictional form of Netsex is virtual teledildonics, in which a complete sensory field is simulated via electronics.

BACKGROUND

Prevalence of Netsex

Netsex is reputedly quite common. Hamman (1997) states that in his experience, about half of all AOL (America Online) chat rooms "have sex related names," and he believes that a large number of AOL users, if not the majority of them, "have at least experimented with having cybersex" (Hamman, 1996). Sannicolas (1997) looked at the chat rooms open on the Microsoft Network (MSN) over a 2-week period and discovered that in the nonregistered rooms, "an average of 98 (21.2%) listed sexual topics." A survey posted to an MSN notice board gained over 9,000 responses and revealed that 45% of respondents claimed to spend over an hour a week on "sexually related activities," with more than 7% reporting they spent 11 or more hours a week on those activities (Cooper, Scherer, Boies, & Gordon, 1999).

Netsex shades into online dating and into using the Internet to meet potential partners (Elias, 1997; Olson, 1999-2000). Obviously, the intensity of any person's use of the Internet for sexual or romantic purposes may vary over their life online.

Explanations for Netsex

Netsex has been explained in terms of sex drive, male dominance, difficulties in finding partners, psychoanalytic projection, addiction, and liberation, as will be described below.

Common motives are that technological development is driven by sex, people will use new technology for sex, and the easier the technology is to use, then the quicker it will be embraced for sex (Dery, 1996). Such theories seem exaggerated, assume a ubiquity and uniformity of sex drives, and have little to do with the actual uses or forms of Netsex as actually employed between people.

Sometimes the prevalence of sex online is explained as the way the Net is marked as a male domain. In this view, women are excluded and harassed by sex talk, and by males trying to pick them up or render them sexual beings alone. Sherman (1995) has even suggested that the publicity given to harassment online was a deliberate attempt on the part of the male-controlled media to frighten women away from the Net. However, harassment is not uncommon for women on the Internet, particularly in chat environments (see Brail, 1997; Branwyn, 1993), and males can also feel harassed by female demands for online sex (Tober, 1995). One of the most famous tales of online life is Julian Dibbell's "A Rape in Cyberspace" (1999), in which the takeover of a woman's avatar and her sense of self is described and given the almost mythic function of originating formal social control and civilisation.

However, online groups may disapprove of sexual harassment and clearly distinguish it from Netsex, and it seems that women participate in Netsex with as much enthusiasm or ambiguity as men (Marshall, 2003). Given this, it is necessary to separate Netsex, a usually private, mutual activity, from the use of sex talk and harassment in public.

Netsex and online pairing is often explained by claiming that people find it extremely difficult to meet potential sexual partners in modern Western society. Albright and Conran (n.d.) write, "Online communities accelerate the expansion of opportunities for relationships begun by personals, video dating, and telephone chat line," while Hamman (1997) claims that a "lot of these people are isolated, either geographically [or] socially." In this view, Netsex can be seen as part of wider social processes that have resulted in increasing isolation, with the suggestion that pairing and sexuality are extremely important in the construction of gender and self-identity (particularly as sexual activity is usually gendered), and in ensuring survival in the contemporary world.

In slight contradiction to this theory of meeting potential partners, it seems that many such affairs involve people who are married and claim their marriages are happy. Over half the cyberromance stories commented upon by Vixen (n.d.) involved at least one married person. Olson (1999-2000) claims that in his survey, despite almost half his respondents claiming to have been in love online and one third of them dating people met online, over 70% claimed to be married (Olson, 1999-2000). The Internet can be seen as a safe place to have affairs and thus even to save marriages as a result (Ben-Ze'ev, 2004).

The prevalence and intensity of sex online can be explained by the supposed blankness of the computer screen, psychoanalytic projection or transference, and the easy activation of fantasy (Albright & Conran, n.d.; Bednarcyk, 1994; Elias, 1997; Hamman,

n.d.; Odzer, 1997; Vixen, n.d.). This explanation can be formulated in terms of escape as when Hering (1994) writes, "Simply stated, the internet is a place where men, women, and children can exercise their fantasies, as well as escape the realities of their boring and pathetic lives; or maybe they're escaping their exciting and overly burdening lives." As it stands, this explanation is little more than a restatement of what is observed: that people often find it easy to have online sex and that it is powerful. We still do not know why projection should so easily take a sexual form, and fantasy is part of the standard explanation for why online sexual relationships, which may sustain people for years, often fail on meeting off-line (Adamse & Motta, 1996; Hamman, n.d.).

A medical-like model phrases Internet usage in terms of addiction, with Netsex seen as either reinforcing this addiction or as a special subcategory of addiction. Cooper, Scherer et al. (1999, p. 154) state, "The first position to emerge was that Internet sexuality is pathological." One therapist quoted by Shachtman (2000) claimed that "cybersex is the crack-cocaine of sex addiction." Online therapist Kimberly Young (n.d.-a) argues that "Cybersexual addiction has become a specific sub-type of Internet addiction" and estimates that 1 in 5 Internet addicts are engaged in online sexual activity. Young (n.d.-a) goes on to remark on the rapid engagement of those with "no prior criminal or psychiatric history" in such behaviour (see also Delmonico, 1997). This is pathologising with a vengeance and guides attention away from events into morals.

Cooper, Putnam, Planchon, and Boise (1999, p. 77) distinguish three types of users of Netsex. First, there are the "recreational or nonpathological users." Second, there are "[i]ndividuals who exhibit sexually compulsive traits and experience a fair amount of trouble in their lives." Third, there are users without histories of sexual compulsivity, but whose "online sexual pursuits have caused problems in their lives" (p. 80). The latter group is held to contain "depressives" who withdraw from off-line social interaction, and "stress reactive types" who use Internet sex to cope with stress or to escape from certain feelings. For another division, see Leiblum (1997). Ferree (2003) claims that women are overrepresented among Internet sex addicts, basing her claims almost entirely on Cooper, Scherer, et al.'s (1999) MSN survey. While it would be foolish to deny people can become addicted to Netsex, it does not seem realistic to make addiction the prime explanation.

In contrast, other views suggest that Netsex is therapeutic or liberatory. Hamman (1996) writes that his informant Rebecca feels that "cybersex has allowed her to become more comfortable with her body as well as her own sexuality." Cooper and Sportolari (1997) write, "When usually hidden parts of the self are seen and accepted by other(s) the experience can be healing, allowing for the gradual integration of that split-off part of the self into the overall personality" (see also Adamse & Motta, 1996). Likewise, online intimacy supposedly frees people from the tyranny of physical appearance. Wallace (1999) describes some research on the importance of physical attractiveness in relationships off-line and its lack of effect online. However, if people are freed from the necessity of looking attractive, attraction can still be increased by fluent writing styles, and can become important again when people meet.

These therapeutic arguments suggest that the external society is repressive of either sex or vital aspects of the personality, and that online life is a compensation or relief valve for this suppression (Hamman, 1996). This liberation may be seen as gendered: According to an MSNBC (1998) article, the psychologist Marlene Maheu said,

Women online are not constrained by the traditional mores we have to struggle with. Since half the sexual population is females, it would make sense that given a chance to speak up, this repressed minority would have something to say.

In other ways, this liberation may also make people nervous as it also liberates those whose sexual behaviour may not be approved and allows them to team up with others of the same ilk. Wallace (1999), writing about online groups and who apparently focused on more extreme sexual behaviours, states that there has been little research to "determine who actually participates, what their motives are and how the availability of these anonymous and introduced outlets for consensual deviance affects them" (p. 155).

Some writers claim that Netsex is safe sex, without the risks of disease, violence, or pregnancy

(Hamman, 1997; Sagan, 1995; Sannicolas, 1997), and that this allows people to express themselves with greater freedom (Benedikt, 1995).

Although one set of dangers may decrease, this does not prevent people fearing other dangers such as falling for a fantasy, being uncertain of a partner's gender, or being subject to electronic harassment or to discredit in their online life through having pictures or transcripts posted to family or employers. People seem to find Netsex as fraught or complicated as off-line sex. In fact, complexity and failure might increase because of the absence of immediate and verifiable feedback (Marshall, 2003).

Potential anonymity shares in this ambivalence. Sometimes, in keeping with the liberation or therapeutic argument, removing constraining social roles and appearances is thought to be beneficial, thus enhancing

other factors such as propinquity, rapport, similarity, and mutual self-disclosure, thus promoting erotic connections that stem from emotional intimacy rather than lustful attraction ... (Cooper & Sportolari, 1997)

People are also held to be free to develop without fear of social consequences (Branwyn, 1993).

On the other hand, anonymity is supposed to increase the ease of irresponsibility, of fraud and deceit, and of socially reprehensible behaviour. Schnarch (1997) argues that anonymity may prevent depth of both self-knowledge and knowledge of the other. However, people generally attempt to work at reducing the anonymity of those they interact with; the problems of identity and gender become problems to be solved, not to be suspended.

Gendered Responses to Netsex

Without ignoring harassment, there is surprisingly little discourse about gender-based differences in the use of Netsex. Young (n.d.-b) writes,

Men tend to seek out dominance and sexual fantasy on-line, while women seek out close friendships, romantic partners, and prefer anonymous communication in which to hide their appearance ... As men tended to look more for

Cybersex, women tended to look more for romance in Cyberspace ... I should note that it is not unusual for women to engage in random Cybersex, but many times they preferred to form some type of relationship prior to sexual chat.

In Cooper, Scherer et al.'s (1999) study of Netsex on MSN, it appeared that:

men most prefer[red] Web sites featuring visual erotica (50% men to 23% women) ... and women favor[ed] chat rooms (49% women to 23% men) ... It is significant that 51% of women reported they never download sexual material. Women, on-line as elsewhere, prefer more interaction and the development of relationships.

Ferree (2003) agrees with both these claims, adding that women are more likely to seek off-line contact. This is an area that needs further research.

CONCLUSION

The positions described above oscillate around issues of liberation or escape, safety or danger, conflicting realities of the off-line and online, affection vs. distance, therapy vs. dissipation, certainty vs. uncertainty, and anonymity vs. authenticity. Usually one of these factors is selected as dominant, though all of them may be involved to a degree, however conflicting. These contradictory interpretations are not just evidence of confusion, but also of the complexities experienced.

Rightly, Cooper and Sportolari (1997) emphasise the conceptually paradoxical nature of the Net: "[O]n the one hand, it seems to epitomize the alienation of the modem world, and yet it also leads to the development of supportive and sometimes intensely intimate, even deeply erotic, relationships." This paradoxical nature cannot be ignored or reduced by saying the Net is either beneficial or not (Civin, 2000). It must be explored.

Explanations should therefore include the interaction between off-line and online lives. Netsex almost certainly is influenced by the difficulties people have in forming successful sexual relationships and pairings off-line, which (whether cross-gendered or intergendered) are important for estab-

lishing their self-identities and survival in the off-line world. However, there are also specific features of online life that need to be considered. In this often-ambiguous environment with weak boundaries, with no markers of presence beyond dialogue and with little elaborated ritual code that can be imported from off-line life, some people might need to fall in love to prove an online relationship actually exists and to sustain the sense of the other's presence. In Netsex, a person can maintain the presence of the other before them via narration and by reference to a commonly available and sustainable bodily reaction. Netsex can be used to restore contact between people when the dialogue slides out of areas of mutual interest. As such, it can be like sharing a drink, or watching a film off-line.

As Western society appears to value authenticity, the conventions of authenticity, especially around intimacy, become part of the way people conduct Netsex. Thus, mistyping can be seen as evidence of genuine excitement, and as Deuel (1996) remarks, preprogrammed MOO (MUD [multiuser domain] object oriented) actions are used less the more that Netsex or the relationship becomes real. However, authenticity generates problems as the use of exaggerated gender symbols to enable the performance of Netsex may also appear to simultaneously delete the presence of real gender or a real self, which might be expressed in uncertainties and hesitations. Given that people seem to fear they could be falling for fantasy images, they hence need to bring the relationship into the off-line reality to check its truth. If the Internet has been liberating them from pressures, at this moment the constraints of off-line life appear again. This makes online relationships fraught, and the person becomes caught in a contradiction between an intimacy that is supposedly only confirmed off-line, and an equally supposed ability to only be who they really are online. Again, the contradictions are a vital part of the experience.

FUTURE TRENDS

It is likely that Netsex will continue to be an ambiguous and multifaceted feature of online life. Research needs to focus on its complexities, the ways it functions in different kinds of groups, the ways that people relate it to their daily lives, and the way it is used to express, subvert, and constrain gender. This may well differ with the cultural bases of the people involved.

REFERENCES

Adamse, M., & Motta, S. (1996). *Online friendship, chat-room romance and cybersex.* Deerfield Beach, FL: Health Communications.

Albright, J., & Conran, T. (n.d.). *Online love: Sex, gender and relationships in cyberspace.* Retrieved February 15, 2006, from http://www.eff.org/pub/Net_culture/Virtual_community/online_love.article

Bednarcyk, M. (1994). Prince princess syndrome: How to avoid pitfalls in Net relationships. *The Arachnet Electronic Journal on Virtual Culture, 2*(3). Retrieved February 15, 2006, from http://www.mith2.umd.edu/WomensStudies/Computing/Articles+Resea rchPapers/ArachnetJournal/netlove

Benedikt, C. L. (1995). Tinysex is safe sex. *Infobahn Magazine, 1*(1), 13-14.

Ben-Ze'ev, A. (2004). *Love online: Emotions on the Internet.* Cambridge: Cambridge University Press (CUP).

Brail, S. (1997). Take back the Net! In G. E. Hawisher & C. L. Selfe (Eds.), *Literacy, technology, and society* (pp. 419-423). NJ: Prentice-Hall.

Branwyn, G. (1993). Compu-sex: Erotica for cybernauts. *South Atlantic Quarterly, 92*(4), 779-791.

Civin, M. (2000). *Male, female, email.* New York: Other Press.

Cooper, A., Putnam, D., Planchon, L. A., & Boise, S. C. (1999). Online sexual compulsivity: Getting tangled in the Net. *Sexual Addiction and Compulsivity, 6*(2), 79-104.

Cooper, A., Scherer, C. R., Boies, S. C., & Gordon, B. L. (1999). Sexuality on the Internet: From sexual exploration to pathological expression. *Professional Psychology Research and Practice, 30*(2), 154-164.

Cooper, A., & Sportolari, L. (1997). Romance in cyberspace: Understanding online attraction. *Journal of Sex Education and Therapy, 22*(1), 7-14. Retrieved February 12, 2001, from http://www.sexcentre.com/index2/internet_and_sex.htm

Delmonico, D. L. (1997). Cybersex: High tech sex addiction. *Sexual Addiction and Compulsivity: The Journal of Treatment and Prevention, 4*(2), 159-167.

Dery, M. (1996). *Escape velocity: Cyberculture at the end of the century.* New York: Grove Press.

Deuel, N. (1996). Our passionate response to virtual reality. In S. Herring (Ed.), *Computer mediated communication: Linguistic and cross-cultural perspectives* (pp. 129-146). Amsterdam: John Benjamin.

Dibbell, J. (1999). *My tiny life.* London: Fourth Estate.

Elias, M. (1997). Some find love online, but cyberdating has its risks. *USA TODAY.* Retrieved August 15, 2001, from http://www.usatoday.com/life/cyber/tech/ctb077.htm

Ferree, M. (2003). Women and the Web: Cybersex activity and implications. *Sexual and Relationship Therapy, 8*(3), 385-393.

Hamman, R. (n.d.). *The role of fantasy in the construction of the on-line other: A selection of interviews and participant observations from cyberspace.* Retrieved February 15, 2006, from http://www.socio.demon.co.uk/fantasy.html

Hamman, R. (1996). *Cyborgasms: Cybersex amongst multiple-selves and cyborgs in the narrow-bandwidth space of America Online chat rooms.* Master's thesis, University of Essex, Colchester, UK. Retrieved February 15, 2006, from http://www.socio.demon.co.uk/Cyborgasms_old.html

Hamman, R. (1997). Interview with Ken Garber on South Africa's Radio 702 Internet show. *Cybersociology, 1.* Retrieved February 15, 2006, from http://www.socio.demon.co.uk/702transcript.html

O

Hering, P. (1994). Internet romances don't work. *The Arachnet Electronic Journal on Virtual Culture, 2*(3). Retrieved February 15, 2006, from http://www.mith2.umd.edu/WomensStudies/Computing/Articles+ResearchPapers/ArachnetJournal/netlove

Leiblum, S. R. (1997). Sex and the Net: Clinical implications. *Journal of Sex Education and Therapy, 2*, 21-28.

Marshall, J. (2003). The sexual lives of cyber-savants. *The Australian Journal of Anthropology, 14*(3), 229-248.

MSNBC. (1998, June 9). *Modem love: Sex and the Net.* Retrieved February 15, 2006, from http://news.zdnet.com/2100-9595_22-510708.html

Odzer, C. (1997). *Virtual spaces: Sex and the cyber citizen.* New York: Berkely Books.

Olson, M. V. (1999-2000). *Internet: Behavioral impacts. How online impacts the lives of users: Findings.* Retrieved February 15, 2006, from http://members.tripod.com/martyman53/survey/impacts.htm

Sagan, D. (1995). Sex, lies and cyberspace. *Wired, 3.01*, 78-83.

Sannicolas, N. (1997). Erving Goffman, dramaturgy, and on-line relationships. *Cybersociology, 1.* Retrieved February 15, 2006, from http://www.socio.demon.co.uk/magazine/1/is1nikki.html

Schnarch, D. M. (1997). Sex, intimacy, and the Internet. *Journal of Sex Education and Therapy, 22*(11), 15-20.

Shachtman, N. (2000). NY gets hip to cybersex woes. *Hotwired.* Retrieved February 15, 2006, from http://www.wired.com/news/culture/0,1284,36289,00.html

Sherman, E. (1995). Claiming cyberspace: Five myths that are keeping women off-line. *Ms, 7*(1), 26-29.

Taylor, C. (1991). *The ethics of authenticity.* Cambridge, MA: Harvard University Press.

Tober, B. (1995). Dames distress dudes. *.Net, 3*, 56-57.

Trilling, L. (1974). *Sincerity and authenticity.* Oxford: Oxford University Press (OUP).

Vixen. (n.d.). *Shared stories.* Retrieved February 15, 2006, from http://members.tripod.com/~VixenOne/index.html

Wallace, P. (1999). *The psychology of cyberspace.* Cambridge: Cambridge University Press (CUP).

Young, K. (n.d.-a). *Cybersexual addiction.* Retrieved February 15, 2006, from http://www.netaddiction.com/cybersexual_addiction.htm

Young, K. (n.d.-b). *Men, women, and the Internet: Gender differences.* Retrieved February 15, 2006, from http://www.netaddiction.com/gender.htm

KEY TERMS

AOL (America Online): The largest commercial provider of Internet services. It is now part of Time-Warner.

Authenticity: The conventional assumption that there is an underlying truth to people's actions, that they should, ideally, display this truth, and that this truth is uncoverable (see Taylor, 1991; Trilling, 1974).

Avatar: A graphic representation of a person's character in a multiplayer computer game or environment. By extension, it is often used to refer to the character itself.

Chat Room: An Internet site that contains the participants in a real-time interaction. Those within the chat room can read what each other is typing. People enter the room to participate and can be banned from it. Some suppliers may allow people to temporarily, or permanently, set up their own rooms.

MOO (MUD Object Oriented): A way of communicating with others over the Internet. It could be thought of as an interconnected set of spaces set up on a single computer. Communication tends to be synchronous and unarchived, so it is not usually possible to access conversations that occurred while one was not there. A person using one of these Internet forums logs in and plays a specific named character, usually with a text description, and can write or accumulate programmed objects, rooms, or devices that add to the environment. This means that he of she has some prop support for playing the character.

MSN (Microsoft Network): A commercial supplier of Internet connection, shopping, and chat rooms.

Teledildonics: The hypothesised use of complete virtual reality to have sex at a distance. The imagined form usually involves the participants wearing bodysuits that stimulate areas of the body and convey the results of the actions of one person to the bodysuit of the other.

O

Online Life and Online Bodies

Jonathan Marshall
University of Technology Sydney, Australia

INTRODUCTION

Bodies are often claimed to be irrelevant to online activity. Online space, or activity, is frequently described as if disembodied, and often this absence of visible bodies is said to contribute to freedom from social pressures around gender, race, and body type (Reid, 1996). However, without bodies, people could not access the Internet, and online there are continual references, directly and indirectly, to bodies, so the term disembodied references a particular type of "ghost" body. Therefore, rather than accepting ideas that naturalise dislocating life online from bodies, it is necessary to explore the situations in which this occurs. Another commonly used body metaphor is the cyborg: the melding of human with machine. In both cases, the body is usually taken as underlying what is happening and as a referent for authenticity.

BACKGROUND

Bodily Referents

In online discourse, signs of the body, such as emoticons (or "smileys" as they are also called), written emotions (*kiss*, *hug*, <smile>, or acronyms such as ROFL [roll on the floor laughing]), and so forth are quite common. These emoticons substitute for gestures and seem more usual in informal situations. However, out of the vast range of emoticons depicted in dictionaries of such things, only three are commonly used in the West: the facial representatives of :), the smile, indicating good humour; :(, the frown, indicating sadness or disappointment; and ;), the wink or wry face, an indicator of knowingness or irony, which is sometimes taken as flirtatious. :-), :-(, and ;-) are equivalents. In Japanese communication, it appears that far more signs are used, as well as "innovative punctuation" (Nishimura, 2003). Stories suggest that Western

emoticons were not originally understandable, or easy to use, in Japan (Pollack, 1996), and neither were Japanese emoticons in other Asian cultures (Koda, 2004). As Katsuno and Yano (2002) write, these emoticons (*kaomoji*) reassert "bodily presence on the computer screen thereby readdressing what has often been called the cybernetic condition of 'leaving the meat behind'" (p. 206).

This variation indicates a need to investigate why particular body signs are deployed in particular cultures and abandoned in others. The most obvious problems that Western smileys deal with involve aggression or the resolution of irony (Dibbell, 1994). Folklore claims that the :-) was directly invented by Scott Fahlman (n.d.) after a joke about a physics experiment in a lift was taken seriously. Crystal (2001) also points out that Western emoticons may be simply expressing rapport, or indicating worries about the effect the text might have.

Appropriate use of these textual gestures may not only vary from culture to culture, but from group to group, and part of fitting in and building group identity involves learning to use them appropriately. Some people, mainly Western males, seem to be quite hostile to their use altogether, seeing people who use them as being linguistically lazy.

These indications of emotional or bodily states are commonly associated with gender in that Western women are widely supposed to be more at ease with emotions and use more emoticons (Witmer & Katzman, 1998). Women may also be required or need to use them to express bodily deference or attenuation in dealing with men (Gurak, 1997). Krohn (2004) suggests that along with gender differences, there may also be generational or authoritative boundaries.

Online communication is often hard to resolve; there is no immediate feedback or body language that might frame conversation, and as a result, references to the body off-line are taken as referents for private emotional states and truth states. In support of this, it seems that messages given off-list or off-line (in private) are more likely to be considered authentic than messages received on-list or online. As Kendall (1998) writes, people "privilege offline identity information over information received online ... This allows them to continue to understand identity in the essentialised terms of a persistent and consistent self, grounded in a particular physical body" (p. 130).

As well as the more temporary emotions, moods such as anger, mourning, sex, or so on can be generated by repeated postings of the same type. These moods then act as a way of framing and resolving communication, either deliberately as in Netsex, or sometimes accidentally when a whole group may seem overwhelmed by flame.

History of Ghosts

In the West, the person and his or her typing body are often alone, cramped, and restrained, although there is nothing inevitable about this particular bodily usage of computers. As Haraway (1989) remarks, "our machines are disturbingly lively and we ourselves frighteningly inert" (p. 152). This off-line constraint can contrast strongly with personal boundaries online that can appear fluid as long as there is no pain. Bodies are sometimes described as extended through the wires, which could easily contribute to a sense of disembodiment (Marshall, 2004).

However, the use of metaphors of disembodiment may have some other function or cause, and it is possible this is interconnected with the history of ideas of the nonphysical, the spiritual, or the ghostly, which might in turn be connected with the gendered history of the Net. A common point made in feminist critiques of Western philosophy and ideologies (e.g., Goldenberg, 1990) has been the tendency of male theorists either to denigrate the body and to praise transcendence, or to derive the world from a set of disembodied categories or processes, while simultaneously constructing the female as an inferior, passive, and physical body.

Whatever the precise history of this split, the ghost becomes more ethereal as this process becomes more pronounced. During the 20th century, the boundary was loosened and the ghost became more solid (Finucane, 1982), while now it seems to be ethereal again. Minds are often described in terms of computer programs, and people use these models to suggest that immortality can be achieved by downloading one's mind and memory into a machine (Moravec, 1990). In science fiction, there are many examples of personalities or spirits being active in computers or computer networks. Such theories tend to imply bodies are discardable and not part of our being.

This emphasis on people as minds could also be linked with the constant attempts to characterise the new elite supposedly dealing with immaterial information as knowledge workers rather than physical service workers or the valueless unemployed (who tend to be ghosted from political action). In some ways the idea of the information economy also ghosts, or displaces, the material basis of power.

If such seems plausible, then as the Internet was built and first colonised by Western males seeking dominance or proficiency via the supposed excellence of their mental, creative, or administrative abilities, it might be expected that they used the Internet to emphasise etherealization as part of their construction of their male, or elite, identity. Hall (1996) writes,

Bodyless communication, then, for many men at least is characterised not by a genderless exchange but rather by an exaggeration of cultural conceptions of masculinity—one realised through the textual construction of conversational dominance, sexual harassment, heterosexism, and physical hierarchies. (p. 158)

It could be expected that some women would be apprehensive with these exclusionary constructions, and Taylor and Saarinen (1994) wrote that their female students using e-mail were:

much more uneasy about the "out-of-body" experience they are having than the men. Cynthia and Kaisu are obsessed with email and yet are deeply disturbed by the evaporation of the material and the absence of face-to-face. The men in the class are much less bothered by all of this.

People could also fight against the disembodying through the use of emoticons or the expressiveness of their bodies.

Contradictions arise as the off-line body can be seen as feminine if compared to the online spirit (true expression of authentic being), but as masculine if seen as active rather than ineffective (with computer use as an escape from, or abandonment of, real life). The representations of bodies can depend on the context of invocation.

Deletion can also occur by reference to gender, particularly when it is conceived as a polarity as it then tends to delete aspects of a person's self that they actually share with the contrasting gender. The resultant tendency magnifies the possibility of distancing the body, which can only be grounded by reference to an excess of symbolic gender (making an ideal body), effectively distancing the body even further. There is no pause in which to explore the underlying feelings or the kinesthetic that might be held to render us present. The cyberbody (particularly on a MOO [MUD (multiuser domain) object oriented] or IRC [Internet relay chat]) becomes constructed as a dyad: ghost and irrelevant matter. Sexuality bridges the gap between these separated poles and charges them with their apparent energy. As Reid (1996) writes, "Cyborgs are born of virtual sex. At the moment of orgasm the line between player and character is the most clouded and the most transparent" (p. 341).

Cyborgs

The cyborg, a melding of human and machine, is also a common, although more deliberate, metaphor. As is well known, Donna Haraway (1989) proposed that the cyborg was a way of dealing with the "border" wars between animal and human, human and machine, and the physical and nonphysical, and was postgendered without being seduced by a desire for wholeness. As gender is vital to Western self-regulation and online behaviour, and people often seem regulated by their machines, there is cause to be sceptical of these assertions.

Cyborgs also have histories and origins despite Haraway's (1989) assertion that "the cyborg has no origin story in the Western sense" (p. 150), her dismissal of this descent as "illegitimate" (p. 151), or her arbitrary separating of it from the android or robot (Haraway & Goodeve, 2000). In many novels and films, the line is vague, and Sawdy (2002) argues that early modern mechanical men grew out of a vision of human nature as purified from irrational (feminine?) emotions. The Renaissance cyborg was "a refuge, a place of sanctuary, a hardened carapace into which the battered psyche might flee ... Only to be an engine is to be free" (p. 174). We might also think of the Luddites as protesters against being made slaves to the machines that overrode their own rhythms of work. Nowadays, if we are employed, we are conceivably in constant computerised touch with work; there are no boundaries to employment. As Wood (2002) points out, an automaton can be either a machine simulating the human, or a human who acts like a machine.

Many theorists involve gender explicitly in "cyborgization." Becoming cyborg can not only be seen as a way of avoiding gender, but of reinforcing it or making symbolic armour for a threatened male ego. Cyborgization, when it is not seen as radical, is usually seen as avoidance of the tender or fleshly feminine and hence as an exaggeration of ideas of masculinity and its transcendence of the (gendered) flesh. Bukatman (1993) summarises the ideas of Springer, Foster, and Dery, suggesting that under a system of technological control, where boundaries and command are drawn and imposed from above, men can identify with the machine and release their fear of dissolution in aggression against outsiders. When women are heroes in cyborg movies, they are also hard (Bukatman). Silvio (1999) demonstrates that an apparently radical cyborg anime actually portrays the control of a female body by a male spirit. For various other views and histories of cyborgs, see Gray (1995).

This leads to the question of whether more aggressive Net users see themselves as cyborgs with greater ease than less aggressive, or whether cyborgization primarily functions in Net discourse to indicate one is an insider and deeply implicated in knowing cyberspace.

Cyborg theory, when used to elucidate online life, tends to be so lax in its definition of cyborgization that there is little difference between being a cyborg and tool use. Although this may be useful in deconstructing ideas of the artificial as opposing the natural, it does not seem to help in analysing actual online experience.

For example, one version of this theory implies that computers are tools that enable self-expression that would otherwise be socially impossible:

Without computer mediated communication, Rebecca would be cut off from a part of herself. Without computers, she could not reach her potential as a human. She could not be fully human ... Rebecca is part human, part machine, and without the machine, she would remain only partly human. The boundary between the human and the machine has blurred. Rebecca has become a cyborg. (Hamman, 1996)

It is doubtful that this tells us much more than Rebecca can use a tool to express herself. We could add that this tool probably has effects on this expression—it both enables and restricts—without using the term cyborg. Similarly, the term can be used to refer to human incompleteness, as can be done without its use. Ito (1995) writes, "[T]o borrow Donna Haraway's imagery, I would like to look at mudders as cyborgs that are never whole." Sometimes it can inadvertently lead to separations, as in the following:

Here it is being suggested that email cyborg intimacy is quite different from bodily-based intimacy; that, in fact, the email cyborg self is a very different self from the body-based self. The implication is not that one shouldn't have email affairs, but that email affairs are distinct things from bodily affairs, and that the two should not be confused. (Stratton, 1997)

The cyborg, in this formulation, almost pathologises the very ambiguity that makes the e-mail affair powerful and possible. Likewise, Reid's (1996) assertion that MOO characters, as cyborgs, "redefine gender, identity and the body" (p. 329) claims too much. People can play with identifiers, but ultimately want to know the "reality" when it shifts out of play mode, and their awareness of online and off-line differences and off-line cultural meanings drives their social dynamics.

Research into the actual use of cyborgs as images shows that portrayals of cyborgs on the Internet tend to be thoroughly gendered (DeVoss, 2000), and Wilkerson (1997) writes that "from the standpoint of feminist bisexual identity ... I contend that this [cyborg] myth evades the very issues of race and sexuality which it seems to be addressing" (p. 164) and suggests that "it is vitally important to keep tensions of race and sexuality present rather than to blur the boundaries" (p. 172).

Cyborgization seems to contrast with the discourse of disembodiment, but may be implicated in it. Cybermachines, according to Haraway (1989), are not haunted, yet they appear more lively than us and hence haunted. When she claims that:

[o]ur best machines are made of sunshine; they are all light and clean because they are nothing but signals, electromagnetic waves, a section of a spectrum, and these machines are eminently portable, mobile ... Cyborgs are ether, quintessence. (p. 153)

it suggests that cybermachines are already homes for disembodiment as is implied by the immortalism of Moravec (1990). The connection of the disembodied intellect with etherealization is further suggested by Haraway's admission that her cyborg has no disruptive unconsciousness. She explains this by her reluctance to embrace the totalising explanations of Freudian psychoanalysis as if that included the only theory of the unconscious (Haraway, 1991). In a way, the ghost and the cyborg are related for they both have the tendency to delegate our bodies to the scrap heap and make them secondary to machine transcendence. This may not be supposed to happen, but the question is whether it does or not, not whether it should or should not.

The cyborg and the ghost further intersect in the virtual body of the records kept about a person, which may be treated as more important than the off-line body by corporate or government agencies without the person themselves being aware of it. Cyborg power may always be displaced elsewhere.

CONCLUSION

People continually refer to their bodies and to processes associated with the body in order to frame and resolve communication and to decide appropriate behaviour and responses. Metaphors of ghosts and cyborgs play their own roles in this process, but they are not neutral and feed into previous off-line histories of gender and technology. Rather than on abstract conception, research about life online needs to focus on the culturally specific ways that references to the body are used, and the off-line history of these references.

REFERENCES

Bateson, G. (1972). *Steps to an ecology of mind.* Chandler Publishing.

Bukatman, S. (1993). *Terminal identity: The virtual subject in post-modern science fiction.* Duke University Press.

Crystal, D. (2001). *Language and the Internet.* CUP.

DeVoss, D. (2000). Rereading cyborg(?) women: The visual rhetoric of images of cyborg (and cyber) bodies on the World Wide Web. *Cyber-Psychology and Behaviour, 3*(5), 835-846.

Dibbell, J. (1994, October 4). 2 cute 4 words: In defense of the smiley. *Village Voice.* Retrieved from http://www.juliandibbell.com/texts/smiley.html

Fahlman, S. (n.d.). *Smiley lore :-).* Retrieved February 20, 2006, from http://www-2.cs.cmu.edu/~sef/sefSmiley.htm

Finucane, R. C. (1982). *Appearances of the dead: A cultural history of ghosts.* London: Junction Books.

Goffman, E. (1974). *Frame analysis.* Harper and Row.

Goldenberg, N. R. (1990). *Returning words to flesh: Feminism, psychoanalysis and the resurrection of the body.* Boston: Beacon Press.

Gray, C. H. (Ed). (1995). *The cyborg handbook.* New York: Routledge.

Gurak, L. (1997). *Persuasion and privacy in cyberspace: The online protests over Lotus market place and the clipper chip.* Yale University Press.

Hall, K. (1996). Cyberfeminism. In S. Herring (Ed.), *Computer mediated communication: Linguistic, social and cross-cultural perspectives.* Amsterdam: John Benjamin.

Hamman, R. (1996). *Cyborgasms: Cybersex amongst multiple-selves and cyborgs in the nar-row-bandwidth space of America Online chat rooms.* Master's thesis, University of Essex, Colchester, England. Retrieved February 20, 2006, from http://www.socio.demon.co.uk/Cyborgasms_old.html

Haraway, D. (1989). Cyborg manifesto. In *Simians, cyborgs and women* (pp. 149-181). New York: Routledge.

Haraway, D. (1991). Cyborgs at large: An interview. In C. Penley & A. Ross (Eds.), *Technoculture.* Minneapolis, MN: University of Minneapolis Press.

Haraway, D., & Goodeve, T. N. (2000). *How like a leaf.* New York: Routledge.

Ito, M. (1995). *Cyborg couplings in a multi-user dungeon.* Retrieved February 20, 2006, from http://www.koni.ch/cyborg/couplings.html

Katsuno, H., & Yano, C. R. (2002). Face to face: On-line subjectivity in contemporary Japan. *Asian Studies Review, 26,* 205-232.

Kendall, L. (1998). Meaning and identity in cyberspace: The performance of gender, class, and race online. *Symbolic Interaction, 21*(2), 129-153.

Koda, T. (2004). Interpretation of emotionally expressive characters in an intercultural communication. In *Lecture notes in artificial intelligence: Eighth International Conference on Knowledge-Based Intelligent Information & Engineering Systems (KES2004)* (Vol. 3214, Pt. 2., pp. 862-868). Retrieved February 20, 2006, from http://www.lab7.kuis.kyoto-u.ac.jp/publications/04/koda-kes2004.pdf

Krohn, F. B. (2004). A generational approach to using emoticons as non-verbal communication. *Journal of Technical Writing and Communication, 34*(4), 321-328.

Marshall, J. (2004). The online body breaks out? Asence, ghosts, cyborgs, gender, polarity and politics. *Fibreculture Journal, 3.* Retrieved February 20, 2006, from http://www.journal.fibreculture.org/issue3/issue3_marshall.html

Moravec, H. (1990). *Mind children: The future of robot and human intelligence.* Harvard University Press.

Nishimura, Y. (2003). Linguistic innovations and interactional features of casual online communication in Japanese. *JCMC, 9*(1). Retrieved February 20, 2006, from http://jcmc.indiana.edu/vol9/issue1/nishimura.html

Pollack, A. (1996, August 12). Happy in the East (—) or smiling :-) in the West. *New York Times*, p. 5.

Reid, E. (1996). Text based virtual realities: Identity and the cyborg body. In P. Ludlow (Ed.), *High noon on the electronic frontier* (pp. 327-345). Cambridge, MA: MIT Press.

Sale, K. (1995). *Rebels against the future.* Reading, MA: Addison Wesley.

Sawdy, J. (2002). "Forms such as never were in nature": The Renaissance cyborg. In Fudge, Gilber, & Wiseman (Eds.), *At the borders of the human.* Houndsmills: Palgrave.

Silvio, C. (1999). Refiguring the radical cyborg in Mamoru Oshii's *Ghost in the Shell. Science Fiction Studies, 77*. Retrieved February 20, 2006, from http://www.depauw.edu/sfs/backissues/77/silvio77.htm

Stratton, J. (1997). Not really desiring bodies: The rise and rise of email affairs. *Media Information Australia, 84*, 28-38. Retrieved February 20, 2006, from http://web.archive.org/web/20011218005627/http://www.newcastle.edu.au/department/so/stratten.htm

Tannen, D. (1993). *Framing in discourse.* Oxford: OUP.

Taylor, M. C., & Saarinen, E. (1994). *Imagologies: Media philosophy.* London: Routledge.

Wilkerson, A. (1997). Ending at the skin: Sexuality and race in feminist theorizing. *Hypatia, 12*(3), 164-174.

Witmer, D. F., & Katzman, S. L. (1997). Smile when you say that: Graphic accents as gender markers in computer-mediated communication. In Sudweeks, McLaughlin, & Rafaeli (Eds.), *Network and Netplay: Virtual groups on the Internet* (pp. 3-11). Menlo Park: AAAI/MIT Press.

Wood, G. (2002). *Living dolls: A magical history of the quest for mechanical life.* London: Faber.

KEY TERMS

Android: An artificial humanoid. They may be robotic, biological, or both. In folklore, the distinction between humans and androids seems to be that humans have emotions, or some kind of irrational mentation, which makes them superior and able to overcome the androids.

Cyborg: A cybernetic organism, or usually a melding of human and machine. The idea originally seems to have been to replace parts of humans by machines so that they could survive in space or other harsh environments. The ambiguities of the cyborg became clear in the British children's science-fiction program *Dr. Who*, in which the cybermen who replaced their human parts with machines became strictly hierarchical galactic warriors in the process, and only ambiguously human.

Emoticon: An icon depicting emotions. The term is usually used to refer to what are called "smileys," which are glyphs that are, in the West, read sideways. Japanese smileys are read vertically rather than horizontally. Often the term is extended to cover other textual expressions of emotion or bodily expression such as <OWWWW!!!>.

Framing: The theory that meaning is decided by the context or frame. Thus, the same set of words can have different meanings depending on the frame brought to them. Framing can be contested, and much politics is about framing things in such a way that one's own political discourse makes sense and the other's does not (see Bateson, 1972; Goffman, 1974; Tannen, 1993).

Luddite: The Luddites were workers in the 19th-century British cloth industry who objected to the new looms because the looms destroyed their self-determined patterns of work, left the tools in the hands of managers, and de-skilled their craft-based activities, leaving them to work harder for less income. They smashed some looms and in turn were smashed by the state (see Sale, 1995).

Outsourcing to the Post–Soviet Region and Gender

Elena Gapova
European Humanities University, Belarus

INTRODUCTION

The purpose of this article is to analyze the outsourcing of information technology (IT) jobs to a specific world region as a gendered phenomenon. Appadurai (2001) states that the contemporary globalized world is characterized by objects in motion, and these include ideas, people, goods, images, messages, technologies and techniques, and jobs. These flows are a part of "relations of disjuncture" (Appadurai, 2001, p. 5) created by an uneven economic process in different places of the globe and involving fundamental problems of livelihood, equality or justice. Outsourcing of jobs (to faraway countries) is one of such "disjunctive" relationships. Pay difference between the United States (U.S.) and some world regions created a whole new interest in the world beyond American borders. Looking for strategies to lower costs, employers move further geographically; and with digital projects, due to their special characteristics, distribution across different geographical areas can be extremely effective. First, digital networks allow reliable and real-time transfer of digital files (both work in progress and final products), making it possible to work in geographically separated locations. Second, in the presence of adequate mechanisms for coordination through information exchange, different stages of software production (conceptualization, high-level design and low-level analysis, coding) are also separable across space (Kagami, 2002).

In the Western hemisphere, the argument for outsourcing is straightforward and powerful. It is believed that if an Indian, Chinese, Russian or Ukrainian software programmer is paid one-tenth of an American salary, a company that develops software elsewhere will save money. And provided that competitors do the same, the price of the software will fall, productivity will rise, the technology will spread, and new jobs will be created to adapt and improve it. But the argument against outsourcing centers on the loss of jobs by American workers. Although there is no statistics on the number of jobs lost to offshore outsourcing, the media write about the outcry of professionals who several years ago considered themselves invulnerable.

BACKGROUND

With digital networks, global enterprises gained access to a skilled labor force like never before. Roughly at the same time, former Soviet high-tech professionals became available for the global employment market after the disintegration of the USSR in 1991 and the decay of its science-military-industrial complex, previously their main employer. Though the bigger nations of India and China continue to be main outsourcing destinations, Eastern Europe has been noted as a promising region. Science and technological development used to be a part of the Soviet pride, and *Whitepaper on Offshore Software Development in Russia* (2001) admits that:

Russia's major advantage over other common offshore software development locales is the technical skills and education of its workforce. Russia has more personnel working in R&D than any other country, and ranks 3rd in the world for per capita number of scientists and engineers. Many of these engineers have solid experience and accomplishments in advanced nuclear, space, military, energy and communications projects. (p. 4)

The same is true for Ukraine, Belarus and, to some extent, Kazakhstan, which used to be highly technological areas during the Soviet period.

CURRENT TRENDS

The mechanism of the region's interaction with international IT jobs providers was (and still is, in part) shaped by the specific trends peculiar to the period after 1991, termed as "transition." The reconfiguration of the fundamentally important social institutions and economic restructuring, to an extent, took the form of "privatization of the state" by those best positioned in the old society and now becoming capital owners. At the same time, national output declined drastically (and still has not reached the level of prior to 1991 in any post-Soviet countries), and the new businesses were emerging in the climate of downward movement and the absence of strong governing and market institutions. Initially, much of the subcontracting into the region arose within informal economy. First, IT groups that developed projects for Western customers were growing on the basis of friendly networks in research laboratories or departments whose members would establish contacts with a western IT contractor, often through a colleague who had relocated, and were paid in cash for short-term projects. With time, transnational employers working with semi-formal or informal groups were complemented with foreign-owned/offshore ventures, establishing permanent offices in the country, but hiding their real outputs from state fiscal agencies. Much of the outsourcing into the region is involved with shadow economy, and "many programmers are paid in cash," the *Whitepaper* (2001, p.12) recognized. According to my data, based on interviews conducted in Minsk in 2005, employees may get about 30% of their salary "officially," while 70% is provided "in an envelope," which implies tax evasion. Cash flows to thousands of people employed in offshore software services via global banking systems, as IT (electronic communications and computer technologies in general) is at the core of these systems. Some virtual employers, which may be "non-existent" in the legal space of post-socialist countries, might be interested in this arrangement as well, when the very "concept of a 'job'—a working place with a contract, employees' rights, sick leave, retirements, working hours— is being abolished" (Rotkirch, 2001). What matters is the final product: its cost, quality and availability to a deadline.

In the 1990s, post-Soviet societies seemed to have lacked the political will to make use of the opportunities for national economies, arising from a strong base in science and technology, while universities continued to provide, as a legacy of socialism, nominally free education for IT (and most other) specialists (alongside with new "commercial" educational opportunities). Currently, 200,000 engineers (of all types and specializations) are prepared annually in Russia alone. Many of these find jobs with transnational IT employers, while benefiting from national social resources: free education, almost free healthcare, extensive systems of affordable public transportation or housing, to name a few. Largely functioning within shadow economy, employing specialists on a temporary basis, withholding from involving with institutions and paying their virtual employees non-taxable cash, some of the outsourcing is similar to the "hijacking" of public resources and has smaller beneficial effects on host societies than they potentially might be. The trade-off, involving a restricted social group and in the absence of methods of administration of the technological-social system as a whole, does not yield adequate investments into national social needs nor spreads extensively across industries: It rather implies buying, at discount prices, its intellectual resources.

The local computer lobby, though, normally resorts to the rhetoric of the good of the nation or even national salvation when justifying the place of honor for IT. The logic of IT as national salvation in the long run is about economic interests of a certain group of people and of particular companies, and thus, is a part of the discourse over wage/economic inequality that emerged in post-socialism. Economists tend to explain higher wages of IT professionals in most national economies by the growing demand in skilled labor (James, 1999). This is only partially the case with post-Soviet high techs, whose salaries tend to be higher because they participate in the global, not national, employment market. The role that IT may play in national economies depends on the social context. It may tend to accentuate, rather than ameliorate, economic and technological differences. Gains from IT accrue mainly to economic agents that form part of the modern technological system in respective countries, as distinct from agents who belong to the traditional system

(James, 1999); small business, unable to introduce high-end products; or state-funded sectors (teachers and doctors). These latter groups may be worse off than before new IT products were introduced, which they are unable to buy (James, 1999).

IT elites (predominantly male), arguing their corporate interests of "world-scale" pay, are among the youngest, most energetic and well-off Russians, as well as Ukrainian or Belarusian urbanites, who, according to the sociological polls, are also the group most favorably looking at tax evasion (Klyamkin, 2000), because they are well positioned to take advantage of it. Young professionals possess expert capital, rooted in certain competencies that, in the era of global labor force, has greater value than before and becomes a basis for moral capital. Their class situation might be viewed as that of a distinct group based on their common unique relationship to the new technological base of modern capitalism. The claim of national salvation through new technologies veils the fact that in the neo-liberal economic restructuring, those benefiting from it are largely those involved with it.

In the post-Soviet job market, corporations granted some high-tech specialists an increase in wages; in return, they demanded a different type of work contract. The new arrangement is based on the idea of an autonomous and competitive, risk-taking "global" worker whose private matters are not supposed to interfere with the pursuit of profit in a market-driven system, while someone else just takes care of them. As corporations were moving overseas in search of skilled labor, outsourcing became embedded in the general process of reconfiguration of gender relations in the post-Soviet region. Under socialism, power relationships were structured differently, as Soviet women were more dependent on the socialist state than on men for their livelihood. Historically, USSR had the highest rates of female labor participation in the world: Nine out of 10 women of reproductive age worked. They entered the workforce massively in the early 1930s, when the socialist industrialization project was launched, and by the 1980s were better educated than men (in terms of their ratio among those with college degrees). Their numbers in science and technology were equally significant and, probably, higher than in any other country: According to some estimates, they made 40%.

As women were very intensively involved in the workforce, the birthrates were falling, and eventually a system of extended benefits was worked out to support motherhood through social policy. It embraced paid parent leaves (e.g., partially paid leave of 3 years to take care of a new child, or fully paid leaves to take care of a sick child, available for any family member, but usually taken by mothers, were counted as work time for retirement), free childcare and healthcare, and so forth. The Soviet gender arrangement was mostly a dual earner/state career contract (Rotkirch, 2001) that did not radically reverse the traditional belief that family obligations are more women's work than men's. The benefits were especially good with richer employers, like the military/science/industrial complex, which provided its workers with holiday packages, sanatoria, children's camps, daycare centers, sports facilities and so forth. Thousands of women were employed there as researchers and engineers and especially programmers, while men were more involved with hardware development (believed to be a male thing). Programming was viewed as less prestigious.

Post-Soviet radical economic and social reforms, carried with the advice of such proponents of neo-liberal economics as the World Bank and International Monetary Fund, included partial dismantling of the system of social services. Sperling (1999) states that when receiving Western loan money, enterprises were required to turn their social services elsewhere, as these were impending economic efficiency. The employees most dependent on leaves and benefits because of family responsibilities turned out as inconvenient in the new system and were massively rejected (under various pretexts) by the market-oriented companies. In the post-communist neo-liberal discourse, social issues began to be largely seen as individual, not collective, responsibility and largely associated with women. As the social policy changed, caring work, without which no economic or social system can function (Jochimsen, 2003), was relegated from the state to the private sphere. In households with perceived lack of time, care may be provided by hired domestic workers, but most often it comes from wives (Rotkirch, 2001).

Meanwhile, as hardware development, which requires investments into facilities and equipment, was largely gone from the job market, and programming became relatively well paid, it turned into a prestigious and mostly male occupation. Ahuja (2002) states that recent trends towards globalization hamper women's chances of hiring or retention IT positions, due to structural and cultural factors. Women were drifting out if IT, as high-tech jobs in the state sector disappeared, and corporations, not concerned with any social provisions, became main employers. New software groups and joint ventures, started through the old boys' net, were either reluctant to hire women or organized in a way that did not imply the private sphere and demanded what was perceived as "male qualities" but, in fact, implied a certain lifestyle, characterized by long working hours and excluding participation in caring work, which is viewed as distracting employees from exclusive concentration on their task. Work-family conflict, social expectations and occupational culture together produce barriers in women's professional lives.

CONCLUSION

Outsourcing to the post-Soviet region became possible after the disintegration of the USSR in 1991 and the decay of its science/military/industrial complex. It became involved with the economic restructuring (transition from central administrative economy to the market economy) and post-socialist emergence of a new systemic superiority of men over women, as the system of social security was reshaped. In this situation, outsourcing was benefiting smaller groups of knowledge workers involved with it. Recently, post-socialist governments that became aware of the potential of outsourcing began developing some policy decisions to use it in the interests of local societies.

REFERENCES

Ahuja, M. K. (2002). Women in the information technology profession: A literature review, synthesis, and research agenda. *European Journal of Information Systems, 11*, 20-34.

Appadurai, A. (2001). Grassroots globalization and the research imagination. In A. Appadurai (Ed.), *Globalization.* Durham; London: Duke University Press.

Information Technologies and Telecommunications Committee of The American Chamber of Commerce in Russia. (2001, April). *Whitepaper on Offshore Software Development in Russia.* Information Technologies and Telecommunications Committee of The American Chamber of Commerce in Russia.

James, J. (1999). *Globalization, information technology and development.* New York: St. Martin's Press.

Jochimsen, M. A. (2003). *Careful economics. Integrating caring activities and economic science.* Boston; Dordrecht; London: Kluwer Academic Publishers.

Kagami, M., & Masatsugu T. (Eds.). (2002). *Digital divide or digital jump: Beyond 'IT' revolution.* Institute for Developing Economies, Japan External Trade Organization.

Klyamkin, I., & Timofeev, L. (2000). *Tenevaya Rossiya. Economico-sociologicheskoe issle dovanie.* Moscow: RGGU.

Rotkirch, A. (2001). The internationalization of intimacy: A study of the chains of care. In *Proceedings of the 5th Conference of the European Sociological Association Visions & Divisions.* Retrieved August, 2005, from http://www.valt.helsinki.fi/staff/rotkirch/ESA%20paper.htm

Sperling, V. (1999). *Organizing women in contemporary Russia. Engendering transition.* Cambridge, MA: Cambridge University Press.

KEY TERMS

Military-Science-Industrial Complex: The aggregate of a nation's armed forces (military establishment), the industries that supply their equipment, materials and armaments, and research units that conduct military-funded research and development.

Neo-Liberal: Neo-liberalism is a political-economic philosophy that de-emphasizes or rejects government intervention in the economy, focusing instead on achieving progress and even social justice by encouraging free-market methods and fewer restrictions on business operations and economic development. The main points of neo-liberalism include the rule of the market; cutting public expenditures for social services; reducing government regulation; privatization; and replacing the concept of "public good" with "individual responsibility."

Outsourcing: Work done for a company by people other than the company's full-time employees. Offshore outsourcing in IT involves the "relocation" of jobs oversees to areas with noticeable pay difference.

Transition: The period of political and economic change in the former communist (socialist) states after the fall in 1989 (1991 in the former USSR) of communist-led governments. The changes involve the transition from the administrative (centrally planned) economy to the market economy, and from the one-party system to liberal democracy. Sometimes termed "the transition to capitalism."

Pair Programming and Gender

Linda L. Werner
University of California, Santa Cruz, USA

Brian Hanks
Fort Lewis College, USA

Charlie McDowell
University of California, Santa Cruz, USA

INTRODUCTION

Studies of pair programming both in industry and academic settings have found improvements in program quality, test scores, confidence, enjoyment, and retention in computer-related majors. In this article we define pair programming, summarize the results of pair programming research, and show why we believe pair programming will help women and men succeed in IT majors.

BACKGROUND

Traditional undergraduate introductory programming courses generally require that students work individually on their programming assignments. In these courses, working with another student on programming homework constitutes cheating and is not tolerated. The only resources available to help students with problems are the course instructor, the textbook, and the teaching assistant. They are not allowed to work with their peers, who are struggling with the same material. This pedagogical approach teaches introductory programming students that software development is an individual activity, potentially giving students the mistaken impression that software engineering is an isolating and lonely career. Gender studies suggest that such a view will disproportionately discourage women from pursuing IT careers (Margolis & Fisher, 2002).

Cooperative or collaborative learning models involve two or more individuals taking turns helping one another learn information (Horn, Collier, Oxford, Bond, & Dansereu, 1998). Some researchers differentiate between cooperative and collaborative methods by stating that cooperative learning involves students taking responsibility for subtasks, whereas collaborative learning requires that the group works together on all aspects of the task (Underwood & Underwood, 1999). The consensus from numerous field and laboratory investigations is that academic achievement such as performance on a test is enhanced when an individual learns information with others as opposed to individually (O'Donnell & Dansereu, 1992; Slavin, 1996; Totten, Sills, Digby, & Russ, 1991).

Cooperative activities have been taught and practiced for other software system development tasks such as design and software engineering but not for programming (Basili, Green, Laitenburger, Lanubile, Shull, Sorumgard, et al., 1996; Fagan, 1986, Sauer, Jeffrey, Land, & Yetton, 2000; Schlimmer, Fletcher, & Hermens, 1994). Often cooperative methods are used in upper division computer science (CS) courses such as compiler design and software engineering in which group projects are encouraged or required. In these courses, the group projects are split up by the group members and tackled individually before being recombined to form a single solution. Sometimes a software engineering instructor offers assistance to the student groups regarding techniques for cooperation but these topics are rarely discussed in other CS courses.

The benefits of collaboration while programming in both industrial and academic settings have been discussed by Flor and Hutchins (1991), Constantine (1995), Coplien (1995), and Anderson, Beattie, Beck, Bryant, DeArment, Fowler, et al. (1998). However, the recent growth of extreme programming (XP) (Beck, 2000) has brought considerable attention to the form of collaborative programming known as pair programming (Williams & Kessler, 2003). Extreme programming is a software development method that differs in a number of ways from generally accepted prior software development methods. These differences include writing module tests before writing the modules, working closely with the customer to develop the specification as the program is developed, and an emphasis on teamwork as exemplified by pair programming, to name just a few. The emphasis on teamwork is an aspect of extreme programming that may be particularly appealing to women.

With pair programming, two software developers work side-by-side at one computer, each taking full responsibility for the design, coding, and testing of the program under development. One person is called the driver and controls the mouse and keyboard; the other is called the navigator and provides a constant review of the code as it is produced. The roles are reversed periodically so that each member of the pair has experience as the driver and navigator. Studies have shown that pair programming produces code that has fewer defects and takes approximately the same total time as when code is produced by a solitary programmer (Nosek, 1998; Williams, Kessler, Cunningham, & Jeffries, 2000). Any code that is produced by only one member of a pair is either discarded or reviewed by the pair together prior to inclusion into the program.

PAIR PROGRAMMING IN THE CLASSROOM

Early experimental research with pair programming using small numbers of students or professional programmers found that pairs outperformed those who worked alone (Nosek, 1998; Williams, Kessler, Cunningham, & Jeffries, 2000). Pairs significantly outperformed individual programmers in terms of program functionality and readability, reported greater satisfaction with the problem-solving process, and had greater confidence in their solutions. Pairs took slightly longer to complete their programs, but these programs contained fewer defects.

A series of experiments conducted at the University of California at Santa Cruz (UCSC) (Hanks, McDowell, Draper, & Krnjajic, 2004; McDowell,

Werner, Bullock, & Fernald, 2003a; Werner, Hanks, & McDowell, 2005) found that students who pair programmed in their introductory programming course were more confident in their work. They were also more likely to complete and pass the course, to take additional computer science courses, to declare computer-related majors, and to produce higher quality programs than students who programmed alone.

Naggapan, Williams, Ferzli, Yang, Wiebe, Miller, et al. (2003) report that pair programming results in programming laboratories that are more conducive to advanced, active learning. Students in these labs ask more substantive questions, are more productive, and are less frustrated.

To ensure that paired students enjoy these benefits, it is important that they have compatible partners. Researchers at the University of Wales (Thomas, Ratcliffe, & Robertson, 2003) investigated issues regarding partner compatibility for pair programming students. They asked more than 60 students to indicate their self-perceived level of expertise and confidence in their programming abilities, and used these rankings to evaluate pairing success. It is important to note that self-reported ability and actual ability are different measures, as 5 of the 17 students who felt that they were highly capable did very poorly in the course.

Thomas et al. found some evidence that students do their best work when paired with students with similar confidence levels. Students with less self-confidence seem to enjoy pair programming more than those students who reported the highest levels of confidence. As there were only seven women in the class, no conclusions about how pairing affected them can be made.

Researchers at North Carolina State University investigated factors that could affect student pair compatibility. Out of 550 graduate and undergraduate students, more than 90% reported being compatible with their partner (Katira, Williams, Wiebe, Miller, Balik, & Gehringer, 2004). Factors such as personality type, actual skill level, and self-esteem appear to have little, if any, effect on partner compatibility. The authors do not discuss any relation between gender and compatibility. Students reported they were more compatible with partners who they perceived to have similar levels of technical competence as themselves; unfortunately, there is no pro-active way that instructors can use this factor to assign partners.

The benefits of pair programming are enjoyed by all students, but women students appear to benefit more. Research conducted at UCSC (McDowell, Werner, Bullock, & Fernald, 2003, Werner, Hanks, & McDowell, 2005a) indicates that although students who pair are more confident in their work and are more likely to declare computer-related majors, these increases are greater for women than men.

The UCSC study looked at four sections of an introductory programming course over three academic terms, one in the fall, two in the winter, and one in the spring. The fall and winter sections required pair programming; the spring term required students to work alone. The fall and spring sections were taught by the same instructor; the winter sections were taught by two additional instructors. Students enrolled into sections without knowing about the pair programming experiments. There were no differences in SAT Math scores or high school GPA between the pairing and non-pairing groups.

Among the UCSC introductory programming students who indicated intent to declare a computer-related major, 59.5% of the paired women had declared a CS-related major one year later, compared with only 22.2% of the women who worked alone. Men who paired were also more likely to declare a CS-related major within one year of taking the introductory course; 74.0% of the paired men had declared a CS-related major compared with 47.2% of the men who worked alone. This is an instance of a possible positive impact on the gender gap due to pair programming. Without pairing, men are 2.13 times more likely than women to declare a CS-related major. When pairing is used, men are only 1.24 times more likely than women to declare a CS-related major.

In this same study, the confidence of students was also measured. To assess student confidence levels, students in the study responded to the question, "On a scale from 0 (not at all confident) to 100 (very confident), how confident are you in your solution to this assignment?" when they turned in each of their programming assignments.

Overall, students who were paired reported significantly higher confidence in their program solutions (89.4) than students who worked independently (71.2). Although men as a group were significantly more confident (87.0) than women (81.1), there was a significant interaction between pairing and gender with regard to reported confidence. Simple effects follow-up tests of the interaction indicated that pairing resulted in increased confidence for both women (86.8 vs. 63.0) and men (90.3 vs. 74.6). Women's confidence increased by 24 points when they paired compared with a 15-point increase for men. Pairing had a statistically significantly greater effect on confidence levels for women and, therefore, may have a visible positive impact on the gender gap. Unpaired men reported 1.18 times greater confidence than unpaired women, while paired men reported 1.04 times greater confidence than paired women. Pairing seems to close the confidence gap between women and men.

FUTURE TRENDS

The Pipelines column of Computing Research News published a short article on pair programming (Werner, Hanks, McDowell, Bullock, & Fernald, 2005b). Computing Research News is a publication of the Computing Research Association whose mission is to influence policy. The appearance of this topic in a publication directed at computing related department chairs of PhD granting institutions is indicative of its importance to the future of introductory computer programming instruction.

One drawback of pair programming is its collocation requirement. We are investigating techniques for extending pair programming to situations where it is difficult or impossible for students to physically meet (Hanks, 2004, Hanks, 2005).

Recommendations for next steps with this research include: the study of pair programming in high school and middle school, additional study about pair compatibility and what is needed in order to increase performance, and to determine what strategies or instructional support is needed to create effective pairs.

CONCLUSION

Why does pair programming hold promise for closing the gender gap regarding the field of information technology? The American Association of University Women Education Foundation Commission on Technology, Gender, and Teacher Education's report in 2000 gives four reasons for the decline in enrollment of women in computer science (CS) programs. These reasons are:

1. A perception that a career in computing is not conducive to family life
2. A belief that work in the information technology field is conducted in a competitive rather than a collaborative environment
3. A perception of CS as a solitary occupation
4. Concern about safety and security about women working alone at night and on weekends in computer laboratories

The use of pair programming combats at least three of these four reasons. A typical beginning programming course requires individual work; with the use of pair programming, women may view programming as a collaborative exercise. Williams and Kessler suggest that "peer pressure" may be at work as a possible explanation for higher completion rates among paired vs. solo programming students (Williams & Kessler, 2000). It may be the collaborative aspect of pair programming that is a major reason that the students remain in the class. The increased levels of confidence that can be attributed to pairing are probably also a factor in improved retention. This same collaboration could help combat the perception of CS as a solitary occupation. Additionally, an outcome of pair programming is that no one works alone late at night or on weekends in a computer laboratory. Partners work together. We hypothesize that these reasons cause pair programming to contribute to persistence of women in CS.

REFERENCES

American Association of University Women Education Foundation Commission on Technology, Gender, and Teacher Education. (2000). *Tech-savvy educating girls in the new computer age*. Retrieved January 1, 2005, from http://www.aauw.org/2000/techsavvy.html

Anderson, A., Beattie, R., Beck, K., Bryant, D., DeArment, M., Fowler, M., et al. (1998). Chrysler goes to extremes. *Distributed Computing* (pp. 24-28). Retrieved January 1, 2005, from http://www.xprogramming.com/publications/dc9810cs.pdf

Basili, V. R., Green, S., Laitenburger, O., Lanubile, F., Shull, F., Sorumgard, S., et al. (1996). The empirical investigation of perspective-based reading. *Journal of Empirical Software Engineering*, *1*(2), 133-164.

Beck, K. (2000). *Extreme programming explained: Embrace change*. Reading, MA: Addison-Wesley.

Constantine, L. L. (1995). *Constantine on peopleware*. Englewood Cliffs, NJ: Yourdon Press.

Coplien, J. O. (1995). A development process generative pattern language. In J. O. Coplien & D. C. Schmidt (Eds.), *Pattern languages of program design* (pp. 183-237). Reading, MA: Addison-Wesley.

Fagan, M. E. (1986). Advances in software inspections. *IEEE Transactions on Software Engineering*, *12*(7), 744-751.

Flor, N. V., & Hutchins. E. L. (1991). *Analyzing distributed cognition in software teams: A case study of team programming during perfective software maintenance*. Empirical Studies of Programmers: Fourth Workshop.

Hanks, B. (2004). Distributed pair programming: An empirical study. In *Proceedings of the 4th Conference on Extreme Programming and Agile Methods—XP/Agile Universe* (LNCS No. 3134, pp. 81-91). Springer.

Hanks, B. (2005). Student performance in CS1 with distributed pair programming. To appear in *Proceedings of the 10th Tenth Annual Conference on Innovation and Technology in Computer Science Education* (pp. 316-320), Caparica, Portugal.

Hanks, B., McDowell, C., Draper, D., & Krnjajic, M. (2004). Program quality with pair programming in CS1. In *Proceedings of the 9th Annual SIGCSE conference on Innovation and Technology in Computer Science Education* (pp. 176-180).

Horn, E. M., Collier, W. G., Oxford, J. A., Bond, C. F., & Dansereu, D. F. (1998). Individual differences in dyadic cooperative learning. *Journal of Educational Psychology*, *90*(1), 153-160.

Katira, N., Williams, L., Wiebe, E., Miller, C., Balik, S., & Gehringer, E. (2004). On understanding compatibility of student pair programmers. In *Proceedings of the 35th SIGCSE Technical Symposium on Computer Science Education* (pp. 7-11).

Margolis, J., & Fisher, A. (2002). *Unlocking the clubhouse: Women in computing.* Cambridge, MA: MIT Press.

McDowell, C., Werner, L., Bullock, H., & Fernald, J. (2003). The impact of pair programming on student performance, perception, and persistence. In *Proceedings of the 25th International Conference on Software Engineering*, Portland, Oregon (pp. 602-607).

Naggapan, N., Williams, L., Ferzli, M., Yang, K., Wiebe, E., Miller, C., & Balik, S. (2003). Improving the CS1 experience with pair programming. In *Proceedings of the 34th SIGCSE technical symposium on computer science education* (pp. 359-362).

Nosek, J. T. (1998). The case for collaborative programming. *Communications of the ACM, 41*(3), 105-108.

O'Donnell, A. M., &. Dansereu, D. F. (1992). Scripted cooperation in student dyads: A method for analyzing and enhancing academic learning and performance. In R. Hartz-Lazarowitz & N. Miller (Eds.), *Interactions in cooperative groups: The theoretical anatomy of group learning* (pp. 120-141). London: Cambridge University Press.

Sauer, C., Jeffrey, D. R., Land, L., & Yetton, P. (2000). The effectiveness of software development technical review: A behaviorally motivated program of research. *IEEE Transactions on Software Engineering, 26*(1), 1-14.

Schlimmer, J. C., Fletcher, J. B., & Hermens, L.A. (1994). Team-oriented software practicum. *IEEE Transactions on Education, 37*(2), 212-220.

Slavin, R. E. (1996). Research on cooperative learning and achievement: When we know, what we need to know. *Contemporary Educational Psychology, 21*, 43-69.

Thomas, L., Ratcliffe, M., & Robertson, A. (2003). Code warriors and code-a-phobes: A study in attitude and pair programming. In *Proceedings of SIGCSE Technical Symposium on Computer Science Education* (pp. 363-367).

Totten, S., Sills, T., Digby, A., & Russ, P. (1991). *Cooperative learning.* New York: Garland.

Underwood, J., & Underwood, G. (1999). Task effects on cooperative and collaborative learning with computers. In K. Littleton & P. Light (Eds.), *Learning with computers* (pp. 10-23). New York: Routledge.

Werner, L., Hanks, B., & McDowell, C. (2005a). Female computer science students who pair program persist. *Journal on Education Resources in Computing, 4*(1).

Werner, L., Hanks, B., McDowell, C., Bullock, H., & Fernald, J. (2005b, March). Expanding the pipeline: Want to increase retention of your female students? *Computing Research News*, p. 2. Retrieved March 21, 2005, from http://www.cra.org/CRN/issues/0502.pdf

Williams, L. A., & Kessler, R. R. (2000). The effects of "pair-pressure" and "pair-learning" on software engineering education. *The 13th Conference on Software Engineering Education and Training.* Austin, TX: IEEE Computer Soc.

Williams, L., & Kessler, R. (2003). *Pair programming illuminated.* Addison-Wesley.

Williams, L., Kessler, R., Cunningham, W., & Jeffries, R. (2000, July/August). Strengthening the case for pair-programming. *IEEE Software, 17*.

KEY TERMS

Active Learning: Based on a theory of learning where people are able to set goals, plan, and revise. This is to be contrasted with the theory of learning based on Piaget where newborns are born with a blank slate on which learning is placed.

Collaborative Learning: A learning model where students work together in a group and the group works together on all aspects of the task.

Cooperative Learning: A learning model where students work together in a group but individual students take responsibility for various subtasks.

Extreme Programming: A method of software development that emphasizes customer involvement and teamwork. One component of the teamwork is the use of pair programming for all code development.

Module: An independent part of a computer program. Different computer languages typically have their own terminology for module. Some of these are: function, procedure, subroutine, and method.

Pair Programming: A method of software development where two software developers work side-by-side at one computer, each taking full responsibility for the design, coding, and testing of the program under development. One person is called the driver and controls the mouse and keyboard; the other is called the navigator and provides a constant review of the code as it is produced.

Software Development: The development of software goes through the steps of requirements specification, design, code, test, and maintenance. The development method that is used determines the order, length, and specific details for each of these steps.

Parental Support for Female IT Career Interest and Choice

Peggy S. Meszaros
Virginia Tech, USA

Anne Laughlin
Virginia Tech, USA

Elizabeth G. Creamer
Virginia Tech, USA

Carol J. Burger
Virginia Tech, USA

Soyoung Lee
Virginia Tech, USA

INTRODUCTION

Although adolescents become progressively independent from their parents in the high-school years, they continue to depend heavily on parents in the area of career development (Peterson, Stivers, & Peters, 1986; Sebald, 1989). The role of parental support in children's career choice has been demonstrated empirically in the career-development literature (Altman, 1997; Fisher & Griggs, 1994; Ketterson & Blustein, 1997; Kracke, 1997; Way & Rossman, 1996). Researchers have found that parents impact career choice more than counselors, teachers, friends, other relatives, or people working in the field of interest (Kotrlik & Harrison, 1989), but are not adequately informed about how to help (Young, Friesen, & Borycki, 1994). Although parents hold a powerful role in the career advising of both their male and female children, most of the reported studies use a male model and focus. Researchers are beginning to develop a knowledge base for the career development of girls and the unique issues they face in deciding on a career. Greater understanding of these issues is urgent, especially as females are recruited into nontraditional fields like information technology. This article will review research on parental support for female career choice, including the research findings from the Women and Information Technology (WIT, 2002-2005) project funded by the National Science Foundation.

BACKGROUND

Early Research Findings of Parental Support for Females

Schulenberg, Vondracek, and Crouter (1984) provided the only existing review of the early published literature as they examined how families influence the vocational development of both females and males using family variables (e.g., demographic variables of the family, family configurations, and process-oriented features of the family) that have been shown to influence different aspects of vocational development. This review, covering studies conducted before the 1980s, found that the early parental-support research for both females and males demonstrated links between career development and socioeconomic status, parents' educational and occupational attainment, and cultural background

More specifically, this review of early findings identified a substantial number of studies conducted in the 1970s regarding possible associations between family process variables and women's career development. Much of this research focused on identifying family characteristics of women who had entered nontraditional careers and suggested that women who pursed nontraditional paths tended to perceive themselves as being similar to their fathers (Tangri, 1972), felt supported by their mothers (Standley & Soule, 1974), and came from families that valued educational and occupational pursuits (Standley & Soule; Trigg & Perlman, 1976).

When examining the early research, Schulenberg et al. (1984) found that except for the work of Roe (1956), who studied parenting styles and career orientation, family interaction-pattern influences on career development had virtually been ignored. Newer research was needed that investigated family interactions such as attachment, psychological separation, conflict, and enmeshment with more sophisticated research methodologies. Given the changes in the world of work, the increased participation of females in the workforce, and changes in the family, families likely influence the career development of females in different ways than in earlier generations.

A different body of research that considers the effects of family functioning is now emerging. Family functioning, a broader concept that encompasses parenting style, parental support and guidance, positive or negative environmental influences, and family members' interaction styles, has been found to exert a greater influence on career development than earlier research that examined family structure or parents' education and occupational status (Fisher & Griggs, 1994; Trusty, Watts, & Erdman, 1997). This newer approach to parental support for career development also includes the effects of parent-child attachment (Ketterson & Blustein, 1997), parent-child communication (Middleton & Loughead, 1993), parental support, guidance, positive and negative environmental influences, and family members' interaction styles (Altman, 1997).

New Studies of Parental Support for Females

A comprehensive review (Whiston & Keller, 2004) of research published since 1980 is related to family support influences on career development and occupational choice. It gives us a picture of family influences on children, adolescents, college students, and adults over the life course. While few studies of childhood and female careers were found, several findings related to communication and perceived family power were cited. Birk and Brimline (1984) studied children enrolled in kindergarten, third grade, and fifth grade and found that parents who talked to their children about their occupational goals had children who aspired toward more gender-traditional occupations. Lavine (1982) asked children aged 7 to 11 years what they wanted to be when they grew up; whether boys, girls, or both could have certain jobs; and which parent made the decisions in their homes. Findings revealed that girls who viewed their mother as having significant power within the family perceived more careers as being open to both men and women and aspired to less feminine-stereotyped careers compared with girls who viewed their mother as having little power.

In the Whiston and Keller 2004 review of research, a theme of mothers and daughters emerged, and studies related to adolescent females found that freshmen and sophomore girls were more likely than boys to report that their mothers provided positive feedback, supported their autonomy, and were open to discussions about their career decisions (Paa & McWhirter, 2000). Continuing the mother-daughter bond of influence, Fassinger's (1990) model of females' career development proposed that a complex set of relationships among agency, ability, and gender-role attitudes influence women's career orientation and choice. O'Brien and Fassinger (1993) found that the relationship with the mother contributed to the model, and their model reflected that a combination of an attachment to the mother and a healthy movement toward individuation contributes to adolescent girls' career orientation. Rainey and Borders (1997) concluded that the career orientation of adolescent females is influenced by a complex interplay of their abilities, agenting characteristics, gender-role attitudes, and relationships with their mothers. The mother-daughter relationship may be significant in adolescent girls developing a career orientation and may play a pertinent role in their feeling efficacious about career decision making.

At the college level, researchers have also found that parental attachment is positively associated with vocational exploration (Ketterson & Blustein, 1997). Felsman and Blustein (1999), and Ryan, Solberg, and Brown (1996), however, found that maternal attachment was more salient than paternal attachment. Finally, a qualitative study by Schultheiss, Kress, Manzi, and Glasscock (2001) examined family influences on both vocational exploration and career decision making. The majority of participants felt their mothers, fathers, and siblings had played a positive role in their career exploration by indirect means such as providing emotional esteem and informational support, and by more tangible means such as providing educational materials. Furthermore, 36% of participants indicated that their mother was the most influential person in their career exploration process, while 21% indicated this was true of their father.

Research Findings from the Women in Information Technology Project Related to Parental Support for Female IT Career Choice and Interest

The next section summarizes key findings from a research project funded by the National Science Foundation to explore women's interest in science, technology, engineering, and math (STEM) fields, particularly information technology (Women in Information Technology, 2002-2005). The larger project was composed of several substudies involving qualitative and quantitative research methods, and samples from a variety of populations. The analytical framework for the project was the developmental theory of self-authorship, defined as "the ability to collect, interpret, and analyze information and reflect on one's own beliefs in order to form judgments" (Baxter Magolda, 1998, p. 143). Self-authorship influences how individuals make meaning of the advice they receive from others, how susceptible they are to feedback, and the extent to which the reasoning they employ to make a decision reflects an internally grounded sense of self (Baxter Magolda, 1998, 1999, 2001; Baxter Magolda & King, 2004).

Mother and Daughter Career Conversations

One of the substudies of this larger project involved qualitative interviews with 11 matched pairs of mothers and high-school daughters (Meszaros, Creamer, Burger, & Matheson, 2005). This study explored answers to the questions of how career decisions are made and who influences them. The daughters were high-school sophomores, aged 16 to 18, from 10 urban and suburban high schools in Virginia. When asked whom they talked to about their future, 7 of the 11 girls identified their mother first, followed by friends, grandmothers, parents (both mother and father), and teachers. The daughters described career conversations with their mothers as being mostly nondirective, often including messages of support and encouragement for making good decisions about a career. Rather than directing their daughters to a specific career, the mothers served more as sounding boards and guides to other resources including the Internet and libraries. Most of the mothers reported that they did not want to influence the specific occupational choice of their children. Several mothers served in the role of providing an active information resource beyond just encouraging their daughters' good decisions. This exploratory study indicates that mothers actively support the career-decision processes of their daughters. If this is the case, mothers could benefit from additional resources about information technology and other nontraditional careers so that the guidance they provide to their daughters supports the consideration of a wide range of career options.

Sources of Influence for College Women's Career Interests

Another substudy involved interviews with college women (n=40) who were asked to identify people who had a significant influence on their career interests (Creamer & Laughlin, 2005). Nearly all (n=39, 98%) of the women identified one or both parents as influencers, while other family members, particularly siblings, were identified next most frequently (n=13, 33%). Participants were also asked why they considered opinions from these people important. The most common reply had to do with the participant's sense that the people giving advice (in most cases, parents) cared for them and would know what was best for them. For example, one woman stated, "Just because I trust them [parents and sister]—I know that they are looking out for my best interests; they're not going to tell me something that is going to hurt me." Another participant explained, "Because they're my parents and I think they know what's best for me sometimes." These findings suggest that many college women turn to parents for advice about career decisions, adding support to past studies showing the primary impact of parents on career choice (Kotrlik & Harrison, 1989). The trust placed on parents to know what is best may override the authority of others, like advisors and/or faculty members, who are better acquainted with a wide range of career options, especially in highly technical fields, but are less trusted because they do not know a student personally.

Predicting women's interest in IT. A theoretically driven path model was developed to predict women's interest in careers in information technology (Creamer, Burger, Meszaros, Lee, & Laughlin, 2005). The model is based on responses to the *Career Decision-Making Survey* (Women in Information Technology, 2005), administered between 2001 and 2005 in three waves to high-school and college women in rural and urban locations in the mid-Atlantic region (*n*=1,621). The findings presented here are from 373 high-school and college women who completed the questionnaire in the fall of 2004 and spring of 2005, resulting in a model that predicted 27% of the variance in the interest and choice of computer-related fields.

Parental support was one of three mediating variables (along with computer use and positive attitudes about IT workers) that had a significant direct effect on the dependent variable IT career choice and interest. Parental support is defined by the perceptions that parents support the importance of a career and encourage career exploration, as well as agreement with the belief that parents have an idea of what would be an appropriate career choice. Students who expressed an interest in an IT career believed that their parents support this choice. This is consistent with interview data (noted above) indicating that female college students trust parents to provide guidance about suitable majors and career options, and to know what is best for them. Parental support had a direct effect on how likely the respondent was to seek and listen to advice about career options, and how positively they viewed individuals working in IT fields. It also impacted how respondents made judgments about whose advice was considered credible. Participants whose mother had completed a college degree were significantly more likely than those whose mother had only completed a high-school degree to report parental support for a career, career exploration, and computer use.

FUTURE TRENDS

Although researchers have agreed that parental involvement in career decisions is very important, and there is a growing body of research evidence, more investigation of how and to what extent parents influence their female children's career decisions is needed (Kerka, 2001). The research on female choice of nontraditional careers such as IT is notably missing. More research is also needed for particular groups of women and girls. For instance, some research shows that for African-American females, early gender-role socialization is less sex typed, and African-American girls often experience more crossovers between traditionally male and female roles and duties in the household, perhaps making them more open to considering nontraditional careers (Hackett & Byars, 1996). There is also a need to examine both positive and negative effects of parental involvement on career decisions.

CONCLUSION

Parental support for female IT career choice and interest emerged as a key variable in the theoretically driven path model of our research. Few studies have been conducted using a family-relations approach. Viewing parental support through a family-relations lens provides insights into Baxter Magolda's (2004) theory of self-authorship with its emphasis on parental challenge and support. The parent-child literature using attachment theory suggests close relationships provide experiences of security promoting exploration and risk taking. Furthermore, parents who are willing to discuss issues openly and promote independent thinking in their children encourage more active career exploration (Ketterson & Blustein, 1997).

REFERENCES

Altman, J. H. (1997). Career development in the context of family experiences. In H. S. Farmer (Ed.), *Diversity and women's career development: From adolescence to adulthood* (pp. 229-242). Thousand Oaks, CA: Sage.

Baxter Magolda, M. (1998). Developing self-authorship in young adult life. *Journal of College Student Development, 39*(2), 143-156.

Baxter Magolda, M. (1999). *Creating contexts for learning and self-authorship: Constructive-developmental pedagogy.* Nashville, TN: Vanderbilt University Press.

Baxter Magolda, M. (2001). *Making their own way: Narratives for transforming higher education to promote self-development.* Sterling, VA: Stylus.

Baxter Magolda, M., & King, P. M. (Eds.). (2004). *Learning partnerships: Theory and models of practice to educate for self-authorship.* Sterling, VA: Stylus.

Birk, J. M., & Brimline, C. A. (1984). Parents as career development facilitators. *The School Counselor, 31*, 310-317.

Creamer, E. G., Burger, C. J., Meszaros, P. S., Lee, S., & Laughlin, A. (2005, February). *Predicting young women's interest in information technology careers: A statistical model.* Paper presented at the Third International Symposium on Gender and ICT: Working for Change, Manchester, UK.

Creamer, E. G., & Laughlin, A. (2005). Self-authorship and women's career decision-making. *Journal of College Student Development, 46*(1), 1-15.

Fassinger, R. E. (1990). Causal models of career choice in two samples of college women. *Journal of Vocational Behavior, 36*, 225-248.

Felsman, D. E., & Blustein, D. L. (1999). The role of peer relatedness in late adolescent career development. *Journal of Vocational Behavior, 54*, 279-295.

Fisher, T. A., & Griggs, M. B. (1994). *Factors that influence the career development of African-American and Latino youth.* Paper presented at the Annual Meeting of the American Educational Research Association, New Orleans, LA.

Governor's Commission on Information Technology. (1999). *Investing in the future: Toward the 21st century technology workforce.* Richmond, VA: Office of the Governor.

Hackett, G., & Byars, A. (1996). Social cognitive theory and career development of African American women. *Career Development Quarterly, 44*, 322-340.

Kegan, R. (1994). *In over our heads: The mental demands of modern life.* Cambridge, MA: Harvard University Press.

Kerka, S. (2001). *Parenting and career development.* Retrieved December 2, 2005, from http://www.ericdigests.org/2001-1/career.html

Ketterson, T. U., & Blustein, D. L. (1997). Attachment relationships and the career exploration process. *Career Development Quarterly, 46*(2), 167-178.

Kotrlik, J. W., & Harrison, B. C. (1989). Career decision patterns of high school seniors in Louisiana. *Journal of Vocational Education, 14*, 47-65.

Kracke, B. (1997). Parental behaviors and adolescents' career exploration. *Career Development Quarterly, 46*(2), 341-350.

Lavine, L. A. (1982). Parental power as a potential influence on girls' career choice. *Child Development, 53*, 658-663.

Meszaros, P., Creamer, E., Burger, C., & Matheson, J. (2005). Mothers and millennials: Career talking across the generations. *Kappa Omicron Nu Forum, 16*(1). Retrieved June 7, 2005, from http://www.kon.org/archives/forum/16-1/meszaros.html

Middleton, E. B., & Loughead, T. A. (1993). Parental influence on career development: An integrative framework for adolescent career counseling. *Journal of Career Development, 19*(3), 161-173.

O'Brien, K. M., & Fassinger, R. E. (1993). A causal model of the career orientation and career choice of adolescent women. *Journal of Counseling Psychology, 40*, 456-469.

Paa, H. K., & McWhirter, E. H. (2000). Perceived influences on high school students' current career expectations. *The Career Development Quarterly, 49*, 29-44.

Peterson, G. W., Stivers, M. E., & Peters, D. F. (1986). Family versus nonfamily significant others for the career decisions of low-income youth. *Family Relations, 35*, 417-424.

Rainey, L. M., & Borders, L. D. (1997). Influential factors in career orientation and career aspiration of early adolescent girls. *Journal of Counseling Psychology, 44*, 160-172.

Roe, A. (1956). Early determinants of vocational choice. *Journal of Counseling Psychology, 4*, 212-217.

Ryan, N. E., Solberg, V. S., & Brown, D. D. (1996). Family dysfunction, parental attachment, and career search self-efficacy among community college students. *Journal of Counseling Psychology, 43*, 84-89.

Schulenberg, J. E., Vondracek, F. W., & Crouter, A. C. (1984). The influence of the family on vocational development. *Journal of Marriage and Family, 46*, 129-143.

Schultheiss, D. E. P., Kress, H. M., Manzi, A. J., & Glasscock, J. M. J. (2001). Relational influences in career development: A qualitative inquiry. *The Counseling Psychologist, 29*, 214-239.

Sebald, H. (1989). Adolescents' peer orientation: Changes in the support system during the past three decades. *Adolescence, 24*, 937-946.

Standley, K., & Soule, V. (1974). Women in male-dominated professions: Contrasts in their personal and vocational histories. *Journal of Vocational Behavior, 4*, 245-258.

Tangri, S. S. (1972). Determinants of occupational role innovation among college women. *Journal of Social Issues, 28*, 177-199.

Trigg, L. J., & Perlman, D. (1976). Social influences on women's pursuit of a nontraditional career. *Psychology of Women Quarterly, 1*, 138-150.

Trusty, J., Watts, R. E., & Erdman, P. (1997). Predictors of parents' involvement in their teens' career development. *Journal of Career Development, 23*(3), 189-201.

Way, W. L., & Rossman, M. M. (1996). *Lessons from life's first teacher: The role of the family in adolescent and adult readiness for school-to-work transition.* Retrieved December 7, 2004, from http://vocserve.berkeley.edu/abstracts/MDS-807/MDS-807.2.html

Whiston, S. C., & Keller, B. (2004). The influences of the family of origin on career development: A review and analysis. *The Counseling Psychologist, 32*(4), 493-568.

Women in Information Technology. (2002-2005). *Research project funded by the Program for Gender Equity in Science, Mathematics, Engineering, and Technology (PGE) at the National Science Foundation* (Grant No. 0120458). Retrieved February 21, 2006, from http://www.clahs.vt.edu/WIT/

Women in Information Technology. (2005). *Career decision-making survey.* Retrieved February 21, 2006, from http://www.clahs.vt.edu/WIT/

Young, R., Friesen, J. D., & Borycki, B. (1994). Narrative structure and parental influence in career development. *Journal of Adolescence, 17*, 173-191.

KEY TERMS

Information Technology: This refers to a variety of jobs that involve the development, installation, and implementation of computer systems and applications. Careers in IT encompass occupations that require designing and developing software and hardware systems, providing technical support for computer and peripheral systems, and creating and managing network systems and databases (Governor's Commission on Information Technology, 1999).

IT Career Interest and Choice: This is the dependent variable in the Women in Information Technology path model presented in this article. It reflects an expressed interest in a career in IT or a choice to pursue a career in IT.

Nontraditional Career: This refers to a career in which less than 25% of the workforce is of one's gender.

Parental Support: In the Women in Information Technology path model presented in this article, this variable reflects the respondent's perceptions that her parents support the importance of a career and encourage career exploration, as well as agreement with the statement that parents have an idea about what would be an appropriate career choice.

Path Model: This is a statistical model that shows directional relationships between variables in the form of a diagram. The variables in the diagram are arranged according to theoretical assumptions regarding their causal relationships and are con- nected by paths or directional arrows that display regression weights.

Self-Authorship: "The ability to collect, inter- pret, and analyze information and reflect on one's own beliefs in order to form judgments" (Baxter Magolda, 1998, p. 143).

Traditional Career: This refers to a career in which more than 75% of the workforce is of one's gender.

Participation of Female Computer Science Students in Austria

Margit Pohl
Vienna University of Technology, Austria

Monika Lanzenberger
Vienna University of Technology, Austria

INTRODUCTION

The situation of women in computer science education has been a major topic of feminist researchers. It has received widespread attention in many countries all over the world. In general, it can be said that in almost all the countries of the world women are underrepresented in computer science education at university level. This phenomenon is, however, a very complex one. In many industrialized countries, there was a peak in women's participation in computer science studies in the middle of the 80s of the 20th century. After that, the number of women who studied computer science in these countries decreased again. This development has been discussed, for example, by Behnke and Oechtering (1995) for Germany, by Kirkup (1992) for Great Britain, and by the EECS Women Undergraduate Enrollment Committee (1995) for the USA. In recent years, there is some indication that the percentage of women who choose Computer science at universities is rising again, at least in Germany (Kompetenzzentrum, 2003) and in Austria (Österreichisches Statistisches Zentralamt, 1971-2001). Apart from that, some cultural differences can be observed. It has been mentioned in several publications that the percentage of Asian women studying computer science is often higher than that of women in Western industrialized countries (Greenhill, von Hellens, Nielsen, & Pringle, 1997).

In the following text, we want to discuss possible reasons for the increase in female computer science students in the last years in a few countries. We want to analyze the reasons for this increase. Detailed information about the motivation of women who study computer science at universities should be helpful in formulating strategies to overcome the under-representation of women in this area. Such strategies should take differences between countries into account. Case studies for single countries could provide relevant information in this context. The following text describes the situation in Austria.

BACKGROUND

The under-representation of women in computer science has been discussed quite extensively. Many reasons have been given why women apparently avoid computer science despite the fact that computer systems do influence all aspects of working conditions and the private life of people quite heavily. A higher percentage of women who take part in shaping computer technology would, therefore, be very desirable. Nevertheless, women do not feel motivated to participate in this process. Several publications have discussed the reasons for this in great detail (see e.g., Gürer & Camp, 2001; Margolis & Fisher, 2003; Schinzel, 1997).

Gürer and Camp (2001) developed a very comprehensive framework of reasons for the under representation of women in computer science consisting of 13 different issues, which have to be considered. Among these issues, Gürer and Camp mention that a positive attitude towards computers is necessary, that computer games developed specifically for boys exert a negative influence, and that equal access of girls and boys to computers is important. In general, the environment in the family, in school, at work, and in society plays a crucial role.

Schinzel (1997) discussed a few additional reasons why women do not go into computing anymore. She points out that (at least in Germany) there was "a shift in the definition of computer science from discrete applications of mathematics to an orientation towards applied and engineering sciences" (p. 368). She argues that this made it more difficult for women to identify with computer science. In her view, computer science strongly influenced by mathematics is more appealing to women. This argument is probably quite specific for Germany. There is empirical evidence that many female German computer scientists originally studied mathematics or feel very attracted by the mathematical side of computer science (Erb, 1996). This contradicts Margolis' and Fisher's (2003) view who assume that the similarity of computer science to mathematics makes it especially difficult for women to get interested in this subject. Schinzel also emphasizes the concept of highly diverse socially created gender-based interests of girls and boys.

Margolis and Fisher (2003) assume that an important reason for the under representation of women in computer science is the "geek mythology" prevailing among computer scientists. Computer science students are seen as persons who are obsessed with their machines and who are rarely ever communicating with normal people. This image is supposed to deter many women from computer science.

Margolis and Fisher also point out that there is a difference between female American students and students from other countries (especially from Asia) who study at U.S. universities. In their interviews, they found out that these students often had no previous computer experience but persisted because they experienced financial pressure either from companies, which granted them scholarships or from their families. In some cases, these women acquired an increased self-confidence towards computer science. This is also supported by the work of Greenhill et al. (1997). They investigated female Asian students of computer science in Australia who also seem to be very job-oriented.

Given that the under-representation of women in computer science still persists several authors tried to formulate measures to overcome this problem (see e.g., Gürer & Camp, 2001; Margolis & Fisher, 2003). In addition, it seems to be necessary to analyze in more detail which conditions promote the access of women to computer science studies and which do not. In Austria, for example, there has been an increase in the percentage of female computer science students during the last few years (since 1998).

Several reasons might explain this development. Especially in the year 2000, the shortage of IT experts was widely discussed in Austria and other European countries. It should be noticed that the increase of female computer science students coincides with this discussion. It might, therefore, be argued that these students were motivated by the demand for IT specialists on the labor market. In the literature, there is some indication that an increased demand for experts in specific areas can lead to a higher degree of women in the workforce (see e.g., Roloff, 1989). The economic crisis and the problems with the so-called new economy led to a decrease in the demand for IT experts but there is still a shortage of IT experts with specific skills. Another development which might be responsible for the increase in female students of computer science might be the fact that there is a tendency in Austria that young girls and women use computers and the Internet as often as boys/young men do. In the age group between 16 and 24 years, 90.9% of all women have used a computer and 68.5% the Internet. The corresponding figures for males are 90.6% and 68.3%. In contrast to that, there is a pronounced gender difference in all other age groups (Statistik Austria, 2003). This indicates that the problem of access for girls and young women is not extremely relevant anymore. A third explanation might be that in the year 2001 five different computer science bachelor studies were introduced at Austrian universities. The disciplines of media informatics and medical computer science were especially attractive for women. The introduction of such fields of specialization might aid female students to choose computer science as a subject. To find out whether there is some empirical evidence for these explanations we conducted a survey at an Austrian technical university concerning attitudes of computer science students towards their discipline. Selected results from this survey will be discussed in the next section.

RESULTS OF A SURVEY CONDUCTED IN 1993 AND 2004

In 2004, we conducted a survey, which was based on 41 females and 247 males, who are students of the BSc in computer science at the Vienna University of Technology. Most of them (256) enrolled in October

Table 1. Enrollments of the subjects, Survey, 2004

	Females	F %	Males	M %	Sum
Data Engineering & Statistics	1	**2**	4	**2**	5
Media Informatics	28	**65**	94	**37**	122
Medical Computer Science	5	**12**	23	**9**	28
Software & Information Engineering	5	**12**	82	**32**	87
Computer Engineering	2	**5**	43	**17**	45
B.Sc. Studies in Computer Science	41	**95**	246	**96**	287
Other Studies	2	**5**	9	**4**	11
Sum	43	**100**	255	**100**	298

2004—40 females and 216 males study in the first semester. Since a small number of participants study more than one BSc, the figures in Table 1 show 298 enrollments in total. The percentages in Table 1 describe the distribution among women and among men. Obviously, the bachelor studies in computer science at the Vienna University of Technology are still dominated by male students; only 14% of the subjects are female. However, there is a significant difference related to the type of bachelor study. Whereas male students tend to choose computer engineering and software and information engineering, female students prefer media informatics. Although data engineering and statistics show a participation of 20% female students, we do not consider this as relevant information because the data sample of five subjects is too small. This gender gap might indicate that the type and labeling of the Bachelor Studies activates a gender-stereotyped context. A focus on engineering seems to attract men and discourage women.

Feminist research offers three major approaches to explain this tendency (Collmer, 1997). The first model assumes a radial difference between women and men and their adoption and relation to nature and technology (Mies, 1980; Jansen, 1986). Another explanation claims shortcomings in the socialization and education of girls and women, which cause a distance to technology (Metz-Göckel, 1990; Schorb, 1990). The third approach asserts an ambivalent role of female engineers (Wagner, 1991). Women experience a double bind situation and alienation caused by contrariness between the concept of technology

and the concept of female. The first approach was criticized for biological reasoning and questionable concepts of female and male. The other approaches differ in the consideration of structural and individual aspects of women. However, they seem to represent the current research results in a more appropriate way.

In 1993, a similar survey was conducted in order to investigate the motivation and approaches of students in computer science. Table 2 shows the enrollments of the survey conducted in 1993 at the Vienna University of Technology.

At this time only two types of studies existed: computer science on the one hand and information systems on the other hand. Both studies correspond to a combination of BSc and MSc requiring a minimum duration of study of eight semesters. Table 2 shows the distribution among women and among men. In the year 2001, the bachelor studies were introduced. Computer science has been split up into five different types of studies listed in Table 1. Since 1993, the number of students in information systems dropped dramatically and was not considered in the survey in 2004 because the situation of these studies offered by three different universities in Vienna is very specific. Therefore, there are limitations when comparing the numbers of 1993 and 2004. However, both investigations yield relevant statements on Austrian computer science students in the respective period.

In the following, we describe two results, which indicate a significant change between 1993 and 2004. The first question addresses the availability of

Table 2. Enrollments of the subjects, Survey, 1993

	Females	F %	Males	M %	Sum
Computer Science	27	**64**	149	**73**	176
Information Systems	15	**36**	55	**27**	70
Sum	42	**100**	204	**100**	246

the computer (PC), as a tool broadly associated with computer science. In particular, the subjects were asked whether they had owned a computer before enrolling at the university. In 1993, 48% of the female students and 83% of the male students had owned a computer before their enrollment. In 2004, the outstanding number of 100% of the female students and 97% of the male students had already owned a computer before their enrollment. Whereas there was a significant difference in 1993, the values are nearly the same for women and men in 2004. Today having a computer of one's own seems to be an important and equally distributed precondition for students in computer science. However, the small number of women (14%) studying computer science in 2004 might be caused by this requirement. Therefore, it would be interesting to ask the pupils in Austrian schools whether they owned a computer.

The second question asks for the importance of extensive interest in computers. This question also covers the activities in leisure time. Particularly, subjects were asked whether they think interest in computers—also in leisure time—is a pre-condition for studying computer science. In 1993, 68% of the female and 73% of the male students agreed that private interest is very important. There is an explicit increase in 2004 especially in the answers of female students: 95% of the female students and 89% of the male stated this as a precondition. Although there is no significant difference between men and women, for female students the engagement with computers is nowadays much more important. Therefore, owning a computer and using it also in spare time could ease the entry into computer science, today even more than a decade ago.

FUTURE TRENDS

Austria is a fairly small country and its IT industry is not as developed and vital as in other countries. The organization of Austrian universities certainly differs from the one in other countries. It is, therefore, difficult to generalize our results of a moderate increase in female students of computer science since 1998. It seems to be plausible that the Austrian trend is at least partly due to specific conditions like, (e.g., the introduction of the bachelor studies). It

would be valuable to compare our results with data from several other countries to find out whether the trend we describe is a general international trend or a specific Austrian phenomenon.

CONCLUSION

The under representation of women in computer science seems to be quite a complex phenomenon. Many different reasons have been offered to explain the lack of participation of women in computing but we still do not know enough about the relative importance of these reasons. In Austria, some of the explanations given for the under-representation of women in computer science do not hold anymore, as, for example, lack of access to computer technology. Our own questionnaire and the micro census data both indicate that there is a tendency to more equal access among younger age groups. The future will show whether these changes will lead to a higher percentage of female students of computer science in the long run.

Another problem is the bachelor studies with a high percentage of female students. Women apparently feel attracted by media informatics and medical computer science and deterred by computer engineering and information and software engineering. This conforms to general stereotypes. From a feminist point of view, the introduction of subjects like medical computer science is contradictory because on the one hand, it attracts women to computer science but on the other hand, it reinforces gender differences.

The higher demand for (female) IT specialists is also difficult to interpret. It is highly probable that this demand will fluctuate in the future in accordance with the development of the economy. Consequently, the proportion of female IT specialists will probably fluctuate as well. Again, it is an open question whether there will be an increase in female IT specialists in the long run.

In general, there is some empirical evidence that in Austria the situation for female computer scientists is improving but that this development is not unequivocal. Only long-term experience can show whether this trend will be only temporary or not.

ACKNOWLEDGMENTS

We want to thank Selva Ardic, Emine Kara, Thomas Lederer, Stephan Sykacek, Andreas Fritz, and Hannes Windisch for their assistance with the survey.

REFERENCES

Behnke, R., & Oechtering, V. (1995). Situations and advancements measures in Germany. *Communications of the ACM, 38*(1), 75-82.

Collmer, S. (1997). Frauen und Männer am Computer, Deutscher Universitätsverlag, Wiesbaden, 48-63.

EECS Women Undergraduate Enrollment Committee. (1995). *Women undergraduate enrollment in electrical engineering and computer science at MIT.* Final Report, January 3, 1995.

Erb, U. (1996). Frauenperspektive auf die Informatik, Münster.

Greenhill, A., von Hellens, L., Nielsen, S., & Pringle, R. (1997, May 24-27). Australian women in IT education: Multiple meanings and multiculturalism. Women, work and computerization. Spinning a Web from past to future. *Proceedings of the 6ᵗʰ International IFIP-Conference* (pp. 387-397), Bonn, Germany.

Gürer, D., & Camp, T. (2001). *Investigating the incredible shrinking pipeline for women in computer science.* Final Report—NSF Project 9812016.

Jansen, S. (1986). Magie und Technik. Auf der Suche nach feministischen Alternativen zur patriarchalen Naturnutzung. Christa Lippmann (Ed.), *Technik ist auch Frauensache* (180-197). Hamburg.

Kirkup, G. (1992) The social construction of computers: Hammer or harpsichord. In G. Kirkup & L. S. Keller (Eds.), *Inventing women* (pp. 267-281). Cambridge.

Kompetenzzentrum Frauen in Informationsgesellschaft und Technologie. (2003). Frauen in Ingenieur- und Naturwissenschaften an deutschen Hochschulen 2003 "At a Glance", Bielefeld.

Margolis, J., & Fisher, A. (2003). *Unlocking the clubhouse. Women in Computing.* Cambridge, MA: MIT Press.

Metz-Göckel, S. (1990). Von der Technikdistanz zur Technikkompetenz, *Frauen leben Widersprüche—Zwischenbilanz der Frauenforschung*, Sigrid Metz-Göckel, & Elke Nyssen (Eds.), Weinheim & Basel, 140.

Mies, M. (1980). Gesellschaftliche Ursprünge der geschlechtlichen Arbeitsteilung. *Beiträge zur feministischen Theorie und Praxis, 3,* 61-78.

Österreichisches Statistisches Zentralamt. (1971-2001). Österreichische Hoch-schulstatistik, Wien.

Roloff, C. (1989). Von der Schmiegsamkeit zur Einmischung. Professionalisierung der Chemikerinnen und Informatikerinnen. Pfaffenweiler.

Schorb, B. (1990). Frauen und Computer: eine problematische Beziehung? Modelle der Erklärung und der pädagogischen Praxis. *Basic für Eva? Frauen und Computerbildung,* Bernd Schorb and Renate Wielpütz (Eds.), Opladen, 10.

Schinzel, B. (1997, May 24-27). Why has female participation in German Informatics decreased? Women, work, and computerization. Spinning a Web from past to future. *Proceedings of the 6ᵗʰ International IFIP-Conference* (pp. 365-378), Bonn, Germany.

Statistik Austria. (2003). IKT-Einsatz in Haushalten. Ergebnisse der Europäischen Erhebung über den Einsatz von Informations- und Kommunikationstechnologien in Haushalten, Wien.

Wagner, I. (1991). Organisierte Distanz? Frauen als Akteurinnen im Handlungsfeld Technik. *Proceedings of "Wer Macht Technik? Frauen zwischen Technikdistanz und Einmischung"*, Frauenakademie München (F.A.M.) 3.

KEY TERMS

BSc: Bachelor Studies in Austria are characterized by a duration of six semesters and require general qualification for university entrance (Matura). Graduates receive the degree Bakkalaurus/Bakkalaurea (Bakk.).

Coeducation: Coeducation is the education of both sexes in the same institution. It is supposed to be beneficial for the equality of women and men. There is some empirical evidence, however, that at least in some subjects, it might be better to have single-sex classes, so that female students get more attention than they get in mixed classes.

Computer Science (at Austrian Universities): A general label for academic studies in theoretical informatics, human-computer interaction, computer and software engineering and similar subjects.

Media Informatics: Media Informatics is a sub-discipline of Computer Science, which deals with all aspects of multimedia systems (technological and programming aspects, design, usability).

Medical Computer Science: A sub-discipline of Computer Science, which deals with applications in medicine.

Mentoring: Mentoring means providing support, counsel, friendship and constructive example. In the context of universities, mentors could be professors but also advanced students. Mentors for female students are necessary because they get less feedback and encouragement than their male counterparts and are not integrated into networks of students.

MSc: Austrian Masters' Studies are either a full study (10 semesters with the BSc included) or graduate studies (4 semesters). Graduates receive the degree Dipl.-Ing.

New Economy: Small and innovative companies which deal with all kinds of information technology form a sector of the economy. This sector might be called New Economy. It has been claimed that the organizational structure of these companies and their business strategies are fundamentally different to more traditional sectors.

NOTE

Part of this work was done while the author was an ERCIM Research Fellow at IDI, Norwegian University of Science and Technology (NTNU), Trondheim, Norway.

Participation of Women in Information Technology

Tiffany Barnes
University of North Carolina at Charlotte, USA

Sarah Berenson
North Carolina State University, USA

Mladen A. Vouk
North Carolina State University, USA

INTRODUCTION

Our nation's continued global competitiveness is widely believed to depend upon the United States maintaining its leadership in the development and management of new information technologies (Freeman & Aspray, 1999; Malcom, Babco, Teich, Jesse, Campbell, & Bell, 2005; Sargent, 2004). Rapidly changing technologies have pervaded every sector of American society, infusing nearly everyone's work and personal lives. Over the long term, we may face a shortage of highly educated IT workers who are needed to maintain and increase the economic productivity of the United States. Interestingly, according to Freeman and Aspray, if women were represented in the IT workforce in equal proportion to men (assuming the percentage of men in IT vis-à-vis other professions remained constant), this impending shortage and its potentially economically devastating consequences could be prevented.

We identify the pipeline of potential female IT workers as beginning in the middle grades, with the girls who take college-prep algebra by the eighth grade and elect college-bound courses in math, science, and computer science through high school. These girls are then prepared to complete a bachelor of science degree in computer science, computer engineering, or electrical engineering and become creative future IT workers.

In this article, we examine some of the factors that, as suggested by the literature, influence the low participation of women in IT. We also discuss the open research issues in understanding and modeling the (educational) persistence of young women in IT-related disciplines, and we outline some results from Girls on Track, an intervention program for middle-school girls. We end with some suggestions for making IT more appealing to this currently underrepresented population.

BACKGROUND

While the enrollment of girls in advanced science and mathematics courses in high school continues to increase, their enrollment in high-school computer-science courses is extremely low (Congressional Commission on the Advancement of Women and Minorities in Science, Engineering and Technology Development [CCAWM], 2000). With women's increased participation in advanced high-school mathematics and science, the achievement gap is closing between men and women (National Science Board, 2000). However, our research seems to indicate that these academic gains may not translate into future career gains in IT.

In the later stages of this pipeline, undergraduate women continue to be underrepresented in computer-science, electrical-engineering, and computer-engineering majors (American Association of Undergraduate Women's [AAUW's] Educational Foundation, 2000; Vesgo, 2005). While in recent years, women's representation in the U.S. undergraduate population has risen to more than 50%, their overall numbers in computer-science programs have in fact declined (Freeman & Aspray, 1999; Vesgo, 2005). Figure 1 illustrates this trend using publicly available North Carolina State University (NCSU, 2005) data.

Figure 1. Recent gender trends at an engineering
school (CSC=computer science)

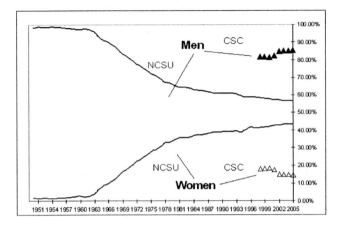

The National Science Foundation (NSF, 2000) has identified several related issues.

- Women are prevalent in fields such as psychology and biology.
- Women are less likely to choose science and engineering.
- Women are more likely to work part time.
- Women holding doctorates in science or engineering are less likely to be tenured or to hold the rank of full professors at educational institutions.
- Women scientists and engineers tend to receive lower salaries.

Positive developments are that the number of bachelor's degrees earned by women in all major science and engineering fields, except mathematics and computer science, are increasing, and the number of younger women engineers in management positions seems to be increasing as well (NSF, 2000).

On the other hand, the number of undergraduates seeking computer-science degrees is down sharply since 2000, and the percentage of women has also declined (Malcom et al., 2005; Vesgo, 2005; Zweben, 2005).

Hence, we continue to be concerned with the declining numbers of women in computer science, particularly as many researchers have reported that this problem has roots in girls' decisions, dispositions, and experiences as early as elementary school.

We now focus on these trajectories of personal and academic development among college-bound females aged 12 to 20. In this context, the term college bound implies middle-grade students that take algebra by the eighth grade, achieve in the top third of their class, and have a predisposition and preparation to take calculus later in their studies.

Underrepresentation of Females in Information Technology Fields

Women's underrepresentation in science, engineering, mathematics, and technology courses and careers has been studied extensively (e.g., AAUW's Educational Foundation, 2000; Malcom et al., 2005; Vesgo, 2005). While the achievement gap in mathematics and science is closing as more women select advanced courses in high-school science and mathematics (National Science Board, 2000), the enrollment of young women in CSC courses and advanced-placement classes in high school continue to remain low (AAUW's Educational Foundation).

A number of hypotheses have been generated to explain the declining enrollment of women in CSC as a function of girls' experiences from ages 12 to 18. For example, Freeman and Aspray (1999) cited the following issues.

- Lack of appropriate equipment in high school
- Lack of computer experiences
- Nature of computer games
- Lack of career guidance
- Perception of competitive environment
- Gender differences in socialization
- Perception of solitary occupation, requiring long hours in unsafe working environments
- Lack of women role models

We add to this the possibility of a very strong influence of parents of the girls (Berenson, Howe, & Vouk, 2005). These hypotheses are supported by an ethnographic study of 20 female CSC undergraduates that found that prior class experiences, as well as interest in computers and the promise of the field, were primary motivators for majoring in CSC (Margolis & Fisher, 2001).

Personal Factors

Previous research by Kerr (1997) sought to identify common patterns in attitude change among early adolescent girls and how these changes could affect the girls' achievement and motivation. In 1991, the AAUW conducted a landmark study documenting a steep decline in self-esteem among Caucasian adolescent girls, with a lesser decline for African-American girls. Other findings indicated a circular relationship among girls who enjoyed mathematics and science in that they had higher self-esteem and aspirations. Conversely, girls who had higher aspirations enjoyed mathematics and science. Family and school had a greater impact on self-esteem than the peer group (AAUW, 1991). Kerr noted that "pride in schoolwork, the belief that one is able to do many things well, and the feeling of being important in one's own family were the major contributors to self-esteem in this study" (p. 169).

Adolescent Girls

In elementary school, gifted girls demonstrate excellent social knowledge and achieve better grades than gifted boys (Kerr, 1997). By high school, however, while girls continued to attain high grades and were highly involved in extracurricular activities, many took less rigorous courses and suffered declines in their IQs (intelligence quotients), self-esteem, and confidence. Researchers have reported conflicts between conformity and achievement among gifted high-school girls (e.g., Arnot, David, & Weiner, 1999). Kerr reported that, throughout adolescence, girls tend to lower their expectations, choosing moderate over high prestige careers, attending less selective postsecondary institutions, and dropping out of graduate programs and professional training more often than men.

Parents, Teachers, and Mentors

Evidence exists that some teachers and parents have different expectations for girls and boys, and these expectations can impact children's achievement (e.g., Leder, 1992). While Lareau (1992) noted that mothers bear primary responsibility for their children in schools, Stevenson and Baker (1987) found that a majority of mothers spend more time and effort in helping their sons with schoolwork and are less likely to accept poor grades from their sons than their daughters. Hanna, Kundiger, and Larouche (1988) found that countries with high support for learning mathematics had fewer gender differences in achievement, but that in countries with low support for learning mathematics, the achievement gap increased in favor of males.

MODELING EDUCATIONAL PERSISTENCE INTO IT CAREERS

A general open issue is modeling educational persistence leading to undergraduate study in IT careers among young women who take college-prep algebra by eighth grade. Some specific issues include the following.

- Identification of school, social, and personal factors associated with young women's decisions to pursue and persist in undergraduate study in IT fields
- Creation and testing of models using the above factors to predict young women's decisions to pursue and persist in IT undergraduate study
- Development of appropriate tools and interventions to increase young women's interest in IT careers based on the Women in Information Technology (WIT) model

Approach and GoT Activities

From 1999 to 2003, NSF funded Girls on Track (GoT; NSF 9813902) to provide a year-round enrichment program for more than 200 talented girls in Grades 7 and 8 who were selected to take Algebra I on the fast track. The project has been so successful that the GoT camp still runs every summer. All GoT information and deliverables are online at http://ontrack.ncsu.edu.

The ages of the girls in GoT range from 11 to 13, with 60 to 65% of them being Caucasian, 25 to 30% being African American, and 10% being Asian. The girls attend a 2-week summer camp where they investigate community problems using mathematics and information technologies. In addition, girls in the program may receive tutoring in the fall and math mentoring in the spring. In the first few years, GoT also incorporated a professional-development component for middle-school math teachers, preservice teachers, and guidance counselors. As part of GoT and WIT (NSF 0204222), we have collected 7 years of quantitative and qualitative data from these girls, teachers, counselors, and parents. We are currently building models to assess and predict the factors that influence the decisions of our participants to pursue IT careers.

Types of quantitative data about the girls over time include standardized test results in mathematics and computer literacy; mathematics, science, and computer-science course selections; confidence in mathematics and information technologies; proportional reasoning scores; and career interests. Qualitative data include individual interviews, reflections, Web pages, and focus-group discussions. Additionally, data were collected from camp counselors and parents.

Completed analyses disclose several interesting findings. First, the data still indicate that proportional reasoning appears to be an important indicator of success in Algebra I for these talented girls. Overall, the correlation between proportional-reasoning scores and the aptitude and achievement scores seems to indicate that an understanding of proportional reasoning is an important contributor in both standardized measures of aptitude and achievement for Algebra I and for staying on track in math.

Another finding is that, although the girls enjoyed working with information technologies, many of them had not had extensive prior experience using IT tools. Before coming to the GoT summer camp, many girls did not use IT for academic purposes. For example, 82% percent of the Year 2 girls had either rarely or never used a spreadsheet. Similarly, 65% of these girls had rarely or never used IT tools to solve math or science problems. During the camp, girls were given opportunities to use the Internet and spreadsheets to investigate and research community problems. They created graphs, Web pages, and PowerPoint presentations to showcase their findings to parents, camp counselors, and other girls. In their postcamp surveys, 100% of the girls rated their PowerPoint experiences positively, and 96% rated their Web-page construction positively. Prior to

Girls on Track, girls tended to use the Web often, but in service of personal rather than scholarly interests.

In terms of mathematics attitudes, survey results for the girls from the first 3 years of camp reported high or very high levels of confidence in their abilities to do math-related activities. A postcamp survey highlighted that for Year 2 girls, 93% acknowledged that Girls on Track helped them "understand that math is a part of everyday life." One year after camp, the Year 2 girls reported statistically significant increases in their readiness to study advanced mathematics and increased confidence in their abilities.

It should be noted that GoT subjects were turning 14, the age at which a plunge in self-confidence has been found by a number of researchers. In terms of career choices, 66% of the Year 1 (1999) girls planned on entering math- or science-related careers. For Year 2 girls, survey results indicated that 81% of these girls expressed interest in working with computers and mathematics in their future careers. During the camp, they were given the opportunity to examine the relationship between various careers, salaries, and the mathematics needed to succeed in those careers. Over half of the girls agreed that the program helped them to "think of new ideas about careers, especially with technology." Results were similar for Year 3 girls. Unfortunately, recent telephone interviews indicate that most of the girls that are on track with respect to algebra are not thinking of continuing in IT careers (Berenson et al., 2005). We suspect that one of the reasons could well be the way we teach and deliver technology in schools and colleges.

Information Technology

What can be done about the issue? The next generation of IT users should, and do, expect not only the provision of effective, high-quality computing engines, but also equally good educational, training, and outreach (ETO) services. These diverse users will require differentiated support that is smoothly integrated with advanced computational and networking frameworks, and with the users' day-to-day operations. Next-generation users (including students and teachers) expect IT to come to them in the form of an appliance or service that aids their work flows (e.g., computational genetics, or the provision of state-of-the-art training in a remote school) rather than hampers them with excessive overhead.

Unfortunately, the seamless and widespread integration of new technologies into everyday operational and educational work flows is still to come. The situation is particularly acute in the following areas.

a. In very rapidly moving fields, such as bio and medical sciences, where users must keep pace with both rapid advances in their own field and the latest developments in computing and networking. These users often suffer from technological overload and ETO-service deficiency.

b. In the case of groups that are traditionally underrepresented in IT, for whom technological and ETO obstacles are exacerbated by economic, social, and other factors (e.g., women, minorities, rural school districts, smaller universities). These groups face a daunting catch-up task at best, and a continuously widening and dangerous technological and skills gap, with all that it implies, at worst.

What is needed is the development of methods and approaches for effectively reaching communities at technological risk, especially those concerned with math and sciences education, via facilitation, mentoring, and training programs. We saw the GoT effort as a major opportunity for the exploration and piloting of IT appliances for (a) the teaching of math and sciences and (b) the on-track steering and retention of underrepresented student populations in IT. We see WIT as doing that by assessing methods for the following.

1. Reducing the technological overload through the introduction of appliance-like high-technology solutions that enhance user activities and allow users to concentrate on their work flows

2. Promoting and increasing exposure to state-of-the-art ETO services in appropriate communities

An appliance-like solution does not mean just technological leveling of the field, but also the development of community- and group-appropriate pedagogical, training, and social interventions that in-crease the technological awareness of the community, reduce its aversion to technological change, ease that change, and advance its workforce into a state where it can sustain an influx of innovation through a combination of stable remote and local resources. This means easier access to state-of-the-art equipment (through network-based solutions), a better trained and continuously upgraded local instruction cadre, and an active technology assistance program that makes new technologies readily accessible and a source of eager anticipation rather than frustration for both teachers and students.

In addition to making IT more accessible, it is also important to involve young children in understanding and exploring the uses of IT for careers and real applications, and to carry this emphasis through the college level. We may also need to involve both parents and counselors early on to engage and interest girls in IT.

FUTURE TRENDS

As indicated by Blum and Frieze (2005) in their analysis of women in computer science at Carnegie Melon, the field of IT itself changes as the population of IT workers changes. Highlighted differences between men and women, such as choice of topic for study, are diminished when the number of women passes a certain threshold. When the workforce becomes more balanced, it is easier to recruit women, and opportunities for leadership and full participation become much more available (Blum & Frieze; Cohoon, in press). We also foresee that this broadening of participation will introduce new innovations and improved working conditions for all IT workers (CCAWM, 2000).

CONCLUSION

Although the demand for IT jobs continues to grow, the percentage of women in IT-related fields continues to decline. Some of the possible reasons for this decline include gender socialization, a lack of experience with and access to computers, a lack of career guidance, and perceptions of IT through the nature of computer games and work environments.

Findings from our Girls on Track and Women in Technology programs indicate that parental influence may also be a very strong factor in girls' career choices. We find it disconcerting that even girls who are excellent in math may not choose to enter IT careers. To address these issues, we believe it is important to make IT more accessible and to involve young children and their parents, teachers, and counselors early in the discovery of what makes IT exciting and useful.

NOTE

This work was supported in part by NSF Grant No. 9813902 and No. 0204222, and by IBM Corporation.

REFERENCES

American Association of Undergraduate Women (AAUW). (1991). *Shortchanging girls, short-changing America.* Washington, DC: Author.

American Association of Undergraduate Women's (AAUW's) Educational Foundation. (2000). *Tech-savvy: Educating girls in the new computer age.* Washington, DC: American Association of Undergraduate Women.

Arnot, M., David, M., & Weiner, G. (1999). *Closing the gender gap: Postwar education and social change.* Malden, MA: Blackwell.

Berenson, S., Howe, A., & Vouk, M. (2005). Changing the high school culture to promote interest in IT careers among high achieving girls. In *Proceedings of the Crossing Cultures, Changing Lives International Research Conference,* Oxford, UK.

Blum, L., & Frieze, C. (2005). In a more balanced computer science environment, similarity is the difference and computer science is the winner. *Computing Research News, 17*(3), 2-16.

Cohoon, J. M. (2005). Just get over it or just get on with it. In *Women and information technology: Research on under-representation* (pp. 205-238). Cambridge, MA: MIT Press.

Congressional Commission on the Advancement of Women and Minorities in Science, Engineering and Technology Development (CCAWM). (2000). *Land of plenty: Diversity as America's competitive edge in science, engineering and technology.* Washington, DC: Author.

Freeman, P., & Aspray, W. (1999). *The supply of information technology workers in the United States.* Washington, DC: Computing Research Association.

Hanna, G., Kundiger, E., & Larouche, C. (1988). *Mathematical achievement of grade 12 girls in fifteen countries.* Paper presented at the Sixth International Congress on Mathematical Education, Budapest, Hungary.

Kerr, B. S. (1997). *Smart girls: A new psychology of girls, women and giftedness.* Scottsdale, AZ: Gifted Psychology.

Lareau, A. (1992). Gender differences in parent involvement in schooling. In J. Wrigley (Ed.), *Education and gender equality* (pp. 207-224). Washington, DC: Falmer.

Leder, G. C. (1992). Mathematics and gender: Changing perspectives. In D. Grouws (Ed.), *Handbook of research on mathematics teaching and learning* (597-622). New York: Macmillan.

Malcom, S., Babco, E., Teich, A., Jesse, J. K., Campbell, L., & Bell, N. (2005). *Preparing women and minorities for the IT workforce: The role of nontraditional educational pathways.* Washington, DC: American Association for the Advancement of Science & Commission on Professionals in Science and Technology.

Margolis, J., & Fisher, A. (2001). *Unlocking the clubhouse: Women in computing.* Cambridge, MA: MIT Press.

National Science Board. (2000). *Science & engineering indicators.* Arlington, VA: Author.

National Science Foundation (NSF). (2000). *Women, minorities, and persons with disabilities in science and engineering.* Arlington, VA: Author.

North Carolina State University (NCSU). (2005). *NCSU planning and analysis. Enrollment history.* Retrieved October 4, 2005, from http://www2.acs.ncsu.edu/UPA/enrollmentdata/index.htm

P

Sargent, J. (2004). An overview of past and projected employment changes in the professional IT occupations. *Computing Research News, 16*(3), 1-21.

Stevenson, D. L., & Baker, D. P. (1987). The family-school relation and the child's school performance. *Child Development, 58*, 1348-1357.

Vesgo, J. (2005). CRA Taulbee trends: Female students & faculty. *Computing Research Association Taulbee survey 2005.* Retrieved October 4, 2005, from http://www.cra.org/info/taulbee/women.html

Zweben, S. (2005). 2003-2004 Taulbee Survey: Record Ph.D. production on the horizon. Undergraduate enrollments continue in decline. *Computing Research News, 17*(3), 7-15.

KEY TERMS

ETO Services: Educational, training, and outreach services. IT used for these services is often the only exposure that underrepresented populations may have to computers and IT.

Girls on Track: An intervention program designed to keep talented middle-school girls on the "fast math track." (http://ontrack.ncsu.edu)

IT Appliance: Software that can be readily used by novices in a natural way, without technical training, much like a refrigerator or toaster.

IT Career: A career requiring an electrical-engineering, computer-science, or computer-engineering degree. Emphasis is placed on technical and creative roles rather than support roles.

Network-Based Education: The use of tools over a network for education and training.

Pipeline: It identifies sources of potential IT workers, including preparatory courses such as algebra and calculus. Particular focus is paid to places and issues where people, and particularly women, leave the pipeline, such as the choice of less advanced math classes in high school.

Women in Technology (WIT): A longitudinal study of the Girls on Track program designed to model the educational persistence of young women in IT-related fields. (http://wit.ncsu.edu)

Personality Characteristics of Established IT Professionals I: Big Five Personality Characteristics

Ronald A. Ash
University of Kansas, USA

Joshua L. Rosenbloom
University of Kansas & National Bureau of Economic Research, USA

LeAnne Coder
University of Kansas, USA

Brandon Dupont
Wellesley College, USA

INTRODUCTION

Women are under represented in the information technology (IT) workforce relative to the overall labor force, comprising about 35% of the IT workforce and 45% of the overall labor force (Information Technology Association of America, 2003). A basic question to be addressed is whether this under representation is a function of barriers to employment of women in this career field, or a function of career-related choices that a majority of women make during their lives. The research reported here is part of a series of studies attempting to better understand the reasons underlying this under representation of women in this reasonably lucrative profession. Through a grant provided by the National Science Foundation (NSF 29560) and in partnership with Consulting Psychologists Press, we have been able to design and conduct an extensive survey of professional workers, IT professionals and a comparable set of non- IT professionals. The non-IT professionals included individuals who are similar to the IT sample in terms of education level (but not specific degree fields) and who work in jobs with comparable human attribute demands, including written comprehension, oral comprehension, oral expression, written expression and deductive reasoning. The survey items include measures of Big Five personality constructs (NEOAC) and Core Self-Evaluations (CSE). The purpose of this article is to document similarities and differences between established IT and non-IT professionals and between males and females on these variables, thereby establishing a benchmark for comparisons with future samples of IT professionals.

Why is this worth doing? Because in the last decade of the 20th century, a critical mass of knowledge related to personality in work organizations has developed. Personality contributes to all that happens during a person's career, and informs our understanding of things like work motivation, job attitudes, citizenship behavior, leadership, teamwork, well-being and organizational culture. Increasingly, we have realized that personality plays an important role in determining who is hired and fired (see Schneider & Smith, 2004).

BACKGROUND

The Survey Sample

Data were obtained from individuals who voluntarily responded to an online survey prepared and managed by the Policy Research Institute at the University of Kansas between December 2003 and September 2004. Participation in the survey was solicited from employees at several large organizations with offices in the central United States (U.S.), and from business school and computer science alumni of a large mid-western university. Note that this business school offers several management information systems (MIS) courses.

Each survey respondent was asked to indicate his or her current career field (one of 13 categories or "Other") and specific job title (open-ended). The researchers used this information to classify respondents as either an IT or non-IT professional. The sample consists of 703 working professionals who completed the survey. Seventy-three percent (510) are non-IT professionals; 27% (193) are IT professionals. Fifty-eight percent (405) are male; 42% (298) are female. The non-IT professionals include accountants, auditors, CEOs, CFOs, presidents, consultants, engineers, managers, administrators, management analysts, scientists, technicians, nurses, teachers and so forth. The IT professionals include application developers, programmers, software engineers, database administrators, systems analysts, Web administrators and Web developers.

Respondent Demographic Information

Table 1 shows means and standard deviations on several demographic variables for the sample.

The age range of the respondents is 22 to 70 years. The mean age is 38.7 years, with a standard deviation of 9.8. Twenty-two percent of the respondents are in their 20s, 33% in their 30s, 30% in their 40s, 13% in their 50s, and 2% are 60 years of age or older. On average, IT respondents are 1.8 years older than non-IT respondents.

The respondents are highly educated. Ninety-three percent hold 4-year college degrees, and 45% have graduate school degrees. Six percent have some college, and less than 1% report having only completed high school. Mean years of formal education is 16.8, with a standard deviation of 1.5. On average, the non-IT respondents have 0.8 year more formal education than the IT respondents.

IT professionals in the sample report having worked for pay 20.1 years on average, almost 2 years more than the non-IT professionals. This is consistent with the average age of the IT respondents being almost 2 years older than that of the non-IT respondents. The IT professionals report having worked in their current career field 13.8 years on average, 3 years more than the non-IT professionals. Respondents report having worked for their current employer for an average of 7 years (standard deviation = 6.5 years), and having held their current positions for an average of 4.3 years (standard deviation = 4.4 years). They report having held an average of 3.3 jobs in their current career field (standard deviation = 2.2).

Male respondents report being exposed to computers an average of 1.6 years earlier than female respondents (at 15.4 vs. 17.0 years of age). The majority (55%) took no computer science courses in high school, 21% took one computer science course,

Table 1. Demographic information for the professional worker sample (NEOAC & CSE)

Demographic Variable	Means (top) and Standard Deviations (bottom)								
	Total Sample N = 703	Non-IT N = 510	IT N = 193	Male N = 405	Female N = 298	Non-IT Male N = 273	Non-IT Female N = 237	IT Male N = 132	IT Female N = 61
Age	38.7	38.2	40.0	38.6	38.9	38.3	38.1	39.2	41.8
	9.8	9.7	10.0	9.8	9.8	9.7	9.8	10.2	9.3
Years of formal education	16.8	17.0	16.2	16.9	16.6	17.3	16.8	16.2	16.0
	1.5	1.4	1.5	1.4	1.5	1.2	1.5	1.6	1.3
Years worked for pay	18.7	18.2	20.1	18.7	18.6	18.3	18.0	19.6	21.1
	9.5	9.4	9.7	9.7	9.2	9.6	9.1	9.9	9.2
Years in current career field	11.6	10.8	13.8	11.8	11.4	11.0	10.4	13.3	15.1
	8.1	7.8	8.6	8.2	8.1	7.9	7.6	8.4	8.7
Years with current employer	7.0	6.8	7.5	7.0	7.1	6.9	6.8	7.1	8.3
	6.5	6.6	6.3	6.4	6.8	6.5	6.8	6.2	6.4
Years in current position	4.3	4.2	4.6	4.4	4.1	4.4	3.9	4.4	5.2
	4.4	4.4	4.3	4.5	4.2	4.7	4.1	4.2	4.6
Number of jobs held in current career field	3.3	3.2	3.4	3.2	3.3	3.2	3.2	3.3	3.7
	2.2	2.2	2.3	2.3	2.1	2.3	2.1	2.3	2.2
Age first exposed to computers	16.1	16.2	15.9	15.4	17.0	15.6	16.8	15.0	17.8
	6.8	7.0	6.4	6.2	7.5	6.3	7.8	6.1	6.5
Number of computer science courses taken in high school	0.8	0.8	0.9	0.9	0.7	0.9	0.7	0.9	0.7
	1.2	1.1	1.3	1.2	1.1	1.2	1.0	1.2	1.4
Number of computer science courses taken in college	5.3	3.5	10.1	6.2	4.0	4.2	2.6	10.5	9.1
	7.1	5.3	9.1	8.0	5.6	6.1	3.9	9.6	7.8

P

15% took two, and less than 10% report taking three or more computer science courses in high school.

Differences show up in terms of the number of computer science courses taken in college. The total sample mean is 5.3 courses with a standard deviation of 7.1, indicating substantial variability. IT professionals took an average of 10.1 computer science courses, while non-IT professionals took 3.5 computer science courses in college. Males took 6.2 computer science courses in college, while females took an average of 4.0 computer science courses.

The Big Five Personality Constructs

During the 1980s, after some four to five decades of research, development and elaboration, the five factor model (FFM) of personality—also called the "Big Five" model—was recognized as representing the five most basic dimensions underlying the traits identified in both natural languages and psychological questionnaires (Digman, 1990). Essentially, five synonym clusters appear to account for the majority of differences between individual personalities. These five personality traits reflect the physiological activities of different underlying arousal systems, and represent predispositions to behave in certain ways when in the presence of particular stimuli (Howard & Howard, 2001). The five traits of this model are explained briefly in the following paragraphs. These descriptions are paraphrased largely from Howard and Howard (2001), because their descriptions use less psychological terminology and are more accessible to the broader spectrum of working professionals.

Factor N, or Neuroticism, refers to one's need for stability. A person high in N is very reactive and prefers a stress-free work environment. A person low in N is typically very calm and relatively unaffected by stress that might result in ineffective behavior in others. In general, women score higher than men on measures of N.

Factor E, or Extraversion, refers to one's positive emotionality or sociability. A person high in E likes to be in the thick of the action, typically interacting with other people, while a person low in E likes to be away from the noise and hubbub, crowds and so forth. In general, there are no systematic differences between women and men on measures of E.

Factor O, or Openness to Experience, refers to one's originality or imagination. A person scoring high in O has a voracious appetite for new ideas and activities, and is easily bored with routine or highly familiar situations. A person low in O prefers familiar territory and tends to be more practical, conventional and conservative. In general, there are no systematic differences between women and men on measures of O.

Factor A, or Agreeableness, refers to one's accommodation or adaptability. A person high in A tends to accommodate or adapt to the wishes and needs of others, and is often viewed as cooperative. A person low in A tends to focus on his or her own personal needs and priorities, and is often described as competitive or critical. In general, women score higher than men on measures of A.

Factor C, or Conscientiousness, refers to one's will to achieve, or consolidation. A person high in C tends to focus or consolidate his or her energy and resources on accomplishing one or more goals, and typically appears to be well organized, ambitious and strong-willed. A person low in C prefers a more spontaneous work style, is more comfortable switching from one task to another, is typically lackadaisical in working toward his or her goals, and often appears to be less organized, less punctual and so forth. In general, there are no systematic differences between women and men on measures of C.

Core Self-Evaluations

CSE is a broad personality trait that has been shown to be a significant predictor of job satisfaction and job performance (Judge, Erez, Bono, & Thoresen, 2003). It is a combination of four primary personality traits that have been featured prominently in psychological research for decades. These include *self-esteem*, the overall value one places on oneself as a person; *generalized self-efficacy*, an evaluation of how well one can perform across a variety of situations; *neuroticism* (Factor N of the Big Five), the tendency to have a negativistic cognitive/explanatory style and to focus on negative aspects of the self; and *locus of control*, beliefs about the causes of events in one's life—locus is internal when individuals see events as being contingent upon their own behavior, and external when they see events as caused largely by forces and events outside themselves and not under their control. CSE is a basic, fundamental appraisal of one's worthiness, effectiveness and capability as a person. Individuals high in CSE are generally more satisfied with their jobs, their work and their lives than are individuals low in CSE. Individuals high in CSE also tend to perform their work and their jobs better than those low in CSE. Judge, Erez, Bono, and Thoresen (2003) have suggested that existing measures of Neuroticism are too narrow to capture self-evaluations, perhaps due to the origin of Neuroticism measures in psychopathology, and hence appear to be less valid predictors of work-related outcomes as compared to CSE. Judge and his colleagues have developed and convincingly demonstrated both the reliability and multi-faceted construct validity of a 12-item direct measure of CSE—the *Core Self-Evaluations Scale* (CSES). There are no systematic differences between women and men on this measure.

COMPARISON OF IT AND NON-IT MALE AND FEMALE PROFESSIONALS ON BIG FIVE AND CSE PERSONALITY VARIABLES

Table 2 contains the means and standard deviations for 273 non-IT males, 237 non-IT females, 132 IT males and 61 non-IT females on measures of N, E,

O, A, C and CSE. The results are expressed in standardized score (T-score) format where the norm group mean = 50 and the norm group standard deviation = 10. The measures of N, E, O, A and C are derived from the 12-item scales of the NEO Five Factor Inventory (NEO-FFI), and standardized using combined gender norms derived from a sample of 500 men and 500 women selected in a stratified manner designed to match U.S. census projections for 1995 in the distribution of age and race groups (Costa & McCrae, 1992). The measure of CSE is the 12-item CSES mentioned above, standardized using norms derived from four different samples yielding CSES results on 841 individuals (Judge et al., 2003).

Table 2 also contains the results of a two-factor (gender X career field) analysis of variance used to test for significant differences in the personality variables as a function of gender (Male/Female), career field (IT/non-IT) and the interaction of gender and career field. This analysis reveals whether there are differences on the respective personality variables between males and females in the total sample (gender effect) and between IT and non-IT professional workers (career field effect), and whether there is an interaction effect (gender by career field interaction) indicating that males or females within either IT or non-IT are different from male and female professional workers in general.

Table 2. Big Five and Core Self-Evaluations personality scale results by gender (M/F) and career field (IT/Non-IT)

| Personality Construct | Means (top) and Standard Deviations (bottom) | | | | Two Factor ANOVA Results Effects (p < .05) | | |
| | Non-IT Career Field | | IT Career Field | | | | Gender by Career Field Interaction |
	Male N = 273	Female N = 237	Male N = 132	Female N = 61	Gender	Career Field	
N Neuroticism	45.8[a] 10.9	49.2[b] 11.0	46.4[ab] 9.1	48.7[ab] 11.3	Yes	No	No
E Extraversion	55.3 10.7	55.5 11.0	50.6[a] 10.6	56.5 11.2	Yes	Marginal	**Yes**
O Openness to Experience	52.1[ab] 11.0	51.9[a] 11.4	55.0[b] 10.7	53.9[ab] 10.3	No	Yes	No
A Agreeableness	48.4[a] 11.2	52.6[b] 10.7	48.3[a] 11.1	52.3[ab] 11.5	Yes	No	No
C Conscientiousness	52.2[a] 9.9	54.5[b] 10.3	49.6[a] 9.6	53.4[ab] 10.5	Yes	Yes	No
CSE Core Self-Evaluations	48.6 10.8	47.4 10.5	47.3 8.6	47.5 11.4	No	No	No

[ab]Means in each row with common superscripts are **not** reliably different from each other.

There is an overall gender effect for N. Females score higher on Neuroticism than males, a common finding for this personality construct. There is neither a career field effect nor an interaction effect for N. IT professionals are similar to non-IT professionals in the need for stability.

For E there is a gender effect, a marginal career field effect and a gender-by-career field interaction effect. The Extraversion mean for IT males is about one-half standard deviation lower than those for IT females, non-IT females and non-IT males. While IT females are significantly more extraverted than IT males, IT females are similar in Extraversion to non-IT male and female professionals. IT males are different—lower—in Extraversion compared to other professional workers, including IT females.

For O there is a career field effect. On average, IT professionals score higher on Openness to Experience than do non-IT professionals, indicating that they are somewhat more original and imaginative, and probably more easily bored. There are no gender differences on O, and there is no interaction effect.

There is an overall gender effect for A. Females score higher on Agreeableness than males, a common finding for this personality construct, indicating that female professionals are more accommodating, helpful and cooperative than male professionals. There is neither a career field effect nor an interaction effect for A. IT professionals are similar to non-IT professionals in accommodation and adaptability.

For C there is a gender effect and a career field effect, but no interaction effect. On average, female professionals score higher on Conscientiousness than males, and non-IT professional score higher than IT professionals. Detailed analysis (post hoc comparisons of individual group means) shows that mean Conscientiousness for non-IT females is significantly higher as compared to the means for both non-IT and IT males.

For CSE, there are no differences between males and females or IT and non-IT professionals, and no interaction effect, either.

FUTURE TRENDS

For the early part of the 21st century, at least, it appears that the FFM of personality and its "Big Five" personality constructs will be the predominant broad measures of personality applied by social science practitioners in studying and helping people understand and adjust to work and other social situations (see Smith & Schneider, 2004). This article documents similarities and differences between established IT and non-IT professionals and between males and females on these variables at this point in time, and presents a benchmark against which to measure and evaluate potential changes in the future. For example, it is possible that recent trends for businesses to outsource code writing jobs may result in changes in the nature of the work of IT-professionals remaining in these businesses, leading to some different skill set requirements—possibly higher levels of problem solving or managerial skills (see Darais, Nelson, Rice, & Buche, 2004; Gallivan, 2004; Todd, McKeen, & Buche, 2004). This, in turn, could eventually be reflected in changes in broad personality characteristics of established professionals in the IT field.

CONCLUSION

In this article, we report descriptive results comparing established professionals in IT and non-IT career fields by gender on six major personality constructs. We found significant differences between IT and non-IT professionals on three of the six constructs, and significant differences between male and female professionals on four of the six constructs. However, we found significant career field by gender interaction effects for only one of the constructs.

IT professionals are higher than non-IT professionals on Openness to Experience, indicating that they are somewhat more original and imaginative, and more easily susceptible to boredom with routine.

IT professionals are lower than non-IT professionals on Extraversion, indicating that they prefer to be away from the noise, hubbub, crowds and so forth. IT professionals are also lower than non-IT professionals on Conscientiousness, indicating that they tend to be less organized, less punctual, more spontaneous and so forth.

Female professionals are higher than male professionals on Neuroticism, indicating that they are more reactive and have a stronger preference for stress-free environments. Female professionals are also higher in Extraversion, indicating stronger preferences for being in the thick of the action and interacting with others. Female professionals are higher than male professionals in Agreeableness, indicating higher levels of accommodating or adapting to the wishes and needs of others. Finally, female professionals are higher than male professionals on Conscientiousness, indicating that on average they are more organized, ambitious, goal-directed and focused.

While there are several notable personality differences between IT professionals and other professionals, and between male and female professionals, there are very few differences (only one uncovered in this study) between male and female IT professionals that do not also exist between male and female professionals in general. Males in the IT profession tend to be significantly lower in extraversion relative to females in IT and other non-IT male and female professionals. Females in IT are quite similar to other male and female professionals in terms of positive emotionality and sociability, whereas males in IT have a significantly stronger preference to be away from noise, crowds, social stimulation and so forth.

Is the nature of IT work such that individuals lower in extraversion have a significantly better chance of being successful and happy in this career field? Other research on occupational personality variables suggests that relative to other professionals, IT professionals tend to be more realistic, more investigative, less social and less enterprising, prefer to work with ideas/data/things rather than people, prefer to work alone, prefer to accomplish tasks independently and prefer to minimize risks (Ash, Rosenbloom, Coder, & DuPont, 2005). This certainly sounds like a profession dominated by individuals lower in extraversion.

These findings raise the possibility that a higher proportion of males relative to females are attracted to work in IT due to a better match of that work with personality differences underlying preferences for aspects of IT work. In general, women are more social and more cooperative relative to men, but given potential significant changes in the nature of IT work, the status of underlying broad personality characteristics of established professionals in this field may be evolving. Of course, we have examined only one aspect of the potential set of causes for the notable current difference in proportions of males

and females in the IT career field, and at this point, the role of gender in career choice remains an open question.

REFERENCES

Ash, R. A., Rosenbloom, J. L., Coder, L., & DuPont, B. (2005). *Gender differences and similarities in personality characteristics for information technology professionals.* Retrieved September 21, 2005, from www.ku.edu/pri/ITWorkforce/pubs/GenderDifferences.shtml

Costa, P. T., & McCrae, R. R. (1992). *NEO PI-R: Professional manual.* Lutz, FL: Psychological Assessment Resources.

Darais, K. M., Nelson, K. M., Rice, S. C., & Buche, M. W. (2004). Identifying the enablers and barriers of IT personnel transition. In M. Igbaria & C. Shayo (Eds.), *Strategies for managing IS/IT personnel* (pp. 92-112). Hershey: Idea Group Publishing.

Digman, J. (1990). Personality structure: Emergence of the Five-Factor Model. *Annual Review of Psychology, 41,* 417-440.

Gallivan, M. J. (2004). Examining IT professionals' adaptation to technological change: The influence of gender and personal attributes. *Database for Advances in Information Systems, 35*(3), 28-49.

Howard, P. J., & Howard, J. M. (2001). *The owner's manual for personality at work.* Marietta: Bard Press.

Information Technology Association of America. (2003). *Report of the ITAA blue ribbon panel on IT diversity.* Retrieved December 14, 2004, from www.itaa.org/workforce/studies/diversityreport.pdf

Judge, T. A., Erez, A, Bono, J. E., & Thoresen, C. J. (2003). The core self-evaluations scale: Development of a measure. *Personnel Psychology, 56,* 303-331.

Schneider, B., & Smith, D. B. (2004). *Personality and organizations.* Mahwah: Lawrence Erlbaum Associates.

Smith, D. B., & Schneider, B. (2004). Where we've been and where we're going: Some conclusions regarding personality and organizations. In B. Schneider & D. B. Smith (Eds.), *Personality and organizations*. Mahwah: Lawrence Erlbaum Associates.

Todd, P. A., McKeen, J. D., & Gallupe, R. B. (1995). The evolution of IS skills: A content analysis of IS job advertisements from 1970 to 1990. *MIS Quarterly, 19*(1), 1-27.

KEY TERMS

Agreeableness: One's accommodation or adaptability.

Conscientiousness: One's will to achieve, or consolidation.

Core Self-Evaluation (CSE): A broad personality trait that is a combination of four primary personality traits—self-esteem, generalized self-efficacy, neuroticism and locus of control.

Extraversion: One's positive emotionality or sociability.

Generalized Self-Efficacy: An evaluation of how well one can perform across a variety of situations.

Locus of Control: Beliefs about the causes of events in one's life—locus is internal when individuals see events as being contingent upon their own behavior, and external when they see events as caused largely by forces and events outside themselves and not under their control.

Neuroticism: One's need for stability.

Openness to Experience: One's originality or imagination.

Personality: A set of scores or descriptive terms that describe the individual being studied in terms of the variables or dimensions that occupy a central position within a particular theory; the most outstanding or salient impression that one creates in others.

Self Esteem: The overall value one places on oneself as a person.

Personality Characteristics of Established IT Professionals II: Occupational Personality Characteristics

Ronald A. Ash
University of Kansas, USA

Joshua L. Rosenbloom
University of Kansas & National Bureau of Economic Research, USA

LeAnne Coder
University of Kansas, USA

Brandon Dupont
Wellesley College, USA

INTRODUCTION

Women are underrepresented in the information technology (IT) workforce relative to the overall labor force, comprising about 35% of the IT workforce and 45% of the overall labor force (Information Technology Association of America, 2003). A basic question to be addressed is whether this underrepresentation is a function of barriers to employment of women in this career field or a function of career-related choices that a majority of women make during their lives. The research reported here is part of a series of studies attempting to better understand the reasons underlying this underrepresentation of women in this reasonably lucrative profession. Through a grant provided by the National Science Foundation (NSF 29560) and in partnership with Consulting Psychologists Press, we have been able to design and conduct an extensive survey of professional workers, IT professionals and a comparable set of non-IT professionals. The non-IT professionals included individuals who are similar to the IT sample in terms of education level (but not specific degree fields) and who work in jobs with comparable human attribute demands, including written comprehension, oral comprehension, oral expression, written expression and deductive reasoning. The survey items include measures of occupational personality constructs (RIASEC) and *Personal Style Scales* (*PSS*). The purpose of this article is to document similarities and differences between established IT and non-IT professionals and between males and females on these variables, thereby establishing a benchmark for comparisons with future samples of IT professionals.

Why is this worth doing? Because in the last decade of the 20th century, a critical mass of knowledge related to personality in work organizations developed. Personality contributes to all that happens during a person's career, and informs our understanding of things like work motivation, job attitudes, citizenship behavior, leadership, teamwork, well-being, and organizational culture. Increasingly we have realized that personality plays an important role in determining who is hired and fired (cf. Schneider & Smith, 2004), as well as who voluntarily stays in and leaves organizations (cf. Harmon, Hansen, Borgen, & Hammer, 1994; Holland, 1997).

BACKGROUND

The Survey Sample

Data were obtained from individuals who voluntarily responded to an online survey prepared and managed by the Policy Research Institute at the University of Kansas between December 2003 and September 2004. Participation in the survey was solicited from employees at several large organizations with offices in the central United States (U.S.), and from business school and computer science alumni of a large Midwestern university. Note that this business school offers several management information systems (MIS) courses.

Each survey respondent was asked to indicate his or her current career field (one of 13 categories or "Other") and specific job title (open-ended). The researchers used this information to classify respondents as either an IT or non-IT professional. The sample consists of 523 working professionals who completed the survey and the revised *Strong Interest Inventory* (SII) (Donnay, Morris, Schaubhut, & Thompson, 2005)—the measures of occupational personality and personal style scales used in this study. Seventy-three percent (382) are non-IT professionals; 27% (141) are IT professionals. Fifty-four percent (285) are male; 46% (238) are female. The non-IT professionals include accountants, auditors, CEOs, CFOs, presidents, consultants, engineers, managers, administrators, management analysts, scientists, technicians, nurses, teachers, and so forth. The IT professionals include application developers, programmers, software engineers, database administrators, systems analysts, Web administrators, and Web developers.

Respondent Demographic Information

Table 1 shows means and standard deviations on several demographic variables for the sample.

The age range of the respondents is 22 to 70 years. The mean age is 39.3 years, with a standard deviation of 10.0. Twenty percent of the respondents are in their 20s, 33% in their 30s, 30% in their 40s, 14% in their 50s, and 3% are 60 years of age or older. On average, the IT respondents are 1.8 years older than the non-IT respondents.

The respondents are highly educated. Ninety-two percent hold 4-year college degrees, and 45% have graduate school degrees. Five percent have some college, and less than 1% report having only completed high school. Mean years of formal education is 16.8, with a standard deviation of 1.5. On average, the non-IT respondents have 1 year more formal education than the IT respondents.

IT professionals in the sample report having worked for pay 20.6 years on average, almost 2 years more than the non-IT professionals. This is consistent with the average age of the IT respondents being almost 2 years older than that of the non-IT respondents. The IT professionals report having worked in their current career field 14.2 years on average, 3 years more than the non-IT professionals.

Respondents report having worked for their current employer for an average of 7.4 years (standard deviation = 6.9 years), and having held their current positions for an average of 4.5 years (standard deviation = 4.8 years). They report having held an

Table 1. Demographic information for the professional worker sample (RIASEC & PSS)

Demographic Variable	Means (top) and Standard Deviations (bottom)								
	Total Sample N = 523	Non-IT N = 382	IT N = 141	Male N = 285	Female N = 238	Non-IT Male N = 190	Non-IT Female N = 192	IT Male N = 95	IT Female N = 46
Age	39.3	38.8	40.6	39.3	39.2	39.1	38.5	39.9	42.0
	10.0	9.9	10.0	10.1	9.8	10.0	9.9	10.4	9.1
Years of formal education	16.8	17.1	16.1	17.0	16.6	17.4	16.7	16.2	15.8
	1.5	1.5	1.5	1.5	1.6	1.3	1.6	1.6	1.2
Years worked for pay	19.2	18.7	20.6	19.4	19.0	19.0	18.5	20.3	21.0
	9.7	9.6	9.8	10.1	9.2	10.0	9.2	10.1	9.2
Years in current career field	11.9	11.1	14.2	12.2	11.6	11.5	10.8	13.7	15.1
	8.3	8.0	8.6	8.3	8.2	8.1	7.9	8.6	8.8
Years with current employer	7.4	7.3	7.5	7.3	7.4	7.5	7.2	7.1	8.2
	6.9	7.1	6.4	6.8	7.0	7.0	7.1	6.3	6.4
Years in current position	4.5	4.4	4.7	4.6	4.3	4.8	4.0	4.2	5.5
	4.8	4.9	4.5	5.1	4.4	5.5	4.3	4.3	4.9
Number of jobs held in current career field	3.3	3.2	3.5	3.3	3.3	3.3	3.2	3.4	3.7
	2.3	2.2	2.4	2.3	2.2	2.2	2.2	2.5	2.2
Age first exposed to computers	16.4	16.4	16.3	15.7	17.2	15.8	17.0	15.4	18.1
	7.1	7.4	6.3	6.6	7.6	6.7	8.0	6.4	5.9
Number of computer science courses taken in high school	.8	0.8	0.8	0.9	0.7	0.9	0.6	0.9	0.7
	1.1	1.1	1.2	1.2	1.0	1.3	0.9	1.2	1.1
Number of computer science courses taken in college	5.0	3.2	10.0	6.0	3.8	4.0	2.5	10.3	9.4
	7.2	5.1	9.5	8.3	5.3	6.2	3.4	10.3	7.7

average of 3.3 jobs in their current career field (standard deviation = 2.3).

Male respondents report being exposed to computers an average of 1.5 years earlier than female respondents (at 15.7 vs. 17.2 years of age). The majority (56%) took no computer science courses in high school, 22% took one computer science course, 14% took two, and less than 10% report taking three or more computer science courses in high school.

Differences show up in terms of the number of computer science courses taken in college. The total sample mean is 5.0 courses with a standard deviation of 7.2, indicating substantial variability. IT professionals took an average of 10.0 computer science courses, while non-IT professionals took 3.2 computer science courses in college. Males took 6.0 computer science courses in college, while females took an average of 3.8 computer science courses.

Occupational Personality and the General Occupational Theme (GOT) Scales

In 1927, E.K. Strong introduced the *Strong Vocational Interest Blank (SVIB)* (now *SII*) (Most, 1993). This measure was used to determine the degree of similarity between a person's interests and those of workers in an occupation. Strong realized in the late 1930s that a systematic clustering of the scales was necessary, but he was unable to find a system that had reliable psychometric qualities. In 1959, Holland introduced six basic occupational interest categories that closely resembled the dimensions found in research on vocational interests using the SVIB. Holland's classification system was an extension of the trait and factor theory from the 1920s and implied that the main goal of vocational counseling is to match people and jobs. In 1974, Strong's empiricism and Holland's theory were combined to develop the GOT (Harmon et al., 1994). The six vocational types of the GOT model are described below. The descriptions are paraphrased from Harmon et al. (1994) and Holland (1997).

The Realistic Theme, or R, refers to a person's preference for activities that entail the explicit, ordered or systematic manipulation of objects, tools, and machines. Realistic types enjoy jobs and activities that involve mechanical manipulations or repairs and construction. They are interested in action rather

than thought and prefer concrete problems to ambiguous, abstract problems. Sample Realistic occupations include auto mechanic, gardener, plumber, and engineer.

The Investigative Theme, or I, refers to a person's preference for activities that entail the systematic or creative investigation of physical, biological, and cultural phenomena. Investigative types enjoy gathering information, uncovering new facts or theories, and analyzing and interpreting data. They prefer to rely on themselves rather than on others in a group project. Sample Investigative occupations include college professor, physician, psychologist, and chemist.

The Artistic Theme, or A, refers to a person's preference for activities that are ambiguous, free, non-systematic and that entail the manipulation of materials to create art forms or products. Artistic types have a great need for self-expression. They are also comfortable in academic or intellectual environments. Sample Artistic occupations include artist, lawyer, librarian, musician, architect, reporter and English teacher.

The Social Theme, or S, refers to a person's preference to lead others or for activities that entail the manipulation of others to inform, train, develop, cure, or enlighten. Social types enjoy working with people, sharing responsibilities, and being the center of attention. They also like to solve problems through discussions of feelings and interactions with others. Sample Social occupations include elementary school teacher, nurse, social worker, and occupational therapist.

The Enterprising Theme, or E, refers to a person's preference for activities that entail the manipulation of others to attain organizational goals or economic gain. Enterprising types seek positions of power, leadership, and status. They like to take financial risks and participate in competitive activities. Sample Enterprising occupations include traveling salesperson, buyer, realtor, sales manager, and marketing executive.

The Conventional Theme, or C, refers to a person's preference for activities that entail the explicit, ordered, systematic manipulation of data. Conventional types often enjoy mathematics and data management activities. These individuals work well in large organizations but do not show a distinct preference for or against leadership positions.

Sample Conventional occupations include book-keeper, accountant, banker, actuary, and proof-reader.

Occupational Personality and the Personal Style Scales (PSS)

The *PSS* were added to the *SII* in 1994. The *PSS* measure a person's broad styles of living, learning, playing and working. They complement the traditional vocational interest scales (i.e., RIASEC) that measure preferences for more specific aspects of the work itself. A distinguishing characteristic of the *PSS* is that they are constructed as bipolar scales, with a distinctive style (or preference) associated with both the right and left pole of each scale (Harmon et. al, 1994). There are five *PSS* attached to the *SII*: work style, learning environment, leadership style, risk-taking/adventure and team orientation. Descriptions for the first four were taken from Harmon et al., (1994).

The Work Style Scale distinguishes individuals who prefer to work with ideas, data or things (left pole or low scores) from those who prefer to work with people (right pole or high scores). The "works with people" pole links strongly to the Enterprising and Social types. The "works with ideas/data/things" pole ties strongly to the Realistic and Investigative types. Occupations whose members prefer to work with ideas, data or things include biologist, chemist and computer programmer. Occupations whose members prefer to work with people include high school counselor, flight attendant and human resources director.

The Learning Environment Scale differentiates people who prefer more practically oriented, hands-on learning situations (left pole or low scores) from those who prefer academic learning environments (right pole or high scores). Occupations whose members prefer an academic learning environment include college professor, lawyer, psychologist and physicist. Occupations whose members prefer a practical learning environment include auto mechanic, dental assistant and nurse.

The Leadership Scale contrasts those who lead by example and prefer to work alone (left pole or low score) from those who enjoy meeting, directing, persuading and leading other people (right pole or

high score). Occupations whose members prefer a "leads by example" leadership style include auto mechanic, chemist, farmer and mathematician. Occupations whose members prefer a "directs others" leadership style include elected public official, minister, broadcaster and realtor.

The Risk Taking/Adventure Scale differentiates those who like to "play it safe" (left pole or low scores) from those who like to take a chance or be spontaneous (right pole or high scores). Occupations whose members prefer a "play it safe" approach include librarian, mathematician and dental hygienist. Occupations whose members prefer the "take a chance" approach include an athletic trainer, police officer and electrician.

In 2004, a new *PSS*, Team Orientation, was added to the *SII*. This construct distinguishes those who prefer to accomplish tasks independently (low scores or left pole) from those who prefer to accomplish tasks as part of a team (high score or right pole). Occupations whose members prefer to accomplish tasks independently include artist, graphic designer, medical illustrator and musician. Occupations whose members prefer to accomplish tasks as part of a team include operations manager, school administrator, sales manager, and rehabilitation counselor (Donnay, Morris, Schaubhut, & Thompson, 2005).

COMPARISON OF IT AND NON-IT MALE AND FEMALE PROFESSIONALS ON OCCUPATIONAL AND PERSONAL STYLE PERSONALITY VARIABLES

Table 2 contains the means and standard deviations for 190 non-IT males, 192 non-IT females, 95 IT males, and 46 IT females on the RIASEC occupational personality measures. The results are expressed in standardized score format (T-Scores) where the norm group mean and standard deviation are 50 and 10, respectively. The measures of R, I, A, S, E and C are derived from the 20-item scales of the *SII* and are standardized using combined gender norms derived from a sample of 9,484 men and 9,467 women (Harmon et al., 1994). Table 2 also contains the results of a two-factor (gender X career field)

Table 2. RIASEC occupational personality scale results by gender (M/F) and career field (IT/non-IT)

| Occupational Personality Variable | Means (top) and Standard Deviations | | | | Two-Factor ANOVA Results Effects (p < .05) | | |
| | Non-IT Career Field | | IT Career Field | | | | Gender by Career Field Interaction |
	Male N = 190	Female N = 192	Male N = 95	Female N = 46	Gender	Career Field	
R Realistic	54.8^a 8.4	46.6^b 8.2	56.2^a 8.5	48.8^b 7.9	Yes	Yes	No
I Investigative	54.0^a 9.2	50.7^b 9.7	55.1^a 9.3	54.7^{ab} 9.5	Yes	Yes	No
A Artistic	47.5^a 9.2	50.5^b 10.5	47.7^{ab} 9.2	51.4^{ab} 9.1	Yes	No	No
S Social	46.4^{ac} 9.1	52.3^b 9.4	44.2^a 8.9	50.0^{bc} 8.9	Yes	Yes	No
E Enterprising	51.6^a 10.9	52.0^a 10.7	44.2^b 9.4	46.0^b 10.3	No	Yes	No
C Conventional	54.2 9.6	55.0 11.2	51.8 8.3	55.8 10.1	Yes	No	No

^{abc} *Means in each row with common superscripts are **not** reliably different from each other.*

analysis of variance used to test for significant differences in the RIASEC variables as a function of gender (Male/Female), career field (IT/non-IT) and the interaction of gender and career field. This analysis reveals whether there are differences on the respective RIASEC variables between males and females in the total sample (gender effect), between IT and non-IT professional workers (career field effect), and whether there is an interaction effect (gender by career field interaction) indicating that males or females within either IT or non-IT are different from male and female professional workers in general.

There is an overall gender effect for R. Males score higher on the Realistic Theme than females, a common finding for this GOT. There is also an overall career effect for R. IT professionals scored significantly higher than non-IT professionals on the Realistic Theme. There is not an interaction effect for R.

For I there is an overall gender effect: Males scored higher on the Investigative Theme than did females. In addition, there is an overall career effect for the Investigative Theme: IT professionals scored significantly higher than non-IT professionals on I. There is not a significant interaction effect for I.

There is an overall gender effect for A: Females scored significantly higher than males on the Artistic Theme. There is neither a career effect nor an interaction effect for A.

For S there is an overall gender effect: Females scored significantly higher than males on the Social Theme. In addition, there is a career effect for S: Non-IT professionals scored significantly higher on S than did IT professionals. There is not a significant interaction effect for S.

There is an overall career effect for E: Non-IT professionals scored significantly higher on the Enterprising Theme than did IT professionals. There is not an overall gender effect or an interaction effect for E.

For C there is an overall gender effect: Females scored higher than males on the Conventional Theme. There was neither an overall career effect nor an interaction effect for C.

Table 3 contains the means and standard deviations for 190 non-IT males, 192 non-IT females, 95 IT males and 46 IT females on measures of the *PSS* of the *SII* occupational personality inventory. The results are expressed in standardized score format (T-Scores) where the norm group mean and standard deviation are 50 and 10, respectively. The

Table 3. Personal style scale results by gender (M/F) and career field (IT/non-IT)

| Personal Style Scale | Means (top) and Standard Deviations | | | | Two-Factor ANOVA Results Effects (p < .05) | | |
| | Non-IT Career Field | | IT Career Field | | | | |
	Male N = 190	Female N = 192	Male N = 95	Female N = 46	Gender	Career Field	Gender by Career Field Interaction
Work Style	44.9[a] 8.1	54.5[b] 9.2	39.8[c] 6.8	48.9 8.2	Yes	Yes	No
Learning Environment	53.9 7.6	52.2 10.4	52.3 8.0	52.5 7.8	No	No	No
Leadership	50.5[a] 9.6	50.3[a] 10.1	46.0[b] 9.4	46.5[ab] 9.0	No	Yes	No
Risk Taking	55.2[a] 8.8	47.2[b] 8.8	52.1[c] 9.3	45.1[b] 7.9	Yes	Yes	No
Team Orientation	50.0[a] 10.1	53.2[b] 9.5	48.1[a] 8.8	50.6[ab] 12.1	Yes	Yes	No

[abc] *Means in each row with common superscripts are **not** reliably different from each other.*

measures of work style, learning environment, leadership, risk taking and team orientation are derived from the 20-item scales of the *SII* and are standardized using combined gender norms derived from a sample of 9,484 men and 9,467 women (Harmon et al., 1994). Table 3 also contains the results of a two-factor (gender X career field) analysis of variance, similar to the one described for Table 2.

There is an overall gender effect for work style: Females scored substantially higher than males, meaning that in general, females prefer to work with people and men prefer with data, ideas and things. There is also a significant career effect for work style: Non-IT professionals scored substantially higher than IT professionals, meaning that non-IT professionals prefer to work with people while IT professionals prefer to work with data, ideas and things. There is not an interaction effect for work style.

For the learning environment scale, there are not any significant differences between males and females or between IT professionals and non-IT professionals.

There is an overall career effect for leadership: Non-IT professionals scored higher than IT professionals. In general, this means that non-IT professionals enjoy meeting, directing, and persuading others to a greater extent than IT professionals, who tend to prefer to lead by example and work alone. There is neither an overall effect for gender nor an interaction effect for leadership.

For risk taking, there is a significant gender effect: Males scored significantly higher than females, meaning that males in general are more likely to take risks and live spontaneously as compared to females in general. There is also an overall career field effect for risk taking: Non-IT professionals scored higher than IT professionals, meaning that non-IT professionals are somewhat more likely to take risks than are IT professionals. There is no interaction effect for risk taking.

There is an overall effect by gender for team orientation: Females scored higher than males in team orientation, meaning that females have a stronger preference than males for accomplishing tasks as a team, whereas males show a stronger preference for accomplishing tasks individually. There is also a career effect for team orientation: Non-IT professionals scored higher than IT professionals, meaning that non-IT professionals have a stronger preference for accomplishing tasks as part of a team whereas IT professionals show a stronger preference for accomplishing tasks individually.

FUTURE TRENDS

For the early part of the 21st century, at least, it appears that the GOT model, as operationalized by the *SII* with its RIASEC variables and *PSS* for measuring occupational personality, will be among the most prominent measures used by social science practitioners and career counselors in helping match people to careers and jobs they find interesting, meaningful and satisfying (cf. Donnay, Morris, Schaubhut, & Thompson, 2004; Walsh, 2004). This article documents similarities and differences between established IT and non-IT professionals and between males and females on these variables at this point in time, and presents a benchmark against which to measure and evaluate potential changes in the future. For example, it is possible that recent trends for businesses to outsource code writing jobs may result in changes in the nature of the work of IT-professionals remaining in these businesses leading to some different skill set requirements—possibly higher levels of problem solving or managerial skills (cf. Darais, Nelson, Rice & Buche, 2004; Gallivan, 2004; Todd, McKeen, & Gallupe, 1995). This, in turn, could eventually be reflected in changes in occupational personality characteristics of established professionals in the IT field.

CONCLUSION

In this article we report descriptive results comparing established professionals in IT and non-IT career fields by gender on 11 major occupational personality constructs. We found significant differences between IT and non-IT professionals on 8 of the 11 constructs, and significant differences between male and female professionals on 8 of the 11 constructs. However, we found no significant career field by gender interaction effects for any of these occupational personality constructs.

IT professionals are higher than non-IT professionals on the Realistic and the Investigative general occupational themes. This means that, in general, IT professionals have a preference for ordered and systematic manipulation of data and things, and for working on concrete (rather than ambiguous) problems. Furthermore, IT professionals have a preference for gathering information, analyzing and interpreting data, and to rely on themselves rather than on others.

IT professionals are lower than non-IT professionals on the Social occupational theme, indicating in general a lower preference for work requiring interactions and discussions. IT professionals are also lower than non-IT professionals on:

- **Work Style:** Indicating a preference for working with ideas, data, or things as opposed to people
- **Leadership:** Indicating a preference to work alone and to lead by example
- **Risk Taking:** Indicating a preference to "play it safe"
- **Team Orientation:** Indicating a preference to accomplish tasks independently.

Relative to male professionals, female professionals in general are higher on the following occupational personality constructs:

- **Artistic:** Prefer ambiguous, free, non-systematic activities
- **Social:** Prefer working with people, sharing responsibilities, having discussions and interactions with others
- **Conventional:** Prefer data management activities
- **Work Style:** Prefer working with people as opposed to ideas, data or things
- **Team Orientation:** Prefer to accomplish tasks as part of a team.

Relative to male professionals, female professionals in general are lower on the following occupational personality constructs:

- **Realistic:** Lower preference for ordered and systematic manipulation of data and things, and for working on concrete (rather than ambiguous) problems

- **Investigative:** Lower preference for systematic or creative investigation
- **Risk Taking:** Prefer to "play it safe."

While the results of this research reveal a number of notable occupational personality differences between IT professionals and other professionals, and between male and female professionals, the results also show that *there are no occupational personality differences between male and female IT professionals that do not also exist between male and female professionals in general.* This is an important finding.

The current research suggests that relative to other professionals, IT professionals tend to be more realistic, more investigative, less social, and less enterprising, prefer to work with ideas/data/things rather than people, prefer to work alone, prefer to accomplish tasks independently and prefer to minimize risks. This is consistent with other research on broad (Big Five) personality constructs, suggesting that in general the nature of IT work is such that individuals lower in extraversion have a significantly better chance of being successful and happy in the IT career field (Ash, Rosenbloom, Coder, & DuPont, 2005).

These findings raise the possibility that a higher proportion of males relative to females are attracted to work in IT due to a better match of that work with occupational personality differences underlying preferences for aspects of IT work. In general, women are more social and more cooperative relative to men. However, given potential significant changes in the nature of IT work, the status of occupational personality characteristics of established professionals in this field may be evolving. This research effort examines only one aspect of the potential set of causes for the notable current difference in proportions of males and females in the IT career field, and at this point, the role of gender in career choice remains an open question.

REFERENCES

Ash, R. A., Rosenbloom, J. L., Coder, L., & DuPont, B. (2005). *Gender differences and similarities in personality characteristics for information technology professionals.* Retrieved September 21, 2005, from www.ku.edu/pri/ITWorkforce/pubs/GenderDifferences.shtml

Darais, K. M., Nelson, K. M., Rice, S. C., & Buche, M. W. (2004). Identifying the enablers and barriers of IT personnel transition. In M. Igbaria & C. Shayo (Eds.), *Strategies for managing IS/IT personnel* (pp. 92-112). Hershey: Idea Group Publishing.

Donnay, D. A. C, Morris, M. L., Schaubhut, N. A., & Thompson, R. C. (2005). *Strong interest inventory manual: Research, development, and strategies for interpretation.* Mountain View, CA: CPP, Inc.

Gallivan, M. J. (2004). Examining IT professionals' adaptation to technological change: The influence of gender and personal attributes. *Database for Advances in Information Systems, 35*(3), 28-49.

Harmon, L. W., Hansen, J. C., Borgen, F. H., & Hammer, A.L. (1994). *Strong Interest Inventory: Applications and technical guide.* Stanford, CA: Stanford University Press.

Holland, J. L. (1997). *Making vocational choices: A theory of vocational personalities and work environments* (3rd ed.). Lutz: Psychological Assessment Resources.

Information Technology Association of America. (2003). *Report of the ITAA Blue Ribbon Panel on IT diversity.* Retrieved December 14, 2004, from www.itaa.org/workforce/studies/diversityreport.pdf

Most, R. (1993). Edward K. Strong: A thoroughly applied psychologist. *The Industrial Organizational Psychologist, 30*, 37-40.

Schneider, B., & Smith, D.B. (2004). *Personality and organizations.* Mahwah: Lawrence Erlbaum Associates.

Todd, P. A., McKeen, J. D., & Gallupe, R. B. (1995). The evolution of IS skills: A content analysis of IS job advertisements from 1970 to 1990. *MIS Quarterly, 19*(1), 1-27.

Walsh, W. B. (2004). Vocational psychology and personality. In B. Schneider & D. B. Smith (Eds.), *Personality and organizations.* Mahwah: Lawrence Erlbaum Associates.

KEY TERMS

Artistic: A person's preference for activities that are ambiguous, free, non-systematic and that entail the manipulation of materials to create art forms or products.

Conventional: A person's preference for activities that entail the explicit, ordered, systematic manipulation of data.

Enterprising: A person's preference for activities that entail the manipulation of others to attain organizational goals or economic gain.

General Occupational Theme (GOT): One of six broad constructs that reflect an individual's overall orientation to work (Realistic, Investigative, Artistic, Social, Enterprising, Conventional).

Investigative: A person's preference for activities that entail the systematic or creative investigation of physical, biological and cultural phenomena.

Leadership Scale: Distinguishes individuals who lead by example and prefer to work alone from those who enjoy meeting, directing, persuading and leading other people.

Learning Environment Scale: Differentiates individuals who prefer more practically oriented, hands-on learning situations from those who prefer academic learning environments.

Occupational Personality: A set of scores or descriptive terms that describe the individual being studied in terms of the variables or dimensions that occupy a central position within a particular theory of career or vocational choice; the most outstanding or salient impression that one creates in others with respect to his or her orientation to work.

Personal Style Scales (PSS): A set of scales that measure aspects of the style with which an individual likes to learn, work, assume leadership, take risks and participate as part of a team.

Realistic: A person's preference for activities that entail the explicit, ordered or systematic manipulation of objects, tool, machines and so forth.

Risk Taking/Adventure Scale: Differentiates individuals who like to "play it safe" from those who like to take a chance or be spontaneous.

Social: A person's preference to lead others or for activities that entail the manipulation of others to inform, train, develop, cure or enlighten.

Team Orientation Scale: Differentiates individuals who prefer to accomplish tasks independently from those who prefer to accomplish tasks as part of a team.

Work Style Scale: Differentiates individuals who prefer to work with ideas, data or things from those who prefer to work with people.

A Perspective of Equality and Role for Women in IT

Susan E. George
University of South Australia, Australia

INTRODUCTION

There are many disciplines and professions where women are not well represented, are paid less than male counterparts, and rise less quickly to leadership positions. IT is one such field, encompassing a broad range of topics from software development to telecommunications. This "inequality" has created a sense of injustice among some, leading to more aggressive stands for rights, for positive discrimination, and cries for all manner of "equality" within the workplace—specifically that male and female peers are able to play the same roles and indeed should have fair opportunity to play the same roles.

This article questions the "equality" that is pursued by the "equal opportunity" agenda. In many instances demanding women, given opportunities to take traditional male dominated positions in the workplace underlines the male-dominated world, what it values, and what it requires. A worldview that rejects male domination at its core may do more to help the "equality" of women and men. Moving toward this entails (1) recognising the roles played by women in the workplace and improving remuneration—rather than forcing women to take more male orientated roles, (2) couching the well renumerated roles that males play in more female friendly language to change perceptions of who is suitable for the role, and (3) recognising the female skills that many male roles require and not failing to give women novel workplace arrangements that permit pursuing roles outside the workplace.

In each of these suggestions the importance of male and female differences are recognised. This represents an understanding of "personhood", that is, not forcing all people to be equal regardless of gender, but recognising the intrinsic worth of people above gender—and that there may be gender differences. The idea of intrinsic worth of people is based upon one theological perspective of personhood drawn from the Christian tradition. It asks for equality of personhood to be recognised over and above gender issues and gender "differences" to be actually incorporated into professional environments.

GENDER EQUALITY

Women = An Unequal Discriminated Group

Data in the 1982 and 2000 reports from the National Science Foundation of America (NSF, 1982) indicate that women are a minority group in science and engineering fields. To this we can add IT (information technology). Relatively small percentages of women earn degrees in these fields; women are more likely than men to be employed part time (if the opportunity exists) and to be unemployed. Women doctoral scientists and engineers employed in educational institutions are less likely than men to be tenured or have the rank of full professor, and women scientists and engineers receive lower salaries than men. Many other reports would reveal the same basic inequality between genders, inequality that society is increasingly unable to tolerate. There are two basic approaches we may take to this inequality.

First, we may conclude that women are not treated in the same favourable way as men and conclude that there is discrimination. We would seek to focus on strategies to restore some moral order to society that rectifies the injustice done to the female gender. For example, the UN Convention on the Elimination of All Forms of Discrimination Against Women (UN, 1979) can be understood as a statement on what the principle of gender equality of opportunity should mean. Other definitions of gender equality point to the "discrimatory" nature of inequality. For example:

Gender equality means that there is no discrimination on grounds of a person's sex in the allocation of resources or benefits, or in the access to services. Gender equality may be measured in terms of whether there is equality of opportunity, or equality of results. (www.bigpond.com.kh/users/gad/glossary/gender.htm)

Second, we may more carefully consider what equality means and what the basis of equality is; we may question whether women are discriminated against, even whether efforts to alleviate the discrimination are actually only reinforcing the problem. Another definition of gender equality states, "Gender equality means that women and men have equal conditions for realizing their full human rights and potential to contribute to national, political, economic, social and cultural development, and to benefit from the results" (www.socialpolicy.ca/g.htm). This definition is slightly improved since it focuses on realising human rights. It enables the possibility that men and women may realise those "rights" in different ways and so touches upon the main point of this article, the meaning of equality.

To some extent there has been recognition of a tendency to consider men's characteristics as the norm and women's characteristics as different from the norm, but it is not widespread. The United Nations Population Fund (UNFPA) recognises "there are differences between the roles of men and women, differences that demand different approaches" (www.unfpa.org/gender/faq_gender.htm). In many places equality is taken to be women permitted to enter the male world on a level footing; not a recognition that there are "differences". In many instances the battles fought concern arguments over women "doing the same job". All these lines of reasoning and argument implicitly force women to "be the same" as men. A more equal opportunity might be to actually create valued roles that are designed for women.

Equality Based on Personhood

The deficiency in understanding equality has largely stemmed from the corresponding focus on gender that accompanies notions of equality. This article looks at a more fundamental equality of person. The perspective resonates to some extent with the idea of human rights and that there is some more fundamental concept involved over gender, that of "being human". Equal opportunity based on personhood rather than role may sound like the antithesis to equal opportunity, based on recognising that those who can do the same job should be given the same chance. But it is an equal opportunity that would actually create roles that are perfectly suited to women, and just as valued in the workplace. It is actually based on an understanding of person that sees all people as equal, while recognising there are gender differences. These differences do not make people unequal, just different and suited to different tasks/roles. And it recognises that many workplaces are traditionally male dominated to the extent that the female role is eliminated.

The concept of equality that will be expanded here is a particular Christian perspective of personhood, in which all human beings are equal, regardless of gender, ethnic status, age, and so forth. It stems from a spiritual perspective of the person gained from Christian Scripture. Genesis 1:27 states, "So God created man in his own image, in the image of God he created him; male and female he created them." The traditional Christian belief is that all persons are equal because they are made in the divine image, which was additionally the pinnacle of creation. This image bearing reality, which all humans possess, defines our personal worth, not the role we play. It is also the basis for true equality, but equal human beings are also different because they are male and female. Thus equality should not be defined by role played, but instead on "intrinsic worth". The Christian perspective does actually go on to suggest that God gave men and women different roles to play. Some Christian perspectives (e.g., Christian feminism) argue against the interpretation of these different roles and seek to make men and women play the same role (but there are a wide variety of theologies and types of feminism within Christianity) (Australian Catholic University, 2005). In arguing against different roles, the dignity of personhood and value that comes from simply being in God's image, equal but different (Ortlund, 1995) is somehow lost.

What are some of those differences of personhood that may underpin gender differences, without undermining personhood equality? In very general terms women are more nurturing (at least many have the biological capacity to nurture developing life); they are generally more adept with language at an earlier age, more empathic, less aggressive, and are likely to create networks of support and be supportive (helping) rather than competitive. Women who have succeeded in the business world report they have qualities that are traditionally male; they are independent, "tough", and have generally fought a lone battle to achieve the role that they do play. Many women may not want to "prove" themselves in this way, especially if it means they must deny their instinctive role. They are forced to by the way the role is defined and worth perceived.

A MALE-FEMALE WORLD

When role and worth are so tightly bound together, women in male dominated fields must always play exactly the same role as men to have the same value. In many ways the equity movement supports inequity by basing worth on role and not intrinsic human value. Recognising the different roles and not making women subject to male domination is perhaps the true way to achieve equal opportunity, although it may be practically difficult if not impossible to implement. Here we outline some ways that differences in role may be accommodated in the secular workplace, enabling all persons to be equally valued.

A New World

There is almost nothing we can say within the current male-dominated worldview that would not simply support and reinforce its message. If we talked about how to encourage women into male dominated fields, we would simply be reinforcing those valued roles played by people in those fields, those desirable roles that women have been denied for a long time. We deny women the right of being valued, regardless of what they do. Certainly incentives to support women in both workplace roles and more traditional family roles come under criticism as sending the message that it is the woman's responsibility to juggle different roles, not the male's (PersonnelToday.com, 2005). Certainly there are problems when roles within and outside the workplace are valued differently. How-

ever, here we confine ourselves to roles within the workplace and make three observations:

1. Traditional women's roles in the workplace need to be valued and financially rewarded. Often the workplace values technical roles more highly than administrative roles. It is not necessarily the case that the technical role requires more skill, just different skills, skills that women may be more suited for. The existing roles that attract the most prestige, financial advantage, and respect do not need to be male roles. Forcing women into male roles, rather than addressing the under-valuing that occurs in women's roles, does not address the need to base equality on personhood and intrinsic worth.

2. Most roles within the workforce are framed in such a way as to make men the most suitable candidates. Tasks are often defined in male terms (e.g., "provide leadership" rather than "nurture and facilitate growth"). Traditional male roles in the workplace need to be described and considered in female terms. This may genuinely open the door for female candidates to undertake a job, appropriate to their skills, while not forcing women into roles that may be more suited to men because of the job description.

3. New non-stereotypical, complimentary roles for people in the workforce need to be considered and male roles augmented with skills that women inherently own. For example, the technical world is realising that while computer programming skills may be useful, communication is equally important in a team-oriented workplace. Traditional male roles augmented with these other skills may open up avenues for women.

These three observations are expanded more fully in the following chapters. In each case there is a call to value the intrinsic worth of a person, while respecting that there are differences between people created by gender, differences that may controversially influence the task performance or preference.

Financially Rewarding Traditional Female Roles

Where women are contributing to the paid workforce, the tasks they do are often less valued than those men do. For example, the workplace often values technical roles more highly than administrative roles; universities value research more than teaching. In Australia the Premier's Department on Remuneration and Work found that (end 1999) "over the whole public sector, the average female full-time equivalent remuneration rate ($42,613 p.a.) was 8.6 % lower than the average rate for males ($46,619)" (http://rrd.premiers.nsw.gov.au/rrd/public/1999/remun.html). There are numerous instances where females could receive better remuneration but do not. This situation represents a failure to recognise the intrinsic worth of a person and starts to rate some skills as more desirable than others. It is at the root of the problem of a technological 'modern' society where productivity and economics is primary.

Changing the Language of Males Roles

Many traditional male roles are described in male language, emphasising the traditional male qualities and minimising the (traditional) female aspects required in a certain role. Simply describing positions in more female-friendly language has lots of potential to make women (and others) change their perceptions of the task—and hence who is appropriate for the role. Female-friendly language, acknowledging female-friendly roles within an organisation, would do females more justice since it would no longer hide the fact that women's qualities are actually needed. (The alternative is for the woman to mould herself to fit with the male stereotype and perpetuate the male worldview.) Changing a job description would not revolutionise the world overnight, but it would start to shift perceptions; it would identify roles women must play, qualities (typically female) needed and valued at the very top of an organisation (and indeed, all through it). The first step in changing a self-perpetuating male dominated world is to accurately describe the roles. This will cause a shift in perceptions of who is appropriate for the job.

For example, the description of specific duties at the professoriate level at the University of South Australia include: "*fostering* the research of other groups and individuals within the unit, playing an *active* role in the maintenance of academic standards and participating in and providing *leadership* in community affairs". Such descriptions actually call for roles that women traditionally play, often more frequently than men but this is hidden in the language used. A quiet, young woman is highly unlikely to be seen as suitable professor material. Language in job descriptions tends to value males and male traits, and indeed creates a situation where only a male is seen as suitable for fulfilling certain roles; whereas in reality a female may actually be what the organisation is looking for. In the previous example the words "leadership", "active", and "foster" colour perceptions of what is required and who is suitable.

First, the candidate must provide "leadership" in the community. The young woman is probably not going to be regarded as providing traditional male leadership (which at its worst dominates, forces its agenda, and imposes). However, she could be perceived as *supporting, working with, helping, cooperating with,* and *building*. Describing the role as "*service* to the community" or "*support* of community building activities", or "*cooperation* with community" may radically change the perception of whether the young female candidate is suitable for the job or not. Using language of a male-dominated world would make her seem unsuitable; revising the language and shifting perceptions of what is actually required makes the position accessible to her.

Second, the candidate must also play an "active" role in the maintenance of academic standards. Active implies up-front, visible, and dominating. We can imagine an articulate, forceful male putting forth his perspective at a meeting, but we may not be so quick to recognise the female (or indeed male) who performs the task in a less outspoken way. Language in a job description, such as *innovate, create ethical basis,* or *foster growth of* academic standards, would help to broaden perceptions of suitable candidates.

New Expectations of Male Roles

Many traditional male roles emphasise male qualities. The essential criteria are those skills and traits that men are likely to possess, while the desirable criteria are often more female traits. Adding in desirable female components to the job description would also help make the workplace accessible to women. The desirable criteria for a job may include "tactful negotiator", "supportive member", or "ability to empathise", but it is unlikely that these (more female) desirable qualities would make their way into the essential criteria in a world where there is already a well-defined perception of who should to be at the top.

We also note that there are some expectations accompanying traditional male positions—for example, that the role is undertaken full-time, the employee shows full commitment to it and demonstrates that it is a priority in his or her life and not secondary to other roles the person may play—whether in family or study or community. Unfortunately, women are more likely to be involved in these other roles and often unwilling to give them up to follow the male-dominated model where the one working or professional role is the most important thing in life. Some women may want to sacrifice family or community involvement in order to play male roles, but why should women be forced to do so before they are taken seriously? Why should the majority of part-time work available be in low-paid unskilled jobs, and why should women have to relinquish family or community in order to qualify for those worthy male jobs that must be full-time?

Naturally flexibility in terms of job-sharing, part-time work, particular duties undertaken, and so forth would have serious implications for employers (who for many reasons prefer full-time staff), but if the male-dominated worldview is to be challenged and changed, making positions truly accessible to all (and not imposing the traditional view) is necessary. It is a complicated issue, how the benefits of full-time salaried staff (e.g., in sick leave, superannuation, etc.) could be extended to employees who operate on a more casual basis (and naturally valuing the roles being played by each type of employee, would require similar benefits)—complicated, but this flexibility is surely part of the agenda; ambitious, but would ultimately make the workplace available to more than a few "determined" individuals. It is reasonable to expect that outside the male-dominated worldview, there are people who require and want greater flexibility and have just as much to offer the workforce, albeit on a part-time basis. They do not want to sacrifice their other roles to contribute to the workforce.

CONCLUSION

This article questions the equality that is pursued by the equal opportunity agenda. In many instances demanding women are given opportunity to take traditional male dominated positions in the workplace. This merely reinforces the male-dominated world, what it values, and what it requires. A worldview that rejects male domination at its core and acknowledges that there are many valid roles to play within the workforce—not just the stereotypical male role that receives the most credit—may benefit women more, rather than help them to chase after male-dominated creations. Thus this article questioned whether enabling women to play *exactly* the same roles as men is in fact the most beneficial way to address inequality. If equality has to be based on role at all, then let the roles be redefined and women allowed to play roles which are not *exactly* the same as male-defined roles, yet equally esteemed.

REFERENCES

Australian Catholic University. (2005). *Introduction to theology*. Retrieved December 30, 2005, from http://dlibrary.acu.edu.au/research/theology/theo102/MODULE11.htm

NSF. (2000). *Women and minorities in science and engineering: 1982* (NSF 82-302). Washington, DC.

Ortlund, R. C., Jr. (1995). Male-female equality and male headship. In J. Piper & W. Grudem (Eds.), *Recovering biblical manhood and womanhood: A response to evangelical feminism*. Wheaton, IL: Crossway Books. Retrieved December 29, 2005, from http://www.leaderu.com/orgs/cbmw/rbmw/index.html

PersonnelToday.com. (2005). *Pro-family policies damage women's equality*. Retrieved December

29, 2005, from http://www.personneltoday.com/Articles/2005/03/07/28387/Pro-family+policies+damage+women's+equality.htm

UN. (1979). Convention *on the elimination of all forms of discrimination against women*. Retrieved December 29, 2005, from http://www.pch.gc.ca/progs/pdp-hrp/docs/cedaw/cn_e.cfm

KEY TERMS

Discrimination: The illegal treatment of a person or a group of persons based on a prohibited factor, including gender.

Equal Opportunity: The right of all persons to be accorded full and equal consideration on the basis of merit or other relevant, meaningful criteria.

Equality: The quality of being the same in quantity or measure or value or status.

Personhood: Person is normally equated with human although there are often certain disputes about whether certain humans are persons; in some instances divinity may be attributed personhood and the concept cannot so easily be equated with being human.

Role: The actions and activities assigned to, required, or expected of a person or group.

Workplace: The sector of society that receives financial remuneration for the activities undertaken, excluding most students, volunteers, homemakers, the retired, and pensioners.

The Pipeline and Beyond

Martha Myers
Kennesaw State University, USA

Janette Moody
The Citadel, USA

Catherine Beise
Salisbury University, USA

Amy Woszczynski
Kennesaw State University, USA

INTRODUCTION

Women have been involved with IT since the 19th century, when Ada the countess of Lovelace was the first programmer for Charles Babbage's analytical engine. Grace Murray Hopper's contributions to COBOL and computing several decades ago are considered so significant that an annual conference is held in her honor (see http://www.grace hopper.org). In fact, the earliest computer programmers tended to be women more often than men (Panteli, Stack, & Ramsay, 2001). As the IT field progressed, however, it evolved into what many still view as a male-dominated domain, some say due to its increasing association with power and money (Tapia, Kvasny, & Trauth, 2003). Today, women make up at least half of World Wide Web users (Newburger, 2001), but this has apparently not translated into a proportionate participation in IT careers.

IT managers must recruit and retain a skilled and diverse workforce in order to meet the needs of increasingly global enterprises where cross-cultural, heterogeneous work groups are the norm. However, numerous sources (Information Technology Association of America [ITAA], 2003; Zweben, 2005) agree that the proportion of females to males selecting and completing degrees in IT-related fields is declining. Not only are women missing out on career opportunities, but the IT profession is also missing potentially valuable alternative perspectives on system design (Woodfield, 2002).

Worldwide, the digital divide is more extreme for women than men (Hafkin & Taggart, 2001), with the result that in many developing countries, women's access to computers is more limited than men's access. However, IT is an important driver for economic development and should provide women with new opportunities to better their circumstances, provided that a variety of challenges, such as technical education and social and political norms, can be addressed (Hafkin & Taggart, 2001).

Even in more developed countries, females face well-documented (Margolis & Fisher, 2002; von Hellens, Nielsen, & Beekhuyzen, 2004) obstacles all along the pipeline beginning as early as middle school and continuing through college, graduate school, and the career. Developing solutions to recruit and retain women in IT may serve other underrepresented groups as well, making IT classrooms and IT workplaces more inviting and ultimately more productive environments for everyone.

BACKGROUND

Part of the challenge of recruiting and retaining women in IT stems from a lack of knowledge by the public in general about the changing nature of IT work. The original focus of IT in the 1950s was on writing code to create computer programs. Unfortunately, many today, including the media, still see programming as the primary IT job (Denning, 2004). Earlier investigations into women and computing suggested that IT work by its nature was a poor fit for females, seen as solitary and boring, a double-edged stereotype that apparently still exists today (American Association of University Women [AAUW], 2000a; Galt, 2002; Symonds, 2000). It is double edged because it perpetuates myths about IT, as well as about women, and so restricts their access to the field.

Another part of the challenge of recruiting and retaining women in IT is related to the definition of IT work, which is difficult to pin down (Gallivan, 2004). The field has evolved, and IT has become more integrated into most business organizations and into the work and home lives of many individuals, creating a wide variety of IT jobs. Today, IT work includes not only job titles such as programmer, systems analyst, system administrator, and software designer, but also software engineer, business analyst, database designer, database administrator, network analyst, network administrator, Web developer, Web engineer, human-interface designer, project manager, applications developer, security administrator, and help-desk technician.

Concurrent with the expansion of IT job titles and responsibilities, there has been an expansion of the venues in which IT is taught. For example, the computer-science (CS) curriculum focuses primarily on technical and related theoretical concepts with emphasis placed on software creation. The information-systems (IS) curriculum integrates technical skills and knowledge with applied business and organizational concepts. IS programs are sometimes found in business schools, other times in schools of science, engineering, or even stand-alone IT units. Variations include business information systems (BIS), computer information systems (CIS), and management information systems (MIS). The information-technology curriculum may focus on a specific subset of technology such as fourth-generation languages and maintenance. For the purposes of this article, IT is defined as an umbrella term that encompasses a variety of job categories that continue to evolve as hardware, software, and methods continue to increase in sophistication.

EARLY INFLUENCES

A growing body of educational research documents many factors that influence female attitudes, perceptions, and behaviors toward computers in K-12 (Ahuja, 2002; AAUW, 2000b; von Hellens et al., 2004; Young, 2000). In addition to general sociocultural trends that appear to dampen girls' enthusiasm and self-efficacy regarding math and sciences around middle school, girls seem to be influenced by the following:

- Low interest in computer games
- Teacher, parental, and peer attitudes
- Lack of access to and experience with computers
- Perceived usefulness, or lack thereof, of computers to themselves and to society
- Lack of IT role models and mentors
- Media images

These variables may be grouped into two main categories: environmental and individual. Environmental variables are those that make up the context within which career decisions are made, such as school or work, while individual variables are characteristics of individuals, such as aptitudes and preferences (Woszczynski, Myers, & Beise, 2003). Both interact to influence the choices and behaviors of individual girls and women (Trauth, 2002).

COLLEGE, GRADUATE SCHOOL, AND THE IT WORKPLACE

Some of these same factors apply as women move into college and graduate school. In a survey of Canadian college freshmen, both male and female students' perceptions of a career's prestige, required education, and starting salary were inversely related to their estimate of the percentage of women employed in it (Harris & Wilkinson, 2004). Often, due to less previous experience with computers and less preparatory coursework, women continue to experience ambivalence about their interest and abilities in IT, in spite of equal performance regarding computer skills. They often encounter hostile academic environments (Margolis & Fisher, 2002; McGrath-Cohoon, 2001), and their career choices are often influenced by work-family balance concerns.

Women are discouraged from the IT workplace in a number of ways. Women are often relegated to IT positions with less prestige, reduced rewards, and lower expectations of technical capability. Earlier studies reported that women are employed at lower levels, make less money, and are more likely to leave their organization than men (Baroudi & Igbaria, 1994-1995; Igbaria, Parasuraman, & Greenhaus, 1997),

Women are often channeled into "softer" positions that are coincidentally lower in status, generate reduced compensation, and lead to less visibility (Panteli et al., 2001). Female IT workers were found to be disproportionately assigned to end-user support or help desks, positions less favored by male employees. Also, females preferred project-management tasks while males preferred network design and maintenance work (Martinsons & Cheung, 2001). In a large survey of the IT workplace, Dattero and Galup (2004) reported that women are more often assigned to legacy systems using COBOL skills than they are to engineering new software using Java. In general, they report that women are assigned to tasks that are considered less challenging. Managers in one large company viewed women as having less technical expertise (Gallivan, 2004).

Critics have rationalized such findings by stating that the differences are due more to variances in age, experience, and education than to gender. When studies have controlled for these potentially confounding variables, the results are mixed (Baroudi & Igbaria, 1994-1995; Igbaria et al., 1997). However, more current research is needed to update these studies, which do not reflect recent industry changes, such as the dot-com boom and bust at the end of the 1990s.

Academic institutions and business organizations alike are realizing that they need to focus on retaining as well as recruiting women in school and in the workplace (Tapia et al., 2003). As with recruitment, both environmental and individual variables will interact to determine retention outcomes for women.

INTERVENTIONS AND SOLUTIONS

A growing body of literature provides a range of useful approaches to facing these challenges (Margolis & Fisher, 2002; Wardle & Burton, 2002). One way of addressing the problem is to focus on individual factors, that is, to change the individuals by changing attitudes, dispelling stereotypes, improving preparation, and increasing experience with computers. Suggestions aimed at K-12 levels and beyond include the following.

- Providing more (and more equal) access to computing resources.
- Designing girl-friendly games and applications to encourage more experience with computers, which leads to higher self-efficacy.
- Creating videotapes and other positive media images that demonstrate women in professional IT roles, thus changing attitudes.
- Encouraging girls to take more courses that adequately prepare them for IT-related college majors.

Taken to an extreme, this approach implies that in order to succeed in this male-dominated field, women must become more like men. An alternative perspective, then, is to change the environment by making it less hostile, less masculine, more family-friendly, and more accepting of diversity. Interventions that have been suggested and implemented (AAUW, 2000b; Ingram & Parker, 2002; Werner, Hanks, McDowell, Bullock, & Fernald, 2005) to address environmental factors, at multiple academic levels, include the following.

- Train teachers to provide more equal access and to reduce the stereotyping of computing as a male domain.
- Provide students with female mentors and role models, including female faculty members and mentors from industry.
- Create communities, study groups, clubs, and other social supports for female students.
- Broaden the range of computing activities for younger students.

- Develop programs that do not depend on a substantial mathematical and scientific background prior to beginning college. Provide bridge opportunities to increase experience, build competency, and improve self-efficacy
- Consider the use of pair programming and other learning strategies in IT courses to provide opportunities for more teamwork

Business organizations have also implemented programs that target many of the individual and environmental factors listed above (Bentsen, 2000; Ingram & Parker, 2002; McCracken, 2000; Taggart & O'Gara, 2000). Suggested initiatives include the following.

- IT training (intraorganizational and community outreach)
- Mentoring programs, support networks, and general diversity training
- Visible top-management support and an external advisory council to promote cultural change
- Promotion of work-life-family balance values and programs for women and men, such as continuing education, flextime, day and elder care, and concierge services (often viewed by women in particular as more important than on-site game tables and sports outings)
- Examination of explicit and implicit reward systems, which may evaluate men differently than women, and which may not reward teamwork and communication as valuable leadership skills as much as more traditional definitions of achievement
- Staffing project teams for more gender balance to promote productive communication styles

FUTURE TRENDS

When IT workers are compared to other categories of workers, two interesting findings emerge. First, persons with a formal education in IT are less likely to pursue work outside of IT than are persons with formal education in other areas. Second, the IT workforce contains a large number of people without formal IT credentials or even traditional under-

graduate education (Wardle & Burton, 2002). This situation has likely arisen because IT is a relatively new field, because it has grown rapidly, and because there is a growing need to apply IT to other functional areas from which these employees come.

One frequently cited source of enrollments in computer-science programs (Zweben, 2005) focuses solely on research institutions (defined as those that offer a PhD in computer science). PhD-granting programs clearly play an important role in the pipeline by providing female IT academics to serve as role models for women enrolled in college. However, a recent study in Georgia (Randall, Price, & Reichgelt, 2003) suggests (a) that more women (all majors) attend non-PhD-granting state universities and colleges than research institutions, (b) that CS, IS, and IT degree programs at these state universities and colleges attract a higher percentage of women than do the research institutions, and (c) that IS and IT programs attract more women than CS programs. The applied nature of IS and IT programs is likely to be part of the reason why these programs are more attractive than CS programs, given women's apparent preferences for work that involves social interaction and social benefit (AAUW, 2000a).

The picture of why women leave the IT field, or choose not to get into it in the first place, is still incomplete and fragmented. Further research is needed, particularly in workplace settings, in order to better understand the problem and apply effective solutions. Furthermore, although a number of interventions have been suggested and even implemented, little is known about their effectiveness over time. Researchers need to engage in longitudinal studies in order to further enrich the body of knowledge.

CONCLUSION

Interventions for recruiting and retaining women in the IT workforce need to address self-confidence regarding computing; the related lack of experience, precollege preparation, mentors and role models, and community and study groups; and the importance of valuing both family and work priorities. More systemic solutions would transform the masculinized IT academic departments and workplaces into friendlier, more supportive environments for all workers. A summary of recommendations for IT practitioners, IT educators, and IT researchers toward addressing both individual and environmental factors includes the following.

1. **Advisors to Young Women:** Educate the public. Share knowledge about the evolving nature of IT work, particularly with parents, counselors, teenagers, and other influential groups. Narrow stereotypes need to be replaced by the realities of an exciting and socially fulfilling IT career.

2. **IT Educators:** Educators should get involved in all levels of education to identify, attract, support, and develop well-qualified women for IT positions. Broaden the definition of IT to include not just CS, but also IS and related majors, and contribute to continued curriculum development that balances important theoretical foundations with applied, practical application. Such curricula are more likely to appeal to women, who tend to view computers more as useful tools than fun toys.

3. **IT Practitioners and Educators:** The business climate should be examined to identify ways to make IT more attractive to women. Rather than on-site games and sports outings, many women find more appealing practical programs that support time management and work-family concerns (Bentsen, 2000). Such programs are likely to increasingly appeal to men as well, as youthful IT workers age, marry, and have children.

4. **IT Practitioners:** Initiatives that have been successful in IT education should inform managers seeking qualified women for IT positions. These initiatives include the development of mentoring programs, support networks, and general training on respect for diversity and multiculturalism for all workers.

Finally, many organizations are beginning to appreciate the value of a newer, more facilitative leadership style that is often associated with women, which includes teamwork, participatory decision making, and interpersonal communication skills. If such behavior is recognized and rewarded, this could help attract and promote more women into the managerial ranks of IT, eventually transforming the IT workplace and perhaps leading to broader perspectives in software and interface design.

REFERENCES

Ahuja, M. K. (2002). Women in the information technology profession: A literature review, synthesis and research agenda. *European Journal of Information Systems, 11*(1), 20.

American Association of University Women (AAUW). (2000a). Girls see computer jobs as lonely, boring. *Women in Higher Education, 9*(6), 3.

American Association of University Women (AAUW). (2000b). *Tech-savvy: Educating girls in the new computer age.* Washington, DC: American Association of University Women Educational Foundation.

Baroudi, J. J., & Igbaria, M. (1994-1995). An examination of gender effects on the career success of information systems employees. *Journal of Management Information Systems, 11*(3), 181-201.

Bentsen, C. (2000, September 1). Why women hate IT. *CIO Magazine.* Retrieved December 10, 2004, from http://www.cio.com/archive/090100/women.html

Dattero, R., & Galup, S. D. (2004). Programming languages and gender. *Communications of the ACM, 47*(1), 99-102.

Denning, P. (2004). The field of programmers myth. *Communications of the ACM, 47*(7), 15-20.

Gallivan, M. (2004). Examining IT professionals' adaptation to technological change: The influence of gender and personal attributes. *The DATA BASE for Advances in Information Systems, 35*(3), 28-49.

Galt, V. (2002, May 15). IT image a turnoff for girls, group finds. *The Globe and Mail,* p. C1.

Hafkin, N., & Taggart, N. (2001). *Gender, IT, and developing countries: An analytic study.* Washington, DC: Academy for Educational Development.

Harris, R., & Wilkinson, M. A. (2004). Situating gender: Students' perceptions of information work. *Information Technology and People, 17*(1), 71-86.

Igbaria, J., Parasuraman, J., & Greenhaus, J. H. (1997). Status report on women and men in the workplace. *Information Systems Management, 14*, 44-53.

Information Technology Association of America (ITAA). (2003). *ITAA report of the Blue Ribbon Panel on IT Diversity.* Arlington, VA: Author.

Ingram, S., & Parker, A. (2002). The influence of gender on collaborative projects in an engineering classroom. *IEEE Transactions on Professional Communication, 45*(1), 7-20.

Margolis, J., & Fisher, A. (2002). *Unlocking the clubhouse: Women in computing.* Cambridge, MA: MIT Press.

Martinsons, M. G., & Cheung, C. (2001). The impact of emerging practices on IS specialists: Perceptions, attitudes and role changes in Hong Kong. *Information and Management, 38*, 167-183.

McCracken, D. M. (2000). Winning the talent war for women: Sometimes it takes a revolution. *Harvard Business Review, 78*(6),159-167.

McGrath-Cohoon, J. (2001). Toward improving female retention in the computer science major. *Communications of the ACM, 44*(5), 108-115.

Newburger, E. (2001). *Home computers and Internet use in the United States.* Washington, DC: U.S. Census Bureau.

Panteli, N., Stack, J., & Ramsay, H. (2001). Gendered patterns in computing work in the late 1990s. *New Technology, Work, and Employment, 16*(1), 3-17.

Randall, C., Price, B., & Reichgelt, H. (2003). Women in computing programs: Does the incredible shrinking pipeline apply to all computing programs? *Inroads: The SIGCSE Bulletin, 35*(4), 55-59.

Symonds, J. (2000). Why IT doesn't appeal to young women. In E. Balka & R. Smither (Eds.), *Women, work and computerization: Charting a course to the future* (pp. 70-77). Boston: Kluwer Academic Publishers.

Taggart, N., & O'Gara, C. (2000, September/October). Training women for leadership and wealth creation in IT. *TechKnowLogia*, 40-43. Retrieved from http://www.techknowlogia.org

Tapia, A., Kvasny, L., & Trauth, E. (2003). Is there a retention gap for women and minorities? The case for moving in vs. moving up. In M. Igbaria & C. Shayo (Eds.), *Strategies for managing IS/IT personnel* (pp. 143-164). Hershey, PA: Idea Group Publishing.

Trauth, E. M. (2002). Odd girl out: An individual differences perspective on women in the IT profession. *Information Technology and People, 15*(2), 98-118.

Von Hellens, L., Nielsen, S., & Beekhuyzen, J. (2004). An exploration of dualisms in female perceptions of IT work. *Journal of Information Technology Education, 3*, 103-116.

Wardle, C., & Burton, L. (2002). Programmatic efforts encouraging women to enter the information technology workforce. *Inroads: The SIGCSE Bulletin, 34*(2), 27-31.

Werner, L. L., Hanks, B., McDowell, C., Bullock, H., & Fernald, J. (2005). Want to increase retention of your female students? *Computing Research News, 17*(2), 2.

Woodfield, R. (2002). Woman and information systems development: Not just a pretty (inter)face? *Information Technology and People, 15*(2), 119-138.

Woszczynski, A., Myers, M., & Beise, C. (2003). Women in information technology. In M. Igbaria & C. Shayo (Eds.), *Strategies for managing IS/IT personnel* (pp. 165-193). Hershey, PA: Idea Group Publishing.

Young, B. (2000). Gender differences in student attitudes toward computers. *Journal of Research on Computing in Education, 33*(2), 204-216.

Zweben, S. (2005). 2003-2004 Taulbee survey. *Computing Research News, 17*(3), 7-15.

KEY TERMS

COBOL: COmmon Business Oriented Language; a programming language used for business applications on mainframe and minicomputers.

Computer Science (CS): A more traditional IT curriculum whose focus is technical and theoretical rather than applied, with emphasis on software creation.

Environmental Variables: The context within which career decisions are made, such as the school and work environments.

ERP: enterprise resource planning; an integrated information system to serve all departments of an organization.

Fourth-Generation Language: Business application languages and tools such as database- and decision-support tools like SQL, ACCESS, and EXCEL; ERP and other reporting tools; and Web development environments such as Cold Fusion and Frontpage.

Individual Variables: Characteristics of individuals, such as attitudes and preferences.

Information Systems (IS): A curriculum that integrates technical skills and knowledge with applied business and organizational knowledge. It is sometimes found in business schools, and other times in schools of science or engineering, or in stand-alone IT academic units. Variations include business information systems, computer information systems, and management information systems.

Information Technology (IT): (a) An umbrella term that encompasses a range of professional positions requiring at least a baccalaureate degree in computer science, information systems, or closely related majors. (b) A major that focuses mainly on fourth-generation language application development and maintenance.

SQL: Structured Query Language; used to query and process data in relational databases.

Postcolonial ICT Challenges

Birgitta Rydhagen
Blekinge Institute of Technology, Sweden

Lena Trojer
Blekinge Institute of Technology, Sweden

INTRODUCTION

This article has a particular interest in the introduction of ICT in the postcolonial parts of the world. The fundamental arguments for investing in ICT all over the world rest on the view of ICT as a necessity for successful integration into the world economy. ICTs are regarded as having great potential to promote development in key social and economic areas where a shortage of capital, knowledge and local capacity obstructs progress. However, "information itself does not feed, clothe or house the world" (Main, 2001, p. 96), and it remains to be seen whether ICTs in developing countries will create wealth among the poor in those countries or among the already wealthy.

In the promotion of ICTs for development, the introduction of these technologies is mainly discussed in technical terms, considering the problems of electricity, telephone access, and expensive computers. The argument for introduction is also rather instrumental, expecting income generation and economic improvement. At the same time, ICTs are sometimes referred to as revolutionary, but they will travel on existing technologies, modes of communication and (post) colonial relationships.

The introduction of new technologies will not only be regarded as a technical issue. It may also be politically sensitive, if the technology shows signs of disrespect for the local culture, if it promotes only specific groups and ways of life in the local society, or if it bypasses the local society when reaching out for a specific target like a company (see e.g., Redfield, 2002). As for example Weckert and Adeney (1997) argue, the spread of ICTs in diverse cultural settings might very well be regarded as cultural imperialism, given the unequal access to resources for alternative technologies or content. The directions that ICTs lead towards, for example distant communication, may be interpreted as unifying and networking on a global scale between interest groups to their own and society's benefit. ICTs may also lead to an increased spread of (androcentric) American and western ideals and commercial products, increasing the global dominance of the U.S. and other western nations. These examples show the impossibility in treating technologies as neutral tools.

The aim of this article is to develop postcolonial and feminist technoscience requests for context sensitive and distributed ICT processes in relation to the development of ICTs for Tanzania at the University of Dar es Salaam.

BACKGROUND

The position of "having never as much" (Redfield, 2002, p. 810) will for a long time be the position from which people in the Third World will receive ICTs. In his study, Redfield showed what reactions and tensions this position may create. Are ICTs yet another way of imposing control, of deciding what is important to know and to have, of showing who is in charge of globalisation? Are they yet another demand on transfer from national to private and commercialisation of common goods? A tool "to make the poor dream the same dreams as the rich" (Martín-Barbero, 1993, p. 165)?

Mörtberg (2000) raises the issue of equal access to ICT in a time when we see less of arguments for "technology in a democratic society" and more of arguments for "democracy in an information society". Equal access, referring to gender, class, race, religion, language etc. is by no means inherent in the ICTs. The gender dimension in the case of ICT in the postcolonial context relates to a double burden of men's supposed supremacy in technological matters, and women's specific barriers in the developing world, including illiteracy, unfamiliarity with English (that dominates the Internet), domestic work load, lack of valuable information on the Internet, and lack of connectivity in rural areas where women primarily live (Gurumurthy, 2004).

The links between equal-level[1] participation and ICT development or ICT policy development are created by means of hard work and tedious dialogues, multidimensional partnership co-evolution with developed and working sensitivity and awareness of diverse interests, gender dimensions and cultural—ethnic pluralism, among other components in an increasingly complex world.

Suchman (2002) argues that the *design from nowhere* is a result of the idea that technical systems could be constructed with a minimal cultural connection "as commodities that can be stablized and cut loose from the sites of their production long enough to be exported en masse to the sites of their use" (p. 140). Suchman also points out that the distinction between designer and user is not straight forward. The designers are users of their own products, and that invisible design-in-use often takes place without rigorous documentation. "Even to keep things going on 'in the same way' in practice requires continuous, mundane forms of active appropriation and adaptation of available resources" (p. 143).

Requests for access to communication (not only information; Colby, 2001), relevant material (e.g., Morley & Robins, 1995) and appropriate modes of communication practices (oral/literal, face-to-face or over distance; Mejias, 2001) highlights the borderline between ICTs supporting imperialism or pluralism. "Our challenge lies in theorizing exactly this interstitial space between agency and the lack thereof, between being constructed within structures of domination and finding spaces of exerting agency" (Shome & Hegde, 2002, p. 266).

These issues make it necessary to investigate and de-naturalise the discussion of former colonies as nations in need of ICT *transfer*. As Rwandan ICT expert Albert Nsengiyumva has stated[2], all electronic technologies have been brought into the African countries from outside. The new ICTs are often referred to as a sign of the jump from the modern into a postmodern age. Hess (1995) is very critical of the reference to a global postmodern age,

before claiming that "we" are living in a postmodern age, it is worth remembering that not everyone is included in that we. Cyberspace is an elite space ... There is a glass ceiling, and for many in the world a large part of postmodern technoculture lies well above it. (p. 116)

THE ROLE OF THE UNIVERSITY IN TANZANIAN ICT DEVELOPMENT

Feminist technoscience with emphasis on ICT is certainly motivated by transformation goals. The needs for transformation are not only seen in the ongoing difficulties of achieving appropriate ICT system solutions especially in low income countries, but also in a more general process of knowledge and technology development (Gibbons et al., 1994; Nowotny, Scott, & Gibbons, 2001). The latter urge for transformation not the least within academy and technical faculties (Etzkowitz & Leydesdorff, 1997). Feminist technoscience within technical faculties is a driving force for the transformation processes required (Trojer, 2002). The transformation on a deeper level is vital to address appropriateness, access and utilisation not only for women within the academia, but for the majority of women in the local society (Gurumurthy, 2004).

In order to be able to understand and learn about distributed knowledge and technology production you have to be situated in a very concrete, day to day practice as well as achieve broad contextual knowledges. The postcolonial situation carries the potential for distributed knowledge production that are of particular interest in this sense. Experiences from Tanzania and the role of the main university of the country will be used to elaborate on these negotiation processes.

Relevance and Transformation

The University of Dar es Salaam (UDSM) is the main university out of five in the country and the only university holding a technical faculty. The challenge for the university as an actor in societal development is huge. High expectations are placed on the implementation of ICT, which can be recognised in strategic documents of UDSM:

As part of the ongoing transformation programme, the UDSM has initiated a number of reforms aimed at improving its main outputs (teaching, research and services to the society) through ICT. The improvement of ICT aims to suit the needs of the students and staff, the working environment and establish linkages with both industry and government. The new ICT developments are also expected to contribute to income generation in order to complement government and other funding sources to ensure sustainable academic programmes. (University of Dar es Salaam, n.d., para. 1)

The vice chancellor emphasizes that within the larger transformation activities of the university the issue of *relevance* becomes central. As far as possible a public university in a very poor country must aim to be relevant to the developmental aspiration of the people in all knowledge areas.

The transformation should go deeper in the academic organisation culture, the vice chancellor argues.

I must say it is not easy. If you want to bend a fish you bend it while it is still alive, before it is dry. If dry you crack it. We have come to learn that it is a bit difficult. We are still struggling with it. (vice chancellor, interview September 12, 2003)

Resource for Society and Government

The experience of approving ICT at the university started in 1993. Responsible people at UDSM put in a 2 MBite wireless line to the university main campus. UDSM even brought Internet to the Tanzania telephone company (TTCL) and not the other way around. Now, twelve years later, Tanzania has Internet backbone in every region. In order to reduce the costs and ensure connection for the ministries, the university also connected eight government ministries to the wireless internet line at UDSM. As a result of this process, expert people from UDSM are now managers at TTCL.

Today, when the university competes with several other Internet service providers (ISP), a number of governmental bodies are still connected through the university link. The impact of the initiative coming from the university was an increased motivation for the university staff to keep on with ICT development, as the university staff members were the only ICT experts within Tanzania at that time. For the content development for the government (eGov) the process is both ways. The governement as well as the university are looking to find the easiest way to implement the governement's own processes and demands, which are monitoring, evaluation and easy communication. UDSM is trying to provide that kind of solutions.

ICT Politics and Borders in Question

The University of Dar es Salaam (UDSM) played a key role in the national ICT policy process. The policy draft was developed in a very broad and open process to reduce the dominance of the academy. The role of UDSM in the policy process can be viewed as part of a sensitive technopolitical agreement between the university and the governement. We have to keep in mind that the knowledge experts of ICT in Tanzania were and are mostly located at UDSM or trained at the same place as the only institution having a technical faculty in the country.

The national ICT policy gives a substantial understanding of the status of ICT in Tanzania as well as strategic areas for ICT and development. One of the central statements concerns the needs for Tanzania to move from being mere consumers of technology to being the designers and manufacturers of ICT.

At UDSM, the issue of how to achieve the dreams like poverty reduction, more education, gender equality and so forth is on the agenda. ICT can provide tools for this, but how much is really Tanzanian? The academic staff regards ICT as more promising than other technological fields in this sense:

We have a kind of technology where we can provide significant content of products, more than 60% as equal partners in the provision of products and services. This is mainly knowledge based. We have an opportunity to do that (provision) much more than in for example nuclear physics. (interview with academic staff, September 2003)

A department director at Tanzania Commission for Science and Technology (COSTECH) stressed that,

it is very unfortunate that computers came to Africa as prestigious tools, as elite, sophisticated tools and not as non rocket signs. This is a myth that came with them. Computers are just ordinary technology, much easier than automobile and more powerful than automobiles, because they are all knowledge based. Knowledge based technologies transform individuals. Many have a lot of interest in them. The West pushed computers as tools for private sector. That this is not true was not understood by the governement ... It all depends on how you look at things within your own country ... This element of articulation is what we need to do.

FUTURE TRENDS

Experiences from international feminist research closely linked to dominant areas of technology (information technology, biotechnology, and material engineering) imply recognition of techno- and research politics deeply rooted in understandings of knowledge and technology production as processes which occur in distributed systems. In other words, knowledge creation today takes place on the boundaries between universities, private sector, public sector and the political spheres.

We can recognize ICT as one of the technological science fields most evidently challenging the borders between academic research and politics/society (Gulbrandsen, 2000) and experience how the negotiations (Aas, 2000) about the character of academic research take place in society. Academic ICT and its applications in society and every day life force our attention towards the relation between dominating actors, of which the university is one. It stresses relevant knowledge about its prerequisites, which in turn results in transformation challenges within the traditional universities. One model explored for these processes has been the *triple helix model* stating that the three institutional bodies university, industry and governement are increasingly working together (Uhlin & Johansen, 2001). The triple helix model focuses more on the outer

frame for the processes. The actual knowledge and development processes are more explicitly discussed within the concept sphere of *mode 2* (Nowotny, Scott, & Gibbons, 2001). Mode 2 knowledge is created in a broad and transdisciplinary social and economic context involving varying actors and participants in the research process. This is seen in contrast with the traditional scientific knowledge, produced in separate, academic institutions with efforts to *reduce* influence from the society.

In a developing country like Tanzania in particular, the process of mode 2 knowledge production in a triple helix formation will present an example for the traditional academic institutions in the industrialised parts of the world. We see that the ICT development in Tanzania and other postcolonial countries will have an advantage in this sense of intensive interaction between different actors in society.

However, as already stated, equal access for people in rural areas, women in particular, and disadvantaged groups, will require active participation also from local organisations and NGOs with feminist agendas (see Gurumurthy, 2004, p. 42ff).

CONCLUSION

From our perspective, the situation at UDSM carries potentials of a contextual awareness that opens for a benign triple helix knowledge and technology production. A more broadly defined group of stakeholders in the early phases of ICT development may increase the robustness of the choices that are made along the road.

At UDSM in Tanzania, the priority of collaboration with institutions outside the university shows a potential for an ICT development drawing from a more conscious technopolitical work in a postcolonial situation. The recognition of the necessary efforts to enter into technology *development* has a potential to bring about a more domestic and context aware ICT development process. Feminist technoscience perspectives are supporting these processes.

If the ICT priorities of African nations shall become directed towards the population, internal expertise needs to develop the technologies and to utilise them. As the quotation from the department director at COSTECH indicates, the way ICT was introduced by foreign companies and nations can be criticised for attempting to retain an unnecessary control, carrying colonial marks, over the use of technology in the postcolonial context. The people at UDSM and COSTECH, however, have made conscious efforts to change the situation of "having never as much" into a situation where the control over the ICTs lies with the domestic expertise. As the interviewed academic staff member acknowledged, this is easier to achieve with ICT, which is knowledge based, than in other technological fields which are more technology based.

The issues of software and content have not been addressed as thoroughly as the technical infrastructure. What we here regard as knowledge production within a triple helix or mode 2 system involves the technical expertise at the Tanzanian universities and in the Tanzanian society. Continuous efforts to address equality issues and access to ICT for diverse user groups, including women and men, rural and urban and so forth are needed.

ACKNOWLEDGMENT

The authors wish to acknowledge support from the Swedish International Development Cooperation Agency.

REFERENCES

Aas, G. H. (2000). Kvinneforskningens samfunnskontrakt. In L. Trojer, M. Eduards, M. Glass, E. Gulbrandsen, B. Gustafsson, S. Björling, et al. (Eds.), *Genusforskningens relevans (The relevance of gender research)*. Report from the expert group of the Swedish Research Councils for Integration of Gender Research, Stockholm. Retrieved October 15, 2005, from http://www.bth.se/tks/teknovet.nsf

Colby, D. (2001). Conceptualizing the "digital divide": Closing the "gap" by creating a postmodern network that distributes the productive power of speech. *Communication Law and Policy, 6*(1), 123-173.

Etzkowitz, H., & Leydesdorff, L. (1997). *Universities in the global knowledge economy. A triple helix of university–industry–government relations*. London: Pinter.

Gibbons, M., Limoge, C., Nowotny, H., Schwartzman, S., Scott, P., & Trow, M. (1994). *The new production of knowledge*. London; Thousand Oaks, CA; New Dehli: SAGE Publications.

Gulbrandsen, E. (2000). Integrering av kvinne- og kjönnsforskning i Norges forskningsråd. *Genusforskningens Relevans*, Stockholm.

Gurumurthy, A. (2004). *Gender and ICTs. Overview report*. Brighton: Institute of Development Studies.

Hess, D. (1995). *Science and technology in a multicultural world: The cultural politics of facts and artifacts*. New York: Columbia University Press.

Main, L. (2001). The global information infrastructure: Empowerment or imperialism? *Third World Quarterly, 22*(1), 83-97.

Martín-Barbero, J. (1993). *Culture and hegemony. From the media to mediations*. London; Newbury Park, CA; New Delhi: SAGE Publications.

Mejias, U. (2001). Sustainable communicational realities in the age of virtuality. *Critical Studies in Media Communication, 18*(2), 211-228.

Morley, D., & Robins, K. (1995). *Spaces of identity. Global media, electronic landscapes and cultural boundaries*. London; New York: Routledge.

Mörtberg, C. (2000, April 14). *Information technology and gender challenges in a new millennium*. Paper presented at the Women and the Information Society Conference, Reykjavik. Retrieved from http://www.simnet.is/konur/erindi/christina_iceland2.htm

Nowotny, H., Scott, P., & Gibbons, M. (2001). *Rethinking science. Knowledge and the public in an age of uncertainty*. Cambridge, UK: Polity.

Redfield, P. (2002). The half-life of empire in outer space. *Social Studies of Science, 32*(5-6), 791-825.

Shome, R., & Hegde, R. (2002). Postcolonial approaches to communication: Charting the terrain, engaging the intersections. *Communication Theory, 12*(3), 249-270.

Suchman, L. (2002). Practice-based design of information systems: Notes from the hyperdeveloped world. *The Information Society, 18*(2), 139-144.

Trojer, L. (2002). *Gender research within technoscience* (Genusforskning inom Teknikvetenskap–en drivbänk för forskningsförändring). Stockholm: Högskoleverket.

Uhlin, Å., & Johansen R. (2001, May). *Innovation and the post-academic condition*. Paper presented at the 2nd Research Conference on University and Society Co-operation (HSS01), Halmstad University, Sweden.

University of Dar es Salaam. (n.d.). *Sida-Sarec ICT Cooperation.* Retrieved February 15, 2005 from http://www.sida-sarec.udsm.ac.tz

Weckert, J., & Adeney, D. (1997, June 20-21). Cultural imperialism and the Internet. Technology and society at a time of sweeping change. In *Proceedings of the International Symposium on Technology and Society,* IEEE, University of Strathclyde, Glasgow (pp. 288-295).

KEY TERMS

Cyberspace: A world-wide computer network that allows people to communicate with each other.

eGov: The governement's information and communication with citizens via the use of the Internet.

Internet Backbone: A larger transmission line that carries data gathered from smaller lines that interconnect with it. On the Internet or other wide area network, a backbone is a set of paths that local or regional networks connect to for long-distance interconnection. The connection points are known as network *nodes* or telecommunication data switching exchanges (DSEs).

Mode 2 Knowledge Production: Some characteristics of mode 2 knowledge productions are situated in the context of application; distributed knowledge processes; development of robust knowledge; subject to multiple accountabilities.

Postcolonial: The period after the independence for colonized states in Africa, Latin America and Asia. Postcolonial also refers to a discursive space that has opened up for diverse positionings, discussions and practices after independence.

Technoscience: In the new fields of ICT, biotechnology and material sciences in particular, science and technology are so intimately related that they have merged into one. The concept of technoscience signals that the boundaries between science, technology, politics and society are about to weaken.

Triple Helix: Knowledge production taking place in the collaboration between the university, the government and the industry.

Wireless Technology: Wireless is a term used to describe telecommunications in which electromagnetic waves (rather than some form of wire) carry the signal over part or the entire communication path. Wireless technology is rapidly evolving, and is playing an increasing role in the lives of people throughout the world. In addition, ever-larger numbers of people are relying on the technology directly or indirectly.

ENDNOTES

[1]   The authors have borrowed Jan Åhlander's concept of the equal-level perspective in order to overcome the dichotomy of the top-down / bottom-up perspective (lecture notes, Jämshög folk high school, 1991).

[2]   Workshop held at Blekinge Institute of Technology October 23, 2003.

Postmodern Feminism

Clancy Ratliff
University of Minnesota, USA

INTRODUCTION

Since the 1970s, researchers have been using gender as an analytic category to study information technology (IT). In the decades since then, several questions have been raised on an ongoing basis, such as: How is gender constituted and reproduced in electronic spaces? Can the Internet be a place where there is no gender, a place where gender becomes fluid and malleable? How are identity and the politics of identity constructed online? Some scholars studying these questions have relied on feminist standpoint theory to frame and inform their inquiries into these issues, which foregrounds the differences between men's and women's experiences in electronic spaces and computing in general. However, others, particularly throughout the 1990s, have found postmodern feminist theory to be not only more accurate for explaining the actual practices of electronic communication and behavior, but also more conducive to the achievement of feminist political goals. The sections that follow will explain the general principles of postmodern feminist theory and its use in studies of gender and computer-mediated communication.

BACKGROUND

What is known as postmodern feminism is often associated with the work of Judith Butler (1990, 1993) and is marked, in part, by a "linguistic turn," a view of gender as a discursive construction and performance rather than a biological fact. These theorists criticize the conflation of sex and gender, essentialist generalizations about men and women, and the tendency to view gender as fixed, binary, and determined at birth, rather than a fluid, mobile construct that allows for multiple gender expressions. The gender dichotomy of man/woman so pervasive in Western culture can be understood in terms of the

cultural imperative to be heterosexual and a history of biological determinism in Western philosophy. Postmodern feminism rejects a dualistic view of gender, heteronormativity, and biological determinism, pointing to the inseparability of the body from language and social norms. Medical professionals can, for example, conform to and reinforce social norms by surgically transforming an infant with ambiguous genitalia into a culturally intelligible girl proper whose clitoris is a socially acceptable size (Butler, 1990). Medical technological intervention is also responsible for sexual reassignment surgery, making the materiality of gender malleable and blurring the boundaries between "man" and "woman." Postmodern feminists argue against the assumption that all women share a common oppression; this assumption has, unwittingly totalized and naturalized the category of "woman" into a white, heterosexual, middle-class, able-bodied, young- to middle-aged norm. Moreover, avowing political categories such as "woman" or "queer" as part of one's identity, what is called "identity politics" is both intellectually and politically misguided. Identitarian terms, such as "transgender," according to the postmodern school of thought, emerge into discourse at certain points in history, and it is important to keep this point in the foreground. Ignoring a term's history can end up reifying the term and reinforcing its place in a discursive hierarchy.

From this body of work, the theorist whose work has been particularly influential to scholars of gender and IT is Donna Haraway. Haraway (1985) argues that in a culture of high technology, the boundaries are no longer clear between human and animal, animal and machine, or human and machine. While not a new observation, Haraway recasts it as a windfall for feminist theory; hierarchical dualisms such as man/woman, heterosexual/homosexual, and white/black are no longer stable in high-tech culture. High technology is embedded so deeply in politics and knowledge (examples include artificial intelligence, genetic modification of organisms, and reproductive technologies) that the technologies are no longer tools deployed by agents in positions of power but now, to a great extent, they construct those agents. New technologies prompt redefinition of such concepts as literacy, work, nature, reproduction, and culture. Haraway argues that taking the cyborg, a figure without boundaries that is both human and machine, as a metaphor for socialist feminist theoretical interventions can be useful for feminist theory because it can help feminist theorists imagine a world that is not seen in or confined to hierarchical dualisms. The cyborg resists and eludes final definitions, as should feminist theory to avoid totalizing the category of "woman."

Braidotti (2003) suggests three potential ways to use the cyborg metaphor as an intellectual tool. First, the cyborg as an analytical tool "assists in framing and organizing a politically invested cartography of present-day social and cognitive relations" (p. 209). Second, the cyborg functions in a normative mode to offer a more complex and nuanced evaluation of social practices (see Selfe & Selfe, 1996). Third, we can use it as a "utopian manifesto" for imagining ways to "[reconstruct] subjectivity in the age of advanced technology" (p. 209). Also, with its focus on the organic and technological body, the cyborg metaphor keeps the body in view; one charge against postmodern feminism is that the materiality of the body "on the ground" gets lost in theorists' preoccupation with discourse.

POSTMODERN THEORY AND IDENTITY IN CYBERSPACE

When the World Wide Web became popular and commonly used, some wondered if the Web could become a truly democratic place, where discrimination on the basis of race, class, or gender could be eliminated. As much research has shown, however, and indeed as anyone who happens upon racist and misogynistic Web sites can attest to, the Web is not a utopia. Feminists have responded to gender inequalities online in several fashions, but Hall's (1996) study of women's experiences online makes a useful distinction between what she calls "liberal cyberfeminism" and "radical cyberfeminism" in online discussion practices (see also Wolmark, 2003). What

Hall terms liberal cyberfeminism is "influenced by postmodern discussions on gender fluidity by feminist and queer theorists, imagines the computer as a liberating utopia that does not recognize the social dichotomies of male/female and heterosexual/homosexual" (p. 148). Radical cyberfeminists, on the other hand, are concerned with everyday online problems: homophobia, harassment of women and pornographic representations of women, and they seek to create safe spaces for women only (see Herring, 1996).

Liberal cyberfeminism corresponds with a postmodern feminist view of gender as mobile and performative, not necessarily tied closely to identity. In online spaces, identity is constructed in communities with certain discursive norms, and identity is based on conversations and credibility established in those conversations; as such, only the community decides whether they accept the user as a woman, a disabled person or the like. Turkle (1995) and Stone (1995) use postmodern theories that problematize the humanist subject to show that online heightens the sense that identity is shifting, fluid, de-centered, and multiple; online, identity is a series of fictions and textual play—"personae all the way down" (Stone, 1995, p. 81). Turkle (1995) claims that computing is taking us "from a modernist culture of calculation toward a postmodernist culture of simulation," from "centralized structures and programmed rules" to "a postmodern aesthetic of complexity and decentering" (p. 20). Turkle (1995) agrees with Haraway that "the computer is an evocative object that causes old boundaries to be renegotiated" (p. 22). One such boundary is that between "man" and "woman."

Turkle (1995) cites netsex as one such simulation that allows for the flexibility of identitarian categories, with what she suggests is rampant "virtual gender-swapping" (p. 212; see also Bruckman, 1993). Turkle describes several cases of gender swapping and finds that "a virtual gender swap gave people greater emotional range in the real" (p. 222). Not only does this kind of gender play give users a space in which to express masculine and feminine aspects of their personalities, virtual gender-swapping also lets users explore their sexuality. For example, women can play men to have netsex with other women, and men can play women to have netsex with other men. Heterosexual women can play lesbian and bisexual women, and heterosexual men can play gay and bisexual men (Turkle, 1995). Gender-swapping in online spaces disrupts traditional gender hierarchies, which is desirable for postmodern feminism. Butler (1990) argues that subverting gender norms through parodizing traditional gender roles, cultural unintelligibility (e.g. rejecting gender as do gender queers), and gender proliferation is good for feminism: "The loss of gender norms would have the effect of proliferating gender configurations, destabilizing substantive identity and depriving the naturalizing narratives of compulsory heterosexuality of their central protagonists: 'man' and 'woman' " (p. 187). Hall (1996) claims that a postmodern feminist approach is "identified by an insistence on equality rather than oppression" (p. 151; see also Plant, 1996). Playing with gender or abandoning it altogether allows women and men to reject traditional roles and avoid occupying assigned positions in the hierarchical dualism of man/woman.

FUTURE TRENDS

Much of the work that draws upon postmodern feminism has used multi-user domains (MUDs) and MUD object-oriented (MOOs) as its objects of study. With their encouragement of role-playing, gaming, and other forms of creativity, including imaginative work such as writing detailed descriptions of objects, rooms, sensations, and people, MOOs and MUDs provide ample opportunity for spontaneity and experimentation. However, the technology is changing; people do not use MUDs and MOOs much anymore. Instead, asynchronous publication tools, such as weblogs, wikis, and audio software for podcasts are becoming more popular. While gender-swapping is not entirely absent on weblogs (Sorgatz, 2004), it is less common, and the studies of gender and communicative practices using these new tools reflect this. As a result, many researchers in computer-mediated communication interpret gender online on the practical basis of self-identification and gender cues. Comstock (2001), in her study of grrrl zines, approaches gender from a third wave feminist perspective, carefully pointing out that she is not using the term "women" (or grrrls) to mean one group with common attributes, which often is hegemonic, taking white middle-class women's agendas and experi-

ences to be the norm. Rather than stating her theoretical assumptions about gender, she only writes that she is not using the term "women" to mean one set of attributes, and she seems to rely on self-identification only as a gender indicator. Comstock discusses grrrl zine authorship, which is highly political and helps young women to participate in a public sphere. Discourse on grrrl zines "is often articulated at the site of the traumatized, adolescent female body" and embodied in narratives of abuse and body image (2001, p. 388). Comstock's research relies in a tacit manner on postmodern feminist theory, with a focus on gender as discursively produced.

In their study of gender and weblogs, Herring, Kouper, Scheidt, and Wright (2004) clearly explain how they classified bloggers as men or women:

Gender of blog authors was determined by names, graphical representations (if present), and the content of the blog entries (e.g., reference to "my husband" resulted in a "female" gender classification, assuming other indicators were consistent). Age of blog authors was determined by information explicitly provided by the authors (e.g., in profiles) or inferred from the content of the blog entries (e.g., reference to attending high school resulted in a "teen" age classification). The gender of the blog author was evident in 94%, and the age of the author in 90%, of the blogs in the combined samples. (online)

Herring et al. acknowledge heterosexual gender norms and how bloggers present a particular gender online, and while they do not discuss specific feminist theories that inform their interpretation, one can see that they interpret gender based on self-identification but also on discursive performance, which speaks to Butler's (1990, 1993) theories of gender as performance.

What weblogs, podcasts, and zines have in common is a desire to reach an audience. Whereas with synchronous conversation, in which role-playing is common, it might be easy to try on various identities, new publication technologies encourage the maintenance of a stable Web presence, the garnering of a readership, the interaction with a public. It is more difficult to inhabit another gender category for an evening than it was before, and it is also more difficult to use theories that argue for shifting and multiple gender identities to study online activity when the actual practices are so divergent from the theories' claims. Also, postmodern feminist theorists who study science and technology, such as Plant (1996) and Haraway (1985), have been criticized for an overly utopian view of technology. Because the emphasis has shifted from MOO and MUD activity to citizen media, future feminist work with computer-mediated communication will make more use of feminist theories that deal with notions of public, private, and personal narrative, including Fraser (1992) and Benhabib (1992).

CONCLUSION

There is now a large body of research and theory of gender and IT. Gender and IT is a field with a history, but more research remains to be done, and lingering questions and problems exist. One such problem is postmodern feminism's concern with the continued assumption (and potential reification) of a man/woman, masculine/feminine gender binary (LeCourt, 1999). LeCourt (1999) observes that much feminist research in computer-mediated communication relies on essentialist or constructivist models of gender, both of which reify a masculine/feminine, man/woman gender binary. However, while it might be theoretically sophisticated to approach IT from the perspective of postmodern feminism, in the everyday practices of IT—education, work, and communication—most people do identify themselves as either men or women, however problematic those identitarian processes may be. Moreover, much astute policy research has been done on behalf of women in IT, research that assumes a male/female dichotomy. Recovery work on the history of technology, especially that of Sadie Plant (1997) and Cheris Kramarae (1988), reveals that women have always been part of the production and use of machines. The underrepresentation of women in IT careers and education is still a problem, and while the research done so far on women's experiences in IT education and the IT workplace has been helpful, more needs to be done to assess the industries' and educational institutions' efforts to create a more egalitarian workplace.

REFERENCES

Benhabib, S. (1993). Models of public space: Hannah Arendt, the liberal tradition, and Jürgen Habermas. In C. Calhoun, (Ed.), *Habermas and the public sphere* (pp. 72-98). Cambridge: The MIT Press.

Bruckman, A. S. (1993). *Gender swapping on the Internet*. Retrieved August 1, 2005, from www.cc.gatech.edu/~asb/papers/gender-swapping.txt

Butler, J. (1990). *Gender trouble*. New York: Routledge.

Butler, J. (1993). *Bodies that matter: On the discursive limits of "sex."* New York: Routledge.

Comstock, M. (2001). Grrrl zine networks: Recomposing spaces of authority, gender, and culture. *JAC, 21,* 386-409.

Fraser, N. (1992). Rethinking the public sphere: A contribution to the critique of actually existing democracy. In C. Calhoun, (Ed.), *Habermas and the public sphere* (pp. 109-142). Cambridge: The MIT Press.

Gerrard, L. (1999). Feminist research in computers and composition. In K. Blair & P. Takayoshi (Eds.), *Feminist cyberscapes: Mapping gendered academic spaces* (pp. 377-400). Stamford: Ablex.

Hall, K. (1996). Cyberfeminism. In S. C. Herring (Ed.), *Computer-mediated communication: Linguistic, social and cross-cultural perspectives* (pp. 147-170). Amsterdam: John Benjamins.

Haraway, D. (1985). A manifesto for cyborgs: Science, technology, and socialist feminism in the 1980s. *Socialist Review, 80,* 65-108.

Herring, S. C. (1996). Two variants of an electronic message schema. In S. C. Herring (Ed.), *Computer-mediated communication: Linguistic, social and cross-cultural perspectives* (pp. 81-108). Amsterdam: John Benjamins.

Herring, S. C., Kouper, I., Scheidt, L. A., & Wright, E. L. (2004). Women and children last: The discursive construction of weblogs. In L. J. Gurak, S. Antonijevic, L. Johnson, C. Ratliff, & J. Reyman (Eds.), *Into the blogosphere: Rhetoric, community, and culture of weblogs.* Retrieved August 13, 2004, from http://blog.lib.umn.edu/blogosphere/women_and_children.html

Kramarae, C. (Ed.). (1988). *Technology and women's voices: Keeping in touch.* New York: Routledge.

LeCourt, D. (1999). Writing (without) the body: Gender and power in networked discussion groups. In K. Blair & P. Takayoshi (Eds.), *Feminist cyberscapes: Mapping gendered academic spaces* (pp. 153-175). Stamford: Ablex.

Plant, S. (1996). On the matrix: cyberfeminist simulations. In R. Shields (Ed.), *Cultures of the Internet: Virtual spaces, real histories, living bodies* (pp. 170-183). London: Sage.

Plant, S. (1997). *Zeroes + ones: Digital women + the new technoculture.* London: Fourth Estate Limited.

Selfe, C. L., & Selfe, R. J. (1996). Writing as democratic social action in a technological world: Politicizing and inhabiting virtual landscapes. In A. H. Duin & C. J. Hansen (Eds.), *Nonacademic writing: Social theory and technology* (pp. 325-358). Mahwah: Erlbaum.

Sorgatz, S. (2004, June). Girl, interrupted. *City Pages, 25.* Retrieved August 1, 2005, from www.citypages.com/databank/25/1230/article12271.asp

Stone, A. R. (1995). *The war of desire and technology at the close of the mechanical age.* Cambridge: The MIT Press.

Turkle, S. (1995). *Life on the screen: Identity in the age of the Internet.* New York: Simon & Schuster.

Wolmark, J. (2003). Cyberculture. In M. Eagleton (Ed.), *A concise companion to feminist theory* (pp. 215-235). Malden: Blackwell.

KEY TERMS

Essentialism: The belief that all members of a group share common attributes. For gender, men might be seen as naturally strong, aggressive, logical, and independent, women as naturally weak, passive, emotional, and dependent on social relationships.

Grrrl Zines: Not-for-profit publications (print or online) containing art, creative writing, political rants, essays, collages, or anything else the author can imagine. Often associated with punk rock and the Riot Grrrl movement, a feminist movement situated in the early 1990s in Olympia, Washington and Washington, DC.

MOO: An abbreviation for MUD, Object Oriented. Unlike a basic synchronous chat space, a MOO contains rooms and objects. Both MOOs and MUDs are text-based, meaning that paragraphs of description serve in the stead of images of objects.

MUD: An abbreviation for Multi-User Dungeon, Multi-User Domain, or Multi-User Dimension. MUDs are spaces where multiple users are logged in at the same time, often to participate in role-playing games.

Netsex: Also called cybersex, netsex refers to the act of participating in a role-playing sexual situation. Netsex usually takes place in synchronous chat venues. Turkle (1995) also uses the term "TinySex" to denote netsex that takes place in TinyMOOs.

Performativity: A term associated with poststructuralist feminism. The idea that gender is a learned, daily act grounded in social norms of heterosexuality, femininity, and masculinity rather than biological sex.

Weblog: Also called "blog." A frequently updated Web site consisting of timestamped posts in reverse chronological order.

Wiki: A Web site that runs on software that enables any reader to add or edit content on the site. As a result of this affordance, wikis are highly collaborative.

Predicting Women's Interest and Choice of an IT Career

Elizabeth G. Creamer
Virginia Tech, USA

Soyoung Lee
Virginia Tech, USA

Peggy S. Meszaros
Virginia Tech, USA

Carol J. Burger
Virginia Tech, USA

Anne Laughlin
Virginia Tech, USA

INTRODUCTION

Research has supported the need to develop separate models for predicting men's and women's career interests. Women's career interests, particularly in nontraditional fields in science, engineering, and technology (SET), are considerably more difficult to predict than are men's (O'Brien & Fassinger, 1993). A number of factors have a significant impact on women's career interests and choices but have little effect in predicting men's career interests (O'Brien, Friedman, Tipton, & Linn, 2000). One of the most striking gender differences is that there is a much weaker connection for women than for men between interests, enjoyment, and career choice (O'Brien & Fassinger). The failure to make this connection is one explanation for the troubling finding that the majority of young women express interest in sex-typical careers that do not match their skills and are far below their ability (O'Brien & Fassinger).

Gender differences in the factors that predict career interest apply to the field of information technology as well. There are significant gender differences in all aspects of the IT pipeline, from how women become interested in the computing field to how they enter and remain in it, as documented by Almstrum (2003).

BACKGROUND

Understanding women's interest in IT careers cannot be reduced to a single factor or to a cluster of factors. It requires the consideration of a broad palette of environmental, social, and personal characteristics that generally are beyond the means of a single instrument to capture. Parental characteristics and support are environmental factors often recognized as central to women's career orientation and choice. Attitudes about technology and computer use are two additional factors that are frequently linked to interest in IT in the research literature. These factors are reviewed briefly in the following section.

Parental support. A fairly substantial body of empirical research documents the instrumental role of parents in the career orientation and choices of high-school and college women (Altman, 1997; Fisher & Griggs, 1994; Ketterson & Blustein, 1997). Parents have a greater impact on career choice than do counselors, teachers, friends, other relatives, and people working in the field (Kotrlik & Harrison, 1989).

An instrumental role played by parents in career decision making is in their support of career exploration. Parental attachment is positively associated with vocational exploration among college women

(Ketterson & Blustein, 1997). Parents who discuss issues openly and promote independent thinking in their children encourage more active career exploration (Ketterson & Blustein). Mothers, fathers, and siblings play a positive role in promoting career exploration by indirect means such as providing emotional esteem and informational support, and by more tangible means such as providing educational materials (Schultheiss, Kress, Manzi, & Glasscock, 2001).

A fairly large body of research provides support for the role of mothers in women's career orientation (O'Brien & Fassinger, 1993) and vocational choice (Felsman & Blustein, 1999). Adolescent girls were more likely than boys to report that their mothers provided positive feedback, supported their autonomy, and were open to discussions about career decisions (Paa & McWhirter, 2000). The career orientation of adolescent females is influenced by a complex interplay of ability, agenting characteristics, gender-role attitudes, and the relationship with their mothers (Rainey & Borders, 1997).

Computer use. Experience with computers is associated with positive attitudes toward computers (Dryburgh, 2001; Lips & Temple, 1990) and interest in computers (Lips & Temple; Shashaani, 1997). High-school programming experience has also been shown to be a significant predictor of women's success in computer science at the college level (Bunderson & Christensen, 1995). Enjoyment with using computers is associated with an interest in majoring in computer science (Lips & Temple).

Attitudes. Research is somewhat mixed about the connection between attitudes about the nature of work in computer-related fields and women's interest in IT and related fields. Some research suggests that negative stereotypical views deter women from enrolling in computer-related fields (Breene, 1993; Fountain, 2000), while other research indicates that women have more positive views than do men about computer technology (Ray, Sormunen, & Harris, 1999). Gender differences in attitudes about computer technology have probably narrowed as the gender gap in access to computers has virtually disappeared in the last decade.

MAIN THRUST OF THE ARTICLE

This article summarizes key findings of a theoretically driven, causal model that the Women and Information Technology (WIT) team developed to predict women's interest and choice of careers in information technology. This model reflects the results of a path analysis and predicts 27.4% of the variance in women's interest and choice of computer-related fields. The model was refined through analysis of the responses to three revisions of a questionnaire, *The Career Decision-Making Survey*, administered between 2002 and 2005 in three waves to high-school and college women in rural and urban locations in the mid-Atlantic region ($N=1621$). The findings discussed here are from 373 high-school and college women completing the questionnaire in the fall of 2004 and spring of 2005. The theoretical implications of the model were further extended through the analysis of interview data with high-school and college women ($N=151$). The analysis of data collected through one-on-one interviews with female high-school ($N=53$), community college ($N=39$), and university ($N=59$) women attending public schools in urban and rural locations throughout the mid-Atlantic region added to the theoretical development of the model.

Characteristics of questionnaire respondents. The sampling technique of purposefully targeting students enrolled in a balance of rural and urban institutions produced a diverse pool of respondents. Slightly more than half of our respondents (50.7%) are racial minorities. Nearly 40% have parents whose highest level of education is high school or less (37.3% mothers, 38.1% fathers), reflecting that more than one third of our sample is probably from middle- and low-income families. More respondents were enrolled in high school ($N=293$) than in college ($N=80$).

One exogenous variable (race) and four mediating variables (parental support, computer use, positive attitudes about the attributes of IT workers, and sources of career information) have direct effects on the dependent variable: IT career choice and interest. Each of these variables consists of a number of questionnaire items that were confirmed through factor analysis. All of the variables had a reliability index (Cronbach's alpha) of 0.60 or better, with most exceeding 0.70.

KEY VARIABLES IN THE MODEL

The model captures cognitive and social dimensions of how high-school and college women make career decisions. In most cases, respondents indicated their agreement to questionnaire items using a four-point Likert scale (1=*disagree*, 2=*slightly disagree*, 3=*slightly agree*, 4=*agree*).

The dependent variable, IT career interest and choice, contains seven questionnaire items. Students indicating interest in careers in computer-related fields were likely to agree that they had some familiarity with the nature of IT work, knew people working in the field, derived satisfaction from using computers, and perceived that their parents would consider IT a good choice of a career field. When all factors were controlled, minority women expressed significantly more interest in careers in computer-related fields than did Caucasian women.

The four mediating variables in the model that directly impact interest in a career in a computer-related field are the following.

1. Parental support
2. Computer use
3. Positive attitudes
4. Sources of career information

Four variables related to the process young adults use to make career decisions are at the center of our model. One of these variables, the number of people consulted about career options, had a direct and significant impact on an interest in and the choice of an IT career, but the direction of the relationship was not what we expected. In general, women who expressed interest in IT as a potential career choice perceived that their parents supported the choice, but the choice was not significantly impacted by information from other sources. Most surprisingly, the fewer contacts respondents had made with various types of people to discuss career options, the more likely they were to express interest in careers in IT. This indicates that individuals are expressing interest in IT

with minimal information from people outside their immediate family. This supports the conclusion that one of the biggest challenges facing educators who want to promote women's interest in IT is to develop a portfolio of strategies that engage young women in thoughtful reflection about career options that are good matches for their values, skills, and interests.

Key elements of each of the four principal variables in our statistical model are discussed in the following section. Each of these offers insight into the types of interventions educators might design to promote interest in IT-related majors.

Parental Support

Our findings support previous research that documents the central role of parents in career decision making for high-school and college women. In our model, parental support includes nine questionnaire items relating to perceptions that parents support the importance of a career and encourage career exploration, as well as agreement with the belief that parents have an idea of what would be an appropriate career choice. Students who express interest in IT careers believe that their parents support their career choices.

Parental support had a direct and positive impact on career information-seeking behavior, a central part of our model. Parents directly influenced three attitudes and behaviors related to seeking career information. These were, first, how likely respondents were to seek input from others when making an important decision; second, the credibility they awarded to the several different groups of individuals as sources of career information (parents, family members, friends, teachers, counselors, and employers); and, third, how often career options had been discussed with the same groups of people. Not surprisingly, the level of education achieved by the respondent's mother and father had a positive and significant impact on parental support.

The direct relationship between parental support and interest in an IT career supports the idea that sharing career information with parents or involving them, particularly mothers, in educational activities is likely to have a positive impact on women's IT career interest.

Computer Use

It is no surprise to discover that computer use has a direct and positive impact on women's interest in careers in computer-related fields. What is unusual about our findings, however, is that for women, interest in IT careers is significantly related to their amount of computer use, but not necessarily their type of computer use.

The respondents to our questionnaire gauged how often they used different types of computer applications, ranging from simple communication through e-mail and instant messaging to more sophisticated purposes, such as the development or design of Web pages. Regardless of the application, the more time respondents spent using the computer, the more likely they were to have positive attitudes about the attributes of IT workers and to consider IT as a possible career option.

Our findings add empirical support for interventions that provide opportunities for hands-on use of many kinds of computer applications. It is very likely that experience and comfort with more sophisticated computer applications are associated with the ability to persist in a computer-related major or career, but they are not, according to our findings, a prerequisite for preliminary interest in a computer-related major.

Positive Attitudes

Previous research supports the idea that negative or stereotypical attitudes about workers in the IT-related fields explain some of the reluctance women express about computer-related careers (Ahuja, 2002). Findings from our model reveal a somewhat more nuanced pattern. Positive, not-stereotypical views had a direct positive impact with an interest in a career in IT. Women with positive attitudes were significantly more likely to express interest in IT careers.

Women who expressed an interest in computer-related fields were significantly more likely than other women to believe that IT workers are interesting, hard working, smart, and creative. The importance of this finding is magnified because not only did positive attitudes about the attributes of IT workers have a significant positive impact on interest in a career in IT, they also had a significant direct

effect on two other variables in the model: computer use and receptivity to advice about IT careers. Respondents with positive attitudes about the attributes of IT workers used computers more frequently and were more receptive to career advice than those who used computers less frequently.

Educational programs that effectively communicate the idea that working in a computer-related field is interesting and creative are likely to translate into gains in the number of women expressing interest in careers in IT.

CONCLUSION

Our model demonstrates that high-school and college women who express interest in careers in computer-related fields share three central characteristics, all of which can be addressed through educational programs.

1. They perceive that parents support IT as an appropriate career choice.
2. They use computers frequently and in various ways.
3. They have positive attitudes about the attributes of IT workers.

Despite the fact that women responding to our questionnaire were significantly more likely than their male counterparts to seek career information from several sources, this was, much to our surprise, negatively related to IT career interest and choice. Most surprisingly, the fewer interactions our respondents reported about career options, the more likely they were to express interest in IT careers. This supports findings from our qualitative data and suggests that an expression of interest in a career in the IT field is often made with little concrete information from sources outside of the immediate circle of trusted friends and family members. This finding does not bode well for the likelihood of long-term persistence in the field.

It is our conclusion that one of the biggest challenges facing educators who want to promote women's interest in SET fields is not simply to develop strategies designed to ensure widespread dissemination of career information about sex-atypical careers. Ensuring women's engagement in a deliberative process about choosing a career requires the creation of a portfolio of developmentally appropriate strategies that engage young women in thoughtful reflection about career options that match their values, skills, and interests.

FUTURE TRENDS

Some researchers have struggled to understand why even proactive efforts to recruit women to degree programs in computer science are frequently not successful (e.g., Cohoon, Baylor, & Chen, 2003). Findings from this research project suggest that recruiting women to IT majors and careers requires considerably more ingenuity than simply delivering information in an engaging way. Recruiting efforts are most likely to be successful when they include a long-enough period of engagement during which participants can begin to feel a sense of affinity and trust for those that guide them.

NOTE

This research has been supported by the National Science Foundation (#0120458).

REFERENCES

Ahuja, M. K. (2002). Women in the information technology profession: A literature review, synthesis, and research agenda. *European Journal of Information Systems, 11*, 20-34.

Almstrum, V. L. (2003). What is the attraction to computing? *Communications of the ACM, 46*(9), 51-55.

Altman, J. H. (1997). Career development in the context of family experience. In H. S. Farmer (Ed.), *Diversity and women's career development: From adolescence to adulthood* (pp. 229-242). Thousand Oaks, CA: SAGE.

Breene, L. A. (1993). Women and computer science. *Initiatives, 55*(2), 39-43.

Bunderson, E. D., & Christensen, M. E. (1995). An analysis of retention problems for female students in university computer science programs. *Journal of Research on Computing in Education, 28*(10), 1-15.

Cohoon, J. M., Baylor, K. M., & Chen, L. Y. (2003). *Continuation to graduate school: A look at computing departments.* Paper presented at the Association for Institutional Research Forum, Tampa, FL.

Dryburgh, H. (2000). Under-representation of girls and women in computer science: Classification of 1990s research. *Journal of Educational Computing Research, 32*(2), 181-202.

Felsman, D. E., & Blustein, D. L. (1999). The role of peer relatedness in late adolescent career development. *Journal of Vocational Behavior, 54*, 279-295.

Fisher, T. A., & Griggs, M. B. (1994). *Factors that influence the career development of African American and Latino youth.* Paper presented at the Annual Meeting of the American Educational Research Association, New Orleans, LA.

Fountain, J. E. (2000). Constructing the information society: Women, information technology, and design. *Technology in Society, 22*, 45-62.

Governor's Commission on Information Technology. (1999). *Investing in the future: Toward the 21st century technology workforce.* Richmond, VA: Office of the Governor.

Ketterson, T. U., & Blustein, D. L. (1997). Attachment relationships and the career exploration process. *Career Development Quarterly, 46*(2), 167-178.

Kotrlik, J. W., & Harrison, B. C. (1989). Career decision patterns of high school seniors in Louisiana. *Journal of Vocational Education Research, 14*(2), 47-65.

Lips, H. M., & Temple, L. (1990). Majoring in computer science: Causal models for women and men. *Research in Higher Education, 31*(1), 99-113.

O'Brien, K. M., & Fassinger, R. E. (1993). A causal model of the career orientation and career choice of adolescent women. *Journal of Counseling Psychology, 40*, 456-469.

O'Brien, K. M., Friedman, S. M., Tipton, L. C., & Linn, S. G. (2000). Attachment, separation, and women's vocational development: A longitudinal analysis. *Journal of Counseling of Psychology, 47*(3), 301-315.

Paa, H. K., & McWhirter, E. H. (2000). Perceived influences on high school students' current career expectations. *Career Development Quarterly, 49*, 29-44.

Rainey, L. M., & Borders, D. (1997). Influential factors in career orientation and career aspirations of early adolescent girls. *Journal of Counseling Psychology, 44*(2), 160-172.

Ray, C., Sormunen, C., & Harris, T. (1999). Men's and women's attitudes toward computer technology: A comparison. *Information Technology, Learning, and Performance Journal, 17*(1), 1-8.

Schultheiss, D. E. P., Kress, H. M., Manzi, A. J., & Glasscock, J. M. (2001). Relational influences in career development: A qualitative inquiry. *The Counseling Psychologist, 29*, 214-239.

Shashaani, L. (1997). Gender differences in computer attitudes and use among college students. *Journal of Educational Computing Research, 16*(1), 37-51.

KEY TERMS

Direct Effect: In a statistical model, it is a variable that impacts the dependent variable, or the variable being predicted, in a direct and statistically significant way.

Exogenous Variable: A variable assumed to be outside of the scope of the statistical model to predict.

Indirect Effect: In a statistical model, it is a variable that impacts the dependent variable, or the variable being predicted, indirectly through another variable.

Information Technology: Refers to a variety of jobs that involve the development, installation, and implementation of computer systems and applications. Careers in IT encompass occupations that require designing and developing software and hardware systems, providing technical support for computer and peripheral systems, and creating and managing network systems and databases (Governor's Commission on Information Technology, 1999).

IT Career Interest and Choice: The dependent variable in our statistical model that identifies the characteristics of respondents who either express an interest in a career in IT or have already made a choice to pursue a career in IT.

Parental Support: Parental support for a career and for career exploration.

Path Diagram: Represents results of a path analysis by showing the relationship between variables derived from theory in the form of a diagram. The variables in the model are connected by paths or directional arrows that display regression weights.

Reliability: Measures that reflect the consistency or dependability of a variable or construct.

A Psychosocial Framework for IT Education

Janice A. Grackin
State University of New York at Stony Brook, USA

INTRODUCTION

The most recent U.S. national statistics available indicate that among those earning degrees in engineering in 2000-2001, women made up only 18% of bachelor's degrees, 21% of master's degrees, and 17% of doctorates (NCES, 2003). A similar pattern emerges among those earning degrees in computer and information sciences, with women awarded only 28% of bachelor's degrees, 34% of master's degrees, and 18% of doctorates in those areas in 2000-2001 (NCES, 2003).

These and related statistics suggest a continuing gender imbalance in engineering and computer and information science education, academic pathways that lead to careers which are among those traditionally accorded higher prestige and greater financial reward than traditionally "female" occupations (Kennelly, Misra, & Karides, 1999). The situation is particularly dire in computer and information science education. According to testimony at a recent congressional hearing, although the proportion of computer science graduates who were women increased steadily from 14% in 1972 to 37% in 1984, from 1984 to 2000 those numbers began to steadily decline again and are currently at less than 28% (Borrego, 2002).

If computer and information technology education draws only from the 49% of the population which is male, the resulting gender imbalance is bound to translate into a shortage of trained IT personnel to fill existing positions. The aging IT workforce means that employers will need to fill not only new positions but those vacated by retiring personnel over the next twenty years (Jackson, 2004). The sheer number of technical professional positions to be filled now and in the foreseeable future makes it imperative that we tap the entire pool of young talent through early implementation of formal and informal strategies that encourage girls and young women to develop technical interests and skills and to enter technical training and post-secondary computer and information science education programs.

BACKGROUND

Although the factors associated with educational and career choices are complex, the relatively small numbers of young women choosing to pursue education and careers in technology and engineering may be directly related to *psychosocial* factors, such as a lack of professional *role models* (Smith, 2000). In simplest terms, a role model is someone who shares substantial characteristics of the observer and by extension is doing something the observer could do. The presence or absence of same-sex role models may transmit to individuals a powerful message regarding the *gender congruity* of various pursuits, including education and careers. The absence of female role models in computer and information science therefore limits the number of young women entering these education pathways, resulting in a situation where neither the academic presence nor the corporate representation of women increases.

As mentioned in the previous section, the number of women in computer and information science fields is not increasing and has, in fact, decreased over the last twenty years (Women Yield High-Tech Field, 1998). Insufficient numbers of women IT academics and field practitioners means that newcomer access to senior women who can provide *psychosocial* and *career mentoring* (Johnston, 2002) is adversely impacted. Of course, young women coming into the IT education and career pathways can and do find mentors among academics and field practitioners of both sexes. However, in light of the minority status of women as a group in computer and information science, female mentors may be better equipped to guide new female entrants through the social and professional vagaries of the educational and career process (Smith, 2000). Senior women who have successfully weathered the process may be able to impart specialized knowledge regarding *coping* and adapting, especially important for newcomers. Minimal presence of female mentors may be one cause of the previous decade's female exodus from computer and information science fields.

GENDER IDENTITY, CULTURAL EXPECTATIONS, AND COGNITIVE SCHEMAS

That there are proportionately fewer women currently working in or choosing to enter computer and information science fields may be due in large part to gendered cultural expectations (Smith, 2000) and the *gender schemas* associated with them.

Kohlberg's (1966) theory of development describes the acquisition of *gender constancy* as a process not completed before children reach the age of five or six. It is at this point in psychosocial development that children understand that being male or female is immutable, just as they begin to integrate the gendered cultural expectations that have swirled around them since before birth and to internalize a *gender identity*, that is, a strong sense of what it is to be male or female.

From these pervasive expectations, culturally derived cognitive *schemas* are built. At this point, children begin to categorize their world in more constrained ways, using schemas or frameworks into which information can be sorted automatically in order to efficiently organize and process the huge amount of incoming information about the world. Gender schemas are comprised of experienced and culturally defined elements of human "femaleness" and "maleness," including aptitudes and behaviors, for comparison to anything that might be defined as or characteristic of female or male.

Perceptions of *gender roles* are culturally driven (although there is a fair amount of cross-cultural correspondence) and so the resulting gender schemas for "maleness" and "femaleness" are generally shared by members of the same culture. In the case of a particular educational or career path, a culture defines the skills required for associated pursuits, and these skills are often associated with aptitudes believed to be inherently and dispositionally "male" or "female." In this way, certain careers come to be perceived within a culture as traditionally appropriate for women (e.g., "nurse") or for men (e.g., "engineer"). These culturally defined *gender role schemas* are internalized by individuals over the course of their development, reinforced along the way by the popular media and by the attitudes and behaviors of parents, teachers and peers (Smith, Jussim, & Eccles, 1999).

The cultural expectations regarding various groups can rise to the level of *stereotypes*, setting the stage for individual members of that group to experience what is known as *stereotype threat* (Steele, 1995, 1997). In performance situations where individuals are aware that a negative group stereotype exists, the anxiety produced can adversely affect performance for a variety of reasons unrelated to ability (Steele, 1995, 1997; Threats Within, 2004). Girls and young women find themselves in a stereotype threat situation any time they are performing with technology in general and computers in particular (Cooper & Weaver, 2003). The anxiety produced may negatively affect cognition and performance, resulting in performance that does not truly reflect abilities. Girls may come to doubt their own abilities and out of a need to preserve their own self-esteem they may then dissociate from technology, embracing the prevailing gender schemas that inform us that this is a male domain and that technology competence is unimportant for females.

Impact of Gender Role Schemas on Educational Choices

Given the strength and pervasiveness of cultural expectations, it comes as no surprise that gender-related schemas become quite rigid over time. Such appears to be the case with computer and information science. In western cultures particularly, skills and aptitudes associated with these educational and career paths have come to be perceived as traditionally male, and as a result, girls and young women may not consider computer and information science appropriate pursuits for females (Colley, Gale, & Harris, 1994). The very "culture" of computers has become associated with male values (AAUW, 2000), forcing girls to choose between technology pursuits and their basic gender identity or "femininity."

While young people can and do choose "non-traditional" education and career paths, it is clear that cultural expectations drive and constrain such choices. It can be extraordinarily difficult to successfully travel a path that defies the culture's expectations regarding gender appropriateness. Some "non-traditional" education and career paths may be retained as viable options for males and females through high school and even beyond college entry. A student's prior lack of interest does not substantially preclude becoming a biologist or chemist. However, the choice of post-secondary computer and information science education may be largely contingent on long-term interest and skills development, which familiarize the individual with IT culture and make these pathways recognized and viable options.

Unlike traditional sciences such as biology and chemistry, computer and information science courses, even basic ones, are not offered in all high schools; in schools that offer such classes students are not required to take them. Given that these courses are often elective, the students who choose them are more likely to be those who have had a longstanding interest in computers and technology. As noted above, gendered expectations around education and careers virtually guarantee that high school students with a history of interest in computers are overwhelmingly male and white. Male students are more likely than female students to have a computer in their room at home, and to have participated in pre-high school and pre-college extracurricular computer activities such as camps, clubs and competitions (Clarke & Teague, 1996). As a result of gender differences in early technology-related experiences outside of school those taking basic and advanced placement computer and information science courses in high school are overwhelmingly male (Bitten by the Tech Bug, 2000; Clarke & Teague, 1996; Cooper & Weaver, 2003; NCES, 2004). The gender discrepancy persists after high school graduation, with many fewer women than men enrolling in computer and information science courses when they reach college (Cooper & Weaver, 2003).

Early Intervention in IT Education

The psychosocial framework as outlined in the previous sections, including gender role expectations and gender differences in early technology experiences,

makes it desirable that interventions to increase the presence of women in the information technology workforce be focused on influencing the developmental process at the point at which girls are acquiring gender identity and an awareness of gender role expectations. Given the young age at which these processes occur, to be effective such interventions must be made at the earliest possible point in girls' formal and informal education.

Perusal of the 2003 and 2004 proceedings of the National Science Foundation's ITWF & ITR/EWF Principal Investigator Conference suggests that many current programmatic approaches to increasing gender equity in computer and information science, particularly initiatives involving institutions of higher education, focus on young women of high school and college age. There appear to be many fewer such programs targeting girls of middle or junior high school age, and still fewer programs that are structured to reach girls in elementary school and that could truly be described as early interventions. Absent entirely appear to be programs that offer a continuum of technology experiences for girls from elementary school to high school. The lack of early, long-term programmatic interventions may be due in part to the challenges and complexities of working with very young populations, as well as to the challenges inherent in building and sustaining partnerships with school districts, families, and community organizations over a period of more than a few years. However, these are obstacles that can be and must be surmounted because early intervention is essential to achieving equity in IT.

The overall goal of early intervention projects must be to influence girls' perceptions of gendered cultural expectations, and to support the development of gender role schemas that include female traits as compatible with computer and information science. Recent research suggests that young women may steer away from careers in technology not because they lack interest in technology-related activities as a whole, but because they perceive the culture of technology to be incongruent to their gender identities. That is, they perceive technology to be a male culture, unappealingly boring and antisocial, peopled with "computer geeks" working in isolation on projects that have little relevance in real terms to the advancement of the human condition (AAUW, 2000; Bitten by the Tech Bug, 2000). A *gender normative* technology environment, one in which a community of girls and women intersects with technology on a continuing long-term basis, offers girls an alternate gender schema that links being female with being technical.

In addition, the well-established importance of role models and mentors makes it clear that if we are to attract more young women to the technology and engineering fields, we must provide very early and ongoing exposure to female role models and mentors in these fields (Bitten by the Tech Bug, 2000). For girls, early and ongoing exposure to female role models and mentors illustrates that the pairing of women and technology is both natural and desirable.

FUTURE TRENDS

Given the developmental implications of the technology gender gap, future research must focus more on enhancing gender equity in IT education through development of early interventions within a psychosocial framework. Many gender equity in IT research projects incorporate potentially effective strategies, but without explicitly setting them in a psychosocial framework that focuses on influencing construction and acquisition of girls' gender schemas. Given the similarities of approach among projects, it is clear that we have learned through our assessments that certain components "work," that is, we can engage girls and young women in computer and information technology activities at certain points in their development, and they express interest in and enjoyment of those activities. Future research might do well to start from broader theoretical frameworks which will enable assessment to go beyond participants' apparent, but often fleeting, interest and enjoyment to not only identify what is "working," but also to define from a psychosocial perspective what "working' means, and why and how some strategies appear to "work" better than others. In this way, we may move toward a higher degree of confidence in identifying strategies that can be generalized across populations to enhance equity in IT education.

Given a psychosocial framework, we must direct our IT education research toward exploring intervention at critical early developmental junctures (Brown, 2001). We have often focused on develop-

ing projects that engage girls in middle and high school in technical activities, which they clearly enjoy, but we have not paid as much attention to girls' relationships with technology. In other words, many of our interventions focus on technical learning, unlike the *Girl Power 21st Century* project, which focuses on girls' evolving relationship with technology in the context of their gender identity development.

We have seen that no matter how enjoyable technical experiences are and no matter how female-friendly the environment in which they take place, merely exposing girls to technical experiences does not with any degree of certainty translate into more women entering IT education and career pathways. In the future, it might be wise for researchers to utilize a more developmental approach, intervening early and providing a continuum of experiences that may influence the development of gender schemas, as well as attitudes regarding what is and is not a gender appropriate academic or career interest. A necessary parallel focus on influencing attitudes of family members and other influential individuals in girls' social networks, such as educators and peers, falls quite naturally within this framework (Smith, Jussim, & Eccles, 1999).

CONCLUSION

Early intervention set in a psychosocial framework is a promising direction for future research. We have long accepted that early intervention is necessary to achieve our goal of equal educational opportunity (e.g., Head Start programs), and equity in IT education requires us to think in similarly developmental terms. We know also that achieving a "critical mass" of women in any field facilitates change, as demonstrated by changes in the practice of medicine and, to a lesser degree, law, and academia. Only at the point of critical mass do women have the power to begin influencing a work culture from within. By intervening early in the psychosocial process that limits perceived career choices for girls, we will be helping girls to develop into young women who have incorporated different expectations into the way they view the world of work. In this way, we will enable them to introduce new cultural norms, and to create change from within the IT education and work culture, even before women have achieved that "critical mass" presence.

REFERENCES

AAUW (American Association of University Women). (2000). *Tech-savvy: Educating girls in the new computer age.* Washington, DC: American Association of Women Educational Foundation.

Bitten by the Tech Bug. (2000, Fall). *AAUW Outlook,* 28-34.

Borrego, A. M. (2002). Witnesses differ on progress of women in science and engineering since Title IX's passage. *Chronicle of Higher Education.*

Brown, B. L. (2001). *Women and minorities in high-tech careers.* ERIC Digest No. 226.

Clarke, V. A., & Teague, G. J. (1996). Characterization of computing careers: Students and professionals disagree. *Computers and Education, 26,* 241-246.

Colley, A. M., Gale, M. T., & Harris, T. A. (1994). Effects of gender identity and experience on computer attitude components. *Journal of Educational Computing and Research, 10,* 129-137.

Cooper, J., & Weaver, K. D. (2003). *Gender and computers: Understanding the digital divide.* Mahwah, NJ: Lawrence Erlbaum Associates.

Jackson, A. J. (2004). The beauty of diverse talent. In American Association for the Advancement of Science (AAAS) & National Action Council for Minorities in Engineering (NACME) (Eds.), *Standing our ground: A guidebook for STEM educators in the post-Michigan era.*

Johnston, W. B. (2002). The intentional mentor: Strategies and guidelines for the practice of mentoring. *Professional Psychology: Research and Practice, 33,* 88-96.

Kennelly, I., Misra, J., & Karides, M. (1999). Historical context of gender, race, and class in the academic labor market. *Diversity Folio (Race, Gender, & Class), 6,* 125-141.

Kohlberg, L. (1966). A cognitive-developmental analysis of children's sex-role concepts and attitudes. In E. E. Maccoby (Ed.), *The development of sex differences* (pp. 82-173). Stanford, CA: Stanford University Press.

National Center for Educational Statistics (NCES). (2003). *Digest of education statistics* (NCES 2003-060) (pp. 336-337). Washington, DC: U.S. Dept. of Education, Institute of Education Sciences.

National Center for Educational Statistics (NCES). (2004). *Trends in educational equity of girls and women: 2004* (NCES 2005-016) (pp. 45-46). Washington, DC: U.S. Dept. of Education, Institute of Education Sciences.

National Science Foundation (NSF). (2003, October 26-28). Proceedings of the Information Technology Workforce & Information Technology Research/Education and Workforce (ITWF & ITR/EWF) *Principal Investigator Conference.* Albuquerque, NM: University of New Mexico.

National Science Foundation (NSF). (2004, October 24-26). Proceedings of the Information Technology Workforce & Information Technology Research/Education and Workforce (ITWF & ITR/EWF) *Principal Investigator Conference.* Philadelphia: Pennsylvania State University School of Information Sciences and Technology.

Smith, A. E., Jussim, L., & Eccles, J. (1999). Do self-fulfilling prophecies accumulate, dissipate, or remain stable over time? *Journal of Personality and Social Psychology, 77,* 0022-3514.

Smith, L. B. (2000). Socialization of females with regard to a technology-related career: Recommendations for change. *Meridian, 3, Summer 2000.* Retrieved from www.ncsu.edu/meridian/sum2000/career/

Steele, C. M. (1995). Stereotype threat and the intellectual test performance of African-Americans. *Journal of Personality and Social Psychology, 69,* 797-811.

Steele, C. M. (1997). A threat in the air: How stereotypes shape intellectual identity and performance. *American Psychologist, 52,* 613-629.

Threats Within. (2004, November). *APA Monitor on Psychology*, p. 101.

Women Yield High-Tech Field. (1998, August 26). *Newsday*, pp. A2, A41.

KEY TERMS

Career Mentoring: Mentoring that is in the form of coaching, protection, and sponsorship and that advances the mentee's career development and which prepares the mentee for advancement.

Coping: Refers to individual behaviors in response to environmental stimuli and can be individually adaptive (resulting in better functioning) or maladaptive (resulting in unchanged or worse functioning).

Gender Congruity: The degree to which a behavior or set of behaviors is perceived to align with the culturally defined female or male gender schema.

Gender Constancy: The transition between knowing the labels "girl-boy" and recognizing that the labels are immutable which occurs around the ages of 5-6.

Gender Identity: An individual's awareness of her/his own gender and its implications, i.e., what it means, in a context that includes culture as well as biology, to be male or female.

Gender Normative: Describes activities and behaviors that are perceived to be "normal" for females or males, and which serve to reinforce gender roles within a given culture.

Gender Role: A set of behaviors and attributes accepted within a cultural context and internalized by the individual as being appropriately linked with one sex or the other.

Gender Role Schemas: Organized sets of culturally derived beliefs and expectations about males and females that provide a framework for efficient cognitive processing of information.

Psychosocial Mentoring: Mentoring that is in the form of role modeling, counseling, and friendship, and that enhances the mentee's sense of competence, identity, and work-role effectiveness.

Psychosocial: Pertaining to the psychology of social interaction.

Role Model: Individual perceived as an exemplar to be emulated in a specific area.

Schema: Organized set of beliefs and expectations that guide information processing about a particular thing.

Stereotypes: Oversimplified beliefs about a group of people, generalized to individual members of the group based on uncritical judgments.

Stereotype Threat: Anxiety and apprehension experienced by individual members of minority groups in a setting where the individual's performance or behavior has the potential to validate an existing cultural stereotype.

Pushing and Pulling Women into Technology–Plus Jobs

Chris Mathieu
Copenhagen Business School, Denmark

INTRODUCTION

This article discusses the causes and implications of an empirically observed tendency to channel a disproportionate number of female computer professionals working in IT companies into what we term *technology-plus* positions. Technology-plus positions are positions requiring technological knowledge and skills but also containing a significant "non-technological" component. The most common such positions are project and group management, but also some sales/business development tasks, technical and specifications writing, and positions entailing substantial client contact can also be included in this category. Channeling a disproportionate number of female computer professionals into technology-plus positions is seen as evidence of gendered "segregation"[1] at the occupational sub-specialization level in the high end of the IT industry. This process is primarily based on horizontally differentiated positions and tracks rather than vertically hierarchical positions, though as argued below, a particular status hierarchy plays a central role in this process. Space constraints mandate sacrificing depth for breadth in making the argument here (see also Davies & Mathieu, 2005).

BACKGROUND

Studies on occupational segregation in the IT industry have documented the general gap between men in the upper and women in the lower echelons of the industry (Millar & Jagger, 2001; Pantelli, Stack, Atkinson, & Ramsey, 1999). Less attention has been paid to differentiation and segregation processes among computer professionals in its higher reaches. There are notable exceptions. Wright (1997) has quantitatively studied the issue in the U.S., concluding that some, but insufficient inroads have been made. Wright and Jacobs (1994) strike an optimistic tone regarding the prospects for sex integration, noting that males are not fleeing the occupation as more women move in. Fondas (1996, p. 284) shares this optimism. Woodfield (2000), however argues that sex-integration of the branch has been optimistically foreseen in three periods in the comparatively brief history of computing, but never realizing the optimistic projections. Some qualitative research has also been conducted in the field. Woodfield's (2000) case study in England found that even in an organization with progressive gender policies and that praises hybrid skills, men were still systematically advantaged and advanced past women. Tierney's (1995) study of an Irish software unit displays how the informal contacts between senior and junior men in the organization open avenues for advancement via informal access to information and currying favoritism based on in-group membership revolving around selectively male interests (football, drinking). In national statistics and four case studies, Pantelli et al. (1999) also find evidence of barriers for the advancement of women in the Scottish IT sector, with few women moving into management. To a certain extent, we find the same with regard to *senior* management. However, we found a different phenomenon at the middle management level—where women were consciously moved into technology-plus positions.

The empirical conclusions presented here derive from a research project on sex and gender equality/inequality in the IT sectors of Sweden and Ireland carried out from 2001-2004.[2] A total of 84 interviews were conducted with male and female employees and managers in eight companies in Sweden and five companies in Ireland. All but one of the companies fall into the NACE 72 category and ranged in size from seven to several thousand employees. In another part of the study, a total of 49 telephone-interviews were conducted with two cohorts of *women* who had studied computer science at a major Swedish university in the early and mid-1990s. The qualitative career choice and career history data was used to uncover processes and mechanisms which lie behind outcomes that are often construed as "free" choices (see Bertaux & Thompson, 2003; Evetts, 1993; Reskin, 2000, 2003).[3]

WOMEN IN TECHNOLOGY-PLUS POSITIONS: THE EVIDENCE

National occupational statistics are not sufficiently differentiated to see if observations from our companies are representative for the branch as a whole in Ireland and Sweden. This is because the technology-plus positions that actually are *positions*, such as group, unit, or line manager are often grouped together with other comparable positions for statistical purposes, and that others are *roles* or *functions* within a broader undifferentiated category, such as project leaders among "computer specialists." Even when statistical material is available down to the *positions* of group and project leader, the "lack of a standardization of job titles in IT" (Pantelli et al., 1999, p. 173) makes comparison difficult.

Evidence for the preference for and actual over-representation women in technology-plus positions comes from three sources in our study. First, our Irish and Swedish informants, male and female, employees and managers, reported an over-representation of women in technology-plus positions in their companies and in the branch general. A corollary to this is when asked if men and women tend to move into the same or different specializations, it was stated that women tend to move into project or group management, and seldom if ever into leading-edge technical specialist positions.

Secondly, data we collected about employees, titles, jobs and from organizational charts corroborated our informants' observations. Group leadership in Sweden was distributed fairly consistently on a 50-50 basis in Sweden (compare with a general presence of women in these companies at about the national average of 25%); while in Ireland it was slightly less. Project leadership appeared as both a formal *position* and a *role* and thus was more difficult to gauge, but the rough estimate in our

companies ranged from 40-60% women, significantly higher than the proportion of women in qualified computer jobs in the companies in general.

The third piece of evidence is the rhetorical support for the recruitment and "appropriateness" of women in such positions. A characteristic comment was made by a female Irish engineer by training, currently working as a project staffer: "To be honest, what I've seen here [at company X] is that women are much better at project leadership than men are…Because project leadership requires you to be very disciplined and organized…women are more systematic. If you look here most of the women here are project leaders. They [senior management at company X] like women to do project leader roles." The "fit" between women and technology-plus roles is based on "matching" skills associated with central tasks in these roles with certain basic characteristics commonly ascribed to men and women. In addition to the culturally based notion that women have better "soft" or communication skills, a number of more specific attributions were ascribed to women—having a holistic orientation; a client/user's perspective; being more systematic and organized and preferring simple, functional solutions to "technological overkill;" and men—jumping into solving particular problems without seeing the "big picture," being oblivious to client/user needs, and being so fascinated by technology that "the more sophisticated the better" was the guiding theme in their work. It was also contended that women tolerated ambiguity or "fuzziness" better.

WOMEN IN TECHNOLOGY-PLUS POSITIONS: SEGREGATION, INTEGRATION, FEMINIZATION?

Following on the point made directly above, technology-plus positions are conceptually *feminized*, but empirically or de facto integrated, as there are roughly as many men as women in such positions. This may be due to the fact that some males are attracted to such positions, or the relatively low number of women in the branch, but it shows that despite the conceptual feminization of these positions, males are not denied access to these positions. Technology-plus is constructed as feminine but it is not a female sphere. This is possibly due to the gender elasticity of the concept of management, which Wajcman (1998) argues is prototypically masculine. Technology-plus *isn't* gender ambiguous–it was presented as highly gendered. However, positions that involve management and sales can alternatively be constructed as masculine by drawing upon general, branch-external gendered associations.

Structural factors facilitate this phenomenon. One is the tripartite sub-specialization process in the branch: (1) becoming a (leading-edge) technical *specialist*; (2) becoming a well-rounded technical *generalist*; or (3) going into project or group management. While these are not irreversible choices, practitioners in the field see them as slippery slopes— once one had embarked on one of them, it is difficult to change tracks. The *de facto* rigidity of the specialization process makes the channeling of women into technology-plus pernicious; a fluid flow between the tracks would be less troublesome. Another factor is that technology-plus positions are usually filled by internal labor markets as they often require knowledge of a company's products, previous projects and internal resources, or industry-wide platforms or environments.

Another important "structural" factor is that this is where "holes" or openings exist in the companies studied. These holes are created by few (men and women) in the branch being interested in these positions. Personal interests tend overwhelmingly to lie in working with technology, and this is buttressed by the predominant cultural, social and organizational arrangements in the branch. These holes result from a *"reverse hierarchization"* of practical/technical and managerial/leadership roles and positions in the branch. We found a *strong* opinion in the branch that *technology* is what is important; especially the forefront of technological creation and innovation is what is *prestigious* and accords *status*. Moving away from technology, even if it means acquiring planning and managerial "power" entails a drop in status. In a branch where technical knowledge, innovation and creation are the highest values, "repetitive" and "non-productive," non-technological tasks such as management were seen as necessary evils—and referred to derogatively as "administration" and "overhead"—and were to be avoided if at all possible, even if one could earn slightly more. In questioning whether money compensates for what is lost, one Swedish

female group leader with 15 years branch experience stated: "You get paid the same, so why take the responsibility?…One gets paid a little bit more, but it doesn't make a difference." Thus, though these are "management" and "leadership" and organizationally vital, well-paid positions, they are culturally "peripheral" and thus it is less surprising to find women there (Bagilhole, 2002; Reskin & Roos, 1990).

PUSHING AND PULLING INTO TECHNOLOGY-PLUS POSITIONS

Causes

Both push and pull factors are at work here, and sometimes it is difficult to separate them. Push causes are generally factors that operate negatively, consist of barriers, limit choices, and push individuals in certain directions. Pull causes operate via attraction. Several push causes were identified in our study. Some women reported not being able to get onto the specialist track due to not being given the opportunity to show what they were capable of. One female Swedish programmer from our cohort study stated: "As a woman it isn't easy to get other [more technologically advanced—author] tasks. They often remain sitting there with the routine jobs, while guys are more aggressive and get to do the more exciting things. Men are more self-assured and display themselves as more "clever" even if they really aren't." While it is difficult in our (non-ethnographic) study to specify why far more males (though far from all males) get on the pure and advanced technology tracks, its clear that processes of accretive, cumulative advantage and disadvantage (Bielby & Bielby, 1996) shape outcomes with women frequently being left with more routine tasks. Those who initially obtain more advanced tasks tend to "prove" their capability and continue to acquire such tasks in the future and vice versa (Trauth, 2002, p. 111). The likely gendered origins of these personal trajectories are unacknowledged and quickly eradicated as all assessments become seen as based on what the advantaged and disadvantaged have "actually" proven or (not) done. As large numbers of women are condemned to routine tasks, the opportunity to develop and "rise" by going into technology-plus positions becomes an attractive option. Push begins to appear as pull.

Pure technology positions were often described as being set in unattractive work environments: socially and physically isolated, sedentary and monotonous—"churning out code on one's own all day" and the description "asocial" arose in several interviews.[4] While some factors were ascribed to "branch pressures," most were frequently seen as both unattractive and unnecessary by a number of women and men. Why these environments were not changed was explained as due to a lack of demand for change by those (primarily men) already there. Another push cause is the absence of women already in these positions. This operates as a social reproduction factor (Blackburn, Brown, Brooks, & Jarman, 2002)—sustaining a barrier by not providing women with "social leads" into the areas where they are underrepresented and many social leads into areas where they are over-represented.

On the pull side, these holes or organizational needs open up "opportunities." These "opportunities" are portrayed as attractive for personal and professional development; an opportunity to engage in more "social" activities than endless coding, an escape from unstimulating routine tasks, and to attract positive attention by doing something important for the company as a whole. The conceptual feminization of technology-plus positions leads to the argument that women are best suited for these positions, and can make a greater contribution here. Based on company needs and the gendered conception of skills, women are steered towards "opportunities" in technology-plus positions. Most of the women (and men) we interviewed in technology-plus positions were aware of the trade-offs entailed in going this route, and *still* expressed their preference for technical work, while reconciling themselves to the position they have or keep their dream of "returning to the technical side" later alive.

Another pull factor reported was that in technology-plus positions it's easier to plan one's time as one isn't hit by enormous workload waves as project deadlines approach to the same extent as those with purely technical tasks. This can be understood as an industry-specific reversal of Crompton's (2002; Crompton & Harris, 1998) practitioner-managerial career argument, where in the IT sector project and

group management resemble in central characteristics "practitioner" positions in other branches and vice-versa.

Consequences

From a sex-integration perspective, several questions can be raised. On the positive side, profiling technology-plus positions may open inroads into the branch that are more attractive to a broader range of women. Carving out a niche within the branch may serve to crack the "technical" façade of the branch, something that is frequently portrayed as scaring girls and women away from the branch and educational programs leading to it.

However, there are disturbing effects associated with this phenomenon. Shunting women into technology-plus appears to play a role in facilitating the exit of skilled, experienced women from the sector. One of the "mysteries" among our informants was "why are there so few women over 40 in the sector?" As one Swedish female consultant put it "I entered the sector with 10 female classmates from university ... and I am the only one left in the branch" after almost 20 years.[5] As women move into technology-plus positions, the "plus" activities afford them the opportunity to develop and document more broadly applicable skills. Whereas programming and systems work is a rather narrow skill set, management and leadership, client relations, etc., are applicable in a wider variety of settings. Simultaneous to developing and documenting these skills, technology-plus positions often entail contact with other organizations (read: potential employers) frequently outside of the IT branch. Losing senior women also means losing social leads into and through the branch for other women.

FUTURE TRENDS AND CONCLUSION

Future action depends upon how this phenomenon is documented in further research and the significance and meaning its implications become associated with. Multiple scenarios are possible. One sees this "natural tendency" for women to gravitate towards these functions as unproblematic. Even if recognized as a distinguishable pattern, if its causes are attributed to individual choices or preferences (Hakim, 2000), then it is not problematic and should be left alone. Another scenario entails exploitation of this phenomenon to draw more women into the sector and educational programs oriented towards specifically filling technology-plus positions–a strategy of "integrating more women into the branch" via greater and more explicit occupational or role segregation. More critical perspectives might see the problematic aspects of the phenomenon. From an individual choice perspective it can be argued that shunting women via gendered processes into these jobs and tasks denies them the opportunity to truly make free choices about career paths. If one's primary concerns are primarily with increasing the numbers of women in the branch, the contribution of this process to exit might be the utmost concern. If one's interest is in the prospective (business and/or social) benefits of mixed/balanced workplaces and workgroups, the phenomenon may or may not cause alarm. If having female project leaders leads to more openness in the project group, a greater willingness to flag when problems are encountered, collective solution seeking, an increased sensitivity to client and user needs, etc, then the current division of labor might be fine. However, if one is interested in the prospective benefits of diversity/integration at all levels and in all groups and communities within the organization, the current occupational segregation fails to realize these ambitions.

REFERENCES

Bagilhole, B. (2002). *Women in non-traditional occupations: Challenging men.* New York: Palgrave Macmillan.

Bertaux, D., & Thompson, P. (2003). *Pathways to social class: A qualitative approach to social mobility.* Oxford: Clarendon.

Bielby, D., & Bielby, W. (1996). Women and men in film: Gender inequality among writers in a culture industry. *Gender and Society, 10*(3), 248-270.

Blackburn, R., Brown, J., Brooks, B., & Jarman, J. (2002). Explaining gender segregation. *British Journal of Sociology, 53*(4), 513-536.

Crompton, R. (2002). Employment, flexible working, and the family. *British Journal of Sociology, 53*(4), 537-558.

Crompton, R., & Harris, F. (1998). Explaining women's employment patterns: "Orientations to Work" revisited. *British Journal of Sociology, 49*(1), 118-136.

Davies, K., & Mathieu, C. (2005). *Gender inequality in the IT sector in Sweden and Ireland.* Stockholm: Arbetslivsinstitutet

Evetts, J. (1993). Women in engineering: Educational concomitants of a non-traditional career choice. *Gender & Education, 5*(2), 167-179.

Fondas, N. (1996). Feminization at work. In M. Arthur & D. Rousseau (Eds.), *The boundaryless career: A new employment principle for a new organizational era* (pp. 282-293). Oxford: Oxford University Press.

Hakim, C. (2000). *Work-lifestyle choices in the 21st century.* Oxford: Oxford University Press.

Korvajärvi, P. (2004). Women and technological pleasure at work? In T. Heiskanen & J. Hearn (Eds.), *Information society and the workplace* (pp. 125-142). London: Routledge.

Millar, J., & Jagger, N. (2001). *Women in ITEC courses and careers.* Nottingham: DfES Publications.

Pantelli, A., Stack, J., Atkinson, M., & Ramsey, H. (1999). The status of women in the UK IT industry: An empirical study. *European Journal of Information Systems, 8*(3), 170-182.

Reskin, B. (1993). Sex segregation in the workplace. *Annual Review of Sociology, 19*(1), 241-270.

Reskin, B. (2000). The proximate causes of discrimination. *Contemporary Sociology, 29*(2), 319-29.

Reskin, B. (2003). Including mechanisms in our models of ascriptive inequality. *American Sociological Review, 68*(1), 1-21.

Reskin, B., & Roos, P. (1990). *Job queues, gender queues: Explaining women's inroads into male occupations.* Philadelphia: Temple University Press.

Tierney, M. (1995). Negotiating a software career: Informal work practices and "The Lads" in a software establishment. In R. Gill & K. Grint (Eds.), *The gender-technology relation: Contemporary theory and research* (pp. 192-209). London: Taylor & Francis.

Tilly, C., & Tilly, C. (1994). Capitalist work and labor markets. In N. J. Smelser & R. Swedberg (Eds.), *The handbook of economic sociology* (pp. 283-312). Princeton: Princeton University Press.

Trauth, E. M. (2002). Odd-girl out: An individual differences perspective on women in the IT profession. *Information Technology and People, 15*(2), 98-118.

Wajcman, J. (1998). *Managing like a man: Women and men in corporate management.* University Park, PA: Penn State Press.

Woodfield, R. (2000). *Women, work, and computing.* Cambridge: Cambridge University Press.

Wright, R. (1997). *Women computer professionals: Progress and resistance.* Lewiston, NY: Edwin Mellen.

Wright, R., & Jacobs, J. (1994). Male flight from computer work: A new look at occupational resegregation and ghettoization. *American Sociological Review, 59*(4), 511-536.

KEY TERMS

Hybrid Skills: Woodfield (2000) defines hybrid skills as combining technical and social skills and the hybrid worker as: "as much a conduit of information and catalyst of group dynamics as a deployer of purely technical skills" (p. 35).

Internal Labor Markets: Recruitment to positions that occurs within specific parameters, such as a firm, sector or profession.

IP: Intellectual property: An idea or design that can be sold or leased to another company for exploitation or development.

NACE Codes: NACE is a pan-European system for classifying companies or organizations based on their area of business activity. The categories relating primarily to computing activities are K.72.

Occupation: A "set of jobs in different firms that employers and government officials consider equivalent" (Tilly & Tilly, 1994, p. 289).

Occupational Segregation: The predominance of some distinguishable group in a given occupation. "Predominance" is a variable term, ranging from a marginal majority (for example 60%) to total dominance (100%). See Reskin, (1993, p. 244-245).

Technology-Plus Positions: Positions or roles where technological skills and knowledge are combined with non-technology tasks such as management, leadership, sales, client relations, technical and specifications writing.

ENDNOTES

[1] The "segregation" process results in integration or sex balance in these positions–see below.

[2] Funded by the Swedish VINNOVA foundation (project number 18327-1). Karen Davies and Katja Bierlein also worked on the project.

[3] See Davies and Mathieu (2005) for a deeper description of the project, methodology and results.

[4] However, our female informants reported and interest in technology and a desire for stimulating work tasks on par with or exceeding our male informants, in concurrence with Pantelli et al. (1999) and Korvajärvi (2004).

[5] Nine of the ten were still in full-time employment in other branches.

Questioning Gender through Deconstruction and Doubt

Cecile K. M. Crutzen
Open University of The Netherlands, The Netherlands

Erna Kotkamp
Utrecht University, The Netherlands

INTRODUCTION

Questioning gender can lead to a reformulating research into: "Why did the hard core of methods, theories and practices of the informatics discipline and domain become a symbol for masculinity?" and "Why is femininity constructed as situated only in the discipline's soft border of the interaction with the users of ICT-products?" In the view of Judith Butler, questioning gender is a strategy to disrupt the obvious acting of every actor, designers and users in the informatics domain:

The abiding gendered self will then be shown to be structured by repeated acts that seek to approximate the ideal of a substantial ground of identity, but which in their occasional discontinuity, reveal the temporal and contingent groundlessness of this "ground." The possibilities of gender transformation are to be found precisely, in the arbitrary relation between such acts, in the possibility of a failure to repeat, a deformity, or a parodic repetition that exposes the phantasmatic effect of abiding identity as a politically tenuous construction. (Butler, 1990, p.141)

In every interaction world, there is a continuity of ongoing weaving of a complex web of meanings in which we live, constructed by the interactions that take place in that world.s In that web of meanings, gender is a web of meanings on women and men, masculinity and femininity, which is connected to other webs of dualistic meanings. Gender is a process[1] in which the meaning of masculinity and femininity are mutually constructed, situated at symbolic, individual and institutional levels of a domain.

All social activities, practices, and structures are influenced by gender. The meaning of gender is thus embedded in social and cultural constructions and is always dynamically linked to the meaning of many concepts, such as technology or the relation between use and design. The performances of gender are the symbols for power relations in a domain (Harding, 1986; Scott, 1988).

RE-GENDERING THE INFORMATICS DOMAIN

Gender is covered by the unquestioned habits of the domain and discipline of informatics. The performance of gender can become visible through questioning and doubting: What has been overvalued, what has been undervalued and what has been ignored? The deconstruction[2] of the opposition "use-design" will function as a source for doubts on the discourse and the acting, methods and theories in the informatics discipline and the application of information and communication technologies (ICTs) in the informatics domain. Analyzing these kinds of power oppositions, such as use-design, could prevent the risk of reducing masculinity and femininity to fixed attributes based on biology and sex. The hierarchical opposition "use-design" is linked to other oppositions, such as "technical-human," "hard-soft" and "secure-doubtful." These gendered symbolic links are established and reinforced through the military, mathematical and technological traditions of the informatics discipline and through concepts of female informatics based on essentialist and deterministic views on femininity and technology. Strategies to destabilize this matrix of links are not easily found and executed for female ICT professionals. To accept the established horizon of the informatics discipline means to lose the potential of doubt, because socialization demands a commitment to the practices of the discipline. To oppose could be interpreted as a reinforcement of the link between the technical-social and male-female oppositions.

Use and Design of ICT Representations

Deconstruction of the opposition "use-design" in the informatics domain reveals that use and design are treated as activities in different worlds—the world of senders and the world of receivers—while ICT products are seen as the exclusive links between these worlds. ICT representations are perceived as the products of a design process if the product is new and innovative in the receiver world whether that the process of making was only a process of applying obvious methods and routines of the informatics discipline.

The symbolic meaning of use and design is constructed as an opposition in which "design" is active and virtuous and "use" is passive and uncreative. Designers see themselves and are seen as makers of a better future and working in a straightforward line of progress. Designers follow the ideal of making ICT products that cause no disturbances and fit completely within the assumed expectations of the users. The concept of "user friendliness" is based on this notion of non-problematic interaction, doubtlessness and reliability of interaction. "Good" design is defined as making a product for users that should not create disharmony or doubt in the life of the users. Easiness is equal to progress and "user friendliness" (Markussen, 1995).

There is a dominant belief in the objectivity of values: a belief that qualities as "good," "innovative," "friendly," "secure," and "reliable" can be measured objectively and that their achievement can be planned in advance before sending the product into the users' world. The design of ICT products is characterized as decision making, problem solving, optimizing, controlling, prescribing and predicting, and therefore has become an activity of displaying power. Design is focused on generalized and classified users. Users are turned into resources, which can be used by designers in the process of making ICT products.[3] The announcement of new products often is performed like a religious proclamation. The use of expert languages and methods within the closed-interaction world of informatics also establishes the dominance of design over use.

Cause, Doubt, and Change

One of the main causes of the hierarchical opposition between use and design is that oversimplified models for interaction and communication are used in the informatics domain. For instance, "use-cases" in UML are presented in simple action-reaction diagrams. In models such as the transmission model and the impulse-response-model, there is no room for processes of meaning construction. "Communication" is defined as the transmission of representations from a sender to a receiver through a neutral channel. Transmissive models of communication do not have "a message to the message." The meanings of a message, the role of sender and receiver, are fixed and separated. The sender has the active role and the receiver has the passive role.

The channel of communication is conceived as neutral. It cannot influence the interaction of sender and receiver. There is no room in the models for negotiation or doubt. Interaction and communication are only defined on a technical and syntactical level but then are used on semantic and pragmatic levels to construct planned and closed interaction. The semantic and pragmatic ambiguities that occur in "being in interaction" are ignored. Ambiguity is seen as troublesome and inconvenient and, thus, has to be prevented and "dissolved" at the technical and syntactical level (Crutzen, 1997, 2000).

Those models of interaction are frozen into the behavior of computer scientists and into the ICT representations they themselves use and apply and force back onto the informatics domain by ICT products ready-made for users. Design in informatics is seen as making a product for a remote world, whose interaction can be modeled from a distance and without being experienced. In the process of making ICT representations, professionals are mostly not designing but using established methods and theories. They focus on security and non-ambiguity, and are afraid of the complex and the unpredictable. Meaning construction processes have disappeared in processes of doubtless syntactical translation.

Users are not given enough opportunities to intertwine use and design. They are not subjects, but mere objects in the ICT representations.

FUTURE TRENDS

Transformative Critical Rooms

The creation of transformative critical rooms is necessary for making the gendering of the informatics domain visible and present. Interdisciplinary interaction and deconstruction are helpful strategies to search for places of interaction where transformative critical rooms can be created in a discipline. By deconstructing the "use-design" opposition in the informatics discipline and domain, the vanishing of the critical "subject-position" and the vanishing of design as a changing activity focused on an openness of the future can be "disclosed." Changing the frozen habits can start with the disclosure and the repair of a variation of "transformative critical rooms" that were closed in the past. These "rooms" should be reopened and redecorated with differences. However, that redecoration is only possible in interdisciplinary fashion.

Creating "transformative critical rooms" needs actors who have a habit of causing doubt. The discourse in Gender Studies on "subject-object" relations, "subjectivity-objectivity" and possible constructions of truth and reality in three main feminist tendencies toward generating new theories of knowledge—feminist empiricism, feminist standpoint theory, and feminist postmodernism—are developed out of critical positions in and towards these three tendencies. They have in common that they reject the claim of universal truths; truths are always particular and situated. Actors socialized in gender studies can cause doubt in the informatics discipline by presenting their critical way of acting and exploiting the play of differences.[4]

The Interaction Between Human Actors and ICT Representations

One transformative critical room, where redecoration is urgent, is that of the interaction between human actors and ICT representations. Much of people's life consists of interaction with themselves and interaction with others: people, machines, animals, objects and so forth. In the future, people will live in ICT-based webs of connections, in ICT-based webs of interaction systems. Webs will be in the people, and at the same time, people will be nodes in several webs of interaction. They will become "Cyborgs" and live in "cyborg worlds" (Haraway, 1991). So, a lot of interaction will be influenced by ICT. This influence should not be a deterministic one, because people themselves should construct the meaning of the technology. In the view of Heidegger, the essence of technology is *"disclosing something, for bringing it forth, for letting it be seen"* (Zimmerman, 1990, p. 229).[5] It is the opening of "Dasein" itself, even to the discovery that human actors will become *standing reserve within the global technical system"* (Heidegger, 1962, pp. 21-28; 1936, pp. 39, 41; Zimmerman, 1990, pp. 215, 229).

Redecoration means to reconstruct the meaning of "use": Using ICT representations means always designing and redesigning a flexible world of interactions between human and non-human actors in which the connections can always be disconnected by the actors involved. Doubting the obvious use of ICT representations can uncover this projective acting into the future. "Being-in-interaction" means that the activities of use and design are always intertwined in a process of learning. In this view, designing can be conceptualized as changing and changed acting as a projection to future acting.

CONCLUSION

The dichotomy between design and use has no traditional gender roots. This dichotomy is present in the discipline and the domain of informatics. It is only in a continuous repeated meaning construction process in our culture that this dichotomy is linked to masculinity and femininity, in which people have established the meanings that design activities are more consistent with masculinity, while use is more consistent with femininity. It is unjustly to repeat the questions "Are males better qualified to tackle design tasks, while females are better equipped to understand "use"? Question gender means a redecoration of the empty rooms between the binary oppositions use and design. These rooms cannot be found at the border of the informatics discipline but only within the discipline itself. Through creating transformative critical rooms, which are leavable, women can cultivate an erotic relation to ICT representations, feeling attraction and antipathy simultaneously. In these rooms, it can be proved that simple communication models are not sufficient enough to represent the way people are interacting. In this redecoration process, the meaning of use and design will change and it will be seen as intertwining activities. The empty room between use and design will be filled with many different meanings on use and design and can lead to a more critical use of methods and theories in informatics.[6]

REFERENCES

Adam, A. (1998). *Artificial knowing, gender and the thinking machine*. London: Routledge.

Babich, B. (1999). The essence of questioning after technology: Techne as constraint and the saving power. *British Journal of Phenomenology, 30/1*, 106-124. Available at www.fordham.edu/philosophy/lc/babich/tech.htm

Biesta, G. (1998). The right to philosophy of education: From critique to deconstruction. In *Yearbook of philosophy of education*. Chicago: University of Illinois, the Philosophy of Education Society. Available at http://www.ed.uiuc.edu/PES/1998/biesta.html

Booch, G. (1994). *Object-oriented analysis and design, with applications* (2nd ed.). Redwood City: Benjamin/Cummings.

Brunick, E. (1995/1996). *Introduction to linguistics and critical theory, 1995/1996*. Retrieved December 17, 2004, from http://tortie.me.uiuc.edu/~coil/contents.html

Butler, J. (1990). *Gender trouble: Feminism and the subversion of identity*. New York: Routledge.

Crutzen, C. K. M. (1997). Giving room to femininity in informatics education. In A. F. Grundy, D. Köhler, V. Oechtering, & U. Petersen (Eds.), *Women, work and computerization: Spinning a Web from past to future* (pp. 177-187). Berlin: Springer-Verlag.

Crutzen, C. K. M. (2000). *Interactie, een wereld van verschillen. Een visie op informatica vanuit genderstudies* (dissertatie). Open Universiteit Nederland, Heerlen.

Crutzen, C. K. M., & Kotkamp, E. (2006). Questioning gender through transformative critical rooms. In E. M. Trauth (Ed.), *Encyclopedia of gender and information technology*. Hershey, PA: Idea Group Reference.

Culler, J. (1983). *On deconstruction. theory and criticism after structuralism*. London: Routledge and Kegan.

Faulconer, J. E. (1998). *Deconstruction*. Retrieved February 19, 2005, from http://jamesfaulconer.byu.edu/deconstr.htm

Figal, G. (2000). Martin Heidegger, *Phänomenologie der Freiheit*. Weinheim: Beltz Athenäum

Haraway, D. J. (1991). A cyborg manifesto: Science, technology and social-feminism in the late twentieth century. In D. J. Simians Haraway (Ed.), *Cyborgs, and women. The reinvention of nature* (pp. 149-181). London: Free Association Books.

Harding, S. (1986). *The science question in feminism*. Ithaca: Cornell University Press.

Heidegger, M. (1926). Heidegger, Martin: *Sein und Zeit*. Tübingen, Niemeyer, 17. Auflage, 1993.

Heidegger, M. (1936). *Der Ursprung des Kunstwerkes* (1936). Used edition: (1960). Stuttgart: Philipp Reclam jun., 1960.

Heidegger, M. (1962). *Die Technik und die Kehre*. Stuttgart: Günther Neske.

Inwood, M. (1991). *Heidegger dictionary*. Oxford: Backwell Publishers.

Jacobson, I., Christerson, M., Jonsson, P., & Övergaard, G. (1992). *Object-oriented software engineering. A use case driven approach*. Reading: Addison Wesley Publishing.

Mallery, J. C., Hurwitz, R., & Duffy, G. (1987, revised 1994). Hermeneutics: From textual explication to computer understanding? In S. C. Shapiro (Ed.), *The encyclopedia of artificial intelligence* (pp. 362-376). New York: John Wiley & Sons. Retreived December 17, 2004, from www.ai.mit.edu/ people/jcma/papers/1986-ai-memo-871/memo.html

Markussen, R. (1995). Constructing easiness—Historical perspectives on work. In S. L. Star (Ed.). *The cultures of computing invisible work und silenced dialogues in knowledge pepresentation* (pp. 158-180). London: Blackwell Publishers.

Meijer, M. (1991). Binaire ppposities en academische problemen. *Tijdschrift voor Vrouwenstudies 45*(12), 1, 108-115.

Rumbaugh, J., Blaha, M., Premerlani, W., Eddy, F., & Lorensen, W. (1991). *Object-oriented modeling and design*. Englewood Cliffs: Prentice-Hall.

Scott, J. W. (1990). Deconstructing equality-versus-difference: Or, the uses of poststructuralist theory for feminism. *Feminist Studies, 14*(1), 35-50. Reprint in: M-K. Hirsch & E. Fox (Eds.), *Conflicts in Feminism* (pp. 134-148). New York: Routledge, Chapman and Hall.

Suchman, L. (1994). Working relations of technology production and use. *Computer Supported Cooperative Work (CSCW), 2*(1-2), 21-39.

Suchman, L. (1994). Do categories have politics? The language/action perspective reconsidered. *Computer Supported Cooperative Work (CSCW), 2*(3), 177-190.

Woolgar, S. (1997). Configuring the user: The case of usability trials. In J. Law (Ed.), *A sociology of monsters. Essays on power, technology and domination*. London: Routledge.

Zimmerman, M. E. (1990). *Heidegger's confrontation with modernity. technology, politics, art*. Bloomington: Indiana University Press.

KEY TERMS

Change: Every interpretation and (re-)presentation will influence future action. Acting is a representation of the interpretation of the phenomenal features of the world. It includes the interpretation of the results of our action. Not only the actual behavior but also the actions, which are not executed in the interaction (actions in deficient mode), are presentable and interpretable because these absent actions influence the ongoing interpretation and representation processes. Every action and interaction causes changes. However, if changes caused by interaction are comparable and compatible with previous changes, then they will be perceived as obvious. They are taken for granted.

Deconstruction of Gender: Deconstruction is a method to evaluate and analyze implicit and explicit aspects of binary gendered oppositions, such as "use-design." The meaning of the terms of oppositions, constructed as a weave of differences and distances, can be traced throughout the discourse of a discipline and its domain. By examining the seams, gaps and contradictions, it is possible to disclose the hidden meaning on gender and the gendered agenda. Identifying the positive valued term, reversing and displacing the dependent term from its negative position will reveal the gendering of the opposition and create a dialog between the terms in which the differences within the term and the differences between the terms are valued. It uncovers the obvious acting in the past and how it has been established.

Informatics Domain: The informatics domain is a world of actors in which ICT representations are designed and used, presented and interpreted. ICT representations are present in this world not only as hardware and software. Methods and theories used for designing and making ICT products are representations within this world. The informatics discipline is a part of and an actor in this world of interactions.

Interaction: The concept "interaction" is seen as an exchange of representations between actors. All acting of an actor is a representation of the actor and, through acting, the world of the actor changes. This exchange of representations is not a simple transmission process from sender to receiver. Interaction is a process of constructing meaning through repeated interpretation and representation, which is always situated in the interaction itself and depends on the horizons and the backgrounds of the actors and representations involved. Heidegger calls this mutual action projected in the future Sorge (care),

Fürsorge (solicitude) and Besorgen (concern) (Heidegger 1926, §12,15, 26; Figal 2000, pp. 81, 144; Inwood, 1999, pp. 35-37; Mallery, 1994).

Interaction World: An interaction world is constructed by the interactions that take place in that world by human, non-human and artificial actors. It is a process of constructing meaning through repeated interpretation and representation of actors. It develops by and stabilizes through the continuity of ongoing weaving of a complex web of meanings in which humans live. Being in an interaction world is being with others, sharing each others' meaning through acting. Interaction is an ongoing process of mutual actions from several actors in a (series of) situation(s).

Transformative Critical Rooms: A "transformative critical room" is a place of negotiation between interpretation and representation. In rooms in which differences can be present, truth can be seen as an ongoing conversation and a process of disclosure, and not as correspondence to reality. Truth is, then, merely a construction of actors in interaction. In transformative critical rooms, actions that cause doubt are seen as fruitful. In such rooms, doubts on representations are possible and can be effective in a change of the acting itself and in a change of the results of this acting: the interpretations and representations. The "preferred reading" of representations can be negotiated. There is space between interpretation and representation. Differences and different meaning construction processes are respected. The act of doubting is a bridge between obvious acting and a possible change of habitual acting. Doubt is always situated in the interaction itself. Doubt cannot only occur by the visible in the interaction but also by the invisible. Doubt is a meaning given to a situation in the interaction, which could lead to the change of meaning, to changed acting and, in the end, to a change of the established routines in an interaction world. Actors and representations are present in an interaction world that functions as a critical transformative if they are willing or have a potential of creating doubt and if they can create disrupting moments in the interaction.

Unified Modeling Language (UML): UML is a general-purpose notational language for specifying and visualizing complex software, especially large, object-oriented projects. UML is also used as a standard notation for the modeling of real-world objects as a first step in developing an object-oriented design methodology. Its notation is derived from and unifies the notations of three object-oriented design and analysis methodologies:

- Grady Brooch's methodology for describing a set of objects and their relationships
- James Rumbaugh's Object-Modeling Technique (OMT)
- Invar Jacobson's approach that includes a use case methodology.

Use-Case: A use case is a methodology used in system analysis to identify, clarify and organize system requirements. The use-case is made up of a set of possible sequences of interactions between systems and users in a particular environment and related to a particular goal. A use-case is a collection of possible scenarios related to a particular goal.

A use-case (or set of use cases) has these characteristics:

- Organizes functional requirements
- Models the goals of system/actor (user) interactions
- Records paths (called *scenarios*) from trigger events to goals
- Describes one main flow of events
- Is multi-level, so that one use-case can use the functionality of another one.

Use cases are employed during several stages of software development, such as planning system requirements, validating design, testing software and creating an outline for online help and user manuals.

ENDNOTES

[1] Judith Butler sees gender as a daily performance of each individual: "… rather, gender is an identity tenuously constituted in time, instituted in exterior space through a stylized repetition of acts." (Butler, 1990, pp.140)

[2] On deconstruction, see Brunick, 1995-1996, II. D. Deconstruction; Meijer, 1991; Culler, 1983, p.155, pp. 213-215, p.228; Faulconer, 1998; Biesta, 1998; Crutzen, 2000.

[3]   Steve Woolgar tells us about the opinion on users of a company that develops a PC: "The user's character, capacity and possible future actions are structured and defined in relation to the machine. ... This never guarantees that some users will not find unexpected and uninvited uses for the machine. But such behavior will be categorized as bizarre, foreign, perhaps typical of mere users." (Woolgar, 1991, pp. 89)

[4]   See Adam (1998); Suchman (1994a, 1994b).

[5]   Heidegger expresses the essence of modern technology as a challenging-forth or challeng-ing-revealing. "This challenging sets upon what is, nature, the genetic profile of the individual human being, the graphic imagination of the human relationship to the cybernetic domain, and so on and reveals it on the terms of that same technical challenge or set up" (Babich, 1999). According to Babich, Heidegger's questioning of technology reveals that questioning is more than a "calculative convention (namely that of question and answer)." It is "an open-ended or attentive project."

[6]   See Crutzen and Kotkamp (2006).

Questioning Gender through Transformative Critical Rooms

Cecile K. M. Crutzen
Open University of The Netherlands, The Netherlands

Erna Kotkamp
Utrecht University, The Netherlands

INTRODUCTION

Using the discourse of Gender Studies (Harding, 1986), proves to be a fruitful strategy to question methods, theories and practices of the Informatics discipline (Suchman, 1994a, 1994b). It shows the problematic notion of the binary opposition of use-design and it uncovers the objectification of both users and designers in ICT-representations in the designing process (Crutzen, 1997, 2000a, 2000b).

To further this analysis of the informatics discipline the concept of the transformative critical room is a very important one. A transformative critical room creates space where the interpretation of ICT-representations can be negotiated and where doubt can occur as a constructive strategy. Creating these rooms require actors who already have a habit of causing doubt and who accept that truths are always situated. Within gender studies these concepts of situated knowledge's and the critical assessment of subject-object relations are at the core of many feminist theories (Crutzen, 2003; Crutzen & Kotkamp, 2006).

A transformative critical room where a feminist analysis is of great importance is the room where interactions take place between human actors and ICT-representations. In this interaction, the meaning of "use" needs to be reconstructed. Using ICT representations imply the (re)design of a flexible environment where the connection between human and non-human actors can always be disconnected. When introducing this possible disruption in these ICT-representations it shows that the activities of use and design occur simultaneously with a process of learning. This means that designing is always an ongoing process where change takes place and where actability becomes an important condition.

MAIN THRUST

Disclosing ICT-Representations

Open ICT-representations are "mutual actable" for an actor. Actability is not a condition of the ICT-presentation. Mutual actability is the process in which the intertwining process of use and design can be based on doubting the obvious way of interacting and the ready-to-hand routines of the ICT-representation. Mutual actability is a process between an actor and a representation and depends on the presence of an ICT-representation for an actor. The process of intertwining design and use is always individual and situated in the interaction. It depends on the affective disposition and the state of mind of the actor.

Therefore, the intertwining of use and design needs the presence-at-hand of the ICT representations. Their readiness-to-hand should not be fixed. ICT-representations are present in a world of actors if they cause doubts and if the representation is at the

Figure 1. The intertwining of use and design

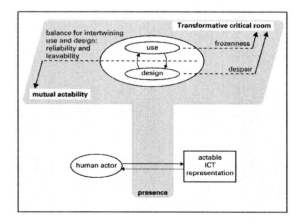

same time "leavable" and reliable. The doubt in acting should be possible but should not lead to desperation or to a forced routine acting. Despair caused by the ready-made acting of the ICT-representations, which allows only acting in a pre-given closed way or by their breakdowns, which leads to doubting your own acting and not the ready-made acting of the ICT-product. ICT-representations have a presence of leavability if representations allow the user to use the ICT-representations as a routine but also give the users the opportunity of learning in which situations the ICT-representations are adequate and in which situations they should be abandoned.[1]

The acting and interacting of people will be influenced by the acting of the ICT-representations that are made ready. Processes of negotiation and construction are necessary not only with the contents of the representations but also with the behavior and memory of ICT-representations to make the range between desperation and obvious acting leavable, useful, and reliable. Translations and replacements of ICT-representations must not fit smoothly without conflict into the world they are made ready for. A closed readiness is an ideal, which is not feasible because in the interaction situation the acting itself is ad-hoc and, therefore, unpredictable. The ready-made behavior and the content of ICT-representations should be differentiated and changeable to enable users to make ICT-representations ready and reliable for their own situated use.

FUTURE TRENDS

The Method OO as a Transformative Critical Room

The object-oriented approach (OO)[2] is used in the Informatics discipline as a method for interpretation and representation, for analyzing worlds of interaction, representing design models, and producing hardware and software systems. OO as it is used for the representation of the dynamics of interaction worlds leads us beyond the data-oriented approach and makes room for the opportunity to discuss the character of human behavior. Knowing that the essence of human behavior is not predictable and is situated in the interaction itself we can discover that OO will only disclose planned action. With abstraction tools in OO such as classification, separation, and inheritance, they colonized real world analysis processes.

This colonization from ICT-system realization into world analysis is dictated by the analyzing subjects' focus of avoiding complexity and ambiguity by selecting the most formalized documents, texts, tables, schemes in the domain which are close to the syntactical level of object oriented programming languages and by transforming natural language into a set of elementary propositions. This results in hierarchical structures and planned behavior to be enlightened, and in ad hoc actions and interactions to be darkened.

This use of OO in Informatics is exemplary for the ontological and epistemological assumptions in the discipline: not only is it possible to "handle the facts" but also to handle and therefore control real behavior itself. The expert users of the object-oriented approach suggest very heavily that OO can objectively represent the total dynamics of reality with its method to create OBJECTs: artificial representations.

Feminist theories can give arguments for doubting the assumptions within the OO approach because these approaches are always based on the same illusions of objectivity and neutrality of representation, the negation of power and dominance by its translation into something "natural and obvious." Leaving OO means to use it only for the purpose it was originally meant for: the production of software. OO-based software, which consists in predictable and planned interaction, cannot be the fundament for the representation of humans.

However, a total rejection of OO cannot be the answer to the doubts. The presence of OO-based products enforces the disclosure of some unwanted consequences of OO. In OO, ambiguity and doubt are hidden, but they are not absent. As a starting point for the use OO-based systems, a comparison with the theatre metaphor is useful for changing the position of the user (see Figure 2).

The OBJECT[3] is the basic unit in an OBJECT world description: the SCRIPT for an "interaction play" of cooperating OBJECTs. In the position of Audience or ACTOR (the intended roles of the user)

Figure 2. OO as theatre

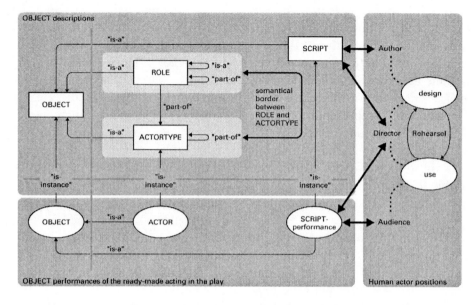

humans can enjoy the OBJECTs playing in the OO world. The OBJECT's ready made acting can be a useful tool integrated in the activities of our daily life. In this aspect the performance of an OBJECT play should be reliable. However if we are, as users, just ACTORs in an OO play, then we are determined to play as OBJECTs, with no doubt, and follow the prescriptions of the pregiven SCRIPT without thinking. Functioning as an OBJECT, we cannot escape the life cycle of states and transition rules of the ACTORTYPE "user" that is embedded in the script. Intertwining use and design activities is possible if we could change our position of only being a passive "Audience" member or a forced ACTOR to the position of being the "Director" or "Author" of the play. Humans should have the possibility to rehearse and to experiment with ROLE's and to test differentiated and situated leavability of the OO-software. As a "Director" or "Author", we could create out of ROLEs new ACTORTYPEs or aggregate old ones to a new surprising play. We could solve the conflicts within the aggregated ACTORTYPE so that they can cooperate in a way that is suited for special situations and act in a way that gives priority to our purposes, even if this could make our self-created ACTORs unpredictable or unreliable. As humans, we like to create new interaction worlds out of the present-at-hand ROLEs. Moreover, plays of which we do not know the plot we enjoy the most. Open and connectable OBJECTs can give us the opportunity to

edit the OO play and to replace parts with our own (inter)actions. And that is precisely the essence of their attraction. OBJECTs are aggregated ROLEs in a play or are things to use and integrate in acting. Humans are Authors and Directors creating the play; rewriting the ontology of the OO-approach to make it possible to look at OO realizations as plays of artificial ACTORs directed by users.

Free/Libre Open Source Software as Open ICT-Representation or Transformative Critical Room

An example of a transformative critical room is Free/Libre Open Source Software (F/LOSS). Software under an open source licensing form destabilizes the power relation at work when designing an ICT product. It deconstructs the before mentioned dominant believe of objectively measurable user friendliness, reliability etc. that can be planned in advanced. By merely providing the opportunity to adapt, the product in its base level of source code after distributing the software it again acknowledges different interpretations of use/design. This renders it impossible to hold on to a belief in objective measuring.

The available source code itself is a representation of the designer and the designer creates the opportunity of doubt to the expert user. In the case of OO-software it comes close to the possibility of (the almost literally) rewriting the ontological assumptions of the OO approach. By providing (expert) users access to and insight into the source code it opens up a space to negotiate the different performances involved. It gives the expert user the opportunity to create and or alter the artificial representation (OBJECT) as well as direct the aggregated ROLEs of the OBJECTs. Expert users are given the position of an author, because they can rewrite the software script.

By creating this opening, the objectivity of the representation is automatically questioned since the starting point of the creation of OBJECT, ACTORTYPE, ROLE, and SCRIPT is keeping it visible to the different human actor positions. In doing so, it shows the necessity of acknowledging the need and possibilities for different representations. F/LOSS does not set out to represent humans and all human interactions. The underlying assumption is that ambiguity and doubt is not only present but facilitates different ways of dealing with it by different actor positions. Furthermore, F/LOSS offers the most effective way of using OO: by being open source it provides the opportunity to combine models and reuse class definitions from other ICT products that might not seem to be an effective combination for one user/designer but might for another. In providing this, it encourages discovery and creativity by both designer/users and prevents user/designer/ OBJECT to turn into pre-defined resources. By preventing this user-designer, OBJECT cannot fit into a pregiven class since the pregiven class is changeable and adaptable.

For the non-expert user F/LOSS creates the possibility to choose between the different available software. For the non-expert user it can create a position of a director since the software producer cannot determine anymore the mutual actability of software. Furthermore, in using open source software, users have more opportunities for direct contact to designers and expert-users, to discuss changed performances of the open source software.[4]

Overall, the concept of F/LOSS products is an example of a critical room within the domain Informatics. Apart from being a strategy to challenge the ontological assumptions of OO, F/LOSS development also provides an opportunity to challenge the much-used dichotomy between user and designer. Just as feminist theories are a theoretical

strategy to deconstruct the binary opposition of user-designer; open source software provides a practical strategy to deconstruct this binary opposition. By literally making it possible for the user to become a designer and for the designer to be a critical user of its own product, it shows the impossibility of maintaining this opposition.

CONCLUSION

To ignore ICT-products is impossible. Therefore, one should be pragmatic and live between the borders of the binary oppositions and assess the construction of subject-object relations.

By creating leavable transformative critical rooms, a feminist approach can blow up the separation of use and design, and intertwine use and design through doubting the ready-made interactions. Through the creation of an opening and a redecoration in this cleared room between use and design, processes of intertwining use and design, and of changing interactions and representations can be started.

Transformative critical rooms are the necessary condition for making visible the gendering of the Informatics domain and for presenting and allowing a mutual dialogue between the female and the male in which differences can continue to exist. This answer is not a closed solution. It is the designing behavior of women and men, which can vivify the differences in future worlds of interactions.

REFERENCES

Booch, G. (1994). *Object-oriented analysis and design, with applications* (2nd ed.). Redwood City: Benjamin/Cummings.

Crutzen, C. K. M. (1997). Giving room to femininity in informatics education. In A. F. Grundy, D. Köhler, V. Oechtering, & U. Petersen (Eds.), *Women, work, and computerization: Spinning a Web from past to future* (pp. 177-187) Berlin: Springer-Verlag.

Crutzen, C. K. M., & Gerrissen, J. F. (2000a). Doubting the OBJECT world. In E. Balka & R. Smith (Eds.), *Women, work, and computerization:*

Charting a course to the future, (pp. 127-36). Boston: Kluwer Academic Press.

Crutzen, C. K. M. (2000b). *Interactie, een wereld van verschillen. Een visie op informatica vanuit genderstudies.* Dissertatie, Open Universiteit Nederland, Heerlen.

Crutzen, C. K. M. (2003). ICT-representations as transformative critical rooms. In G. Kreutzner & H. Schelhowe (Eds.), *Agents of change: Virtuality, gender, and the challenge to the traditional university* (pp. 87-106). Opladen, Germany: Leske+Budrich.

Crutzen, C. K. M., & Kotkamp, E. (2006). Questioning gender, deconstruction, and doubt. In E. M. Trauth (Ed.), *Gender and information technology Encyclopedia.* Hershey, PA: Idea Group Reference.

Harding, S. (1986). *The science question in feminism.* Ithaca, NY: Cornell University Press

Heidegger, M. (1936). *Der Ursprung des Kunstwerkes.* Used edition: (1960). Stuttgart: Philipp Reclam jun., 1960.

Jacobson, I., Christerson, M., Jonsson, P., & Övergaard, G. (1992). *Object-oriented software engineering. A use case driven approach.* Reading, MA: Addison Wesley Publishing Company.

Rumbaugh, J., Blaha, M., Premerlani, W., Eddy, F., & Lorensen, W. (1991). *Object-oriented modeling and design.* Englewood Cliffs, NJ: Prentice-Hall.

Suchman, L. (1994a). Working relations of technology production and use. In *Computer Supported Cooperative Work (CSCW), 2*(1-2), 21-39.

Suchman, L. (1994b). Do categories have politics? The language/action perspective reconsidered. In *Computer Supported Cooperative Work (CSCW), 2*(3), 177-190.

Wijnants, M., & Cornelis, J. (2005). *How open is the future. Economic, social, and cultural scenarios inspired by free and open source software.* Brussels, VUB Brussels University Press. Retrieved from http://crosstalks.vub.ac.be/documents/howopenisthefuture_CROSSTALKS BOOK1.pdf

KEY TERMS

Actability: Human actors can experience other actors as "actable" if these actors present themselves in a way, which is interpretable out of their own experiences. That does not mean that this is the intended interpretation because each actor has her or his horizon of experiences and expectations. Tools and things are actable, if humans can give meaning to them by drawing them into their interactions. A necessary condition of actability of artificial tools is that humans can perceive the performance of the non-human actor.

Free/Libre Open Source Software: Also referred to as open source software or free software. F/LOSS is a licensing type of software that has 4 main characteristics as defined by the Free Software Foundation:

- The freedom to run the program, for any purpose (freedom 0).
- The freedom to study how the program works, and adapt it to your needs (freedom 1).

Access to the source code is a precondition for this.

- The freedom to redistribute copies so you can help your neighbor (freedom 2).
- The freedom to improve the program, and release your improvements to the public, so that the whole community benefits (freedom 3). Access to the source code is a precondition for this.

The free in F/LOSS is not be confused with the price of the software but refers to the freedom as in freedom of speech. This is the reason the Libre is added; to clarify this. There are many different types of licensing that have variations on the 4 characteristics; one of the most important ones is the GNU General Public License (GNU-GPL) that was instated by Richard Stallman, who is with Thorvald Linus, the most important developer of the GNU/Linux operating system. There is much debate over advantages and disadvantages of open source software from different angles ranging from economical arguments to technical arguments to academic to socio-political arguments. See for more information

How open is the future. Economic, Social and Cultural Scenarios inspired by Free and Open Source Software" edited by Marleen Wynants and Jan Cornelis, published in 2005 and downloadable here: http://crosstalks.vub.ac.be/documents/ howopenisthefuture_CROSSTALKSBOOK1.pdf (last checked February 18th, 2005)

Mutual Actability: Mutual actability is a process of interactions that creates a transformative critical room between actors. Mutual actability of a tool and user can develop if the actor can intertwine use and design activities. The borders of the transformative critical room are on the one side, the frozenness of use and on the other side, the despair of a continuous forced design, for instance the tool causes doubt continuously by mal functioning.

Object-Oriented Programming: Object-oriented programming (OOP) is a programming language model organized around "objects." Objects are the things you think about first in designing a program and they are also the units of code that are derived from the design process. Each object is made into a generic class of object and even more generic classes are defined so that objects can share models and reuse the class definitions in their code. Each object is an instance of a particular class or subclass with the class's own methods or procedures and data variables. An object is what actually runs in the computer. Examples of objects range from human beings (described by name, address, and so forth) to buildings and floors (whose properties can be described and managed) down to the little widgets on your computer desktop (such as buttons and scroll bars).

ENDNOTES

[1] Heidegger calls this "Verläßlichkeit". He used it in two meanings: leavable and trustworthy (reliable) (1936, pp. 28-29)

[2] For the construction of such transformative critical rooms on OO in Education see: Crutzen 2000a, Crutzen 2000b, pp. 368-391. For the OO approach see: Booch 1994, Jacobson 1992, Rumbaugh 1991

[3] With the word "OBJECT" in uppercase is meant a constructed artificial entity within the ontology of the object-oriented approach. The word "object" in lowercase is an entity in reality, which can be observed and represented by a subject.

[4] Even though access to these resources and knowledge are determined by in- and exclusion mechanisms that can be shown through a gender analysis of these processes.

Race and Gender in Culturally Situated Design Tools

Ron Eglash
Rensselaer Polytechnic Institute, USA

INTRODUCTION

In theer study of equity issues in information technology (IT), researchers concerned with workforce diversity often utilize the metaphor of a "career pipeline." In this metaphor a population full of gender and race diversity enters the pipeline in kindergarten, but its delivery at the pipeline outflow in the form of software engineers and other IT workers is disproportionately white and male. While we might question the metaphor—its lack of attention to economic class or social construction, its illusion of rigid boundaries, etc.—the phenomenon it describes is well established by a broad number of statistical measures. For example, the U.S. Bureau of Labor Statistics' Current Population Surveys shows that between 1996 and 2002 the percentage of women in the overall IT workforce fell from 41% to 34.9%; during the same period the percentage of African Americans fell from 9.1% to 8.2%. Not only are women and certain minority groups under-represented, but the gap is in some cases getting worse. Returning to the pipeline, we might ask what barriers are encountered by women and minorities that act as impediments to this flow. Some of these barriers can be attributed to economic status; in particular the impact that poor educational resources have on low-income minority student academic success (Payne & Biddle, 1999). But other barriers appear to be more about cultural identity, including both race and gender identity. This essay describes Culturally Situated Design Tools (CSDTs), a suite of Web-based interactive applets that allow students and teachers to explore mathematics through the simulation of cultural artifacts, including Native American beadwork, African American cornrows, ancient Mayan temples, urban Graffiti, and Latin percussion rhythms (see http://www.rpi.edu/~eglash/csdt.html). Our preliminary evaluation indicates that some of the identity barriers preventing women and minori-

ties from participating in IT careers can be mitigated by the use CSDTs in classroom and out of class learning environments.

BACKGROUND

Cultural Identity Conflict in Science and Technology Curricula

While many problems in minority student performance can be directly attributed to economic circumstances, other barriers are more cultural. Fordham (1991) and Ogbu (1998) document the ways in which African American students perceive a forced choice between Black identity and high scholastic achievement (e.g., the accusation that they are "acting white," which Fordham connects to "peer-proofing"). Although some researchers (Ainsworth-Darnell & Downey, 1998) have critiqued this framework for its conflict with the positive view of education reported on minority attitude surveys, Mickelson (2003) has shown that there is a difference between what she terms "abstract" conceptions of education, which all racial groups respond to positively, and "concrete" conceptions of education which differ across racial groups and correlate with disparity in academic achievement. Martin (2000) reports a similar finding in African American conceptions of "cultural ownership" of mathematics. Powell (1990) found that pervasive mainstream stereotypes of scientists and mathematicians conflict with African-American cultural orientation. Eglash (2002a) describes conflicts in the identity of the "black nerd" in both popular imagination and reality. Similar assessment of cultural identity conflict in education has been reported for Native American students (Moore, 1994), Latino students (Lockwood & Secada, 1999), and Pacific Islander students (Kawakami, 1995). In addition to peer-proofing and identity conflict, the recent NSF-sponsored study of how minority students are lost in the science and technology career "pipeline" (Downey & Lucena, 1997) is also consistent with these results: many minority students with good math aptitude reported that they dropped out because they did not see how science or technology could have the connections to their cultural background offered by arts, humanities and social studies. Thus, perceived social irrelevance is a third obstacle to minority flow through the pipeline.

Mathematics is a key gate-keeper to the science and technology career pipeline, and the above cultural barriers—peer-proofing, identity conflict, and social irrelevance—can all be observed in minority barriers to mathematics achievement (Martin, 2000; Moore, 1994). In addition to these conflicts between cultural identity and mathematics education, another component for the poor mathematics performance in these minority groups is suggested by the work of Geary (1994). His review of cross cultural studies indicates that while children, teachers, and parents in China and Japan tend to view difficulty with mathematics as a problem of time and effort, their American counterparts attribute differences in mathematics performance to innate ability. This myth of genetic determinism then becomes a self fulfilling prophecy, lowering expectations and excusing poor performance.

Similar barriers are found for gender. Stipek and Heidi (1991) for example found that girls in their study expressed a persistent belief that they lacked an innate math ability. Tiedemann (2000) found that teachers in his study tended to attribute poor performance in girls to genetic causes, and poor performance in boys to lack of effort. The recent controversy over comments by the president of Harvard University proclaiming his belief that the underrepresentation of women in the sciences could be attributed to biological differences shows that this mythology persists at the highest levels of education. In addition to biological determinism, the kinds of cultural determinism that minority students must resist—peer-proofing, identity conflict, and social relevance—also operate as barriers to girls. For example, Armstrong (1979) found that sexist stereotypes of parents, peers, and teachers influenced girls' decisions not to participate in math. The previously cited study of Downey and Lucena (1997)

showed that some female students, like minority male students, also reported that they dropped out because they did not see how science and technology could give them the focus on beneficial social change offered by arts, humanities, and social studies.

Ethnomathematics: An Alternative View of Mathematical Knowledge

Ethnomathematics (Ascher, 1992; D'Ambrosio, 1990) is the study of mathematical practices in various cultures. While many studies have reported on variations in numeration systems (the Mayans used base 20, etc.) ethnomathematics takes a much broader view. It also considers the mathematical practices that are embedded in designs such as architecture, baskets, beadwork, divination, navigation, sculpture, textiles, etc. What distinguishes ethnomath from the broader category of "multicultural mathematics" is that ethnomathematicians strive to *translate* between indigenous concepts and the corresponding representations in standard (Western) mathematics. In my own work (Eglash, 1999) for example, interviews with traditional African artisans showed that they used specific geometric algorithms to construct recursive scaling structures (fractals). Other researchers have applied the same translation process to vernacular knowledge rather than indigenous knowledge. Lave (1998) for example considers the algebraic properties of knitting, which is generally more identified by gender than race. Other examples of gender-based ethnomathematics analyses can be found in Gilmer 1998.

Cultures characterized as "primitive" by colonialists can be shown to utilize sophisticated mathematical ideas, and vernacular knowledge that was dismissed by sexist stereotypes as trivial "women's work" can be appreciated in a new light. Thus ethnomathematics analyses are useful in opposing harmful myths of cultural and biological determinism in both race and gender.

Given the race and gender barriers to math achievement in pre-college education, it might seem like an obvious step to move from this academic research on ethnomathematics to direct application of ethnomathematics in the K-12 classroom. But this is more difficult than it sounds. First, many of these mathematical topics, such as fractals, are not part of the standard K-12 math curriculum. Second, many children are unaware of their own heritage culture, or uninterested in traditional activities. For these two reasons we have developed a suite of Web-based applets called "Culturally Situated Design Tools." Each design tool allows students to simulate a cultural practice such as Native American beadwork, African sculpture, etc. Their grade levels range from 4 to 12. Each site includes a cultural background section, a tutorial, the software itself, and a "teaching materials" section of lesson plans, evaluation instruments, samples of student work, and other support. The design tools can be integrated into standards-based curricula, including a variety of specific math topics as well as state and national standards for more general areas such as technology use and understanding patterns. At the same time, this expressive computational medium affords new opportunities for researchers to explore the relationships between youth identity and culture, including that of gender identity.

MAIN THRUST OF ARTICLE

From Ethnomathematics to CSDTs: Examples from Cornrow Curves

Our first software application began with the "African Fractals" material, using graphical simulations of recursive scaling patterns in architecture, textiles, sculpture, etc. Discussions with teachers from inner city areas in which there were large numbers of African American students lead to the following three critiques: First, they mentioned that fractal geometry is not part of the K-12 standard curriculum. Second, they noted that the computer design was primarily carried out with mouse movements, not numeric or symbolic inputs. Finally, they noted that not only were their students doing poorly in math, but that they also knew little about Africa.

The cultural challenge of making the connection with African American students who had little knowledge of Africa was solved by three innovations. First, following teachers suggestions, we focused on the example of fractal patterns in cornrow hairstyles, and titled the software "Cornrow Curves" (Figure 1).

Typically each criss-cross ("plait") of the hair diminishes in size, creating several iterations of scale in a single braid; to that extent it matches the other African traditions of recursive scaling. But cornrows are part of the African mathematical heritage that not only made it through the middle passage, but are still an on-going area of innovation in contemporary culture. Because many students we spoke to were unaware of this legacy ("Cornrows were invented in the 1970s" was a common statement, we developed an historical background page for cornrows, which included images and information covering the indigenous

Figure 1. Cornrow curves

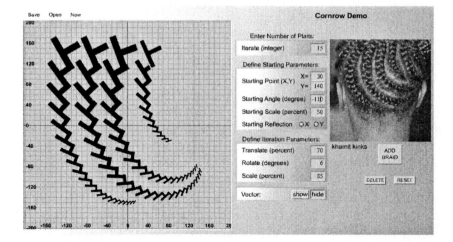

styles and meaning in Africa, their appearance in the pre-civil war era, and their new significance in post-civil rights and hip-hop cultures.) Finally, we also created a series of "goal images": photos of styles for students to simulate, including both professional styles and photos the students took of themselves and their friends. This allowed students to make the connection from contemporary vernacular identity to heritage identity, which then opened up a wide selection of indigenous African fractal patterns.

It was a simple matter to replace the mouse-driven interface with numeric input, as can be seen in Figure 1. Separating these parameters numerically also addressed the problem of fitting fractal geometry into the curriculum: now it was about geometric transformations, not fractals, and transformational geometry is a component of the state and national standards. In addition to several years of software development and cultural studies (ranging from history of black culture to interviews with stylists) the following discussion makes use fieldwork carried out from 2001-2004 in summer and after-school workshops in Albany, NY, and Troy, NY, with students ranging from middle school to high school. The classes ranged from all-minority to majority white. Most minorities were African American and Puerto Rican (which included students with African heritage).

Figure 2. "tis-abay"

We found that students quickly became adapt at using the software to not only create simulations of the hairstyles, but also generate patterns of their own choosing. Figure 2, for example, shows a pattern one girl named after a waterfall in Ethiopia (her father's place of birth), "tis-abay."

Another advantage of choosing cornrow patterns was that both genders typically found it to be readily accessible. During discussions girls were just as engaged as boys, and even went so far as to suggest that African American males had "stolen" their hairstyle. We found little difference in performance on mathematical, cultural, or aesthetic outcomes across genders, although this was not tracked statistically. Both male and female students generally understood the fundamental ethnomathematics concept—that this was a claim for mathematical knowledge, albeit one that is embedded in a practice rather than represented through symbols. The strongest reaction we heard from any participant was from an African American girl—"I never realized there was math in cornrows!!" The only hesitation in using the software came from two white males; neither had any difficulty once they started working with it. Not large enough for statistical significance, but it does suggest at least a minor role in race and gender interactions in the users' relationship to the software.

For all the design tools, including the Cornrow Curves software, there are three categories of use. First, the software can be used to simulate the original designs. Second, it can be used as an artistic tool in itself. And third, it can be used to conduct mathematical explorations. Some of the mathematical exercises are chosen by teachers. For example, how many plaits does it take to make a braid form a complete circle with no dilation and a 10 degree rotation? The answer is 36, because $36 * 10 = 360$. Now try the same with a 7 degree rotation. Now give us a general formula for any degree rotation. Other mathematical exercises are chosen by the students themselves. Several Latino students, for example, have tried to produce an image of the Puerto Rican flag in the virtual bead loom; a task which raises several mathematical challenges (as well as exemplifying the playful hybridity which students enthusiastically bring to these projects).

Other Design Tools

Space does not permit us to describe any other design tools in detail, but we can briefly describe each of them, and make some notes about associated gender issues.

1. The **Native-American** design tools:
   - **The Virtual Bead Loom:** This has been our most popular tool with math teachers. Native students on and off the reservation schools have expressed great enthusiasm for it, but so have white, black, and Latino students. Like Cornrow Curves, it makes use of indigenous practices which are now more female-associated. However, both genders do well with this software. Our graphic designer (co-PI Audrey Bennett) suggests that girls' designs tend to make better use of color. See Eglash, 2002b for more details
   - **Yupik Star Navigation:** Used with 5th grade students in Alaska. The indigenous practice is male-associated, but we did not discern any gender difference. It was interesting to note here the strong interest in cornrows shown by native Yupik girls, but not by Yupik boys, indicating that gender may play a strong role in some cross-cultural experiences
2. The **African/African-American** design tools:
   - **Cornrow Curves:** See description in previous section
   - **Mangbetu Design:** This software is very similar to Cornrow Curves, but based on recursive scaling patterns and other geometric symmetries in the artifacts of the Mangbetu society. The artisans themselves tend to be male, but the subject matter was both male and female, and use of the artifacts is by both genders. We obtained similar results to the Cornrow Curves software, with little gender differences observed
3. The **Latino** design tools:
   - **Virtual Temple Builder:** This software allows the users to construct three-dimensional simulations of ancient mayan temples. There was little gender difference observed, although the software is relatively new and we do not have as many samples
   - **Rhythm Wheels:** This software is also relatively new; again we found little gender difference
4. The **Youth Subculture** design tools:
   - **Graffiti Grapher:** We found the reactions to Graffiti Grapher to be profoundly multicultural—for example the two white male students who were hesitant about Cornrow Curves were immediately attracted to it, but so were African American and Latino students. Girls were also interested in it, but seemed less "obsessed." On the other hand, in a recent classroom competition for best Graffiti Grapher design, the winning entry was from a girl. Discussions indicated that girls saw graffiti as a more male-oriented activity.

In summary, the students do seem to have both race and gender associations for the cultural activities that form the basis of CSDTs, but these do not appear to be limiting factors for participation, and in some cases do appear to encourage participation.

Quantitative Results

Our quantitative assessment did not separate results by gender, but both male and female students seemed equally engaged with most activities. Using the Bath County Computer Attitudes Scale we found statistically significant improvement for attitudes toward information technology careers in comparison to a baseline measure in the classes with 90% minority students. There was no significant difference from baseline measures in the sample with majority white students. We did find statistically significant improvement in pre-test/post-test comparisons for mathematics performance for classes of all ethnic compositions, with one exception (where there was a positive increase but small class size prevented statistical significance).

FUTURE TRENDS

We are constantly expanding both the number of design tools, the utility of their interface, and their associated classroom materials, as well as evaluation. We hope to further the qualitative aspects of gender analysis with a more ethnographic approach. One of the most important questions we would like to ask is about the comparison with ostensibly "neutral" software. For minority girls with enthusiasm about engaging software marked as female and non-white, does that contrast with less enthusiastic engagement of software that is marked (and marketed) as neutral? Can this kind of software make visible the currently invisible semiotics of race and gender currently hiding under a mask of generic universalism?

In their excellent anthology on gender in computer games, Cassell and Jenkins (1998) note the wide variety of feminist perspectives. In the question of violence, for example, these range from an "essentialist" positions that would posit an inherent non-violence in female gender constructions, to strongly anti-essentialist positions which applaud girl gamers who can tackle "first person shooter" games and their ilk. In my own experience teaching female engineering students in a design studio, where we read selections from the Cassell and Jenkins volume, I have also found that these women draw their academic and technological strength from a wide variety of positions. For every design studio in which I have seen a cluster of women working together in one group, there has always been one or two others who elbow their way into the most aggressive male group (and often insist on being the one who operates the power tools). It is our hope that the wide variety of experiences offered by the design tools can do justice to this valuable spectrum of gendered identities, and provide the young girls with the positive technological experiences that can bring them through the other side of the IT pipeline.

CONCLUSION

We have seen a broad spectrum of reactions to these design tools in students. While there are some indications that both race and gender associations remain relevant (the enthusiasm of minority girls for the cornrows, the hesitation of the two white males when working with the same, the preference of males for Graffiti Grapher, etc.), what is much more striking is the enthusiasm by students of all ethnic and gender identities for the design tools in general. We believe that one of the reasons for this is the active role that students themselves are able to take in design creation when using these tools. They name their designs after everything from dogs to musicians, and often students have a story to tell with their work.

That is not to say we can overlook the complexity of race and gender identity in our efforts to improve female and minority participation in IT. As Nakano Glenn (1992) argues for the case of service workers, gender and race cannot be reduced to "additive oppressions," and must be seen as the site of an interlocking or relational dynamic. If, for example, a middle school text book says that "women won the right to vote in 1920," is that statement helpful for celebrating a feminist victory, or harmful for ignoring the Jim Crow laws that prevented minority women from voting? There has been a debate in both feminist and anti-racist literature concerning the advantages and disadvantages of tradition. Should women re-evaluate knitting and embrace it as a valued heritage to be recovered, or should they distain its patriarchal roots in sexist division of labor and "busywork"? If we are attempting to respectfully represent traditional cultural practices, should we also represent their gendered divisions of labor?

We posit that the main advantage of the design tools is their ability to facilitate new forms of cultural hybridity, new translations between formally disparate domains (both socially as well as mathematically), and thus new ways of allowing students to create their own identities in relation to science and technology. Expressive media are natural aids to identity formation, and offering students new ways to link identity to math and technology can be empowering across both gender and race.

ACKNOWLEDGMENT

This material is based upon work supported by the National Science Foundation under Grant No. 0119880.

REFERENCES

Ainsworth-Darnell, J. W., & Downey, D. B. (1998). Assessing the oppositional culture explanation for racial/ethnic differences in school performance. *American Sociological Review, 63,* 536-553.

Armstrong, J. M. (1979). *Achievement and participation of women in mathematics* (Report No. NIE-G-7-0061). Denver, CO: Education Commission of the States. (ERIC Document Reproduction Service No. ED 184878).

Ascher, M. (1990). *Ethnomathematics: A multicultural view of mathematical ideas.* Pacific Grove: Brooks/Cole Publishing.

Cassell, J., & Jenkins, H. (1998). *From Barbie to Mortal Kombat: Gender and computer games.* Cambridge: The MIT Press.

D'Ambrosio, U. (1990). *Etnomatematica.* Sao Palulo: Editora Atica.

Downey, G. L., & Lucena, J. (1997). Weeding out and hiring in: How engineers succeed. In G. L. Downey & J. Dumit (Eds.), *Cyborgs & citadels: Anthropological interventions in emerging sciences and technologies.* Santa Fe, NM: School of American Research Press.

Eglash, R. (1999). *African fractals: Modern computing and indigenous design.* New Brunswick: Rutgers University Press.

Eglash, R. (2002a, Summer). Race, sex, and nerds: From Black geeks to Asian-American hipsters. *Social Text, 20*(2), 49-64.

Eglash, R. (2002b, June 21). A two-way bridge across the digital divide. *Chronicle of Higher Education,* p. B12.

Fordham, S. (1991). Peer-proofing academic competition among black adolescents: Acting white Black American style. In C. Sleeter (Ed.), *Multicultural education and empowerment* (pp. 69-94). Albany, NY: SUNY Press.

Gilmer, G. (1998, September 6-11). Ethnomathematics: An African American perspective on developing women in mathematics. In *Proceedings of the 1ˢᵗ Mathematics Education and Society Conference (MEAS1),* Nottingham University, UK. Retrieved from http://www.nottingham.ac.uk/csme/meas/papers/gilmer.html

Kawakami, A. J. (1995). *A study of risk factors among high school students in the Pacific region.* Honolulu, HI: Pacific Resources for Education and Learning.

Lave, J. (1998). *Cognition in practice.* New York: Cambridge University Press.

Lockwood, A. T., & Secada, W. G. (1999). *Transforming education for Hispanic Youth: Exemplary practices, programs, and schools.* NCBE Resource Collection Series, No. 12.

Martin, D. (2000). *Mathematics success and failure among African-American youth: The roles of sociohistorical context, community forces, school influence, and individual agency.* Mahwah, NJ: Lawrence Erlbaum Associates.

Mickelson, R. A. (2003). When are racial disparities in education the result of racial discrimination? A social science perspective. *Teachers College Record, 105*(6), 1052-1086.

Moore, C. G. (1994). Research in Native American mathematics education. *For the Learning of Mathematics, 14*(2), 9-14.

Nakano Glenn, E. (1992). From servitude to service work: Historical continuities in the racial division of paid reproductive labor. *Signs, 18,* 1-43.

Ogbu, J. (1998). Voluntary and involuntary minorities: A cultural-ecological theory of school performance. *Anthropology and Education Quarterly, 29*(2), 155-188.

Payne, K. J., & Biddle, B. J. (1999). Poor school funding, child poverty, and mathematics achievement. *Educational Researcher, 28*(6), 4-13.

Powell, L. (1990). Factors associated with the under-representation of African Americans in mathematics and science. *Journal of Negro Education, 59*(3).

Stipek, D., & Granlinski, H. (1991, September). Gender differences in children's achievement-related beliefs and emotional responses to success and failure in mathematics. *Journal of Educational Psychology, 83*(3), 361-71.

Tiedemann, J. (2000). Gender-related beliefs of teachers in elementary school mathematics. *Education Studies in Mathematics, 41*, 191-207.

KEY TERMS

Acting White: Phenomenon in which minority students perceive a forced choice between "authentic" ethnic identity and high scholastic achievement, largely as the result of peer pressure.

Cultural Hybridity: Mixing of cultures, usually in terms of the identity of an individual who partakes in both "parent" cultures (such as Texas and Mexican combining to form "Tex-Mex."

Culturally Situated Design Tools: User-controlled computer programs which allow simulation of indigenous or vernacular designs in visual or auditory media.

Ethnomathematics: The study of mathematical concepts and practices as they occur in various cultures, with particular attention to the translation from mathematics embedded in indigenous or vernacular designs or practices to their analogous representations in "mainstream" (i.e., Western) mathematics.

Heritage Culture: The culture that an individual envisions as their heritage; often nostalgic or mythic in comparison to the actual culture referenced.

Indigneous Culture: Typically refers to the culture of small-scale (non-state) societies that existed previous to colonial invasion or third-world nationalization.

Vernacular Culture: The culture of every-day life; often used to invoke "street smarts" or non-elite knowledge.

Race and the IT Workforce

Elaine K. Yakura
Michigan State University, USA

INTRODUCTION

As the existence of the present volume attests, gender is crucial in understanding the IT workforce. To stop there, however, would be to miss many other aspects of identity that influence issues of satisfaction, recruitment, retention and attrition in IT organizations.[1] In this article, I will focus on race[2]—one of the most salient identity characteristics for today's workforce. The goal is to summarize some of the research approaches from sociology, psychology, and management that have furthered our understanding of race. These perspectives are presented as possibilities for extending the repertoire of strategies for enriching our research on IT women, especially women of color.

BACKGROUND

As shown in Table 1, African Americans, Hispanic Americans, and Native Americans are still underrepresented in the U.S. IT workforce (Niederman & Mandviwalla, 2004); women of color in the IT workforce are even scarcer. While the Information Technology Association of America ("ITAA") and the National Science Foundation (Wardle & Burton, 2002) have funded initiatives focusing on increasing the diversity of the IT workforce, there is still a dearth of relevant research. There is evidence that U.S. IT organizations have problems attracting and retaining minorities (Tapia & Kvasny, 2004), not to mention promoting them (Igbaria & Wormley, 1992,

1995). These problems are exacerbated by differential access to technology in general—a phenomenon referred to as the "digital divide" (Compaine, 2001; Kuttan & Peters, 2003; Mack, 2001; Servon, 2002).

The demography of the IT workforce is also affected by characteristics of IT organizations. We know race as a cultural factor is important in the shaping of technology—in other words, technology and technology organizations are not race neutral. Nor are they colorblind, despite the majority of whites in the U.S. who believe otherwise (Gallagher, 2003). Research often treats race as a static individual trait, yet this is hardly reflective of the experiences of IT professionals. Researchers often separate out one identity element at a time, such as gender, or race, or age, and the literature has tended to emphasize simple, monolithic categories, such as "black" and "white." Of course, there are many other categories and combinations. The 2000 U.S. census was redesigned to allow citizens to check multiple boxes to identify themselves as multiracial. Nationwide, approximately 2.4% of the population, over 6.8 million Americans, checked two or more races, in dozens of different combinations (http://www.censusscope.org/us/chart_multi.html). By 2050, the rate of multiracial identification could increase to 1 in 5 (Lee & Bean, 2004). In typical organizational research, however, members of these groups are often too few in number to constitute statistically significant subgroups and must be removed from the population of research subjects (Cox, 2004). Hence, Asian American, Hispanic American, and dozens of multi-racial groups are

Table 1. Percentage by race, 2002

	White	African-American	Asian/Pacific Islander	American Indian	Hispanic
IT occupations	77.7	8.2	11.8	0.6	6.3
All occupations	83.5	10.9	4.0	0.9	12.2

Source: ITAA, 2003

excluded from research data. Individuals who do not "fit" the major categories become invisible. This problem is particularly acute for women of color.

The potential multiplicity of racial or ethnic self-identification can be affected by computer mediated communication technology, such as the Internet. Kolko, Nakamura, and Rodman (2000) note that in textual "cyberspace," visual cues that can identify race are invisible, but users can present themselves in a wide variety of different ways using textual cues and signifiers. Thus, people can construct identities in ways that were not possible IRL ("in real life"). However, Kendall (1998, 2002) has also found that "Gendered, classed, and raced identities continue to have salience in online interactions, with power relations often operating in much the same ways as they do offline, even when participants understand that people's online identities might differ from their offline identities" (1998, p. 150). Thus, while online identities can be fluid in a way that is different from offline identities, this does not necessarily extend to the power hierarchies in which identities (such as gender, class, race, or age) are embedded.

While novel in some respects, the fluidity of race in cyberspace is just the one of the more recent instances of the recognition of the socially constructed nature of identity in the research literature. Social psychologists have described the individual's identification with different groups:

Tajfel first defined social identity as "the individual's knowledge that he belongs to certain social groups together with some emotional and value significance to him of this group membership. (Hogg, Abrams, Otten, & Hinkle, 2004, p. 249)

Different group identities (such as age or ethnicity) might become more salient given the particular context, or at different times. Thus, it is misleading to think of a particular individual as having a single social identity group (such as "female"), since an individual will belong to more than one group, and the significance of these identifiers depends on the context. In practice, different sets of identities become salient at various times.

Perhaps the most challenging development concerns the acknowledgement that multiple identities related to race, class, or gender, should not be explored as separate elements. In people's actual experiences, aspects of identity are experienced together rather than severally. For example, Kvasny (2003) offers an analysis of the "triple jeopardy" that arises from the intersection of race, gender, and social class in the world of information technology through participant observation at an inner-city computer technology center. Such a theoretical (as well as methodological) approach can provide us with a far more nuanced understanding of the experiences of women of color in IT than approaches that rely solely on a limited number of racial checkboxes on a survey.

However, there are many hurdles that race researchers must overcome before their research is published (Cox, 2004). For instance, focusing on a non-white group can cause journal reviewers to reject such studies. Race studies can give rise to a number of methodological dilemmas, including sampling issues, or overlaying research constructs onto the accounts of the research participants (Cuadraz & Uttal, 1999). As He and Phillion (2001) have noted, during the research process, "previously reified formalistic notions of race, gender, and class" can be "shattered" (p. 47) since research participants do not fit nicely into the categories that have been created for them. Race researchers who strive to preserve the narratives and experiences of their research participants must also contend with issues of their own identity in the research process (McCorkel & Myers, 2003). Smith (2002) called for broader level qualitative analyses of race and the workplace as well as longitudinal studies, but added the caveat that the challenges of this type of research are myriad, including the problem of employer reluctance to participate and release information about the organization. However, organizational or occupational level studies (House, Rousseau, & Thomas-Hunt, 1995) are critical in understanding the dynamics of the workplace, as discussed in the next section.

MESO LEVEL: ORGANIZATIONAL CULTURE AND OCCUPATIONAL SUBCULTURES

Issues of race and identity take on significance against a background of organizational cultures and occupational subcultures. Organizational cultures can play an important role in fostering an environment that is "warm" rather than "chilly" toward minorities and other underrepresented groups (Roldan, Soe, & Yakura, 2004). An important part of this effect is created through sheer numbers (Kanter, 1977). As the ratio or percentage of a particular group increases, the comfort level generally increases for members of that group. Because numbers are important for comfort, under-representation of minorities creates a self-perpetuating problem for the IT workforce.

Similarly, the culture of an occupation can be chilly towards women and minorities. For example, the ITAA report (2003) notes that women and minorities might not consider IT as a profession: "Females and minorities may feel isolated or unaccepted in the IT profession" (p. 6). Consider the stereotypical white male "geek," who represents a "prestigious stigma" (Moore & Love, 2004) in IT subculture, but may appear less than prestigious to women of color (see also Camp, 1996). Within an organization, occupational subcultures (Trice, 1993) can exist which foster attitudes that are implicitly or explicitly exclusionary.

Even in an organization with a high percentage of women, the comfort level for women of color may be quite low. Attempting to address problems of attraction and retention of minority IT professionals using policies that target a single identity factor will not always help. For example, any single initiative targeted for women may not help women of color. Furthermore, the process of attempting to accommodate women runs the risk of alienating men (as well as some of the women). To address these concerns, we need to foster cultures that are less stereotypic and simplistic in attitude.

Stereotyping operates in many different ways in the IT workplace. In addition to the obvious effect of generating bias, it can set up invidious comparisons among workers. For example, consider the impact of the "model minority" (Cheng, 1997) stereotype on the IT workplace. Asian Americans are often considered "excellent" in IT-related work, a thesis that sets up an implicit hierarchy of stereotypical expectations vis-à-vis other ethnic groups within the organization. As researchers (Cheng, 1997; Woo, 2000) have indicated, depending on how data on Asian Americans in the workforce are analyzed, arguments

can be made that either support or defeat the thesis that Asian Americans are socioeconomically successful. For example, aggregating the data on Asian Americans as a whole "perpetuates a picture of high achievement" (Woo, 2000, p. 32). Stereotypes that reproduce and perpetuate these kinds of misunderstandings are damaging not only to those stereotyped but also to other dominant or non-dominant groups (see also Frankenberg, 1993).

Stereotyping is not the only area of concern regarding perceptions. We tend to think of racial stereotyping as a potential source of bias in hiring or performance appraisal, but in some respects, double standards are more insidious. Foschi (2000) notes that, in contrast to direct evaluation bias, "double standards ... occur when evaluations have already been made and are accepted to be objective (i.e., exempt from evaluation bias). The use of double standards is thus a subtle exclusionary practice ..." (p.27). Thus, even where the inputs to an evaluation process are fair and unbiased, the outcomes may not be. Different cross-groups can experience different outcomes in downsizing (Spalter-Roth & Deitch, 1999), working shifts (Presser, 2003), workplace authority (Elliot & Smith, 2004), or promotional practices (Maume, 2004; Wilson, Sakura-Lemessy, & West, 1999).

FUTURE TRENDS

Projections indicate that the North American population and labor market will continue to diversify (Fong & Shibuya, forthcoming; Grieco & Cassidy, 2001). Surveys by organizations such as the ITAA will continue to highlight this issue, and focus attention on the need to do more in terms of attracting and retaining a diverse set of employees. NSF funding initiatives will also aid in providing necessary resources for research that has been neglected in the past.

Our research must continue to deepen our understanding of the different identity groups that abound in the workforce. In particular, research that illuminates the complexities of the interactions and intersections of multiple identities (McCall, 2005) can enhance our ability to attract and retain more and more diverse sets of people in the IT workplace. Of course, understanding issues at the intersection of gender and race involves novel and complicated challenges for researchers as well as managers. While statistical research can provide useful information for examining changes in the IT workforce, qualitative and interpretivist research strategies will be required to address these challenges. While these approaches are not always easier or simpler, the additional efforts will be repaid with better outcomes for IT professionals and organizations.

CONCLUSION

Ideally, our research should reflect the complexity of individual identities against the backdrop of organizational, occupational, and even societal levels of culture. Monolithic, single-factor models of identity fail to capture key aspects of the experience of women of color in the IT workplace. Researchers are just beginning to explore the nuances of the gendered (Wilson, 2002) and raced (Vallas, 2003) spaces of U.S. organizations, and a consistent and continuous effort is required to address the negative and insidious aspects of issues of culture and climate as they affect women of color (Tapia & Kvasny, 2004; Taylor, 2002). For researchers committed to this deeper understanding of the experiences of women of color, this presents a difficult but ultimately more satisfying research process and outcome. This is not just an academic issue, but a practical problem as well that confronts IT organizations on a daily basis.

REFERENCES

Camp, L. J. (1996). We are geeks, and we are not guys: The systers mailing list. In L. Cherney & E. R. Weise (Eds.), *Wired women: Gender and new realities in cyberspace*. Seattle, WA: Seal Press.

Cheng, C. (1997). Are Asian American employees a model minority or just a minority? *Journal of Applied Behavioral Science, 33*, 277-290.

Compaine, B. M. (Ed.). (2001). *The digital divide: Facing a crisis or creating a myth?* Cambridge, MA: MIT Press.

Cox, Jr., T. (2004). Problems with research by organizational scholars on issues of race and ethnicity. *Journal of Applied Behavioral Science, 40*(2), 124-145.

Cuadraz, G. H., & Uttal, L. (1999). Intersectionality and in-depth interviews: Methodological strategies for analyzing race, class, and gender. *Race, Gender, & Class, 6*(3), 156-171.

Elliott, J. R., & Smith, R. (2004). Race, gender, and workplace power. *American Sociological Review, 69*(3), 365-386.

Fong, E., & Shibuya, K. (forthcoming). Multiethnic cities in North America. *Annual Review of Sociology, 31*.

Foschi, M. (2000). Double standards for competence: Theory and research. *Annual Review of Sociology, 26*, 21-42.

Frankenberg, R. (1993). *White women, race matters: The social construction of whiteness*. Minneapolis, MN: University of Minnesota Press.

Gallagher, C. A. (2003). Color-blind privilege: The social and political functions of erasing the color line in post race America. *Race, Gender & Class, 10*(4), 22-37.

Grieco, E., & Cassidy, R. (2001). *Overview of race and Hispanic origin: Census 2000 brief*. Retrieved May 25, 2005, from http://www.census.gov/population/www/cen2000/briefs.html

He, M. F., & Phillion, J. (2001). Trapped in-between: A narrative exploration of race, gender, and class. *Race, Gender & Class, 8*(1), 47-56.

Hogg, M. A., Abrams, D., Otten, S., & Hinkle, S. (2004). The social identity perspective: Intergroup relations, self-conception, and small groups. *Small Group Research, 35*(3), 246-276.

House, R., Rousseau, D. M., & Thomas-Hunt, M. (1995). The meso paradigm: A framework of the integration of micro and macro organizational behavior. *Research in Organizational Behavior, 17*, 71-114.

Igbaria, M., & Wormley, W. (1992). Organizational experiences and career success of MIS professionals and managers: An examination of race difference. *MIS Quarterly, 16*(4), 507-529.

Igbaria, M., & Wormley, W. (1995). Race differences in job performance and career success among IS people. *Communications of the ACM, 38*(3), 82-92.

Information Technology Association of America. (2003). *Report of the ITAA blue ribbon panel on IT diversity.* Presented at the National IT workforce convocation, Arlington, VA. Retrieved February 5, 2005, from http://www.itaa.org/workforce/docs/03divreport.pdf

Kanter, R. M. (1977). *Men and women of the corporation.* New York: Basic.

Kendall, L. (2002). *Hanging out in the virtual pub: Masculinities and relationships online.* Berkeley, CA: University of California Press.

Kendall, L. (1998). Meaning and identity in "cyberspace": The performance of gender, class, and race online. *Symbolic Interaction, 29*(20), 129-54.

Kolko, B. E., Nakamura, L., & Rodman, G. B. (2000). *Race in cyberspace.* New York: Routledge.

Kuttan, A., & Peters, L. (2003). *From digital divide to digital opportunity.* Lanham, MD: Scarecrow Press.

Kvasny, L. (2003). Triple jeopardy: Race, gender and class politics of women in technology. In *Proceedings of the 2003 ACM SIGMIS Computer Personnel Research Conference: Leveraging Differences and Diversity in the IT Workforce,* Philadelphia (pp. 112-116).

Lee, J., & Bean, F. D. (2004). America's changing color lines: Immigration, race/ethnicity, and multiracial identification. *Annual Review of Sociology, 30,* 221-242.

Mack, R. L. (2001). *The digital divide: Standing at the intersection of race & technology.* Durham, NC: Carolina Academic Press.

Maume, Jr., D. (2004). Is the glass ceiling a unique form of inequality? Evidence from a random-effects model of managerial attainment. *Work & Occupations, 31*(2), 250-274.

McCall, L. (2005). The complexity of intersectionality. *Signs, 30*(3), 1771-1800.

McCorkel, J. A., & Myers, K. (2003). What difference does difference make? Position and privilege in the field. *Qualitative Sociology, 26*(2), 199-231.

Moore, J. E., & Love, M. S. (2004). An examination of prestigious stigma: The case of the technology geek. In *Proceedings of the 2004 ACM SIGMIS Computer Personnel Research Conference,* Tucson, AZ (pp. 103).

Niederman, F., & Mandviwalla, M. (2004). The evolution of IT (computer) personnel research: More theory, more understanding, more questions. *Database, 35*(3), 6-8.

Presser, H. B. (2003). Race-ethnic and gender differences in nonstandard work shifts. *Work & Occupations, 30*(4), 412-439.

Roldan, M., Soe, L., & Yakura, E. (2004). Perceptions of chilly IT organizational contexts and their effect on the retention and promotion of women. In *Proceedings of the 2004 ACM SIGMIS Computer Personnel Research Conference,* Tucson, AZ (pp. 108-113).

Schein, E. H. (1992). *Organizational culture and leadership* (2nd ed.). San Francisco: Jossey-Bass.

Servon, L. J. (2002). *Bridging the digital divide: Technology, community, and public policy.* Malden, MA: Blackwell.

Smith, R. A. (2002). Race, gender, and authority in the workplace: Theory and research. *Annual Review of Sociology, 28,* 509-542.

Spalter-Roth, R., & Deitch, C. (1999). I don't feel right sized; I feel out-of-work sized: Gender race, ethnicity, and the unequal costs of displacement. *Work & Occupations, 26*(4), 446.

Tapia, A., & Kvasny, L. (2004). Recruitment is never enough: retention of women and minorities in the IT workplace. In *2004 ACM SIGMIS Conference Proceedings on Computer Personnel Research* (pp. 84-91).

Taylor, V. (2002). Women of color in computing. *SIGCSE Bulletin, 34*(2), 22-23.

Trice, H. (1993). *Occupational subcultures in the workplace.* Ithaca, NY: Cornell University Press.

Vallas, S. P. (2003). Rediscovering the color line within work organizations: The "knitting of racial groups" revisited. *Work and Occupations, 30*(4), 379-400.

Wardle, C., & Burton, L. (2002). Programmatic efforts encouraging women to enter the information technology workforce. *SIGCSE Bulletin, 34*(2), 27-31.

Wilson, E. (2002). Family man or conqueror? Contested meanings in an engineering company. *Culture and Organization, 8*(2), 81-100.

Wilson, G., Sakura-Lemessy, I., & West, J. P. (1999). Reaching the top: Racial differences in mobility paths to upper-tier occupations. *Work & Occupations, 26*(2), 165-186.

Woo, D. (2000). *Glass ceilings and Asian Americans: The new face of workplace barriers.* Walnut Creek, CA: Altamira Press.

KEY TERMS

Model Minority Stereotype: The stereotype that Asian Americans have achieved economic success.

Multiracial Identification: Identifying oneself as having an ancestry that includes more than one race or ethnicity.

Occupational Subcultures: Subgroups within an organization that share assumptions as members of an occupational group.

Organizational Culture: A pattern of shared basic assumptions that the group learned as it solved its problems of external adaptation and internal integration, that has worked well enough to be considered valid and, therefore, to be taught to new members as the correct way to perceive, think, and feel in relation to those problems (Schein, 1992, p. 12).

Race/Ethnicity: Groups that can be distinguished on the basis of ancestry and/or color (Lee & Bean, 2004, p. 223).

Social Identity: An identity which an individual has as a result of their acknowledgement of membership in a particular social group; an individual can have multiple social identities.

ENDNOTES

[1]   This discussion is limited to the U.S. context, since other contexts, such as more homogeneous societies of Japan or Korea, would have a very different workforce demographic.

[2]   As Lee and Bean (2004, p. 225) have noted, "Today, social scientists generally agree that race is a social rather than biological category and have documented the processes by which ethnic and racial boundaries have changed throughout our nation's history." In their review, they (p. 223) chose to use term "race/ethnicity" to refer to "groups that distinguish themselves on the basis of ancestry and/or color." In this article, for the sake of brevity, I use the term "race" as shorthand for their term, "race/ethnicity."

Reasons for Women to Leave the IT Workforce

Peter Hoonakker
University of Wisconsin-Madison, USA

Pascale Carayon
University of Wisconsin-Madison, USA

Jen Schoepke
University of Wisconsin-Madison, USA

INTRODUCTION

Turnover has been a major issue among information technology (IT) personnel since the very early days of computing as well as nowadays (Moore, 2000; Niederman & Summer, 2003). IT personnel have a strong tendency to frequently switch employers. Annual turnover in the information systems (IS) field ranged between 15% and 20% during the 1960s and the early 1970s. In the late 1970s, the turnover was as high as 28% annually and around 20% in the early 1980s. By the 1990s, the turnover rate reached 25 to 33% annually (Jiang & Klein, 2002). Many large American companies had a 25 to 33% turnover rate among their IS personnel in the late 1990s (Hayes, 1998). Although *women* represent an increasingly important segment of the labor force, their turnover rate can exceed 2½ times the turnover rate of men (Chusmir, 1982; Cotton & Tuttle, 1986; Davis & Kuhn, 2003; Giacobbe Miller & Wheeler, 1992; Schwarz, 1989). A meta-analysis by Cotton & Tuttle (1986) of 120 datasets showed strong evidence for gender differences in turnover: women are more likely to leave their job than men. Gender differences in turnover are less consistent among nonmanagerial and nonprofessional employees, and are stronger among professional (Cotton & Tuttle, 1986). However, recent evidence suggests that educated women start resembling men with regard to turnover rate and pattern (Griffeth, Hom, & Gaertner, 2000; Royalty, 1998). Educated women are more likely to leave to take on another job, while less educated women are more likely to abandon the labor force (Royalty, 1998). Furthermore, part of the higher turnover rates for women can be explained by individual variables that turnover studies conducted by economists and focused on industry do not consider, such as age, tenure, marital status, occupation and salary (Giacobbe Miller & Wheeler, 1992). In this chapter, we look at gender differences in reasons why IT personnel want to leave their job, and in their intentions once they have left their job.

BACKGROUND

Age, tenure, and number of dependents are negatively related to intention to leave one's organization (Cotton & Tuttle, 1986; Griffith et al., 2000). Married employees are less likely to quit than unmarried persons (Cotton & Tuttle, 1986). The effect of education on turnover is ambiguous. Results from meta-analysis show education to be positively related to turnover (Cotton & Tuttle, 1986). However, some studies have found a negative relationship between education and turnover (Cotton & Tuttle, 1986; Porter, Steers, Mowday, & Boulian, 1974). Salary is negatively related to turnover (Cotton & Tuttle, 1986). We also know that demographic variables have direct effects on work-related attitudes such as job satisfaction (Compton, 1987; Igbaria & Greenhaus, 1992). Age and organizational tenure are positively related to satisfaction and involvement (Cotton & Tuttle, 1986; Igbaria & Greenhaus, 1992). Education has been found to be negatively related to satisfaction (Igbaria & Greenhaus, 1992; Parasuraman, 1982), and organizational involvement (Mottaz, 1988). Demographic variables have direct effects on turnover intention beyond their effects on turnover intention through satisfaction and involvement (Igbaria & Greenhaus, 1992; Parasuraman, 1982).

An important question rarely addressed in the literature is: where do women go after they leave their job? In a study on gender differences in turnover intention, Giacobbe Miller and Wheeler (1992) found that turnover intention among women was twice as high as men in comparable occupations. However, after controlling for age and job dissatisfaction, the gender effect disappeared. The researchers also found that meaningfulness of work was a strong predictor of intention to leave for women. We found similar results in a study on gender differences in job and organizational factors as predictors of quality of working life (Hoonakker, Marian, & Carayon, 2004b). For female employees in the IT department of a large public organization, task identity was one of the most important factors explaining gender differences in quality of working life (job strain, job commitment, and job satisfaction). A study by Allen, Drevs, and Ruhe (1999) looked at reasons why college-educated women change employment. The top three reasons were promotion, better pay/opportunities, and relocation (marriage/family). When asked what the employer could have done to make the respondents stay in their position of employment, the three most important suggestions were: to provide more pay or recognition (25%), to change working conditions (19%) and to move to another position (16%). The study by Giacobbe Miller and Wheeler (1992) showed that for both men and women, promotional opportunities predict intention to leave. In our study, we found similar levels of turnover intention among women and men (Hoonakker, Carayon, Schoepke, & Marian, 2004a; Schoepke, Hoonakker, & Carayon, 2004). Female IT employees perceived job and organizational characteristics and quality of working life in a manner similar to men. There were no gender differences in job satisfaction, organizational involvement, tension, fatigue and burnout. However, we found important differences in the factors that predict turnover (i.e., pathways to turnover) (Hoonakker et al., 2004a). For men, three pathways played an important role in predicting turnover: (1) the pathway from IT demands to emotional exhaustion to turnover (partly mediated by job satisfaction); (2) the pathway from challenge, career opportunities and rewards to job satisfaction to turnover; and (3) the pathway between rewards and turnover intention (also partly mediated by job satisfaction). Supervisory support did not play a significant role for men, but played a significant role in turnover of female IT employees. Supervisory support is related to nearly all the job and organizational characteristics, has a direct effect on turnover intention, as well in the pathways that are mediated through emotional exhaustion and job satisfaction. Job satisfaction also plays a central role for women: it is highly related to turnover intention, and many of the pathways to turnover intention are mediated by job satisfaction (Hoonakker et al., 2004a).

MAIN THRUST OF THE ARTICLE

Methods

We used a Web-based survey to collect the data (see Barrios (2003) for a detailed description of the Web based survey management system). A total of five IT organizations participated in the study: one large organization (N>500), one medium sized company (N=200) and three small companies (N<100). A total of 624 respondents responded to the survey (56% response rate). Twenty-seven cases in the sample had missing data on gender and were not used in the analysis. For the analysis reported in this chapter, we used data of 324 male employees (54%) and 273 female employees (46%). Respondents vary in age from 20 to 68 years (mean=39.7 years). The majority of the sample is married (61%); 9% is living with a partner; 1% is separated; 6% is divorced, 1% is widowed and 9% is single. Fifty-six percent of the respondents have children; 83% of the respondents who have children have children that still live at home. Forty-three percent of the respondents have one or more children younger than 7 years. Ten percent of the women and 2% of the men have a part-time job. Fifty percent of the men and 38% of the women telecommute or work remotely from their office.

Turnover intention was measured using a single item: "How likely is it that you will actively look for a new job next year?" on a 7-point scale (1: not at all likely-2-3: somewhat likely-4-5: quite likely–6-7: extremely likely) (mean=2.87, sd=1.83). Twenty questions were asked about reasons why respondents would leave their job. Five questions were asked about intention after leaving the job: (1) intention to look for another job at the same company; (2) looking for a similar job at a different company; (3) looking for another job at a different company; (4) no more job in the IT industry; and (5) no intention to look for a job at all. The questionnaire has been shown to be reliable and valid (Carayon, Schoepke, Hoonakker, Haims, & Brunette, 2006).

Results

Overall, there are no gender differences in turnover intention (see Figure 1).

Figure 1. Turnover intention

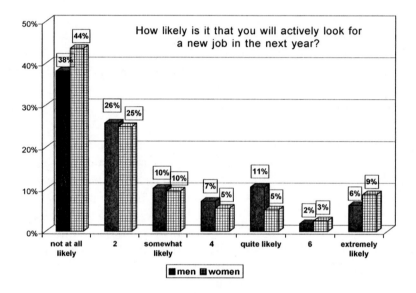

Table 1. Reasons for leaving one's job (percentages and rank)

	Men	Women	Total
High job demands	13%	16%	14%
Long working hours	15%	18%	16%
Lack of challenge or boredom	32%	31%	30%
Lack of social support	7%	11%	9%
Inadequate flexible work practices**	12%	20%	15%
Conflicts between work and family***	14%	25%	18%
Ineffective management	32%	33% (5)	31%
Feeling discriminated against**	5%	10%	7%
Feeling of not fitting in	7%	12%	9%
Lack of training	20%	15%	17%
Lack of development opportunities	29%	25%	26%
Inadequate rewards/reviews/raises	36% (4)	29%	32% (5)
Inadequate career advancement	31%	27%	28%
Want a higher job status	27%	21%	24%
Want to advance my career*	44% (2)	36% (3¹)	39% (3)
Want a higher salary**	55% (1)	44% (2)	48% (1)
Want a better compensation plan**	27%	18%	22%
Want to obtain more expertise	35% (5)	36% (3¹)	34% (4)
Want the opportunity to learn new things	41% (3)	45% (1)	42% (2)
Want more challenge in my job	26%	28%	26%
Other reasons...	10%	12%	11%

Notes: ¹ Tied; Difference between men and women significant at the 0.05 level (*), 0.01 level (**) or 0.001 level (***)

There are gender differences in reasons why respondents would leave their job (see Table 1). The most important reason for leaving one's job is the wish for a higher salary: this reason is significantly more important for men than for women. For women, the most important reason to leave one's job is the need to learn new things in the job: this is the third reason why men want to leave their job. Other important reasons to leave one's job include: wish to advance one's career; wish to obtain more expertise; inadequate rewards/reviews/raises; and ineffective management. Conflicts between work and family, inadequate flexible work practices, and feeling discriminated against are more important reasons for women than for men to leave one's job.

There are gender differences in intention after leaving one's job. Table 2 shows the results for the following question: "If you'd decide to leave your current job, what would be your intentions? (Please check all the intentions that apply)". Men are significantly more often than women interested to stay in a similar type of job, but more likely to move to a different company or to move to a different IT job in a different company (47% vs. 38%, p=0.02). Women are more likely than men to look for another job within the same company (42% vs. 36%, p=0.006). Women are more likely than men not to look for another job at all (10% vs. 6%, p=0.02).

We conducted a series of analyses to predict turnover intention. First, we looked at objective variables: age, tenure, education, formal and informal IT education, salary, job type (managerial vs. professional), job title, marital status, children, number of children still at home, and number of children age six or younger. In general, both men and women are *less* likely to look for another job when they still have children at home. In a second analysis, we entered the objective variables in step 1. In step 2 we added reasons for leaving one's job (see Table 3).

Table 3 presents the Odd's ratios for seriously looking for another job. An Odd's ratio represents the likelihood for an event to occur, in this case looking for another job. An Odd's ratio greater than 1 means that the chance for an event to occur is higher. For example, the Odd's ratio for looking for another job for men who feel they do not fit in is 4.44: men who feel they do not fit in are four times more likely to look for another job. Odd's ratios less than 1 mean that the chance employees will look for another job is lower than average. For example, the Odd's ratio for female employees with children at home will look for another job is eight times (1/0.12=8.3) lower than average the changes that women without children will look for another job.

Results show that age, marital status, education, and years of tenure with employer and in current job do not predict turnover intention. Male employees with children at home are less likely to look for another job; but male employees with children younger than six years are slightly more likely to look for another job. Female employees with children at home are also less likely to look for another job. Employees who want a higher job status, who want to advance their career and who want more challenge in their job are more likely to actively look for another job. Employees who complain about ineffective management and employees who feel not fitting in are also more likely to look for another job. Employees who experience work/family conflict are less likely to look for another job.

Results also show some interesting gender differences. For women, the wish to have a higher job status is one of the most important reasons to look for another job; while for men the wish to advance

Table 2. Intentions after leaving one's job

	Men	Women	All
I would intend to look for a different type of job in my same company	36%	42%	37%
I would intend to stay in a similar type of job, but move to a different company*	47%	38%	42%
I would intend to be in a different IT job in a different company**	28%	19%	23%
I would intend to no longer work in the IT field	20%	17%	18%
I would intend not to look for another job at all*	6%	10%	7%

Note: Difference between men and women significant at the 0.05 level (*) or 0.01 level (**)

Table 3. Predictors of intention to leave one's job

		Men	Women	Total
Step 1	Age	1.14	0.86	0.98
	Marital status (1 = non-single, 2 = single)	0.98	1.13	1.07
	Children (0 = no, 1 = yes)	1.97	4.72	2.59*
	Children at home (0 = no, 1 = yes)	0.17*	0.12*	0.24*
	Children < 6 year (0 = no, 1 = yes)	3.49*	1.35	1.94
	Highest level of education	1.21	1.12	1.06
	Formal IT education	0.95	0.84	0.92
	Informal IT education	1.22	1.07	1.12
	Tenure (in years for employer)	0.73	0.82	0.69
	Experience (in years in current job)	1.40	1.16	1.44
	Job type (1=professional, 2 = managerial)	1.32	0.69	1.02
	Job title	0.93	0.97	0.96
Step 2	High job demands	0.68	0.85	0.74
	Long working hours	2.76	0.69	1.27
	Lack of challenge or boredom	1.33	0.88	1.06
	Lack of social support	1.01	0.37	0.72
	Inadequate flexible work practices / options	1.16	0.84	0.96
	Conflicts between work and family	0.19*	0.63	0.39^
	Ineffective management	2.16^	2.37	2.41*
	Feeling discriminated against	1.92	2.73	2.20
	Feeling of not fitting in	4.44*	2.49	3.34*
	Lack of training	0.95	1.95	1.58
	Lack of development opportunities	0.62	0.58	0.55
	Inadequate rewards / reviews / raises	0.96	3.08	1.25
	Inadequate opportunities for career advancement	1.32	1.74	1.64
	Want a higher job status	1.58	3.27^	1.98*
	Want to advance my career	2.57^	1.08	1.95^
	Want a higher salary	0.70	1.01	0.74
	Want a better compensation plan	0.67	0.37	0.57
	Want to obtain more or different expertise	0.77	0.42	0.54
	Want the opportunity to learn new things	1.72	2.17	1.82
	Want more challenge in my job	2.29	2.92	2.37*

Note: Difference between men and women significant at the 0.10 level (^) or 0.05 level (*)

their career is one of the most important reasons to look for another job.

FUTURE TRENDS

High turnover of skilled IT personnel, especially among women, remains a problem for IT organizations. Part of the high turnover rates for women can be explained by a number of variables such as age, tenure, marital status, occupation and salary. Moreover, recent studies suggest that educated women start resembling men in turnover rate and pattern (Griffeth et al., 2000; Royalty 1998). Women in our study are higher educated than men, although they still lag behind men with regard to specific IT education. They are less likely to hold computer-related bachelor's and graduate degrees (see also the article by Carayon, Hoonakker, & Schoepke on "Gender Differences in Education and Training in the IT Work Force" in this encyclopedia).

CONCLUSION

In our study we do not find gender differences in turnover intention, but we find gender differences in intention after leaving one's job and in reasons for leaving one's job. We also found some gender differences in reasons that predict turnover intention. Our results suggest that men follow a different strategy than women in order to achieve their professional and personal goals. Men are more likely than women to look for another job with *another company* when they leave their job. Women tend to look more often for another job within the *same company*. The reasons for leaving one's job vary between men and women. This result confirms our earlier findings that show that different job and organizational factors predict turnover intention for men and women (Hoonakker et al., 2004a). Therefore, IT organizations need to adapt their retention strategies to the specific needs of men and women.

NOTE

Funding for this research is provided by the NSF Information Technology Workforce Program (project #EIA-0120092, Pl: P Carayon).

REFERENCES

Allen, W. R., Drevs, R. A., & Ruhe, J. A. (1999). Reasons why college-educated women change employment. *Journal of Business and Psychology, 14*(1), 77-93.

Barrios, E. (2003). *Web survey mailer system* (WSMS 1.1) (No. 186). Madison, WI: Center for Quality and Productivity Improvement.

Carayon, P., Schoepke, J., Hoonakker, P. L. T., Haims, M., & Brunette, M., (2006). *Evaluating causes and consequences of turnover intention among IT users: The development of a questionnaire survey.* Paper accepted for publication by Behaviour and Information Technology.

Chusmir, L. H. (1982). Job commitment and the organizational woman. *Academy of Management Review, 7*(4), 595-602.

Compton, T. R. (1987). Job satisfaction among systems personnel. *Journal of System Management, 38*(7), 28-31.

Cotton, J. L., & Tuttle, J. M. (1986). Employee turnover: A meta-analysis and review with implications for research. *Academy of Management Review, 11*(1), 55-70.

Davis, J., & Kuhn, S. (2003, April 10-12). What makes Dick and Jane run? Examining the retention of men and women in the software and Internet industry. In *Proceedings of the 2003 ACM SIGMIS CPR Conference*, Philadelphia.

Giacobbe Miller, J., & Wheeler, K. G. (1992). Unraveling the mysteries of gender differences in intentions to leave the organization. *Journal of Organizational Behavior, 13*(5), 465-479.

Griffeth, R., Hom, P., & Gaertner, S. (2000). A meta-analysis of antecedents and correlates of employee turnover: Update, moderator tests, and research implications for the next millennium. *Journal of Management, 26*(3), 463-488.

Hayes, F. (1998). Labor shortage is real. *Computer World, 32*, 8.

Hoonakker, P. L. T., Carayon, P., Schoepke, J., & Marian. A. (2004a). Job and organizational factors as predictors of turnover in the IT work force: Differences between men and women. In H. M. Khalid, M. G. Helander, & A. W. Yeo (Eds.), *Working with computing systems 2004* (pp. 126-131). Kuala Lumpur, Malaysia: Damai Sciences.

Hoonakker, P. L. T., Marian, A., & Carayon, P. (2004b, September 20-24). The relation between job characteristics and quality of working life: The role of task identity to explain gender differences. In *Proceedings of the Human Factors and Ergonomics Society (HFES) 48th Annual Meeting* (pp. 1571-1575). New Orleans, LA.

Igbaria, M., & Greenhaus, J. H. (1992). Determinants of MIS employees' turnover intentions: A structural equation model. Association for Computing Machinery. *Communications of the ACM, 35*(2), 34-51.

Jiang, J. J., & Klein, G. (2002). A discrepancy model of information system personnel turnover. *Journal of Management Information Systems, 19*(2), 249-272.

Moore, J. E. (2000). One road to turnover: An examination of work exhaustion in technology professionals. *MIS Quarterly, 24*(1), 141-168.

Mottaz, C. J. (1988). Determinants of organizational commitment. *Human Relations, 41*(6).

Parasuraman, S. (1982). Predicting turnover intentions and turnover behavior—a multivariate—analysis. *Journal of Vocational Behavior, 21*(1), 111-121.

Porter, L. W., Steers, R. M., Mowday, R. T., & Boulian, P. V. (1974). Organizational commitment, job satisfaction, and turnover among psychiatric technicians. *Journal of Applied Psychology, 59*, 603-609.

Royalty, A. B. (1998). Job-to-job and job-to-nonemployment turnover by gender and educational level. *Journal of Labor Economics, 16*, 392-443.

R

Schoepke, J., Hoonakker, P. L. T., & Carayon, P. (2004, September 20-24). *Quality of working life among women and men in the information technology workforce*. Paper presented at the Proceedings of the Human Factors and Ergonomics Society 48th Annual Meeting, New Orleans, LA.

Schwarz, J. (1989). Management women and the new facts of life. *Harvard Business Review*, 65-76.

KEY TERMS

IT Organization: An organization is a group of people intentionally organized to accomplish an overall, common goal or set of goals. An IT organization is an organization where information technology is used to accomplish the common goal or set of goals. This definition is much larger than the definition for an IT company: a company that produces information technology (hardware and/or software). Estimates show that 90% of IT employees are employed in non-IT companies, such as hospitals, banks and insurance companies.

Job Satisfaction: Describes how content an individual is with his or her job. There are a variety of factors that can influence job satisfaction, such as the level of pay and benefits, perceived fairness of the promotion system within a company, the quality of working conditions, leadership and social relationships, and the job itself (i.e., variety of tasks, challenge in the job, and the clarity of the job description/requirements).

Meta-Analysis: A statistical technique for combining and comparing results of many studies.

Quality of Working Life (QWL): Represents the quality of the relationship between employees and their total working environment, with human dimensions added to the usual technical and economic considerations.

Task Identity: The extent to which employees do an entire piece of work (instead of small parts) and can clearly identify the results of their effort.

Turnover: The movement of workers in and out of employment with a particular firm. Turnover intention is an employee's intention to leave the organization.

Web-Based Survey: In the broad sense of the notion of a survey, any Hypertext Markup Language (HTML) form that solicits input from respondents can be considered a survey. In our definition, a Web-based survey is a well-defined questionnaire, that has been proven to be reliable and valid in research and that, with the use of HTML, is put on the Web and solicits responses from specifically sampled respondents.

A Reflexive Analysis of Questions for Women Entering the IT Workforce

R

Valerie Pegher
Booz Allen Hamilton Inc., USA

Jeria L. Quesenberry
The Pennsylvania State University, USA

Eileen M. Trauth
The Pennsylvania State University, USA

INTRODUCTION

There are many resources available for young college graduates entering the workforce. Colleges and universities have entire departments and buildings dedicated to the process of moving students into the "real world." Questions such as "what should my salary be?" "which firm is rated the best in the country?" and "how do I fit into the corporate environment?" are typically asked by both male and female students and are answered by the staff. Yet given that business is generally a male dominated field, questions such as "have you encountered a glass ceiling in your career?" are less likely to be answered with the whole truth. Hence, this article seeks to answer some of the questions that women may have upon entering the information technology (IT) workforce.

As a woman who is graduating from college and preparing to enter the IT workforce, I[1] constantly ask myself questions about what it means to be a minority in a male dominated industry. In order to be prepared for my future career, I synthesized my questions into three central issues of coping strategies, social networking and gender identity:

1. **Coping Strategies:** How do women cope with being minority, and what do women do when treated unfairly because of their gender?
2. **Social Networking:** When should social networking begin, and how does a woman form a personal network?
3. **Gender Identity:** Do women have to display more masculine traits to get ahead in the IT workforce, and does business attire matter?

These questions are of importance because they are typical of the kinds of questions that a woman entering the IT field may have. Hence, the purpose of this article is to address these questions through a reflexive analysis in order to better prepare myself and others for careers in the IT workforce.[2]

MAIN THRUST OF THE ARTICLE

In order to address the questions raised previously, two primary sources of data are included in this article. First, a literature review is included that identifies the main themes of social networking, coping, and gender identity. Incorporated in this review are other resources that readers can use for additional information on the topic. Second, a reflexive analysis of the first author is included that details personal reflections from transcribing 25 qualitative interviews with American women working in the IT workforce[3] and from internships at a local school district and a large financial services institution.

Coping Strategies

According to Merriam-Webster Online (2006), *coping* is defined as "deal[ing] with and attempt[ing] to overcome problems and difficulties." This definition is interesting because it addresses two aspects of coping, both dealing with and overcoming problems. With regard to the IT workforce, it is important to have strategies to deal with and accept issues. Since coping is not always the easiest thing to do, it is important for women to consider coping strategies they might utilize as minorities in a male dominated industry.

Menaghan and Merves (1984) argue that most coping studies look at direct action on environment or self, interpretive reappraisal regarding environment or self, or emotional management. Dornbusch and Scott (1975) suggest that three types of workplace coping exist for employees: (1) leave an organization; (2) lower expectations; or (3) communicate dissatisfaction to the authority system and suggest changes in conditions. Fennell, Rodin, and Kantor (1981) argue that responses that create change are considered constructive, yet occur infrequently. Likewise Lim and Teo (1996) add that most coping does not require change, but primarily support from others.

These authors' arguments align with Trauth's (2004) research on coping strategies used by women in the IT workforce. Trauth explains that women in the IT workforce utilize three forms of coping: assimilation, accommodation, and activism. Assimilation accounts for the lack of stress because one is not aware of gender issues or uses selective perception to deal with hostile situations. Accommodation is the management of gender issues by recognizing the unequal treatment, but not taking action regarding the environment. Activism is the heightened awareness and the attempt to alleviate stress of gender issues through an active tendency toward changing the situation.

Based on a reflexive analysis of the interview transcriptions, it appears that some women in the IT workforce use group support as a primary means of coping. Perhaps people tend to feel better being around those who have similar beliefs, problems, and conflicts. Many of these women stated that they value social networks that allow for issue discussion and resolution. Other women in the interviews attempted to change their behaviors and actions to fit those of the group, such as taking on an activity that others in work are participating in, so they can have something to talk about. Furthermore, some women tended to take matters into their own hands. When faced with workplace issues, these women said

something to their manager or co-workers in order to bring the situation to the attention of others. At times they found that by speaking up, others in their organizations realized the problems that were occurring, and progress was made in attempts to correct them.

Social Networking

There are two main social networking theories. The first considers the strength of weak and strong ties. This theory was created by Mark Granovetter in his 1973 article, "The Strength of Weak Ties." The article is one of the most cited works in social network literature. His main argument is: Our acquaintances *(weak ties)* are less likely to be socially involved with one another than are our close friends *(strong ties)*. Thus the set of people made up of any individual and his or her acquaintances comprises a low-density network (one in which many of the possible relational lines are absent) whereas the set consisting of the same individual and his or her close friends will be densely knit (many of the possible lines are present). Put another way, actors gain novel information from less intimate ties than close ties because actors who are strongly connected share information directly; therefore, they possess the same knowledge. New information comes from external connections which are likely to be weak.

Another social networking theory builds upon the weak tie relationships of Granovetter and discusses structural holes within social networks. Developed by Robert Burt in 1992, structural hole theory means that an actor is in a more advantageous position to gain and control novel information if s/he is connected to others who themselves are not directly connected to one another. The more non-redundant connections an actor has, the more information will flow with greater efficiency and with little constraint (Burt, 1992). The theory was developed to explain interpersonal communication within a competitive environment.

Based on a reflexive analysis of the interview transcriptions, it seems as though several women had their own ideas of how to build their social networks, including taking up new hobbies. One woman even went to the extreme. Her colleagues all took flying lessons and they would talk about it at work. She went out and took lessons, too, and could then join in their conversations. While it is not recommended for women to change their lifestyles to fit that of those they work with, adding some additional activities could be beneficial. If coworkers take a half day on Friday or meet Saturday mornings to play golf, perhaps a woman should think of taking up lessons. If not, do not feel as if you must keep at it to impress anyone. The next step would then be to find people in the company who do things that you are interested in. This is where the strong social networking skills come in. [4]

When transcribing the interviews, I also found it interesting that several participants considered why woman would want to be part of the "old boy's club." The participant viewpoints on male social networking were numerous and diverse. For example, a few women seemed to perceive female networks as less effective and lacking the power of male networks. There was also the belief that many women did not want to network with other women because they were afraid to share their secrets. Some women enjoyed being the minority, especially if that uniqueness earned them certain privileges. Other women did not like to stand out, and said that blending in with men is the easiest way to not draw attention to oneself in an organization.

Gender Identity and Attire

Sex or gender roles in our society are characteristics that actually differentiate the sexes, are stereotypically believed to differentiate the sexes, or are considered to be differentially desirable in the two sexes (Lenney, 1991). Saunders and Stead (1986) argue that manner of dress has increasingly been used by women to overcome unfavorable gender stereotypes and to improve their employment opportunity and advancement. Likewise, Forsythe, Darke, and Cox (1985) explain that female attire is an important factor in hiring decisions and career advancement recommendations. Furthermore, Malloy (1976) argues that the appropriate business uniform for men is more narrowly defined than it is for women.

In the transcripts, the women interviewed expressed a number of themes about masculine and feminine gender identity and attire. As much as we may hate to admit it, clothing is very important to perception within a company. Although the days of

dressing like a feminized version of a man are over (e.g., Annie Hall), we still have to be conscious of what we wear to work. Because the styles of clothes change with the seasons and years for us, there are more choices. But, are those choices correct for entering the workplace? The Milan outfits from Ralph Lauren or Dior do not showcase shirts, pants, or skirts that are appropriate for the office. Most of the clothes that come off the runways are not ready to wear. In the age of *Sex and the City* and *Project Runway*,[5] what is a girl to do about shopping for work? Not so long ago, it was socially unacceptable for women to wear pants to work. It was all about the power suit, the navy blue or red blouse/jacket with the mid-length or long skirt. Pants were seen as inappropriate for a woman, leading to other stereotypes about women. Now, pants suits and separates are the most common business attire for women. There are many stores now that specialize in clothing for the professional woman.

In my experiences, what I wore at work was very important. During my job working for a school district, when school was not in session, the technology staff was allowed to wear casual clothes. It was brought to my attention that I was not allowed to wear shorts that were "too short" or any other articles of clothing that would be revealing. I felt offended because this was brought up before I was even able to wear casual clothing. Nothing like it was mentioned to the males I was working with, and I thought that it was unfair. I also paid close attention to what I wore when I worked for a financial institution. Historically, the banking industry is very conservative in the way employees are supposed to dress. The IT section of the company was "business casual."[6] For men it is easy—khaki slacks and a polo or a dress shirt without a tie. However, I found it interesting when I received the dress code policy, that the number of things that a woman could or could not wear was much longer. In my situation, I found myself dressing more towards business professional than business casual because I felt that my co-workers would take me more seriously that way. I noticed that the executives, both male and female, still wore the power suits, even when others around them looked like they were going golfing. I learned that clothing creates an image, and looking polished and professional is important.

There is also more to gender identity than clothing. The biggest example of gender identity that I have encountered is the way that women act in the professional IT world. During my internship experience, the functional manager for my group was a woman. Several teams besides mine reported to her, and she had only been with the company for six years. There were several team members who had been there much longer. She told me that she still felt like a newcomer in the organization, and I found that interesting, because to me, six years is a long time to be somewhere. My team members and I were discussing an issue that was handed down from the functional manager, and they were saying that she liked to have everything a certain way, and that they did not like it that she was that strict and demanding. I made note of the fact that our direct manager did not expect anything less than that from us, and the only difference was that he was a man. I said that maybe she felt that she had to act that way for them to take her seriously or that they were being overly critical because she was a woman. They all said that they had not thought of that before, and that makes sense.

FUTURE TRENDS

The Center for Women and Information Technology (CWIT) at the University of Maryland, Baltimore County, recommends the following courses of action for those women entering the IT field looking for avenues for discussing issues such as those discussed in this article:

- **Network by Joining an E-Mail Discussion List for Women in IT:** If you join one that serves the area where you live, you should be able to make some helpful personal contacts. To find a local list for women in IT, check out the listings at WorldWIT (http://www.worldwit.org) or the chapters of the Association of Women in Computing (http://www.awc-hq.org/chapters.html). Many of these discussion groups have e-mail lists and monthly face-to-face meetings. If regional dis-

cussion lists are not available in your area, try a national list such as SYSTERS (http://wwwsysters.org).

- **Learn More about Women and the IT Workforce through Gender-Related Electronic Forums:** A list of international and national forums on women related issues in science and technology can be found at CWIT's Gender-Related Electronic Forums (http://research.umbc.edu/~korenman/wmst/forums.html).
- **Attend Regional Events Offered through Area Organizations:** The Association for Women in Computing (http://www.awc-hq.org) is a national organization with local chapters that usually offer regular meetings, e-mail lists, and workshops.
- **Find a Mentor:** To find out more about what mentors do and how they can help, visit MentorNet (http://www.mentornet.net/).

CONCLUSION

Social networking seems to be the biggest method of coping. This can mean that a woman either accepts the problems and talks to someone about them, or that a woman looks to change them. Finding a mentor in the organization is an excellent method for coping with an uneven playing field. Women should prove what they can do. There are several examples of women not being taken seriously, and then performing well on a task, and gaining the respect of their coworkers. Most of the time, unfair treatment is the result of lack of respect. Women should pit themselves in positions to make a difference and to be noticed.

With regard to social networking, maybe the answer is not breaking into the "old boy's club," but creating a women-centered club. Other women in a company who feel the same way you do. Maybe there already is a network, and you just need to join. Get to know your coworkers and the women in higher-ranking positions. Have lunch with them, or seek them out on a coffee break. Knowing others within the organization can help when it comes time for promotions and reviews. It also helps to look outside of your direct company. There are lots of resources on the Internet for building networks of women. There are several conferences that offer opportunities for growth of both your business and networking skills. When you first start with a company, inquire as to the types of programs offered for women. "Is there an on-site daycare for children?" is often a good question to start off with, because if a company does this for its employees, it generally means that there are other opportunities as well.

With regard to gender identity and attire, before you start working, ask what the dress is like for the company. Most IT departments of large corporations have moved towards business casual, but that does not mean that your look needs to be casual. It is written more for the men, letting them know that they do not have to wear a tie and jacket around the office. In the more conservative fields, such as banking and some consulting firms, business professional is still active. If you are unsure of what to wear, dress on the conservative side until you have learned what is appropriate.

ACKNOWLEDGMENT

This article is from a study funded by a National Science Foundation Grant (EIA-0204246).

REFERENCES

Burt, R. (1992). *Structural holes.* Chicago: University of Chicago Press.

Dornbusch, S. M., & Scott, R. W. (1975). *Evaluation and the exercise of authority.* San Francisco: Jossey-Bass.

Fennell, M. L., Rodin, M. B., & Kantor, G. K. (1981). Problems in the work setting, drinking, and reasons for drinking. *Social Forces, 60,* 114-132.

Forsythe, S., Darke, M. T., & Cox, C. E. (1985). Influence of applicant's dress on interviwer's selection decisions. *Journal of Applied Psychology, 70,* 374-378.

Granovetter, M. (1973). The strength of weak ties. *American Journal of Sociology, 78,* 1360-1380.

Lenney, E. (1991). Sex roles: The measurement of masculinity, femininity, and androgyny: Measures of personality and social psychological attitudes. In J. P. Robinson, P. R. Shaver, & L. S. Wrightsman (Eds.), *Measures of personality and social psychological attitudes, Volume 1* (pp. 573-660). San Diego: Academic Press.

Lim, V. K. G., & Teo, T. S. H. (1996). Gender differences in occupational stress and coping strategies among IT personnel. *Women in Management Review, 11*(1), 20-29.

Malloy, J. T. (1976). *Dress for success.* New York: Warner Books.

Menaghan, E. G., & Merves, E. S. (1984). Coping with occupational problems: The limits of individual efforts. *Journal of Health and Social Behavior, 25,* 406-423.

Merriam-Webster Online. (2006). *Cope.* Retrieved February 22, 2006, from http://www.m-w.com/dictionary/coping

Morgan, A. J., Quesenberry, J. L., & Trauth, E. M. (2004). Exploring the importance of social networks in the IT workforce: Experiences with the "Boy's Club". In E. Stohr & C. Bullen (Eds.), *Proceedings of the 10th Americas Conference on Information Systems* (pp. 1313-1320). New York.

Saunders, C. S., & Stead, B. A. (1986). Women's adoption of a business uniform: A content analysis of magazine advertisements. *Sex Roles, 15*(3/4), 197-205.

Trauth, E. M. (2002). Odd girl out: An individual differences perspective on women in the IT profession [Special Issue on Gender and Information Systems]. *Information Technology and People, 15*(2), 98-118.

Trauth, E. M. (2004, November 26). *Seeing the forest and the trees: The role of the human dimension in the interpretation of computer-supported analysis of qualitative data.* Keynote Address at the Qual/IT Conference, Brisbane, Australia.

Trauth, E. M., Quesenberry, J. L., & Morgan, A. J. (2004). Understanding the under representation of women in IT: Toward a theory of individual differ-

R

ences. In M. Tanniru & S. Weisband (Eds.), *Proceedings of the 2004 ACM SIGMIS Conference on Computer Personal Research* (pp.114-119). New York: ACM Press.

KEY TERMS

Coping Strategies: Mechanisms used to address or overcome issues, problems, and/or difficulties.

Gender Roles: Characteristics that actually differentiate the sexes, are stereotypically believed to differentiate the sexes, or are considered to be differentially desirable in the two sexes (Lenney, 1991).

Individual Differences Theory of Gender and IT: A social theory developed by Trauth (2002, Trauth et al., 2004) that focuses on within-group rather than between-group differences to explain differences in male and female relationships with information technology and IT careers. This theory posits that the underrepresentation of women in IT can best be explained by considering individual characteristics and individual influences that result in individual and varied responses to generalized environmental influences on women.

"Old Boy's Club": An informal social network where men are able to share information in an informal setting, in order to build trust, personal relationships, and career advantage.

Social Networks: The web of personal and professional relationships that people utilize to exchange resources, information, and services.

ENDNOTES

[1] Use of first person in this article refers to the first author whose voice is reflected in this article. The second author contributed to the writing, and the third author collected the empirical data used in this article.

[2] This article is based on a research paper that was a requirement of a directed study course with Eileen M. Trauth, PhD, in the final year of the first author's undergraduate degree program.

[3] The transcribed interviews are a part of a National Science Foundation study on gender and the IT workforce (grant number: EIA-0204246; principal investigator: Eileen M. Trauth, PhD). The purpose of this study is to engage in development of the Individual Differences Theory of Gender and IT (Trauth, 2002; Trauth et al., 2004).

[4] Additional information on social networking in this dataset can be found in Morgan, Quesenberry, and Trauth (2004).

[5] *Sex and the City* and *Project Runway* are two American television shows.

[6] *Business casual* refers to office clothing that is not as professional as a suit, but more professional than jeans. An example of business casual is dress slacks with a button down shirt.

Retaining Women in Undergraduate Information Technology Programs

Tona Henderson
Rochester Institute of Technology, USA

INTRODUCTION

While the experiences of women in computer science (CS) are well documented (Cohoon, 2001, 2002; Computing Research Association, 2002; Margolis & Fisher, 2001), information technology is a relatively new discipline (Denning, 2001; Mitchell, 2002) and does not enjoy the same level or scope of inquiry. This study focuses on women in undergraduate IT programs and attempts to identify the factors involved in the attrition of women from these programs. In Phase 1 of this study, all freshman IT and CS women as well as a random sample of IT men at an eastern university (15,000 students) were interviewed and asked about their experiences in the IT program. These interviews were qualitatively analyzed, and the results are currently being used to develop a national survey of women in undergraduate IT programs. The primary research question of this study is, What factors are most influential in the decision of female students in IT undergraduate programs to enter these programs, and, where applicable, what factors most influence their decision to leave the programs during their first year of study?

BACKGROUND

Attrition in both undergraduate and graduate programs has been shown to be significantly different across academic disciplines; Cohoon (1999) demonstrated the differences between gendered attrition rates in CS and Biology. As a result, IT programs will not necessarily have attrition rates that are comparable to (or based on the same factors as) those in CS. While the literature contains numerous examples of research into gendered differences in CS, these examples are not sufficient to identify issues in the IT population, but they are more than adequate as a foundation for this inquiry.

Literature Review

In one of the earliest large-scale studies of differences in educational attainment by sex, Alexander and Eckland (1974) confirmed what many women already knew then and now: "a relatively strong and unmediated depressant sex effect remained for the educational attainment of women ... " Based on a longitudinal study of 2,077 women who were high school sophomores in 1955 and a follow-up in 1970, the results indicated that even with controls for a wide variety of variables (background, performance, self-esteem, etc.), evidence suggested that sex was negatively related to college attendance and achievement in school.

To explore a more generic model of academic persistence, Tinto (1974) suggested a model predicting the "degree of fit" between an individual and college as strongly related to persistence and retention. Tinto based his model on the continuum of changing commitments and experiences of students based on the assumption that students entered school with specific backgrounds and varying levels of commitment to completion. His development of five different factors causally related to persistence provided the basis for subsequent research into gendered attrition: (a) background (family, gender, high school), (b) initial commitment (commitment to graduation or academic major), (c) academic and social integration, (d) subsequent goals and commitment, and (e) withdrawal decisions (persistence).

Pascarella and Terenzini (1983) conducted a longitudinal study from 1976 to 1978 of 1,457 students to examine Tinto's theory. Students were surveyed at three data points to collect information about their freshman-year experience. Interesting differences between men and women appeared in the effect of initial goal commitment. While both men and women demonstrated that initial goal commitment positively influenced subsequent goal commitment, women also demonstrated a strong relationship between initial goal commitment and both social integration and persistence. Significantly, social integration appeared to have a stronger influence on female persistence than did academic integration (with the reverse being true for men).

Stage (1989) followed suit with an exploration of the Tinto model using another longitudinal study conducted between 1984 and 1985. Focusing on motivational orientation as related to initial commitment, Stage recognized seven categories (certification, cognitive, community service, change, social, recommendation, and escape), with the majority of respondents being classified into the three largest subgroups: certification (i.e., the goal is to earn a degree or get a job), cognitive (the goal is to learn and grow), and community service (the goal is to gain skills and experience in helping people). In the certification and cognitive categories, gender appeared as significantly correlated to social integration.

In 1997, with the publication of "The Incredible Shrinking Pipeline" (Camp, 1997), an examination of the decline of female enrollment in CS was underway with urgency. In 1997, Fisher, Margolis, and Miller also examined the experiences of women in CS using factors like persistence and motivation. Interestingly, Fisher et al. decided to let the students speak as "expert witnesses in their own world." West (2002) also talked with women (in this case, in an introductory programming course). Additionally, *Unlocking the Clubhouse* (Margolis & Fisher, 2002) was highly influential in both discovering and describing the experience of women in an undergraduate CS program.

Other studies like Beyer et al. (2003), Liu and Blanc (1996), and Scragg and Smith (1998) have also employed survey methodologies to gain data and insight into the experiences of women in CS.

MAIN THRUST

Methodology

The underlying technique for data collection and analysis of interview results in Phase 1 is the sense-making model developed by Brenda Dervin (1992). This technique is based on situating individual deci-

sions and choices along a continuum of time and space. In other words, as students journey through their academic experiences, they encounter what Dervin calls "gaps" that force the successful bridging of these gaps or decisions to pursue completely different paths (thus skirting the gaps). Troublesome situations (Dervin & Clark, 1987) are defined as "any situation ... where a person faces some kind of gap preventing a movement ahead." Conversely, helps are defined as those experiences, people, activities, thoughts, ideas, and/or resources that successfully bridge any gaps.

The interview technique allowed for asking open-ended questions that focus on what Dervin (1992) considers the significant questions. For example, what stopped this person from accomplishing goals, what information or bridges were sought, and what assistance was ultimately necessary or lacking? Based on the situation, the gap, and the help, a sense-making triangle encircles experiences and provides a context for examination and discussion. This sense-making triangle forms the basis of the time-line interview. In the interview, students are asked about their experiences and, using open-ended questions, encouraged to talk at length about the situation, the gaps, and the helps.

Study Design

In Phase 1 of the study, 33 respondents were invited to participate (10 IT women, 13 CS women, and 10 IT men). One hundred percent of freshman women in both the IT and CS programs agreed to participate. The 10 men who agreed to participate were randomly selected from the entering freshman class of 200 freshman men. Each participant agreed to sit down for a face-to-face interview in the fall and spring quarter, as well participating in e-mail exchange during the winter quarter. This design allowed for three data-collection points. Respondents were given a cash incentive for the fall and spring interviews. All interviews were taped and transcribed for later analysis.

For the second phase of the study, two surveys are underway. Beginning with the ACM SIG-ITE membership list, schools offering a BS in information technology were identified and solicited to answer a demographic survey about their IT programs. Questions in this survey included faculty, student, and institutional descriptors. This survey also requested data about the number of women entering the IT program over the past 2 to 5 years, current FTE enrollment, the number of women known to have left the program (transfers out of the program or departures from the institution), and graduation rates.

Running concurrently during Phase 2, an individual survey will be administered to individual women enrolled in the identified IT programs; it is scheduled to be distributed at the beginning of the 2005 to 2006 academic year. Survey questions, under development, are based on the qualitative analysis of interview transcripts conducted in Phase 1 of the study.

Analysis

Coding the transcripts was based on the key components of sense making (situation, gaps, and helps). Trees of coding terms (nodes) were created so that passages could be broadly or narrowly identified as appropriate. One example is Helps/People, or more narrowly, Helps/People/School/Counselor. Two passes were used through the transcripts: The first pass coded each appropriate passage with terms, while the second pass combined terms as required. For example, Gaps/No Money and Gaps/Cost were combined into Gaps/Financial Considerations.

The fall and spring transcripts were lengthy and frequently ran over 40 pages in length. Students spoke freely and openly, but often diverged from the structure of the interview instrument. Extracting the key information from the transcripts was difficult and required focus and attention to detail. Conversely, the winter e-mail exchanges were more structured and required less sifting and winnowing. Comparable amounts of relevant data were obtained despite the difference in length. In retrospect, the e-mail exchanges forces a certain focus on the question at hand and reduced the judgment calls of the interviewer significantly. The use of e-mail allowed the interviews to be iterative, and allowed both the researchers and the participants to take time for reflection between responses.

After two passes through the transcripts, nodes were reviewed for frequency and distribution in each population (i.e., how many men, IT women, and CS women mentioned this topic in their interview). Differences in the conceptual framing of the topic were also apparent as in the case of women who identified counselors as playing a dual role as help and gap (depending on the situation).

CONCLUSION

Some of our preliminary findings confirm work by previous researchers in gendered attrition. For example, while the impact of guidance counselors and teachers is important to both men and women, women more frequently identify counselors as negative factors in pursuing an IT major. And, in keeping with other research into gender differentials in retention, none of the men interviewed cited concerns about proper preparation for their studies, while two of the IT women and four of the CS women indicated that was a concern in their decision-making process. This concern persisted for the women throughout their first year.

Of unique interest is an early indicator that while female students in both CS and IT described frustration with programming classes (as opposed to men who infrequently indicated issues with programming), the IT women also expressed a sense of irrelevance about programming and doubts about how programming fit with their career goals. In contrast, the CS women more frequently accepted programming as both relevant and important to their career goals.

Analysis of the Phase 1 transcripts is ongoing and is expected to finish by the summer of 2005. These results will then be used to identify and develop questions for the Phase 2 national surveys. Upon completion of the national survey, quantitative analysis of the results will provide a basis for programmatic activities and future directions for retaining women in undergraduate IT programs.

REFERENCES

Alexander, K., & Eckland, B. (1974). Sex differences in the educational attainment process. *American Sociological Review, 39*, 668-682.

Beyer, S., et al. (2003). Gender differences in computer science students. In *Proceedings of the 34th SIGCSE Technical Symposium on Computer Science Education*, 49-53.

Camp, T. (1997). The incredible shrinking pipeline. *Communications of the ACM, 40,* 10.

Cohoon, J. M. (1999). Departmental differences can point the way to improving female retention in computer science. *ACM SIGCSE Bulletin, 31*(1), 198-202.

Cohoon, J. M. (2001). Toward improving female retention in the computer science major. *Communications of the ACM, 34*(5), 108-114.

Cohoon, J. M. (2002). Recruiting and retaining women in undergraduate computing majors. *ACM SIGCSE Bulletin, 34*(2), 48-52.

Computing Research Association. (2002). *2000-2001 CRA Taulbee study.* Retrieved September 26, 2002, from http://www.cra.org/statistics/

Denning, P. (2001). The IT schools movement. *Communications of the ACM, 44*(8), 19-22.

Dervin, B. (1992). From the mind's eye of the "user": The sense-making qualitative-quantitative methodology. In J. D. Glazier & R. R. Powell (Eds.), *Qualitative research in information management* (pp. 61-84). Englewood, CO: Libraries Unlimited.

Dervin, B., & Clark, K. (1987). *ASQ: Asking significant questions. Alternative tools for information need and accountability assessments* (ERIC Document Reproduction Service No. ED286519). Belmont, CA: Peninsula Library System.

Fisher, A., Margolis, J., & Miller, F. (1997). Undergraduate women in computer science: Experience, motivation and culture. In *Proceedings of the Twenty-Eighth SIGCSE Technical Symposium on Computer Science Education* (pp. 106-110).

Liu, M.-L., & Blanc, L. (1996). On the retention of female computer science students. *Proceedings of the Twenty-Seventh SIGCSE Technical Symposium on Computer Science Education,* 32-36.

Margolis, J., & Fisher, A. (2002). *Unlocking the clubhouse: Women in computing.* Cambridge, MA: MIT Press.

Mitchell, W. (2002). *Information technology education: One state's experience.* Retrieved September 26, 2002, from http://www.ualr.edu/~wmmitchell/itcp.htm

Pascarella, E., & Terenzini, P. (1983). Predicting voluntary freshman year persistence/withdrawal behavior in a residential university. *Journal of Educational Psychology, 75*(2), 215-226.

Scragg, G., & Smith, J. (1998). A study of barriers to women in undergraduate computer science. In *Proceedings of the Twenty-Ninth SIGCSE Technical Symposium on Computer Science Education* (pp. 82-86).

Stage, F. (1989). Motivation, academic and social integration, and the early dropout. *American Educational Research Journal, 26*(3), 385-402.

Tinto, V. (1987). *Leaving college: Rethinking the causes and cures of student attrition* (1st ed.). Chicago: The University of Chicago Press.

West, M. (2002). Retaining females in computer science: A new look at a persistent problem. *Journal of Computing Science in Colleges, 17*(5), 1-7.

KEY TERMS

Attrition: A reduction in membership, number, or strength.

Coding Nodes: The process of examining interview transcripts and assigning classification information and/or descriptive terms to excerpts of the textual information.

Motivation: Any inducement or incentive that improves persistence.

Persistence: Continuation or continuance.

Qualitative Analysis: The process of discovering and/or assigning meaning based on descriptive narratives.

Sense Making: A theoretical approach to interviewing that examines the continuum of decisions and choices along time and space.

Troublesome Situation: Any situation that involves a gap or some other problem that prevents a person from moving ahead in time and space.

APPENDIX: THE INTERVIEW INSTRUMENT

Thanks again for agreeing to participate in this study! I'd like to start by gathering some basic information about you. As we discussed, this information will be kept completely confidential; only broad characteristics, such as your age and gender, will be linked with your replies in the research results. Your name will not be stored with any of the replies you provide. Where necessary in the reporting of results, even this demographic information below may be masked or "genericized" in order to protect your identity.

Age:
Hometown:
Other members of home household (parents, siblings, etc.):
Parent(s) occupation:
Computer in your house? If yes, what years? What kind?
High School
 Public/Private?
 Size of graduating class?
 Did your high school offer programming or computer classes?
 If yes, did you take any of those classes? Which one(s)?

Now that we have the general framework of information in place, I want to ask you some specific questions about the events in your life leading up to your decision to go to college, and particularly your decision to come to school here. As we proceed, you may have some particularly strong memories, and I'd like to hear all of them. I have a questionnaire structure which guides how and when I ask you about different parts of your memories. By using this structure, we can compare your experiences with others', while still allowing you to share details that are unique to your particular experiences.

What I'd like for you to do now is to choose a situation that occurred in the past that related to your decision to attend college, and where you felt that you were in some way blocked or hindered from accomplishing your goals or getting the information you required. The "block" could take the form of a person, a bureaucracy, a rule, a lack of a specific item, or any number of things. This situation could be a discussion with a family member, friend, or colleague; visiting a campus; reading an article or book; or simply an occasion when you were particularly focused on the idea of going to college. It's important that you be able to recall the situation clearly, and reconstruct the questions and thoughts that you were having at the time. I'll give you a minute or two to choose the situation and recall the details. Feel free to write it down.

(Here's an example: Suppose I'm in a grocery store, and have just wheeled my cart into the produce section. That is my specific situation, one that I can picture in my mind. I'm blocked because I can't find anyone to answer my questions about the produce.)

When did this situation occur? (An approximate date is fine.)

Can you describe the situation for me?

In the situation you just described, how did you see yourself as blocked?

Now I'd like you to go back in your mind and try to identify what the questions you had in your mind at the time were. By questions, I mean things that you wanted to find out, learn about, and come to understand, unconfuse, or make sense out of. You need not have asked the question out loud, nor found an answer; we simply want to identify gaps in understanding that you faced at the time. These may not have been in your mind as questions, but rather as unclear aspects of the situation or your feelings. In these cases, what I need you to do is to translate that aspect into a question, or to simply talk about that aspect so that together we can translate it into a question. Take a few minutes to think about the questions. Feel free to write them down.

(In the grocery store situation I used as an example, my questions might be: Where are the avocados? Is that man in the bright green plants an employee? Are mushrooms still $3.00 a pound? I wonder if the corn is as good as it looks? Etc.)

Now, thinking about your situation, what questions did you have in your mind?

Let's talk about the specific questions that you listed. As we go through this process, you may think of other questions that were in your mind at the time. If so, feel free to add them to the list as we go, and we'll talk about them as well.

APPENDIX: THE INTERVIEW INSTRUMENT (CONT.)

We're going to start by looking at the motivations behind each of your questions. By that, I mean what you were trying to accomplish or understand by asking the question.

(Going back to the grocery store example that we used before, when I entered the produce department and asked if the mushrooms were still $3.00 per pound, what I might have been trying to do was decide whether or not to get mushroom for myself because I like them, but I was worried about paying too much. Or I might have been curious because even though I don't like mushrooms, it is interesting to me that some people will pay up to $3.00 per pound for them, like maybe that guy in the green pants.)

Given that, think back to when you asked [insert question here]. What was it that you were trying to do by asking this question?

When you had this question in your mind, did you see yourself as blocked or hindered in some way?
 If yes, how did you see yourself blocked?

Were you able to find an answer to this question?
  If yes:
 Was it a complete or a partial answer?
 How did you get the answer?
 How did the answer help you?
  If no:
 What do you see as having prevented you from getting an answer?
 How do you think the answer could have helped you?

Repeat for each question.

Okay, that's all for this interview.

Is there anything you'd like to add at this point, or questions that you have about what we just discussed?

Thanks again for your participation. We'll be back in touch with you in January for the second part of the interview. As a "thank you," here's the $25 that you were promised. You'll receive the $25 at the end of the school year, after our third interview with you.

If you have any questions about the research process, please feel free to contact us—

Schema Disjunction Among Computer Science Students

Rebecca L. Crane
University of Colorado at Boulder, USA

Liane Pedersen-Gallegos
University of Colorado at Boulder, USA

Sandra Laursen
University of Colorado at Boulder, USA

Elaine Seymour
University of Colorado at Boulder, USA

Richard Donohue
University of Colorado at Boulder, USA

INTRODUCTION

In her book *Why So Slow?: The Advancement of Women*, Virginia Valian describes a schema as "a set of implicit, or nonconscious, hypotheses about … differences." (Valian, 1998). Individuals use schemas about particular social groups to guide their interpretations of and behavior toward members of those groups. However, problems can arise when multiple conflicting schemas are applied to the same person. This phenomenon, *schema disjunction*, is particularly well illustrated by the situation of female undergraduate computer science majors.

Extensive interviews with introductory computer science students of both genders reveal a significant discontinuity between their schema of women and their schema of successful computer scientists. Despite professing conscious egalitarian beliefs about the ability of women to do computer science, many students unconsciously hold disjunct schemas that help facilitate an environment hostile to novice women and may deter them from pursuing computer science careers (Pedersen-Gallegos, Laursen, Seymour, Donahue, Crane, DeAntoni, et al., 2004).

Valian argues that, starting in childhood, we acquire *schemas* through observation of adult behavior toward others. Schemas are generally more inclusive than stereotypes and carry fewer negative connotations: They are not necessarily unfair or pejorative. In fact, schemas are a normal way that humans use categorization to negotiate our environments. However, Valian also explains that schemas can become unjustly misrepresentative of individuals due to errors that creep in during their development. These errors are then reinforced during maintenance and application of those erroneous schemas. These generalized beliefs about certain types of people are often unarticulated, and may be even consciously disavowed by those who hold them. Yet people can still operate unconsciously on the basis of ingrained schemas while remaining unaware of them.

Because schemas color our interpretations of people we interact with, they also shape how we behave towards those people. We treat each other, and ourselves, in accordance with our schematic expectations. When these expectations are unfairly pejorative, they can have a damaging impact on the self-concepts and lives of the people to whom they are applied, often resulting in a self-fulfilling prophecy. Echoing Cooley's (1902) classic notion of the "looking-glass self," Valian describes this phenomenon with a focus on gender schemas:

All of us—boys and girls, men and women—become in part what others expect us to become, thereby confirming hypotheses about the different nature of males and females. While no one is infinitely malleable, no one is completely indifferent to others. One way we learn who we are is through others' responses to us. As men and women, we also develop expectations for our own behavior, based on characteristics we believe we possess. We then explain our successes and failures in terms of those abilities and traits. (Valian, 1998)

BACKGROUND

The report *Attracting and Retaining Women in Information Technology Programs* reports on a study done for the National Science Foundation by Ethnography and Evaluation Research at the University of Colorado at Boulder. To illuminate some of the obstacles to achieving gender equality in the field of computer science, we undertook in-depth qualitative interviews with a wide variety of computer science (CS) students and faculty (Pedersen-Gallegos et al., 2004). Interview subjects included students who were just beginning their studies and those who were nearing completion of a CS major or minor. All interviews were tape-recorded and transcribed *verbatim* into *The Ethnograph*, a computer program that allows for multiple, overlapping and nested coding of a large volume of transcribed documents, and supports analysis to a high degree of complexity. This discussion draws primarily on data from 70 introductory students, including both under-

graduates who had persevered in the discipline and those who had left it or chosen alternative programs. All uncited quotations in this article come directly from those interviews.

THE INTERVIEWS

We asked the students we interviewed to tell us what they thought made a successful computer scientist. We also gleaned their impressions about what women are like through their conjectures about why there are so few women in the discipline. Questions about gender issues were asked only at the end of interviews with students, so as to make a clear distinction between gender issues and more generic learning issues. One of the most striking findings, especially among students at the introductory levels, was that although most students professed egalitarian beliefs about women's rights and abilities to be computer scientists, the characteristics they considered schematic of a computer scientist were often antithetical to those included in their schema of a woman.

Comparing the novice students' descriptions of computer scientists to their implied beliefs about women, we can see how the schemas of these two groups stand in opposition to each other, as illustrated in Table 1.

Table 1. Students' schema for computer scientists and for women

Students' Schema for Computer Scientists:	Students' Schema for Women:
Highly intelligent	Find CS difficult or intimidating
Inherent passion for CS	Natural interest in arts and humanities
Early experience with math, science, computers	Inexperienced with math, science, computers
Self-motivated, independent learners	Communal and social learners
Logical, analytical thinkers	Less spatial and abstract thinkers
Poor social skills - "Nerd" image	Concerned with social image
Competitive and driven	Less competitive, more nurturing
Focused on prestigious, challenging careers	Less concerned with career success
Can handle the long, stressful hours of a programming job	Want more balanced career, opportunity to raise a family
Represented among faculty/discipline	Underrepresented among faculty/discipline

Students we interviewed described traits that they considered as characterizing good computer scientists or facilitating success in the discipline. Their answers provided a picture of the computer scientist schema held by introductory level students. Both men and women described a typical computer scientist as fitting the traditional "computer geek" image —brainy but shy, scatterbrained and lacking in social skills. "Even other engineers make fun of computer science students for their nerdiness," one student admitted. Students also described computer scientists as members of an extremely intelligent, prestigious elite, with access to high-status careers, such as "being a rocket scientist." Most students considered this reputation a key element of what it was to be a computer scientist, priding themselves on their prestige and often on the very "geekiness" associated with it. This echoes the finding of Seymour and Hewitt that engineering students prided themselves on the "hardness" of their discipline as a source of status (Seymour & Hewitt, 1997).

Students also identified a variety of specific experiences, skills and personality traits that they saw as constituting the "CS type." They told us that success in CS depended on a prior interest in computers and programming, especially a passion to pursue the subject. This desire was seen by many as something that could not be learned, but rather was inherent to CS types. Another strongly held image of the CS type was that of the self-motivated, independent programmer. Several students related archetypical CS anecdotes, combining both passion and independence, about how they or students they knew had taught themselves much of the material needed for success in CS courses, often before coming to college:

None of the people I know started coding in high school, they started coding when they were like in middle school or something, and then they kept coding in high school, and they probably would ace [the introductory CS course].

In addition to being self-motivated learners with strong drives to master the material, students' computer scientist schemas often included intellectual aptitudes, such as having a natural facility for math, science or foreign languages. They cited more general ways of thinking as well, such as having a logical, linear or analytical mind. Not only did students list traits linked to programming ability itself, they also considered certain personal characteristics necessary for surviving the rigors of CS *education*. These included an ability to learn quickly and independently, a willingness to challenge oneself and an early development of fluency with CS basics. Finally, they saw successful CS students and computer scientists as capable of handling a high level of stress, using it as a motivating factor rather than as a hindrance. This also corresponds to Seymour and Hewitt's finding that engineering students valued their own ability to "hack it" by tolerating large amounts of stress.

We asked students, in addition to delineating the characteristics of any successful computer scientist, to explain why women might be underrepresented or struggle in the discipline. Regardless of the gender of the respondent, their hypotheses illuminated similar pictures about what women are like. Quite a few students of *both* genders believed that women have inherent biological or psychological traits that limit their potential for success in the CS major, while others saw them as being unjustly but no less insurmountably limited by their socialization and cultural location.

Supporting Valian's assertion that both women and men hold detrimental gender schemas, nearly equal numbers of men and women made statements in interviews about women's "natural" aptitudes or inclinations. Some students suggested that these characteristics arose through a process of socialization, or had been culturally reinforced. However, they were nonetheless seen as inescapable and deterministic of women's beliefs and choices. For example, one student suggested that the toys and games little boys play with could engender neurobiological development different from that of little girls.

The most common assumption was that women's enrollment in CS was low because women are "naturally less interested" in technical fields, preferring arts, humanities and the social sciences. Further claims reinforcing a schematic disjunction were that women are less able than men to think spatially and abstractly, that women are naturally less competitive, and that women tend to have less focus and drive toward their careers. They also believed that women were more concerned with their social lives and ability to make friends, and suggested that some might be put off from CS and other engineering disciplines for fear of being stigmatized as anti-social "nerds."

Not all students believed that women suffered from in-born disadvantages when it came to CS. In almost equal numbers, they suggested that women were at a disadvantage due to cultural factors or were limited by aspects of their upbringing and roles imposed upon them by society. Regardless, students retained an image of women as being disadvantaged, lacking the social and cultural resources necessary to succeed. The most common attribution among this group was that men made better CS students because they tended to have more experience with math and science, due to encouragement at an early age:

You go back to elementary school, and you know, the teachers are concerned when little boys don't do well in math, but when it's little girls, it's, "Oh, okay, that's all right, ... girls aren't good at math."

Men were seen as often being more conditioned than women into a mindset that facilitated success in CS. Their early encouragement in math and science gave them more exposure to frustrating, detail-oriented tasks and fast-paced learning environments similar to those they would later encounter in CS. Meanwhile, students suggested, women's socialization might lead even those with a potential talent for CS to be intimidated by the math needed to pursue a CS degree.

These schematic characteristics correspond with Valian's description of the schema of women, which includes being expressive, communal, social and aesthetic, as opposed to men, who are seen as agentic (capable of individual, autonomous action), assertive, task-oriented and analytic (Valian, 1998). Due to these two schemas being largely comprised of diametrically opposed characteristics, women can often have trouble simultaneously reconciling their gender identity with their identity as a computer scientist. As one student told us, "... even just in engineering, a lot of girls won't tell people immediately that they're engineers. Especially when they're trying to meet guys." Jane Roland Martin (1994) also paradigmatically makes this point with her assertion that women in science face a "double bind."

CONCLUSION

Although most students claimed that the knowledge and skills needed for CS could be learned by anyone, frequently this belief was held concurrent with the view that only a certain type of person inherently "has what it takes"—in terms of personality, perspective and drive, as well as technical ability—to be a successful computer scientist. Regardless of the potential of any real individual woman do CS, introductory CS students demonstrated beliefs about women as a category that were in clear opposition to their beliefs about computer scientists. The images students had of female attributes did not resemble those they saw as needed by successful CS majors.

The persistence of this schema disjunction is understandably inimical to women's inclusion and success within CS programs. It is not the accuracy of the ascribed traits that is of critical concern, but the damage these expectations do to women who take them to heart. These women may exclude themselves from entering the major, or enter but then struggle more than their male peers with feeling accepted and capable.

Of course, not all women are intimidated out of the discipline by a feeling of not belonging. The schema disjunction is exacerbated and reinforced by a complex system of social factors, some unique to CS and some more universal, for which there is not room to address here. However, this interplay of diverse factors may occur in ways that are more or less detrimental to particular women. Furthermore, some women do develop strategies for persevering in the discipline. Although this article cannot describe these strategies in full depth, they generally involve the individual privileging one or the other of the schematic groups as a more fundamental part of her identity: either somewhat rejecting her identity as a woman in favor of seeing herself as a computer scientist, or else strongly and consciously identifying her gender as an especially relevant element of her CS experience.

One hopeful and illuminating finding: More advanced female students, given several years of personal experience with the realities of what success at CS actually entails, tended to see fitting the CS schema as less crucial to the identity of a computer scientist than their less experienced peers. These students had a much broader view of what CS

S

is about, and emphasized that most useful traits are learnable rather than innate characteristics of an elite "type." It appears that, through experience in CS, they eventually found ways to reconcile their identities by altering their schemas rather than their self-concepts.

However, a strong belief in the standard schemas was still prevalent among students at the introductory levels, and thus more likely to impact novice students' experience of belonging. Consequently, those who did not initially feel like they "fit in" sometimes dropped an introductory course or switched out of the major before they had a chance to truly gauge their own individual potential and desire to succeed at CS.

Sustained effort continues to research and understand the relative dearth of women and other underrepresented groups in STEM disciplines such as CS. A complex web of interrelated factors operates to limit access to and success in CS by members of these groups. Although untangling that web necessitates simultaneous work in many areas, *schema disjunction* is one relevant contributing factor. Even if CS students do not consciously hold that particular women necessarily lack the capacity to *do* CS, they still express a background belief that *people like women*—those who exhibit traits deemed characteristic of *real* women—do not fit the necessary description to *be* real computer scientists. The persistence of this view, especially among students at the novice level, makes introductory CS an understandably discouraging environment for any young woman exploring her options.

REFERENCES

Cooley, C. H. (1902). *Human nature and the social order*. New York: Scribner's.

Martin, J. R. (1994). *Changing the educational landscape: Philosophy, women, and curriculum*. New York: Routledge.

Pedersen-Gallegos, L., Laursen, S., Seymour, E., Donahue, R., Crane, R., DeAntoni, T., et al. (2004). *Attracting and retaining women in information technology programs: A comparative study of three programmatic approaches*. A report to the National Science Foundation, IT Workforce Grant NO. 0090026.

Seymour, E., & Hewitt, N. (1997). *Talking about leaving: Why undergraduates leave the sciences*. Boulder, CO: Westview Press.

Valian, V. (1998). *Why so slow?: The advancement of women*. Cambridge: The MIT Press.

KEY TERMS

Boys' Club Mentality: A form of elitist atmosphere that is male-oriented to the extreme of implicitly rejecting and excluding women.

Computer Geek: As used here, the stereotype of a single-minded programmer, who is brilliant with computers but socially inept. Computer science students used the term affectionately, admitting that, while an exaggeration, it has some truth to it, and is even considered a badge of honor to some extent.

In-Group: A collection of people defined as close to and similar to oneself. Fellow members of a social set.

Out-Group: A collection of people defined as distant from and other to oneself. Not a member of a particular social set.

Schema: A set of implicit assumptions or generalized hypotheses about members of a given social group.

Schema Disjunction: An incompatibility between two or more schemas when they are applied to the same person or group.

Self-Fulfilling Prophecy: A situation in which certain expectations held by a person or persons end up becoming true only because the people in question become (consciously or unconsciously) aware of the expectations and adjusts their actions to meet them.

STEM: Science, technology, engineering and mathematics. A cluster of disciplines from which women and many ethnic minorities have traditionally been excluded, and in which they are currently underrepresented.

The Shrinking Pipeline in Israeli High Schools

Larisa Eidelman
Technion – Israel Institute of Technology, Israel

Orit Hazzan
Technion – Israel Institute of Technology, Israel

INTRODUCTION

Worldwide surveys indicate that the number of women studying undergraduate-level computer science (CS) has been constantly decreasing in the last 20 years (Camp, 1997, 2002; Camp, Miller, & Davies, 1999; National Center for Education Statistics, 2004). According to Galpin (2002), the low participation of women in the computing studies is recognized worldwide. As it turns out, the situation is similar among high-school students as well (Davies, Klawe, Nyhus, Sullivan, & Ng, 2000). However, while many studies are carried out at the university level and programs are implemented in order to change the situation, high-school students do not attract such attention. In Israel too, as far as we know, no research has ever been performed that focused on female high-school students studying CS. This article presents such a study. Specifically, it focuses on high-school female students studying advanced-level CS.

Based on data collected in Israel, significant differences were found in the percentages of female high-school students studying advanced-level CS among different sectors. More specifically, while the percentage of female high-school students studying advanced-level CS is about 50% for the Arab minority sector, the percentage of female students studying CS at the same level among the Jewish majority sector is only about 25%. Different studies around the world identified various factors that discourage women from studying CS and from persisting in the field. By focusing on the Israeli high-school female students studying CS at the highest level and coming from two sectors, we suggest that the research presented in this article may partially explain the above-mentioned phenomenon. Further findings are presented in Eidelman and Hazzan (2005).

BACKGROUND

Margolis and Fisher (2002) suggest that the underrepresentation of women in the computing fields is important on two levels: on the personal level and on the societal-cultural level. Therefore, Margolis and Fisher suggest that the significant differences between the representation of women and men in the CS fields in general, and in high-school CS classes in particular, should not be ignored. This underrepresentation has a special significance in Israel, a small country in which the efficient utilization of its human resources is of great importance.

Underrepresentation of Women in the Computing Fields: A Worldwide Perspective

As mentioned in the introduction, the underrepresentation of women in the computing fields is recognized worldwide (Galpin, 2002). However, recent in-depth analysis of this phenomenon reveals that the problem is not universal, but rather is restricted to specific countries and cultures (Adams, Bauer, & Baichoo, 2003; Galpin; Lopez & Schulte, 2002; Schinzel, 2002). More specifically, in certain countries and cultures, such as Greece, Turkey, Spain, Portugal, Mauritius, Romanic countries (e.g., France and Italy), North African countries, Arabic countries, and South American countries, the representation of women in CS is high and constant in contrast to the United States, Israel, Anglo-Saxon countries, Scandinavian countries, and German- and Dutch-speaking countries, in which the representation of women in CS is relatively low and decreasing. Accordingly, it is reasonable to assume that cultural factors play an important role in encouraging or discouraging women from studying CS.

Underrepresentation of Women in the Computing Fields: An Israeli Perspective

As mentioned previously, a significant difference exists in the percentages of female high-school students studying advanced-level CS between the Arab and Jewish sectors in Israel. As it turns out, high school is a critical point in the CS pipeline, at which many female students, mainly in the Jewish sector, are lost. This situation encouraged us to initiate the research described in this article, which examines factors that influence the enrollment and persistence of Israeli female high-school students (of two populations, Arab and Jewish) in advanced-level CS studies.

The Israeli education system has a unique characteristic that may be useful for research works of this kind; that is, the Arab and the Jewish pupils learn in separate educational systems according to the same curriculum. One solution that has been suggested for countries in which women are underrepresented in CS (and that would like to change this situation) is to visit countries in which this problem does not exist, and to identify the cultural differences, as well as actions taken to encourage women to study CS, that may explain why women in such countries find CS an attractive field (e.g., Adams et al., 2003). From this point of view, Israel is a perfect place for such research. Specifically, in order to understand the low participation of Jewish female high-school students in CS, there is no need to visit another country. It is sufficient to investigate the differences that exist between the two populations, which live in the same country and, as has been mentioned before, study CS according to the same curriculum, one of which (the majority) suffers from this underrepresentation, while the other population (the minority) does not.

MAIN THRUST OF THE ARTICLE

This section describes the research setting and its results.

The research population consisted of 12th-grade CS students from nine typical high schools from both sectors (five schools from the Jewish sector, four schools from the Arab sector). Table 1 describes the

Table 1. Distribution of research population

	Total	Male	Female
Number of students from the Jewish sector	90	65 (72%)	25 (28%)
Number of students from the Arab sector	56	22 (39%)	34 (%61)
Total	**146**		

distribution of the students according to gender and sector. CS teachers were included in the research population as well.

Three comparisons were conducted in the research: Jewish female students vs. Jewish male students, Arab female students vs. Arab male students, and Jewish female students vs. Arab female students. This article focuses on the differences between Jewish and Arab female students.

The research applied both quantitative and qualitative approaches. Data were gathered using the following research tools: comprehensive questionnaires that included closed and open questions completed by all students; ethnographic, semistructured interviews with 18 Jewish and Arab female students; and classroom observations during CS lessons (both lab lessons and traditional classroom lessons). In addition, interviews were conducted with CS teachers.

In what follows, we present the analysis of data, gathered by questionnaires and interviews, with respect to three topics: support and encouragement, future and success orientation, and the perception of CS.

Support and Encouragement

Several questions in the questionnaire addressed the support and encouragement to study CS that students receive from different sources. One of the questions was "Who encouraged you to choose CS studies?" for which the pupils could choose from a given list of figures more than one figure. Table 2 presents the distribution of answers to this question.

Table 2 reflects an unequivocal conclusion: Arab female high-school students receive much more encouragement to learn CS than do their Jewish counterparts.

In another question, the students were asked to rate their agreement with the following statement: "Our school encourages its students to study advanced-level CS." The difference between the two

Table 2. Percentages of females' encouragement by others

Source	Jewish Female Pupils	Arab Female Pupils
Mother	40%	56%
Father	40%	44%
Siblings	16%	44%
Friends	20%	40%
Acquaintances who had studied CS	20%	50%
Teachers	8%	56%

populations was significant (p=0.000): 91% of the Arab female high-school students agreed with the statement compared to only 28% of the Jewish female high-school students.

Additional evidence can be found in students' answers to a question in which students were asked to rate the influence of different factors on their choice to study advanced-level CS. The results are presented in Table 3.

A similar picture emerges also from students' answers to open questions. For example, a Jewish female student explained in an open questionnaire that:

[p]arents don't encourage the girls enough to begin studying CS and since the female students are influenced by their female friends' attitudes

towards computing, they don't turn to CS studies (and not because of a lack of required skills, since they do have them).

The difference in the participation of female high-school students in CS studies between the two sectors is reflected not only in the number of female students attracted to study CS, but also in the number of female students that persist in their CS studies. As it turns out, Jewish female students are more likely, compared to their Arab counterparts, to abandon their CS studies during the high-school years (especially at the beginning of 12th grade, when the material becomes more complicated). In the interviews, the female students were asked about their own and their friends' persistence in CS studies. The Arab female students' attitudes to this

Table 3. Factors influencing the choice to study advanced-level CS

Factors	Jewish Female Students	Arab Female Students	
Supporting and helpful CS teachers	1.2	2.4	*
School recommendation (teacher, counselor, principal)	1.2	2.0	*
Family recommendation (parents, siblings, uncles/aunts)	1.9	2.3	**
Friends' recommendation	1.4	2.3	*

*Notes: * (p<.01), ** (p<.03)*
Values range from 1 (no influence at all) to 3 (much influence). The table presents the averages.

Table 4. Factors influencing the choice to study advanced-level CS

Factors	Jewish Female Students	Arab Female Students	
A matriculation certificate with CS will help me find a job	1.8	2.3	**
CS is an essential subject for academic studies	1.9	2.5	*

*Note: * (p<.01), ** (p<.04)*
Values range from 1 (no influence at all) to 3 (much influence). The table presents the averages.

Table 5. Intended academic major

Major	Jewish Female Students	Arab Female Students
CS related	31.8%	21.2%
Undecided	45.5%	9.1%
CS unrelated	22.7%	69.7%

issue further highlighted the importance of the support that they receive from their environment.

Future and Success Orientation

Interesting outcomes were revealed in students' answers to a question in which they were asked to rate the influence of future- and success-oriented factors on their decision to study advanced-level CS. Table 4 presents two such factors.

Table 5 presents answers to a question that asked about the field the students plan to major in for their academic studies. The results presented in Table 5 will be further discussed in the conclusion.

When asked about gender-related considerations involved in the choice of majors, some of the female Arab students stressed their duties as wives and mothers: "The boys consider how to earn a living, but the girls consider both how to earn a living and be at home."

At the same time, however, at the high-school level, female Arab students conceive CS as a way to increase their self-esteem. An Arab female student said:

CS is like I have a job. People will regard me as ... if I have a higher status ... There is a different perspective on a girl who is studying CS. There is more appreciation ... When my parents tell their friends that I'm studying CS it gives them more pride and appreciation.

Attitudes Toward CS Studies

A difference in the Arab and the Jewish female students' perspectives was observed also with respect to their conception of the field of CS. As can be seen in Table 6, the female students from the two sectors view CS in a significantly different way. Only 32% of the Jewish female students who are already studying advanced-level CS think that CS is important; a higher percentage of Arab female

students think that CS is difficult and frightening, but yet a much higher percentage of Arab female students like and enjoy CS.

One of the questions in the questionnaire asked to rate factors related to attitudes toward CS studies according to their influence on the students' decision to study advanced-level CS. Table 7 presents several factors and strengthens the previous results.

Currently, the Jewish and Arab sectors in Israel study in separate educational systems. As mentioned previously, both educational systems have a similar structure in terms of the basic curriculum. In what follows, we explain the above findings through some of the structural differences that do, nevertheless, exist between the two educational systems.

High-school students can choose to specialize (advanced-level studies) in specific subjects taken from two groups: Group A includes the traditional scientific fields of mathematics, physics, chemistry, and biology, while Group B includes CS, economy, communication, psychology, sociology, languages, law, art, drama, music, tourism, and theater. As it turns out, the most prominent difference between the two educational systems is expressed in the diversity of the subjects offered at the advanced level. Both systems offer the same variety of Group A subjects; however, while most Jewish schools offer a variety of Group B subjects, most Arab schools offer very few subjects from Group B. As a result, when 10th-grade students are required to choose specialization subjects for their high-school years, Jewish students have many options to choose from in Group B compared to Arab students who have less choices. Specifically, Jewish female students have the option of choosing traditional "feminine" subjects (like psychology) while Arab female

Table 6. Attitudes toward CS among Jewish and Arab female students

Attitude toward CS	Jewish Female Students	Arab Female Students
Interesting	68%	85% *
Difficult	16%	35% *
Essential	56%	38% *
Enjoyable	28%	53% *
Frightening	8%	21% *
Important	32%	77% *
Liked	8%	50% *
Surprising	0%	47% *

*Note: * (p<.01)*

Table 7. Factors influencing the choice to study advanced-level CS

Factors	Jewish Female Students	Arab Female Students	
CS is a prestigious subject	2.0	2.4	*
CS studies is a challenge for success	1.9	2.4	*
I enjoy studying CS	2.1	2.4	*
Friends' recommendation	1.4	2.3	*

*Notes: * (p<.05)*
Values range from 1 (no influence at all) to 3 (much influence).

students are, in most cases, restricted to a choice of more "masculine" subjects (such as CS). Moreover, most of the other Group B subjects are considered by students to be easier and to require less effort compared to CS. Thus, for Jewish female students, the alternatives seem much more attractive.

This situation is reflected clearly in the interviews, in which the female students were asked, "Which subjects did you have doubts about when you had to choose your majors?" A typical answer given by a Jewish female student was:

I didn't choose physics, because I didn't really see myself there ... I didn't feel the connection in any way. And versus chemistry I had drama, so I preferred drama, because I felt a connection. Later I changed drama to social sciences and instead of physics I took CS.

Such answers clarify the idea that because of the alternatives that exist in the Jewish schools, the choice of CS is not trivial or obvious.

Ultimately, we see that the limited options in the Arab sector actually benefit Arab female students and expose them to prestigious areas, such as CS. At the same time, the existing diversity offered in Jewish high schools, while aiming to enable all students to study subjects according to their capabilities and areas of interest, draws female students away from CS. Well, subject diversity—is it good or bad?

FUTURE TRENDS

In this section, we briefly present several of the solutions suggested in the literature for the encour-

agement of women to learn CS, which may fit to the Israeli female high-school students: providing girls with early experiences with computing; conducting meetings or workshops for parents, teachers, and counselors to increase their awareness with respect to their influence on their daughters and female students to choose CS studies; carrying out direct attempts to recruit more female pupils into advanced CS studies; implementing mentoring programs in high schools; conducting classroom visits and talks by successful women in computing positions; including activities to increase female students' self-confidence; establishing diverse learning and working environments; creating all-female CS classes; and reversing the image of CS and refuting its stereotypes. This list of activities is based on the following resources: Gürer and Camp (2001, 2002), Jepson and Perl (2002), Leever, Dunigan, and Turner (2002), Margolis and Fisher (2002), and Verbick (2002). The potential influence of each of these activities may be a subject of future research.

CONCLUSION

As mentioned above, noticeable differences exist in the extent of encouragement Arab female students receive from various agents, especially from teachers, in comparison to Jewish female students. Even in light of the renowned, greater appreciation and respect given to teachers in the Arab sector (Sherer & Karnieli-Miller, 2004), the differences, we suggest, are still enormous. One possible explanation for such an extent of encouragement is based on findings of other studies that explored cultural and familial differences between Arab and Jewish adolescents. According to these studies, since Arab students are part of an Eastern, collective culture, as well as a minority group in Israel, it is likely that they are pushed by their parents to higher scholastic achievement in order to improve their social status (Peleg-Popko, Klingman, & Abu-Hanna Nahhas, 2003).

In addition, Arab students perceive their family environment as more authoritarian than do their Jewish counterparts. The hierarchical structure of the Arab family is based on age and traditionally requires the young to obey the old members and adhere to their expectations (Peleg-Popko et al., 2003). Furthermore, it was found that peer influence in the Arab sector is much more positive than it is in the Jewish sector, possibly because of the relative independence from family and friends that exists in the Jewish sector (Azaiza & Ben-Ari, 1997). This might explain the lower influence of parents and peers in the Jewish sector.

The picture painted by the results of our study is that Arab female students perceive CS studies as a way to provide themselves with increased professional opportunities and especially higher social status. This assumption is reinforced by results of general research regarding the future orientation of Arab adolescents, which concluded that Israeli-Arabs perceive high-school education as a crucial element in the opening up of employment opportunities and in achieving a higher economic status (Azaiza & Ben-Ari, 1997; Seginer & Vermulst, 2002). In order not to be inferior in the eyes of their family in particular and their society in general, it seems that Arab female students are highly motivated to study CS since they consider these studies as a way to prove their skills and capabilities.

Despite the fact that Arab female students are about half of the students in advanced-level CS classrooms in high schools, according to their future orientations (Table 5), this will probably not help to expand the shrinking pipeline in the Arab sector. Most of the female students have already decided on their future professions, and only a small percent of female Arab students consider majoring in CS. Thus, the better starting point might not carry over to higher education and industry.

Arab female students also hold relatively positive attitudes toward CS compared to Jewish female students. Since it was found that positive attitudes of the female students toward computing influenced the success of the students and their continued enrollment in computer courses (Charlton & Birkett, 1999; Gürer & Camp, 2002), the attraction and retaining levels of CS studying among the Arab female students is clear. By creating an atmosphere that supports the development of positive attitudes toward CS, we can probably attract more female students to study advanced levels of CS.

Furthermore, as has been discussed previously, different social and cultural characteristics stimulate the extensive encouragement the Arab female students receive. Naturally, we can conclude that encouragement may be one solution for attracting female students to study CS and keeping them there.

ACKNOWLEDGMENT

We would like to thank the Samuel Neaman Institute for Advanced Studies in Science and Technology and the Technion Fund for the Promotion of Research for their generous support in this research.

REFERENCES

Adams, J., Bauer, V., & Baichoo, S. (2003). An expanding pipeline: Gender in Mauritius. *Inroads-SIGCSE Bulletin, 35*(1), 59-63.

Azaiza, F., & Ben-Ari, A. T. (1997). Minority adolescents' future orientation: The case of Arabs living in Israel. *International Journal of Group Tensions, 27*(1), 43-57.

Camp, T. (1997). The incredible shrinking pipeline. *Inroads-SIGCSE Bulletin, 40*(1), 103-110.

Camp, T. (2002). Message from the guest editor. *Inroads-SIGCSE Bulletin,* 6-8.

Camp, T., Miller, K., & Davies, V. (1999). *The incredible shrinking pipeline unlikely to reverse.* Retrieved March 8, 2005, from http://www.mines.edu/fs_home/tcamp/new-study/new-study.html

Charlton, J. P., & Birkett, P. E. (1999). An integrative model of factors related to computing course performance. *Journal of Educational Computing Research, 20*(3), 237-257.

Davies, A. R., Klawe, M., Nyhus, C., Sullivan, H., & Ng, M. (2000). Gender issues in computer science education. In *Proceedings of the National Inst. Science Education Forum*, Detroit, MI. Retrieved from http://www.wcer.wisc.edu/archive/nise/News_Activities/Forums/Klawepaper.htm

Denning, P. (1989). Computing as a discipline. *Communications of the ACM, 32*(1), 9-23.

Eidelman, L., & Hazzan, O. (2005). Factors influencing the shrinking pipeline in high schools: A sector-based analysis of the Israeli high school system. In *Proceedings of SIGCSE 2005: The 36th Technical Symposium on Computer Science Education,* St. Louis, MO (pp. 406-410).

Galpin, V. (2002). Women in computing around the world. *Inroads-SIGCSE Bulletin*, 94-100.

Gürer, D., & Camp, T. (2001). *Investigating the incredible shrinking pipeline for women in computer science* (Final Report NSF 9812016). Retrieved from http://www.acm.org/women/

Gürer, D., & Camp, T. (2002). An ACM-W literature review on women in computing. *Inroads-SIGCSE Bulletin*, 121-127.

Jepson, A., & Perl, T. (2002). Priming the pipeline. *Inroads-SIGCSE Bulletin*, 36-39.

Leever, S., Dunigan, M., & Turner, M. (2002). The power to change is in our hands. *The Journal of Computing in Small Colleges, 18*(2), 169-179.

Lopez, A., & Schulte, L. (2002). African American women in the computing sciences: A group to be studied. *Inroads-SIGCSE Bulletin, 34*(1), 87-90.

Margolis, J., & Fisher, A. (2002). *Unlocking the clubhouse: Women in computing.* MIT Press.

National Center for Education Statistics. (2004). *Trends in educational equity of girls & women.* Retrieved from http://nces.ed.gov/pubsearch/pubsinfo.asp?pubid=2005016

Peleg-Popko, O., Klingman, A., & Abu-Hanna Nahhas, I. (2003). Cross-cultural and familial differences between Arab and Jewish adolescents in test anxiety. *International Journal of Intercultural Relations, 27*, 525-541.

Schinzel, B. (2002). Cultural differences of female enrollment in tertiary education in computer science.

In *Proceedings of the IFIP 17th World Computer Congress: TC9 Stream/6th International Conference on Human Choice and Computers*, 283-292.

Seginer, R., & Vermulst, A. (2002). Family environment, educational aspirations, and academic achievement in two cultural settings. *Journal of Cross-Cultural Psychology, 33*(6), 540-558.

Sherer, M., & Karnieli-Miller, O. (2004). Aggression and violence among Jewish and Arab youth in Israel. *International Journal of Intercultural Relations, 28*, 93-109.

Verbick, T. (2002). Women, technology, and gender bias. *Journal of Computing Sciences in Colleges, 17*(3), 240-250.

KEY TERMS

Computer Science: The systematic study of algorithmic processes that describe and transform information: their theory, analysis, design, efficiency, implementation, and application (Denning, 1989). The *Computing Curricula 2001: Computer Science* volume, published on December 15, 2001 (http://www.computer.org/education/cc2001/cc2001.pdf), presents the following areas as topics in the computer-science body of knowledge: discrete structures, programming fundamentals, algorithms and complexity, architecture and organization, operating systems, Net-centric computing, programming languages, human-computer interaction, graphics and visual computing, intelligent systems, information management, social and professional issues, software engineering, computational science, and numerical methods.

Israeli High-School Computer-Science Syllabus: The syllabus includes the core of the discipline and is considered to be relatively advanced in comparison to the CS syllabi of other countries.

Shrinking Pipeline: The pipeline represents the ratio of women involved in computer science from high school to graduate school. The pipeline shrinkage problem focuses on several exit junctions: from high school to undergraduate school, at the bachelor's level and at the senior levels both in academia and the industry.

Skills of Women Technologists

Maria Elisa R. Jacob
DePaul University, USA

INTRODUCTION

Women technologists practice careers in various fields of information technology. They traditionally are educated and trained to acquire primarily technical skills. However, in response to organizational change and industry shifts, today's women technologists are acquiring a multitude of diverse skill sets—on top of their conventional technical skills—to excel and succeed in the workplace.

This article delves into various skill sets in today's IT workplace and how women technologists have adopted and updated their skill sets to redefine their role to align with today's industries.

BACKGROUND

Skills

A skill is defined as a proficiency, talent, or ability that is developed through training, education, or experience (Beckhusen & Gazzano, 1993, p. 10). Throughout a woman technology professional's career, a unique combination of on-the-job training, technical apprenticeships, internships, job shadowing, and formal education results in acquiring basic skills. This basic foundation later develops into specialized expertise through years of work experience (Gordon, 2000, p.133).

Skills vary in use and purpose. Beckhusen's et al. work (1993) on strategic skills assessment divides skills into five main categories. This assessment is an excellent baseline, which covers numerous skill sets of technology professionals. Some skills are transferable and portable, and thus are listed under multiple categories. The following are the seven main skills categories.

Creative Expression Skills

Skills in this category are designing, developing, authoring, composing, displaying, inventing, performing, and producing (Beckhusen et al., 1993). Skills of creative expression serves key for technology professionals involved in such roles as architecting software, designing user interface or developing new hardware. They use their artistic inclinations to develop innovative solutions and designs. Examples of technical professionals who have highly developed creative expression skills are software developers, Web designers, graphic user interface (GUI) designers, and research/development specialists.

Communication Skills

Skills in this category consists of consulting, facilitating, explaining, speaking, writing, interviewing, persuading, selling and motivating (Beckhusen et al., 1993). Communication skills are critical for successful interaction between various stakeholders of projects—from the client and all the way to the project team. Communication is vital throughout all phases of project management, planning, and development processes. Gill (2002) states that communication is a major determinant for the success or failure of a project. Brooking (1999) adds that in an organization, it is important to know "who needs to know what" and make sure they get the information they need at the right time they need it. Examples of these technical professionals with highly developed communication skills are project managers, functional analysts, business analysts, technical writers, documentation specialists, and system diagrammers.

Mental Creative Skills

Skills in this category consist of intuitive, conceptualizing, brainstorming, improvising, memorizing, synthesizing, and visualizing (Beckhusen et al, 1993, p. 10). Advances in technology that are results of using mental creative skills are expert systems that capture deep knowledge to derive conclusions based on if-then analysis (Brooking, 1999, p. 89). Examples of technology professionals that rely heavily on mental creative skills are software engineers, architects, project managers, quality assurance testers, systems assurance analysts, and performance capacity analysts.

Mental Analytic Skills

Skills in this category consist of analytical, budgeting, categorization, editing, investigating, observing, monitoring, researching, and problem solving (Beckhusen et al., 1993, p. 10). Mental analytical skills are used extensively during information management efforts that are both the responsibility of business experts as well as the information technology functions. Information is seen as having a key role in business processes (Evernden, 2003, p. 6-8). Examples of these technical professionals heavily utilizing these skills are database administrators, programmers, functional leads, business analysts, and systems analysts.

Leadership/Management Skills

Skills in this category consist of initiating, coordinating, deciding, delegating, implementing, organizing, mediating, negotiating, and supervising (Beckhusen et al., 1993, p. 11). These skills are used by developers and testers when coordinating a team, organizing a meeting and supervising progress. Furthermore, according to Cohen (2002), exceptional negotiation skills are at play when the team ends up mutually committed to fulfilling the agreement they have reached (Cohen, 2002, p. 3). Technical professionals having strong leadership skills are project managers, senior architects, lead system administrators, technical leads, and Web directors.

Physical Skills

Skills in this category consist of building, constructing, operating, and restoring (Beckhusen et al., 1993, p. 11). Physical skills are directly used by profes-

sionals dealing with computer hardware in performing troubleshooting, configuration, set-up and installation tasks. Examples of technology professionals requiring strong physical skills are computer technicians, PC operators, help desk support specialists, customer support representatives, telecommunications specialists, and infrastructure support personnel.

Humanitarian Skills

Skills in this category consist of advocacy, coaching, mentoring, counseling, instructing, listening, and training (Beckhusen et al., 1993, p. 11). Humanitarian skills include empathizing and building rapport with fellow project team members. Empathy is the art of relationship building and used by project team members to show their support for team performance. Rapport is insightfulness regarding other's feelings, motives, and concerns (Johnson, 1997, p. 230). These skills are exhibited by all technical professionals such as trainers, managers, team leads, and LAN administrators as they do their daily jobs.

SKILLS OF WOMEN TECHNOLOGISTS

Women technology professionals possess diverse and multiple skill sets that enable them to successfully work in their field. As industries and businesses evolve, today's organizations now need fewer yet far better educated and skilled workers due to technical advances in the workplace (Gordon, 2000, p. 2). Fields (2001) calls people who have a diverse set of skills "indispensable employees".

Field generalizes that most organization's ultimate goal is to recruit and retain these competent indispensable employees. In order to achieve this status, women technology professionals strive to have most or all of the skill categories above either as their core and/an secondary skill. The following sections differentiates core from secondary skills.

Core Skills

Core skills comprise of mental analytical skills. They are acquired through intensive technical training.

Core skills are specialized expertise in the areas of software, hardware, telecommunications, network, server, telecommunications, security, and Web services. Advanced math, logic, software engineering, CASE implementation, programming languages, data flow diagramming, decomposition diagramming, entity relationship modeling, data normalization, prototyping, structure charting, walk-through, testing are some examples of techniques, and strategies used that take heavily from mental analytical skills (Gordon, 2000, p. 9; Fletcher & Hunt, 1993, p. 154-155).

Secondary Skills

Secondary skills comprise the rest of the other skill categories—creative expression, communication, mental creative, leadership/management, physical, and humanitarian skills. Other secondary skills come about through interaction with the project team and intercultural collaboration. Skills in project management is gaining wider acceptance as IT departments become more projectized. This skill includes competencies in such techniques as critical path scheduling, PERT charting and GANTT charting (Fletcher, 1993, p. 154-155). Intercultural skills are also emerging to be an important secondary skill as multicultural and global teams increasingly being formed (Laroche, 2003, p. 58).

MAIN THRUST OF THE ARTICLE

Women technologists exhibit a strong sense of adapting multiple skills as their careers progress in the workforce. A single or combination of a variety of methods is employed by women to acquire these skill sets at various points in their career. The acquisition of skill sets is divided into two stages that directly correspond to the stage of a woman's career.

Skills Specialization Stage

This stage occurs during the initial 1-3 years of a woman's work experience. As a newly grad or a junior technical professional, she seeks to define her expertise in her specific line of work, function or environment. During this part of her career, successful women technologists specialize to carve their niche and build a reputation for themselves based on solid technical core skills. If success is not achieved immediately, she may seek specialized technical training to augment and/or update existing skills for her to move up in the corporate ladder. She may also consider acquiring professional certification, licensure, advanced degrees, joining professional organizations, and networking (Laroche, 2003, p. 57).

At this stage, successful women technologists have a strong belief in their ability to control their careers. Very often successful women have legitimate power to control their destiny, which would reinforce a personal sense of internality and career drive. Many of them say that they have the tenacity and perseverance which enabled them to strive hard consistently throughout their careers (White, Cox, & Cooper, 1992, p. 85, 62).

Skills Diversification Stage

This stage occurs after a woman has completed the skills specialization stage. Skills diversification usually occurs during the next 4-10 years of a woman's career. Upon mastery of specialized skills, she sets about to complete this next stage that extends her specialized skill set to include diverse and complementary general skills. Tanton (1994) assesses effective women technologists have a good mix of both their technical specialization and have strong analytical, financial, marketing, planning, project management, decision making, and people skills. Tanton refers to people skills as communicating, negotiating, motivating, listening, involving, counseling, and appraising and delegating skills. In this stage, women technologists usually act the role of technology knowledge workers, sources, and gatekeepers.

Women perform the role of technical knowledge workers by gaining an intrinsic knowledge about people and how they can work together to become better teams (Brooking, 1999, p. 38). Brooking added that since job titles often mask past knowledge about people, proficient women technologists are competent and have experience in identifying team resources based on networking relationships and working with them in previous teams. This skill adds value to products and services. Thus, women technologists who fulfill this role become key business assets that link jobs, knowledge, and profit in every organization (Gordon, 2000, p. 6-7).

Women also perform the role of sources by being identified as a person with more knowledge to solve problems and be creative than the rest of the group (Brooking, 1999, p. 102). Women technologists fulfilling this role are not only experienced in managing an IT environment, but also on working on custom application development on all levels. A source may also be characterized as a technologist who is multilingual and can speak the language of developers and the language of business experts to serve as the bridge across that gap (Gill, 2002, p. 135).

Lastly, women perform the role of gatekeepers by knowing how to get things done, how to solve a particular problem or remembering corporate historical literature (Brooking, 1999, p.105). A woman technologist's technical competence becomes less important, while her social skills become more important as her career progresses. The social skills of self-constraint and compassion are apparent in the ability to organize groups, negotiate solutions, empathize, and create rapport (Gordon, 2000, p. 50). These abilities taken together create skills that are seen as interpersonal intelligence. Women who possess high interpersonal intelligence can connect with people quite smoothly, be astute in reading the reactions and feelings, lead and organize, handle the disputes that result from human interactions (Johnson, 1997, p. 223-224).

ISSUES AND CONTROVERSIES

Quantifying the return of investment (ROI) on skills training is an issue management and organizations constantly face. Organizations admit that there is enormous value in investing in training of women technologists, however there a controversy on how to quantify the results of skills training and effectively measure it back to the increased efficiency and dollar savings. Although many strategies have been proposed through the years, there is still difficulty in comparing skills knowledge management principles skills versus improvement of the bottom line.

A common complaint about skills training is it does not always bring the return or benefits that one might expect, given the huge sums invested espe-

cially when analyzing the ROI (Evernden, 2003, p. 61; Gordon, 2000, p. 49). ROI is an industry standard for how an organization weighs the cost of education versus compensation. It answers the question of what does the organizations get back in return for its investment (Mingus, 2002, p. 50). The results of skill training are not necessarily easily quantifiable, direct or concrete (Gordon, 2000, p.3).

Furthermore, skills training are unfortunately considered as just lip service among some organizations. It is sometimes largely dismissed for most employees as a burdensome cost that must be cut. Some CEOs, presidents and business owners still see no connection between company profits and investing in their human capital because they believe you cannot measure it (Gordon, 2000, p. 3). Organizations have continuously sought to quantify the increase in skill set by setting benchmarks and metrics to measure skills advancement. Although not concrete, some case studies show a relationship between training and profit increase (Laroche, 2003, p. 30).

Assimilation is also an issue for new women technologists. Assimilation is the process of adapting to the organization's culture, customs and attitudes. It involves the risk of bringing in a new person and finding out how well they blend in and fit with the corporate culture. If a new woman technologist finds it difficult to assimilate for whatever reasons, her colleagues might discount her knowledge no matter how valuable it is (Brooking, 1999, p. 95). This issue can be addressed by proper employer/employee matching and screening for a potential fit during the organization's recruitment process.

PROBLEMS

A shortage of mentors in their specific field willing to contribute to their skills development is a problem faced by women technologists. Mentoring is the process of giving advice on education, developing job related knowledge, advance competency and proficiency (Brooking, 1999, p.96). Due to limited number of practicing women technologists, there is also a shortage of available and willing women technical mentors. This is still prevalent today. However, trends show that efforts and awareness campaigns are underway to lessen the gap.

Aside from a shortage of woman mentors, there is also a general labor force shortage of specialized technical professionals. There is a huge skills gap between the technical proficiency necessary in the workforce and the labor force's actual general lack of know-how and skills. This skills deficiency has been identified for both female and male technologists (Fields, 2001, p. 20). A potential solution to this problem would be integrating technology into educational curriculum early on to encourage developing talent.

FUTURE TRENDS

Women technologists will continue to display a diverse set of skill set as their roles in the organization evolves and changes. The trend towards mastering their core skills first, then consistently refining their secondary skills will be prevalent in the future.

There will also be a continued emphasis on "techno-education", wherein educational curriculum will be heavily infused with technical subjects in order to expose students, increase awareness, and generate interest early on to information technology.

There will also be potentially increasing numbers of women entering the technology field as traditional job descriptions evolve. Women who possess a good balance of diverse skill sets and adaptable to change will thrive in future careers.

Additional areas for research include defining information technology skill sets from male technologists. It is also recommended to do a comparative study between skill sets and attitudes of technology professionals and methods for acquiring diverse skills.

CONCLUSION

Women technologists have adapted, and will continue to adapt, their skill sets to organizational change and industry shifts. Successful women technologists first specialize, then constantly diversify their skill sets to enhance employability. The most responsive woman technologists in today's organizations must truly possess skills beyond their technical training.

REFERENCES

Beckhusen, L., & Gazzano. (1993). Strategic skills assessment. *Skillscan Professional Pack—Level I*, 1-15.

Brooking, A. (1999). *Corporate memory. Strategies for knowledge management.* International Thompson Business Press.

Cohen, S. (2002). *Negotiating skills for managers.* New York: Mc-Graw Hill.

Evernden, R., & Everden, E. (2003). *Information first. Integrating knowledge and information architecture for business advantage.* Elsevier Butterworth-Heinemann.

Fields, M. R. A. (2001). *Indispensable employees. How to hire them. How to keep them.* NJ: Career Press.

Fletcher, T., & Hunt, J. (1993). *Software engineering and CASE: Bridging the culture gap* (pp. 150-160). New York: McGraw-Hill, Inc.

Gill, T. (2002). *Planning smarter: Creating blueprint-quality software specifications.* NJ: Prentice Hall.

Gordon, E. E. (2000). *Skill wars: Winning the battle for productivity and profit.* Boston: Butterworth-Heinemann.

Johnson, H. E. (1997). *Mentoring for exceptional performance.* CA: Griffin Publishing.

Laroche, L. (2003). *Managing cultural diversity in technical professions.* Amsterdam: Butterworth Heinemann.

Mingus, N. (2002). *Alpha teach yourself project management in 24 hours.* CWL Publishing.

Nielsen, J. (2000). *Designing Web usability: The practice of simplicity.* New Riders Publishing.

Tanton, M. (1994). *Women in management. A developing presence.* Routledge.

White, B., Cox, C., & Cooper, C. (1992). *Women's career development. A study of high flyers.* Blackwell Publishers.

S

KEY TERMS

Data Dictionary: Establishes a clear and unambiguous vocabulary for bookkeeping and documentation purposes. This may consist of precise definitions of data elements in order for forms and databases to be constructed.

Information Map: Can be any diagram, from a simple graphic to a complex software model, depicting corporate information.

Knowledge Management: To manage institutional intelligence as they would any other precious asset. It defined how well an organization develops its people and their capabilities to make a profit.

Knowledge Worker: A person with certain amounts of specialized training, based on prior, sound basic education.

Mentoring: To facilitate, guide, encourage continuous innovation, learning and growth to prepare the business for the future

Scope Creep: Describes the phenomenon in which the features are added or changed after the specification is finalized so they increase the time required to deliver the product.

User Interface: The object used to interact between a human and a physical device. It is a set of commands or menus through which a user communicates with a program.

The Social Construction of Australian Women in IT

Sue Nielsen
Griffith University, Australia

Liisa von Hellens
Griffith University, Australia

INTRODUCTION

The declining participation of women in IT education and professional work is now a well-documented research area (Adam, Howcroft, & Richardson, 2004), but the causes and remedies remain puzzling and complex. Studies have indicated that there are signs of the "shrinking pipeline" (Camp, 1997) even in the years between junior and senior high school (i.e., Meredyth, Russell, Blackwood, & Thomas, 1999) when girls' interest and confidence in the use of computers declines markedly.

A lack of clarity as to what constitutes the IT industry and the rapid rate of change complicate attempts to understand the reasons for the declining participation of women in the IT industry, as well as the declining interest in IT degrees. This is despite the fact that IT salaries compare well with other professional salaries and are superior to most traditional female occupations (Megalogenis, 2003). Our research also demonstrates that many people—especially women—enter the IT workforce via other qualifications indicating that traditional IT education is not very successful in attracting either the quantity or quality of students required to meet workforce needs. Furthermore, IT has not matched the rise in female participation in the traditionally male-dominated professions of science, engineering, and medicine.

AUSTRALIAN SITUATION

By the start of the last decade, the proportion of females in the IT workforce was beginning to decline in most western countries including Australia (von Hellens, Pringle, Nielsen, & Greenhill, 2000).

About 20% of the members of the Australian Computing Society are women. The Australian Bureau of Statistics show that the percentage of women working in IT occupations decreased in 1996-2001 despite the total number of people (men and women) with IT qualifications increasing significantly, to around 70% (Byrne & Staehr, 2003). IT education suffered one of the highest drop-offs in students' numbers between 2003 and 2004 and the IT schools' downturn has continued in 2005 (O'Keefe, 2005). Figures from the Australian Department of Education, Science, and Training reveal a 6.7% drop in IT, from the 68,271 students (domestic and overseas) in 2003 to 63,651 in 2004. The decline was even higher with women: a 14.7% decrease. The percentage of female students in IT declined from 24.02% to 21.97%. The information systems area of IT in the business schools has traditionally had more female students than the core IT programs. However, the downturn of students and female students in particular is apparent there too (Business/Higher Education Round Table, 2004).

WOMEN IN INFORMATION TECHNOLOGY PROJECT

To deal with this complexity, the Women in Information Technology (WinIT) research project has taken several perspectives, reflecting the multidimensional nature of the problem (von Hellens & Nielsen 2001; Trauth, Nielsen, & von Hellens, 2003). The project has surveyed and interviewed male and female high school students as well as undergraduate and postgraduate IT students. Interviews have also been carried out with high school teachers, vocational guidance counselors, IT academics, and professionals in the IT industry. A complete list of published research can be found at http://www.cit.gu.edu.au/~jenine/WinITProject/. This article provides an overview of the WinIT research and the findings to date.

The perspective taken in the WinIT research is based on widespread views of gender in IT:

- **"Sameness":** That women are capable of entering and succeeding in male domains by adapting to those domains.
- **Social Construction:** That in order to understand IT as a field of education and work, it is worthwhile to view it in terms of its social and political construction, in the same way that "female" domains such as child care and nursing may also be viewed.

Although the WinIT focus has been on female's perception, male students have also been interviewed and surveyed to further clarify the female experience. We have also focused on Confucian Heritage (CH) students from a range of South-East Asian countries who represent a major non-English speaking background ethnic group among IT students in the Australian education system including high school and universities, especially their female contingent.

HIGH SCHOOL STUDENTS

Male and female high school students in the IT entry-level mathematics course were surveyed and interviewed (Nielsen, von Hellens, Pringle, & Greenhill, 1999). Both Asian and non-Asian females had similar views about the work of an IT professional, who they characterized as a person working alone with a computer and thus not requiring any communication or people skills. Females perceived computing as boring, requiring logic or mathematics skills, and involving little contact with other people. There was a strong impression that women preferred different types of work that would require personal contact and communication. Female students expected boys to be more interested in computers than girls, as computing was perceived to be a masculine pastime. There was a false perception among female students that the participation of women in IT was improving rather than declining.

The use of computers outside school was equal between Asian and non-Asian girls, but Asian females were more inclined to choose computing and IT subjects at school, despite the presence of negative perceptions. It appeared that the usefulness of computing and the favorable prospects for employment, rather than personal interest, were stronger motivating factors for Asian students than for other female students.

FIRST YEAR UNIVERSITY STUDENTS

Research revealed that the IT degree was much harder than expected, and female students were either ignored or harassed (Nielsen et al., 1999). Males dominated the study environment physically. Although not all female students perceived this as a problem, some expressed resentment at other female students' reliance on male students and also of bad behaviour by male students in computer labs. Condescending behavior towards female students took the form of stereotyping based on intellectual achievement and physical appearance. Inclusion in the "pretty and smart female" group meant a privileged acceptance into the male-dominated setting.

There was a notable difference in the way Asian and non-Asian female students experienced gender-based discrimination. The former felt their opinions were not valued as highly as those of non-Asian students and all non-Asian people largely ignored them. As a result, they felt isolated and ignored and had to band together for guidance and assistance. On the other hand, non-Asian female students felt they were the focus for sexual harassment (e.g., via remarks and uninvited e-mails). They received unwanted positive discrimination in the form of easier marks, which was offensive to their sense of achievement. There were subtle differences in staff behavior towards female students, particularly if they were Asian. Interviewees claimed that lecturers preferred male students. Competition between students did not worry female students. Their social interactions were comfortable and they found it easy to initiate collaborations, however there was some confusion about the differences between working together, and plagiarism.

It was found that information about the IT degree programs was inadequate and poorly distributed. Teachers in high schools had provided misleading information about IT programs at university. Students were unclear about the skills required in the IT industry and what job opportunities the degree would provide. The Asian female students, in particular, were pessimistic about future job prospects. The reasons for choosing IT education were often arbitrary, including inability to qualify for a preferred course of study.

There was an overall perception of the IT course as difficult and demanding, and all students expressed reliance on networks for troubleshooting, group study, and support. Elements of racism and sexism operated against successful group work. Students in the university avoided working with other students due to racially defined differences, such as language, assuming it would result in a disadvantage for oral group presentations.

The applications of IT education were perceived and valued differently by the Asian and non-Asian females. Whereas the non-Asian females considered computing as having limited career options, the Asian females emphasized further career opportunities from IT education, although this did not necessarily indicate a better understanding of the IT industry. Asian girls experienced pressure from home to pursue IT and were aware of a family responsibility to provide for elderly parents.

Although many students valued IT as providing "flexible" work arrangements, not many students were sure what was meant by "flexible"; in terms of location, time, amount of work or wider job opportunities. Younger students with lack of work experience seemed confused about IT jobs in general. Mature-aged students with more experience in the workforce were able to acknowledge both the advantages and disadvantages of job flexibility.

CULTURE AND GENDER INFLUENCES

Existing research on perceptions of IT job in Asian and Westerns countries informed our research on high school and university students (Burn, Ma, & Ng Tye, 1995: Couger, 1996; Hofstede, 1980, 1994;

Igbaria & McClosky, 1996). The conceptualization of the factors of culture and gender affecting students' perception of IT careers (Figure 1) summarizes our findings. This model (discussed in detail in Nielsen et al., 1999) attempts to display how factors of culture, gender and life history shape the students' perceptions of IT and suggests a categorisation of students according to why Computing and IT skills are valued. The extreme cases are students who value computing and IT skills for escapism for leisure and study at one end of the spectrum and those who prefer collaborative practices at the other end of the spectrum. The categories in the middle are those who value computing and IT skills for flexible work arrangements and those who value computing and IT skills in order to secure employment and to achieve an occupation of status. The model suggests that combining these two sets of values allows for a broader perception of IT (as social systems technically implemented).

Our research also suggests students did not fully understand the new organizational forms that create, and are created by, IT innovations. Also, IT educators may not be conveying these changes and their relevance very effectively. First-year students were surprised at the amount of mathematics and programming required, especially as it seemed to contradict their experience of what was expected in the real world. Most students seemed to have a very narrow view of IT and were more concerned with understanding about computers, rather than the tasks to be supported by technology and the changing context in which IT is used. The conceptual model illustrates the changing environment of IT professionals.

Most females reinforced the expectation that females and males have equal ability, although some females believed men were smarter and better at the sort of study required for IT.

RESEARCH AMONG WOMEN IT PROFESSIONALS

Our research among female IT professionals examines the perceptions of women in the IT industry as expressed in their discourse about IT work. The research data was collected primarily through in-

Figure 1.

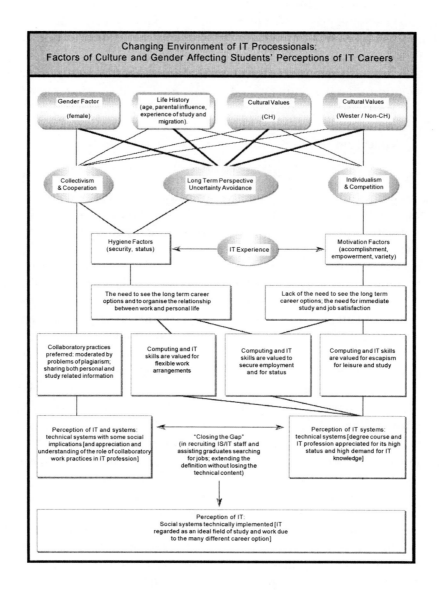

terviews with female IT Professionals, and a small number of male IT Professionals to further understand the declining female participation in IT education and work (von Hellens & Nielsen, 2001).

Our interest is in identifying patterns in women's discourse about IT work, which both reinforce and challenge the institutional discourse. Especially evident are discursive patterns in relation to skills required in the IT industry. There is a very large body of literature on the demand for IT skills, however our research differs from the primarily positivist approach taken in IT skills research. For example, Todd, McKeen, and Gallupe (1995) reported that the skills that are required from IT

professionals are business, technical and systems skills. We have adopted Giddens' view (1984, 1991) and challenge the strictly objective/subjective approach to the formation and maintenance of institutions. From a structuration theory point of view, demand for IT skills cannot be treated as a matter of social fact, but is implicated in the structuration of the IT industry, and reinforced or challenged by industry participants.

The female IT professionals saw themselves as task oriented, client focused, and hard working. They also emphasized the importance on mentoring and parental support and the element of chance that triggered their entry into the IT industry. They did

Table 1. Dualisms of skills in IT work.

Dualisms of skills in IT work
Home/work
IT industry/career certainty
IT work/emotion
Intuition/analysis
Programmer/people focused
Soft skills/technical education
Solving problems/talking to people
Technology/business problems
Technical/management
Technical/people skills
Technology/communication skills

not view their gender as making a difference, although successful women adopted ways of working that were not common to women, in order to adapt to different IT environments.

These perceptions and their representations may not have a straightforward relationship to the types of categorisations (business/technical, etc) made by studies of IT skills (Nielsen, von Hellens, & Beekhuyzen, 2004). In order to examine this relationship we use the notion of "dualisms"—which represent IT works and skills as a set of either or propositions (see Table 1)

One of the most interesting dualisms is the separation of soft and hard skills, not only in terms of their association with women and men, but also in terms of how they are learned and valued. This is expressed by one interviewee as follows: "Skills like communication and organisational ... they're things that you can train into yourself from a little child—your parents can help you with those. But things like technical skills like IT would need ... you will only get that at a tertiary institution."

The women viewed themselves as being able to switch between "male" and "female" skills, such as "people" and "technical skills" but challenged the conventional wisdom that women have superior communication and organizational skills (von Hellens et al., 2000). The interview data showed a number of examples of gender specific dualisms, as listed in Table 2.

One interviewee commented why political skills are perceived as uncommon among women "I think one of the key skills for large corporate organisations

is the networking skills and if I use the term 'office politics' I mean that's a more negative term but its understanding the culture of the organisation, what do you need to do, who do you need to talk to. Women traditionally I think are not good at talking about what they do and how they do it and their success."

Our findings among women IT professionals are not very different from other professions where also the importance of mentors and encouragement is emphasized, as well as the chance factor, being in the right place at the right time. Traditional wisdom argues that women are valued for their "female" skills in the IT industry. However, the contradictions in the discourse show skills attributed to gender are difficult to define, and that it is not clear why some women are attracted to and successful in the IT industry.

FUTURE TRENDS

Countries like Australia face unpredictable shortages of qualified IT personnel. The IT industry is one of fastest growing, with more than 2.5 times the average growth of the Australian economy as a whole (Business/Higher Education Round Table, 2004), but interest in IT education is declining, particularly amongst women. Strategies to improve the gender balance and to make IT education more appealing to adolescents are not uniformly successful and the situation has not improved.

Table 2. Gender specific dualisms in IT work

Issue	Men	Women
Attention to detail	Uncommon	Common
Broad perspective	Uncommon	Common
Continuous learning and rapid change	Common	Uncommon
Assertiveness	Common	Uncommon
Communication	Uncommon	Common
Noisy vocalism	Common	Uncommon
Political skills	Common	Uncommon
Networking	Common	Uncommon
Left brain	Common	Uncommon
Programming	Common	Uncommon
Technical skills	Common	Uncommon

Low female participation in IT education and work is a complex social problem, reflecting general confusion about the identity of the IT industry itself. More research is required to understand how this identity is constructed. Ongoing research on how female identities and gender differences in IT work are constructed (Anthias, 1999; Nielsen et al., 2004) will assist in clarifying the identity of the IT industry. To extend the use of Giddens' structuration theory beyond what has been done in the WinIT research to date is recommended, including work practice observations and the examination of documentation.

CONCLUSION

The WinIT research has taken a social construction view of female participation in the IT industry, assuming that social institutions are constructed by as well as constraining the activities of their participants, and that unhelpful constructions may be deconstructed and rebuilt. The IT industry was earlier perceived as a "level playing ground," a new industry which would offer broad participation to all interested parties. Many of the institutions deriving from the IT industry, such as the Internet, have had a liberalizing influence. However, to date, the IT industry appears to be constructed as a masculinised domain, unattractive to most women and many men. Considering its importance to modern society, this construction of the IT industry needs to be challenged so that all stakeholders can participate in the design as well as the utilization of IT products and services.

REFERENCES

Adam, A., Howcroft, D., & Richardson, H. (2004). A decade of neglect: Reflecting on gender and IS. *New Technology, Work, and Employment, 19*(3). 222-240.

Anthias, F. (1999). Theorising identity, difference, and social divisions. In M. O'Brian, S. Penna, & C. Hay (Eds.), *Theorising modernity: Reflexivity, environment, and identity*. In Giddens Social Theory (pp. 156-178). London: Addison Wesley Longman Limited.

Business/Higher Education Round Table. (2004). *The gender gap in the ICT industries: Failing to fully utilise a national resource.* Women in ICT Task Force Report, August 2004.

Burn, J. M., Ma, L. C. K., & Ng Tye, E. M. W. (1995). Managing IT professionals in a global environment. *Computer Personnel, 16*(3), 11-19.

Byrne, G., & Staehr, L. (2003). The participation and remuneration of women in the Australian IT industry: An exploration of recent census data. In S. Spencer (Ed.), In *Proceedings of the 2003 Australian Women in IT Conference: Participation, Progress and Potential, School of Information Systems* (pp. 109-116). University of Tasmania.

Camp, T. (1997). The incredible shrinking pipeline. *Communications of the ACM, 40*(10), 103-110.

Couger, J. D. (1996). The changing environment for is professionals: Human resource implications. In M. J. Earl (Ed.), *Information management* (pp. 426-435). Oxford: The Organizational Perspective.

Giddens, A. (1991). *Modernity and self identity: Self and society in the late modern age.* Cambridge, Polity Press.

Giddens, A. (1984). *The constitution of society: Outline of the theory of structuration.* Cambridge, Polity Press.

Hofstede, G. (1980). *Culture's consequences: International differences in work-related values.* Beverly Hills, CA: Sage.

Hofstede, G. (1994). Management scientists are human. *Management Science, 40*(1), 4-13.

Igbaria, M., & McClosky, D. W. (1996). Career orientations of MIS employees in Taiwan. *Computer Personnel, A Quarterly Publication of the SIGCPR, ACM Press, 17*(2), 3-24.

Megalogenis, G. (2003, April 26-27). Women (not) on top. *The Weekend Australian, 17.*

Meredyth, D., Russell, N., Blackwood, L., Thomas, J., & Wise, P. (1999). *Real time: Computers, change, and schooling. National Sample Study of the IT Skills of Australian School Students, a project funded by the Commonwealth Depart-*

ment of Education, Training and Youth Affairs. Retrieved from http://www.detya.gov.au/archive/schools/Publications/1999/realtime.pdf

Nielsen, S., von Hellens, L., Pringle, R., & Greenhill, A. (1999). Students' perceptions of information technology careers: Conceptualising the influence of cultural and gender factors for it education. *GATES (Greater Access to Technology Education and Science), 5*(1), 30-38.

Nielsen, S., von Hellens, L., & Beekhuyzen, J. (2004, June 14-17). The discursive divide: Women in the IT industry. In *Proceedings of the 12th European Conference on Information Systems*, Turku, Finland.

O'Keefe, B. (2005). *Women lead IT course exodus. The Australian*, 18. Retrieved January 12, 2005 from http://news.com.au

Todd, P. A., McKeen, J. D., & Gallupe, R. B. (1995). The evolution of IS job skills: A content analysis of IS job. *MIS Quarterly, 19*(1), 1-28.

Trauth E. M., Nielsen, S. H., & von Hellens, L. A. (2003). Explaining the IT gender gap: Australian stories for the new millennium. *Journal of Research and Practice in Information Technology, 35*(1), 7-20.

von Hellens, L., & Nielsen, S. (2001). Australian women in IT. *Communications of the ACM 44*(7), 46-52.

von Hellens, L. A., Pringle, R., Nielsen, S. H., & Greenhill, A. (2000, April 6-8). People, business, and IT skills: The perspective of women in the IT industry. In W. Nance (Ed.), *Proceedings of the 2000 ACM SIGCPR Computer Personnel Research Conference: Electronic Commerce and Internet Business: Roles, Relationships, Skills and Strategies for the New Millennium* (pp. 152-157), Chicago.

KEY TERMS

Asian: In our study "Asian" is not defined categorically, but refers to a range of East and South-East Asian Confucian-heritage (CH) culture, which represent a major non-English speaking background ethnic group among IT students in Australian.

Confucian-Heritage (CH) Culture: CH refers to a range of East and South-East Asian Confucian-heritage (CH) cultures, including Singapore, Hong Kong, Taiwan, PRC and to some extent Vietnam, Korea, Japan, emphasizing self-control, adherence to social hierarchy and social and political order. CH culture provides a background against which personal life histories can be considered.

High School: Also called Secondary School; grades 7 or 8 to 12. Students must attend school in Queensland, Australia until 14 and 9 months (usually Grade 10). Students may choose to continue into senior secondary school, Grades 11 and 12, and specialize in their choice of subjects. Completion of year 12 is usually required for entry to university degree programmes.

Information Technology Education: The IT degree programmes referred to in this article are undergraduate Australian university degrees, commonly called the Bachelor of Information Technology (BIT). They are professionally accredited and provide core IT education.

Information Technology Field: The technical and people oriented area of information technology development and utilisation in organisations and in society. Information technology education provides core technical skills for the analysis, design, development and use of computer-based information systems.

Private School: Independent school that receives only a small amount of government funding.

Public School: Government funded school.

Social Construction of Gender and Sexuality in Online HIV/AIDS Information

Jing Chong
The Pennsylvania State University, USA

Lynette Kvasny
The Pennsylvania State University, USA

INTRODUCTION

HIV (human immunodeficiency virus) and AIDS (acquired immunodeficiency syndrome) represent a growing and significant health threat to women worldwide. According to the United Nations (UNAIDS/WHO, 2004), women now make up nearly half of all people living with HIV worldwide. In the United States, although males still accounted for 73% of all AIDS cases diagnosed in 2003, there is a marked increase in HIV and AIDS diagnoses among females. The estimated number of AIDS cases increased 15% among females and 1% among males from 1999 through 2003 (Centers for Disease Control and Prevention, 2003). Looking closer at HIV and AIDS infections among women in the United States, Anderson and Smith (2004) report that HIV infection was the leading cause of death in 2001 for African-American women aged 25 to 34 years, and was among the four leading causes of death for African-American women aged 20 to 24 and 35 to 44 years, as well as Hispanic women aged 35 to 44 years. The rate of AIDS diagnoses for African-American women (50.2 out of 100,000 women) was approximately 25 times the rate for white women (2 out of 100,000) and 4 times the rate for Hispanic women (12.4 out of 100,000; Centers for Disease Control and Prevention). African-American and Hispanic women together represented about 25% of all U.S. women (U.S. Census Bureau, 2000), yet they account for 83% of AIDS diagnoses reported in 2003 (Centers for Disease Control and Prevention).

Women's vulnerability to HIV and AIDS may be attributed to gender inequalities in socioeconomic status, stereotypes of AIDS as a gay-male disease, and cultural ideology around sexual practices such as abstinence, monogamy, and condom use. Be-cause of cultural mores and socioeconomic disadvantages, women may consequently have less access to prevention and care resources. Information is perhaps the most important HIV and AIDS resource for women, and the Internet provides a useful platform for disseminating information to a large cross-section of women. With the flourishing use of e-health resources and the growing number of public-access Internet sites, more and more people are using the Internet to obtain health-care information. Over two thirds of Americans (67%) are now online (Internet World Statistics, 2005). On a typical day, about 6 million Americans go online for medical advice. This exceeds the number of Americans who actually visit health professionals (Fox & Rainie, 2002). Studies also show that women are more likely to seek health information online than are men (Fox & Fallows, 2003; Fox & Rainie, 2000; Hern, Weitkamp, Hillard, Trigg, & Guard, 1998). HIV and AIDS patients are among the health-care consumers with chronic medical conditions who increasingly take the Internet as a major source of information (Kalichman, Weinhardt, Benotsch, & Cherry, 2002).

As more Americans go online for health information, the actual efficacy of the information consumption becomes salient. Recent digital divide studies call for shifting from demographic statistics around technological access to socially informed research on effective use of technology (Gurstein, 2003; Hacker & Mason, 2003; Kvasny & Truex, 2001; Payton, 2003; Warschauer, 2002). Although the Internet provides a health information dissemination platform that is continuous, free, and largely anonymous, we should not assume that broader access and use will be translated into positive benefits. We must begin to critically examine the extent to which e-health content meets the needs of an increasingly diverse population of Internet users.

To combat the AIDS pandemic, it is necessary to deliver information that is timely, credible, and multisectoral. It has to reach not just clinicians and scientists, but also behavioral specialists, policy makers, donors, activists, and industry leaders. It must also be accessible to affected individuals and communities (Garbus, Peiperl, & Chatani, 2002). Accessibility for affected individuals and communities would necessitate targeted, culturally salient, and unbiased information. This is a huge challenge. For instance, health providers' insensitivity and biases toward women have been documented in the critical investigation of TV programs (Myrick, 1999; Raheim, 1996) and printed materials (Charlesworth, 2003). There is a lack of empirical evidence to demonstrate the extent to which and the conditions by which these biases are reproduced on the Internet. In what follows, we provide a conceptual framework for uncovering implicit gender biases in HIV and AIDS information. This framework is informed by the role of power in shaping the social construction of gender and sexuality. We conclude by describing how the framework can be applied in the analysis of online HIV and AIDS information resources.

BACKGROUND

Gupta (2000) has explored the determining role of power in gender and sexuality. Gender, according to Gupta, concerns expectations and norms of appropriate male and female behaviors, characteristics, and roles shared within a society. It is a social and cultural construct that differentiates women from men and defines the ways they interact with each other. Distinct from gender yet intimately linked to it, sexuality is the social construction of a biological drive, including whom to have sex with, in what ways, why, under what circumstances, and with what outcomes. Sexuality is influenced by rules, both explicit and implicit, imposed by the social definition of gender, age, economic status, ethnicity, and so forth (Dixon Mueller, 1993; Zeidenstein & Moore, 1996).

What is fundamental to both sexuality and gender is power. The unequal power balance in gender

relations that favors men translates into an unequal power balance in heterosexual interactions. Male pleasure supersedes female pleasure, and men have greater control than women over when, where, and how sex takes place (Gavey, McPhillips, & Doherty, 2001). Therefore, gender and sexuality must be understood as constructed by a complex interplay of social, cultural, and economic forces that determine the distribution of power. As far as HIV and AIDS, "the imbalance in power between women and men in gender and sexual relations curtails women's sexual autonomy and expands male sexual freedom, thereby increasing women's and men's risk and vulnerability to HIV" (Gupta, 2000, p. 2; Heise & Elias, 1995; Weiss & Gupta, 1998).

Based on this feminist approach to theorizing gender and sexuality, Gupta (2000) categorized HIV and AIDS programs in terms of the degree to which historical power dynamics in gender and sexuality were maintained. The categories summarized in Table 1 are depicted in Figure 1 ranging from the most damaging to the most beneficial ones.

In the theory of social construction, HIV and AIDS are represented as a set of social, economic, and political discourses that are transmitted by media (Cullen, 1998). In symbolic interactionism's theory of gender, mediated messages in advertising, TV, movies, and books tell quite directly how gender is enacted (Ritzer, 1996). As the latest platform for computer-mediated communication, the Internet may also adhere to these gendered representations. We theorize that online HIV and AIDS information follows a similar pattern of power reconstruction, and that these categories could be applied to empirically determine how and why online HIV and AIDS information reproduces these power relations.

FUTURE TRENDS

This theoretical framework could be employed in empirical studies that deconstruct online materials to demonstrate how HIV and AIDS gain their social meanings at the intersection of discourses about gender and sexuality. Prior studies in this area have focused on the cultural analyses of AIDS (Cheng, 2005; Sontag, 1990; Treichler, 1999; Waldby, 1996) rather than structural determinants of risk such as political policy, globalization, industrialization, and

Table 1. Categories of HIV and AIDS programs based on gender and sexuality

Category	Description
Stereotypical	The damaging stereotypes of men are reinforced with terms like "predator, violent, irresponsible," and the role of women is given as "powerless victims" or "repositories of infection" (Gupta, 2000, p. 5).
Neutral	The target is the general population instead of either gender or sex. Despite there being no harm done and the data being better than nothing, the different needs of women and men are ignored. Very often the basis is research that only has been tested on men or works better for men.
Sensitive	The different needs and constraints of individuals based on their gender and sexuality are recognized and responded to. One example is in the provision of female condoms. Thus, women's access to protection, treatment, or care can be improved, but little is done to change the old paradigm of imbalanced gender power.
Transformational	The aim is to transform gender relations to make them equitable. The major focus is on the redefinition of gender roles at the personal, relationship, community, and societal levels.
Empowering	The central idea is to "seek to empower women or free women and men from the impact of destructive gender and sexual norms" (Gupta, 2000, p. 6). Women are encouraged to take necessary actions at personal as well as community levels to participate in decision making. One misunderstanding that needs to be corrected is that empowering women is not equal to disempowering men. The fact is more power to women would eventually lead to more power to both since empowering women improves households, communities, and entire nations.

Figure 1. Continuum of the social construction of gender and sexuality

Stereotypical   Neutral   Sensitive   Transformational   Empowering

the economy. Cultural analysis is based upon the belief that this disease operates as an epidemic of signification based on largely predetermined sexual and gendered conventions. The female has now become socially constructed as a body under siege in AIDS discourse. This gendered body is not, however, a stable signifier. Previously, the body was constructed as white, gay, and male. Now the global discourses on HIV and AIDS have constructed the body as third world, heterosexual, and female. Thus, we see a feminization of HIV and AIDS.

Analysis of the social construction of AIDS using this framework could occur at different levels of analysis and with various populations. We conclude with a few examples.

- Garbus et al. (2002) provide a categorization of HIV and AIDS Web sites that could be used for a cross-category or within-category analysis of the representation of gender and sexuality.
- Cultural ideologies around condom use for AIDS prevention and reproductive health could be studied.

- Given the wide disparities in HIV and AIDS infections among women in the United States, research is needed to examine the discursive practices surrounding HIV and AIDS, socio-economic status, geographic region, and ethnicity or race.
- The absence of lesbians in the HIV-AIDS and women discourse can be analyzed.
- The social construction of the female body in the HIV and AIDS discourse can be studied.
- Discursive practices surrounding HIV and AIDS, gender, and development in developing countries are a potential research subject.
- The tension in the social construction of women as both vulnerable receivers and immoral transmitters of this deadly disease can be deconstructed.

CONCLUSION

HIV and AIDS are a complex and pressing issue. It is not just an issue of health, but has also been framed as an issue of personal responsibility, economics, development, and gender equity. It impacts every nation and individual across the globe. In this article, we argue that the increasing epidemic of HIV and AIDS among women is also an issue of information. We propose a framework for unpacking discursive practices that construct women as the new face of HIV and AIDS. We also provide examples of problem domains in which the feminist analysis informed by this framework can be conducted.

REFERENCE

Anderson, R. N., & Smith, B. L. (2003). *Deaths: Leading causes for 2001. National Vital Statistics Reports, 52*(9), 32-33, 53-54. Retrieved October, 2005, from http://www.cdc.gov/nchs/data/nvsr/nvsr52_09.pdf

Bull, S. S., McFarlane, M., & King, D. (2001). Barriers to STD/HIV prevention on the Internet health education research. *ProQuest Psychology Journals, 16*(6), 661-670.

Centers for Disease Control and Prevention. (2003). Cases of HIV infection and AIDS in the United States. *HIV/AIDS Surveillance Report, 15,* 1-46. Retrieved February 10, 2005, from http://www.cdc.gov/hiv/stats/hasrlink.htm

Charlesworth, D. (2003). Transmitters, caregivers, and flowerpots: Rhetorical constructions of women's early identities in the AIDS pandemic. *Women's Studies in Communication, 26*(1), 60-87.

Cheng, S. (2005). Popularising purity: Gender, sexuality and nationalism in HIV/AIDS prevention for South Korean youths. *Asia Pacific Viewpoint, 46*(1), 7-20.

Cullen, J. (1998). The needle and the damage done: Research, action research, and the organizational and social construction of health in the "information society." *Human Relations, 51*(12), 1543-1564.

Dixon Mueller, R. (1993). The sexuality connection in reproductive health. *Family Planning, 24*(5), 269-282.

Fox, S., & Fallows, D. (2003). Internet health resources: Health searches and email have become more commonplace, but there is room for improvement in searches and overall Internet access. *Pew Internet & American Life Project: Online Report.* Retrieved July 18, 2003, from http://www.pewinternet.org/reports/toc.asp?Report=95

Fox, S., & Rainie, L. (2000). The online health care revolution: How the Web helps Americans take better care of themselves. *Pew Internet & American Life Project: Online Report.* Retrieved November 28, 2000, from http://www.pewinternet.org/reports/pdfs/PIPHealthReport.pdf

Fox, S., & Rainie, L. (2002). Vital decisions: How Internet users decide what information to trust when they or their loved ones are sick. *Pew Internet and American Life: Online Report.* Retrieved April 2005 from http://www.pewinternet.org/PPF/r/59/report_display.asp

Garbus, L., Peiperl, L., & Chatani, M. (2002). AIDS in the digital age. *British Medical Journal, 324*(7331), 246.

Gavey, N., McPhillips, K., & Doherty, M. (2001). "If it's not on, it's not on": Or is it? Discursive constraints on women's condom use. *Gender and Society, 15*(6), 917-934.

Gupta, G. R. (2000, July 12). Gender, sexuality, and HIV/AIDS: The what, the why, and the how (Plenary Address). *Proceedings of the 13th International AIDS Conference,* Durban, South Africa (pp. 1-8).

Gurstein, M. (2003). Effective use: A community informatics strategy beyond the digital divide. *First Monday, 8*(12). Retrieved May 2005 from http://www.firstmonday.org/issues/issue8_12/gurstein/

Hacker, K., & Mason, S. (2003). Ethical gaps in studies of the digital divide. *Ethics and Information and Technology, 5*(2), 99-115.

Heise, L., & Elias, C. (1995). Transforming AIDS prevention to meet women's needs: A focus on developing countries. *Social Science and Medicine, 40*(7), 933-943.

Hern, M. J., Weitkamp, T., Hillard, P. J. A., Trigg, J., & Guard, R. (1998). Promoting women's health via the World Wide Web. *Journal of Obstetric, Gynecologic, and Neonatal Nursing, 27*(6), 606-610.

Internet World Statistics. (2005). *Internet usage statistics for the Americas.* Retrieved May 2005 from http://www.internetworldstats.com/stats2.htm

Kalichman, S. C., Weinhardt, L., Benotsch, E., & Cherry, C. (2002). Closing the digital divide in HIV/AIDS care: Development of a theory-based intervention to increase Internet access. *AIDS Care, 14*(4), 523-537.

Kvasny, L., & Truex, D. (2001). Defining away the digital divide: A content analysis of institutional influences on popular representations of technology. In B. Fitzgerald, N. Russo, & J. DeGross (Eds.), *Realigning research and practice in information systems development: The social and organizational perspective* (pp. 399-415). New York: Kluwer Academic Publishers.

Myrick, R. (1999). Making women visible through health communication: Representations of gender in AIDS PSAs. *Women's Studies in Communication, 22*, 45-65.

Payton, F. C. (2003). Rethinking the digital divide. *Communications of the ACM, 46*(6), 89-91.

Raheim, S. (1996). The reconstruction of AIDS as a women's health issue. In R. L. Parrott & C. M. Condit (Eds.), *Evaluating women's health messages: A resource book* (pp. 402-413). Thousand Oaks, CA: Sage Publications.

Ritzer, G. (1996). *Modern sociological theory* (4th ed.). New York: The McGraw-Hill Companies, Inc.

Sontag, S. (1990). *Illness as metaphor and AIDS and its metaphors*. New York: Farrar Straus and Giroux.

Treichler, P. (1999). *How to have theory in an epidemic: Cultural chronicles and AIDS*. Durham, NC: Duke University Press.

UNAIDS/WHO. (2004). *AIDS epidemic update: 2004*. Retrieved May 2005 from http://www.unaids.org/wad2004/EPI_1204_pdf_en/EpiUpdate04_en.pdf

U.S. Census Bureau. (2000). *Census brief: Women in the United States: A profile*. Retrieved March, 2000, from http://www.census.gov/prod/2000pubs/cenbr001.pdf

Waldby, C. (1996). *AIDS and the body politic*. London: Routledge.

Warschauer, M. (2002). Reconceptualizing the digital divide. *First Monday, 7*(4). Retrieved from http://www.firstmonday.org/issues/issue7_7/warschauer/index.html

Weiss, E., & Gupta, G. R. (1998). *Bridging the gap: Addressing gender and sexuality in HIV prevention*. Washington, DC: International Center for Research on Women.

Weiss, E., Whelan, D., & Gupta, G. R. (2000). Gender, sexuality and HIV: Making a difference in the lives of young women in developing countries. *Sexual and Relationship Therapy, 15*(3), 233-245.

Zeidenstein, S., & Moore, K. (Eds.). (1996). *Learning about sexuality: A practical beginning*. New York: Population Council.

KEY TERMS

Digital Divide: Unequal access to and use of computers and the Internet resulting from such socioeconomic gaps as income, education, race, and age.

E-Health: The applications of the Internet and global networking technologies to medicine and public health.

Empowerment Theory: The study of how perceptions of power affect behaviors and how individuals can increase their power through social interaction.

Feminist Theory: Women-centered theory that treats women as the central subjects, seeks to see the world from the points of women in the social world, and seeks to produce a better world for women.

Gender: Expectations and norms of appropriate male and female behaviors, characteristics, roles, and ways of interaction that are shared within a society.

Sexuality: Social construction of a biological drive, including whom to have sex with, in what ways, why, under what circumstances, and with what outcomes.

Social Construction of Information: Information is examined not as objective missives, but rather as data inextricably intertwined with the social settings in which they are encountered.

The Social Impact of Gender and Games

Russell Stockard, Jr.
California Lutheran University, USA

INTRODUCTION

An increasingly important area of gender and information technology is that of Internet, computer, and video games. Besides women increasingly playing conventional entertainment-oriented or role-playing games, there are a number of pertinent developments in gaming. They are adver-games, casual games, games for change or "serious" games, and games aimed at women and/or developed by women.

Computer and video games are a significant area of interest for a number of reasons. In the United States, games generate substantially more annual revenue than motion picture exhibition, totalling over $11 billion for three consecutive years from 2002 to 2004 (Hollywood Game Daemon, 2004; Traiman, 2005). Research by the Entertainment Software Association indicates that:

half of all Americans play computer and video games, with women making up the second largest (demographic) group of gamers. Games are steadily becoming a dominant way that people spend their leisure time, often stealing time away from traditional media, like television. (Games for change mentioned at NYC Council Hearing, 2005)

In addition, games often reinforce traditional gender roles (Cassells & Jenkins, 2000) and reproduce negative racial and ethnic stereotypes, even as male players comfortably assume female identities (Baker, 2002). As greater numbers of consumers spend time gaming, the advertising industry has taken notice and is following the population into the game world with advertising. The game enthusiasts comprise a desirable target, freely spending on games and other products. Gamers spend an estimated $700 a year per capita on games (Gamers are spending 700 dollars a year, 2005).

CONTROVERSIAL ASPECTS OF ENTERTAINMENT-ORIENTED GAMES

While many games are designed to entertain players involve team and individual sports, car racing, and life simulators such as "Sim City" and "The Sims" do not fundamentally disturb the guardians of public morality. "The Sims" and "Sims 2" permit players to guide simulated beings through their daily lives in cyberspace (Surette, 2005). Another segment that has traditionally attracted the attention of the majority of players is marked by violence and sexual stereotypes. The impact of this category of game has stimulated such controversy that laws have been passed to ban the sale of games depicting violence against law enforcement officials, and guidelines governing game sales to minors have been imposed by retailers (Carlson, 2005; Morris, 2003; Muir, 2004). At least one study indicates that exposure to violent video games even leads to increased short term aggressive behavior in young women (Anderson & Murphy, 2003). Another study, an online survey, tested the hypothesis that an aggressive personality is attracted to aggressive video games, and that women are less likely to play computer games because they are socialized to be less aggressive. Women who used the computer but did not play games and women gamers were subjects of the survey. Women who played computer games perceived their online environments as "less friendly but experienced less sexual harassment online, were more aggressive themselves, and did not differ in gender identity, degree of sex role stereotyping, or acceptance of sexual violence" (Norris, 2004) when compared against the non-gaming women.

Many IT professionals trace their interests in the field to their childhood exposure to games (Kaji, 2002). Gorski (2001) finds this link unfortunate given that most entertainment-oriented games depict women as "damsels in distress or sideshow prostitutes."

More than half of girls and women do not find in games a hospitable environment and, thus, miss an early opportunity to enter the computer science education and career pipeline enjoyed by their male counterparts. A game blogger site reported that an IT manager's game recently released by Intel had to be withdrawn because it did not give players the option to hire women (Water Cooler Games, 2004).

WOMEN'S USE OF CONVENTIONAL COMPUTER AND VIDEO GAMES

In a very real sense, a kind of gender "digital divide" has tended to exist in the gaming environment. Specifically, the rift between boys and girls begins to widen as early as kindergarten (Agosto, 2004). The action-orientation of hard-core gaming has favored boys, male adolescents, and young men. However, this situation has begun to change. Despite the often unwelcoming environment historically posed by the gamescape, many girls and women are increasingly drawn to a number of games. ABC News (2005) recently reported that women enjoy video games for the same reasons men do—for excitement and competition. The report went on to cite video game experts who credited a single game title, "The Sims," for helping to change the gaming industry "virtually overnight." Earning $3 billion last year, half of the game's players are women.

Other gender issues in computer and video games have to do with marketing and content. The Women's Game Conference scheduled for October 2005 includes the site and process of women's purchase of games and the message—"marketing can hurt as well as help." Girls learn early that games are marketed and designed for boys. A number of studies indicate that games are perceived as "boys' toys" and "the disconnect between many computer games available today and girls' game content and design preferences" (Agosto, 2004). Another topic planned for the conference is the representation of women in ads. The literature indicates that the portrayal of characters in games can influence girls' interest in games. Most of the characters are male, and female characters are portrayed negatively. Content issues planned include the importance of art for female players and female entertainment criteria (Women's Game Conference, 2005).

CASUAL GAMES AND GENDER

BBC News recently reported that a research firm found that while hardcore online gaming continues to be dominated by young men, "bored housewives" are stimulating the growth of other game categories available on the Internet (BBC News, 2004). Female players constitute two thirds of the growing market in such skill games as cards, solitaire, and puzzles. This segment is termed casual games.

In contrast to hard-core gamers, casual gamers are classified as those who have played online games within the last three months (Twist, 2005). Even though men spend more time on the Internet each week than women (23.2 vs. 21.6 hours), female gamers over 40 spend the greatest time per week playing online games (9.1 hours or 41% of their online time vs. 6.1 hours—26% of their online time—for men).

In order to serve this growing market, women are being recruited to create games attractive to women. Women gamers are viewed as intelligent players who like a challenge and strategy (ABC News, 2005). At the same time, the future of growth in video gaming depends on the development of easier user interfaces. One of the key players in Canadian digital entertainment is Ana Serrano, director of Habitat New Media Lab, the interactive think tank at the Canadian Film Centre (Seguin, 2005). Among her responsibilities is growing the video game industry.

THE GAMES FOR CHANGE OR SERIOUS GAMES MOVEMENT AND GENDER

The games for change or "serious" game movement consists of video and computer games being used as tools for social change rather than as mere entertainment. The serious games initiative "focuses on uses for games in exploring management and leadership challenges facing the public sector" (Muir, 2005, p. 4). Further, the initiative seeks to link the electronic game industry and projects relating to the use of games in education, training, health, and public policy. The Web site for Social Impact Games carries a quote from Will Wright, Designer of "Sim City" and "The Sims" that notes that "Many game designers today are looking to maximize the social impact of their work." (Social Impact Games, 2005)

The Social Impact of Games site lists a number of game categories. They range from public policy and political games to education and learning games to health and wellness games. One job learning title is set in a simulated high-tech company where students learn job skills as part of a training course. A particularly salient topic is food security. The World Hunger—Food Force game addresses world hunger and targets young people. The United Nations World Food program establishes a presence in the gamescape, where a major crisis has developed in the Indian Ocean. Clearly, this undertaking could attract girls or adolescents who want to use computers and/or IT to solve a simulated real world problem.

At the opposite end of the spectrum another, decidedly more commercial game category includes the adver-game, that is, a type of game with a promotional or advertising purpose. The United States Army has developed such an adver-game to brand the particular service branch and to attract recruits into the service (Oser, 2005). Other marketers have followed the Army's lead. One of them is Daimler-Chrysler with its Race the Pros Games for Dodge and another Coca-Cola Company's NCAA Championship Run.

THE FUTURE OF SERIOUS GAMES AND GENDER

The increase in advertising in video games played online will parallel the rise in adver-games. In 2004, $280 million was spent on advertising in games. By 2008, that amount will increase to $1.05 billion. While the serious games movement will not be completely buoyed by swelling advertising dollars, the rise in such spending may well sustain the trend of more women as players and as designers of computer and video games. Cell phones, MP3 players, PDAs, set-top boxes, children's toys, and even exercise equipment are among the expanding number of platforms

for games (Seguin, 2005). Cell phones are proliferating in the hands of men and women alike, so the movement of games to mobile phones may provide an impetus to further progress in the diversity of the gaming industry. While mobile phone gaming may be plagued in the short run by relatively slower response times (Roman, 2005), the rise of this platform is certainly likely to enable playing of skill games by women casual gamers.

The issue of equity in information technology careers may be addressed as more girls are exposed to a more diversified gaming environment. With casual, serious/games for change, and even adver-games offsetting the traditional hard-core, more girls and women may play games and enter the computer science pipeline. The advantage of serious games is that the sector addresses the oft-expressed interest of girls and young women in using IT not as an end in itself, but as a tool to resolve actual real-world problems. As the slogan of the Water Cooler Games states, it is a "site about video games with an agenda." This phrase should appeal to girls and young women developing their own agendas in education, advocacy, and even advertising.

REFERENCES

ABC News. (2005, March 27). *Girls got (video) game.* Retrieved January 10, 2006, from http://abcnews.go.com/GM.A/Technology/story?id=617384&page=1

Agosto, D. (2004). Girls and gaming: A summary of the research with implications for practice. *Teacher Librarian, 31*, 8-15.

Anderson, A., & Murphy, C. (2003). Violent video games and aggressive behavior in young women. *Aggressive Behavior, 29*, 423-429.

BBC News. (2004). *Net games lure "bored housewives."* Retrieved February 25, 2006, from http://news.bbc.co.uk/go/pr/fr/-/1/hi/technology/4151431.stm

Carlson, S. (2005, May 20). Best Buy posts video game policy. *Knight Ridder Business News*, p. 1.

Cassells, J., & Jenkins, H. (1998). *From Barbie to Mortal Combat: Gender and computer games.* Cambridge, MA: MIT Press.

Gamers are spending 700 dollars a year on games. (2005, March 28). *Businesswire.*

Games for Change mentioned in NYC Council Hearing. (n.d.). Retrieved January 10, 2006, from http://www.seriousgames.org/gamesforchange/archives/000046.html

Gorski, P. (2001). *Understanding the digital divide from a multicultural education framework.* Retrieved January 10, 2006, from http://www.edchange.org/multicultural/net/digdiv.html

Hollywood Game Daemon. (2004, May 16). *Hollywood interest in video games grows.* Retrieved January 10, 2006, from http://m3.typepad.com/hla/game_daemon/

Kaji, J. (2002). *Dream job: Computer game guru.* Retrieved January 10, 2006, from http://www.salary.com/salary/layoutscripts/sall_print_article.asp

Morris, C. (2003). *Wash. bans violent video games.* Retrieved January 10, 2006, http://money.cnn.com/2003/05/20/technology/gaminglaw/

Muir, H. (2004). The violent games people play. *New Scientist, 184,* 26.

Norris, K. (2004). Gender stereotypes, aggression, and computer games: An online survey of women. *Cyber Psychology & Behavior, 7,* 714-727.

Oser, K. (2005). Big guns fall in line behind Army's adver-game. *Advertising Age, 76,* 83.

Roman, D. (2005, February 16). Cell phone games still play for time. *Wall Street Journal,* p. 4E.

Seguin, D. (2005, April 11-24). Search and employ. *Canadian Business, 78,* 67-68.

Social Impact Games. World hunger—food force. (n.d.). Retrieved January 10, 2006 from http://www.socialimpactgames.com

Surette, T. (2005, July 22). Sims 2 content "worse than hot coffee." *The Game Spot.* Retrieved January 10, 2006, from http://www.gamespot.com/news/2005/07/22/news_6129609.html

Traiman, S. (2005, February 12). Retailers chase gaming dollars. *Billboard, 117,* 43-44.

Twist, J. (2005). Casual gaming to take of in 2005. *BBC News.* Retrieved February 25, 2006, from http://news.bbc.co.uk/go/pr/fr/-/1/hi/technology/4151431.stm

Water Cooler Games. (2004). Retrieved January 10, 2006, from http://www.watercoolergames.org

Women's Game Conference. (2005). Retrieved January 10, 2006, from http://womensgameconference.com/program.html

KEY TERMS

Adver-Games: Games designed and published for specific advertisers, including the U.S. Army and Coca-Cola.

Branding: The application of a name, term, symbol, design, or a combination of all of these to differentiating a firm's product from another.

Casual Games: Online skill games such as solitaire played by gamers within the last three months.

Digital Divide: The condition of unequal access to computer-related resources, varying along the demographic dimensions of age, gender, race and ethnicity, education, income, and nationality.

Diversity: In a social and cultural context, the presence in a population of a wide variety of cultures, opinions, ethnic groups, socio-economic backgrounds, disabilities, and sexual preferences.

Equity: The process of achieving fairness among women and men while the term "gender equality" speaks to equality in status and equal enjoyment of fundamental human rights.

Gender: The sex of a person or organism, or of a whole category of people or organisms.

Serious Games: Computer and/or video games with a primary purpose other than entertainment.

Strategies of ICT Use for Women's Economic Empowerment[1]

Sonia N. Jorge
Consultant in Communications Policy and Regulation, Gender, and Development, USA

INTRODUCTION

Information and communication technologies (ICT) provide a great development opportunity by contributing to information dissemination, providing an array of communication capabilities, and increasing access to technology and knowledge, among others. Access to and the cost of ICT continue to be a major development obstacle, particularly in the developing world. Despite the growth in mobile telephony, peri-urban[2] and rural areas—home to a great majority of women and poor populations—continue to lack infrastructure and ICT services in general. For ICT to become meaningful development tools, ICT policy and programs must address the needs of women and the poor in general. This article discusses the main challenges and obstacles faced by women, suggests practical strategies to address those challenges and provides recommendations on how to proceed to improve the conditions leading to women's economic empowerment.

CHALLENGES OF ICT USE FOR WOMEN'S ECONOMIC EMPOWERMENT

Just as in many other areas of development (e.g., agriculture, health, and education), women face enormous challenges to use ICT for their own economic empowerment. Using and benefiting from ICT requires learning, training, affordable access to the technology, information relevant to the user and a great amount of support (to create enabling environments). The challenges are many and they fall in a few categories. The following is a discussion of some of these challenges and how they hinder ICT use for women's economic empowerment.

Affordable Access and Availability of Infrastructure

Access to affordable services and availability of infrastructure is, without a doubt, a major requirement if ICT are to be used for women's economic empowerment. While this discussion focuses on access to telecommunications and ICT infrastructure, it is important to note that other infrastructure and service-related factors may influence the use of ICT, such as availability of electricity, transportation means and security, among others.

Access to Telecommunications Infrastructure

Telecommunications infrastructure is limited in most developing countries and costs are exceedingly high. Whatever little infrastructure is available is concentrated in the larger urban areas, and services provided are only affordable to a few. Bandwidth costs as well as transmission costs incurred by Internet Service Providers (ISP) are high and passed on to users. In rural areas, where women make up the majority of the population, infrastructure is almost non-existent, and services are generally too expensive to poor populations.

The rapid expansion of wireless technologies as well as decreasing costs provide great opportunities in rural areas and areas with little or no infrastructure. For example, in the Dominican Republic, fixed wireless public telephones (using satellite technology) were installed in rural areas without service. These telephones, operated using phone cards, not only provided the community with greatly needed telephone service, but also provided rural women with an additional source of income, as many sell the phone cards that they buy in bulk from the telephone company.

Figure 1.

> **Grameen Phones, Bangladesh**
> Evidence shows that when provided with access and business opportunity, women have become owners and frequent users of cellular telephone services. The Grameen Phones program in Bangladesh, which sets Village Phone Operators to resell wireless telephone services, is an example of an initiative that has contributed greatly to women's economic empowerment and has increased cellular telephone use by women tremendously. In fact, "where women are operators, 82% of the users were women; with men operators, women comprised only 6.3% of Grameen phone users." (Hafkin & Taggard, 2001).

Access to ICT

Access to ICT is highly dependent on telecommunications infrastructure, particularly if one is focusing on telephone service, faxes, e-mail and the Internet. However, the use of ICT is not only based on these services. Radio, for example, provides a great source of information dissemination in many areas of the world, and so does television. Where available, computers may be used as a source of information and a tool for training without the use of telecommunications. The use of CD-ROMs, such as in the case of the IDRC-IWTC CD-ROM for illiterate women in Uganda, *Rural Women in Africa: Ideas for Earning Money* (both in English and Luganda), illustrate that ICT can be used in creative ways and in ways that are more effective and affordable than other solutions (such as browsing or obtaining the information via the Internet) (Mijunbi, 2002).

Radio and television, as the widest form of communication, provide one-way solutions for information dissemination. Women's radio clubs are increasing in Africa, Latin America and the Caribbean, and provide a means to share information on development issues. Recent projects show that radio can be used well beyond a listening-only device, and effectively become a successful two-way communication tool.

In Zimbabwe, some 52 women's radio listening clubs are active in the Development Through Radio (DTR) project, aiming at giving rural women access to radio through participation in production of programmes based on their development needs and priorities. Information exchange is a significant part of the programmes.

Women pose questions and an information intermediary puts the question to a concerned official. The response becomes part of the weekly broadcast. (Hafkin & Odame, 2002, p. 28)

Access to ICT is crucial if they are to be used as a means for women's economic empowerment, and no community should be left short-changed simply because of a few alternatives. The challenge here is that we need to work towards universal access to ICT while actively devising creative solutions to provide alternative access to information to those who need it the most and can immediately benefit from the exchange of information, increased knowledge and diminished isolation.

Cost of Access and Lack of Affordable Solutions

Even when infrastructure is available, affordable access is a concern in most developing countries. Pricing regulation within a privatized and competitive environment (as is the case for most telecommunications operators) has not proven successful in ensuring universal and affordable access. Universal-access policies aim at developing solutions that provide community access at affordable prices, including expansion of public telephones and ICT access points (e.g., in post offices, community centers).

Lack of Gender Awareness in Telecommunications and ICT Policy

Telecommunications and ICT policy lack a gender focus in most countries of the world. With a few exceptions (e.g., South Africa, Korea), there is no emphasis on gender-specific projects or any attempt at ensuring that policies reflect gender equality goals. There are no positive discrimination efforts in place to improve women's access to ICT, increase women's participation in decision-making or project-management positions. Policies that increase women's participation in decision-making and policy-making positions or that, for example, ensure a proportion of funds to be allocated to women's ventures, women's organizations or organizations with a strong gender focus, can contribute to increase use of ICT by women and, consequently, contribute to economic empowerment (Jorge, 2001).

Social, Cultural, and Economic Factors

This discussion focuses on some of the most important social, cultural and economic factors that challenge the use of ICT for women's economic empowerment.

Language and Content Limitations

Lack of local and community-related content as well as content in local languages continue to be a major barrier in women's use of ICT for economic empowerment. Multimedia tools are essential, as they can be developed to provide information both in spoken and written languages. The challenge is to develop content that is relevant and useful to communities in their own language.

Education and Skills

With a great percentage of illiterate women and many speaking local and regional languages, ICT face tremendous challenge to be effectively used by these communities. Again, recent experiences show that it is possible to address these issues (such as the CD-ROM projects in Uganda), and ICT advocates and policy-makers should focus on developing programs that address the development needs and demands of these communities in ways that they can benefit from it. Particularly, it is important to involve community women in the process of deciding what kind of projects will be most useful.

ICT require that users have some skills, and no one should assume that by providing the facilities, everyone in the community will immediately embrace the technology. There are two important aspects here. First, as Eva Rathgeber clearly stated, "The key issue is that the technologies should be adapted to suit women rather than that women should be asked to adapt to suit the technology" (Marcelle, 2002). And second, ICT training is critical if women are to use the technology of choice. Gaining the required skills not only allows women to feel comfortable as ICT users but further empowers women to use ICT in many other ways, such as, for example, increasing their employment choices and their contri-

bution to community development. In Ecuador, where BarrioNet and Chasquinet established telecenters in peri-urban areas of Quito, women are using ICT and the telecenter facilities to assist them in community-organizing efforts, effectively communicating with government officials on issues to improve their community's environment and improving their small business activities.[3]

Addressing Women in the Informal Sector

Women in developing countries are primarily active in the informal economy as handicraft makers, street vendors and so forth. Most of these women are poor and could potentially benefit from ICT in a number of ways. The challenge will continue to be one of reaching women in this sector and consequently providing them with ICT tools that they feel can make a difference in their income generation potential. For example, there are a few organizations (e.g., Development Workshop, ADRA and OMA) working with women in the informal markets in Luanda, Angola; however, with women's lack of time and daily pressure to make ends meet, it is difficult to bring them to resource centers and organize training sessions of any kind. The well-known Self Employed Women's Association (SEWA) in India has done extensive work to assist women in informal markets and has recently established an ICT program aiming to increase efficiency of rural micro-enterprise activities.

PRACTICAL STRATEGIES OF ICT USE FOR WOMEN'S ECONOMIC EMPOWERMENT

Understanding the challenges allows us to better address the problems at stake and devise strategies to consider the complex dimensions of women's lives. Extremely interesting work in this area has been done recently by Nancy Hafkin and Nancy Taggard, who documented the many opportunities for women's economic empowerment through IT use (Hafkin & Taggard, 2001). This discussion identifies some practical strategies that can facilitate women's use of ICT in ways that truly empower women and contribute to their development.

- **Providing Community Access to ICT:** Community access to ICT addresses two of the greatest challenges in ICT use: lack of access and cost of access. Community access can be provided in numerous ways, such as with phone shops, telecenters (with different models for different settings), public phones and libraries, among others, and in strategic locations (e.g., near or at the informal market area, adjacent to health clinics or support organizations, at women's organizations, etc.). Community access can be affordable and based on dependable technology solutions (e.g., wireless and fixed wireless and satellite connections) that can rapidly be installed and effectively utilized. Policies may include, among other things, community access tariffs, subsidized tariffs for areas with extremely low incomes and special incentives for companies that invest in rural areas.[4]

  It is important to view ICT as a tool to meet women's development needs and priorities, and as such, all forms of ICT should be considered to determine which is more appropriate in a particular setting and for the particular program. Despite the fact that the Internet may provide more comprehensive information on a particular topic, it may very well be that a radio program or video produced in the local language will be more effective in the short run in disseminating requested information for women in a rural area. These types of solutions may be accompanied by discussion groups, where women can exchange ideas and share concerns. There is a responsibility to make technology work for the people and, in many cases, that requires a gradual transformation in the use of ICT themselves. For example, women in the informal sector may decide that cellular telephones are all that they need to improve their businesses, but may become more interested in the use of the Internet for business purposes once their businesses grow and as they feel more comfortable using technology.

- **Be Familiar with and Take Advantage of Telecommunications Development Funds (TDF) and Other Universal Access Policies:** There is a great disconnect between universal access policies and resources and the many ICT projects being implemented throughout the developing world. TDF are established and administered by telecommunications regulators to finance the expansion of universal access to ICT in underserved and rural areas. Funds are distributed based on the quality and cost of the proposed projects. Most TDF are established to finance ICT access projects, including telecenters, phone shops, public telephones and libraries. TDF have been successfully developed and implemented in many Latin American countries (e.g., Peru and Chile) and several countries in Africa and Asia are currently working towards developing their own TDF (e.g., Zambia, Uganda, Nigeria, Sri Lanka).

- **Advocate for and Develop Government-Funded Training Programs:** As in all areas of technology use and innovation, training is a crucial factor to ensure that a country is prepared to utilize the technology and increase productivity. Training programs should be offered free of charge or, in fact, be considered a "job," in that participants are paid a certain salary as an incentive to participate and increase their education and qualification level.

- **Develop Special-Interest Content in Local Languages:** Content in local languages is extremely important if ICT are to make a difference in women's lives. In fact, development of content that address priority issues of a particular community may be a major incentive for the use of ICT and to increase community interest in learning more about ICT use for economic empowerment. It is, therefore, extremely important to develop content that addresses local/regional/national needs, to provide information relevant to local/regional/national issues and consequently disseminate the information in appropriate languages. These efforts should be coordinated and supported by those involved in each particular area of interest. For instance, the government and ministry of health of a specific country may have already developed guidelines regarding disease prevention for each region of the country, but lack the capability to properly disseminate the information, particularly in remote rural areas. ICT access points can become major venues for dissemination of public health information and, as a result, introduce women and men to the use of ICT. Some organizations working on health issues may also qualify as hosts for ICT access points and, as a result, become effective users of ICT to improve their own work and consequently the health of the population in their region.

• **Gender-Aware Participatory Methods to Assess the ICT Needs and Demands of Women:** There are numerous potential uses and areas where ICT could contribute to poverty reduction and improved economic opportunities. However, it is important to realize that one woman's use of ICT may be completely irrelevant for another woman in a different setting. Again, the point is to ensure that ICT can meet women's needs rather then having women adjust to ICT (Marcelle, 2002).

Gender and ICT advocates and practitioners must engage in gender-aware participatory methods to assess the needs of women and develop a clear understanding of how ICT can best be used as a tool for women's economic empowerment. As a result, we can develop creative solutions that use ICT to provide programs that promote and facilitate the use of ICT. Using the example of women in the informal sector, it is important to allow women to choose the technology they feel most comfortable with, such as a cellular telephone to call for market prices, even if it may not be the most efficient solution (when the local NGO may provide daily up-to-date price information at no charge).

WHERE DO WE GO FROM HERE?

With access and cost being some of the greatest barriers for ICT use, it is of the utmost importance to engage women and gender advocates in the policy-making process and dialog. Advocates must make an effort to familiarize themselves with the various aspects of ICT policy and understand the gender dimension of these aspects. It is important to engender ICT policy to ensure that women, particularly rural and poor women, benefit from ICT. And gender and ICT advocates are responsible to inform

the ICT debate on gender issues and to ensure that gender analysis becomes an integral part of the policy process. The work done by gender and ICT advocates throughout the WSIS has been critical and reflects some of these lessons. The same is true with respect to ICT project analysis and design. If we want to address gender with ICT projects, gender must be considered from the start of project design (Hafkin & Jorge, 2002). Only then can ICT policy and projects properly address the gender digital divide and further contribute to women's economic empowerment.

It is essential to engage the ITU and other U.N. agencies and programs involved in ICT work in more active training of policy makers and ICT advocates on gender analysis. ITU frequently conducts training seminars and workshops for regulators and policy makers of member states. These training activities should incorporate gender considerations and gender analysis in their plans. In addition, each country's ministry of women affairs or equivalent agency should be involved in the process to mainstream gender among government organizations and should develop specific gender training programs to educate policy makers on gender issues and gender analysis.

As an outcome of WSIS and within the context of the MDGs, the U.N. and all its partners must make a special effort to develop and use gender-disaggregated data and indicators at all levels of ICT development (i.e., from national ICT use to ICT program indicators). This will establish a baseline of information that will be essential to monitor and evaluate access to ICT and the impact of ICT use for women's economic empowerment.

Last, but not least, steps must be taken to ensure that there is greater participation and access to the policy process and to information resulting from policy decision. It is frustrating to see that, even where there are policies and programs in place to improve access (e.g., TDF), few women's organizations or organizations working towards gender equality benefit from the programs. There is no reason for these organizations not to receive funds to establish ICT access points or even to implement telecenter-type programs. These sort of initiatives would certainly contribute to women's economic empowerment.

REFERENCES

Hafkin, N., & Taggart, N. (2001). *Gender, information technology and developing countries.* Washington, DC: Academy for Educational Development. Retrieved from http://learnlink.aed.org/Publications/Gender_Book/Home.htm

Hafkin, N., & Odame, H. (2002). *Gender, ICTs and agriculture. A situational analysis for the 5th consultative expert meeting of CTA's ICT observatory meeting on Gender and Agriculture in the Information Society.* Retrived from www.agricta.org/observatory2002/documents.htm

Hafkin, N., & Jorge, S. (2002). Get in and get in early: Ensuring women's access to and participation in ICT projects. *Women in Action, No. 2, ISIS International-Manila,* 11-15.

Marcelle, G. (2002). *Information and communication technologies (ICT) and their impact on and use as an instrument for the advancement and empowerment of women.* Report from the online conference conducted by the UN Division for the Advancement of Women. Retrieved from http://www.un.org/womenwatch/daw/egm/ict2002/online.html

Mijunbi, R. (2002). *ICT as a tool for economic empowerment of women: Experiences from the use of a CD ROM by rural women in Uganda.* Prepared for the UN-DAW expert group meeting on Information and Communications Technologies and their impact on and use as an instrument for the advancement and empowerment of women, Seoul, Korea. Retrieved from http://www.un.org/womenwatch/daw/egm/ict2002/documents.html

WWW RESOURCES AND SITES CONSULTED

- www.barrioNet.org
- www.chasquinet.org
- www.dw.angonet.org
- www.genderit.org
- www.sewa.org
- www.un.org/womenwatch/daw/egm/ict2002/documents.html
- www.whrnet.org/icts/aisgwg_intro.html
- www.worldbank.org/gender/digitaldivide/digitaldivide20.htm

KEY TERMS

Community Access Centers or Telecenter: Generally refers to a community-owned and -operated initiative with the goal of providing affordable access to ICT and ICT-related services relevant to the specific community it serves (including locally relevant content in local language, a focus on the preferred technology, etc.).

Gender: Gender refers to culturally and socially constructed roles and behaviors of women and men, and identifies the social relations between women and men. It is distinct from the biological differences between women and men.

Gender Analysis: It involves the systematic research and analysis of information based on gender differences and relations, in order to address inequalities and work towards gender equality.

Information and Communication Technologies (ICT): New ICT generally refer to telecommunications technologies (i.e., telephone, fixed and mobile; fax; radio; TV; satellite), computer technology (i.e., to process data) and networking technologies (i.e., the Internet, voice-over Internet platforms, etc.).

Universal Access to ICT: The ability of any individual, regardless of location (geographical), income, gender, race, age, language and so forth, to access ICT at a shared community access point at a reasonable distance (which can be defined by walking distance, number of meters, kilometers or miles) from his or her household. The concept of universal access also encompasses the notion of affordable access (i.e., a price community members can pay).

Women's Empowerment: Refers to the ability to make choices and shape one's environment based on those choices (which can be different in different contexts, societies and cultures). Empowered women are able to challenge and define their environments at all levels of their lives, as individuals, as workers, as mothers, as daughters and so forth.

ENDNOTES

[1] Adapted from an earlier paper by the author: Jorge, S. (2002). The economics of ICT: Challenges and practical strategies of ICT use for women's economic empowerment, prepared for the UN-DAW expert group meeting on information and communications technologies and their impact on and use as an instrument for the advancement and empowerment of women, Seoul, Korea. (http://www.un.org/women watch/daw/egm/ict2002/documents.html)

[2] Peri-urban areas are those areas in the periphery of major cities and where poor populations tend to settle when they migrate closer to urban centers. Peri-urban areas often include large shanty-towns or illegal settlements with little or no infrastructure and high rates of poverty.

[3] See www.barrioNet.org and www.chasquinet.org

[4] For a detailed discussion of such policies, see Jorge, Sonia, "Gender Sensitive ICT Policy: Rethinking Policy Making," comprehensive training workshop, prepared for the Workshop on Equal Access of Women to ICT, Seoul, Korea, October 2001; Marcelle, Gillian, "Getting Gender into African ICT Policy: A Strategic View," in Eva Rathgeber and Edith Ofwona Adera, *Gender and the Information Revolution in Africa*, IDRC, 2000; and African Information Society-Gender Working Group, "Engendering ICT policy: guidelines for action," Johannesburg, 1999. www.whrnet.org/icts/aisgwg_intro.html

S

Student and Faculty Choices that Widen the Experience Gap

Lecia J. Barker
University of Colorado, USA

Elizabeth R. Jessup
University of Colorado, USA

INTRODUCTION

A major teaching challenge for higher education faculty is students' wide differences with respect to experience or knowledge with the subject matter or skill set of a class. In computing education research, this is often referred to as the "experience gap." Research shows that the experience gap contributes to the low participation of women in professional information technology (IT) careers. Women are significantly more likely to enter college-level IT courses with little or no computer programming experience than are their male peers (College Board, 2004). Yet, programming experience is positively associated with success, especially in introductory classes (Taylor & Mounfield, 1994; Bunderson & Christensen, 1995; Brown, 1997; Margolis & Fisher, 2002), and low grades are positively associated with attrition from the major (Strenta, Rogers, Russell, Matier, & Scott, 1994). When women receive low grades due to inexperience, they may be more likely than males to lose confidence and leave the major (Cohoon & Aspray, in press).

Another type of experience gap becomes evident in cross-disciplinary teams, where students encounter others whose areas of expertise and knowledge are substantially different, often to the point where students have difficulty understanding each other. According to IEEE Computer Society/ACM Computing Curricula Task force, "Computing education is also affected by changes in the cultural and sociological context in which it occurs" (IEEE and ACM Joint Task Force, 2001, p. 10). For this reason, both Computing Curricula 1991 and 2001 strongly recommend the integration of experiences and opportunities for student understanding of real-world applications and the people who need them. Courses that provide opportunities for collaborative and interdisciplinary learning are also often recommended to increase retention of women in science, technology, engineering and mathematics (STEM) courses in general (Agogino & Linn, 1992; Felder, Felder, Mauney, Hamrin, & Dietz, 1995) and in computing, in particular (McDowell, Werner, Bullock, & Fernald, 2003; Barker, Garvin-Doxas, & Roberts, 2005). Yet, collaborative learning and, in particular, project-based courses, must be carefully planned and managed for students to have similar learning outcomes. In this article, we demonstrate how students' choices can reinforce and even widen differences in experience and reduce their ability to develop cross-disciplinary understandings.

BACKGROUND

Unlike most assignments in the computer science curriculum, team projects are too complex to be completed by a single student. Team projects involve building practical solutions to substantial problems, requiring that students evaluate alternative designs in terms of cost, performance and so forth. Team members must determine what they will deliver and how—and how to distribute the work. The process of bringing idea to product is part of what gives students the professional experience.

In their review of projects within the traditional computer science curriculum, Fincher, Petre and Clark (2001) characterize project work as a way for students to "show their stuff." To be successful in project work, they note, students must demonstrate mastery of a diverse collection of technical skills acquired over terms or years of study. The authors believe that the most diverse project teams are formed by splitting up "affinity groups" (friendships or cultural groups) and creating a mix of interpersonal and technical abilities. Such teams may increase the potential for peer learning. Teams formed from computer science majors from a single institution, however, are quite likely limited in their diversity of knowledge and may not bring students professional experience in interacting with people substantially different from themselves.

More innovative IT programs present the possibility of more heterogeneous project teams with students of different majors. Ideally, the tasks performed by such groups require that all students share their knowledge and expertise as well as their questions and uncertainties in ways that lead to peer learning (Tinzmann, Jones, Fennimore, Bakker, Fine, & Pierce, 1990). However, this ideal assumes that students take equal responsibility for the roles of teacher and student, and that tasks focus on learning through dialog and hands-on activities (Johnson & Johnson, 1994). Knowledge asymmetry, when one group member is more expert on a topic than another, is to be encouraged and expected in group projects because it creates an opportunity for peer tutoring, benefiting both the more expert and less expert students. Further, in successful learning groups, students alternate between different types of roles and communication: those involving peer tutoring in which the roles of "teacher" and "student" are clear and well defined, and collaborative sequences where students work together in free discussion to create knowledge and understanding with no clear role differences (Haller, Gallagher, Weldon & Felder, 2000).

Yet, simply putting students in project groups does not automatically lead to improved or cross-disciplinary learning through the processes described above, because students' understanding of collaboration may be quite limited by lack of experience and even a belief that collaboration is cheating (Barker, Garvin-Doxas, & Jackson, 2002). Instead, they often divide the work, taking on the part most consistent with their "comfort zone" or most expedient for finishing the project. When students have different levels or areas of knowledge, students of both sexes can take on gendered roles or roles based on experience. Further, students may accept less learning in the interest of getting a product and "showing their

stuff." The case study we present below demonstrates how students take on roles in project groups.

WIDENING THE EXPERIENCE GAP

This case study describes one project team from Technology for Community (T for C), an undergraduate computer science course taught at the University of Colorado at Boulder. In T for C, student teams work with local community service agencies, building computational solutions to problems confronting those agencies. The course has no prerequisites, and participants have diverse backgrounds in terms of educational experience, major and expertise with technology. Although few computer science majors are female, this course has consistently attracted a large proportion of female students. Most of those women, however, come from the technology, arts and media (TAM) certificate program.

The TAM certificate program, open to all undergraduates, requires that students take six courses, three of which require hands-on development in project teams. Students acquire expertise with high-end software packages (e.g., Flash, Photoshop) and some HTML programming, with the goal of designing and producing multimedia materials both for self-expression and to serve clients. Programming courses (e.g., Java) are optional, and most students do not take them. Three of the courses are focused on historical issues related to information and communication technologies, communication theory, implications of media for society and the like. TAM student enrollment is consistently more than half women.

In each T for C project, students are expected to acquire new skills and experience. While most students improve their abilities to interact with a real client with real needs and to design for and test with real users, not all succeed in enhancing their technical skills. Instead, students, succumbing to real or imagined time constraints, fall into their comfort zones. Because of the multidisciplinary nature of the course, students arrive with different levels of technical, communication and design experience. Learning to work across disciplines within project teams is a new experience for many of the students, and different project teams succeed to different degrees in responding to that diversity. The case study below illustrates how one student team responds to both types of experience gap.

The case is based on data collected in an ethnographic study conducted by the first author. Ethnography is a qualitative research method that focuses on articulating the shared—yet often unspoken—rules, beliefs and values produced and made visible in everyday communicative interaction. The second author developed the T for C course and has taught it for four semesters. The case study is based on 35 hours of observation of T for C, a focus group interview with 10 TAM students (including both of the females presented below) and experiences of the instructor.

On one T for C team, Jane (studying TAM, Spanish and education), Maria (studying TAM, journalism and molecular biology) and John (studying computer science) were tasked with creating an educational Web site for middle school children in a bilingual charter school. They attended a workshop including the school principal, conducted interviews with her and visited the school to understand teachers' and children's needs. Building the site required design and content development and HTML coding. Once the parameters of the project were defined, the students decided which of them would carry out which tasks. They then began their work on those tasks, individually or jointly.

All three students were bright, creative and accomplished in their chosen majors, yet early in the project, the differences in those majors led to a severe breakdown in their interactions. The two women complained to the instructor that John was uncommunicative, not available to meet and a poor contributor. The instructor was surprised: She knew John from two previous courses to be conscientious, friendly and well liked by peers. She nonetheless called John in for a discussion about his work in T for C. As it turned out, John was unaware of the dissatisfaction of his teammates. He was working hard on his parts of the project and checking in with Maria and Jane regularly by e-mail. The conflict turned out to be a problem of culture. John was accustomed to solitary, late-night work in the computer lab while the women were expecting regular, in-person project team meetings. The instructor intervened, making John aware of the need for such meetings while

letting Jane and Maria know that the meetings would have to take place late in the day to accommodate John's work schedule. With those simple ground rules in place, the students quickly found common ground in their project, and they came to see that everyone was making good contributions. For the remainder of the semester, theirs was a very successful project team, and lasting friendships among the three resulted.

While Maria, Jane and John ultimately developed a high-quality site for the school, a closer look at their interactions revealed some troubling features with respect to the experience gap. In particular, the students split up the work along disciplinary (perhaps gendered) lines, with Jane and Maria making most design and content decisions, and John doing the HTML coding. It is easy to see why Jane, as a Spanish major and pre-service teacher, would be especially good at providing content for this project. However, John's and Maria's roles were not quite so straightforward and were negotiated through many interactions. For example, at one point, the three were discussing icons and images. As Maria explained to John at length why she saw an image as particularly good, he repeatedly expressed agreement with her, but she did not acknowledge it. It was as if she could not register that he might possibly know enough about images to agree. On another occasion, the team was having a technical problem saving an image as a GIF file. Jane's opinion overrode John's suggestion for solving the problem on the implied basis that she had a better eye for image quality; his solution, she said, would diminish that quality. Maria's view of John's potential contributions to the project was limited, and limiting to John.

Yet John was clearly conscious of design and thoughtful about content and users' needs. For example, when deciding on the approach to helping children find the help they need with research, the purpose of the site, it was his idea to use questions ("Do you want to find a book?") rather than headings ("Research"), because as he said, "kids don't think in those terms." During whole-class discussion about intuitive tasks, John pointed out that in the Macintosh interface, dragging the floppy disk icon to the trashcan to eject can make people afraid that they're somehow deleting files on the disk. In fact, throughout the project, we documented many of John's insightful comments about the site's appearance and content. But with his ideas and knowledge repeatedly ignored or brushed aside, John learned implicitly that design and content were not to be his domain. Jane and Maria seemed to have staked out content and design as their territory. This move may appear to be gendered, yet it was also made by virtue of their unarticulated beliefs about what kinds contributions a computer science major can make.

During the focus group interview, TAM students discussed the image of a computer science major. Maria said, "I've taken a C++ programming class and there's really no creativity there." It is not clear whether Jane shared this belief (several other students objected to this), but Maria's inability to see computer science as "creative" may have impeded her ability to "hear" John's design ideas. Her initial beliefs and the implicit messages embedded in many interactions were likely a factor in John's relegation to HTML coding—at the expense of his obvious desire to participate in design decisions. Also interesting is the revelation that although Maria adopted the role of content and design, she actually had substantial programming experience. When we asked Maria why she and Jane did mainly content and design and John the coding, she narrowly characterized John's contribution, explaining that he was a fast programmer, that she and Jane had made a similar Web site before and that they were under time pressure (more on this later). In spite of her own ability to function in any of the three roles, Maria was unable to see John, a CS major, as functional in the roles of content and design. In fact, all three of those students could have provided more input or support in all three knowledge domains.

T for C is intended as a long-term projects course where students may enroll for several semesters, continuing work on very large-scale projects. Nonetheless, students usually feel pressured to finish their projects within a one-semester time frame. At the end of the semester, when we asked Maria why John mainly coded while she and Jane did content and design, she cited John's speed and the women's prior experience with such a site, adding, "it's just the way the class was set up." Asked what she meant by the class being "set up" that way, she said, "Because of the time pressure. We had to get it all done." Her perception was at odds with the

professor's beliefs and observation data about the structure of the courses; that is, students were told explicitly that they did not have to finish projects to earn a grade, but that learning was more important.

FUTURE TRENDS

Many of the behaviors described in the case study were also displayed by other T for C project teams. Students chose or were subtly assigned roles based on their perceptions of experience and apparent beliefs about what kinds of people should make what kinds of contributions. The ways students conceptualize teamwork and the potential contributions of team members have a profound effect on their ability to extend their skills and knowledge beyond what they already know. What students already know is influenced by years of gendered choices, with female students being more likely to have design experience and male students more likely to have programming experience.

When students are allowed to select their roles based on expediency or comfort, it works against the benefits of collaborative and cross-disciplinary learning. While this approach may seem practical and efficient, it does not provide any of the students with a new learning experience, but instead practice of existing skills. Thus, those with less experience fall into this trap, missing out on the opportunity to advance their experience and knowledge about software development and, as a result, continue to remain behind. At the same time, students like John, with technical experience, lose the opportunity to work on skills that they do not already possess, such as client relation skills and content and design.

Without overt and explicit measures implemented and enforced by instructors, other pressures, such as perceived time constraints, a tendency to allow group members to focus on what they already know how to do well, and gendered and disciplinary beliefs about what is appropriate for people of different categories to contribute to a project, male and female students alike can miss out on the opportunity to add to their skills and will instead complete their projects using skills they have already mastered. The multi-disciplinary understanding that comes from having worked in another knowledge domain is a desired, but not acquired, learning gain.

CONCLUSION

Both men and women are gendered beings, imposing their beliefs about appropriate behavior for men and women on themselves and on others. The literature on women in IT often portrays women as powerless, passive victims of a male-oriented curriculum within a male-dominated academic culture. This case demonstrates that both men and women can be the agents of oppression, imposing their expectations for behavior on each other and themselves in ways that preclude full participation in a project. Research into gender issues in IT must take into account that both women and men are gendered beings who make gendered choices if we are going to have a better understanding of how to bring to parity the male-female composition of the IT workforce. Instructors must also make pedagogical choices to impose and enforce learning objectives as part of group assignments. If all students are assessed for particular learning outcomes, the experience gap is less likely to widen.

REFERENCES

Agogino, A. M., & Linn, M. (1992). Retaining female engineering students: Will design experiences help? *NSF Directions, 5*(2), 8-9.

Barker, L. J., Garvin-Doxas, K., & Jackson, M. J. (2002). Defensive climate in the computer science classroom. *SIGCSE Technical Bulletin, 34*(1), 94-99.

Barker, L. J., Garvin-Doxas, K., & Roberts, E. (2005, February 23-27). What can computer science learn from a fine arts approach to teaching? In *Proceedings of 35th SIGCSE Technical Symposium on Computer Science Education*, St. Louis, MO.

Brown, J., Andreae, P., Biddle, R., & Tempero, E. (1997, February 27-March 1). Women in introductory computer science: Experience at Victoria University of Wellington. In *Proceedings of the 28th SIGCSE Technical Symposium on Computer Science Education*, San Jose, CA.

Bunderson, E. D., & Christensen, M. E. (1995). An analysis of retention problems for female students in university computer science programs. *Journal of Research in Computing in Education, 28*(1), 1-18.

Cohoon, J. M., & Aspray, W. (in press). A critical review of the research on women's participation in postsecondary computing education. In J. M. Cohoon & W. Aspray (Eds.), *Women and information technology: Research on under-representation.* Cambridge, MA: MIT Press College Board.

College Board. (2004). *AP Program summary report 2004.* Retrieved from http://apcentral. collegeboard.com/repository/programsummary report_39028.pdf

Felder, R. M., Felder, G. N., Mauney, M., Hamrin, Jr., C. E., & Dietz, E. J. (1995). A longitudinal study of engineering student performance and retention. III. Gender differences in student performance and attitudes. *Journal of Engineering Education, 84,* 151-174.

Fincher, S., Petre, M., & Clark, M. (2001). *Computer science project work: Principles and pragmatics.* London: Springer-Verlag.

Haller, C. R. Gallagher, V. J. Weldon, T. L., & Felder, R. M. (2000). Dynamics of peer education in cooperative learning workgroups. *Journal of Engineering Education, 89,* 285-293.

IEEE & ACM Joint Task Force. (2001). *Computing curricula 2001.* Retrieved from http://www. computer.org/education/cc2001/final/index. htm

Johnson, D. W., & Johnson, R. T. (1994). *Learning together and alone: Cooperative, competitive and individualistic learning.* Boston: Allyn & Bacon.

Margolis, J., & Fisher, A. (2002). *Unlocking the clubhouse: Women in computing.* Cambridge, MA: MIT Press.

McDowell, C., Werner, L., Bullock, H., & Fernald, J. (2002, February 25-March 2). The effects of pair-programming on performance in an introductory programming class. In *Proceedings of the 33rd SIGCSE Technical Symposium on Computer Science Education*, Covington, KY.

Strenta, A. C., Rogers, E., Russell, A., Matier, M., & Scott, J. (1994). Choosing and leaving science in highly selective institutions. *Research in Higher Education, 35*(5), 513-547.

Taylor, H. G., & Mounfield, L. C. (1994). Exploration of the relationship between prior computing experience and gender on success in college computer science. *Journal of Educational Computing Research, 11*(4), 291-306.

Tinzmann, M. B., Jones, B. F., Fennimore, T. F., Bakker, J., Fine, C., & Pierce, J. (1990). *What is the collaborative classroom?* North Central Regional Educational Laboratory. Retrieved from http://www.ncrel.org/sdrs/areas/rpl_esys/collab.htm

KEY TERMS

Attrition: When students switch out of a major area of study.

Collaborative Learning: When students work together to learn new material. It contrasts with individual learning, where a student learns without talking to or working with other students.

Cultural and Sociological Context: Specific types of social situations; the ways people behave in social situations are influenced by the cultures, societies, institutions and so forth in which they occur.

Experience Gap: In education, the difference among students in a class in prior knowledge and experience with a subject or activity.

Gender: A set of social categories that shapes how males and females behave and the ways that others treat them based on deeply ingrained expectations about how males and females *should* respond. In contrast, sex describes biological categories.

Positive Association: A situation where when one element increases, another increases. This does not mean one causes the other, just that they co-occur.

Project-Based Courses: Courses in which students, working individually or in groups, undertake an activity that goes on over a period of time. The outcome is a product, presentation or performance.

S

Survey Feedback Interventions in IT Workplaces[1]

Debra A. Major
Old Dominion University, USA

Lisa M. Germano
Old Dominion University, USA

INTRODUCTION

Several factors may explain the underrepresentation of women in IT. One reason is the portrayal of the IT workplace as hostile, or at least inhospitable, to women. Long work hours, a frenetic pace, and few family-friendly benefits are believed to characterize many IT work environments (Howard, 1995; Lambeth, 1996; Panteli, Stack, & Ramsay, 1999). Another reason is the perception that IT careers afford little social interaction or support (Misic & Graf, 1999). The stereotype of the IT worker as a "geek" who works in isolation from others may be less appealing to women than men (Spender, 1997). Moreover, white males are most frequently portrayed as IT professionals in the media, are most likely to have role models and support systems, and work in work environments that reflect their values and learning styles (Balcita, Carver, & Soffa, 2002). Subtle biases in stereotyping and language use and working in a white male culture may contribute to feelings of exclusion for women. Finally, male IT supervisors may be less likely to develop supportive relationships with women than men (Ragins, 2002), thereby reducing their bond to the organization and leading to their eventual dissatisfaction and departure from IT work and the organizations that employ them (Lee, 2004).

Our research examines how characteristics of the IT workplace can foster inclusion and equal opportunity for IT employees (see Major, Davis, Sanchez-Hucles, Germano, & Mann, 2006). We are particularly interested in identifying barriers and enablers to the career success of women and minorities in IT departments. During Phase 1 of this three-year project, IT departments completed a Web-based survey designed to understand the factors that shape the access that IT employees have to opportunities in the workplace. During Phase 2 of the project, we provided the IT departments with feedback from our survey, conducted focus groups and structured supervisor interviews, and worked with the organizations to identify and implement changes designed to increase opportunity and inclusion for IT employees. During Phase 3 of this project, we administered another survey to assess the effectiveness of the interventions implemented during Phase 2. The remainder of this chapter describes our sample, survey measures, and research methodology.

BACKGROUND

During Phase 1 of our project, 916 IT employees from 11 companies completed our Web-based survey. Participating organizations varied in terms of industry, size, and location, in order to more broadly represent the diversity of IT work experiences and workplace climates. See Table 1 for a detailed description of participants.

Survey Measures

The measures used in the Web-based survey are described in Table 2.

Inclusion

The 13-item inclusion scale was created from existing measures and original items (Chrobot-Mason & Aramovich, 2002; Mor-Barak & Cherin, 1998). Inclusion was assessed using three subscales: belonging, participation, and influence. An example of an item from the belonging subscale is, "I am included as part of the team by my coworkers." A sample

Table 1. Demographic characteristics of the total sample, N = 916

Characteristic	N	% of Total Sample
Gender		
▪ Males	530	57.9
▪ Females	344	37.5
▪ Gender not specified	42	4.6
Race		
▪ American Indian or Alaska Native	27	2.9
▪ Asian (non-Indian)	47	5.1
▪ Asian Indian	35	3.8
▪ Black or African American	74	8.1
▪ Hispanic	51	5.6
▪ Native Hawaiian or other Pacific Islander	8	0.9
▪ White	617	67.4
▪ Multiple Race	5	0.5
▪ Race not specified	52	5.7
Relationship Status		
▪ Single	140	15.3
▪ Married	639	69.8
▪ Living with Partner	22	2.4
▪ Separated	11	1.2
▪ Divorced	59	6.4
▪ Widowed	4	0.4
▪ Did not specify	41	4.5
Educational Attainment		
▪ High school graduate	71	7.8
▪ Vocational/technical school graduate	59	6.4
▪ Associate's degree	115	12.6
▪ Bachelor's degree	472	51.5
▪ Master's degree	144	15.7
▪ Doctorate degree	6	0.7
▪ Did not specify	49	5.3
IT Degree		
▪ IT related	405	44.2
▪ Non-IT related	404	44.1
▪ Did not specify	107	11.7
IT Position		
▪ Conceptualizer	288	31.4
▪ Developer	169	18.5
▪ Modifier/Extender	80	8.7
▪ Supporter/Tender	302	33.0
▪ Did not specify	77	8.4
Salary ($)		
▪ Less than 30,000	21	2.3
▪ 30,000 - 39,000	46	5.0
▪ 40,000 - 49,000	93	10.2
▪ 50,000 - 59,000	133	14.5
▪ 60,000 - 69,000	119	13.0
▪ 70,000 - 79,000	124	13.5
▪ 80,000 - 89,000	85	9.3
▪ 90,000 - 99,000	86	9.4
▪ 100,000 or more	140	15.3
▪ Did not specify	69	7.5
Characteristic	Mean	Standard Deviation
Number of Children	1.11	1.13
Age of Youngest Child	11.08	6.29
Age of Participant	41.98	8.90
Years Worked at Current Organization	10.44	8.51
Years Worked in IT	14.44	8.67
Hours Worked per Week	46.69	7.96

Table 2. Description of survey measures

Measure	Source	# Items	Coefficient Alpha
Inclusion:			
▪ Belonging	Chrobot-Mason & Aramovich (2002)*	5	.94
▪ Participation	Mor-Barak & Cherin (1998) *	4	.93
▪ Influence	Mor-Barak & Cherin (1998) *	4	.90
Climate for Opportunity	Mor-Barak, Cherin, & Berkman (1998)	6	.88
Workplace Relationships:			
▪ Affective Coworker Support	Ducharme & Martin (2000)	5	.92
▪ Instrumental Coworker Support	Ducharme & Martin (2000)	5	.92
▪ Leader-Member Exchange	Graen, Novak, & Sommerkamp (1982)	7	.92
▪ Satisfaction with Mentoring	Ragins & Cotton (1999)	4	.84
Satisfaction:	Hackman & Oldham (1975)*		
▪ Overall Job Satisfaction		16	.91
▪ Satisfaction with Supervision		3	.92
▪ Satisfaction with Job Security		2	.88
▪ Satisfaction with Pay		3	.84
▪ Satisfaction with Social Environment		3	.74
▪ Satisfaction with Growth Opportunities		5	.85
Organizational Commitment	Mowday, Steers, & Porter (1979)*	9	.91
Career Commitment	Blau (1985)*	8	.86

*Note: *Indicates measure was adapted or supplemented with original items.*

item from the participation subscale is, "My judgment is respected by members of my workgroup." An example of an item from the influence subscale is, "I am able to influence decisions that affect my job." Participants used a five-point Likert-type scale anchored by 1 (*strongly disagree*) and 5 (*strongly agree*) to respond to all inclusion items. Coefficient alpha for the entire scale for the present sample is .94.

Climate for Opportunity

Climate for opportunity was assessed using six items developed by Mor-Barak, Cherin, and Berkman (1998) to tap organizational fairness. An example item is, "Managers here give feedback and evaluate employees fairly, regardless of the employee's ethnicity, gender, age, or social background." Participants responded to the items using a six-point Likert-type scale anchored by 1 (*strongly disagree*) and 6 (*strongly agree*). For the present sample, climate for opportunity yielded a coefficient alpha of .88.

Coworker Support

Coworker support was measured using a 10-item scale developed by Ducharme and Martin (2000). The measure assesses affective and instrumental support. Affective coworker support is a form of social support that coworkers offer by being sympathetic, listening to problems, and expressing care and concern. An example of an item tapping affective coworker support is, "Your coworkers take a personal interest in you." In the present sample, coefficient alpha for the affective coworker support items is .92. Instrumental coworker support is tangible helping behavior offered by coworkers in response to specific needs (e.g., assistance with work responsibilities and switching schedules). An example of an item that assesses instrumental coworker support is, "Your coworkers would fill in when you are absent." In the present sample, coefficient alpha for the instrumental coworker support items is .92. Participants responded to the coworker support items using a five-point Likert-type scale anchored by 1 (*strongly disagree*) and 5 (*strongly agree*).

Leader-Member Exchange

The quality of the relationships between IT employees and their supervisors was assessed using Graen, Novak, and Sommerkamp's (1982) seven-item measure of Leader-Member Exchange, the LMX 7. An example item is, "Regardless of how much formal authority he or she has built into his or her position, what are the chances that your leader would use his or her power to help you solve problems in your work?" The response scales for each item vary, but all are five-point Likert-type scales on which higher numbers indicate greater LMX. For the present sample, the LMX 7 yielded a coefficient alpha of .92.

Mentoring

First, IT employees were asked to indicate whether or not they currently had at least one mentor. For participants who indicated that they did have at least one mentor, they were asked to complete a four-item scale developed by Ragins and Cotton (1999) that assessed employees' satisfaction with their mentor(s). Participants responded to the satisfaction with mentoring questions using a five-point Likert-type scale anchored by 1 (*strongly disagree*) and 5 (*strongly agree*). An example of an item assessing satisfaction with mentoring is, "My mentor is someone I am satisfied with." The satisfaction with mentoring scale yielded a coefficient alpha of .84 for the present sample.

Job Satisfaction

Overall job satisfaction and multiple facets of job satisfaction were measured using 15 items adapted from Hackman and Oldham's (1975) Job Diagnostic Survey. The measure includes subscales that assess five facets of job satisfaction, including satisfaction with supervision, satisfaction with job security, satisfaction with pay, satisfaction with social environment, and satisfaction with growth opportunities. Participants responded to the job satisfaction items using a seven-point Likert-type scale anchored by 1 (*extremely dissatisfied*) and 7 (*extremely satis-*

fied). For the current sample, coefficient alpha for overall job satisfaction is .91. Coefficient alpha for the five subscales ranged from .74-.92 (see Table 2). A sample item for satisfaction with supervision is, "The amount of support and guidance I receive from my supervisor." An example of an item assessing satisfaction with job security is, "The amount of job security I have." The following sample item assesses satisfaction with pay, "The amount of pay I receive." A sample item for satisfaction with social environment is, "The chance to get to know other people while on the job." Lastly, an example of an item tapping satisfaction with growth opportunities is, "The amount of personal growth and development I get in doing my job."

Organizational Commitment

Mowday, Steers, and Porter's (1979) nine-item measure of organizational commitment was used to assess how loyal and attached a participant is to his or her employing organization. Participants used a seven-point Likert-type scale anchored by 1 (*strongly disagree*) and 7 (*strongly agree*) to respond to the nine items. A sample item is, "I am willing to put in a great deal of effort beyond that normally expected in order to help this organization be successful." This measure yielded a coefficient alpha of .91 for the current sample.

Career Commitment

An eight-item measure developed by Blau (1985) was adapted to assess career commitment. Career commitment describes one's attitudes towards one's profession or vocation. An example of an item is, "This is the ideal profession for my life's work." Participants were asked to respond to the career commitment items using a five-point Likert-type scale anchored by 1 (*strongly disagree*) and 5 (*strongly agree*). For the current sample, the career commitment measure yielded a coefficient alpha of .86.

Gender Differences

Phase 1 means for participants' age, number of children, years worked in IT, organizational tenure, and hours worked per week are broken down by

Table 3. Mean differences for men and women

	Men			Women				
Factor	N	M	SD	N	M	SD	df	t
Age	521	42.07	9.02	329	41.84	8.72	848	0.36
Number of Children	518	1.19	1.20	334	0.99	1.00	797.5	2.66*
Years Worked in IT	524	15.55	8.94	338	12.75	7.99	860	4.67*
Organizational Tenure	525	9.79	8.59	338	11.46	8.31	861	-2.81*
Hours Worked Per Week	525	47.52	7.88	338	45.42	7.95	861	3.81*

*Note. *p < .01.*

Table 4. Gender comparisons for race, relationship status, and salary

Characteristic	Percentage Men	Percentage Women	Gender Difference (Yes/No)
Race			No
▪ American Indian or Alaska Native	3.0	3.2	
▪ Asian (non-Indian)	5.1	5.8	
▪ Asian Indian	3.0	2.0	
▪ Black or African American	7.2	10.5	
▪ Hispanic	6.4	4.9	
▪ Native Hawaiian or other Pacific Islander	1.1	0.6	
▪ White	70.2	70.6	
▪ Multiple Race	0.4	0.9	
▪ Race not specified	1.3	1.5	
Relationship Status			Yes
▪ Married	79.7	68.9	
▪ Single	20.4	30.8	
▪ Did not specify	0.0	0.3	
Salary ($)			Yes
▪ Less than 30,000	1.1	4.4	
▪ 30,000 - 39,000	5.5	4.9	
▪ 40,000 - 49,000	10.4	11.0	
▪ 50,000 - 59,000	12.3	19.8	
▪ 60,000 - 69,000	13.2	14.0	
▪ 70,000 - 79,000	14.5	13.7	
▪ 80,000 - 89,000	10.0	9.3	
▪ 90,000 - 99,000	10.4	9.0	
▪ 100,000 or more	20.0	9.6	
▪ Did not specify	2.6	4.4	

Note: N = 530 for men. N = 344 for women. Married described respondents who indicated that they were married or living with a partner at the time of the study. Single described respondents who indicated that they were single, separated, divorced, or widowed at the time of the study.

gender in Table 3. There were significant gender differences on each of these variables, with the exception of age. Compared to women, men reported having more children, working in IT a greater number of years, having shorter organizational tenure, and working more hours per week. Gender comparisons were also made for race, relationship status, and salary, as shown in Table 4. There were statistically significant gender differences for relationship status and salary but not race. Women were disproportionately single, and men were dispropor-tionately married. Generally, men reported higher salaries than women.

FUTURE TRENDS

Upon completion of Phase 1 of the research project, we analyzed the data collected from the Web-based survey and created reports for each organization as part of our survey feedback intervention. Phase 2 of the project began with the distribution of these feedback reports to participating organizations. The primary reports for each organization provided information regarding that company's standing on each of the constructs measured in the survey. In addition, each company was provided with benchmarking comparison data from the other participating companies. Thus, each company was able to see how it scored on each construct relative to the other companies in our project. When an organization's sample was sufficiently diverse, companies also were provided with information on any gender and/or ethnic differences in survey responses. Finally, each company also received work group level survey feedback reports for any group with at least five respondents reporting.

After the reports were disseminated to the leadership at a company, feedback meetings with company executives and feedback presentations to IT employees were given. In addition, we invited IT employees to participate in focus groups where a small group of people convened to discuss concepts and to address issues relating to the results from the survey. The aim of the focus group discussion was to obtain qualitative data to aid in interpretation of the survey results, clarify any ambiguities, and point out any related issues that were not addressed in the Web-based survey. Then, based on the survey findings and focus group feedback, we worked with management to develop action plans that capitalized on their strengths and addressed their weaknesses with regard to opportunity and inclusion for all employees. Our feedback has facilitated both managerial and organizational development. In addition to emphasizing the factors that make an IT work environment more inclusive, we have taken this opportunity to educate managers on how to view diversity as a strategic business value.

We are currently in Phase 3 of the project. Phase 3 involves resurveying participating organizations to assess the effectiveness of the interventions implemented during the second phase of the project. Like Phase 2, we will provide each organization with a feedback report detailing that company's standing on each of the constructs measured in the survey. The report also examines any changes in an organization's scores from the first survey to the second. When an organization's sample is sufficiently diverse, companies also receive information

on any gender and/or ethnic differences in survey responses. When the Phase 3 data collection is complete, each company will again be provided with benchmarking comparison data from the other participating companies.

CONCLUSION

We believe that our focus on climate for opportunity and inclusion is not only applicable to the workplace, but is also relevant to educational environments in which IT professionals are trained (e.g., computer science). Based on our research findings, we propose that an inclusive learning environment is likely to be enhanced when students have supportive relationships with faculty members and peers. Enhancing the inclusiveness of learning environments would likely result in greater satisfaction with IT education, heightened commitment to the IT field, and greater likelihood of completing an IT-related degree, thus increasing the number of women IT graduates who enter the workforce. We describe elsewhere in this volume how inclusive educational environments might be created (see Davis, Major, et al.).

REFERENCES

Balcita, A. M., Carver, D. L., & Soffa, M. L. (2002). Shortchanging the future of information technology: The untapped resource. *SIGCSE Bulletin, 34*(2), 32-35.

Blau, G. (1985). The measurement and prediction of career commitment. *Journal of Occupational Psychology, 58,* 277-288.

Chrobot-Mason, D., & Aramovich, N. (2002, April). Assessing the multi-cultural organization: A comparison of whites and non-whites. In D. Chrobot-Mason (Chair), *Defining, measuring, and creating a positive climate for diversity.* Symposium conducted at the 17th Annual Conference of the Society for Industrial and Organizational Psychology, Toronto, Canada.

Ducharme, L. J., & Martin, J. K. (2000). Unrewarding work, coworker support, and job satisfaction: A test of the buffering hypothesis. *Work Occupations, 27,* 223-243.

Graen, G. B., Novak, M. A., & Sommerkamp, P. (1982). The effects of leader-member exchange and job design on productivity and satisfaction: Testing a dual attachment model. *Organizational Behavior and Human Performance, 20,* 109-131.

Hackman, J. R., & Oldham, G. R. (1975). Development of the Job Diagnostic Survey. *Journal of Applied Psychology, 60,* 159-170.

Howard, N. (1995). Don't treat your employees like a commodity. *Government Computer News, 14*(5), 25.

Hunt, D. M., & Michael, C. (1983). Mentorship: A career training and development tool. *Academy of Management Review, 8,* 475-485.

Lambeth, J. (1996, May 30). Time to come to terms with flexible working? (IT professionals face fixed-term contracts). *Computer Weekly,* 18.

Lee, P. C. B. (2004). Social support and leaving intention among computer professionals. *Information & Management, 41,* 323-334.

Major, D., Davis, D., Sanchez-Hucles, J., Germano, L., & Mann, J. (2006). IT workplace climate for opportunity andinclusion. In E. M. Trauth (Ed.), *Encyclopedia of gender and information technology.* Hershey, PA: Idea Group Reference.

Misic, M., & Graf, D. (1999). The interpersonal environments of the systems analyst. *Journal of Systems Management, 44,* 12-16.

Mor-Barak, M. E., & Cherin, D. (1998). A tool to expand organizational understanding of workforce diversity: Exploring a measure of inclusion-exclusion. *Administration in Social Work, 22,* 47-64.

Mor-Barak, M. E., Cherin, D. A., & Berkman, S. (1998). Organizational and personal dimensions in diversity climate. *Journal of Applied Behavioral Science, 34,* 82-104.

Mowday, R. T., Steers, R. M., & Porter, L. W. (1979). The measurement of organizational commitment. *Journal of Vocational Behavior, 14,* 224-247.

Panteli, A, Stack, J., & Ramsay, H. (1999). Gender and professional ethics in the IT industry. *Journal of Business Ethics, 22,* 51-61.

Ragins, B. R. (2002). Understanding diversified mentoring relationships: Definitions challenges and strategies. In D. Clutterbuck & B. R. Ragins (Eds.), *Mentoring and diversity: An international perspective* (pp. 23-53). Oxford, UK: Butterworth-Heinemann.

Ragins, B. R., & Cotton, J. L. (1999). Mentor functions and outcomes: A comparison of men and women in formal and informal mentoring relationships. *Journal of Applied Psychology, 84*(4), 529-550.

Schneider, B., Wheeler, J. K., & Cox, J. F. (1992). A passion for service: Using content analysis to explicate service climate themes. *Journal of Applied Psychology, 77,* 705-716.

Spender, D. (1997). The position of women in information technology—or who got there first and with what consequences? *Current Sociology, 45*(2), 135-147.

KEY TERMS

Action Plan: Strategy that includes concrete steps to take in order to implement positive organization change.

Best Practices: Policies, procedures, philosophies, and practices employed by top performing IT supervisors that were revealed in structured interviews.

Climate: Consists of employees' perceptions of workplace events, practices, and procedures, including which behaviors are rewarded, supported and expected (Schneider, Wheeler, & Cox, 1992).

Focus Group: A small group of people convened to discuss concepts and to address issues relating to the results of the survey. The aim of focus group discussions is to provide qualitative data to aid in interpretation of the survey results, clarify any ambiguities, and point out any related issues.

Mentor: Individuals with advanced experience and knowledge who are committed to providing upward support and mobility to their protégés' careers (Hunt & Michael, 1983)

Structured Interview: An interview in which the interviewer uses a fixed set of questions and asks them in the same order of all respondents.

Survey Feedback Intervention: An organizational change strategy in which employees provide information about current workplace conditions via survey. Specific plans for organizational change are driven by the strengths and opportunities for improvement identified in the survey.

ENDNOTE

[1] This material is based upon work supported by the National Science Foundation under Grant No. 0204430. The authors would like to acknowledge Thomas D. Fletcher for his assistance with data management and analyses.

S

Teaching Gender Inclusive Computer Ethics

Eva Turner
University of East London, UK

INTRODUCTION

Computer ethics as a subject area is finally being debated in wider computer science and information technology academic circles. In most computer science departments the syllabus is based on publications often written specifically to deliver courses. These texts select and prioritize those computer ethics topics seen by the professional bodies as the most important for a computer professional. Much rarer are courses which analyse questions of access and social exclusion, disability, global and green issues.

What has not yet been included in any systematic or conscientious way in the computer ethics syllabi are the questions of gender and associated ethical issues. Most students and staff are still not aware that all computing and ICT related areas are innately gendered and that a cohesive body of research material is available in the form of feminist or gender research in conference papers, proceedings and book publications.

This article analyses the progress of inclusion of gender in computer ethics and argues that the inclusion of gender issues in computer science curriculum must be accommodated. The article outlines how gender issues can be applied to individual computing disciplines in appropriate forms relevant across the spectrum of students.

BACKGROUND

Computer Ethics

Computer ethics is a field that is now widely recognized as a field of philosophical, political, and social enquiry in the use and construction of computing technology. There are now conferences devoted entirely to this field (e.g., ETHICOMP, CEPE) and in Britain there is The Centre for Computing and Social Responsibility at DeMontford University. The first PhD students graduated in this discipline in 1998.

The computing professional organizations embraced computer ethics quite early in the computer's history. In particular, the Association of Computing Machinery (ACM) and Institute of Electrical and Electronics Engineers (IEEE) began debating ethical issues concerning mainly computer hardware and its construction in the 1960s. In 1993 a task force combining members of the ACM and IEEE created a new computing curriculum, which embraced the social, ethical, professional, and legal issues of computing. The perception of ethical issues in this curriculum was rather simple and issues of gender and equality did not appear; yet it was a very advanced beginning (History of the Joint IEEE Computer Society and ACM Steering Committee for the Establishment of Software Engineering as a Profession, 1999).

The British Computer Society (BCS) published both the *Code of Conduct* and the *Code of Good Practice*. In terms of ethical professional behaviour, both publications now prioritise the user and the public as sites of good practice. The BCS course accreditation criteria concerned with ethical issues have also developed from initially only acknowledging the inclusion of the relevant legislation and the Codes itself in the curriculum, to containing a full appendix called "Legal, Social, Ethical and Professional Issues." While not mentioning equal opportunities explicitly, these make references to the Learning and Teaching Support Network (LTSN) computer ethics resource site where some equal opportunities material appears.

The ACM and the National Science Foundation (NSF) sponsored the first course on how to teach computer ethics to computing students in 1998. This course was aimed at U.S. academics, mainly computer scientists and concentrated mainly on issues of professionalism, codes of conduct, hacking, privacy, legislation and the environment. The course did not deal with equality, gender, race, or disability. Since

1998, there have been two other such courses, results of which were communicated to the NSF and the ACM/IEEE in a form of recommendations. The third course in 2001 accepted a gender scenario and gave space to a debate on gender and computing. The meeting of experts at this get-together contributed to a Special Issue of ACM SIGCSE Bulleting Inroads on Women and Computing (2002).

The British Computer Society, sponsored by the LTSN, have called to date three one-day conferences on delivering computer ethics to computing students. These conferences brought together computing academics, who were already interested in this field and who wanted to exchange their experiences and discuss how best to bring the subject to our students. The first conference (in 2000) resulted in the online LTSN resource.

Computer Ethics and Gender

James More (2001) described the development of computing as having three stages: the introduction stage, the permeation stage and the current power stage, which presents the most serious legal, ethical, and social questions. He describes a "policy vacuum," which is currently present and has resulted in the ever-increasing use of information and communication technology in its many forms and disguises. He urges that our "conceptual muddles" are cleared first, before any policy can be formulated. While More's thoughts were on issues such as data protection, the same applies to gender power relations in computing.

To investigate these conceptual muddles in relation to gender in computer ethics a clarification of feminist ethics is necessary. Feminist ethics accepts the experiences of women of any origin, status, sexual orientation, education etc as valid within the context of their social experiences (Porter, 1999). It includes and interrogates the meaning of traditional ethics in relation to these experiences and proposes alternatives to existing perceptions and social behaviours. Feminist ethics criticize the gender-blindness in traditional ethics and gender bias in all walks of life. In computer ethics specifically it examines all three broad areas of computing: the area of design and creation of computing technology, the area of personal and business uses of it and the area of computing education.

As a theory of moral behaviour, ethics draws on the traditional masculine perceptions and experiences, which inform the social systems creating acceptable standards of behaviour, legislation, and perceptions of equality. Feminism investigates the power relations between men and women and exposes their political nature. Feminist ethics attempts to develop social morality, which puts women's equality and emancipation in the centre of moral prescription (Adam, 2005). This of course calls also for men's recognition of the issues and ultimately for changes in men's attitudes. The complex issues of equality and in particular equality in the working environment (Bednar & Bissett, 2001) should be included in the teaching of computer ethics.

The major ethical issue is the under-representation of women in the computing industry and education. Many have documented the decreasing numbers of women in the last 20 years (Camp, 2002; Martin, Liff, Dutton, & Light, 2004; Mortleman, 2004; Turner, 2001) and argued that the lack of women's participation in the creation of technology is excluding a substantial body of human experience from being used in the process (Crutzen 2005; Schiebinger, 1999; Suchman, 1994). The questions of working conditions (e.g., Adam et al., 2005; Richardson & Richardson, 2001) and the expectations the computer industry has of their employees are almost Victorian. Unionization is non-existent and the workers are often expected to be on call 7 days/week. The lack of opportunities for women to return after career break and gender and race discrimination at the point of entry into the profession (e.g., Camp, 2002; Turner, 2001) have often been blamed for women not choosing to work in the industry.

Women are not equally paid for equal work in many industries and the computing industry is not an exception (Martin et al., 2004). The glass ceiling phenomena in the computing is probably worse than in many other industries. While the number of women managers is increasing in the Western countries, the computer industry's own statistics still indicate that only 8% of top management positions are occupied by women (Ezine, 2004).

It is necessary to remember, that while the numbers of women in computing education are extremely low, according to Martin et al. (2004) there are some "50,000 women with science, engineering and technology degrees (including computing)" in the UK alone, who are not using their qualifications. Arguably bringing more women into education is, in itself, not likely to remedy the situation in the industry.

Equal opportunities as an ethical issue guarantee not only equal rights of access but also rights to equal treatment. The low numbers demonstrate that for a variety of reasons women do not have equal access into the profession. The technology created in their name is thus created without their input. Technology is only useful if it helps to improve the position of those who use it and thus feel an ownership of it. It should exist to empower those whose power is socially suppressed (Everts, 1998; Rathgeber & Adera, 2000; Taylor, 2002). Questions of access to computer technology for women all over the world whether for education, business, information, or leisure are an issue of equal opportunities, power, and democracy for women. We talk of virtual democracy, global village, information age, globalisation, global communication, etc, all concepts loaded with socially based meanings in which women have "globally" little representation precisely because of their exclusion from access to creation and uses of ICT.

There is evidence of segregation of men and women and strong gendering of those working in the computing industry. This is an active product of the development of the industry and its power relations (for literature see IFIP WG 9.1, WiC, and ETHICOMP conferences). Women's achievements are not recognized and the power structures within the occupation keep them alienated (Turner, 2001). Studies have shown that in all we do we are influenced by our own experiences and thus if a gender split exists in the workforce, then the end product cannot reflect the needs of those for whom the technology is being constructed. These are questions of justice and fairness as well as issues of quality (e.g., Greysen, 2005; Thimbleby & Duquenoy, 2001; Turner & Stepulevage, 2000).

FUTURE TRENDS

Computer Ethics in Education

There are a number of papers on teaching computer ethics to computing and IT students. The ETHICOMP (see proceedings 1998, 1998, 2001, 2002, 2004) con-

ferences are one of the main forums for those who teach computer ethics to meet and exchange experiences. There is as yet a debate to be had, where and when and how to include gender in these curricula.

Some ethical content is now delivered in many relevant degree programmes in Europe and the U.S. The perceptions of what are the most important issues in this area are often informed by the need to satisfy the accreditation criteria of professional bodies.

In 2001, no computer ethics courses "included the ethical issues of gender and race (except in questions of pornography and freedom of expression" (Turner & Roberts, 2001, p. 224). Pornography on the Internet is often debated as an issue of freedom of expression and freedom of an individual but without any gender analysis. In 2005, (Turner, unpublished) an Internet pole of 110 mainly UK academics and a short study of UK universities computing curricula posted on the web found that in 20 British universities only two undergraduate and two masters modules included a lecture on gender delivered by individual lecturers interested in the issues. Only a few universities in Europe, USA, Australia, and elsewhere include gender in their computing curricula, these being delivered by lecturers personally interested in the issue. This is a sad indication of how little understanding of gender issues the large numbers future computer professionals take with them into their working lives.

The Gender Inclusive Computer Ethics Curriculum

The computer science and ICT curriculum needs to be re-shaped and re-thought to include gender issues in such a format, that the students will understand it, accept it and take it into their working lives. Perceiving them as an integral part of ethical and social debates around technology needs to become the norm, not the exception. Gender inclusive computer ethics therefore needs to permeate the curriculum throughout the students' study.

The gender issues need to be owned by the school/department and made a priority when the current curriculum is re-structured. They should be discussed at "teaching and learning" meetings so that staff can learn from each other. Gender issues can be documented and related to students on case studies and real life examples and resourced by the literature now available. These additions do not have to reduce the amount of the technical material delivered.

Teaching strategies should be adopted for each curricular subject area. Programming, hardware/ computer architecture, systems analysis/database theory/ software engineering, internet and networks are areas which re-occur in a variety of forms throughout the degree programmes. In each, the lecturers should present debates related to gender whether on products and their use, the nature of gathering data and information or creation of gendered web content. The class should discuss the professional's social responsibility related to gender, women's working conditions in developed and developing worlds, interface design, computer game production and gender power relations, not least in students' group assignments. Staff and students should note how they interact in computer labs, what verbal communication takes place and which student gets the most help, when and how.

Only then can a computer ethics module be delivered at a higher level. If the above suggestions are followed, then by this level the students will have a considerable awareness of issues related to gender. They will have been given case studies to work through and had discussions based of real experiences. This module can thus give all that learning experience some formal theoretical and analytical framework. The students can be introduced to ethical and feminist theories, which can be used to analyse a wide variety of ethical, professional, and cultural issues of computing. Most of these issues have a gender element which students should be encouraged to include in their discussions.

An important contribution to a critical debate on gender politics is the Shadow Report on Equal Opportunities (Pavlik, 2004), which is a unique and welcome publication. In most cases computer dissertations students are neither required to include ethical issues in nor to provide any gender analysis of the topic concerned. Only too often an extended computer program is sufficient for a dissertation, with no theoretical section on social and ethical analysis of the environment for which it is being written. It should be imperative on all dissertations to include such a theoretical section, which would require students to investigate gender issues relevant to the theme of their project.

CONCLUSION

In the year 2000 there were a few computer science departments delivering computer ethics courses to their students. The call among academic staff particularly interested in this area of computer science was for the academic management to consider a module in computer ethics a priority (Turner, 2000).

Five years down the road most computer science departments have introduced some form of professional and ethical issues into their teaching, but only very few have included gender as an important topic in their "professional" delivery.

While the experts debate the future direction of computer ethics, they mostly do not include gender in their discussion (e.g., contrast Weckert, 2001; Tavani, 2002, with Adam, 2004). Where they talk about women, they mention them as victims. However there are now more academic researchers and women practitioners world wide, interested in and talking about gender issues in computing. It is important that we promote an atmosphere of acceptance so that the analysis and the delivery of this material becomes an integral part of our teaching. The responsibility for departmental ownership of the issues and their delivery lies still with the deans, heads of schools and university policy makers as well as the professional organisations themselves.

REFERENCES

ACM SIGCSE Bulleting Inroads. (2002). *Special issue women and computing, 34*(2).

Adam, A. (1998). *Artificial knowing: Gender and the thinking machine.* New York; London: Routledge.

Adam, A. (2005). *Gender, ethics, and information technology.* Basingstroke, UK: Palgrave Macmillan.

Adam, A., Griffith, M., Keogh, C., Moore, K., Richardson, & Tattersall, A. (2005). You don't have to be male to work here, but it helps!—Gender and the IT labour market. In J. Archibald, J. Emms, F.

Grundy, J. Payne, & E. Turner (2005). *Gender politics of ICT*. London: Middlesex University Press.

Bednar, P. M., & Bissett, A. (2001). The challenge of gender bias in the IT industry. In T. W. Bynum, H. Krawczyk, S. Rogerson, S. Szejko, & B. Wiszniewski (Eds.), *ETHICOMP 2001 Proceedings*. Gdansk, Poland: Technical University of Gdansk.

Camp, T. (2002). The incredible shrinking pipeline. ACM SIGCSE Bulleting Inroads (2002), *Special Issue Women and Computing*, *34*(2), 129-134

Crutzen, C. K. M. (2005). Intelligent ambience between heaven and hell: A salvation? In J. Archibald, J. Emms, F. Grundy, J. Payne, & E. Turner (2005). *Gender politics of ICT*. London: Middlesex University Press.

Everts, S. (Ed.). (1998). *Gender & technology, empowering women, engendering development*. London; New York: Zed Books Ltd.

Ezine. (2004). *Women in management*. Retrieved December 20, 2004, from http://www.softworks-computing.com/feb04_ezine/dload_women_in_mgmt.html

Greysen, K. R. B. (2005). An initial investigation of students' self-construction of pedagogical agents. In J. Archibald, J. Emms, F. Grundy, J. Payne, & E. Turner, *Gender politics of ICT*. London: Middlesex University Press.

History of the Joint IEEE Computer Society and ACM Steering Committee for the Establishment of Software Engineering as a Profession. (1999). Retrieved December 17, 2004, from http://www.computer.org/tab/seprof/history.htm and retrieved December 20, 2004, from http://www.ics.ltsn.ac.uk/resources/ethics/

Martin, U., Liff, S., Dutton, W., & Light, A. (2004). *Rocket science or social science? Involving women in the creation of computing*. Oxford, UK: Oxford Internet Institute.

More, J. H. (2001). The future of computer ethics: You ain't seen nothin' yet! *Ethics and Information Technology*, *3*(2), 89-91. Kluwer Academic Publishers.

Mortberg, C. (1994). Computing as masculine culture. In E. Gunnarsson & L. Trojer (Eds.), *Feminist voices on gender, technology and ethics*. Sweden: Luela University of Technology.

Mortleman, J. (2004). *Women in IT still outnumbered by men*. Retrieved December 17, 2004, from http://www.computing.co.uk/news/1155314

Porter, E. (1999). *Feminist perspectives on ethics*. London; New York: Longman.

Rathgeber, E. M., & Adera E. O. (2000). *Gender and the information revolution in Africa*. Canada: International Development Research Centre.

Richardson, H., & Richardson, K. (2001). Customer Relationship Management Systems (CRM) and information ethics in call centres—You are the weakest link. Good bye! In T. W. Bynum, H. Krawczyk, S. Rogerson, S. Szejko, & B. Wiszniewski (Eds.), *ETHICOMP 2001 Proceedings*. Gdansk, Poland: Technical University of Gdansk.

Schiebinger, L. (1999). *Has feminism changed science*? London: Harward University Press.

Suchman, L. (1994). Supporting articulation work: Aspects of a feminist practice of technology production. In A. Adam & J. Owen (Eds.), *Women, work, and computerization, breaking old boundaries: Building new forms. Proceedings of the 5th IFIP Conference, UMIST* (pp. 46-60).

Tavani, H. T. (2002). The uniqueness debate in computer ethics: What exactly is at issue, and why does it matter? *Ethics and Information Technology*, *4*(1), 37-54. Kluwer Academic Publishers.

Taylor, V. E. (2002). Women of color in computing. ACM SIGCSE Bulleting Inroads. *Special Issue Women and Computing*, *34*(2), 22-23.

Turner, E. (2000). *Presentation to LTSN conference*. University of Greenwich. Retrieved December 20, 2004, from http://www.ics.ltsn.ac.uk/pub/ethics01/evaturner.doc

Turner, E. (2001). The case for responsibility of the computing industry to promote equality for women. *Science and Engineering Ethics Journal*, *7*(2), 247-260. Opragen Publications.

Turner, E. (2002). Gendered future of computer profession, establishing and ethical obligation on the computer educators. In I. Avarez, T. W. Bynum, A.

J. Assis Lopes, & S. Rogerson (Eds.), *The transformation of organisations in the information age. Proceedings of the 6th International Conference ETHICOMP2002* (pp. 711-722). Lisbon, Portugal: Universidade Lusiada de Lisbon.

Turner, E., & Roberts. P. (2001, June). Teaching computer ethics to it students in higher education: An exploration of provision, practice and perspectives. In T. W. Bynum, H. Drawczyk, S. Rogerson, S. Szejko, & B. Wiszniewski (Eds.), *The social and ethical impacts of information and communication technologies. Proceedings of 5th International Conference ETHICOMP2001,* Gdansk, Poland (Vol. 1, pp. 223-232).

Turner, E., & Stepulevage, L. (2000). Will ETs understand us if they make contact? In E. Balka & R. Smith (Eds.), *Women, work, and computerization, charting the course to the future* (pp. 155-163). Boston: Kluver Academic Publishers.

Weckert, J. (2001). Computer ethics. *Future Directions, Ethics and Information Technology, 3*(2), 93-96.

KEY TERMS

Association of Computing Machinery (ACM) and Institute of Electrical and Electronics Engineers (IEEE): U.S. professional associations for IT and computing professionals. (http://www.acm.org and http://www.ieee.org/)

British Computer Society (BCS): The British Computer Society (BCS) is the industry body for IT professionals, and a chartered engineering institution for information technology (IT). (http://www.bcs.org/)

Computer Ethics: Computer ethics is a branch of practical philosophy which deals with how computing professionals should make decisions regarding professional and social conduct. (en.wikipedia.org/wiki/Computer_ethics)

Learning and Teaching Support Network (LTSN): A British independent organisation working with academic institutions on teaching and learning initiatives, it is now called The Higher Education Academy. (http://www.heacademy.ac.uk/)

National Science Foundation (NSF): U.S. government organisation which supports science and engineering research and education. (http://www.nsf.gov/)

T

A Techno–Feminist View on the Open Source Software Development

Yuwei Lin
Vrije Universiteit Amsterdam, The Netherlands

INTRODUCTION

Current debate on women in free/libre open source software (FLOSS) tends to fall into the gender stereotype of men and women when coming across to the gender issue. This article stays away from a reductionism that simplifies the gender issue in the FLOSS community to the level of a fight between men and women. Instead of splitting women from men in the FLOSS development, this analysis helps motivate both men and women to work together, reduce the gender gap and improve the disadvantaged statuses of women and a wider users' community in the FLOSS development. More importantly, it addresses not only the inequality that women face in computing, but also other inequalities that other users face, mainly emerging from the power relationships between expert and lay person (namely, developer and user) in software design. In so doing, the issue at stake is not only to create a welcome environment for women to join the FLOSS development, but also to come up with a better way of encouraging both sexes to collaborate with each other.

This article starts from how FLOSS can make a difference for today's information society, and present some successful stories of implementing FLOSS in developing countries and rural areas to empower women and the minority. Consequently, it discusses the problem of including more women and the minority in the FLOSS development through deconstructing the myth of the programming skill.

BACKGROUND

The essential element of FLOSS is "freedom" that allows users to run, copy, redistribute, study, change and improve the software. By having source code made available to the public, interested users or developers can study and understand how the software is written and, if competent, they can change and improve it, as well. In other words, apart from serving as an alternative choice for consumers, FLOSS helps open up the black box of software technologies, facilitate the practice of participatory design and provide an opportunity of breaking down the hierarchy of professional knowledge. And this could lead to improved security and usability, because users can configure software to fulfill their local requirements and secure against vandalism, user errors and virus attacks.

Given these opportunities, FLOSS has been adopted and implemented in several developing countries and rural areas. For instance, believing that FLOSS serves as a better technological tool to bridge the digital divide, Brazil, for example, has also required any company or research institute that receives government financing to develop software to license their work under FLOSS licenses, meaning the underlying software code must be free to all (Benson, 2005). In the wave of localization and customization, a group of volunteers in India has started the IndLinux[1] project to create a Linux distribution that supports Indian Languages at all levels. These examples are just two of the many ongoing projects around the world. These continuously emerging cases demonstrate that FLOSS provides a better basis for more widespread access to information and communication technologies (ICT), more effective uses and a much stronger platform for long-term growth and development compared with scaled-down versions of proprietary software.

However, such FLOSS-based technologies meant to be used widely and to empower users have not yet engaged with a diverse range of people in development and implementation. So far, the freedom of FLOSS seems to be enjoyed only by those who are capable of manipulating the technologies. We see imbalanced population distributions in the FLOSS-based knowledge demography, and the unbalanced gender distribution is among those top ones. We see a strong programming culture in the FLOSS development and implementation nowadays—if one does not program, he or she seems to be left out of the FLOSS movement. In other words, instead of breaking down the hierarchy of professional knowledge, a new boundary and barrier of accessing ICT knowledge seems to be established. Abbreviations such as "RTFSC" (Read The F***ing[2] Source Code) or "RTFM" (Read The F***ing Manual) shows how strong this hegemony of software knowledge is. This article aims to challenge the workship on programming knowledge, which is one of the many reasons that causes gender inequality in FLOSS (see Henson, 2002; Lin, 2006).

I would like to stress that to be involved in the FLOSS development, one needs not be a programmer (see Rye, 2004); one could write documentation, report or triage bugs, improve graphic or text content, translate/localize, submit feature-requests or teach how to use FLOSS. These activities are equally important to programming in the software innovation process, because software is not ready to use just as it is written. It needs many efforts to make it user friendly, implement it in different contexts and maintain it over time (Levesque, 2004). To make FLOSS successful, we need not only Richard Stallman or Linus Torvalds, but also a great amount of volunteers reporting and fixing bugs, writing documentation and, more importantly, teaching users how to use OpenOffice.org and Mozilla Firefox browser. When thinking of an approach of including more women and improving the representation of women in FLOSS, these activities can be considered as essential.

Saying that we should start encouraging women to participate in these activities does not imply that women are not good at programming. Not at all! While it is generally recognized that there is no genuine biological difference between men and women in science (American Sociological Association, 2005), the history and cultural and educational backgrounds in turn lead to the circumstance in which many women nowadays do not have as strong programming experience as men do. Given this, we may need an alternative way of including women in FLOSS. But more importantly, it is because neither of these activities (e.g., documentation and localiza-

tion) are subordinate to programming, nor are they peripheral in any case, and we need to encourage women and other minority user groups to participate in these activities in the FLOSS development.

These efforts on documentation and localization (including translation) are so important that they are the keys to opening the black box of the software technologies and allowing more people (regardless of gender, class, race and disability) to participate in the FLOSS development. While some people try to degrade the skill of writing documentation or translation, an experienced female FLOSS user, Patricia Jung, emphasized the importance and challenge of writing documentation on the Debian-women mailing list:

Documentation can be a means of quality insurance, and this power is far too seldom used, not only in Open Source development. The people who write the best code I know write documentation alongside or even before coding: The code has to follow documentation, otherwise it's a bug :), at least documentation and code are never allowed to get out of sync. Which means documentation _is_ development, not just something subordinate.

In a scenario like this, documentation and usability are not just nice to have but an inherent part of development and equally important as writing code, and it finally leads you to better software, to software that is aware of its users and tasks and not just aware of how things are easiest, smartest to implement. But it requires a paradigm shift: Coders are no longer allowed to see documentation as a nasty add-on, as something subordinate, and documentation people don't simply have to follow the software they get but allowed and required to intervene. Software isn't released as long as the doc people don't give their go: Right now code matches documentation, it does what it is supposed to do, now we can release. (Debian-Women, 2005)

Jung's message demonstrates that coding is neither the only nor the foremost activity in the FLOSS innovation process. Programmers do not play a more important role than other contributors in the FLOSS development. The FLOSS community is comprised of diverse people from different social worlds, and each member should gain equal respect from what they do. FLOSS cannot get widespread without people writing documentation, reporting bugs and mentoring. The value of the FLOSS development is embedded and embodied not only in coding and the resulting code, but also in the process of collaboration, networking with others, sharing knowledge and experiences, learning and helping reciprocally. FLOSS gives us a chance to see the co-construction of social and technical activities in a socio-technical innovation process. With a techno-feminist perspective (Wajcman, 2004; Faulkner, 2000, 2001) on the FLOSS development, the socio-technical complexity in the FLOSS community can be observed more deliberately on their power relationships: the haves and have-nots in programming. Strengthening this gap would bring the problem to all members involved (regardless of gender, race, class and disability) in the FLOSS development, rather than just men and women. And this would also clarify the misunderstanding on feminism: Feminism cares not only about the inequality between men, women and other genders, but all inequalities in the society. In other words, involving more women in the documentation or localization (and internationalization) of FLOSS should not turn these fields into a female domain that it might ironically end up somehow cheapening the work suggested by some old thoughts about certain things being women's work; instead, such a conflicting situation raises the question on non-programming work in software design that is continuously undermined and devalued. To mitigate this unbalance, we need to take a feminist perspective, treating the whole mechanism of software design as a "socio-technical system" (Hughes, 1979, 1987; Latour, 1983, 1999; Pinch & Bijker, 1987), "recognizing the various forms of visible and invisible work that make up the production/use of technical systems, locating ourselves within that extended web of connections, and taking responsibility for our participation" (Suchman, 1999, p. 263). The biased power relationships between men and women, developers and users, experts and lay person inscribed in the strong technological determinism in the FLOSS development can be overcome through valuing heterogeneity and situated knowledge in the FLOSS community (Haraway, 1991; Lin, 2004).

FUTURE TRENDS

As argued, FLOSS has a potential of being a platform providing both men and women, expert and lay person with equal opportunities to develop and implement software together. But realizing this potential still requires more deliberate efforts. Many women-led FLOSS groups have been tackling the knowledge gap between expert and lay person (not only men and women) and challenging the "masculine" culture in the FLOSS community. Three examples below show how women are encouraged to become mobile grass-roots information technology (IT) workers supporting organizations and individuals with advice on technology with non-technical language, rather than just coding.

LinuxChix

LinuxChix[3] is a community for women Linux enthusiasts and for supporting women in computing. The membership ranges from novices to experienced users, and includes professional and amateur programmers, system administrators and technical writers. It aims at creating a more hospitable community in which people can discuss Linux, a community that encourages participation, that does not allow the quieter members to be drowned out by the vocal minority. LinuxChix was aimed at women, and it remains primarily a group for supporting women in computing. LinuxChix now has several branches around the world, including LinuxChix Brazil[4] and LinuxChix Africa[5].

Debian-Women

The Debian-Women[6] project, founded in May 2004, seeks to balance and diversify the Debian[7] Project by actively engaging with interested women and encouraging them to become more involved with Debian. Debian-women promotes women's involvement in Debian by increasing the visibility of active women, providing mentoring and role models, and creating opportunities for collaboration with new and current members of the Debian Project. All people (both men or women) who are interested in increasing the participation of women in Debian are welcome. Now Debian-Women has a mailing list running for discussion of related issues, and an Internet relay chat (IRC) channel for discussion of related issues, technical questions and to allow women who are interested in contributing to Debian to meet each other and some of Debian's current contributors. The members also eagerly give talks at conferences, and organize "birds of a feather" (BOF) discussions at Linux conferences to promote discussion of issues concerning women and their involvement in Debian and Linux. These activities have effectively encouraged and educated the Debian community to increase understanding of these specific issues concerning women who wish to contribute more to Debian.

Women's Information Technology Transfer (WITT)

WITT[8] is a portal site to link women's organizations and feminist advocates for the Internet in Eastern and Central Europe. It aims at providing strategic ICT information to all, and supporting, in a collective way, Central and Eastern European women in developing the Web as an instrument in their social activism. WITT is committed to bringing women's actions, activities and struggles into the spotlight, promoting the use of FLOSS as a way to highlight women's voices. The WITT Web site has been developed for women to share their experiences with ICT, to learn about training events provided by WITT and to develop expertise in advocacy on gender and ICT issues. Women can publish on the Web site in their own language (eight languages are available to be used as the site develops).

CONCLUSION

The features of FLOSS have been said to open a range of opportunities to change the power relationships in society: experts and lay people, developers and users, developed and developing countries, rich and poor and so forth. The feature of low development cost, modularized features and transparent information are particularly celebrated in a knowledge-based society. Whereas FLOSS is represented

as a weapon to fight against proprietary software companies, such as Microsoft, neither have we seen an equal status for all members involved in the FLOSS development, nor have we seen an accessible channel for all interested people to enter this world. Drawing on new perspectives in feminist theory and science and technology studies, I challenge the power emerging from the skills of programming and designing technologies that overlooks the requirements of having user-friendly technologies. And it is exactly because of this misconception on the coding skill that makes the composition and structure in the FLOSS social world imbalanced. This misconception fosters a false impression that FLOSS is too technical and difficult to use. This kind of misunderstanding discourages many people, including women, to participate in the FLOSS development, and subsequently results in an imbalance in gender, race and class. Today, when we criticize women's status in the FLOSS social world, we must not forget that a feminist critique not only applies to gender issues, but it aims to challenge all kinds of power inequalities in the world. I have proposed to take a critical view on the attitude in favor of (if not worshiping) people who own programming skills when examining the reason why there are so few women in FLOSS. Instead of narrowing the gender argument to a fight between men and women, I argue that this is not only about men and women, but about all majority and minority, the powerful and the powerless class. In viewing the problem from a techno-feminist angle, we can overcome many dilemmas, such as whether designing software for women is needed. After all, it is not whether software should be designed for men or for women; it is whether the software is designed for users without taking too much pride of the developers. Three women-led FLOSS groups working persistently in this direction were introduced: LinuxChix, Debian-women and WITT. These groups facilitate networking and provide mutual help among women participants in the FLOSS development and computing. They help maintain a pool of women who will not only promote ICT use but also promote a feminist approach of design and usage of ICT. Although this article has focused on the gender-related issues specifically in the FLOSS development, the analytic concepts introduced here can and should be widely applied to software design and other technological designs to explore the ways in which technologies are gendered in their design and use.

REFERENCES

American Sociological Association. (2005). *Statement of the American Sociological Association Council on the causes of gender differences in science and match career achievement.* Retrieved August 30, 2005, from http://www.asanet.org/page.ww?section=Issue+Statements&name=Statement+on+Summers

Benson, T. (2005, March 19). BRAZIL: Free software's biggest and best friend. *The New York Times.*

Debian-Women. (2005, May). *Software quality and documetation, was: Core KDE member about HIG^W female contributors.* Retrieved from http://lists.debian.org/debian-women/2005/05/msg00116.html

Faulkner, W. (2000). Dualisms, hierarchies and gender in engineering. *Social Studies of Science, 30*(5), 759-792.

Faulkner, W. (2001). The technology question in feminism: A view from feminist technology studies. *Women's Studies International Forum, 24*(1), 79-95.

Haraway, D. J. (1991). Situated knowledges: The science question in feminism and the privilege of partial perspective. In *Simians, cyborgs, and women: The reinvention of nature* (pp. 183-201). New York: Routledge.

Henson, V. (2002). *HOWTO encourage women in Linux.* Retrieved June 18, 2005, from http://www.tldp.org/HOWTO/Encourage-Women-Linux-HOWTO/index.html

Hughes, T. P. (1979). The electrification of America: The system builders. *Technology and Culture, 20,* 124-161.

Hughes, T. P. (1987). The evolution of large technological systems. In W. E. Bijker, T. P. Hughes, & T. J. Pinch (Eds.), *The social construction of techno-logical systems. New directions in the sociology and history of technology* (pp. 51-82). Cambridge, MA: MIT Press.

Latour, B. (1983). Give me a laboratory and I will raise the world. In K. D. Knorr-Cetina & M. Mulkay (Eds.), *Science observed. Perspectives of the social studies of science* (pp. 141-170). London: Sage.

Latour, B. (1999). *Pandora's hope: Essays on the reality of science studies.* Cambridge, MA: Harvard University Press.

Levesque, M. (2004). Fundamental issues with open source software development by Michelle. *First Monday, 9*(4). Retrieved from http://firstmonday.org/issues/issue9_4/levesque/index.html

Lin, Y.-W. (2006). Women in the free/libre open source software development. In E. M. Trauth (Ed.), *Encyclopedia of gender and information technology.* Hershey, PA: Idea Group Reference.

Lin, Y.-W. (2004). *Hacking practices and software development: A social worlds analysis of ICT innovation and the role of open source software* (unpublished doctoral thesis). Department of Sociology, University of York, UK.

Pinch, T. J., & Bijker, W. E. (1987). The social construction of facts and artefacts: Or how the sociology of science and the sociology of technology might benefit each other. In W. E. Bijker, T. P. Hughes, & T. J. Pinch (Eds.), *The social construction of technological systems. New directions in the sociology and history of technology* (pp. 17-50). Cambridge, MA: MIT Press.

Rye, J. B. (2004). *I am not a programmer* (IANAP—JBR Disclaimer). Retrieved June 18, 2005, from htt://www.xibalba.demon.co.uk/jbr/linux/ianap.html

Suchman, L. (1999). Working relations of technology production and use. In D. Mackenzie & J. Wajcman, J. (Eds.), *The social shaping of technology, second edition* (pp. 258-265). Buckingham, UK: Open University Press.

Wajcman, J. (2004). *Techno feminism.* Cambridge, UK: Polity Press.

KEY TERMS

Coding or Programming: The craft of implementing one or more interrelated abstract algorithms using a particular programming language to produce a concrete computer program.

Debian GNU/Linux: Debian, organized by the Debian Project, is a widely used distribution of free software developed through the collaboration of volunteers from around the world. Since its inception, the released system, Debian GNU/Linux, has been based on the Linux kernel, with many basic tools of the operating system from the GNU project. The project Web site is accessible at http://www.debian.org/.

GNU/Linux or Linux: A computer operating system that is one of the most famous examples of free/libre open source software (FLOSS) development. Unlike other major operating systems (such as Windows or Mac OS), all of its underlying source code is available to the public and anyone can freely use, modify and redistribute it.

LinuxChix: A community for women Linux enthusiasts, and for supporting women in computing. It aims at promoting interest in and learning about Linux among women around the world.

Mozilla: The Mozilla Organization was founded in 1998 to create the new suite. On July 15, 2003, the organization was formally registered as a not-for-profit organization, and became Mozilla Foundation. The foundation now creates and maintains the Mozilla Firefox® browser and Mozilla Thunderbird e-mail application, among other products. The project Web site can be accessed at http://www.mozilla.com/.

OpenOffice.org: OpenOffice.org is a free/libre and open source office suite, including a word processor, spreadsheet, presentation, vector drawing, and database components. It is available for many different platforms, including Microsoft Windows, Unix®-like systems with the X Window System including GNU/Linux, BSD, Solaris, and Mac OS X. It is intended to be compatible and complete with Microsoft Office; it supports the OpenDocument standard for data interchange; and it can be used at no cost. The project Web site can be accessed at http://openoffice.org/.

Science and Technology Studies (STS): A field in academic research devoted to studying how scientific knowledge is produced, maintained and used. Studies done in this field usually are interdisciplinary and multidisciplinary, involving mainly anthropology, history, philosophy and sociology.

Socio-Technical Systems: An analytical concept describing how technology is created, maintained and used. Staying away from technological determinism, the term stresses the reciprocal interrelationship between humans and technologies.

Techno-Feminism (or Feminist Technology Studies): Both a branch of feminism and that of science and technology studies. This dualist theory studies the co-construction of gender and technologies, how such mutual-shaping influences both how technologies are designed and used, and how gender identities of users and designers are perceived and articulated.

Women's Information Technology Transfer (WITT): A portal Web site to link women's organizations and feminist advocates for the Internet in Eastern and Central Europe. It aims at providing strategic ICT information to all and supporting, in a collective way, Central and Eastern European women in developing the Web as an instrument in their social activism.

ENDNOTES

[1] http://www.indlinux.org
[2] To show politeness, the word "f***ing" now usually is replaced with the word "fine" or hidden in the abbreviations. But the meaning and the way it is used does not change.
[3] http://www.linuxchix.org
[4] http://www.linuxchix.org.br/
[5] http://www.africalinuxchix.org/
[6] http://www.debianwomen.org
[7] http://www.debian.org
[8] http://www.witt-project.net/

Theorizing Gender and Information Technology Research

Eileen M. Trauth
The Pennsylvania State University, USA

INTRODUCTION

A fundamental consideration when attempting to understand the complex factors leading to the underrepresentation of women in IT is the choice and use of theory. Theories about women and their relationships to information technology and the IT profession guide the conceptualization of the research problem, the methods of data collection, the basis for analysis, and the conclusions that are drawn. However, a criticism of gender and IT research is that the topic of gender and IT is currently undertheorized (Adam, Howcroft, & Richardson, 2001, 2004).

This undertheorization takes on several different forms. First, there are cases in which there is no theory in evidence to guide the conceptualization of the research project or to inform the data collection and analysis. Rather, the focus is typically on compiling and representing statistical data regarding the differences between men and women with respect to technology adoption, use or involvement in the IT profession. This form of undertheorization can be labeled *pre-theoretical research*. Second, other research, while not explicitly articulating a particular theory, nevertheless, is guided by a theory-in-use. For example, quite often a theory of inherent differences between males' and females' relationships to IT is used implicitly to guide data collection and analysis. This form of undertheorization can be labeled *implicit-theoretical research*. This approach is considered to be a type of undertheorization in that the lack of explicit discussion of a theory makes it difficult for others to discuss, challenge or extend the research. Finally, the body of research that reflects explicit theory-in-use has been shown to have gaps in the theoretical landscape (Trauth, 2002). That is, an argument has been made that current theories about gender and IT do not fully account for the variation in men's and women's relationships to

information technology and the IT field. This form of undertheorization can be labeled *insufficient-theoretical research*. It is this third condition that is addressed in this article: the need for new theoretical insights to guide our effort to understand the underrepresentation of women in the IT profession.

BACKGROUND

Two dominant theoretical viewpoints are currently reflected in the majority of literature about gender and IT: essentialism and social construction (Trauth, 2002). Essentialism is the assertion of fixed, unified and opposed female and male natures (Wajcman, 1991, p. 9). The existence of biological difference between the sexes has led to a tendency to assume that other observed differences between men and women are due to biological determinates as well (Marini, 1990). When applied to the topic of gender and IT, the essentialist theory presumes the existence of relevant *inherent differences* between women and men with respect to information technology. It uses the observed differences in the participation of women and men in the IT field as evidence of this view. Thus, the causes of gender underrepresentation in IT are attributed to biology. It turns to observed differences in men's and women's behavior for explanations of what are believed to be inherent, fixed, group-level differences that are based upon bio-psychological characteristics.

Essentialism underlies research on gender and IT that views gender as a fixed variable that is manipulated within a positivist epistemology (e.g., Dennis, Kiney, & Hung, 1999; Gefen & Straub, 1997; Venkatesh & Morris, 2000). Adam et al.'s (2001) analysis of this perspective points out that focusing on a background literature of psychology, alone, places too much emphasis on individual gender characteristics where a form of essentialism may creep in. Looking only to psychological explanations of observations without giving attention to the influence of context[1] results in a determinist stance with respect to gender.

One inference that could be drawn from an essentialist approach to gender and IT research is that women and men should be treated differently. For example, Venkatesh and Morris (2000) recommend that trainers adopt different approaches toward men and women and that marketers design different marketing campaigns for men and women. Trauth's critique of essentialist approaches to gender and IT research suggested that one logical extrapolation from this line of thinking to IT workforce considerations would be the creation of two different workforces: a "women in IT" workforce and a "men in IT" workforce. Thus, policies for addressing the gender imbalance would focus on differences between women and men and the equality issue would focus on "separate but equal," something that was rejected in the arena of racial equality decades ago (Trauth, 2002; Trauth & Quesenberry, 2005; Trauth, Quesenberry, & Morgan, 2004).

The other dominant theoretical perspective focuses on the *social construction* of IT as a male domain. According to this theory, there is a fundamental incompatibility between the social construction of female identity and the social construction of information technology and IT work as a male domain. This explanation for women's relationship to information technology looks to societal rather than biological forces. Thus, the causes of gender underrepresentation can be found in both the IT sector and in the wider society.

The literatures of gender and technology in general (e.g., Cockburn, 1983, 1988; Cockburn & Ormrod, 1993; Wajcman, 1991) and that of gender and information technology, in particular (e.g., Adam et al., 1994; Balka & Smith, 2000; Eriksson, Kitchenham, & Tijdens, 1991; Lovegrove & Segal, 1991; Slyke, Comunale, & Belanger, 2002; Spender, 1995; Star, 1995; Webster, 1996) look to social construction theory (Berger & Luckmann, 1966) rather than biological and psychological theories. According to this view, the social shaping of information technology as "men's work" places IT careers outside the domain of women.

Recommendations for addressing this situation vary. One school of thought based on a multi-year investigation of female underrepresentation in both academe and the workplace in Australia explores the development of strategies to help women fit in to this male domain (e.g., Nielsen, von Hellens, Greenhill, & Pringle, 1998; Nielsen, von Hellens, Pringle, & Greenhill, 1999; Nielsen, von Hellens & Wong, 2000; Pringle, Nielsen, von Hellens, Greenhill, & Parfitt, 2000; von Hellens, von Hellens, Nielsen, & Trauth, 2001; Pringle, Nielsen, & Greenhill, 2000). Another school of thought focuses on the need to reconstruct the world of computing to become more of a "female domain." For example, Webster (1996) focuses on the social shaping of female gender identity and the implication for women's relationship to workplace technologies. Based on analysis of women as a social group in cyberspace, Spender (1995) predicted an influx of "female values" into the virtual world that would accompany increased female presence.

Wajcman's (1991) analysis of the social constructivist perspective on gender and technology reveals several issues. For example, there is no universal definition of masculine or feminine behavior; what is considered masculine in one society is considered feminine or gender-neutral in another. Further, while gender differences exist they are manifested differently in different societies. Hence, addressing the gender gap in IT employment based upon an assumed "woman's perspective" is problematic. This analysis suggests a gap in current theory and motivates the articulation of new theory to help us better understand the underrepresentation of women in the IT field.

MAIN THRUST OF THE ARTICLE

The need for an alternative theory to account for the underrepresentation of women in the IT workforce emerges from consideration of the assumptions underlying the two prevailing theories discussed in the previous section. The initial work on the Individual Differences Theory of Gender and IT resulted from an analysis of this theoretical gap and used empirical data from a study of gender and IT in Australia and New Zealand (Trauth, 2002; Trauth, Nielsen, & von Hellens, 2003) to make the case for an alternative theory to occupy the space between essentialist theory and social constructionist theory. Subsequent work has focused on greater articulation of this theory (Trauth & Quesenberry, 2005, 2006; Trauth et al., 2004, 2006) and empirical testing of it (Morgan, Quesenberry, & Trauth, 2004; Quesenberry & Trauth, 2005; Quesenberry, Trauth, & Morgan, 2006; Trauth, Quesenberry & Yeo, 2005).

The Individual Differences Theory of Gender and IT addresses the undertheorization of gender and IT by offering an alternative theory that focuses on *individual differences* among women as they relate to the characteristics of IT work and the IT workplace. This view finds the causes of gender underrepresentation in the varied individual responses to generalized societal influences. Thus, it represents the middle ground between the essentialist and social constructionist theories. In doing so, it investigates the individual variations across genders as a result of the combination of personal characteristics and environmental influences in order to understand the participation of women in the IT workforce. Hence, the focus is on differences *within* rather than *between* genders. The theory also views women and men as individuals who possess different technical talents and inclinations and respond to the social shaping of gender in unique and particular ways that vary across cultures. This individual differences theory takes into account the uniformity of social shaping messages conveyed in a culture. However, it also takes into account the varied influence of individual background and critical life events that result in a range of responses to those messages.

The individual differences theory is comprised of three general constructs that, together, explain women's decisions to enter and remain in the IT field. The individual identity construct includes both personal demographic items (such as age, race, ethnicity, nationality, socio-economic class, and parenting status) and professional items (e.g., industry, type of IT work, etc.). The individual influence construct includes personal characteristics (e.g., educational background, personality traits, and abilities) and personal influences (e.g., mentors, role models, experiences with computing, and other significant life experiences). The environmental influence construct includes cultural attitudes and values (e.g., attitudes about IT, about women in IT), geographic data (e.g., about the geographical location of one's work) and economic and policy data (e.g., about the region/

country in which one works). The Individual Differences Theory of Gender and IT posits that, collectively, these constructs account for the differences among men and women in the ways they experience and respond to characteristics of IT work, the IT workplace and societal messages about women and men and IT.

CONCLUSION

It is ironic that coincident with a documented need for a deeper understanding of the gender imbalance in the IT field, there is insufficient attention being paid to theorizing gender and IT. Given this need, greater theorization in gender and IT research can contribute in several ways to a better understanding of women's relationship to information technology. First, it can lead to more theoretically-informed treatments of gender in IT research. Wajcman (2000) has observed that gender is seldom considered a relevant factor in socio-technical studies of IT in context. Second, much of the published work that does focus on gender places emphasis on data analysis rather than theoretical implications and linking these results to the existing body of gender, and gender and IT literature (Adam et al., 2001). Hence, greater explicit use of theory can strengthen the existing body of gender and IT research. Finally, insufficient attention has been paid to the differences *among women* rather than *between women and men* with respect to information technology adoption, use and work. The development of the Individual Differences Theory of Gender and IT is intended to address this need by providing additional theoretical insights to help us to better understand the individual and environmental forces that account for the underrepresentation of women in IT. It accomplishes this by focusing on women as individuals, having distinct personalities, experiencing a range of socio-cultural influences, and thus exhibiting a range of responses to the social construction of IT. This, in turn, can facilitate more nuanced studies of gender that explore the multiple identities of women—for example, race and gender, or sexual orientation and gender, or age and gender—and their relationships to information technology.

One stream of future work will explore the role of epistemology and methodology in conducting gender and IT research using the Individual Differences Theory of Gender and IT. Another stream of research will explore the contribution of organizational factors to the underrepresentation of women in IT by focusing on the articulation of workplace factors that enhance and inhibit women's participation in IT work and women's varying responses to them. A third stream of research will apply the individual differences theory of gender and IT to an examination of differences in Internet search behavior across a variety of uses.

ACKNOWLEDGMENT

This article is from a study funded by a National Science Foundation Grant (EIA-0204246).

REFERENCES

Adam, A., Emms, J., Green, E., & Owen, J. (Eds.). (1994). *Women, work and computerization: Breaking old boundaries—Building new forms.* Amsterdam, The Netherlands: North-Holland.

Adam, A., Howcroft, D., & Richardson, H. (2001). Absent friends? The gender dimension in information systems research. In N. Russo, B. Fitzgerald, & J. DeGross (Eds.), *Realigning research and practice in information systems development: The social and organizational perspective* (pp. 332-352). Boston: Kluwer Academic Publishers.

Adam, A., Howcroft, D., & Richardson, H. (2004). A decade of neglect: Reflecting on gender and IS. *New Technology, Work and Employment, 19*(3), 222-240.

Balka, E., & Smith, R. (Eds.). (2000). *Women, work and computerization: Charting a course to the future.* Boston: Kluwer Academic Publishers.

Berger, P. L., & Luckmann, T. (1966). *The social construction of reality: A treatise in the sociology of knowledge.* New York: Doubleday.

Cockburn, C. (1983). *Brothers: Male dominance and technological change.* London: Pluto Press.

Cockburn, C. (1988). *Machinery of dominance: Women, men, and technical know-how.* Boston: Northeastern University Press.

Cockburn, C., & Ormrod, S. (1993). *Gender and technology in the making.* London: Sage.

Dennis, A. R., Kiney, S. T., & Hung, Y. (1999). Gender differences in the effects of media richness. *Small Group Research, 30*(4), 405-437.

Eriksson, I. V., Kitchenham, B. A., & Tijdens, K. G. (Eds.). (1991). *Women, work and computerization: Understanding and overcoming bias in work and education.* North-Holland, Amsterdam.

Gefen, D., & Straub, D. W. (1997, December). Gender differences in the perception and use of e-mail: An extension to the technology acceptance model. *MIS Quarterly, 21*(4), 389-400.

Lovegrove, G., & Segal, B. (Eds.). (1991). *Women into computing: Selected papers 1988-1990.* London: Springer-Verlag.

Marini, M. M. (1990). Sex and gender: What do we know? *Sociological Forum, 5*(1), 95-120.

Morgan, A. J., Quesenberry, J. L., & Trauth, E. M. (2004). Exploring the importance of social networks in the IT workforce: Experiences with the "Boy's Club." In E. Stohr & C. Bullen (Eds.), *Proceedings of the 10th Americas Conference on Information Systems,* New York (pp. 1313-1320).

Nielsen, S., von Hellens, L., Greenhill, A., & Pringle, R. (1998, March 26-28). Conceptualising the influence of cultural and gender factors on students' perceptions of IT studies and careers. In *Proceedings of the 1998 ACM SIGCPR Computer Personnel Research Conference,* Boston (pp. 86-95).

Nielsen, S., von Hellens, L., Pringle, R., & Greenhill, A. (1999). Students' perceptions of information technology careers: Conceptualising the influence of cultural and gender factors for IT education. *GATES, 5*(1), 30-38.

Nielsen, S., von Hellens, L., & Wong, S. (2000, December). *The game of social constructs: We're going to WinIT!* Panel discussion paper, Addressing the IT Skills Crisis: Gender and the IT Profession. International Conference on Information Systems (ICIS 2000), Brisbane, Australia.

Pringle, R., Nielsen, S., von Hellens, L., Greenhill, A., & Parfitt, L. (2000). Net gains: Success strategies of professional women in IT. In E. Balka & R. Smith (Eds.), *Women, work and computerization: Charting a course to the future* (pp. 23-33). Boston: Kluwer Academic Publishers.

Quesenberry, J. L., Morgan, A. J., & Trauth, E. M. (2004). Understanding the "Mommy Tracks": A framework for analyzing work-family issues in the IT workforce. In M. Khosrow-Pour (Ed.), *Proceedings of the Information Resources Management Association Conference* (pp. 135-138). Hershey, PA: Idea Group Publishing.

Quesenberry, J. L., & Trauth, E. M. (2005). The role of ubiquitous computing in maintaining work-life balance: Perspectives from women in the IT workforce. In Y. Sorensen, K. Yoo, K. Lyytinen & J. I. DeGross (Eds.), *Designing ubiquitous information environments: Socio-technical issues and challenges* (pp. 43-55). New York: Springer.

Quesenberry, J. L., Trauth, E. M., & Morgan, A. J. (2006). Understanding the "Mommy Tracks": A framework for analyzing work family balance in the IT workforce. *Information Resource Management Journal, 19*(2), 37-53.

Slyke, C. V., Comunale, C. L., & Belanger, F. (2002). Gender differences in perceptions of Web-based shopping. *Communications of the ACM, 45*(7), 82-86.

Spender, D. (1995). *Nattering on the net: Women, power and cyberspace.* North Melbourne, Victoria: Spinifex Press.

Star, S. L. (Ed). (1995). *The cultures of computing.* Oxford: Blackwell Publishers.

Trauth, E. M. (2002). Odd girl out: An individual differences perspective on women in the IT profession [Special Issue on Gender and Information Systems]. *Information Technology and People, 15*(2), 98-118.

Trauth, E. M., Huang, H., Morgan, A. J., Quesenberry, J. L., & Yeo, B. (2006). Investigating the existence and value of diversity in the global IT workforce: An analytical framework. In F. Niederman & T. Ferratt (Eds.), *Managing information technology human resources* (pp. 333-362). Greenwich, CT: Information Age Publishing.

Trauth, E. M., Nielsen, S. H., & von Hellens, L. A. (2003). Explaining the IT gender gap: Australian stories for the new millennium. *Journal of Research and Practice in IT, 35*(1), 7-20.

Trauth, E. M., & Quesenberry, J. L. (2005). Individual inequality: Women's responses in the IT profession. In G. Whitehouse (Ed.), *Proceedings of the Women, Work and IT Forum,* Brisbane, Queensland, Australia.

Trauth, E. M., & Quesenberry, J. L. (2006). Gender and the information technology workforce: Issues of theory and practice. In P. Yoong & S. Huff (Eds.), *Managing IT professional in the Internet age.*

Trauth, E. M., Quesenberry, J. L., & Morgan, A. J. (2004). Understanding the under representation of women in IT: Toward a theory of individual differences. In M. Tanniru & S. Weisband (Eds.), *Proceedings of the 2004 ACM SIGMIS Conference on Computer Personal Research* (pp. 114-119). New York: ACM Press.

Trauth, E. M., Quesenberry, J. L., & Yeo, B. (2005). The influence of environmental context on women in the IT workforce. In M. Gallivan, J. E. Moore, & Yager, S. (Eds.), *Proceedings of the 2005 ACM SIGMIS CPR Conference on Computer Personnel Research* (pp. 24-31). New York: ACM Press.

Venkatesh, V., & Morris, M. G. (2000). Why don't men ever stop to ask for directions? Gender, social influence, and their role in technology acceptance and user behavior. *MIS Quarterly, 24*(1), 115-139.

von Hellens, L., Nielsen, S., & Trauth, E. M. (2001). Breaking and entering the male domain: Women in the IT industry. In *Proceedings of the 2001 ACM*

SIGCPR Computer Personnel Research Conference.

von Hellens, L., Pringle, R., Nielsen, S., & Greenhill, A. (2000). People, business and IT skills: The perspective of women in the IT industry. In *Proceedings of the 2000 ACM SIGCPR Computer Personnel Research Conference.*

Wajcman, J. (1991). *Feminism confronts technology.* University Park, PA: The Pennsylvania University Press.

Wajcman, J. (2000). Reflections on gender and technology studies: In what state is the art? *Social Studies of Science, 30*(3), 447-464.

Webster, J. (1996). *Shaping women's work: Gender, employment and information technology.* London: Longman.

Wilson, M., & Howcroft, D. (2000). The role of gender in user resistance and information systems failure. In R. Baskerville, J. Stage, & J. I. DeGross (Eds.), *Organizational and social perspectives on information technology* (pp. 453-472). Boston: Kluwer Academic Publishers.

KEY TERMS

Essentialism: The assertion of fixed, unified and opposed female and male natures (Wajcman, 1991, p. 9). This theoretical perspective is used to explain the under representation of women in the information technology field by arguing the existence of "essential differences" between males and females with respect to engagement with information technology.

Individual Differences Theory of Gender and IT: A social theory developed by Trauth (Trauth, 2002; Trauth et al., 2004) that focuses on within-group rather than between-group differences to explain differences in male and female relationships with information technology and IT careers. This theory posits that the under representation of women in IT can best be explained by considering individual characteristics and individual influences that result in individual and varied responses to generalized environmental influences on women.

Social Construction: A theoretical perspective articulated by Peter Berger and Thomas Luckmann (1967) that focuses on social processes and interactions in the shaping of actors. This theoretical perspective is used to explain the under representation of women in the information technology field by arguing that technology—socially constructed as a masculine domain—is in conflict with socially constructed feminine identity.

ENDNOTE

[1] See Wilson and Howcroft (2000) for an example of how context enriches the analysis of observed differences in behavior toward IT based upon gender.

Theorizing Masculinity in Information Systems Research

Ben Light
University of Salford, UK

INTRODUCTION

I want to argue that understanding masculinity is an important part of understanding gender and sexuality as it relates to information and communications technologies (ICTs), specifically those under the lens of the information-systems community. In order to do this, the landscape of gender and sexuality research in general is referred to along with such research in the field of information systems (IS), with reference as necessary to masculinity studies. I will then suggest some possible areas where a more thoroughgoing theorization may prove useful. In sum, future research might focus on the relationship between marginalised masculinities and the construction and consumption of IS in work organisations and society.

BACKGROUND

For my purposes here, gender is seen as a system of social practices that creates and maintains gender distinctions, which are used to organise relations of inequality (Wharton, 2005). The field, which Beasley (2005) links with sexuality studies, can be broadly seen as comprising feminist studies, masculinity studies, and sexuality studies. Feminist studies refute the masculine bias of mainstream Western thinking and practices that render women marginal and distort understandings of men. These studies usually put women at centre stage. Masculinity-studies writers are largely social constructivist and profeminist in nature, and aim to critique and destabilise mainstream conceptualisations of masculinity. These studies usually put men at centre stage. Sexuality writers generally focus on lesbian, gay, and transsexual regimes with the similar aim of social destabilisation, usually from a postmodern perspective. Some feminist studies do consider sexu-

ality, but in general, as with masculinity studies, the emphasis is on sexed regimes. Sexuality studies in turn generally pay less attention to gender. Within IS, gender-theory-informed work usually refers to feminist studies and the literatures on gender and technology (again, mostly feminist in nature). Masculinities and the gender link with sexualities are seldom theorized in any great depth. Here, the focus is mostly on masculinity, although I recognise the inextricable links between gender and sexual regimes.

GENDER, IS, AND THEORIZING MASCULINITY

Within IS, gender is a neglected area of investigation. As A. Adam (2002) points out, given gender is viewed as an important part of behaviour in the social sciences, and IS is often viewed as sociotechnical, with the social being as much a part of the genealogy of the area as the technical; it is then surprising that this aspect of the social is neglected. Certainly, gender relations are one of the few things in life that everyone has a stake in, yet this does not seem to enter the discussions about IS as being relevant. Instead, the focus is on relevance to the management of organisations in terms of such things as improving performance. Indeed, the landscape of gender and IS research is pretty limited. A. Adam, Howcroft, and Richardson's (2004) survey of the top mainstream IS journals unearthed only 19 papers published during the period of 1993 to 2002. They found, with a few exceptions, a lack of reference to the more general literature on gender and technology and a focus on using gender as a fixed variable. Moreover, those articles that theorized gender seemed to be free of a thoroughgoing theorization of masculinity. Looking toward the gender and technology literature, studies of masculinities and technologies are also sparse. Whilst Lohan and Faulkner's (2004) introduction to their special issue on masculinities and technologies states that feminist technology studies have much to learn from masculinities studies, and vice versa, they point out that the articles submitted came from the former, despite soliciting contributions widely. This dearth of research matters because whilst there may be some commonly recognised masculinity characteristics, it is necessary to understand the mutual shaping of these and technology. Indeed, those who study gender and technologies, specifically those in the area of feminist technology studies and gender and IS, subscribe to Wajcman's (1991) view of technology as a masculine culture. Moreover, IS is still recognised to be dominated by men as a field of academic study and in work organisations and society (Panteli, Stack, & Ramsay, 2001; Robertson, Newell, Swan, Mathiassen, & Bjerknes, 2001). So, gender is relevant to IS research, and I would suggest that developing a rigorous understanding of masculinities could contribute to the field.

A FOCUS ON MASCULINITIES STUDIES

It is necessary to distinguish between the men's movement that emerged in the 1970s and the academic study of masculinities. The men's movement tends to shore up stereotypical conceptualisations of Western masculinity and its dominant social status. Implicated in the movement are, for example, antifeminism and homophobia. Thus, the academic study of masculinities has an equivocal relationship with the mainstream politics of the men's movement. Masculinity-studies writers do not take up the cause of masculinity; they seek to understand and critique its role in gendered and sometimes sexual regimes. Additionally, in contrast to the men's movement, there is a more wide-ranging recognition of differences amongst masculinities. Masculinity is presented as involving the dominance of men over women and other men. From this there are two considerations.

First, who are these other men and masculinities? To answer this, it is necessary to direct attention to theorizations of masculinity that recognise multiple differences. In this respect, one of the most pervasive concepts in masculinity studies is that of hegemonic masculinity (Carrigan, Lee, & Connell, 1985; Connell, 1987). Hegemonic masculinity draws on Antonio Gramsci's concept of hegemonic domination. It is concerned with the idea that a particular definition of masculinity is culturally exalted and latently imposed upon those in society. This hegemonic masculinity is not seen as merely held by one person or group; it is institutionalised. In line with Wharton's (2005) more general theorization of gender, this locates the forces of oppression outside the individual and recognises its meshing with wider society. Thus, legal, welfare, and educational institutions and others like them become implicitly and explicitly imbued with the masculinity. This hegemonic masculinity subordinates competing masculinities and, of course, is used as a reference for subordinating femininities and constructing emphasized femininity (Connell, 1987). In terms of the domination of men by men, we can see competing, socially constructed categories of masculinities, such as race, age, ethnicity, sexuality, disability, and class, come into play—categories, of course, that are equally appropriate to women (Trauth, 2002). Indeed, as Lohan (2001) notes in relation to the domestic landline telephone, similarities and differences in the use of technologies and the construction of gender relations around these may vary in terms of age and sexuality. Moreover, she notes that such constructions might be further unpacked within any given group, as I shall illustrate later in relation to gay men.

The second point highlights contradictions in masculinity studies. Even though Connell (1987) states that masculinity is not inherently a man's characteristic, men are usually the foci of investigation. Thus, care has to be taken to avoid drifting into essentialist accounts of masculinity that determine it as the purview of one sex. Indeed, Halberstam (2002) argues that those who do not interpret masculinity as a synonym for men are rare. However, in Western society at least, that which is deemed masculine is constructed as held by men. For example, we could look at the film adaptation of the Marvel comic book *The X-Men* (2000). Ironically, in a fantasy world where anything is possible, the status quo prevails. The team is called the X-Men when there are several women members. We learn that two men built the supercomputer Cerebro and that it is too powerful for Dr. Jean Gray to use; she does not have the "control" Professor Charles Xavier has.

Although I have unpacked the idea of masculinity into masculinities, there are undoubtedly marginalised groups of men such as those who are black, gay, or disabled: They are still men. As such, it is important to recognise that in the main, these groups will still usually retain a relatively more privileged position in society as a whole. Their gender often compensates for the aspects of their identity that otherwise marginalise them. Certainly, it might be argued that masculinities research merely privileges the already privileged and steals the very limited space for discussing women. Moreover, assigning masculinity to women and marginalised groups of men implicates them in their own domination. This maybe goes some way to explaining why masculinity studies is filled with studies of white, middle-class, straight men, written mostly by white, middle-class, straight men; these men do not want to speak for other masculinities (Beasley, 2005). Yet, studying masculinities in a critical fashion, particularly with emphasis on marginalised masculinities, does have use. Of course, it is necessary to face the facts of gender power relations in relation to men, but it is also important to do this without simplifying the situation either. After all, if gender is relational, adequate theorizations of the concepts it comprises are required.

FUTURE TRENDS

So far, I have argued that masculinity would benefit from unpacking in gender and IS research, and have drawn upon work from masculinity studies in an attempt to show some of what is available. Drawing upon the concept of hegemonic masculinity and its associations with multiple masculinities, this section will briefly, and by necessity, arbitrarily, cast an eye over a few IS-related examples to give a flavour of the potential insights that might be gained.

Within the context of work, probably the major focus for IS research in general, the most obvious area for attention is men's dominance of the IT sector and the consequent association of IS with men's work. Drawing upon masculinity studies might, for example, lead to considerations of who, if anyone, meets the current hegemonic masculinity in the industry. What I mean here is that it would be worth-

while identifying what constitutes the ideal-type masculinity that dominates the lives of those who work in IT and the coping strategies they employ in response. This further raises the question of how the contradictions between any given hegemonic masculinity and competing lesser masculinities (those that are gay or disabled, for example) are managed to secure an identity. Moreover, it suggests a need for inquiry about the relative positioning of masculinities and their relation to technology at work. For example, in the same way as some so called women's technologies, such as the domestic landline telephone (Lohan, 2001), are seen as lesser in status than more masculine ones, some masculine technologies and the roles associated with them might be relatively positioned to others. Indeed, it has been argued that IS development is perceived as being at the frontier whilst maintenance is viewed as menial, dull work (Swanson & Beath, 1989). In a recent study, for example, it was reported that women still appear to predominate in such low-status roles as maintenance and reporting whether they are new entrants or highly experienced (Dattero & Galup, 2004). Of course, some men may also have to undertake these roles given the lack of women in the industry overall, so it would be interesting to see how their masculinities are managed where they do so. For instance, it has been argued that IS development is particularly amenable to contradictions in masculinity. Knights and Murray's (1994) study of masculinity in software development illustrates that the role is simultaneously presented as requiring an objective, logical, project-focused approach, but one that also relies upon what they call a much-celebrated "'gung ho' form of 'macho' masculinity" (p. 125). From an end-user perspective, there is also the question of where ICTs and IS-based technologies sit in relation to masculinities when other technologies, which are deemed more masculine than ICTs, are used as part of the job. For example, within the U.S. Navy, supply officers who usually undertake administrative roles are seen as weaker because they control people and IS rather than a jet or ship (Barrett, 2001). Finally, the links made between masculinities and technology, as discussed above, are not reserved for developers: A few studies have pointed to this in relation to the adoption of software. For example, Adam and O'Doherty (2000) talk of how at one organisation, a particular piece of technology was adopted "to show the big boys" (that is, to show off to those in larger organisations), and Light (2005) singles out bravado as a shaping influence in the process of adoption. However, those on the receiving end of new technologies that have been adopted may find themselves on the back foot as they are inextricably associated with existing computer-based systems. The problem here is that in order to make way for the new technology, existing arrangements have to be problematized or delegitimised. This usually means that a story is constructed about how they are failing to perform. Thus, in order to preserve some sense of masculinity and save face, users might reconstruct their identities based on discourses around succeeding in getting failing technologies to work, while others might not (Alvarez, 2001). Indeed, there are similarities between some of Alvarez's findings and those of Thorsby and Gill (2004), who report on how men constructed their experiences of in vitro fertilization (IVF) technology in a way that did not undermine their identities as fertile men.

In wider society, the idea of multiple masculinities might be used to consider how marginalised masculinities shape and are shaped by ICTs and IS. For example, on the fringes of IS, there have been a few studies that consider gay men; but whilst representing a further unpacking of masculinity, they do not draw upon the masculinities-studies field. For example, in one UK report, gay men are treated as a homogenous promiscuous group who cannot live without the supporting tools of the mobile phone and Internet chat rooms (Anderson et al., 2002). In another more rigorous study of gay and lesbian Taiwanese (mostly student-based) communities, this stereotypical view is challenged with Internet chat rooms being seen as a useful political device (Yang, 2000). However, multiple masculinities are not really discussed. Thus, these studies miss what I would term any subhegemonic gay masculinity that dominates the construction of what is masculine and feminine within the gay community, and how technology may be implicated in the shaping of this. For instance, one could take the popular Gaydar community Web site that operates in about 159 countries. Gaydar is a software-enabled, Internet-based com-

munity where mostly gay men, and some other groups of men who share a sexual preference for men (such as bisexuals and transsexuals), socialise. The software used to create user profiles is standard in nature and is configured by the user based on drop-down menus, tick boxes, and some freeform text. In configuring the software, the users are essentially configuring a version of their identity based on a mix of masculine and feminine characteristics. Indeed, the way this is configured itself is a representation of this. Thus, for example, users are configured as sexually dominant or submissive, and as enjoying hobbies such as dancing, clothes shopping, car maintenance, or bodybuilding. The profile created can be very detailed and results in the intended and unintended categorisation of the users into groups with identities that are well known within the gay community. Importantly, these groups are subject to interpretation in terms of their associations with a highly personalised view of what it is to be masculine. The mutual shaping of the software and the users on Gaydar ultimately perpetuate these masculinities and their interpretation. Moreover, Gaydar is further shaped and has a wider shaping effect on society as new members join. For instance, the influence of and influence upon those who are discovering their sexuality, know where they stand and have just found Gaydar, or want to extinguish it through online abuse such as cyberstalking all play a part. Indeed, the site is, in some ways, representative of (white, upper and middle-class) male dominance in wider Western society. Gay men are still men, and Gaydar is for men. The community's name, Gaydar, would have you believe this might involve gay women, too—this is not the case. As Adam (2005) states, "we talk of football and women's football, not men's football and women's football" (p. 7). My earlier argument that gender still works favourably for marginalised men, such as those who are gay, finds support here. There are gendered versions of Gaydar, for example, http://www.gaydar.co.uk and http://www.gaydargirls.co.uk. Gay men are the default option; women get singled out as different and arguably lesser individuals, and are bolted on as an afterthought.

CONCLUSION

Gender is a social construction that is used to construct and maintain inequalities in society. Yet, despite gender being seen as a relational concept, feminist studies have largely, some would say quite rightly, neglected to theorize masculinity. This is further in evidence in the literature relating to gender and technology, and gender and IS despite agreement with Wajcman's (1991) view of the realm of technology being deemed masculine. With this in mind, I have attempted to show how a more thoroughgoing theorization of masculinities, especially with a focus on those that are marginalised by hegemonic masculinities, might yield interesting insights into the gender-technology relation.

REFERENCES

Adam, A. (2002). Exploring the gender question in critical information systems. *Journal of Information Technology, 17*(2), 59-67.

Adam, A. (2005). *Gender, ethics and information technology*. Basingstoke, UK: Palgrave Macmillan.

Adam, A., Howcroft, D., & Richardson, H. (2004). A decade of neglect: Reflecting on gender and IS. *New Technology, Work and Employment, 19*(3), 222-240.

Adam, F., & O'Doherty, P. (2000). Lessons from enterprise resource planning implementations in Ireland: Towards smaller and shorter ERP projects. *Journal of Information Technology, 14*(4), 305-316.

Alvarez, R. (2001). "It was a great system": Facework and the discursive construction of technology during information systems development. *Information Technology and People, 14*(4), 385-405.

Anderson, B., Gale, C., Gower, A. P., France, E. F., Jones, M. L. R., Lacohee, H. V., et al. (2002). Digital living: People-centred innovation and strategy. *BT Technology Journal, 20*(2), 11-29.

Barrett, F. J. (2001). The organizational construction of hegemonic masculinity: The case of the US Navy. In S. M. Whitehead & F. J. Barrett (Eds.), *The masculinities reader* (pp. 77-99). Cambridge, UK: Polity Press.

Beasley, C. (2005). *Gender and sexuality: Critical theories, critical thinkers*. London: Sage Publications.

Carrigan, T., Lee, J., & Connell, R. W. (1985). Towards a new sociology of masculinity. *Theory and Society, 14*(5), 551-604.

Connell, R. W. (1987). *Gender and power*. Cambridge: Polity Press.

Dattero, R., & Galup, S. D. (2004). Programming languages and gender. *Communications of the ACM, 47*(1), 99-102.

Halberstam, J. (2002). An introduction to female masculinity. In R. Adams & D. Savran (Eds.), *The masculinity studies reader* (pp. 355-374). Oxford, UK: Blackwell Publishers Limited.

Knights, D., & Murray, F. (1994). *Managers divided: Organisational politics and technology management*. Chichester, UK: John Wiley and Sons Ltd.

Light, B. (2005). Potential pitfalls in packaged software adoption. *Communications of the Association for Computing Machinery, 48*(5), 119-121.

Lohan, M. (2001). Men, masculinities and "mundane" technologies: The domestic telephone. In A. Adam & E. Green (Eds.), *Virtual gender* (pp. 189-206). London: Routledge.

Lohan, M., & Faulkner, W. (2004). Masculinities and technologies. *Men and Masculinities, 6*(4), 319-329.

Panteli, N., Stack, J., & Ramsay, H. (2001). Gendered patterns in computing work in the late 1990s. *New Technology, Work and Employment, 16*(1), 3-17.

Robertson, M., Newell, S., Swan, J., Mathiassen, L., & Bjerknes, G. (2001). The issue of gender within computing: Reflections from the UK and Scandinavia. *Information Systems Journal, 11*(2), 111-126.

Swanson, E. B., & Beath, C. M. (1989). *Maintaining information systems in organizations*. Chichester, UK: John Wiley and Sons.

Trauth, E. M. (2002). Odd girl out: An individual differences perspective on women in the IT profession. *Information Technology and People, 15*(2), 98-118.

Wajcman, J. (1991). *Feminism confronts technology*. Oxford, UK: Polity Press.

Wharton, A. S. (2005). *The sociology of gender: An introduction to theory and research*. Oxford, UK: Blackwell Publishing.

Yang, C-C. (2000). The use of the Internet among academic gay communities in Taiwan: An exploratory study. *Information Communication and Society*, *3*(2), 153-172.

KEY TERMS

Emphasized Femininity: That which is created and supported to ensure compliance with and complicit affirmation of the hegemonic masculinity at any given time.

Hegemonic Masculinity: Drawing upon Gramsci's hegemonic domination, it is concerned with the idea that a particular definition of masculinity is culturally exalted and latently imposed upon those in society through collective construction and institutional application.

Masculinity: Often recounted as the way men behave, and see and think about themselves. However, more sophisticated views recognise it as a socially constructed category including such characteristics as aggression and technical capability, which are not biologically determined and thus are capable of being held by men and women.

Sexed Regime: A prevailing order of gendered categories such as men and women.

Sexuality: An expression of sexual interest and preference.

Sexual Regime: A prevailing order of identities and practices related to sexuality categories such as gay, bisexual, and straight.

Subhegemonic Masculinity: A masculinity that may dominate a specific group of masculinities within a given wider society. This submasculinity may also refer to the hegemonic masculinity, but it is subordinated by it.

Third World Feminist Perspectives on Information Technology

Lynette Kvasny
The Pennsylvania State University, USA

Jing Chong
The Pennsylvania State University, USA

INTRODUCTION

Historically, information systems (IS) researchers have conducted empirical studies of gender and information technology (IT) in business organizations. These studies cover a wide range of topics such as the under-representation of women in the IT workforce (von Hellens, Nielsen, & Trauth, 2001) and the educational pipeline, which prepares women for careers in computer-related fields (Camp, 1997; Symonds, 1999). IS researchers have generally embraced an essentialist approach to examine gender differences in the adoption and use of IT (Gefen & Straub, 1997; Venkatesh & Morris, 2000), career selection (Joshi & Kuhn, 2001; Nielsen, von Hellens, Greenhild, & Pringle, 1998), employment experiences (Gallivan, 2003; Sumner & Niederman, 2002; Sumner & Werner, 2001), and employment outcomes (Baroudi & Igbaria, 1997). More recently, however, researchers have adopted anti-essentialist stances and extended IS gender studies to include individual differences among women (Trauth, 2002; Trauth, Quesenberry, & Morgan, 2004), as well as race and ethnicity (Kvasny & Trauth, 2002; Tapia & Kvasny, 2004; Tapia, Kvasny, & Trauth, 2004).

In this growing body of scholarship, a few researchers have argued persuasively for the inclusion of feminist epistemologies in IS research (Adam & Richardson, 2001; Henwood, 2000; Kvasny, Greenhill, & Trauth, 2005). These proponents contend that feminist epistemologies provide theoretical and methodological insights for studying gender as a complex and multidimensional construct for understanding the use, management, and regulation of IT in multiple domains such as business organizations, households, reproductive health, built environments, and the military (MacKenzie & Wacjman, 1991;

Ormrod, 1994). Feminist scholars have also called for research that considers not only gender, but also the intersection of racial, ethnic, and class identities (Kvasny, forthcoming).

In this article, we adopt a third world feminist perspective to examine perceptions of IT held by black women in Kenya and the U.S. In what follows, we defining third world feminism, especially as it relates to women in the African Diaspora. Next, we discuss our research methodology, which consists of interviews with women in both settings. We conclude by presenting our findings and implications for future research.

BACKGROUND: THIRD WORLD FEMINISM

The term "third world" captures the discourse that typifies women of color from around the globe as an oppressed group having relatively less formal education, higher birth rates, and lower incomes. These discourses generally employ "emblems of oppression", that is, the use of single practices such as foot binding in China, veiling in the Middle East, and female circumcision in Africa as emblematic of the totality of women's experience in a particular culture. In doing so, women's experiences are collapsed into a single, "victimizing practice" which ignores the multiplicity of ways in which these practices are experienced by women and the ways in which women exercise their agency (Lorde, 1985).

Third world also carries the connotation of colonized populations located in geographically distant nation-states under the economic and political control of so-called developed nations in the West. However, women of color in Western contexts have embraced a third world identity by applying the term "third world" to themselves and their politics to call attention to similarities in locations of, and problems faced by, their communities and communities in third world cultures (Narayan, 1997). It is a call for feminism without the boarders of socio-economic class, race, ethnicity, nationality, and sexual orientation. To silence the voices of diverse women is to deny the opportunity to realize the connections as well as the differences among women. It constructs women of color as voiceless victims who are spoken about and constructed by privileged women in the academy. It is unfair to merely assume that working-class women, middle-class women, lesbians, women in developing countries, and women of color share a common oppression based upon a shared gender. This colonialist stance, according to Narayan (1997), replicates the problematic aspects of Western representations of third world communities, and thus poses an obstacle to the need for feminists to form communities of resistance.

For Smith (1981), "Feminism is the political theory and practice to free all women: women of color, working-class women, poor women, physically challenged women, lesbians, old women—as well as white economically privileged heterosexual women. Anything less than this is not feminism, but merely female self-aggrandizement." Much of the gender and IT research has been about the under-representation of highly paid, college educated women employed in the primary IT sector. There are relatively few IS studies which employ an anti-essentialist epistemology for the study of the lived experiences of economically disadvantaged women of color engaging with IT to improve their life chances (Kvasny, forthcoming).

RESEARCH APPROACH

To gain a third world feminist perspective on IT and to understand what specific differences IT has made socially, politically and economically for black women, we interview 40 black female participants in IT training programs in the U.S. (8) and Kenya (32). The women in the U.S. were participating in a community technology center located in an inner-city neighborhood. The women in Kenya were enrolled in IT bachelor's degree program at a univer-

sity. These training programs provided women with their initial entrée into the domain of IT.

Using Cameron's (1992) notion of empowering research, we conducted interviews to understand their motivations for participating in their respective IT programs, and expectations for outcomes resulting from this training. In what follows, we focus exclusively on the women's motivations for participating in IT training programs by recounting the common themes which emerged from their narratives.

FINDINGS

Even though women in Kenya have traditionally been active in the informal economies around agriculture and local trade, and the women in the U.S. had limited formal educations and held low-paying jobs in the service sector, they both perceived IT as a panacea for acquiring desirable job skills and employment that would lift them out of poverty. For instance, nearly one-half of the women in Kenya participated in the IT educational programs because they perceived substantial job opportunities upon graduation. The IT sector was described as "an upcoming field," and as "a new field in Kenya and a very dynamic field which affects all aspects globally." They also believed that there were few IT professionals and therefore skilled people have a competitive advantage. One woman remarked that "not many people in Kenya have this sort of information [and] this is because currently in Kenya there lacks professionals in this field." Not only were there "job opportunities that come with this vast growth," the jobs were seen as well paying. "I think IT is a field that will provide me with a means of earning good income in future." "IT program have proved to be better paying careers than other technical careers in the country."

Many Kenyan women remarked specifically about acquiring skills which would enable them to integrate IT into business organizations. For instance, "the integration of business in the IT program made it even more attractive for me." "This course is not a technical course. I am not interested in details about technologies...I am interested in how I can use IT more efficiently and a broader view." U.S. women mirrored this belief about leveraging IT skills in the workplace. "I will learn a lot of computer applications when I finish this class. I will be able to get a better job and better opportunities. I will conquer the digital divide. We all need to learn these computer applications. We will need this information to be successful in the business world."

Some women were more entrepreneurial, and saw IT as a way to start their own business. "Since I have the basics of IT and my course provides a grounding I can build up on my own, I could start my own enterprise using this knowledge." For some Kenyan women, business ownership was once a dream that now can potentially be achieved. "Given that I would like to learn IT so that I run my own IT firm in future. If I do not take this chance to learn IT, then my dream will not be accomplished."

We also found that the American narratives were not about gradual movement; they were about rapid escape from oppression and positive outlooks on the future. "I have certainly had a successful computer orientation and beginning. It is truly an exciting journey. My goal is to continue my training with the ultimate goal focusing on certification status. Then, it is look out world as I am on my way!"

Kenyan women, in particular, were motivated by perceived gender inequalities. For them, IT offered an opportunity for overcoming oppression and achieving parity with men. "Gone are the days when there were specific jobs/careers for men and women. Women now want the challenge." "More and more women want to play an active role in their society and in the world…women want to be involved in the IT sector (not to be left behind by their male counterpart)." IT represented a vehicle which would enable them to engage in an activity which has been historically perceived as a male domain. "The simple reason why women participate in this IT program is because men do the same thing. Equality is something that women have all been fighting for and have accomplished their goal. If a man can participate in IT, why shouldn't a woman do the same thing?" Women not only want to do the same thing as men, they want to adopt IT "because it is beneficial to them too as much as it is to men… it will enable us as women to compete fully with men in jobs."

Women in both countries believed that women in IT-related professions are "able to successfully represent other women in our country" and "able to adapt to contribute to society by raising awareness about what IT can do for a nation." Thus the training provided immediate benefits to the recipient, but also external benefits to other women and the entire nation. These were pioneering women who were not content to "stick to the stereotype that certain jobs are for women." They wanted to demonstrate that "they are clever enough to prove that they can master a tough course like IT and do well."

African-American women didn't speak of gender inequities, but they did speak of community solidarity. "[W]e are taking computer classes that have connected us with the great information divide. We are no longer left behind … We are still traveling on the road of information freedom and enjoying every minute of it. There is so much to be learned, and the information is available because we made the first step, receiving information and taking the steps to change our future in the usage of the computer in our everyday life. We now realize that the Internet is the mode of travel for today as well as tomorrow." This woman often uses communal words like "we", "our" and "us" to signify the collective advancement of working-class people in her neighborhood. The sense of freedom and inclusion is important as it signifies a break from the isolation expressed by inner city residents.

FUTURE TRENDS AND CONCLUSION

In an IS discourse community comprised of scholars working primarily from a western European frame of reference, what can a third world feminist perspective offer to the scholarly discourse about gender and IT? Third world feminism provides a lens for analyzing the oppression faced by women of color, and the perceived role of IT in alleviating inequities. This oppression comes not only from gender, but also from race, ethnicity, poverty, and institutional policies that limit their human agency. These factors are interwoven, which clearly suggests the need for scholarship that provides a more nuanced understanding of women's issues and IT.

The women in this study saw IT as a mechanism for gaining access to other people's privileges. Notably, the two words "market" and "job" dominate the women's narratives, which hint that those women are still on the level of satisfying their most basic needs, and are seeking employment opportunities for improving their life chances. When examining employment, it is important to consider that jobs provide more than economic ends. Jobs also provide a sense of belonging to society and to a profession. Employment can improve self-esteem, feelings of accomplishment and independence, as well as provide access to healthcare and education.

The women were also attracted by these programs because popular discourses tend to romanticize IT and lull people into believing that practical computing skills are easily translated into high paying jobs. Here practical can be translated into skills desired by employers. Some women even mentioned directly that "it will guarantee an instant job." Thus, the job market was seen as highly elastic for people with marketable IT skills. Women in both settings greatly privileged hands on training and frowned upon theoretical learning because practical training could be more readily converted to marketable skills.

Given the pervasive history of racial and gendered oppression, women across the African Diaspora have limited mechanisms for representing and demonstrating the fullness of their abilities, aspirations, and accomplishments. These women did, however, generate self-defined perspectives which grew out of their struggle to appropriate IT. The dominant discourse of IT as a mechanism for empowerment and increased workforce participation resonated with the women's deep and justifiable frustrations. Even though they suffered, they believed that IT presented a real opportunity for change.

Gender-as-variable studies which measure differences in IT adoption, use, and employment do little to enact social change; they simply measure the status quo. IS researchers should therefore adopt a feminist or other gender-as-relations approaches to understand the situated nature of IT as experienced by diverse women, and enact praxis oriented methods such as action research and participatory design techniques to create socio-technical interventions that enable women to realize their economic, cultural and political potential. Then, perhaps, third world women can fully realize other people's privilege.

REFERENCES

Adam, A., & Richardson, H. (2001). Feminist philosophy and information systems. *Information Systems Frontiers, 3*(2), 143-154.

Baroudi, J., & Igbaria. M. (1995). An examination of gender effects on career success of information systems employees. *Journal of Management Information Systems, 13*(1), 127-143.

Cameron, D. (1992). Respect, please! Investigating race, power, and language. In D. Cameron, E. Frazer, P. Harvey, M. Rampton, & K. Richardson (Eds.), *Researching language: Issues of power and method* (pp. 113-130). London: Routledge.

Camp, T. (1997). The incredible shrinking pipeline. *Communications of the ACM, 40*(10), 103-110.

Davis, A. (1981). *Women, race, & class.* New York: Random House.

Gallivan, M. (2003, April 10-12). Examining gender differences in it professionals' perceptions of job stress in response to technological change. *Proceedings of the ACM SIGMIS Conference on Freedom in Philadelphia: Leveraging Differences and Diversity in the IT Workforce* (pp. 10-23). Philadelphia, PA.

Gefen, D., & Straub, D. (1997). Gender differences in the perception and use of e-mail: An extension to the technology acceptance model. *MIS Quarterly, 21*(4), 389-400.

Henwood, F. (2000). From the women question in technology to the technology question in feminism: Rethinking gender equality in it education. *The European Journal of Women's Studies, 7*(1), 209-227.

Joshi, K. D., & Kuhn, K. (2001, April 19-21). Gender differences in is career choice: Examine the role of attitudes and social norms in selecting the IS profession. *Proceedings of the ACM SIGCPR Conference on Computer Personnel Research* (pp. 121-124). San Diego, CA.

Kvasny, L. (forthcoming). *Let the sisters speak: Understanding information technology from the standpoint of the other.* The DATA BASE for Advances in Information Systems.

Kvasny, L., & Trauth, E. (2002). The digital divide at work and home: The discourse about power and underrepresented groups in the information society. In E. Wynn, E. Whitley, M. Myers, & J. DeGross (Eds.), *Global and organizational discourse about information technology* (pp. 273-291). Deventer: Kluwer.

Kvasny, L., Greenhill, A., & Trauth, E. (2005). Giving voice to feminist projects in MIS Research. *Journal of Technology and Human Interaction, 1*(1), 1-18.

Lorde, A. (1984). *Sister outsider.* Berkeley: Crossing Press Feminist Series.

Mohanty, C. (2003). *Feminism without borders: Decolonizing theory, practicing solidarity.* Chapel Hill: Duke University Press.

Narayan, U. (1997). *Dislocating cultures: Identities, traditions, and third-world feminism.* New York: Routlege.

Nielsen, S., von Hellens, L., Greenhill, A., & Pringle, R. (1998). Conceptualising the influence of cultural and gender factors on students' perceptions of it studies and careers. *Proceedings of the ACM SIGCPR Conference on Computer Personnel Research* (pp. 86-95). Boston.

Ormrod, S. (1994). Let's nuke the dinner: Discursive practices of gender in the creation of a new cooking process. In C. Cockburn & R. Furst-Dilic (Eds.), *Bringing technology home* (pp. 42-59). Buckingham: Open University Press.

Sumner, M., & Niederman, R. (2002, May 14-16). The impact of gender differences on job satisfaction, job turnover, and career experiences of information systems professionals. *Proceedings of the ACM SIGCPR Conference on Computer Personnel Research* (pp. 154-162), Kristiansand, Norway.

Sumner, M., & Werner, K (2001, April 19-21). The impact of gender differences on the career experiences of information systems professionals. *Proceed-*

ings of the ACM SIGCPR Conference on Computer Personnel Research (pp. 125-131), San Diego, CA.

Symonds, J. (2000, June 8-11). Why IT doesn't appeal to young women. *Proceedings of the IFIP TC9/WG9.1 7th International Conference on Woman, Work, and Computerization: Charting a Course to the Future* (pp. 70-77), Vancouver, British Columbia.

Tapia, A., & Kvasny, L. (2004, April 22-24). Recruitment is never enough: Retention of women and minorities in the IT workplace. *Proceedings of the ACM Conference on Computer Personnel Research: Careers, Culture, and Ethics in a Networked Environment* (pp. 84-91), Tucson, AZ.

Tapia, A., Kvasny, L., & Trauth, E. M. (2004). Is there a retention gap for women and minorities? The case for moving in versus moving up. In M. Igbaria & C. Shayo (Eds.), *Strategies for managing IS/IT personnel* (pp. 143-164). Hershey, PA: Idea Group Publishing.

Trauth, E. M. (2002). Odd girl out: An individual differences perspective on women in the IT profession. *Information Technology and People, 15*(2), 98-118.

Trauth, E. M., Quesenberry, J., & Morgan, A. (2004, April 22-24). Understanding the under representation of women in IT: Toward a theory of individual differences. *Proceedings of the ACM Conference on Computer Personnel Research: Careers, Culture, and Ethics in a Networked Environment* (pp. 114-119), Tucson, AZ.

Venkatesh, V., & Morris, M. (2000). Why don't men ever stop to ask for directions? Gender, social influence, and their role in technology acceptance and usage behavior. *MIS Quarterly, 24*(1), 115-139.

von Hellens, L., Nielsen, S., & Trauth, E. M. (2001, April 19-21). Breaking and entering the male domain: Women in the IT industry. *Proceedings of the ACM SIGCPR conference on Computer Personnel Research* (pp. 116-120), San Diego, CA.

Wacjman, J. (1991). *Feminism confronts technology.* University Park, PA: Pennsylvania State University Press.

KEY TERMS

Anti-Essentialism: The belief that while there are biological differences between men and women, gender is not a biological matter; it is a social construction used to create and perpetuate systems of privilege. The term woman (feminine, weak, submissive), for instance, is socially constructed as a binary opposite from the term man (masculine, strong, assertive).

Essentialism: Essentialism is the belief that women are biologically different from men, and that this biological difference has implications for the ways that we think and act.

Feminism: Feminism is a set of social theories and political practices that are critical of past and current social relations which privilege men as a group. Feminism involves the promotion of women's rights, and the belief that men and women should be politically, economically, and socially equal.

Gender as Relation: This is an anti-essentialist view in which women are believed to have unique experiences. The research aim is not to compare men and women, but rather to center women's needs, behaviors, ways of thinking, and experiences.

Gender as Variable: This is an essentialist view in which gender is seen as an objective, often quantifiable, demographic variable. Women are perceived as a single group with common needs, values, and behaviors. Women are generally compared to men to demonstrate gender differences.

Third World Feminism: Third world feminism is a critical set of theories and political practices which gives voice to the issues of women of color from diverse socio-economic class, race, ethnicity, nationality, and sexual orientation. Third world feminism challenges to mainstream feminism have highlighted the ways in which issues central to the lives of women of color have been misrepresented or rendered invisible, and have demanded recognition of the global imbalances in which mainstream feminist agendas are structured.

UN World Summit on the Information Society

Heike Jensen
Humboldt University Berlin, Germany

INTRODUCTION

The World Summit on the Information Society (WSIS) is a United Nations (UN) conference led by the International Telecommunication Union (ITU). It has unique structural features. First, WSIS is comprised of two summit events: one in Geneva, Switzerland, December 10 to 12, 2003, and the other in Tunis, Tunisia, November 16 to 18, 2005. Second, WSIS is characterized by the so-called multistakeholder approach (Association for Progressive Communications [APC] & Campaign for Communication Rights in the Information Society [CRIS], 2003; Hemmati, 2002; Raboy, 2004). In this approach, civil society and the private sector have an institutionalized basis in the summit process from which to engage with governments and inform the political deliberations. The goal set for WSIS is to develop a global consensus on the features that are to characterize the information society and on ways to bring this society about.

BACKGROUND

Historically, WSIS is part of the unprecedented series of UN conferences and summits that have in particular marked the 1990s and early 2000s. These conferences have taken up pressing issues such as sustainable development and the environment, human rights, women's rights, the abolition of racism, and poverty eradication, and have discussed them within a global framework. The impetus to hold a World Summit on the Information Society came from ITU, which adopted a resolution to this effect at its Plenipotentiary Conference in 1998 (Resolution 73; Minneapolis). When the UN General Assembly finally took up this issue and adopted it in 2001 (A/RES/56/183), much of the attention within the UN had already shifted to the Millennium Summit and its follow-up process, particularly the Millennium Development Goals (MDGs). The ongoing WSIS political process, which culminates in 2005 with the Tunis summit, has been eclipsed by this focus to the effect that the UN Millennium Summit +5 (September 14 to 16, 2005) is now termed the 2005 World Summit.

THEMATIC SCOPE OF WSIS

Thematically, WSIS has focused almost exclusively on the new digital and networked information and communication technologies (ICTs) so that the prime developmental question has been how to bridge the digital divide. Given the importance attached to the Millennium Summit follow-up, this question has in the WSIS process been tied to the assertion that ICTs can and should be employed for reaching the goals laid down in the Millennium Declaration. A major point of political contestation has been which financial mechanisms could be utilized in this respect, from the exploitation of existing development cooperation favored by the North to the establishment of a new digital solidarity fund called for by the South. Further main issues of the political debate, whose discussion has not necessarily stood in relation to the question of how to enable development and a sustainable closing of the digital divide, have been the scope and organization of Internet governance; human rights including freedom of expression and communication rights, as well as the right to privacy vs. national security issues; intellectual property rights (IPRs) vs. knowledge commons and public resources; proprietary vs. free- and open-source software models; and media diversity vs. media monopolies.

GENDER DIMENSIONS IN TERMS OF THE WSIS POLITICAL OUTCOME

The principle of gender mainstreaming has not been applied to WSIS, and the political deliberations have hence been characterized by a gender-blind and male-centered discussion process. A broad reaffirmation of women's human rights and of the commitment to women's empowerment and gender equality has been hard to achieve and has not automatically extended from the first to the subsequent political statements that have been discussed: the Declaration of Principles (WSIS-03/GENEVA/DOC/4-E) and the Plan of Action (WSIS-03/GENEVA/DOC/5-E) that were agreed upon at the Geneva Summit, and the Political Chapeau and Operational Part that are currently negotiated for the Tunis Summit. The one strong commitment to women codified so far is Paragraph 12 of the Declaration of Principles. It states,

We affirm that development of ICTs provides enormous opportunities for women, who should be an integral part of, and key actors, in the Information Society. We are committed to ensuring that the Information Society enables women's empowerment and their full participation on the basis on [sic] equality in all spheres of society and in all decision-making processes. To this end, we should mainstream a gender equality perspective and use ICTs as a tool to that end. (United Nations, 2003a)

In addition to this paragraph, two more paragraphs of the Declaration of Principles contain expressly gendered content: Paragraph 2 reiterates the promotion of gender equality and the empowerment of women as well as the improvement of maternal health as goals of the *Millennium Declaration*, toward which ICTs should be employed. Paragraph 29 refers to the special needs of girls and women regarding literacy and universal primary education.

In the Geneva Plan of Action (PoA, United Nations, 2003b), several paragraphs take up specific concerns of girls and women, while it is indirectly claimed that "special attention will be paid" to girls and women with respect to all objectives, goals, and targets of the PoA (para. B.7). Explicitly, the PoA addresses girls' and women's promotion with respect to ICT education, training, and careers (paras. C4.11.g, C7.19.a, C7.19.c, C7.19.d, C8.23.h, C6.13.l); the integration of a gender perspective in ICT education (para. C4.11.g); a focus on "gender-

sensitive curricula in formal and non-formal education"; and the attainment of "communication and media literacy for women" (para. C8.23.h) as well as the promotion of balanced and diverse media portrayals of men and women (para. C9.24.e). Also, the need to acknowledge "women's role as health providers in their families and communities" is referenced in the context of e-health (para. C7.18.e). Importantly, the PoA acknowledges the requirements to monitor the developments and to devise "gender-sensitive indicators on ICT use and needs" (para. E.28.d).

GENDER DIMENSIONS IN TERMS OF THE WSIS PROCESS AND ENTITIES

The larger WSIS process has constituted a unique focal point and has even been a catalyst both for research on women, media, and ICTs, and for feminist advocacy regarding media and ICT that is based on this research. Various stakeholder entities from inside and outside the UN system have been involved in these endeavors and have promoted them in the WSIS process. Important examples from within the UN system during the Geneva phase are the following. The UN International Research and Training Institute for the Advancement of Women (INSTRAW) held a virtual seminar series on gender and ICTs in the summer of 2002 (Huyer & Sikoska, 2002). The same year, the UN Division for the Advancement of Women (DAW), together with other agencies, held two expert group meetings and two online discussions in preparation for them. One meeting was titled Information and Communication Technologies and their Impact on and Use as an Instrument for the Advancement and Empowerment of Women and was held in Seoul, Republic of Korea, November 11 to 14, 2002 (DAW, 2002a). The other was entitled Participation and Access of Women to the Media, and the Impact of Media on and its Use as an Instrument for the Advancement and Empowerment of Women and took place in Beirut, Lebanon, November 12 to 15, 2002 (DAW, 2002b).

The 47th session of the UN Commission on the Status of Women (CSW) in 2003 brought many of the previous findings together under the agenda item Participation and Access of Women to the Media, and Information and Communications Technologies and their Impact on and Use as an Instrument for the Advancement and Empowerment of Women (New York, March 3 to 14, 2003). By submitting its agreed conclusions (CSW, 2003) to the WSIS process, CSW built a bridge between its task to follow up on the Twelve Critical Areas of Concern codified at the Fourth World Conference on Women in Beijing, China, in 1995, and the then current WSIS negotiations. The UN Inter-Agency Network on Women and Gender Equality (IANWGE) additionally brought together the assessments of many other UN bodies and produced fact sheets on ICTs and women to inform the WSIS debates.

Feminist advocacy during the Geneva phase was characterized by two entities, which in the Tunis phase have merged into one. The WSIS Gender Caucus was formed in the summer of 2002 and has been working since. It mirrors the multistakeholder approach of WSIS itself and is open to all gender-equality advocates, from governmental ones to business and civil-society ones. The Gender Strategies Working Group (Non-Governmental Organizations' NGO # GSWG), which came into being shortly after the WSIS Gender Caucus, was only active during the Geneva phase. As the name implies, it was a civil-society entity. Both groups were instrumental in achieving the recognition of gender issues and accomplishing the codification of the gender-equality provisions and special measures for girls and women cited above. Early governmental motions in this direction, which were only put forth by a few countries such as Canada and South Africa, were overruled up until the negotiations in September 2003.

The WSIS Gender Caucus, the NGO GSWG, and their membership organizations made extensive use of the exposition ICT4D (ICT for Development) held in conjunction with the Geneva summit: They showcased a substantial number of ICT-for-women initiatives and presented a multitude of panels, seminars, and workshops on women and gender in the information society. The NGO GSWG also provided substantial input on women's rights and gender equality to the civil-society declaration that was adopted at the Geneva summit, entitled *Shaping Information Societies for Human Needs* (Civil Society Declaration to the World Summit on the Information Society, 2003).

FUTURE TRENDS

At the time of this writing, the outcome of the Tunis summit is still open, as are the kinds of implementations and follow-up processes that will characterize the post-Tunis phase. It is to be expected that most of the post-Tunis political discussions, planning, implementation, and evaluation will be conducted on the regional or subregional levels. On these levels, many of the research findings about women, media, and ICTs that have been brought together or generated in the WSIS context can be applied. These encompass broad overviews (Gurumurthy, 2004; Kuga Thas, Ramilo, & Cinco, in press; Primo, 2003) and applications of fundamental principles such as human rights (Jensen, in press) as well as sector-specific approaches. The latter include ICT policy and regulation as a central area of feminist concern since the late 1990s (Hafkin, 2002; Jorge, 2000). They also encompass questions of women's participation in e-democracy (Martínez & Reilly, 2002; Ramilo, 2002) and in the economy, as well as girls' and women's needs and opportunities in sectors such as e-education and e-health.

CONCLUSION

The WSIS process has been a focal point and catalyst for gender-sensitive research on media and ICTs in all world regions and a center of advocacy with respect to gender equality, nondiscrimination, and women's empowerment in the information society. As such, it has not only brought back to the global agenda the issue of women and media, which constitutes one of the Twelve Critical Areas of Concern of the Beijing Declaration and Platform for Action, adopted at the Fourth World Conference on Women in Beijing in 1995. It has also broadened the scope of this area of concern to encompass gender dimensions of new pressing issues, from free- and open-source software models to Internet governance. It has hence achieved what the Beijing +10 review process, which culminated in the 49th session of CSW in February and March 2005, failed to achieve: an acknowledgment of the tremendous impact of media and ICTs on women's and girls' lives in the current processes of global restructuring, and hence of the need to carefully engineer these ICT-driven processes from a gender-equality point of view. It is true that the political outcomes of WSIS do not provide a comprehensive set of tools in this regard. But the fundamental commitment to gender equality and women's empowerment that was laid down in the WSIS *Declaration of Principles*, together with the acknowledged need to provide special measures for girls and women and to develop gender-sensitive indicators on ICT use and needs, can and should be employed to shape the WSIS implementation phase and thus the information society itself in a more just and sustainable manner.

REFERENCES

Association for Progressive Communications (APC) & Campaign for Communication Rights in the Information Society (CRIS). (2003). *Involving civil society in ICT policy: The World Summit on the Information Society.* Retrieved February 18, 2006, from http://www.ictdevagenda.org/devlibrary/downloads/apc_civil_society_ict.pdf

Civil Society Declaration to the World Summit on the Information Society. (2003). *Shaping information societies for human needs.* Retrieved February 18, 2006, from http://www.itu.int/wsis/docs/geneva/civil-society-declaration.pdf

Commission on the Status of Women (CSW). (2003). *Participation and access of women to the media, and information and communication technologies and their impact on and use as an instrument for the advancement and empowerment of women: Agreed conclusions.* Retrieved February 18, 2006, from http://www.un.org/womenwatch/daw/csw/csw47/AC-mediaICT-auv.PDF

Division for the Advancement of Women (DAW). (2002a). *Information and communication technologies and their impact on and use as an instrument for the advancement and empowerment of women: Report of the EGM.* Retrieved February 18, 2006, from http://www.un.org/womenwatch/daw/egm/ict2002/reports/EGMFinalReport.pdf

Division for the Advancement of Women (DAW). (2002b). *Participation and access of women to the media, and the impact of media on and its use as an instrument for the advancement and empowerment of women: Report of the EGM.* Retrieved February 18, 2006, from http://www.un.org/womenwatch/daw/egm/media2002/reports/EGMFinalReport.PDF

Gurumurthy, A. (2004). *Gender and ICTs: Overview report.* BRIDGE, Institute of Development Studies. Retrieved February 18, 2006, from http://www.bridge.ids.ac.uk/reports/cep-icts-or.pdf

Hafkin, N. (2002). Gender issues in ICT policy in developing countries: An overview. *UN DAW EGM on information and communication technologies and their impact on and use as an instrument for the advancement and empowerment of women.* Retrieved February 18, 2006, from http://www.un.org/womenwatch/daw/egm/ict2002/reports/Paper-NHafkin.PDF

Hemmati, M. (2002). *Multi-stakeholder processes for governance and sustainability: Beyond deadlock and conflict.* Earthscan. Retrieved from http://www.minuhemmait.net/eng/msp/msp_book.htm

Huyer, S., & Sikoska, T. (2002). INSTRAW virtual seminar series on gender and information and communication technologies (ICTs). *UN DAW EGM on information and communication technologies and their impact on and use as an instrument for the advancement and empowerment of women.* Retrieved February 18, 2006, from http://www.un.org/womenwatch/daw/egm/ict2002/reports/Paper%20by%20INSTRAW. PDF

Jensen, H. (in press). Women's human rights in the information society. In R. F. Jørgensen (Ed.), *Human rights in the information society.* Cambridge, MA: MIT Press.

Jorge, S. (2000). *Gender perspectives in telecommunications policy: A curriculum proposal.* ITU. Retrieved February 18, 2006, from http://www.itu.int/ITU-D/gender/projects/GenderCurriculum.pdf

Kuga Thas, A. M., Ramilo, C. G., & Cinco, C. (in press). *ICT and gender.* Bangkok, Thailand: UNDP-APDIP.

Martínez, J., & Reilly, K. (2002). Looking behind the Internet: Empowering women for public policy advocacy in Central America. *UN/INSTRAW virtual seminar series on gender and ICTs. Seminar 4: Engendering management and regulation of ICTs.* Retrieved from http://www.un-instraw.org/en/docs/gender_and_ict/Martinez.pdf

Primo, N. (2003). *Gender issues in the information society.* Paris: UNESCO.

Raboy, M. (2004). The origins of civil society involvement in the WSIS. *Information Technologies and International Development,* 1(3-4), 95-96.

Ramilo, C. (2002). National ICT policies and gender equality: Regional perspective Asia. *UN DAW EGM on information and communication technologies and their impact on and use as an instrument for the advancement and empowerment of women.* Retrieved February 18, 2006, from http://www.un.org/womenwatch/daw/egm/ict2002/reports/Paper-CRamilo.PDF

United Nations. (1995, September 15). *Beijing Declaration and Platform for Action.* Adopted by the Fourth World Conference on Women. Retrieved February 18, 2006, from http://www.un.org/womenwatch/daw/beijing/platform/

United Nations. (2003a, December 12). *Declaration of Principles.* Adopted by the World Summit on the Information Society. Retrieved February 18, 2006, from http://www.itu.int/dms_pub/itu-s/md/03/wsis/doc/503-WSIS-DOC-0004!!PDF-E.pdf

United Nations. (2003b, December 12). *Plan of Action.* Adopted by the World Summit on the Information Society. Retrieved February 18, 2006, from http://www.itu.int/dms_pub/itu-s/md/03/wsis/doc/503-WSIS-DOC-0005!!PDF-E.pdf

KEY TERMS

Beijing Declaration and Platform for Action: The outcome document of the Fourth World Conference on Women, held in Beijing in 1995. It constitutes a comprehensive analysis of the areas in which women's and girls' human rights are violated, and outlines wide-ranging actions to counter this situation in order to empower girls and women and to achieve gender equality. Among the Twelve Critical Areas of Concern codified in Beijing is one entitled Women and the Media. Two strategic objectives are elaborated in this context. One is to "[i]ncrease the participation and access of women to expression and decision-making in and through the media and new technologies of communication" (United Nations, 1995: strategic objective, p. 1). The other is to "[p]romote a balanced and non-stereotyped portrayal of women in the media" (United Nations, 1995: strategic objective, p. 2).

CEDAW (the Convention on the Elimination of all Forms of Discrimination against Women): Adopted by the UN General Assembly in 1979 and entered into effect in 1981. It constitutes the most comprehensive human-rights instrument in existence for women. In its preamble and 30 articles, CEDAW defines discrimination and maps a broad agenda for governments to end discrimination de jure and de facto in all spheres of society. Over 90% of the member countries of the UN are by now parties to CEDAW, with the notable exception of the USA. While CEDAW does not address media in depth, its provision to counter stereotyped roles of men and women and the social and cultural patterns that perpetuate these roles (Article 5a) has lent itself well for an application in the field of media and ICTs, both with respect to media content and with respect to occupational segregation in the media industries. This illustrates that CEDAW is a living convention whose provisions have been adaptable to diverse and changing circumstances.

Millennium Development Goals (MDGs): The time-bound and quantified targets for combating extreme poverty in several dimensions that came out of the UN Millennium Summit of 2000. There are eight goals: (1) eradicate extreme poverty and hunger, (2) achieve universal primary education, (3) promote gender equality and empower women, (d) reduce child mortality (4) improve maternal health, (5) combat HIV, AIDS, malaria, and other diseases, (6) ensure environmental sustainability, and (7) develop a global partnership for development. Within WSIS, Goal 8 is interpreted as a call for the use of ICTs in order to facilitate the achievement of all other goals.

U

Multistakeholder Approach: A form of political deliberation, and sometimes also decision making, that first identifies the individuals and groups who have a stake in the outcome of certain negotiations, and then brings them together to develop solutions. It is expected that multistakeholder processes lead to well-informed and well-balanced decisions, and that the process of developing these creates trust and commitment among the stakeholders that impacts favorably on the implementation process. In WSIS, the main stakeholder groups are representatives of governments, civil society, and the private sector. These groups, however, do not participate on an equal footing, and decisions are entirely up to the governmental representatives.

NGO Gender Strategies Working Group (NGO GSWG): The civil-society entity that promoted gender equality and the participation of women during the Geneva phase of the WSIS process. Its seven priority issues for gender equality were (1) to take an intersectional approach, (2) to build on the global consensus from previous UN world meetings, (3) to focus on people-centered development, (4) to involve a human-rights framework of analysis, (5) to uphold respect for diversity, (6) to uphold peace and human development, and (7) to support local solutions. Like most other civil-society groups, the NGO GSWG was largely self-financed. It was formed at the first WSIS Preparatory Committee Meeting in Geneva in July 2002. Its initial members were African Women's Development and Communication Network (FEMNET), Agencia Latino Americana de Información, the Association for Progressive Communications Women's Networking Support Program (APC WNSP), the International Women's Tribune Centre (IWTC), and Isis International-Manila.

UN Commission on the Status of Women (CSW): A functional commission of the Economic and Social Council established as early as 1946 to promote the equal rights of women and men. In the wake of the Fourth World Conference on Women in Beijing in 1995, CSW was additionally mandated to conduct the follow-up process to that summit and to mainstream a gender-equality perspective within the UN. In 1996 and again in 2003, CSW reviewed issues concerning women and the media. CSW consists of 45 members who meet for 10 days each year.

WSIS Gender Caucus: A multistakeholder group that has promoted the principles of gender equality, women's empowerment, and women's human rights as well as the participation of women in the WSIS process. It operates with financial assistance from Finland, Norway, Denmark, Sweden, and the UN Development Fund for Women (UNIFEM). The WSIS Gender Caucus was formed at the invitation of UNIFEM in 2002 at the African regional preparatory conference in Bamako, Mali, May 25 to 30. The founding members include Abantu for Development; African Centre for Women, Information, and Communications Technology (ACWICT); African Connection Programme; African Information Society Gender Working Group (AIS-GWG); Women's International Netowork of the World Association of Community Radio Broadcasters (AMARC-WIN); AMARC Africa; Association for Progressi ve Communications Africa Women's Programme; AQ Solutions Association of Yam-Bukri; Environment and Development Action (ENDA); Gender Equity and Equality Project (GEEP); FEMNET; Media Institute of Southern Africa (MISA); Network for the Defence of Independent Media in Africa (NDIMA); Network of African Women Economists; United Nations Development Programme Sub-Regional Resource Facilities West Africa (UNDP/SURF West Africa); UNIFEM; Unite d'appui au programme de la cooperation Canada-Malienne; WomensNet South Africa; Women of Uganda Network (WOUGNET); Zimbabwe Women's Resource Centre and Network (ZWRCN); and the Zimbabwe Ministry of Transport and Communications. By now, the WSIS Gender Caucus has more than 100 registered participants on its Listserv, representing all stakeholder groups.

Understanding the Mommy Tracks in the IT Workforce

Jeria L. Quesenberry
The Pennsylvania State University, USA

Eileen M. Trauth
The Pennsylvania State University, USA

INTRODUCTION

Despite the recent growth in the number of women in the American labor force, women are under represented in the IT workforce. Key among the factors that account for this under representation is balancing work-family issues. Some researchers have speculated that IT work is not an ideal fit for working mothers because of long work hours, increased conflicts with family responsibilities, and the difficulty of returning after maternity leave to an industry with ever evolving technologies (Kuosa, 2000; Webster, 1996). This article reports on an empirical study that explored the influence of work-family balance on American women's participation in the IT workforce by using the Individual Differences Theory of Gender and IT (Trauth, 2002; Trauth, Quesenberry, & Morgan, 2004; Trauth, Huang, Morgan, Quesenberry, & Yeo, 2006). In doing so, we summarize a work-family balance study presented in greater detail in Quesenberry, Morgan, and Trauth (2004) and Quesenberry, Trauth, and Morgan (2006) that articulates the ways in which individual and environmental factors influence female responses to issues of work-family balance.

BACKGROUND

Studies of the IT workforce are mixed on the question of whether the IT workplace is a conducive or an unfriendly environment for working mothers. One stream of research points to the IT industry as having a pragmatic approach to working practices that can have a positive impact on working mothers. These practices include innovations in teleworking, job-sharing and technical advances that allow more flexibility in work-family balance (Quesenberry & Trauth, 2005; Zimmerman, 2003). An alternative stream of research highlights several difficulties associated with work-family balance in the IT workforce. Trauth's studies in Ireland and Australia revealed that women found it difficult to manage work-family conflicts despite shifts in societal views about working mothers (Trauth, 2002; 2000; 1995; Trauth Nielsen, & von Hellens, 2003). Webster (2002) adds that this is "particularly hard to reconcile with the working rhythms of IT work" (p. 6) and may not be conducive for many women.

Researchers also highlight the consequences associated with work-family balance for women in the IT workforce. Mennino and Brayfield (2002) found that female employees in male-dominated occupations make more family trade-offs and fewer employment trade-offs than employees in other occupations. Ahuja (2002) reports that women may have to neglect certain family obligations to be eligible for promotional opportunities similar to those of men. Baroudi and Igbaria (1994) point to family-related responsibilities as partial explanation for the under representation of women in managerial positions. Likewise, Sumner, and Werner (2001) found the burden on family-career balance from overtime and administrative tasks to be a barrier to women in management.

MAIN THRUST OF THE ARTICLE

This article reports on one aspect of a multi-year, multi-site qualitative field study of women working in IT whose goal is to investigate the female under representation in IT. Our objective is to contribute to a deeper understanding of specific factors that influence American women in their working lives as IT professionals by examining the work-family balance issues facing women in the IT workforce and how they respond when making decisions about their personal and professional development.

Fifty-seven open-ended, in-depth, face-to-face interviews with female practitioners in the IT workforce were conducted between October 2002 and August 2004. The participants represent a range of geographical locations, ages, demographic backgrounds, educational backgrounds, levels of management and job classifications, relationship statuses and family compositions. The women work and live in three different geographical regions of the U.S.: the Northeast (Boston, Massachusetts), the Southeast (Research Triangle/Charlotte, North Carolina) and the Mid Atlantic (central Pennsylvania). The women range in age from 21 to 58 with the average age being 40.6 years. Furthermore, 35 of the women are married, 2 are in committed relationships, 14 are single and 6 are divorced/not remarried. Thirty-two of the women have one or more child and 26 of the women do not have children.

The guiding theory for this research is the Individual Differences Theory of Gender and IT proposed by Trauth (2002; Trauth, Huang, et al., 2006; Quesenberry, 2004) that focuses on differences among women in the ways they experience and respond to characteristics of IT work, the IT workplace and societal messages about women and IT. This theory focuses on women as individuals, having distinct personalities, experiencing a range of sociocultural influences, and therefore exhibiting a range of responses to the social construction of IT. Thus, the theory elucidates the differences *within* rather than *between* the sexes and examines issues at an individual rather than a group level of analysis.

Analysis of Work-Family Balance in the IT Workforce

What emerged from the analysis of life histories of women in IT are four categories of women in the IT workforce: the non-parent, the working parent, the "back-on-track" parent and the "off-the-track" parent. The categories are by no means static or limiting. Rather they are dynamic in nature, and were created to analyze data to support theory refinement.

The Non-Parent: Balancing Work-Family Issues without Children

The *non-parent* category is comprised of women employed in the IT workforce who do not have children. The non-parent represents 26 women or 45.5% of the women interviewed. These women are single, married, partnered, and divorced and range in age from 21 to 53, with an average age of 37.8 years. The non-parents consist of two groups of women: women who have *not yet* had children and women who *are not* having children. This is important to note this distinction because not all non-parents are young, single women who have not yet reached a point to have children. Rather, many non-parents are women who have made conscious decisions not to have children.

Despite the range of explanations regarding motherhood, one common theme arose regarding work-family balance: the non-parents acknowledged their ability to more easily balance work-family issues in the IT workforce than their co-workers with children. The non-parents felt that they were more able to adjust to the temporal aspects of IT work, including longer work days and late hours. In addition, many non-parents felt more able to participate in after-hour networking events than co-workers with children. Further, several participants also commented on the freedom they enjoyed by not having to make work-child trade-offs. Although non-parents have chosen to not have children and tend to acknowledge the increased ability to balance work and life, it does not mean that they are all workaholics who are focused exclusively on themselves or their careers. Many of the non-parents talked about their values regarding personal life and time spent away from the office. They spoke of elder care, responsibility for nieces and nephews, other family commitments and pets.

Another theme raised by the non-parents is coping with the societal message they sometimes received about motherhood. This was explored through discussion of regional cultural attitudes towards women and women working. Many non-parents spoke about a cultural message that women's family obligations should take precedence over professional obligations. Thus, they should assume domestic child-care roles and men should assume professional income-earning roles. According to this view, the only acceptable jobs are domestic or traditional female occupations. Francie, a 26 year-old software engineer, summarized this message by explaining:

Typically, the family obligations take precedence over the professional obligation. ... I think typically [the societal view] is, that, when the woman has a child that she should stay home and take care of them. The male would be the financial supporter [Francie].

From a personal development perspective, the non-parents spoke about the difficulties reconciling their own identities with what they perceive to be a societal stereotype of women. For instance, Nancy, a 48 year-old Web consultant, spoke about the pressures she felt to have children despite the fact that it conflicted with her own personal desires. The women also spoke about the difficulties reconciling their professional development with views of women working. The non-parents also spoke of the difficulties reconciling their professional development, particularly in job attainment, with a socially constructed view of women as primarily mothers. Many non-parents discussed the difficulty they faced in obtaining IT jobs because of attitudes towards women working in IT.

The Working Parent: Balancing Work-Family Issues Concurrently

The *working parent* category is comprised of women who have both children and a career in IT. The working parent represents 26 women or 45.5% of those interviewed for this paper. Working parents are women who do not fit the working parent dichotomy found in the literature because they place a value on both family and career. Working parents represent a range of ages and relationship statuses including single, partnered, married, and divorced. Furthermore, the working parents represent a range of motherhood scenarios including raising one or more child (biological, adopted, and foster), and vary in age from 27 to 57 years old.

The working parents are typically motivated by both financial and personal desires to simultaneously work and raise their children. Although, these women acknowledge the financial benefit of working in IT,

an overwhelming sentiment is that they seek employment because of the personal value they place on being active and continuing to grow as a professional. Donna, a 39 year-old quality assurance analyst, was asked how important work was in her life:

I think [having a career] is very important for me... It's important to keep my mind active to keep challenged and to like what I do. When I stop having fun at this job that's probably when I'll decide it's time to move on. I think it's very important to stay active [Donna].

Although, many working parents value a professional career, they also acknowledged the difficulties associated with having one. The women frequently spoke about the work-family balance issues that arise, particularly those with regard to the temporal arrangements of IT work such as a 24/7 work day where employees are always accessible. Candace, a 41 year-old systems developer, felt that there is a growing tendency in the last few years for employees to work extra hours and to be "constantly available by computer or phone."

A recurring theme raised by the working parents related to the importance of work flexibility. The women spoke about the benefits of work programs such as job sharing, part-time work and manageable commutes. For instance, Kimberly, a 38 year-old project manager, has the ability to telecommute and was one of the first women at her IT consulting firm to be promoted to manager while on maternity leave. Another prevalent theme expressed by working parents is the importance of supportive partners and spouses. Many working parents spoke about how their partners and spouses share an active role in child rearing, domestic responsibilities, and community volunteer activities.

Several working parents spoke about the societal pressures and mixed messages they receive about raising children and working outside of the home. The participants highlighted the societal messages that women should be stay-at-home mothers. Rose, a 47 year-old director of IT who is Japanese American, explained that her culture puts "a lot of emphasis on the Japanese woman staying home and taking care of the children" and taking on duties such as finances and keeping certain traditions alive. The working parents spoke of the difficulties of reconciling their professional development particularly in getting a job with societal images of women working in IT. They noted the negative stereotypes associated with being a working parent and the influence it had on their careers. To overcome these societal views of women, the working parents shared accounts of having to work harder and longer hours than coworkers, in order to dispel the negative stereotypes of working mothers.

The "Back-on-Track" Parent: Balancing Work-Family Issues Sequentially

The *back-on-track parent* refers to women who, for a variety of reasons, took time away from work to raise children and then later returned to the IT workforce. The back-on-track parents represent 9% or five women interviewed for this paper. The back-on-track parents are older in age, ranging from 44 to 58 with an average age of 50.2 years. All of these back-on-track parents were in committed relationships during their employment break so that the main source of income came from their partners or spouses.

A common theme arising from the back-on-track parent interviews was the idea that women should take time away from work to stay home because it was the right thing to do. This idea seems to stem from societal messages that the women received during their childhood and adult lives. These messages appear to vary by geographical regions included in the study, and the associated cultural influences. Although, Sue, a 53 year-old IT coordinator, from central Pennsylvania eventually returned to work she felt that she was "supposed" to follow the path in life of going to school, getting married and then staying home with kids:

I wound up going to a two year college because I really didn't think that getting into sports was something I was supposed to do. I felt more pressure to go into business and get married and become a secretary or something like that. ... So I went to a two year associate [degree at a] Catholic college and took business, so I got an associate degree in business and then got married, had kids, the whole [thing] [Sue].

Although, societal pressures about motherhood and careers affect women in non-technical and technical careers alike, there is a common shared experience of the back-on-track parents that relates specifically to the IT workforce. This theme relates to the amount of skill preparation required for reentry into the IT workforce. This difficulty of returning to IT work has caused the back-on-track parents to develop plans of action to ease the transition. For instance, Elsie, a 47 year-old website manager, spoke specifically about the amount of "intense" work that was required to reenter the IT workforce. Consequently, she developed a success strategy that involved diligently working to prove herself as a viable employee despite the fact she is ten years older than her cohort.

The stories of back-on-track parents demonstrate that balancing work-family issues in IT work is a constant challenge that pulls women in several directions at once. Societal messages complicate the pressures women feel in decisions about their professional and personal lives. Likewise, long, irregular work hours associated with IT careers makes it difficult to balance work-family issues. Thus, taking time away from the IT workforce is a temporal solution to these conflicts. Leaving the workforce for an extended period of time allows women to balance family responsibilities during the early years of their children's lives and return to their careers at a point in time when their children are more self-sufficient.

The "Off-the-Track" Parent: Balancing Work-Family Issues by Egression

The fourth type of parent is the *off-the-track parent* comprised of employees who permanently leave the IT workforce upon having children. We have not captured data regarding the off-the-track parent because our participants are drawn from women currently employed in the IT workforce. However, for purposes of conceptual completeness we include this category in our discussion.

CONCLUSION

With regard to theory, our research shows that work-family tradeoff considerations are much more nuanced than what is commonly depicted in the literature. Women experience a range of work-family situations that present varying issues and concerns. The remarks illustrate an identifiable theme that crosses geographical regions and timeframes: *societal messages are complex and difficult to digest, and are processed in different ways by different women, yet they contribute to the decisions women make about their professional and personal lives.* More specifically, this research is an example of the application of the Individual Differences Theory of Gender and IT to go beyond the identification of societal messages that operate at the group level to also understand the response variations among women when examining the topic of work-family balance.

Of utmost importance to practice is the realization that not all women, indeed not all employees, have the same work-family balance issues and therefore, do not have the same needs or concerns with regard to their careers. Likewise, we do not advocate special considerations given to any group of individuals as it is highly likely that such a plan of action would weaken these groups in the labor force. Rather, our findings suggest that a more robust and flexible conceptualization of career tracks with multiple avenues would benefit a wider range of IT workers, both men and women, as the traditional view of a career is one that no longer reflects the needs and concerns of workers.

ACKNOWLEDGMENT

This article is from a study funded by a National Science Foundation Grant (EIA-0204246).

REFERENCES

Ahuja, M. K. (2002). Women in the IT profession: A literature review, synthesis and research agenda. *European Journal of Information Systems, 11,* 20-34.

Baroudi, J., & Igbaria, M. (1994). An examination of gender effects on career success of information systems employees. *Journal of Management Information Systems, 11*(3), 181-201.

Kuosa, T. (2000). Masculine World Disguised As Gender Neutral. In E. Balka & R. Smith (Eds.), *Women, work and computerization charting a course to the future* (pp. 119-126). Norwell, MA: Kluwer Academic Publishers.

Mennino, S. F., & Brayfield, A. (2002). Job-family trade-offs: The multidimensional effects of gender. *Work and occupations, 29*(2), 226-255.

Orlikowski, W., & Baroudi, J. (1991). Studying information technology in organizations: Research approaches and assumptions. *Information Systems Research, 2*(1), 1-28.

Quesenberry, J. L., Morgan, A. J., & Trauth, E.M. (2004). Understanding the "mommy tracks": A framework for analyzing work-family issues in the IT workforce. In M. Khosrow-Pour (Ed.), *Proceedings of the Information Resources Management Association Conference*, New Orleans, LA (pp. 135-138). Hershey, PA: Idea Group Publishing.

Quesenberry, J. L., Trauth, E. M., & Morgan, A. J. (2006). Understanding the "mommy tracks": A framework for analyzing work-family balance in the IT workforce. *Information Resource Management Journal, 19*(2), 37-53.

Quesenberry, J. L., & Trauth, E. M. (2005). The role of ubiquitous computing in maintaining work-life balance: Perspectives from women in the IT workforce." In C. SØrensen, Y. Yoo, K. Lyytinen, & J. I. DeGross (Eds.), *Designing ubiquitous information environments: Socio-technical issues and challenges* (pp. 43-55), New York: Springer.

Schwandt, T. A. (2001). *Dictionary of qualitative inquiry,* (2nd ed.). Thousand Oaks, CA: Sage.

Sumner, M., & Werner, K. (2001). The impact of gender differences on the career experiences of information systems professionals. In *Proceedings of the ACM SIGCPR 2001 Conference,* San Diego, CA (pp. 125-131).

Trauth, E. M. (1995). Women in Ireland's information industry: Voices from inside. *Eire-Ireland, 30*(3), 133-150.

Trauth, E. M. (2000). *The culture of an information economy: Influences and Impacts in the*

Republic of Ireland. Dordrecht, The Netherlands: Kluwer Academic Publishers.

Trauth, E. M. (2002). Odd girl out: The individual differences perspective on women in the IT profession. *Information Technology and People, 15*(2), 98-117.

Trauth, E. M., Huang, H., Morgan, A. J., Quesenberry, J. L., & Yeo, B. (2006). Investigating the existence and value of diversity in the global IT workforce: An analytical framework. In F. Niederman & T. Ferratt (Eds.) *Managing information technology human resources* (pp. 333-362). Greenwich, CT: Information Age Publishing.

Trauth, E. M., Nielsen, S. H., & von Hellens, L. A. (2003). Explaining the IT gender gap: Australian stories for the new millennium. *Journal of Research and Practice in IT, 35*(1), 7-20.

Trauth, E. M., Quesenberry, J. L., & Morgan, A. J. (2004). Understanding the under representation of women in IT: Toward a theory of individual differences. In M. Tanniru & S. Weisband (Eds.), *Proceedings of the 2004 ACM SIGMIS Conference on Computer Personal Research*, Tucson, AZ (pp. 114-119). New York: ACM Press.

Webster, J. (1996). *Shaping women's work: Gender, employment and information technology.* New York: Longman.

Webster, J. (2002). *Widening women's work in information and communication technology: Integrated model of explicative variables.* European Union for Information Society Technologies Programme.

Zimmerman, E. (2003). Parent-to-parent: Using technology to stay connected. *Sales and Marketing Management, 155*(7), 58.

KEY TERMS

Individual Difference Theory of Gender and IT: a social theory developed by Trauth (Trauth, 2002; Trauth, Quesenberry et al., 2004) that focuses on within-group rather than between-group differences to explain differences in male and female relationships with information technology and IT careers. This theory posits that the under representation of women in IT can best be explained by considering individual characteristics and individual influences that result in individual and varied responses to generalized environmental influences on women.

Interpretive Research: Research directed at understanding the deeper structure of a phenomenon within its cultural context by exploring the subjective and intersubjective meanings that people create as they interact with the world around them (Orlikowski & Baroudi, 1991, p. 5).

Qualitative Research: A term used to describe forms of social inquiry that aim at understanding the meaning of human action and that rely primarily on qualitative data (i.e., data in the form of words), including ethnography, case study research, naturalistic inquiry, ethno-methodology, life-history methodology and narrative inquiry (Schwandt, 2001, p. 213).

Work-Family Balance: the act of balancing inter-role pressures between the work and family domains, which are generally mutually incompatible.

Work-Family Balance Categories: a description of decisions about work-family balance of women in the IT workforce, which is comprised of four types: the non-parent, the working parent, the "back-on-track" parent and the "off-the-track" parent.

Unintended Consequences of Definitions of IT Professional

Wendy Cukier
Ryerson University, Canada

INTRODUCTION

Attention to women's low participation in information technology is framed in Canada and elsewhere in terms of concern over availability of well-qualified human resources (ITAC & IDC, 2002) as well as equity issues (Applewhite, 2002; Ramsey & McCorduck, 2005).

In most of these discussions, IT Professional is equated with Computer Scientist or Engineer in spite of the evidence that the profession is more diverse. This article suggests that while those directions are worthwhile, the very definition of "information technology professional" framed in the discourse may have unintended consequences which tend to exclude women. Framed by the literatures on gender and institutionalization of professions, this article applies critical discourse analysis to a variety of "texts" concerning the IT profession in Canada as well as available empirical data. Critical discourse analysis focuses on surfacing the political structures which underlie taken for granted assumptions (Fairclough, 1995). We maintain that while it is critically important to continue to attract females to study computer science and engineering, it is equally important to ensure that multiple paths are available and respected and that narrow definitions are not systemic barriers to the participation of women in the IT profession. In addition, more inclusive definitions which broaden the perspective on information technology (and match the reality of the industry) will promote good technology practices.

BACKGROUND

Women and information technology has been the subject of much attention. Although women account for half the Canadian workforce, they represent only around 22% of the IT profession. Much attention has been focused on increasing the number of women studying computer science and engineering programs and their participation in the industry (Applewhite, 2002; Cohoon, 2001). Programs aim at increasing awareness of IT careers, promoting female role models, increasing young girls' participation in math and science, exploring female-friendly pedagogy, offering alternative entry routes, as well as providing mentors and networking opportunities. However the evidence of the long term impacts is uneven (Cukier & Chauncey, 2004).

Scholars (Ramsey & McCorduck, 2005) have begun to probe beyond the barriers to explore issues related to professional identity in the face of systemic stereotyping, dualism, and devaluation. Studies have shown many women articulated an interest in "computing with a purpose" as opposed to "hacking for hacking's sake". There are also issues of perceptions; both male and female respondents lack information about the nature of the work and overwhelmingly perceived it as a masculinized domain; the females mainly saw IT courses as boring and difficult. Females tend to be more interested in the application of technology than "the technical bits" (Grundy, 1996). Computer science programs that place more emphasis on the application of technology in context and a strong emphasis on teamwork, communication, personal growth, and commitment, such as Purdue's EPIC program, attract a substantially higher percentage of women than traditional computer science programs (Jamieson, 2001).

Research shows the tendency to gender type tasks as male or female (Krefting, Berger, & Wallace, 1978), and there is ample research to show that technology work tends to be framed as mens-work (Perry & Greber, 1990). The masculine gender role may become associated with male-dominant contexts and associated behaviors, and values become institutionalized. The literature also suggests that the proportion of women in a particular context is negatively related to rewards and that men tend to occupy higher status and higher paying jobs and occupations than women do (Konrad, 2004).

The nature of the profession is changing. Denning (2000) has suggested that there are over 40 organized professional groups in computing and information technologies and that interdisciplinary studies are growing. He proposes a redefinition of the profession to include what he terms "IT specific disciplines, IT intensive disciplines, and IT supportive occupations".

This article will explore the intersection of these two notions: (1) that women are more likely to be interested in the application of technology to solve problems than in the "technical" bits and (2) that the IT profession is broader than computer science and engineering. It will also consider (3) the ways in which the institutionalization of the profession itself affects the participation of women.

Industry data show that the shape of the information technology profession in Canada has changed and that the skills it demands are multi-disciplinary. Non-technical and soft skills based positions will be increasingly important. Technical competence will not be sufficient (SHRC, 2003). A survey by the Software Human Resources Council of Canada identifies 27 different segments in the IT workplace (Gunderson, Jacobs, & Vaillancourt, 2005).

MAIN BODY: GENDER IMPLICATIONS OF DEFINING THE IT PROFESSION

U

- **Proposition 1: The under-representation of women in engineering and computer science disciplines is not an appropriate metric for assessing women's participation in the information systems profession.** Women's absence in computing is not a general phenomenon, but rather women are under-represented in particular forms of computing such as software engineering (Clegg & Trayhurn, 2000; Grundy, 1996). Women account for 15% of electrical engineering, 21% of applied computer science, 49% of business administration, 62% of mass communications, and 73% of library science programs. Where the focus is the application and management of information technology, female participation is close to 33%. These patterns also hold for the gender distribution of faculty. Similar findings have been reported for the United States (Cukier, Devine, & Shortt, 2002).

Studies from industry reinforce the gender imbalance in subsets of IT occupations.

A study from the Canadian Software Human Resources Council (SHRC), based on a National Survey of IT Occupations, used a broader definition of IT occupations, but the patterns were similar.

Table 1. The evolution of the IT profession

IT-Specific Disciplines	IT-Intensive Disciplines	IT-Supportive Occupations
Artificial intelligence	Aerospace engineering	Computer technition
Computer science	Bioinformatics	Help desk technician
Computer engineering	Cognitive science	Professional IT trainer
Computational science	Digital library science	Security specialist
Database engineering	E-commerce	System administrator
Computer graphics	Financial services	Web services designer
Human-computer interaction	Genetic engineering	Web identity designer
Network engineering	Information science	Database administrator
Operating systems	Information systems	
Performance engineering	Public policy and privacy	
Robotics	Instructional design	
Scientific computing	Knowledge engineering	
Software architecture	Management information systems	
Software engineering	Multimedia design	
System security	Telecommunications	
	Transportation	

Table 2: Proportion of female IT occupations (Gunderson et al., 2005)

	Percentage Female
Writers	74.3
Graphic designers and illustrators	60.8
Analysts	36.8
Support	29.9
Data	25.7
Project Managers	24.9
Multimedia Developers	21.3
Trainers	21.2
Managers	19.0
Programmers	16.7
Operators	15.5
Technicians	14.8
Software engineers	13.0
Other engineers	9.0

Source: Gunderson et al., 2005

Writers and graphic designers are predominantly female. More than one third of analysts are female. On the other hand, only 16% of computer programmers and an even lower proportion of engineers are female.

Another study used different definitions and a different methodology but still showed that more than twice as many women were analysts compared to engineers in the IT sector (Wolfson, 2003). There is substantial variation among segments in terms of the representation of women.

Similar patterns are found in research institutions and successful research grant recipients; interdisciplinary research into technology, such as the TeleLearning Network Centres of Excellence, tend to have more gender balance (50% female) than centers focused on "the technical bits" such as the Canadian Institute for Telecommunications Research (CITR) with 6% (Cukier, 2002).

Finally, while further study is needed, many women who define themselves as "IT professionals" have not entered the profession via computer science or engineering. A recent survey of members of the "Wired Woman" association revealed that while the majority define themselves as IT professionals (64%), only a fraction (11%) are computer scientists and none are engineers (cited in Cukier et al., 2002).

- **Proposition 2: Qualifications for Information Technology which are not related to performance may become "taken for granted" and institutionalized.** The institu-

tionalization of professions and the role of "actors", including the state, legal entities, and professional associations, have been studied in a variety of contexts (MacDonald, 1995; Meyer, 1994). Less attention has been paid to the *process* of institutionalizing the IT profession. It has been shown that organizations are rewarded for conforming to the requirements, such as corporate training programs, irrespective of whether they support improved performance (Scott & Meyer, 1994). Qualifications may become institutionalized and present unintended barriers to certain segments of the population. These become forms of systemic discrimination (Ontario Human Rights Commission, 1999).

Discursive practices are the taken-for-granted, systematic, and consistent ways of speaking about phenomenon that shape the perception of the phenomenon. Language both reflects and shapes a culture and so has implications for women in the professions (Atkinson & Delamon, 1990).

Discourses of associations, employers, IT professionals, and academics reinforce the notion that an "IT professional" is synonymous with computer scientist (Cukier, 2003). This is at odds with the empirical evidence, described previously, which shows the demand in the industry is for a broader range of skills. John Roth, former president of Nortel and an engineer, emphasized the shortage of engineers, mathematicians, and computer scientists (Roth, 1997). In contrast, Carol Stephenson, former president of Lucent Canada and a psychology graduate, stressed the importance of disciplinary diversity in the IT workforce because "the soft skills are hard" (cited in Brody, Cukier, Grant, Holland, Middleton, & Shortt, 2003). In spite of the emphasis on technical skills, she maintained that interpersonal and communications skills were even more valuable and less commonplace.

Hi-tech companies, dominated by computer scientists and engineers, perpetrate the practice of hiring in their own image. Despite clear evidence that graduates of many technology-related disciplines have the necessary skills and do succeed in the sector, there is clear bias toward hiring from traditional (and also male dominated) disciplines (Cukier, 2003).

The Canadian Association of Advanced Technology (CATA) and the Information Technology Association of Canada (ITAC) pushed the government to invest in "doubling the pipeline" and creating more spaces for computer science and engineering students. There was no consideration of other disciplines or attention to the gender dimensions (Ontario Ministry of Education and Training, 1998). Three years later they announced there was actually not a shortage of computer scientists but a need for soft skills (ITAC & IDC, 2002).

Even companies which profess commitment to promoting women in IT reinforce this discourse. At a recent conference, IBM's vice president of Industry Solutions argued that changes in the industry make it necessary to recruit from a broader range of disciplines, yet the presentation on Women in IT referred only to engineering and computer science programs (IBM, 2004), reinforcing definitions and stereotypes which tend to exclude women.

Requirements which do not link to performance may also have unintended consequences. Grundy (1996) argues that proficiency at mathematics does not necessarily help with non-mathematical abstractions in computing and, conversely, students could be proficient in the areas of non-mathematical abstraction in computing without necessarily being good at mathematics. Yet, the focus remains : "Teach girls how to program. To write a program, like solving a math problem, is to discover a pattern with logic. If girls can do mathematics, and they manifestly can, they can program" (De Palma, 2001). At issue here is not an essentialist question, that is, whether or not men are inherently more mathematical than women. Rather relying on mathematics as a barrier to entry and core skill in IT educational programs may exclude women. Data from the University of Limerick (1998) reinforces the point that not all information technology disciplines rely on math. All students in math and computing had advanced math, as did 80% of students in computer systems compared to only 25% of students in applied computing or 10% of those in business and computing did. Mathematics may serve a symbolic or political function (Hacker, 1990; Schiebinger, 1999). Certainly the emphasis on mathematics is consistent with other tendencies to value skills constructed as "male" and devalue skills constructed as female. Discourses of technology contribute to sustaining gender disparities in technol-ogy-related fields (Hanson, 1997). While technology includes a wide range of practices, the tendency to narrow the definition to "the technical bits" has the effect of excluding women (De Palma, 2001).

- **Proposition 3: These institutions are (a) dominated by men, computer scientists, and engineers; and (b) in spite of the evidence to the contrary, they tend to reinforce the definition of Information Systems Professional as computer scientist or engineer and (c) tend to emphasize skills such as mathematics thought to be associated with these disciplines.** Women are under-represented at senior management levels in the IT industry. They are also under-represented in the professional associations and educational institutions. In 2001, for example, the Canadian Advanced Technology Association's board was 96% male (CATA, 2001) and the Information Technology Association of Canada's Board of Directors was 87% male (ITAC, 2002). More research is needed to explore this, but the women who are prominent tend to be computer scientists and engineers.

Even programs sponsored by industry with the stated intention of encouraging more women to enter IT tend to focus on computer science and engineering. A review of 70 existing Canadian programs aimed at increasing women's participation in IT (Cukier & Chauncey, 2004) reveals that most are based on a relatively narrow definition of "information technology professional". Academics in computer science, even those actively engaged in promoting women in the profession, are often also inclined to reproduce their own image, and consistently define the IT Profession as Computer Scientists and Engineers (Klawe, Cavers, Popowich, & Chen, 2000).

Proposition 4: The current institutional environment, which includes government programs, industry and professional associations, employment practices, granting councils, and educational institutions, tends to (a) exclude and marginalize women entering the profession from other disciplines; (b) reinforce negative stereotypes of the IS profession which are known to discourage women from entering the profession; and (c) devalue skills associated with females (the soft skills).

Engineering and computer science programs have lower representation of women than disciplines such as information systems management or library science. Some segments within the IT profession (multimedia designers and analysts) have higher female representation than others. Consequently, a narrow definition of IT professional tends to privilege male dominated disciplines and marginalize those that are female dominated.

Another unintended consequence of this is to reinforce the negative stereotypes of IT professionals by over-emphasizing the "technical bits". Considerable research has shown that this is a disincentive to female participation.

While studies of employer needs have stressed the critical importance of soft skills in the IT profession, these skills, which are often associated with females, are devalued as qualifications. As Grint and Gill (1995) note, the concept of "skill" is subjective. This suggests the tendency to valorize hard skills (which are masculinized) over soft skills (which are feminized) in the IT profession.

FUTURE DIRECTIONS

This article suggests many avenues for further research regarding discursive practices and definition of IT in industry, professional associations, universities, government programs, and media. A systematic assessment is needed of industry recruitment, hiring practices, promotion, work environment, evaluation, and mentoring systems which defines skills and qualifications for IT workers and also influences government policy, industry associations, and educational institutions.

Increasing participation of women depends on valuing diversity in approaches to information technology and "respecting multiple points of entry" (AAUW, 2000).

Progress is slow and even women in engineering and computer science resist broadening the definition of IT. One reviewer wrote,

I have a degree in engineering, a PhD in business and taught both Management of Information

Systems and Systems Analysis for years. I am also a woman. I found many of the ideas expressed ... to be naïve, unfounded, biased and non-theoretical. I also found their arguments on why IT should not be restricted to computer science and engineering but expanded to include things [like] library science, information systems management and digital media design to be naïve and unfounded. ... Deciding to call the sky green does not make it less blue. (SSHRC, 2003)

CONCLUSION

Practices associated with the institutionalization of the IT profession may present barriers to the participation of women. The narrow definition of IT Professional seems at odds with the increasingly multi-disciplinary nature of the industry and the demand for a workforce with a broad range of skills sets.

The notion that computer science, engineering, and mathematics proficiency are the only entry routes to the IT profession may, in the same way, have unintended consequences for women. In addition, the (mis)representation of the IT Profession in this way reinforces the stereotypes of it as a technocentric profession rather than one with broader appeal, requiring diverse disciplines and skill sets. More study is needed to explore the ways in which our definition of IT professional, the occupational categories, skill requirements, education, and admission requirements may present unintended barriers.

REFERENCES

American Association of University Women (AAUW). (1991). *Shortchanging girls, shortchanging America.* Retrieved December 29, 2005, from http://www.aauw.org/research/sgsa.cfm

Applewhite, A. (2002). Why so few women? *IEEE Spectrum, 39*(5), 65-66.

Atkinson, P., & Delamon, S. (1990). Professions and powerlessness: Female marginality in the learned occupations. *The Sociological Review, 38*(1), 89-110.

Brody, L., Cukier, W., Grant, M., Holland, C., Middleton, C., & Shortt, D. (2003). *Innovation nation: Canadian entrepreneurs from Java to Jurassic park*. Toronto: John Wiley & Sons.

Canadian Advanced Technology Association (CATA). (2001). Retrieved December 29, 2005, from http://www.cata.ca

Clegg, S., & Trayhurn, D. (2000). Gender and computing: Not the same old problem. *British Education Research Journal, 26*(1), 75-89.

Cohoon, J. M. (2001). Towards improving female retention in the computer science major. *Communications of the ACM, 44*(5), 108-114.

Cukier, W. (2003). Constructing the IT skills shortage in Canada: The implications of institutional discourse and practices for the participation of women. In *Proceedings of the Computer Personnel Research ACM SIGCPR/SIGMIS 2003,* Philadelphia (pp. 24-33).

Cukier, W., & Chauncey, C. (2004). Women in information technology initiatives in Canada: Towards fact-based evaluation. In *Proceedings of the Americas Conference on Information Systems (AMCIS),* New York (pp. 1222-1230).

Cukier, W., Devine, I., & Shortt, D. (2002). Gender and information technology: Implications of definitions. *Journal of Information Systems Education, 13*(1), 7-15.

De Palma, P. (2001). Why women avoid computer science. *Communications of the ACM, 44*(6), 27-30.

Denning, P. (2000). The profession of IT: Who are we? *Communications of the ACM, 44*(2), 15-19.

Fairclough, N. (1995). *Critical discourse analysis.* Harlow: Longman Group Ltd.

Grint, K., & Gill, R. (1995). *The gender-technology relation*. London: Taylor and Francis.

Grundy, A. F. (1996). *Women and computers.* Intellect Books: Exeter.

Gunderson, M., Jacobs, L., & Vaillancourt, F. (2005). *The information technology (IT) labour market in Canada: Results from the national survey of IT occupations*. Ottawa: Software Human Resource Council (SHRC).

Hacker, S. (1990). *Doing it the hard way: Investigations of gender and technology*. Boston: Northeastern University Press.

Hanson, K. (1997). *Gender, discourse and technology*. Newton, MA: Gender and Diversities Institute.

IBM. (2004, August 30). *Taking ACTion against declining enrolment in computing disciplines.* Presentation, a meeting conducted and hosted by IBM Canada.

Information Technology Association of Canada (ITAC), & International Data Corporation (IDC). (2002). *Meeting the skills and needs of Ontario's technology sector: An analysis of the demands and supply of IT professionals.* Toronto; Ontario: Authors.

Jamieson, L. (2001, May). Expanding the pipeline: Women, engineering and community. *Computing Research News,* pp. 2-16.

Klawe, M., Cavers, J., Popowich, F., & Chen, G. (2000). ARC: A computer science post-baccalaureate diploma program that appeals to women. In E. Balka & R. Smith (Eds.), *Women, work and computerization: Charting a course to the future* (pp. 94-101). Boston: Kluwer.

Konrad, A. (2004). *Gender context and individual outcomes: A meta-analysis.* Unpublished manuscript.

Krefting, L., Berger, P. K., & Wallace, M. J. T. (1978). The contribution of sex distribution, job content and occupational classification to job sextyping: Two studies. *Journal of Vocational Behaviour, 13,* 181-191.

MacDonald, K. M. (1995). *The sociology of the professions*. London: Sage.

Meyer, J. (1994). Social environments and organizational accounting. In W. R. Scott & J. Meyer (Eds.), *Institutional environments and organizations: Structural complexity and individualism* (pp. 121-136). Thousand Oaks, CA: Sage.

U

Ontario Human Rights Commission. (1999). *Guide to the human rights code.* Retrieved December 29, 2005, from http://www.ohrc.on.ca/English/publications/hr-code-guide.pdf

Ontario Ministry of Education and Training. (1998). *Access to opportunities program: Thousands of new student spaces to be created.* Retrieved December 29, 2005, from http://www.edu.gov.on.ca/eng/document/nr/98.05/tech.html

Perry, R., & Greber, C. (1990). Women and computers: An introduction. *Signs: Journal of Women in Culture and Society, 16*(1), 74-101.

Ramsey, N., & McCorduck, P. (2005). *Where are the women in information technology? Preliminary report of literature search and interviews.* Boulder, CO: National Center for Women and Information Technology.

Roth, J. (1997). University capacity halting the high-tech engine: A crisis for Canada's economy. *The Toronto Star.* Retrieved from http://www.communications.uwaterloo.ca/Gazette/1998/apr01/nortel.doc

Scott, W. R., & Meyer, J. W. (1994). *Institutional environments and organizations: Structural complexity and individualism.* Thousand Oaks, CA: Sage.

Schiebinger, L. (1999). *Has feminism changed science?* Cambridge, MA: Harvard University Press.

Social Sciences and Humanities Research Council of Canada (SSHRC). (2003, May 2). *Personal correspondence with Irene Devine.*

University of Limerick. (1998). *Barriers for women in computing.* Retrieved December 29, 2005, from http://www.ul.ie/~govsoc/barrierstw.html

Wolfson, W. G. (2003). *Analysis of labour force survey data for the IT occupations.* Ottawa: Social Sciences and Humanities Research Council of Canada.

KEY TERMS

Discourse: The aggregate of written and spoken textualizations. "Institutionalized language codes for articulation of the social construction of reality and events" or taken for granted ways of speaking which shape ways of understanding.

Discursive Practices: The "taken-for-granted", systematic, and consistent ways of speaking or behaving that characterize phenomena in particular ways, and that limit descriptions of them in other ways; shaping the perception of the phenomenon is shaped by the discourse.

Hard Skills: Skills that are generally borne out of discipline, knowledge, and training. For example, the ability to write a computer program and engineering the architecture of an information system are hard skills. These are often the result of skill or object specific training.

Institutionalization: "A product of the political efforts of actors to accomplish their ends; and the success of an institutionalization project and the form the resulting institution takes depends on the relative power of the actors who support, oppose, or otherwise strive to influence it" (DiMaggio in Alvarez, 1996, p. 94).

Soft Skills: Skills that are generally the result of talent, intuition, socialization, and practice. Soft skills include people-oriented tasks such as communication skills, sales, coaching, and counseling.

Systemic Discrimination: Found in existing structures, policies and/or practices in the school or workplace impose barriers, both subtle and unsubtle, to some individuals, based on a prohibited ground (race, gender, etc.). This may include the existence of a qualification not required for the position, which has the unintended consequence of excluding groups based on a prohibited ground.

Technocentric: Refers to a focus on the technological features and qualities of an idea, concept, or practice. The term implies that features or qualities that are not of a technical nature are marginalized or disregarded.

Virtual Learning and Teaching Environments

Heike Wiesner
Berlin School of Economics, Germany

INTRODUCTION

Without a well-thought-out didactic concept, the best surface design isn't of any help.

This article contains results from an empirical study I conducted independently. It was developed in the context of the umbrella project "gender and information technologies" in the context of the Vifu (virtual international women university), a project financed by Germany's Federal Ministry of Education and Research (BMBF) and carried out at the Center for Interdisciplinary Research on Women at the University of Kiel. This empirical study provides insight into the field of "virtual learning and teaching". It is based on an expert survey and reflects the specific experience instructors had with virtual seminars. The study focuses on the following question:

What convergences and divergences can be identified in experts' specific experience with forms of virtual teaching?

I proceeded from the assumption that intercultural and gender factors affect the design, structure, and implementation of virtual learning and teaching environments. This assumption shaped the study.

Experts were generally defined as all persons who have practical experience in virtual teaching, especially in the research fields of gender and computer sciences. Since virtual learning and teaching environments is a new, experimental field, none of the interviewed experts had more than two to three years of teaching experience in this area. Some of them have programmed and developed the online module themselves.

Most of the fourteen interviews were conducted face-to-face, three of the interview partners were sent a questionnaire by e-mail. With them I conducted "semi-standardized interviews" (Mayring, 1996) and all were evaluated with the core sentence method. Core sentences are "natural generaliza-tions" used by the interview partners themselves (Leithäuser & Volmerg, 1988; Volmerg, Senghaas-Knobloch, & Leithäuser, 1986). These are state-ments that make a point succinctly, reducing entire paragraphs to a single statement. In contrast to the deductive method, the inductive method I used has the advantage that the meaning of the "spoken word" is not lost. It is particularly useful in the development of hypotheses.

FINDINGS

Virtual Learning and Teaching Environment as Science in Action

It has always been that way here. Each project must have two people pushing it ...

Virtual teaching and learning are still fields of experimentation. It is surely not incorrect to claim that all noted interview partners have done pioneer work emphasizing the area of "virtual teaching." One citation taken of an interviewee can thus be considered representative for the situation found in virtual teaching:

... we were well among the first who at all had a virtual college. There was a college [XXX],[1] a virtual college, two years ago, already three years ago, when the net had just gotten started. So we immediately had a huge project since someone was interested in it here. It has always been that way here. Each project must have two people pushing it, carrying the project, and the other fellow workers then more or less work through the daily loads ...

Against this background, the network of scien-tists involved in virtual teaching is to be classified as 'weak,' even if several experts directly cooperate with each other.

The term *science in action,* coined by Bruno Latour (1987), should thus convey the essence of what the surveyed community expressed. In the end, it is all about "science having not become cold" yet. A research and work area which is still in its preliminary stages has been described as a science in action by Latour. The actors and their related technical apparatus are thus to a large extent still in the so-called hot phases, the "science in the making" or "science in action" stage. The capacities are (still) above all oriented inwards. It is not until the stabilization phases (translating interests) that the activities are outwardly intensified (Latour, 1987). These "translations" consist of translations of one's own interpretations into others' taking place in the course of the establishing of scientific and technical constructions. In a successful interpretation, one's own interests are translated into those of others, in whose interest it is also to support one's own interpretations, approach, or technical innovation. Latour's concept of "translating interests" is not only limited to the phase of stabilization. Since this translation process, however, only unfolds its full impact in the phase of stabilization, the localizing from my point of view seems legitimate. With reference to the virtual teaching environment, the actors can not be analytically separated from their financial, technical, and societal structuring potential. They must rather be situated in their interlaced structural diversity—which Latour would describe as hybridity.

It is to be expected that with the expansion of the field of virtual learning intensified—internationally oriented—networking and differentiation will take place. The interviewees' responses thus provide a similarly detailed, as well as lively, snapshot at a certain point in time of the research field of virtual learning.

Technology development can therefore be seen as a social activity which is influenced by existing social structure, gendered values, and cultural practices. Virtual learning proposals should be given attention to make them more user oriented. Gender mainstreaming and diversity strategies in this context have a big impact, because technology is not a fixed product, but an configurable open process (Schelhowe, 2001; Zorn, 2004).

Gender and Culture as Moments of Selection

… computers are much more gendered in the domestic environment than they are in the work environment …

Internet competency should be mentioned as an important aspect in the context of virtual learning environments (Kreutzner & Schelhowe, 2002). For example, one expert was amazed that some participants "couldn't even manage to enroll in a virtual seminar on the internet." This procedure may be considered standard practice by Western academics, but it poses a problem to individuals from other social backgrounds and cultures. In order to avoid misconceptions, we must emphasize here that an individual's inability to carry out what appears to be a "simple procedure," such as enrolling for a course, can by no means be attributed one-dimensionally to social causes, (e.g., that the person is considered) (in a "deficit-oriented approach") to belong to a group characterized by "deficient internet competency." Quite often, the reason simply is a participant's insufficient computer resources or other infrastructural impediments. Thus it is not at all clear what instructors can be expecting from students as basic "internet competency" or as a "technological standard."

An important aspect of gender has often been overlooked: the existence of a computer in the household by no means guarantees access for women! The following quote sheds light on how closely technical infrastructure conditions for example are related to the category of gender:

… computers are much more gendered in the domestic environment than they are in the work environment, and this is based on empirical evidence from our students, men and women, about what they can do and the access that they have over computers. So, for example, men and women who have access to computers in the work place in general have a very similar kind of access. Women are a little bit more restricted, but not terribly. The restrictions tend to be the same,

they tend to have the same amount of access, but men and women who have access to a computer that belongs to the family or home, have totally different kinds of access, which reflects real gendering in the household ... And governments and funders don't want to hear that ... It sounds too big to address, so instead, they much prefer some kind of solution which is something they can do outside of the family"

Technology, as the quote shows, is by no means an independent entity. Unless the role of gender constructions in societal discourse is rendered transparent, the interconnections between technology and society cannot easily be traced. Gender and the cultural and social background are interrelated and together can have a strong, multiple effect on participation and drop-out rates (Green & Adam, 2001; Wiesner, 2003). For this reason, technological infrastructure and internet competency cannot be treated as independent entities, but must instead be analyzed as contextually embedded and interconnected. In brief, the categories *gender* and *culture* play a considerable role in the context of virtual learning environments.

Communication in Action

The largest problem ... with virtual courses is the people who don't say anything!

Even when conducted synchronously, online discussions occur in a temporally slightly delayed fashion. The impression that the contributions argumentatively build on one another can thus be misleading. Further, some virtual seminars come about without video support and hence without facial expressions and gestures. *Pure listeners* can therefore not be observed. They run the danger of being declared inactive students while they consider themselves to be active course participants.

Chatting in this sense also is something that must be learned! It indeed appears to constitute a new—hybrid—form of communication, one of intermediate quality:

Well, it's not as good, but it is better than nothing ... this is somewhere in-between.

Though assuming very different forms, these hybrid phenomena will gain importance in the future. Haraway—much like Latour (1998)—insists on the unaccustomed, (i.e., "the non-anthropomorphic and non-substantial quality of unexpected action competencies") (Haraway, 1995). In this context, the technological challenge for example consists in integrating technology in a "more malleable fashion" into interactive interpersonal processes.

While most of the experts cited in an exemplary fashion either gender-specific or culturally related different styles of communication, one expert did not make such a clear distinction, but rather gave a response pointing out interrelationships:

... if there is a gap concerning being self-confident ... then this might be related to gender, to cultural differences, but also to the level of education. One can under certain circumstances also apply to the M.A. programme without any first university degree, namely on the basis of many years of work experience. For example, in one of my courses, I had a mid-level manager from a steel company, who started as a craftsman, worked his way up and now wanted to acquire the academic degree in order to have more job security. These people have a lot to contribute in such a learning group ... but at first they are mostly for some time reserved and naturally in the area of academics also less capable of articulating themselves well. What of course does not mean that they know less ... Under certain circumstances this plays a greater role than gender.

Even if this description should not be limited to computer-facilitated communication but can also be true for face-to-face methods, it is evident that the internet in no way provides for an unlimitedly unbiased communication (Herring, 2003). In computer-facilitated communication, the use of *language* is of very special importance and the scope of its potential becomes even greater in the frame of intercultural exchange. In the process of mutual perception and evaluation, the intercultural dimension of active conceptualization and codes is of particular importance. Beside the primary socialized cultural background, these are co-determined by every individual life-world and world perspective. Indeed, before taking off on "cultural differences" much too rashly, one should definitely be wary at this point. The cultural differences can be attributed to an often underestimated aspect: the agreement on one "common language."

Determining one language generally leads to the exclusion of others and thus of other groups within the population. Communication in a language other than one's own requires more consideration of the interaction partners and thus demands more time, in particular in connection with communication in writing. The required transformation of expert terminology in the operative language can moreover make communication more difficult. The most interviewed experts did underestimate this aspect in their own point of view.

Didactic Preparation Important

Without a well-thought-out didactic concept, the best surface design isn't of any help.

As far as the conceptual development of the virtual courses is concerned, the experts' virtual teaching style can roughly be divided into three groups: "highly prestructured," "groupwork-oriented" and "flexible."

In this context, the first group is assigned the term "highly prestructured." Their courses are provided with a considerable amount of instructions, e.g. supervision of each step in the lesson, linear text guidelines, and the keeping up with assignments within the given time frame. According to a respondent, this course structure is particularly appropriate for acquiring factual knowledge.

With the second, "group work-oriented," group of experts, the courses were developed closely with the participants. They often work with different concepts by directly reacting to the participants' manner of working and critiques. The terms most frequently used during the interviews in this context were "group-oriented," "network oriented," and "community-building." It should be emphasized that the teaching staff made the social constructive dimension of teaching (and learning) very important in their courses.

The third group is assigned a position half-way between the above-mentioned positions. One person calls the structure of the course a mix-form, a "hybrid structure." Due to the quite factually-oriented and action-oriented approach, the group chose to describe it as "pre-structured as well as network-oriented." The structured course instructions were made "a little more flexible" from the very beginning, with lesson units that included learning through discovering and group-work activities.

Most instructors in the study initially were of the opinion that it would suffice to use conventional seminars and "simply put them on the net." Yet, according to many experts, precisely the conventional course type has proven generally unsuccessful on the net. Just one example:

You have to be a little bit postmodern and you have to take some risks, and accept that what you offer may not be what people want.

Most interview partners thus lamented that they had not taken enough time to reflect on didactic methods. Precisely this was considered by all experts to be a grave deficit in the design and implementation of virtual lesson plans.

The formation of an academic community, (i.e., interpersonal networking), has had a strikingly positive effect on participation and drop-out rates, above all among women (Wiesner et al. 2004; Zorn, 2004). Seminars in which the participants could work networked or in groups discursively had substantially lower drop-out rates (especially of women) than those in which the participants only had direct contact to the instructor and worked on their own. In this regard, two important aspects can be identified in the answer patterns of the interviewed experts. First, a continual *interest-building process* has a positive effect on the participants of the virtual seminars offered by the experts. Second, the networking process of community-building is supported by the opportunity for participants to meet one another outside of official (online) sessions *in an informal setting*, (e.g., via chat rooms, mailing lists or even informal live encounters). But all communication-tools should be embedded in a (moderated) collaborative learning context, otherwise the communication between the participants of a learning module won't work and the learning outcome is low (Baumgartner, 2005; Luck, 2004).

Long-Term Prospects of Virtual Educational Forms

It depends on the target group ..., if you take the normal students at universities ... they actually don't voluntarily take any virtual courses.

It is not only a purposeful pragmatic optimism that misleads the experts to make statements like: "... I think of the seventeen courses, I am sure some of them, probably half, will come up with some good results." Much more, this quote reflects the state of research on virtual learning: "The philosophers' stone has not yet been found." Experimental enthusiasm is just not rewarded from the start. Success and failure are standing close to one another in this first phase.

... what I would like to do, is just to offer a lot of courses, a variety of courses, and just try to test how it works ... We are trying to discover who will be interested, how many people and how we can make our courses more attractive. So what can I say, we are in a very interesting phase—which is an experiment ...

The question now arising of course is: where should we go from here? There is no reason to believe, contrary to recurrent journalistic pronouncements (Gaschke, 2000; Schönert, 2000), that in the near future university education will only take place in virtual classrooms. The core sentence, "[a] virtual seminar as a supplementary offering is certainly worthwhile, but not as a substitute," summarizes the opinions expressed by the experts as a whole.

Individuals who explicitly opt for virtual seminars—and above all those who remain in them—are far more similar to students in "education at a distance" than to "normal students" in conventional presence universities. I would like to again cite a statement I quoted earlier:

It depends on the target group ... When you take the normal students at universities, ... they actually don't voluntarily take any virtual courses.

This statement is in fact confirmed by all interview partners!

The experts were nearly unanimously of the opinion that individuals who participate in virtual seminars must be highly self-organized: "these are people who are actually 80% sure that they will study on their own at home." At the same time, as another interview partner noted, people who take part in such courses are such "for whom the alternative 'I will learn through reading alone' does not suffice ... There are a great number of people who are capable of learning from books. They have no need for virtual courses."

One can thus begin to speak of a new–hitherto unnoticed—or *"other type of student,"* who can be located midway between the students of a traditional distance university and those of a conventional presence university. This assumption is supported by the efforts of experts to develop more open universities: the providers of virtual seminars appear to no longer orient themselves at traditional educational models. Most experts offer their courses not only in initial training or education, but also simultaneously in advanced and continuing education. They construct the division between initial and continuing education opportunities more permeably and thus make it appear obsolete (Wiesner, 2003, pp. 35-37).

Participants in virtual seminars thus constitute a largely unnoticed type of student. The conclusions we can draw from this are ambivalent. Virtual seminars can be organized relatively independent of time and space. Studies show that this aspect could be a real chance for women (Luck, 2004; Zorn, 2004). This may be also well-suited to some male and female students' needs, for an increasingly large group of students declare their studies to no longer be "the focal point of their life". In addition, occupational demands are rising incessantly. Working adults without "additional qualifications" expect to face "harder times" over the medium-term. Life-long learning bears opportunities *and* risks. Virtual teaching and learning environments are extremely "compatible", in that they (seemingly) match the interests of today's "student community" with those of the economy and of demands related to the organization of work particular to each society. Only against this complex and highly ambivalent background can the emergence of virtual courses be explained and meaningfully integrated into the existing educational landscape.

V

REFERENCES

Baumgartner, P. (2005). *Eine neue Lernkultur entwickeln: Kompetenzbasierte Ausbildung mit Blogs und E-Portfolios*. Retrieved July 27, 2005, from http://www.educa.ch/dyn/bin/131141-131143-1-eportfoliodeutsch.pdf

Gaschke, S. (2000, March 30). Verheißung Internet. *Die Zeit,* (a weekly journal in Germany*), 14*, 7.

Green, E., & Adam, A. (2001). *Virtual gender: Technology, consumption and identity*. London: Routledge.

Haraway, D. (1995). *Die Neuerfindung der Natur. Primaten*. Verlag; Frankfurt; New York: Cyborgs und Frauen Campus.

Herring, S. C. (2003). Gender and power in online communication. In J. Holmes & M. Meyerhoff (Eds.), *The handbook of language and gender*. Oxford: Blackwell Publishers.

Kreutzner, G., & Schelhowe, H. (2002). *Agent of change. Virtuality, knowledge and the challenge to traditional academia*. Opladen: Leske+Budrich.

Latour, B. (1987). *Science in action. How to follow scientists and engineers through society*. Cambridge, MA: Harvard University Press.

Leithäuser, T., & Volmerg, B. (1988). Psychoanalyse in der Sozialforschung, Eine Einführung, Westdeutscher Verlag GmbH, Opladen.

Luck, P. (2004). *Problem based management learning-better online?* Retrieved July 27, 2005, from http://www.eurodl.org/materials/contrib/2004/Luck_Norton.htm

Mayring, P. (1996). Einführung in die qualitative Sozialforschung, 3. überarbeitete Auflage, Psychologie Verlags Union, Weinheim.

Pohl, M., & Michaelson, G. (1997). "I don't think that's an interesting dialogue": Computer-mediated communication and gender. In A. F. Grundy, D. Köhler, U. Oechtering, & H. G. Petersen (Eds.), *Women, work, and computerization. Spinning a Web from past to future* (pp. 87-97). Berlin: Springer-Verlag.

Schelhowe, H. (2001). Offene Technologie—Offene Kulturen. Zur Genderfrage im Projekt Virtuelle Internationale Frauenuniversität vifu, in: FIFF Ko 3/2001, 14-18.

Schönert, U. (2000, February 3). Der Draht zum Prof. *Die Zeit*(a weekly journal in Germany*), 6*, 63.

Volmerg B., Senghaas-Knobloch, E., & Leithäuser, T. (1986). Betriebliche Lebenswelt, Eine Sozialpsychologie industrieller Arbeitsverhältnisse Westdeutscher Verlag GmbH, Opladen.

Wiesner, H. (2003). Virtuelle Lehr- und Lernformen auf dem Prüfstand. In *Forschungszentrum Arbeit-Umwelt-Technik* (artec-Paper Nr. 87). Research Centre Work-Environment-Technology, University of Bremen, January 2002, 1-53.

Wiesner, H., Zorn, I., Schelhowe, H., Baier, B., & Ebkes, I. (2004). Die 10 wichtigsten Gender-Mainstreaming-Regeln bei der Gestaltung von Lernmodulen. In *I-com - Zeitschrift für interaktive und kooperative Medien*, Heft 2/2004, 50-52.

Zorn, I. (2004). Designing for intercultural communication in an international virtual community. Reflections on the technology design process. In Jürgen Bolten (hg.) *Interkulturelles Handeln in der Wirtschaft: Positionen-Modelle-Perspektiven-Projekte* (pp. 159-177). Sternenfels: Wissenschaft und Praxis.

KEY TERMS

Core Sentences: Natural generalizations used by interview partners to reduce entire interview response paragraphs to a single statement.

Internet Competency: The adequate ability and knowledge of the Internet.

Science in Action: Coined by Bruno Latour (1987); conveys the essence of what the surveyed community expresses at the preliminary stages of research.

Virtual Learning Environment: An information system used to facilitate student learning via online educational courses.

Virtual Teaching Environment: An information system used to facilitate teaching of online educational courses.

ENDNOTE

[1]   [XXX] indicates anonymity.

Vulnerability to Internet Crime and Gender Issues

Tejaswini Herath
State University of New York at Buffalo, USA

S. Bagchi-Sen
State University of New York at Buffalo, USA

H. R. Rao
State University of New York at Buffalo, USA

INTRODUCTION

A tremendous growth in the use of the Internet has been observed in the past two decades. More than 75% of Americans participate in online activities (University of Southern California Annenberg School Center for the Digital Future, 2004) such as e-mail, Web browsing, working from home, accessing news stories, seeking information, instant messaging, using the Internet in lieu of the library for school work, playing games, and managing personal finance. For professionals, the Internet is an important medium for networking and building social capital. However, along with all positive impacts, there are also negative outcomes. One such negative outcome includes Internet crimes. Dowland, Furnell, Illingworth, and Reynolds (1999) state that "with society's widespread use of and, in some cases, reliance upon technology, significant opportunities now exist for both mischievous and malicious abuse via IT systems" (p. 715).

Internet crimes (cyber crimes) consist of specific crimes dealing with computers and networks (such as hacking, spreading of viruses, and worms) and the facilitation of traditional crime through the use of computers on the Internet (such as child pornography, hate crimes, telemarketing/Internet fraud). This article focuses on Internet crimes, especially those affecting individual users, and offers a discussion of issues regarding Internet crimes and gender.

BACKGROUND

Cyber Crime

Computer crimes can be categorized by who commits them and what their motivation might be (e.g., professional criminals looking for financial gain, angry ex-employees looking for revenge, hackers looking for intellectual challenge), or by the types of computer security that ought to prevent them (e.g., breaches of physical security, personnel security, communications and data security, and operations security). These crimes can also be understood by how they are perpetrated (e.g., by use of the Internet or by use of physical means such as arson). For the purpose of this article, we will consider the method of perpetration, that is, crime committed via the Internet to hurt individual users, as the focus of the discussion below. Table 1 lists some common types of Internet crimes.

The 2004 Computer Crime and Security Survey, recognizes that Internet crime continues to be a significant threat. The E-Crime Watch Survey conducted by CERT Coordination center notes that nearly 70% of their survey respondents reported at least one intrusion while 43% of survey respondents reported an increase in electronic crimes. Organizations are harmed by insiders, such as employees or contractors, and outsiders (Computer Emergency Response Team Coordination Center, 2004, p. 6). CERT 1988-2004 statistics shows that incidents of

Table 1. Selected examples of Internet crimes

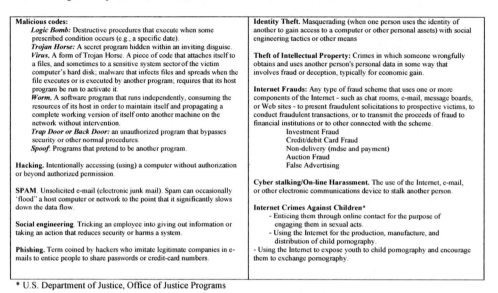

Malicious codes:
    Logic Bomb: Destructive procedures that execute when some prescribed condition occurs (e.g., a specific date).
    Trojan Horse: A secret program hidden within an inviting disguise.
    Virus. A form of Trojan Horse. A piece of code that attaches itself to a files, and sometimes to a sensitive system sector of the victim computer's hard disk; malware that infects files and spreads when the file executes or is executed by another program; requires that its host program be run to activate it.
    Worm. A software program that runs independently, consuming the resources of its host in order to maintain itself and propagating a complete working version of itself onto another machine on the network without intervention.
    Trap Door or Back Door: an unauthorized program that bypasses security or other normal procedures.
    Spoof: Programs that pretend to be another program.

Hacking. Intentionally accessing (using) a computer without authorization or beyond authorized permission.

SPAM. Unsolicited e-mail (electronic junk mail). Spam can occasionally 'flood' a host computer or network to the point that it significantly slows down the data flow.

Social engineering. Tricking an employee into giving out information or taking an action that reduces security or harms a system.

Phishing. Term coined by hackers who imitate legitimate companies in e-mails to entice people to share passwords or credit-card numbers.

Identity Theft. Masquerading (when one person uses the identity of another to gain access to a computer or other personal assets) with social engineering tactics or other means

Theft of Intellectual Property: Crimes in which someone wrongfully obtains and uses another person's personal data in some way that involves fraud or deception, typically for economic gain.

Internet Frauds: Any type of fraud scheme that uses one or more components of the Internet - such as chat rooms, e-mail, message boards, or Web sites - to present fraudulent solicitations to prospective victims, to conduct fraudulent transactions, or to transmit the proceeds of fraud to financial institutions or to other connected with the scheme.
        Investment Fraud
        Credit/debit Card Fraud
        Non-delivery (mdse and payment)
        Auction Fraud
        False Advertising

Cyber stalking/On-line Harassment. The use of the Internet, e-mail, or other electronic communications device to stalk another person.

Internet Crimes Against Children*
    - Enticing them through online contact for the purpose of engaging them in sexual acts.
    - Using the Internet for the production, manufacture, and distribution of child pornography.
- Using the Internet to expose youth to child pornography and encourage them to exchange pornography.

* U.S. Department of Justice, Office of Justice Programs

Table 2. Increase in reported incidents and fraud complaints over time

	1990	1995	2000	2001	2002	2003	2004
Incidents reported at CERT (a)	252	2,412	21,756	52,658	82,094	137,529	-NA-
Complaints received via IFCC website (b)			16,838	49,959	75,063	124,509	190,143

Sources: (a) Computer Emergency Response Team Coordination Center--*CERT/CC Statistics 1988-2004.*
        (b) Internet Fraud Complaint Center–*IC3 2004 Internet Fraud Report*

systems or data intrusions are continuously increasing. In 2003-2004, the reported incidents increased by nearly 67% (see Table 2).

Internet Crimes Targeting Individuals

The Internet Fraud Compliant Center (IFCC) report shows that traditional crimes such as fraud, identity theft, and harassment are on the rise. Furthermore, these crimes are now committed with the use of the Internet (Internet Fraud Complaint Center 2004). IFCC, which deals with Internet fraud, also receives complaints regarding child pornography (redirected to National Center for Missing and Exploited Children), computer intrusion (redirected to National Infrastructure Protection Center), SPAM e-mail, and identity theft (redirected to Federal Trade Commission). Victims of the well-known Nigerian letter fraud, which has been in existence since the early 1980s, have lost millions of dollars. The Nigerian

letter fraud and other credit card frauds are handled by the U.S. Secret Service.

The financial loss incurred through the above fraudulent activities is extensive but the new emergence of crime against children and crime against women using the Internet are even more disturbing (Sones, n.d.). In the recent past, Internet crime against children, such as child pornography, has also been on the increase. Some of the common types of crime against children include enticing them through online contact for the purpose of engaging them in sexual acts, using the Internet for the distribution of child pornography, using the Internet to expose youth to child pornography, and encouraging them to exchange pornography.

Cyber-stalking or online harassment, where one individual harasses another individual on the Internet using various modes of transmission such as electronic mail, chat rooms, newsgroups, mail exploders, and the World Wide Web is also on the rise (National

White Collar Crime Center, n.d.). The Internet makes stalking much easier for the perpetrator than off-line stalking. The anonymity of e-mail and chat rooms hide stalkers well, and stalkers can easily access private information about their victims via the Internet (Violence, Women, & Media, n.d.). Online harassments can range from minor annoyance to deadly outcomes. Ackerman (1997) notes that approximately 1 million women and 370,000 men are stalked each year. In studying off-line stalking, Tjaden and Thoennes (1998) found that women are far more likely to be victims of stalking than men. Such is true also for online stalking.

The growth in crime and unlawful practices connected with Internet use is has resulted in the introduction of regulations or legislations pertaining to Internet use. In the United States, Homeland Security Act of 2003 or the USA Patriot Act has amendments to handle computer crimes. In addition, federal criminal codes have specific provisions to deal with computer intrusion (visit http://www.cybercrime.gov/cclaws.html). At the international level, the EU led International Convention on Cybercrime seeks to set international standards for the policing of electronic networks (Rowe, 2004).

VULNERABILITY TO INTERNET CRIME AND GENDER ISSUES

Gender Issues in Computer Use and Communication

The literature on women in IT has a substantial section focusing on gender differences in computer use, education, and career paths. These studies have shown relevance of gender on how an individual adopts, perceives, and uses technology. In the earlier years of the Internet, the number of male users exceeded female users and created a gender gap. Recent statistics shows a reversal of this trend. The critical difference now is based on the difference in online activities between men and women. For example, women use the Internet more for interpersonal communication, information search or education, while men use the Internet more for one-way communication, entertainment, or personal finance (Banerjee, Kang, Bagchi-Sen, & Rao, 2005).

Gender can play an important role in determining technology use as well (Whitley, Jr. 1997). Considerable research has been conducted to find out whether females are likely to hold negative attitudes toward computers than males resulting in "technological gender gap". For instance, Venkatesh and Morris (2000) found gender difference in individual adoption and sustained usage of technology in the workplace. Ellison (1999) notes that men tend to be more adversarial; they use intimidation tactics to dominate and control online discourse. Elder, Gardner, and Ruth (1987) found that females are more likely to experience physical and emotional burnout caused by inability to adapt to new technology and older workers are more likely to experience "technostress" compared to younger workers.

The gender inequity in information technology can be reasonably explained by different male-female cognitive structures, that is, individual differences in encoding, processing, and organizing information, which leads to differentiated judgments (Bem, 1981). Several studies have attempted to understand the difference in technology use by males and females. In 2001, Jackson, Ervin, Gardner, and Schmitt found that women used e-mails more than males for interpersonal communication while men used Internet for information gathering. Gefen and Straub (1997) found that men and women appear to derive different meaning from the same use of technology and that the women and men differ in their perception and use of e-mail. Specifically, they found that "women perceive networking more useful than do men, and thus have tendency to see support, intimacy and cooperative behavior through e-mail (p. 389)". In their review of sociolinguistic literature, Gefen and Straub (1997) find that, to some extent, women and men use and understand similar messages quite differently.

This article provides a launching ground for exploring whether gender will be one of the factors in understanding vulnerability to Internet crime.

Gender Issues in Vulnerability to Internet Crimes

The Internet enables people to buy, sell, and trade images and videos that portray sexual exploitation of women and enables sex tourism as well as the advertisement of mail order brides. These sex crimes are examples of crimes against women, which use this new uncontrolled media outlet to carry out the crimes that are considered as violation of human rights. Organizations such as "Working to Halt Online Abuse" are working to combat as well get attention of law authorities to this issue.

The attitudinal, emotional, educational, perceptional, and cognitive differences that were discussed in the previous subsection may affect the vulnerability of a user based on gender. For example, perceptional and cognitive differences may put girls and women at more risk to social engineering tactics. As more women go online, the probability of exposure to Internet frauds will be on the rise. Women's need for socialization has been established in several studies confirming that females participate more in e-mail and online communication (Jackson et al., 2001; Banerjee et al., 2005). This may expose women and girls to several other types of Internet crimes. In recent past "phishing scams" are noted to be on rise in which the social-engineering schemes use "spoofed" e-mails to lead consumers to counterfeit websites designed to trick recipients into divulging financial data such as credit card numbers, account usernames, passwords, and social security numbers (http://www.antiphishing.org/phishing_archive.html). Although not much statistics is available on victimization of women to such scams, several news reports document the incidences. For example, the current case in Kansas, USA, (December, 2004) of murder of a pregnant woman and kidnapping of her baby from the womb by a woman who met online to purchase a puppy, sheds gruesome insight into crimes committed against women that are facilitated by the Internet (http://www.nydailynews.com/front/story/263217p-225301c.html).

In their structural-choice model of victimization theory, Meier and Miethe (1993) integrate the current predominant theories of victimization, namely lifestyle-exposure theory and routine-activity theory. They discuss that exposure and target attractiveness as necessary conditions for victimization. Lifestyle characteristics such age, gender, and education as well as routine activities which may predispose individuals to riskier situations may contribute to exposure or target attractiveness of victims.

Gender differences observed and studied in a wide range of samples often disappear in certain select subgroups. For example, Parsuraman and Igbaria (1990) found no male-female differences on attitudes towards computers among highly educated participants holding managerial positions in organizations that are highly dependent on computer technology (e.g., accounting, finance, information systems, and marketing firms). Advanced education and computer training does make users more comfortable with tasks needed to secure the computer and preserve user-privacy. We can argue that relatively younger women who have had education in this era of computing will have more exposure to computer training and hence they will be less vulnerable to these crimes. However, girls with limited exposure to computer training may be more susceptible to specific types of Internet crimes. Therefore, we can postulate that educational level as well as age may influence the level of an individual's vulnerability to Internet crimes which needs to be tested.

CONCLUSION

As more users go online, we need to understand the relative importance of socio-demographic factors that affect an individual's vulnerability to Internet-based crimes. Does gender play a role? If so, within what context is the male-female difference in vulnerability to Internet crime accentuated? Recent trends show that traditional crimes such as harassment, fraud, and identity theft are now committed with the use of the Internet. Privacy and security breach of personal information carried out by virtual criminals pose a significant threat to individuals and to the society. Prior research acknowledges gender differences in the use of information technology. We need to understand if all users benefit or suffer in the same way. If not, who are the victims and how can the victimization be controlled? Most surveys noted in this study focus on computer crimes against organizations. These surveys do not give sufficient insight into crimes affecting individual users. We need to understand what makes an individual vulnerable in order to better prepare task forces and the society as a whole to combat Internet crimes. This article raises questions regarding the role of gender, age, education, and the societal context of the Internet user in predicting vulnerability to Internet crimes. Such understanding will be useful in educating individuals and households about cyber security and will be essential for law enforcement agencies to intervene before a crime is committed.

ACKNOWLEDGMENTS

The authors would like to thank Professor Eileen Trauth for encouragement on this article. The authors were supported by NSF under grant 0420448. The usual disclaimer applies.

REFERENCES

Ackerman, E. (1997). When obsession turns ominous. *U.S. News and World Report, 123*(20), 43.

Banerjee, S., Kang, H., Bagchi-Sen, S., & Rao, H. R. (2005). Gender divide in the use of the Internet applications. *International Journal of E-Business Research, 1*(2), 24-39.

Bem, S. L. (1981). Gender schema theory: A cognitive account of sex typing. *Psychological Review, 88*, 354-364.

Computer Emergency Response Team Coordination Center (CERT/CC). (2004). *2004 e-crime watch survey summary of findings.* Retrieved December 3, 2004, from http://www.cert.org/archive/pdf/2004eCrimeWatchSummary.pdf

Computer Security Institute. (2004). *2004 computer crime and security survey.* Retrieved September 12, 2004, from http://www.gocsi.com/forms/fbi/csi_fbi_survey.jhtml

Dowland P. S., Furnell S. M., Illingworth H. M., & Reynolds P. L. (1999). Computer crime and abuse: A survey of public attitudes and awareness. *Computers and Security, 18*(8), 715-726.

Elder, V. B, Gardner, E. P., & Ruth, S. R. (1987, December). Gender and age in techno stress: Effects on white collar productivity. *Government Finance Review, 3*, 17-21.

Ellison, L. (1999). *Cyberstalking: Tackling harassment on the Internet. The 14th BILETA Conference: "CYBERSPACE 1999: Crime, Criminal Justice and the Internet".* Retrieved on December 23, 2004, http://www.bileta.ac.uk/99papers/ellison.html

Gefen, D., & Straub, D. (1997). Gender differences in the perception and use of e-mail: An extension to the technology acceptance model. *MIS Quarterly, 21*(4), 389-400.

Internet Fraud Complaint Center (IFCC). (2004). *IC3 2004 Internet Fraud Report, January 1, 2004-December 31, 2004.* Retrieved March 15, 2005, from http://www.ifccfbi.gov/strategy/2004_IC3Report.pdf

Jackson, L. A., Ervin, K. S., Gardner, P. D., & Schmitt, N. (2001). Gender and the Internet: Women communicating and men searching. *Sex Roles: A Journal of Research, 44*(5/6), 363-379.

Meier R. F., & Miethe T. D. (1993) Understanding theories of criminal victimization. *Crime and Justice, 17*, 459-499.

National White Collar Crime Center (NW3C). (n.d.). *Cyberstalking.* Retrieved December 11, 2004, from http://www.nw3c.org/research_topics.html

Parsuraman, S., & Igbaria, M. (1990). An extension of gender differences on determination of computer anxiety and attitudes toward microcomputers among managers. *International Journal of Man-Machine Studies, 32*, 327-340.

Rowe, C. J. (2004) *Sexual exploitation of women on the Internet: The council of Europe's convention on cybercrime and Canada's "Lawful Access" discussion paper.* Retrieved December 11, 2004, from http://www.womenspace.ca/policy/research_cybercrime_paper.htm

Sones, M. (n.d.) *Speculations on crime on the Internet.* Retrieved December 23, 2004, http://www.psyctc.org/iafp/nl_3_1/internet_crime.html

Tjaden, P., & Thoennes, N. (1998, April). *Stalking in America: Findings from the national violence against women survey.* Retrieved December 23, 2004, from http://www.ncjrs.org/pdffiles/169592.pdf

V

University of Southern California Center for the Digital Future. (2004, September). *The digital future report, surveying the digital future, year four, ten years ten trends.* Retrieved November 25, 2004, from http://www.digitalcenter.org/downloads/DigitalFutureReport-Year4-2004.pdf

Venkatesh, V., & Morris, M. (2000). Why don't men ever stop to ask for directions? Gender, social influence, and their role in technology acceptance and usage behavior. *MIS Quarterly, 24*(1), 15-140.

Violence, Women, and the Media. Retrieved December 23, 2004, from http://www.mediascope.org/pubs/ibriefs/vwm.htm

Whitley, B. E., Jr. (1997). Gender differences in computer-related attitudes and behavior: A meta-analysis. *Computers in Human Behavior, 13*(1), 1-22.

KEY TERMS

Exploit: Noun: Attacker tool (usually a program) for exploiting a known weakness. Verb: To take advantage of a known vulnerability to attack a system.

Internet Crime (Cyber Crime): Internet crime consists of specific crimes dealing with computers and networks (such as hacking) and the facilitation of traditional crime through the use of computers (child pornography, hate crimes, telemarketing/Internet fraud). In addition to cyber crime, it may cover the use of computers by criminals for communication and document or data storage

Privacy: The protection of individual rights to nondisclosure of information. According to the Calcutt Committee in the United Kingdom, privacy is the right of the individual to be protected against intrusion into his personal life or affairs, or those of his family, by direct physical means or by publication of information. Privacy can then be divided in several related concepts including:

1. **Information privacy,** which involves the establishment of rules governing the collection and handling of personal data such as credit information, and medical and government records. It is also known as "data protection".
2. **Privacy of communications,** which covers the security and privacy of mail, telephones, e-mail and other forms of communications. (http://www.privacyinternational.org)

Security: Policies, procedures, and technical measures used to prevent unauthorized access, alteration, theft, or physical damage to data or computer systems.

Spyware: A software that gathers user information through the user's Internet connection without his or her knowledge, usually for advertising purposes. Once installed, the spyware monitors user activities while on the Internet such as capture your keystrokes while typing the passwords, read and track your e-mail, record what sites you visit, record the credit card numbers; and transmits that information in the background to someone else.

System Vulnerability (Vulnerability): Weakness in an information system, system security procedures, internal controls, or implementation that could be exploited.

Vulnerability to Internet Crimes: Weaknesses that may subject an Internet user to become victim to an Internet Crime (e.g., lack of awareness of current threats and system vulnerabilities), inability or delay in dealing with the system vulnerabilities.

Web–Based Learning and Its Impacts on Junior Science Classes

W

Vinesh Chandra
Curtin University of Technology, Australia

Darrell Fisher
Curtin University of Technology, Australia

INTRODUCTION

The past decade has seen significant improvements in the design and development of information and communication technologies (ICT). The Internet, for instance, has become more efficient, more affordable and more accessible. While the availability of these technologies in classrooms has created new opportunities, it has at the same time presented new challenges for teachers. Teachers have to find innovative methods of implementing these technologies in lessons that are not only effective and efficient but also fair to both sexes.

BACKGROUND

ICT in Education

Technological advances in the past decade have created teaching and learning opportunities of significant proportions that would have been a fantasy a few years ago (Dierker, 1995). Gomory (2001) pointed out that while in the past many were deprived of education due to reasons such as accessibility, affordability, and personal commitments, today technology has addressed some of these issues. Learning can now be blended in with lifestyles and integrated into the daily routines of learners.

Another significant factor is that young people of today are constantly interacting with multimedia, and the new technologies "speak their language." According to Eklund, Kay, and Lynch (2003), many students in Australian schools were more skilled in using computers than their teachers. Stager (2004, para. 12) argued that by correctly harnessing these skills in technology, teachers can "breathe life into

the least effective teaching practices of yore." But is a technology-driven environment a fair learning medium for both sexes?

Gender Issues

Gender issues relating to science subject selections have been an issue for some time. In recent times, the widening gap between the academic performance of boys and girls has emerged as a significant issue for educators across all subjects. According to Biddulph (1997), these days, girls are much more confident, hardworking and motivated than boys. A recent Queensland Government report pointed out that girls were more likely to complete high school than boys (Wenham & Odgers, 2004). Matters, Pitman, and Gray (1997, p. 6) believed that the "original question of whether girls have equal educational opportunities has now been replaced with that of whether boys have equal educational opportunities."

Head (as cited in Cortis & Newmarch, 2000) suggested that the performance of boys was most probably due to their preference for different learning styles. Lerner and Galambos (1996) listed a range of factors, such as motivation, curriculum, student teacher and peer interactions, as some of the possible reasons for the disparity in the performance of the two sexes. Can computers and related technologies provide a level playing field for both sexes? How can the impact of such initiatives on students be studied?

Learning Environments

In the field of learning environments, the impact of such innovations on students can be effectively measured. For more than 30 years, proven qualitative and quantitative research methods associated with learning environments have yielded productive results for educators. In this study, the perceptions of Web-based learning in a blended environment were measured using a modified version of the *Web-Learning Environment Instrument* (WEBLEI) (Chang & Fisher, 1998). The WEBLEI measures students' perceptions across four scales—Access, Interaction, Response, and Results. Theoretically, if students perceived their learning environments favorably, then this was more likely to be transformed into favorable learning outcomes.

THE RESEARCH PROJECT

Design, Development, and Implementation of *Getsmart*

The layout of the *Getsmart* Web site enabled students to engage in learning activities that included opportunities for modeling, coaching, articulation, scaffolding, reflection, exploration, questioning, performance feedback, and direction instruction. These paralleled the instructional methods of "electronic cognitive apprenticeship" (Bonk & Kim, 1998; Collins, Brown, & Newman, 1989; Wang & Bonk, 2001). These learning options were created though Web-based lessons, tests, online chats and interactive activities. Figure 1 shows the general layout of the Web site.

Figure 1. The general layout of Getsmart

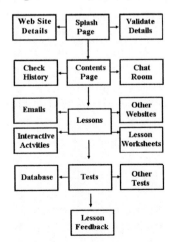

Figure 2. Key features of the lesson page

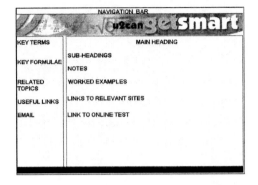

Students accessed the Web site through the *splash page* once their user login and password were verified. Upon a successful login, users were directed to the *contents page,* which listed all available lessons. Each *lesson page* highlighted key aspects of a topic or concept and was closely related to the work done in class. The page layout (Figure 2) was kept uniform throughout the Web site, thus ensuring that students did not have to rediscover the steps of using the Web pages each time they logged in.

Most pages were linked to either a multiple-choice or short-answers test. These tests provided instant feedback to the user. Feedback to the tests did not specify which questions were wrong, for two reasons. First, students had to understand that there was more to an answer than merely choosing an option. Second, a wrong answer was meant to encourage the student to find the correct answer. It was purposely designed in this manner to encourage interaction. The results of these tests were written in a database file that could be accessed by students and teachers.

Students were also issued with or able to download lesson worksheets. These fill-in-the-blanks worksheets served as notes for students. It also facilitated further discussion in the classroom and kept students on task.

Research Sample

This study was conducted at a state high school in Queensland, Australia. Lessons were designed for various topics in Year 10 Science and Advanced

Science. Qualitative and quantitative research data were collected from all these classes. A sample of 315 students in 11 classes participated in this study over a 2-year period.

Internet lessons were scheduled once a week and lasted for one school term at a time. During these lessons, students had the option to access the Internet for up to 35 minutes. Students also had the option of accessing the Internet on their own time.

Data Collection Methods

To obtain a fuller picture of a research initiative, a number of researchers have recommended that both qualitative and quantitative methods should be used (e.g., Jayaratne & Stewart, 1995). Consequently, in this study, data were gathered through both methods. Students' perceptions of their Web-based learning environment were established quantitatively by using a learning environment instrument. A modified version of the WEBLEI (Chang & Fisher, 1998), which was initially developed to measure student perceptions in a tertiary environment, was used in this study. Qualitative data was also gathered through written surveys. Both these surveys were administered at the end of the school term. Students were given the option to respond to the surveys on a voluntary basis.

End-of-term examination results were used to investigate the impact of this initiative on students' academic achievement. Examination results obtained by the sample were compared with their earlier results and also with groups who had completed similar assessments in previous years. In the qualitative survey, students were also asked to specifically answer questions in relation to the extent to which Web-based learning influenced their learning outcomes.

Results and Discussion

Students' Perceptions of the Learning Environment

The 32 WEBLEI items were designed in a five-point Likert response format. The responses were scored as shown in the brackets—*Strongly Agree* (5), *Agree* (4), *Neither Agree nor Disagree* (3), *Disagree* (2) and *Strongly Disagree* (1). The mean and

Table 1. Mean and standard deviations of boys and girls for the four scales of the WEBLEI

WEBLEI Scales	Descriptive Statistics						
	Mean			Standard Deviation		Number of Cases	
	Boys (1)	Girls (2)	Difference in means (1)-(2)	Boys	Girls	Boys	Girls
Access	4.04	3.92	0.12	0.65	0.55	118	90
Interaction	3.65	3.49	0.16	0.70	0.72	114	92
Response	3.86	3.72	0.14	0.68	0.68	118	91
Results	3.98	3.89	0.09	0.61	0.59	116	90

standard deviation of boys and girls in junior science for each scale of the WEBLEI are presented in Table 1.

The results reported in Table 1 are interesting, because both boys and girls perceived the learning environment similarly. While the boys scored higher means for each scale, the difference between the two sexes was not statistically significant ($p < 0.05$). A mean of 4.04 ($SD = 0.65$) for boys and 3.92 ($SD = 0.55$) for girls for the Access scale suggested that both agreed that their learning environment was easily accessible at locations suitable to them. The learning environment was also convenient and it enabled them to work at their own pace. A Web-based learning environment also gave them greater autonomy in achieving their learning objectives. Qualitative data gathered from written surveys suggested that Access characteristics of the WEBLEI were important to students:

It is easier to understand and comprehend because you can read it at your own pace and you don't have to listen to a teacher mumble on.

You can go over the work again as many times as you like. Having the Internet sheets from class lessons help you revise and study. I can go over and over the parts I don't really understand until I do. It is easy to read and understand.

The Interaction scale produced means of 3.65 ($SD = 0.70$) and 3.49 ($SD = 0.72$) for boys and girls, respectively—the lowest of all three scales. An average of three implied that students neither agreed nor disagreed with all the items in the scale. While the WEBLEI's Interaction scale predominantly measured students' perceptions on the use of e-mails, student responses to the written survey suggested that their interaction with the technology was more significant to them than interactions with humans via technology. Hyperlinks, links to other related Web sites, online tests, applets, pictures and graphics were the interactions highly valued.

For the Response scale, mean scores of 3.86 (*SD*=0.68) and 3.72 (*SD*=0.68) for boys and girls were obtained, respectively, which implied that both groups generally agreed that Web-based learning was satisfying and it enabled them to interact with other students and their teachers. They also enjoyed learning in this environment and believed that this approach held their interest in the subject for the whole term.

Qualitative data suggested that students also believed that Web-based learning made learning easier, interesting and, therefore, enjoyable. According to students' responses, one of the reasons why they responded positively to Web-based learning was that it took them away from the usual routine (instead of doing the same thing repeatedly). In this relaxed environment, they did not have to listen to the teacher *carry on*. Their responses also suggested that they did not have to do much writing or copying off the board, and consequently, they believed that they learned more. They were also able to do related activities like *make your own rocket* and they *relaxed* while they were learning.

It is better than sitting in a classroom, bored. You are actually actively doing something.

It was something different and fun to do ... and was better than writing all the time.

No need to hear teacher's voices again (unless you need help) and after everything has been read, it goes straight to your brain.

It allowed girls to move away from certain bigoted boys, thus allowing us to learn without having to hear their very boring life story.

Of the four scales of the WEBLEI, the Results scale is probably the most important, because it gives an indication of whether students felt that they had

gained anything from their online learning environment. For this scale, means of almost four were achieved by both boys (*M=3.98, SD=0.61*) and girls (*M =3.89, SD=0.59*), which suggested that students agreed they could establish the purpose of Web-based lessons. It was also easy to follow, well sequenced and clear. The structure of the Web site kept them focussed and helped them learn better the work that was done in class. The content was presented well and was appropriate for delivery in a Web-based learning environment. The tests at the end of the lessons improved their understanding in the subject.

Responses to questions in the written survey suggested that many students believed that the design and layout of the Web site increased their understanding of the concepts covered in their science lessons. They felt that *Getsmart* not only reinforced content but helped them to look at the content from a different source. One student claimed that such an approach made the class *open-mined* and enabled to *look into new opportunities of learning*. Another student believed that *Getsmart* improved marks, widened knowledge of science and was more than a normal science lesson. These perceptions reflect aspects of the Results scale of the WEBLEI.

The Web pages are presented in a manner that is easy to follow. You can re-read what you do not understand, is put in a way where the content is ... in appropriate categories ... you can find your weaknesses.

Achievement in Science

For many people exam results are regarded as the best indicators of how successful an educational initiative is. Therefore, the success of an innovative approach could be measured by comparing exam data with exams previously completed. However, school assessment instruments are continuously changing which makes reliable data comparisons very difficult. Other important variations also occur such as teachers changing from one year to the next and the variety in their expertise in the subjects can vary.

The students who participated in this project studied units on consumer science and chemistry in

Table 2. Analysis of mid- and end-of-semester results for boys and girls in year 10 science

Assessment type	Difference in the means[#]	Standard deviation (girls)	Standard deviation (boys)
Traditional learning group ($N = 214$)			
Knowledge (mid-semester)	8.73[*]	18.20	22.36
Knowledge (end-of-semester)	3.42	19.70	21.37
Application (mid-semester)	-0.77	24.33	24.70
Application (end-of-semester)	3.61	33.21	34.50
Blended learning group ($N = 244$)			
Knowledge (mid-semester)	2.81	20.02	22.38
Knowledge (end-of-semester)	-2.08	17.87	18.03
Application (mid-semester)	0.92	22.45	24.27
Application (end-of-semester)	-4.01	29.34	30.60

 # Difference in the means equals the mean score of the girls minus the mean score of the boys.
* $p < 0.05$.

term one. In the second term, Web-based learning was introduced over a ten-week period and units on road science and space travel were taught in a blended environment. These units were relatively harder because the focus was on physics related concepts. An assessment of the degree of difficulty in this case was based on teacher observations and student results from previous years.

The tests were designed to measure student's abilities in three performance domains—*Knowledge, Science* and *Application Processes*. The *Knowledge* section of the test examined student's abilities to recall and apply their knowledge to simple situations. *Science Processes* measured their abilities to present and interpret data. The *Application* section measured their abilities to apply themselves in problem solving situations. The results obtained by students in the *Knowledge* and *Application* sections were compared because these were more readily available. The results in the *Science Processes* section included exam and laboratory report marks which made comparison difficult.

In this discussion the changes to student performance possibly because of Web-based learning, are based on the comparison of test results obtained by the students who experienced Web-based learning and those who were taught through traditional teaching methods the year before. Consequently these groups have been referred to as the traditional and blended learning groups. Table 2 presents mid and end-of-semester results for boys and girls in Year 10 science.

The data in Table 2 shows that the overall performance of girls in the traditional learning group was generally better than boys in both *Knowledge* and *Application* sections of mid and end-semester exams. An independent samples t-test showed that the difference in the means of the *Knowledge* results for the two groups in traditional learning group was statistically significant ($p<0.05$) with girls achieving a higher mean than did boys.

In the blended learning group, the girls did better than boys in both sections in their mid-semester exams whereas the boys reversed this difference in the end of semester exams. Even though the boys did better in the end of semester exam, the differences were small and statistically insignificant which suggests that the blended approach was as good as traditional teaching methods. Additionally the results did not show any gender bias. Qualitative data gathered from students suggested that there was an overwhelming belief that Web-based learning did make a difference to their performance in science.

I found that using the online learning technique was of great benefit to me when studying the topics incorporated in science ... I believe this allows students to attain and remain in a state of mind that enables a better understanding of the course material ... In my opinion, the online course taken by the science class was highly beneficial to them and their results.

The website helped me to learn ... I did well in the exam which is better than I expected.

Worksheets completed during the lessons acted like revision sheets for the exam.

Yeah! I only just passed last term, this term I got a B^+ or something. It was great! Internet lessons all the way.

FUTURE TRENDS

New research evidence suggests that the brains of girls and boys develop at different rates (Park, 2004). Consequently, children should be exposed to a variety of learning opportunities and teachers should actively look for innovative teaching approaches. According to Bill Gates (as cited in Levy, 2003, p. 49) "the digital era is far from fading ... it's only now getting interesting." Teachers have to find smart ways of blending these technologies into their lessons. But this should not be achieved at the expense of any group.

CONCLUSION

The findings of this study are significant in the "swinging pendulum debate" where one sex is perceived to be favoured more than the other in classrooms. Palmer (as cited in Bevin, 2002, p. 9), believed that the curriculum "used to be made for boys, but there are now a greater number of female teachers and this has led to discipline, curriculum, going ... more towards girls". In this instance, the Web-based learning environment appears to be a learning medium which is preferred equally by both sexes. Creating such fair environments must be pleasing to education authorities because it has the potential to create equal opportunities for all. The findings of this study warrant further research in this area.

ACKNOWLEDGMENTS

We thank the teachers of the high school in Queensland, Australia where this research was conducted. The first author of this paper is also grateful to Education Queensland for the award of the Premiers Smart State Teacher's Excellence Scholarship in 2002.

REFERENCES

Bevin, E. (2002, March 1). Boys' school crisis. *The NT News*, p. 9.

Biddulph, S. (1997). *Raising boys*. Sydney, Australia: Finch Publishing.

Bonk, C. J., & Kim, K. A. (1998). Extending socio-cultural theory to adult learning. In M. C. Smith & T. Pourchot (Eds.), *Adult learning and development: Perspectives from educational psychological processes*. (pp. 67-87). Mahwah, NJ: Lawerence Erlbaum Associates.

Chang, V., & Fisher, D. (1998). *The validation and application of a new learning environment instrument to evaluate online learning in higher education*. Retrieved July 31, 2003, from http://www.aare.edu.au/01pap/cha01098.htm

Collins, A., Brown, J. S., & Newman, S. E. (1989). Cognitive apprenticeship: Teaching the craft of reading, writing and mathematics. In L. B. Resnick (Ed.), *Knowing, learning and instruction: Essays in honor of Robert Glaser* (pp. 453-494). Hillsdale, NJ: Erlbaum.

Cortis, N., & Newmarch, E. (2000). *Boys in schools: What's happening?* Paper presented at the Manning the Next Millennium—An International Interdisciplinary Masculinities Conference, QUT, Brisbane, Australia.

Dieker, R. A. (1995). The future of electronic education. In E. Boschmann (Ed.), *The electronic classroom: A handbook for education in the electronic environment* (pp. 228-235). Medford, NJ: Learned Information Inc.

Eklund, J., Kay, M., & Lynch, H. M. (2003). *E-learning: emerging issues and key trends*. Retrieved September 25, 2003, from http://flexiblelearning.net.au

Gomory, R. E. (2001, January 11). Internetlearning: Is it real and what does it mean for universities? *Journal of Asynchronous Learning Networks,* *5*(1), 139-146. Sheffield Lecture, Yale University.

Jayaratne, T. E., & Stewart, A. J. (1995). Quantitative and qualitative methods in the social sciences: Feminist issues and practical strategies. In J. Holland, M. Blair, & C. Shelton (Eds.), *Debates and issues in feminist research and pedagogy.* Clevedon, UK: Multilingual Matters Ltd.

Lerner, R. M., & Galambos, N. L. (1998). Adolescent development: Challenges and opportunities for research, programs and policies. *Annual Review of Psychology, 49*, 413-446.

Levy, S. (2003, November 25). He's still having fun. *Newsweek,* 49-51.

Matters, G., Pittman, J., & Gray, K. (1997). *Are Australian boys underachieving?* Paper presented at the 23rd Annual Conference of the International Association for Educational Assessment (IAEA), Durban, South Africa.

Park, A. (2004, May 10). What makes teens tick? *Time,* 46-53.

Stager, G. (2004, July). *The case for computing.* Paper presented at the Australian Computers in Education Conference, Adelaide.

Stager, G. (2004). *A not-so-funny thing happened on the way to the future.* Retrieved 26 July, 2004, from http://www.stager.org/articles/dailyleader.html

Wenham, M., & Odgers, R. (2004, October 20). Glass ceiling begins to crumble. *The Courier Mail,* p. 9.

Wang, F.-K., & Bonk, C. J. (2001). A design framework for electronic cognitive apprenticeship. *Journal of Asynchronous Learning Networks,* *5*(2), 131-150. Retrieved February 10, 2003, from http://www.sloanc.org/publications/jaln/v5n2/pdf/v5n2_wang.pdf

KEY TERMS

Blended Web-Based Learning: This type of learning involves a blend or mixture of traditional and Internet driven teaching methods.

Learning Environment Instrument: A tool designed to quantitatively measure aspects of a learner's learning environment.

Splash Page: This is the introductory page of a Web site. It describes and leads a user to the rest of the Web site.

Web-Based Learning Environment Instrument: An instrument that is designed to quantitatively measure aspects of a Web-based learning environment.

What Women IT Professionals Want from Their Work

Kristine M. Kuhn
Washington State University, USA

K. D. Joshi
Washington State University, USA

INTRODUCTION

Articles in the popular IT press that address the underrepresentation of women often claim that women IT professionals differ from their male counterparts in what they desire from jobs, suggesting that special understanding of women's work-related values is required to improve their recruitment and retention (e.g., Bentsen, 2000; Gotcher, 1999; Paul, 2001). Although a great deal of research has been conducted on possible value differences that may affect women's and girls' attraction to IT as a career choice, there is relatively little empirical research showing that women actually do differ significantly from similarly trained men in the importance they place on particular qualities of work. The tradeoffs IT workers may make across desired attributes and the challenges they face in achieving career goals, however, are likely to show differences along gender lines. More methodologically-rigorous and practically-oriented research in job attribute preferences could help organizations make the changes in job design and personnel policies most likely to increase the representation of women in IT.

BACKGROUND

In general, little is known about the effectiveness of recruitment and retention practices targeted specifically at a particular demographic pool of applicants, such as women (Barber, 1998). It is well established, however, that the fit between what an employee or applicant desires from work and her perception of what she has is an important driver of both attraction to a job and turnover, as shown by both research in human resource management generally and with IT professionals specifically (Jiang & Klein, 2002). Thus, a key question is whether there are job attributes that women, on average, value more (or less) than men.

There is a large work of research literature many decades old comparing men's and women's ratings of how much they desire particular attributes of paid work (often referred to as work values), conducted among the general population, with adolescents, and across various educational and occupational groups (see Konrad, Ritchie, Lieb, & Corrigall, 2000, for a review). Most of these studies have reported gender differences in preferences for at least some job attributes. For example, traditionally it has been widely thought that men value earnings and leadership more, whereas women are especially likely to value social factors such as the opportunity to help others. These beliefs are also found among IT professionals. For example, in a cover story in *Infoworld* (Gotcher, 1999) women IT professionals are described as valuing income less than they value work assignments that allow growth. This article is based on a survey sponsored by Women in Technology International that had no male respondents for comparison purposes, but the underlying assumption is that men *would* respond differently. A software vice-president is quoted to the effect that "it's very different if you're interviewing men or women; I don't get as many questions about money and stock options from women" (Gotcher, 1999, p. 7). However, whether this behavior means that women actually care less about money, or rather that they have been socialized not to ask directly for more (see Babcock & Laschever, 2003), has important implications.

In the most comprehensive review of job attribute preferences, Konrad et al. (2000) conducted a meta-analysis of over two hundred studies, and found statistically significant gender differences for many attributes, generally consistent with gender roles and stereotypes. They note, however, that the effect sizes tend to be small, and that many masculine-typed job attributes, such as prestige and recognition, have become relatively more important to women in recent years, suggesting that aspirations have risen as barriers to equal opportunity have lessened (Konrad et al., 2000).

It is also true that reported job attribute preferences may be shaped at least partially by work and life experiences. Many researchers (e.g., Rowe & Snizek, 1995) have argued that gender differences in job attribute preferences are negligible once such factors as occupation and socioeconomic level are taken into account, which many studies have not done. Konrad et al.'s (2000) meta-analysis did include an analysis of studies of matched adults working in male-dominated occupations. In this subcategory, women rated several feminine-typed items, such as working with people, more highly than men, but they also rated many masculine-typed items, such as earnings, leadership, and challenge, *more* highly than their male counterparts. Most gender effects were small, and suggested a net pattern of greater similarity than difference among men and women currently working in the same predominantly masculine professions.

Research in this area conducted specifically on IT populations by IT researchers is relatively sparse, and the methods used sometimes make comparisons and conclusions difficult. Igbaria, Greenhaus, and Parasuraman (1991) surveyed several hundred members of the Association for Computing Machinery and classified each respondent according to which of eight possible career orientations he or she assigned the average highest score; thus men's and women's average ratings for various job attributes were not given. Igbaria et al. (1991) reported that a much higher proportion of women than men were primarily "life-style integration" oriented, defined as having an emphasis on family concerns and self-development. Conversely, a higher proportion of men were found to be technically oriented (primarily focused on the intrinsic technical content of work). The reported gender differences, however, were not analyzed controlling for type of work within IT or organizational level.

Other research suggests that men and women in IT might not differ significantly in what they desire from a job. Smits, McLean, and Tanner (1993) surveyed recent information systems graduates (IS) from over thirty U.S. colleges and universities and found that the similarities in desired job characteristics between men and women greatly overshadowed the differences. This study, where respondents had similar educational backgrounds and work tenure, also controlled for achievement level, and supports the argument that men and women are less likely to report significantly different preferences when they are similar on relevant factors.

Likewise, in a recent study conducted by the current authors (details available upon request), we surveyed IS alumni from a single program who had received their bachelor's degree within the previous five years. Men and women respondents did not differ significantly on the importance they placed on any of seven job attributes, including income, opportunity to develop technical skills, leadership opportunities, or social climate. Interestingly, both men and women on average rated work-life balance as most important.

After analyzing a broad survey of IT workers within the United Kingdom who rated the importance of several factors when choosing a new job, Panteli, Stack, Atkinson, and Ramsay (1999) concluded that women's responses were essentially similar to men's, with both men and women most likely to rate "job interest" and "challenging" as very important.

FUTURE TRENDS

Methodological Concerns

One problem with research on "what women in IT want" is that it relies on subjective ratings of very generally described job attributes. Typically, survey respondents are asked to rate the importance of each of a series of factors, such as responsibility, job interest, income, and promotional opportunities. However, priming artifacts and social desirability effects may limit the conclusions we should draw from such direct self-report data. Gender socialization and gendered social structures may affect both self-concept *and* self-presentation (see Konrad et al., 2000), and so the tendency for people to respond to such surveys in accordance with perceived social desirability (which, in turn, would be influenced by gender stereotypes) may be problematic.

Nor is it obvious what a respondent who rates "challenge" as more important than "income" would actually do when making career-related decisions. If we do not know how she defines those two attributes, or what she considers an adequate level of income or challenge, we cannot make predictions about the tradeoffs she would make in pursuing one job opportunity over another. It also remains a possibility than men's and women's subjective interpretations or definitions of such terms might differ. For example, it has been observed that women in IT tend to receive lower salaries than men, even when education and experience are controlled for, but that women do not report being any less satisfied with their compensation (see Baroudi & Igbaria, 1995; Sumner & Neiderman, 2003). This does not necessarily mean that women value income less than men, but may mean that their reference points differ from men's, or that their expectations (as opposed to aspirations) may rationally be lower.

Both in-depth qualitative studies and more sophisticated quantitative techniques may be better suited to revealing meaningful, and possibly subtle, gender differences in desired job attributes. For example, policy-capturing (also called conjoint analysis) designs derive the objective decision weights of a variety of independent factors, such as specifically defined job attributes, from an individual's global response to each of a series of combinations of those factors. As applied to job attraction, policy capturing provides a concrete, multi-attribute decision context, reduces social desirability effects, and links attribute information directly to the job choice criterion (see, for example, Zedeck, 1977). Other experimental designs could be used to see, for example, whether men and women IT professionals respond differently to various manipulations of advertised job descriptions. With between-subjects designs, where each individual participant responds only to one version of a job advertisement or description, social desirability and priming artifacts are again less of a problem.

A second major concern in studying gender differences in job preferences is the substantial segre-gation in the types of work men and women perform within the broad category of information technology, with women especially underrepresented in higher-paid IT jobs and more commonly found in categories such as help-desk workers (Igbaria, Parasuraman, & Greenhaus, 1997, Panteli et al., 1999). Whether this results from women choosing different types of work in accordance with their values, or whether self-reported values might be shaped at least partially by the type of work people are engaged in, or the extent to which both are true, are questions that cannot be adequately addressed by single-shot surveys. There is a clear need for more longitudinal research tracking individuals over time.

The Igbaria et al. (1991) study discussed earlier, a one-time survey administration, found that respondents with different career orientations tended to work in different types of IT jobs. For example, technically-oriented employees were especially likely to work in fields such as software engineering, whereas lifestyle integration-oriented respondents were found to be especially likely to work as applications programmers. Although the authors note that an individual's career orientation may not crystallize until she has significant work experience, they also claim that individuals' values cause them to gravitate toward compatible positions. Thus they suggest that because women are more likely to primarily value work-life balance, they may prefer jobs requiring less time commitment. In a later paper, however, Igbaria et al. (1997) noted that differences in career orientations in this sample could not explain women's underrepresentation in IT managerial positions.

Interestingly, Sumner, and Niederman (2003) reported that men and women graduates of IS programs did not tend to differ appreciably in types of work tasks performed. In our recent survey of IS alumni mentioned earlier, we also asked respondents to rate the attractiveness of three potential types of jobs within information systems, assuming they were to search for a new position. Both men and women rated an analyst position as most attractive. Women, however, did rate a service-oriented position somewhat more positively than did men.

Ideally, though, research following individual men and women over time would be best suited to answering the important question of whether inherent preferences and work values, or outside factors, best explain observed differences in career outcomes and occupational segregation within IT. Longitudinal research could also be used productively to assess the impacts of changes in family responsibilities on individuals' reported preferences, as well as their career choices and outcomes, and the related impact of work-life programs.

Research Needs

We suggest that research aimed at finding out what women in IT want from their work could have a much greater impact if it focused on particular aspects of job design and on work-family personnel policies. Rather than just assess subjective ratings of desired attributes, such research would focus on specific, practical options organizations could choose to target recruitment and retention efforts toward women. For example, in our survey of IS alumni mentioned earlier, on average both men and women respondents rated work-life balance as their most important criterion. However, women's actual experience seemed to differ from that of men, in that many more women rated their employing organization as not supportive of work-life balance. This again suggests that subjective ratings of generally described job attributes are insufficient to capture the experienced meaning underlying individuals' job-related preferences and decisions.

Many IT personnel researchers (e.g., Igbaria et al., 1997) have noted that women employees are more likely to carry a heavier burden of domestic labor, in terms of providing care to partners, children, and elders, than are men. Although men are increasingly assuming a greater responsibility for child-rearing, the imbalance is still marked and likely to remain so for some time. Personnel practices such as offering job shares, corporate day care, and part-time consulting work are recommended to organizations seeking to promote work-life balance, with the expectation that these options will be especially appreciated by women. However, there is relatively little hard data on how these policies are viewed by IT employees and applicants, and what impact they have on women in particular.

One common organizational response to work-life concerns is to offer more flexible work arrangements, in terms of either time or the space where the

work is performed. The option to telecommute is particularly relevant to IT workers. It is commonly assumed both that such arrangements will tend to attract women and, because they can better balance their work and family concerns, to retain them. However, the actual effects of such policies may not be so straightforward. Rau and Hyland (2002) examined the effectiveness of flextime and telecommuting in attracting employees by randomly assigning respondents (employed MBA students) to read different versions of a hypothetical job. Those with high role conflict between work and family were more attracted to organizations offering flextime, but were *not* more attracted to work offering telecommuting. Thus, "employers who believe telecommuting will be attractive to job seekers because it will reduce high levels of role conflict may find it ineffective in attracting the very applicants it targets" (p. 133). Similarly, Hill, Miller, Weiner, and Colihan (1998) did not find that telework helped employees balance their work and personal lives.

Research on telecommuting by IT researchers has typically focused on productivity and performance evaluations rather than effects on employees' home lives or attitudes. How telecommuting affects IT professionals' attraction, attitudes, and retention is thus an important topic for future research; a particular need would be the careful comparison of different types of telecommuting arrangements and methods of implementation. For example, the effects of telecommuting full-time vs. only one day a week are likely to be quite different.

As we noted earlier, IT personnel research incorporating the effects of career changes and changes in family responsibilities over time would be especially useful in assessing the practical impacts of work-life policies. As an illustrative example, a recent longitudinal study by Konrad (2003) of MBA students showed that preferences for short, flexible work hours predicted hours of household labor years later. Moreover, this relation was stronger for women than men, supporting the notion that women are especially sensitive to this factor even at "pre-need" stages of their career, in that they may be more likely to make plans for combining work and family. Interestingly, respondents performing a high degree of household labor did not have any corresponding reduction in their desire for high salaries and intrinsically rewarding work (Konrad, 2003).

It would also be useful to research whether there are significant gender differences in the effects of specific alternatives in job design. For example, if women do especially value social connections, then perhaps making work team-based might be especially useful in attracting and retaining women, as well as having other possible positive impacts. A recent article in *CIO Magazine* claimed that "[w]hat women want is a sense of purpose in their work. They also want to feel connected and needed" (Bentsen, 2000). While we would not dispute this, it seems likely that most men would value these factors as well. In fact, task significance—the perception that one's work has an important impact on others—is a key component of the most widely applied motivational-based job design model. Although many job characteristics might be unlikely to differ significantly in terms of their impact on men and women, the widespread belief that men and women in IT are motivated by different factors warrants further investigation, both to establish which differences may exist and which do not.

CONCLUSION

What IT employees most want from their jobs will naturally vary a great deal from person to person, depending upon values, personalities, backgrounds, current circumstances, and expectations. Although many IT academics and practitioners have suggested that men and women employees value job attributes differently, we would agree with other writers on gender and IT (Trauth, 2002) that the essentialist position (i.e., the view that there are essential fixed male and female characteristics) is unsupported in this area. The similarities in desired job attributes between men and women, particularly those with comparable training, probably greatly outweigh any differences. Nevertheless, the tradeoffs IT workers may be willing to accept in achieving their career goals, and the challenges they are most likely to face, may tend to show differences along gender lines. Research geared toward investigating specific policies organizations could actually choose to implement, and which is aware of the methodological limitations of relying solely on subjective ratings of generally described attributes, could there-

fore have great potential in improving the attraction and retention of women in IT.

REFERENCES

Babcock, L., & Laschever, S. (2003). *Women don't ask: Negotiation and the gender divide.* Princeton, NJ: Princeton University Press.

Baroudi, J. J., & Igbaria, M. (1995). An examination of gender effects on career success of information systems employees. *Journal of Management Information Systems, 11*(3), 181-201.

Barber, A. E. (1998). *Recruiting employees: Individual and organizational perspectives.* Thousand Oaks, CA: Sage.

Bentsen, C. (2000). Why women hate IT. *CIO, 13*(22), 81-86.

Gotcher, R. (1999). Building the best. *InfoWorld, September 27,* 7-13.

Hill, E. J., Miller, B. C., Weiner, S. P., & Colihan, J. (1998). Influences of the virtual office on aspects of work and work/life balance. *Personnel Psychology, 51,* 667-683.

Igbaria, M., Greenhaus, J. H., & Parasuraman, S. (1991). Career orientations of MIS employees: An empirical analysis. *MIS Quarterly, 15,* 151-169.

Igbaria, M., Parasuraman, S., & Greenhaus, J. H. (1997). Status report on women and men in the IT workplace. *Information Systems Management, 14,* 44-53.

Jiang, J. J., & Klein, G. (2002). A discrepancy model of information systems personnel turnover. *Journal of Management Information Systems, 19*(2), 249-272.

Konrad, A. M. (2003). Family demands and job attribute preferences: A 4-year longitudinal study of women and men. *Sex Roles, 49*(1), 35-46.

Konrad, A. M., Ritchie, J. E., Lieb, P., & Corrigall, E. (2000). Sex differences and similarities in job attribute preferences: A meta-analysis. *Psychological Bulletin, 126,* 593-641.

Panteli, A., Stack, J., Atkinson, M., & Ramsay, H. (1999). The status of women in the UK IT industry: An empirical study. *European Journal of Information Systems, 8,* 170-182.

Paul, L. G. (2001). Why IT hates women (and the women who stay anyway). *CIO Magazine, 14*(23), 114-116.

Rau, B. L., & Hyland, M. M. (2002). Role conflict and flexible work arrangements: The effects on applicant attraction. *Personnel Psychology, 55,* 111-136.

Rowe, R., & Snizek, W. E. (1995). Gender differences in work values: Perpetuating the myth. *Work and Occupations, 22,* 215-229.

Smits, S. J., McLean, E., & Tanner, J. R. (1993). Managing high-achieving information systems professionals. *Journal of Management Information Systems, 9*(4), 103-114.

Sumner, M., & Niederman, F. (2003). The impact of gender differences on job satisfaction, job turnover, and career experiences of information systems professionals. *Journal of Computer Information Systems, 44*(2), 29-39.

Trauth, E. M. (2002). Odd girl out: An individual differences perspective on women in the IT profession. *Information Technology and People, 15*(2), 98-118.

Zedeck, S. (1977). An information processing model and approach to the study of motivation. *Organizational Behavior and Human Performance, 18,* 47-77.

KEY TERMS

Flextime: Any of a variety of temporally flexible work arrangements in which employees are allowed to schedule their own working hours, generally given specific parameters and limits.

Interrole Conflict: The extent to which the expectations of different roles held by an individual are incompatible, for example work interfering with family life and vice versa.

Job Attribute Preferences: The importance placed on particular qualities and outcomes of paid employment; often referred to as work values.

Job Design: The way the content and tasks of a job are organized, including what tasks are done, as well as how, when, and where.

Targeted Recruiting: Recruiting efforts aimed at a particular group of potential applicants, often members of an underrepresented demographic.

Telecommuting: Performing work at a site away from the office, and transmitting the work electronically.

Work-Life Balance: Finding satisfaction with life both inside and outside paid work, and effectively managing job responsibilities as well as family life and/or other important activities.

The Woman Problem in Computer Science

Vivian A. Lagesen
Norwegian University of Science and Technology, Norway

INTRODUCTION

The longstanding concern about the absence of women in science and engineering in most Western countries has led to much research (see, e.g., Etzkowitz, Kemelgor, & Uzzi, 2000; Zuckerman, Cole, & Bruer, 1991). Today, this is all the more striking since women constitute the majority among students at most Western universities (Quinn, 2003). This article will briefly review research that analyses the lack of women in computer science, with a focus on higher education.

BACKGROUND

The substantial attention given to gender aspects of computer science reflects the significance of computers as a gateway to the emerging information society and the concern about a gendered digital divide. Women should take part in the design of this key technology, not remain users (Bjerknes & Bratteteig, 1987; Green, Owen, & Pain, 1993; Perry & Greber, 1990). The issue has been addressed by many women computer scientists, struggling to get more women into computing. Particularly in the U.S., a range of organisations and private initiatives have dealt with women and computing issues (Henderson & Almstrum, 2002). Australia and many North-European countries have also witnessed many such initiatives, often originating within government bodies.

MAIN THRUST OF THE ARTICLE

The so-called woman problem in computer science was discovered in the early 1980s through observations that young women took less interest in computers than young men (e.g., Dambrot, Watkins-Malek, Marshall, & Garver, 1985). Also, complaints that computer science offered a chilly climate to women students began to surface. Women students and faculty were perceived as women rather than as professionals, they were often treated as invisible, they met with patronising behaviour, and their qualifications were doubted. Also, their social environment was characterised by misplaced expectations, unwanted attention, and even obscenity (Barriers to equality 1983).

The main issue has been the question "why so few?" The explanations invoked may be classified in different ways (Ahuja, 2002; Cronin & Roger, 1999, Dryburgh, 2000, Littleton & Hoyles, 2002, Wilson, 2003). Striking metaphors have been employed to portray the situation, like "the incredible shrinking or leaking pipeline" or "the Silicon Ceiling." For the purposes of this article, contributions have been categorised under the following headlines, which summarize the varied understanding of the woman problem in computer science:

- Women's deficits
- Deficits in the educational practices of computer science and its student culture
- Discriminatory practices and other minority problems
- The masculine image of computer science

In the following sections, I will review research relevant to each of these categories. Please note that some contributions may belong to more than one of the categories, even to all of them.

Women's Deficits

Research has claimed that women have weaker computer knowledge and lesser computer experience, rendering an image of women as more computer anxious (Brosnan, 1998), less confident about their computer skills (Beyer, Rynes, & Haller, 2004; Borge, Roth, Nichols, & Nichols, 1980; Maccoby &

Jacklin, 1974), lacking self-efficacy in relation to computer science (Durndell et al., 2000, Galpin, Sanders, Turner, & Venter, 2003), alienated or just not interested in computers or computer science (Rasmussen, 1997; Siann, 1997; Symmonds, 2000). In this way, an emerging deficit model has become empirically grounded in comparisons with men's use of computers. Women were seen as deviant because men were regarded as the norm (Kramer & Lehman, 1990).

Consequently, a main argument has been that women needed to catch up with men by gaining access to computers and the related set of technical skills (Gansmo, 2004). In this manner, computer science was understood as a neutral set of skills to be acquired. Getting women into the discipline was a matter of "compensatory strategies," like providing better information and encourage girls and women to enter the field (Cronin & Roger, 1999; Henwood, 2000).

Turkle (1988) claimed that women's computer reticence emerged from their unwillingness or even fear to engage with a machine that was seen as intimate. However, her study of the early emerging Internet culture (Turkle, 1996) suggests a change, since women seem as eager as men to perform on the net.

Deficits in Educational Practices

Deficits in educational practices have been identified as obstacles to women to succeed and remain within computer science. Particularly the view that women, more than men, need to learn in a meaningful, goal-oriented and contextualised way, has been a common critique against programming courses. They have been seen as repetitive, playful, and meaningless exercises (Balcita, Carver, & Soffa, 2002; Countryman et al., 2002; Margolis & Fisher, 2002). To raise the quality of computer science teaching seems important to recruit and retain women students. This goal may be achieved by employing faculty that enjoy teaching undergraduates and by maintaining a stable faculty (Cohoon, 2002), by employing better qualified faculty, and by using more resources to hire teaching assistants and to provide better technical facilities and support (Lagesen, 2005; Margolis & Fisher, 2002; Roberts, Kassiandou, & Irani, 2002). To supply more women role models

is generally seen as vital (Richardson & Kavanagh, 1997; Roberts et al., 2002; Townsend, 2002). Also, the importance of networking and support communities has been emphasised (Gabbert & Meeker, 2002).

Discriminatory Practices and Other Minority Problems

As mentioned, being a minority of women in an environment dominated by men is observed to be a major problem for women in computer science (Berg, 2000; Dambrot et al. 1985; Spertus, 1991; Teague 2000). Minority problems are diverse and include discrimination and sexual harassment (Dambrot et al., 1985; Kanter, 1977; Spertus, 1991).

The Masculine Image of Computer Science

Many scholars have explored the assumption that computer science is masculine and consequently a turn-off for women. It has been argued that the discipline has emerged from and is associated with institutions and practices usually perceived as masculine, in particular mathematics. Many recognize this as the parent discipline of computer science (Dambrot et al., 1985, Gressard & Lloyd, 1987; Mahony & van Toen, 1990). When mathematics is unattractive to women, arguably, computer science would be too (Kvande & Rasmussen, 1989; Mahony & Van Toen, 1990; Stepulevage & Plumeridge, 1998). A parallel argument is that computer science technology grew out of the military (Edwards, 1990; Mörtberg, 1987).

Some has advocated to broaden the scope of computer science, from the belief that this would attract more women (Salminen-Karlsson, 1999). To include social and organisational aspects of technology, social issues, and contextualised computer science has been assumed to make computer science more woman-friendly (Clegg & Trayhurn, 1999; Henwood, 2000; Siann, 1997). There are observations that computer science programmes located in social science or humanities environments have a much higher proportion of women (Kvande & Rasmussen, 1989).

A different approach to the presentation of computer science as masculine emerged through the 1990s, when one became concerned with its culture.

The occupational culture of computing has been seen as alienating to everyone who are not enthusiastic about computers (Sproull, Kiesler, & Zubrow, 1987), and it has been characterised as masculine (Wright 1996). Another main finding has been the impact of what was seen as a new form of masculinity, described by reference to hackers, nerds, or geeks (Berg, 2000; Kiesler, Sproull, & Eccles, 1985; Margolis & Fisher, 2002; Rasmussen & Håpnes, 1991).

The image of computer science has been found a vital impediment for women pursuing an IT career. Particularly among young girls, computers, and computer science has been perceived as nerdy (Rasmussen, 1997; Rasmussen & Håpnes, 2003).

Non-IT women graduate students perceived computer science as offering an unwelcoming classroom or workplace environment for women, compared to other fields (Weinberger, 2004). Another U.S. study found that computer science students were seen as unsociable and nerdy, compared to students in other disciplines (Beyer et al., 2004; Jepson & Perl, 2002). A common observation is that computer science is considered more difficult by women than by men, and that women perceive themselves (inaccurately) to be less competent (Corneliussen, 2002; Nordli, 2003). Beyer et al. (2004) found women students more interpersonally oriented than men. Women valued careers that would allow them to help others, to work with people, and to provide opportunities to combine career and family. Computer science was seen to be in conflict with these concerns.

FUTURE TRENDS

In addition, to be a gendered space, computer science is also found to be highly racialised (Galpin, 2003; Katz, Aronis, Albritton, Wilson, & Soffa, 2003; Leggon, 2003; Margolis, Holme, Estrella, Goode, Nao, & Stumme, 2003; Taylor, 2002). In the U.S., African-American women computer scientists constitute less than 3% (Taylor, 2002). Nevertheless, it seems that there are segments of women, based on nationality and/or ethnicity, that do opt for computer science careers. Culley (1986) noted that girls from single sex schools and of Asian origin enjoyed computing more than most other girls. An Australian study show that Asian women students outnumbered non-Asian women students. The Asian women were seen as better able to overcome the apparently hostile "international" computer culture (Greenhill, von Hellens, Nielsen, & Pringle, 1995). Margolis and Fisher (2002) observe that women students most likely to persist in the computer science programme were international students, primarily from Asia and Eastern Europe.

Schinzel (2000) shows that women's participation in computer science courses varies a great deal from country to country, from between 50% in South American, North African, and some East Asian countries, to 10-30% in the U.S., China, Russia, India, and countries in the southern part of Europe. The lowest percentage of 5-10 is found in Scandinavia, German-speaking countries, Japan, most African countries, and Australia. Galpin (2002) provides a statistical account of undergraduate and graduate students in computer science that shows the same pattern. Despite these overwhelming national differences, a common notion that there is a global masculine computer science culture that transcends national cultures, still seems to exist (Galpin, 2002; Suryia & Panteli, 2000; Wright, 1997). Lagesen's study of the situation in Malaysia (2005) shows that here, women constitute about half of the students in computer science at most universities. Also, she shows that computer science is not perceived as masculine subject, but rather considered as a career particularly suitable for women. Here, women computer science students told they wanted to pursue an academic career within computer science because it provided a relatively flexible job situation.

CONCLUSION

The literature reviewed here provides a varied set of explanations of the phenomenon that so few women choose to study computer science. As argued previously, the main reasons given for this problematic situation may be categorised under these four headings: women's deficit, deficits in the educational practice, discriminatory practices and other minority problems, and the masculine image of computer science.

However, it is striking how the woman problem in computer science has been understood mainly as an issue of exclusion. With a few exceptions (see Berg, 2000; Nordli, 2003; Lagesen, 2005), we are left with little knowledge about why women who actually decided to study computer science, have done so. Lagesen (2005) finds that, generally, Norwegian women computer science students are motivated by an interest in science and mathematics, in addition to the prospect of secure and well-paid jobs. The Malaysian counterparts in Lagesen's study have similar motives, but they are also induced to study computer science through outspoken parental advice as well as through a perception that computer science is a well-suited occupation for women.

Future research in the area should pay more attention to what may attract women to computer science, instead of focusing solely on exclusion. Disseminating results from research that has focused on women's positive relationship to computer science may be important to change or modify the widespread notion that women do not belong in this field. Also, an important lesson to be learnt from the case of Malaysia, is that a large share of women among the student population seems to dissolve the otherwise widespread symbolic identity between computer science and masculinity. Thus, we may be less pessimistic in our strive to get more women to into computer science.

REFERENCES

Ahuja, M. K. (2002). Women in the information technology profession: A literature review, synthesis, and research agenda. *European Journal of Information Systems 11*, 20-34.

Balcita, A. M., Carver, D. L., & Soffa, M. L. (2002). Shortchanging the future of information technology: The untapped resource. *SIGSCE Bulletin, 34*(2), 32-36.

Barriers to Equality. (1983). *Barriers to equality in academia: Women in computer science at MIT.* Report prepared by Laboratory for Computer Science and the Artificial Intelligence Laboratory at MIT.

Berg, U. A. L. (2000). *Firkanter og rundinger. Kjønnskonstruksjoner blant kvinnelige dataingeniørstudenter ved NTNU. Skriftserien 3/2000*, Trondheim: NTNU, Centre for feminist and gender studies.

Beyer, S., Rynes, K., & Haller, S. (2004). Deterrents to women taking computer science courses. *IEEE Technology and Society Magazine, 23*(2), 21-28.

Bjerknes, G., & Bratteteig, T. (1987). System development with nurses. In G. Bjerknes, P. Ehn, & M. Kyng (Eds.), *Computers and democracy: A Scandinavian challenge.* Aldershot, UK: Avebury.

Borge, M. A., Roth, A., Nichols, G. T., Nichols, & B. S. (1980). Effects of gender, age, locus of control, and self esteem on estimates of college grades. *Psychological Reports, 47,* 831-837.

Brosnan, M. (1998). *Technophobia: The psychological impact of information technology.* London: Routledge.

Clegg, S., & Trayhurn, D. (1999). Gender and computing: Not the same old problem. *British Educational Research Journal, 26*(1), 75-89.

Cohoon, J. M. (2002). Recruiting and retaining women in undergraduate computing majors. *SIGSCE Bulletin, 34*(2), 48-53.

Corneliussen, H. (2002). Diskursens makt—individets frihet. Kjønnede posisjoner i diskursen om data', *Dissertation,* Faculty of Humanities, University of Bergen.

Countryman, J., Feldman, A., Kekelis, L., & Spertus, E. (2002). Developing a hardware en programming curriculum for middle school girls. *SIGSCE Bulletin, 34*(2), 44-48.

Cronin, C., & Roger, A. (1999) Theorising progress: Women in science, engineering, and technology in higher education. *Journal of Research in Science Teaching, 36*(6), 637-661.

Culley, L. (1986). *Gender differences and computing in secondary schools.* Loughborough: Department of Education.

W

Dambrot, F. H., Watkins-Malek, M. A., Sillings, S. M., Marshall, R. S., & Garver, J. A. (1985). Correlates of sex differences in attitudes toward and involvement with computers. *Journal of Vocational Behaviour, 27,* 71-86.

Dryburgh, H. (2000). Under representation of girls and women in computer science: Classification of 1990s research. *Educational Computer Research, 23*(2), 181-202.

Edwards, P. (1990). The army and the microworld. *SIGNS: Journal of Women in Culture and Society, 16,* 102-107.

Etzkowitz, H., Kemelgor, C., & Uzzi, B. (2000). *Athena unbound. The advancement of women in science and technology.* Cambridge, UK: Cambridge University Press.

Gabbert, P., & Meeker, P. H. (2002). Support communities for women in computing. *SIGSCE Bulletin, 34*(2), 62-66.

Galpin, V., Sanders, I., Turner, H., & Venter, B. (2003). Computer self-efficacy, gender and educational background in South Africa. *IEEE Technology and Society Magazine, 22*(3), 43-48.

Gansmo, H. J. (2004). Toward a happy ending for girls and computing? *STS-report 67/2004,* Trondheim: Centre for Technology and Society.

Gansmo, H. J., Lagesen, V. A., & Sørensen, K. H. (2003). Forget the hacker! A critical re-appraisal of Norwegian studies of gender and ICT. In M. Lie (Ed.), *He, she, and IT revisited.* Oslo, Norway: Gyldendal Akademiske.

Green, E., Owen, J., & Pain, D. (1993). *Gendered by design? Information technology and office systems.* London: Taylor and Francis.

Greenhill, A., von Hellens, L., Nielsen, S., & Pringle, R. (1995). Larrikin cultures and women in IT-Education: Multiple meanings and multiculturalism. In F. Grundy (Ed.), *Women, work, and computerisation: Spinning a Web from past to the future.* Amsterdam: Kluwer.

Gressard, C., & Lloyd, B. (1987). An investigation of the effects of math anxiety and sex on computer attitudes. *School Science and Mathematics, 87*(2), 125-135.

Henderson, P. B., & Almstrum, V. L. (2002). Some resources related to women and computing. *SIGSCE Bulletin, 34*(2), 184-189.

Henwood, F. (1998). Engineering difference: Discourses on gender, sexuality, and work in a college of technology. *Gender and Education, 10*(1), 35-49.

Henwood, F. (2000). From the women question in technology to the technology in question in feminism—Rethinking gender equality in IT education. *European Journal of Women's studies, 7*(2), 209-227.

Jepson, A., & Perl, T. (2002). Priming the pipeline. *SIGSCE Bulletin, 34*(2), 36-40.

Kanter, R. M. (1977). *Men and women of the corporation.* New York: Basic Books.

Katz, S., Aronis, J., Albritton, D., Wilson, C., & Soffa, M. L. (2003). Gender and race in predicting achievement in computer science. *IEEE Technology and Society, 22*(3), 20-27.

Kiesler, S., Sproull, L., & Eccles, J. S. (1985). Pool halls, chips, and war games: Women in the culture of computing. *Psychology of Women Quarterly, 9,* 451-462.

Kramer, P. E., & Lehman, S. (1990). Mismeasuring women: A critique of research on computer ability and avoidance. *SIGNS: Journal of women in culture and society, 16*(1), 158-171.

Kvande, E., & Rasmussen, B. (1989). Men, women, and data systems. *European Journal of Engineering Education, 14*(4), 369-379.

Lagesen, V. A. L. (2005). *Extreme make-over? The making of women and computer science.* PhD-dissertation, Trondheim: NTNU

Leggon, C. B. (2003). Women of color in IT: Degree trends and policy implications. *IEEE Technology and Society Magazine, 22*(3), 36-42.

Littleton, K., & Hoyles, C. (2002. The gendering of information technology. In N. Yelland & A. Rubin

(Eds.), *Ghosts in teh machine*. New York: Peter Lang.

Maccoby, E. E., & Jacklin, C. N. (1974). *The psychology of sex difference*. CA: Stanford University Press.

Mahony, K., & Van Toen, B. (1990). Mathematical formalism as a means of occupational close computing—why "hard" computing tend to exclude women. *Gender and Education, 2*, 319-331

Margolis, J., & Fisher, A. (2002). *Unlocking the clubhouse. Women in computing*. Cambridge, MA: The MIT Press.

Margolis J., Holme, J. J., Estrella, R., Goode, J., Nao, K., & Stumme, S. (2003). The computer science pipeline in urban high schools: Access to what? For whom? *IEEE Technology and Society Magazine, 22*(3), 12-19.

Mørtberg, C. (1987). *Varför har programmeryrket blivit mannligt?* Luleå: Tekniska Högskolan i Lulea, TULEA 1987:042

Nordli, H. (2003). *The net is not enough. Searching for the female hacker* (STS Report 61). Dissertation. Trondheim: Centre for Technology and Society.

Quinn, J. (2003). *Powerful subjects. Are women really taking over the university*? Stoke on Trent, UK: Trentham Books.

Perry, R., & Greber, L. (1990). Women and computers: an introduction. *SIGNS: Journal of Women in Culture and Society, 16*(1), 74-99.

Rasmussen, B., & Håpnes, T. (1991). Excluding women from the technologies of the future? A case study of the culture of computer science. *Futures, 23*, 1107-1119.

Rasmussen, B. (1997). Girls and computer science: It's not me. I'm not interested in sitting behind a machine all day. In F. Grundy et al. (Eds.), *Women, work, and computerisation. Spinning a web from the past to the future*. Berlin: Springer.

Rasmussen, B., & Håpnes, T. (2003). Gendering technology. Young girls negotiating ICT and gender.

In M. Lie (Ed.), *He, she, and IT—revisited* (pp. 173-197). Oslo, Norway: Gyldendal Akademiske.

Richardson, I., & Kvanagh, I. (1997). Positive action: promoting technology and science through female role models. In R. Lander & A. Adam (Eds.), *Women and computing*. Exceter: Intellect.

Roberts, E., Kassiandou, M., & Irani, L. (2002). Encouraging women in computer science. *SIGCSE Bulletin, 34*(2), 84-88.

Salminen-Karlsson, M. (1999). Bringing women into computer engineering. Curriculum reform processes at two institutes of technology. *Dissertation*, Linköping Studies in education and psychology, No. 60.

Schinzel, B. (2000). *Cross country computer science students study*. Paper for Women, Work, and Computerisation Conference 2000, Vancouver.

Siann, G. (1997). We can, we just don't want to. In R. Lander & A. Adam (Eds.), *Women and computing*. Exceter: Intellect.

Spertus, E. (1991). *Why are there so few female computer scientists?* (MIT Artificial Intelligence Laboratory Technical Report 1315). Cambridge, MA: MIT.

Sproull, L., Kiesler, S., & Zubrow, D. (1987). Encountering an alien culture. In S. Kiesler and L. Sproull (Eds.), *Computing and change on campus*. Cambridge: Cambridge University Press.

Stepulevage, L., & Plumeridge, S. (1998). Women taking positions within computer science. *Gender and Education, 10*(3), 313-326.

Suryia, M., & Panteli, A. (2000). The globalisation of gender in IT. A challenge for the 21st century. In E. Balka & R. Smith (Eds.), *Woman, work, and computerization: Charting a course to the future*. Vancouver: Kluwer.

Symmonds, J. (2000). Why IT doesn't appeal to young women. In E. Balka & R. Smith (Eds.), *Woman, work, and computerization: Charting a course to the future*. Vancouver: Kluwer.

Taylor, V. (2002). Women of color in computing. *SIGSCE Bulletin, 34*(2), 22-24.

Teague, J. (2000). Women in computing: What brings them to it, what keeps them in it? *GATES, 5*(1), 45-59.

Townsend, G. C. (2002). People who make a difference: Mentors and role models. *SIGSCE Bulletin, 34*(2), 57-62.

Turkle, S. (1988). Computational reticence: Why women fear the intimate machine. In C. Kramerae (Ed.), *Technology and women's voices*. London: Routledge.

Turkle, S. (1996). *Life on the screen*. London: Phoenix.

Weinberger, C. J. (2004). Just ask! Why surveyed women did not pursue IT courses or careers. *IEEE Technology and Society Magazine, 23*(2), 28-35.

Wilson, F. (2003). Can compute, won't compute: Women's participation in the culture of computing. *New Technology, Work, and Employment, 18*(2), 127-142.

Wright, R. (1996). The occupational masculinity of computing. In C. Cheng (Ed.), *Masculinities in Organisations*. Thousands Oaks, CA: Sage Publications.

Wright, R. (1997). Women in computing: A cross-national analysis. In R. Lander & A. Adam (Eds.), *Women in computing* (pp. 72-84). Exceter: Intellect.

Zuckerman, H., Cole, J. R., & Bruer, J. T. (1991). *The outer circle*. New York: W.W. Norton & Company.

KEY TERMS

Chilly Climate: Environments in class rooms or work places, where individuals are often treated differently because of their gender, race, age, or other "outsider" status. A chilly climate can have a damaging cumulative effect that affects for instance women's self-esteem, aspirations, and participation.

Computer Science: The scientific study of computers and their use.

Education Deficit Model: Weaknesses and flaws in the educational systems and practices in computer science that works to disadvantage women in particular.

Exclusion Perspective: To emphasise the factors that tend to keep women out of computer science.

Hacker: A stereotypical description of a person, usually a man, very much engaged with computers, highly skilled but with low interpersonal qualifications.

Inclusion Perspective: To emphasise the actions that may bring more women into computer science and retain them, including the motives and actions of the women themselves.

Masculine Image: The idea that computer science is somehow better suited for men than for women.

Role Model: A person with skills and appearances that one finds attractive and may try to emulate.

"The Shrinking or Leaking Pipeline": Refer to a pattern where the number of e.g. women in computer science are systematically being reduced as they move to higher levels in the educational system.

"The Silicon Ceiling": A variant of the "glass ceiling", which is a metaphor for an invisible barrier for women's upward career mobility.

Women's Deficit Model: The view that women lack certain qualities, like knowledge, skills or self-esteem, which should be corrected in order to have more women become computer scientists.

Women and Computing Careers in Australia

Gillian Whitehouse
The University of Queensland, Australia

INTRODUCTION

In spite of predictions that the spread of information technology (IT) would help break down the gender segregation that characterized employment in the industrial era, women are under-represented in professional computing occupations throughout the advanced industrialized world, and those who do take up work in the IT sector are most likely to be found in routine and comparatively low paid jobs. The emergence of a "lighter, cleaner, and more sedentary set of occupations than the technologies of iron, oil and steam" (Cockburn, 1985, p.2) has certainly produced new jobs for both women and men, but—as Cockburn argues—gender inequalities have been reshaped rather than eradicated in this process of technological change.

The aim of this article is to extend existing knowledge about gendered employment patterns in professional computing with an examination of the situation in Australia in the early 21st century. Drawing on research conducted as part of a project funded by the Australian Research Council (Whitehouse, Hunter, Smith, & Preston, 2002-5), the analysis illustrates the types of computing jobs that women are most likely to enter, and the extent to which women are ascending career ladders to take up senior technical and/or management positions. While this is primarily a descriptive exercise, it produces a more nuanced picture of gender inequalities in IT employment than observations simply about under-representation, and allows some reflection on strategies to enhance opportunities for women.

BACKGROUND

Feminist analyses have long drawn attention to the "historical and cultural construction of technology as masculine" (Wajcman, 1991, p. 22), with a wide range of studies examining the processes that underpin the reproduction of gender inequalities in IT-related employment. At one end of the spectrum, attention has been paid to the way gender differences are developed in the recreational use of computer technology, as well as in the "educational pipeline" (see, among many, Greenhill, von Hellens, Neilsen, & Pringle, 1996; Henwood, 2000; Margolis & Fisher, 2002). Within the workplace itself, analysis has focused on issues such as masculine cultures and sexual harassment, gender-biased notions of skill, lack of informal networks and role models for women, and demands for skills currency and working-time pressures that restrict the ability to balance work and family (for example, Ahuja, 2002; Webster, 2004; Wright, 1996). Clearly, not all these phenomena are peculiar to IT employment, and within the IT sector there are also contrary propositions, such as an expectation that new IT firms may be less constrained by traditionally gendered culture and practice (see Panteli, Stack, Atkinson, & Ramsay, 1999), and recognition that the technology itself brings potential to maximize time and space flexibilities.

It is not the goal of this article to adjudicate between competing explanations or ascertain the compounding causes of gender differences in IT employment; rather the purpose is to provide additional detail on the shape of gender differences within the IT workforce. Much analytical attention has been focused at the level of occupational choice, with the problem identified broadly as women's under-representation in IT courses and employment. Here, the focus is on patterns within the labor market. In particular, the aim is to illustrate the *horizontal segregation* of men and women across IT occupational categories, as well as *vertical segregation* between status levels within these occupational categories (see definitions at the end of the article). While a global picture of segregation can be identified (for example, with processor assembly work performed largely by women in poor countries), this article is limited to the Australian case and high-skill (or 'professional') computing jobs.

AN AUSTRALIAN OVERVIEW

The material presented below draws on aggregate level survey data to illustrate patterns of segregation in IT employment in Australia. Although a deeper understanding of the development and reproduction of gender inequality requires additional dimensions such as qualitative investigation of workplace culture and practice, it is only survey data that can provide the overview sought in this analysis. The first sub-section outlines the data used, noting its strengths and limitations. The second and third sub-sections address, respectively, patterns of horizontal and vertical segregation. While there is no attempt to establish causal relationships, the implications of the statistical overview are considered briefly in the final sections of the article.

Data

Data are drawn primarily from a commissioned survey of large IT firms operating in Australia (*Survey of Employment and Pay Rates by Gender in the IT Industry*) conducted by Classified Salary Information services (CSi) in November-December 2003. The intent was to produce a gender breakdown of employment and pay rates in a selection of skilled roles relating to the development, configuration and maintenance of computer systems, and for this purpose CSi utilized its comprehensive and regularly updated list of occupational roles and position descriptions in the IT sector. Responses were received from 77 of the 108 companies contacted, and this delivered information on 12,706 employees working in 106 designated occupational roles. The sample cannot be taken as representative of IT employment or computing professionals as a whole, because it only includes information on full-time employees and long-term contractors within the respondent organizations (although these comprise the majority of employees in all cases), and it excludes small IT firms and organizations in which IT is not the primary function. However, the data do provide a relatively comprehensive picture of regular employment in large IT companies in Australia, with an important advantage being the fine level of occupational detail available.

Additionally, data are drawn from the Australian Bureau of Statistics (ABS) 2001 Census of Population and Housing. This provides a useful backdrop for the CSi data, with the main advantage of the census being the statistically reliable picture it provides of the Australian population. However, the standard occupational classification system used in the census lacks the finely detailed categorization of IT jobs in the CSi survey, hence observations from the census are limited to the 4-digit occupational category "Computer Professionals" and some of its constituent sub-groups.

Horizontal Segregation

In Australia, as elsewhere, there is marked horizontal sex segregation within IT-related employment at a broadly inclusive level (that is, where IT employment is defined as ranging from occupations involving routine use of computers, to the design of IT systems and software). For example, data from the 2001 Australian census show that while women made up only 23% of computer professionals at that time, they were over-represented in lower paid areas of IT-related work such as data entry, where they accounted for 85% of employees. As would be expected, a more homogeneous picture was evident *within* the category "computer professionals", although the census data do indicate some variation among the sub-categories, with women more likely to be "systems managers" (close to 30% of this group was female) than "systems designers" (around 18% female).

The degree of variation observed is clearly dependent on the level of occupational definition; thus horizontal segregation is more apparent within the comprehensive list of computing jobs covered in the *Survey of Employment and Pay Rates by Gender in the IT Industry*. Women accounted for 22% of employees covered in this sample, which—as noted earlier—includes skilled jobs associated with the design and maintenance of computer systems, but excludes jobs based solely on the routine use of computers such as data entry. While this overall percentage of women is similar to the figure for computer professionals in the census data, within the occupational roles included, female share varied from under 10% for a range of support engineer roles to over 60% of employees in areas such as technical writing.

It is not possible here to examine differences across all 106 occupational roles included in the survey, hence attention is narrowed to selected groups of occupations, each comprised of a set of associated occupational roles classified by "career level" on the basis of the level of skill and responsibility required. Seven such groups were selected for examination, together accounting for 8,075 (64%) of the employees described in the sample. In ascending order of female share, these groups are: support engineers; consultants; software developers; technical analysts—specialized support; test analysts; and technical writers (definitions are provided at the end of the article).

Table 1 shows the marked variation in female share between these groups—from 9% for support engineers through to 64% for technical writers. Although explanations for this picture of horizontal segregation cannot be assessed within the scope of this article, they are likely to include differences in educational pathways and occupational histories, and gendered assumptions that reinforce women's location in support roles.

The extent to which this level of horizontal segregation is disadvantageous to women is not clear from Table 1, which provides no clear indication of a linear relationship between female share and average pay levels. For example, the groups with the lowest and highest female share—support engineers and technical writers—record very similar average pay rates. However it is also apparent that the two occupational groups with very high female share, test analysts and technical writers, are towards the lower end of the pay distribution. Most importantly for this article,

horizontal segregation figures can conceal disadvantage that becomes apparent once vertical divisions are uncovered.

Vertical Segregation

Three of the groups presented in Table 1 (consultants, software developers, and technical analysts) are examined in more detail in this sub-section to show their constituent career levels and—for the latter two groups—associated management roles. The three groups were selected because they have clear career ladders and sufficient numbers of women to illustrate vertical segregation.

Table 2 presents the data for consultants, showing that although women are relatively well represented in career levels two and three (31%, compared with the survey average of 22%), female share is much lower in career levels four, five, and six. As the table also shows, high career level consultancy jobs are relatively well paid, and thus women's limited career progression has clear ramifications for gender equity.

A slightly different picture is evident for software developers (see Table 3). As with consultants, women's representation is above the survey average at the entry level role—programmer (see also Baroudi & Igbaria, 1995), but the career and pay ladders are not as steep within this group, and there is a separate group of project management positions. Women's representation is fairly consistent within the software development group of roles, with the exception that they are more likely to

Table 1. Female share and average pay in selected computing occupational groups, Australia 2003*

Occupational groups	% Female	Average pay relative to programmer [**]	N (employees)
Support engineers	9	125	2043
Consultants	26	180	2850
Software developers	26	141	2256
Technical analysts, specialized support	28	161	743
Test analysts	44	117	144
Technical writers	64	127	39
TOTAL SAMPLE	22		12,706

Notes: * Data refer to full-time employees in large IT organisations; ** Average pay for the job role "Programmer" = 100. Pay rates include taxable base salary and any salary sacrifice superannuation amounts made by the employee.
Source: Survey of Employment and Pay Rates by Gender in the IT Industry, Australia 2003

Table 2. Female share and average pay for consultants by career level, Australia 2003*

Occupational roles	Career level	% Female	Average pay relative to programmer	N (employees)
Consultants				
Associate Consultant	2	31	105	*346*
Consultant	3	31	150	*1135*
Senior Consultant	4	19	203	*594*
Principal Consultant	5	23	214	*449*
Senior Principal Consultant	6	19	273	*326*

Note: * See notes to Table 1
Source: Survey of Employment and Pay Rates by Gender in the IT Industry, Australia 2003

Table 3. Female share and average pay for software developers and project managers by career level, Australia 2003*

Occupational roles	Career level	% Female	Average pay relative to programmer	N (employees)
Software developers				
Programmer	2	29	100	*345*
Analyst Programmer	3	27	143	*909*
Senior Programmer	3	22	158	*778*
Systems Analyst	4	28	139	*224*
Software project managers				
Project Leader	4	32	145	*523*
Software Project Manager	5	22	190	*166*
Senior Software Project Manager	6	15	249	*79*

Note: * See notes to Table 1
Source: Survey of Employment and Pay Rates by Gender in the IT Industry, Australia 2003

be systems analysts than senior programmers, and although systems analyst is rated at a higher career level than senior programmer, average pay is lower. While these observations suggest the emergence of gendered pathways within the software development function, the degree of vertical segregation is not as marked as for consultants or software project managers.

The software project management roles included in the lower half of Table 3 show that women are much more likely to be represented in project leader than in senior management roles (see also Donato, 1990). Although project leader is classified as career level four, it is not particularly well remunerated (for example, average pay is lower than for level three senior programmers), and there is little evidence that it provides a well-used stepping stone for women into senior management.

The final group to be analyzed includes technical analysts and the management roles within technical support centers. Table 4 shows the marked vertical segregation that exists among technical analysts providing specialized support. While female share among associate technical analysts (career level two) is high at 41%, this declines with each step of the career ladder, reducing to 10% for principal technical analysts (career level five). Although there is not quite as wide a pay disparity between entry and senior level positions as that illustrated for consultants, career progression clearly brings significant financial gain. Among the leadership/management roles, there is again a clear distinction between team leader and senior management roles, both in female share (from 40 to 19%) and level of remuneration (centre team leaders at career level four earn less on average than level three technical analysts).

Table 4. Female share and average pay for technical analysts and technical support centre management roles by career level, Australia 2003*

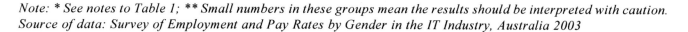

Occupational roles	Career level	% Female	Average pay relative to programmer	N (employees)
Technical analysts-specialized support				
Associate technical analyst-specialized support	2	41	108	*51*
Technical analyst-specialized support	3	26	143	*356*
Senior technical analyst-specialized support	4	14	172	*169*
Principal technical analyst-specialized support	5	10	206	*167*
Technical support centre management				
Technical support centre team leader **	4	40	130	*35*
Technical support centre manager **	6	19	239	*27*

Note: * See notes to Table 1; ** Small numbers in these groups mean the results should be interpreted with caution. Source of data: Survey of Employment and Pay Rates by Gender in the IT Industry, Australia 2003

Overall, the data on vertical segregation have allowed a more detailed assessment of gendered patterns of employment in skilled computing work. They present a less benign picture than the more aggregated horizontal comparisons shown in Table 1, in particular showing that the types of management positions women are likely to occupy are rarely high paid senior roles.

FUTURE TRENDS

Australia, along with a number of other industrialized countries, is experiencing a decline in already very low levels of women in tertiary level IT courses—a trend with clear implications for women's future representation in professional computing work. Women made up a little over one-quarter of commencing students in IT higher education in 2001, but this fell to around 20% in 2004 (Australian Government, n.d.). This trend has been accompanied by a fall in absolute numbers in the IT student cohort. From 2003 to 2004, there was a 15% drop in the number of women studying IT at tertiary level, while at the same time the number of women enrolled in tertiary education overall increased by 2% (O'Keefe, 2005, p. 26). Although the focus of this article is on progression within IT employment rather than tertiary education trends, this decline is not without significance. For example, career prospects for women in areas of professional computing work

where tertiary qualifications are the norm may worsen over time if there are fewer women to challenge masculine working cultures and bring equal employment opportunity issues onto the agenda. Additionally, to the extent that competitive pressures encourage long hours and discourage career breaks and working-time flexibility, and high levels of contracting out allow organizations to sidestep regulatory frameworks, the prospects for improving women's career progression in the sector remain limited.

CONCLUSION

In conclusion, this overview of gendered employment patters in skilled computing work in Australia indicates that the "women in IT problem" is not simply one of access, just requiring strategies to attract more women into the field; it is also one of career progression, underlining the need for workplace level arrangements to facilitate retention and advancement. This is not peculiar to IT employment, although the highly competitive nature of IT work and its strong masculine culture are likely to continue to make advances difficult. The problems are not immutable, however, and ongoing research identifying cross-national variations in segregation patterns and policy frameworks, as well as workplace and individual level studies illustrating the experiences and needs of those engaged in the IT sector, will continue to inform the pursuit of more gender egalitarian employment outcomes.

REFERENCES

Ahuja, M. K. (2002). Women in the information technology profession: A literature review, synthesis and research agenda. *European Journal of Information Systems, 11,* 20-34.

Australian Bureau of Statistics (ABS). (2001). *Census of Population and Housing* (unpublished data).

Australian Government. (n.d.). *Department of Education, Science and Training, online statistics.* Retrieved May 2005, from http://www.dest.gov.au/sectors/higher_education/publications_resources/statistics/

Baroudi, J., & Igbaria, M. (1995). An examination of gender effects on career success of information systems employees. *Journal of Management Information Systems, 11*(3), 181-201.

Cockburn, C. (1985). *Machinery of dominance: Women, men, and technical know-how.* London: Pluto Press.

Donato, K. M. (1990). Programming for change? The growing demand for women systems analysts. In B. F. Reskin & P. A. Roos (Eds.), *Job queues, gender queues: Explaining women's inroads into male occupations* (pp. 167-182). Philadelphia: Temple University Press.

Greenhill, A., von Hellens, L., Neilsen, S., & Pringle, R. (1996). Larrikin cultures and women in IT education. In *Proceedings from the 19th Information Research Seminar in Scandinavia,* Sweden (pp. 13-28).

Henwood, F. (2000). From the woman question in technology to the technology question in feminism: Rethinking gender equality in IT education. *The European Journal of Women's Studies, 7,* 209-227.

Margolis, J., & Fisher, A. (2002). *Unlocking the clubhouse.* Cambridge, MA: MIT Press.

O'Keefe, B. (2005, January 12). Women lead IT course exodus. *The Australian,* p. 26.

Panteli, A., Stack J., Atkinson, M., & Ramsay, H. (1999). The status of women in the UK IT industry: An empirical study. *European Journal of Information System, 8,* 170-82.

Survey of Employment and Pay Rates by Gender in the IT Industry. (2003). Commissioned for G. Whitehouse, R. Hunter, M. Smith, & A. Preston (2002-5) and conducted by Classified Salary Information services (CSi), Melbourne.

Wajcman, J. (1991). *Feminism confronts technology.* University Park, PA: The Pennsylvania State University Press.

Webster, J. (2004, June 17-18). Widening women's work in ICT. In U. Martin & Malcolm Peltu (compilers), *Women in computing professions: Will the Internet make a difference?* Position papers from an Oxford Internet Institute (OII) Policy Forum, Oxford. Retrieved March 2005, from http://www.oii.ox.ac.uk/resources/publications/OIIPP_20040617-WomenInIT_200407.pdf

Whitehouse, G., Hunter, R., Smith, M., & Preston, A. (2002-5) *The production of pay (in)equity: A study of emerging occupations.* Australian Research Council Discovery Project DP0209261.

Wright, R. (1996). The occupational masculinity of computing. In C. Chang (Ed.), *Masculinities in organizations* (pp. 77-96). Thousand Oaks, CA: Sage Publications.

KEY TERMS

All occupational definitions are based on position descriptions used in the *Survey of Employment and Pay Rates by Gender in the IT Industry* conducted in Australia in 2003 and may not be universally applicable.

Consultants: Employees with responsibility for evaluating clients' business needs, developing industry specific systems for clients, and liaising with them during installation and testing.

Horizontal Segregation: Shorthand for the notion of horizontal sex segregation in employment, which involves the separation or uneven distribution of men and women across different occupational groups, such as computer programmers and technical writers.

Software Developers: Employees who create and maintain computer programs, analyse software requirements and existing programs, and monitor new software developments.

Support Engineers: Employees who provide support to customers in the field or on-site, installing and repairing hardware and software.

Software Project Managers: Employees who control software development project schedules and quality standards, and report on costs and progress. Senior management positions may also involve control of budgets and the recruitment and training of staff.

Technical Analysts—Specialized Support: Employees who provide specialized technical support from a remote location by telephone or e-mail.

Technical Writers: Employees who present technical information in forms accessible to users, for example in manuals and online tutorials.

Test Analysts: Employees who prepare testing documentation and test programs.

Vertical Segregation: Shorthand for the notion of vertical sex segregation in employment, which involves the separation or uneven distribution of men and women across jobs of different status levels within the same occupational group, such as consultants and senior consultants.

W

Women and ICTs in the Arab World

Mohamed El Louadi
Université de Tunis, Tunisia

Andrea Everard
University of Delaware, USA

INTRODUCTION

The digital divide manifests itself on the one hand in the lag in Arab world nations vis-à-vis other more developed countries and on the other hand in the existing inequalities between men and women. Although the United Nations and the World Bank publish a variety of reports on the differences between developed and developing nations, very little data is available to fully grasp the meaning of the gap between genders. In terms of information and communication technologies (ICTs), there are two distinct gaps that need to be recognized: the gap between Arab men and Arab women and the gap between Arab women and women from other nations around the world (Figure 1).

Much differs in the lives of men and women. For decades, researchers have published comparative reports, attempting to explain what distinguishes men and women in socio-professional environments. According to Meyers-Levy (1989) men tend to be more comfortable with ICTs and partake more often in gaming and programming. When they use computers, women are more inclined to use them as communication tools. Given women's presumed lack of experience with technology, their upbringing which is different from men's, and that the studies they most often pursue are not technology-oriented, it is not surprising that women are generally less inclined to adopt new technologies. Those who nonetheless have tried their hand at browsing the Web were either witness to or victims of offensive language used during interactive discussion sessions; in some cases, they were harassed via e-mail. In order to avoid this unpleasantness, some women assumed male aliases (Herring, 2003). However, since 2000, when men and women reached parity in Web use (Rickert & Sacharow, 2000), it would appear that using the Internet is presently no more intimidating for females than for males.

An abundance of other differences between men and women exist. The United Nations Development Program (UNDP) acknowledged that there does not exist a society in which women benefit from the same opportunities as men. Everywhere in the world, women are poorer, less educated, and less valued than men. These and other inequalities reduce women's ability to take advantage of the potential benefits of ICTs and to consequently contribute to their nation's economic and social development which is in fact facilitated by these same technologies.

Figure 1. Female unemployment rates by region from 1993 to 2003

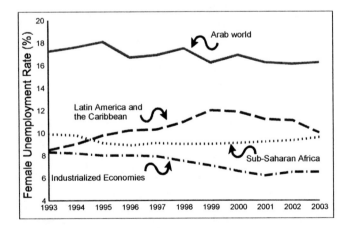

Note: In 2003, 40% of the world's 2.8 billion workers were women, representing a worldwide increase of nearly 200 million women in employment in the past 10 years. In the Middle East and North Africa, the female unemployment rate of 16.5% was 6% higher than that of men.

Source: ILO, 2004

BACKGROUND

If a digital divide is known to exist between northern and southern nations, between developed and developing countries, between knowledge economies and emerging ones, and between the haves and the have nots, it also needs to be recognized that a greater, encompassing-all-of-the-above divide exists based on individuals' gender. In essence, a divide exists between men and women regardless of what other category into which they may fall.

The Gender Digital Divide

Although the gender digital divide may no longer be a concern for a few countries, namely Scandinavian nations, it is very much present in Arab countries where the effect is exacerbated: first, because of the general lag of Arab nations, second because this divide targets the gender which, according to most of the socio-economic-cultural criteria as defined by the UNDP in its 2002 and 2003 reports, is already at a disadvantage.

In December 1998, 34.2% of men and 31.4% of women worldwide used the Internet. In 1999, less women (48%) than men (52%) used the Internet, even in the U.S. In August 2000, it was reported that 44.6% of men and 44.2% of women were Internet users.

A 2002 study undertaken by eMarketer and appearing in the New York Times reported that from a professional perspective men preferred e-mail to the telephone; the opposite was found true for women. For both genders, face-to-face interaction was favored above all (54% of men and 47% of women). These results are in line with Herring's (2001) comments who argued that women prefer the Internet over face-to-face conversations since, traditionally, patterns of male dominance have been observed in face-to-face interaction. Similarly, Consalvo (2002) maintains that the telephone has come to be considered a "female medium" of communication.

In 2003, a study by Nielsen/Netratings revealed that equality between European men and women Internet users was still far from reality. In 2000, this parity had been reached in the U.S. (Figure 2). According to the report, the Internet user population is indeed becoming feminized, albeit slowly. In 2002,

41% of European Internet users were women. In 2003, this number rose to 43%. At this rate, it will take until 2010 for equality in terms of the number of women and men Internet users in Europe to be reached.

Women in the Arab World

Arab countries have exhibited the fastest improvements in female conditions of any region of the world (UNDP, 2002). To illustrate, women's literacy rates have increased threefold since 1970 from 16.6% to 52.5% and female school enrolments have more than doubled. Today, women make up more than 70% of the student population in most Gulf-area universities. They represent more than 25% of judges in Tunisia (as compared with 20% of federal judges in the U.S.) and 10% of members in the Moroccan Parliament (as compared with 13% in the U.S. Congress) (Al-Hamad, 2003).

However, only 32% of women are active participants in their country's labor force, the lowest rate in the world (World Bank, 2004) (Figure 4). Though working women are generally more educated than their male counterparts, female unemployment is often highest among more educated women, who regularly leave the labor force to get married and have children (World Bank, 2004).

Figure 2. In the U.S., parity in Internet use between men and women was reached in 2000

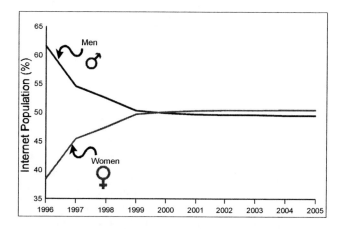

See Rickert & Sacharow, 2000; and eMarketer, 2000

Figure 3. The evolution of the literacy rate for Arab women as compared to that of men and to the world average

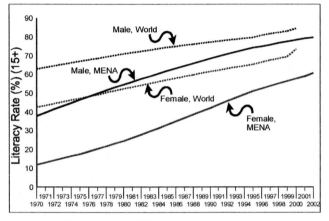

Source: World Bank, 2004

Figure 4. Male and female labor force participation, by region

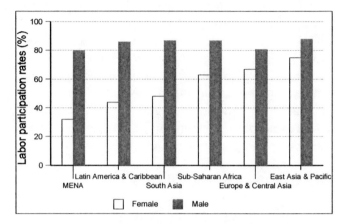

Note: Despite this significant growth and despite the high potential for women to participate in the labor force, actual rates of participation remain among the lowest in the world.
Source: World Bank, 2004

Figure 5. Literacy rates of women and men in selected Arab countries in 2004

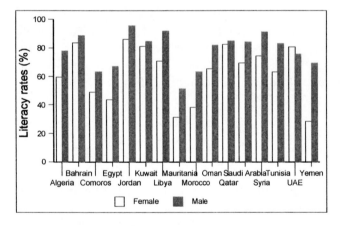

Source: UNESCO, The UIS literacy table (September 2004 release) applying to the reference period 2000-2004, www.uis.unesco.org

The 2003 edition of the UNDP report is a bit more pessimistic in that the quality of higher education was actually found to be declining and enrollments to be down. On closer examination, investments in education have actually declined since 1985. Even more alarming news is the fact that the average expenditure on R&D was a tiny 0.2% of

GNP. Among the 65 million illiterate Arabs (of a total population estimated at 280 million), two-thirds are women (Figure 5).

The 2002 UNDP report already highlighted that the condition of women is one of the top three inadequacies in the Arab world. The 2003 report added that, in general, the utilization of Arab women's capabilities is amongst the lowest in the world both in quantitative and in qualitative terms. This inspired Christiaan Poortman, World Bank Vice President for the Middle East and North Africa (MENA) region, to state that "No country can raise the standard of living and improve the well-being of its people without the participation of half its population."

Arab Women and Information and Communication Technologies

Over the years, changes have also been sensed on the ICTs front, at least for younger generations. The introduction of Internet browsing and electronic messaging through Internet cafés extended across the Arab world. The World Bank estimates that over 8 million Arabs used the Internet by the end of 2002, compared with less than 40,000 in 1995.

When looking at ICTs and the Arab world, the picture is further complicated because gender and culture are confounded. The digital divide is thus doubled, even squared, in such a case.

A number of analyses, including those of the UNDP and the World Bank, report that the proportion of Arab women Internet users was 4% while the European average was 42%. This figure rose to 6% in 2000 and a few years later to 19-20% with peaks reaching 36% in the United Arab Emirates.

One could expect the Internet to be the preferred medium of communication for women in some Arab countries. Indeed, in other parts of the world the Internet has been claimed to lead to greater gender equality (Consalvo, 2002) because text-based computer-mediated communication would allow women and men to participate equally, in contrast with patterns of male dominance observed in face-to-face conversations and because it would allow women to find community in pursuit of their own interests (Herring, 2001). Furthermore, the World Wide Web would allow women to engage in entrepreneurial activities without transgressing socially constructed norms and rules such as those prevailing in the Gulf where men and women are not allowed to mingle.

FUTURE TRENDS

The ICT Skills Deficit in the West

In 2002, it was forecast that Western Europe would lack 1.6 million ICT specialists representing 12% of the demand. The same year, it was expected that France would have a need for 67,000 network specialists, England 31,000 and Germany 188,000. And the outlook, if anything, looks bleaker as demonstrated in a study reported in the French magazine Les Echos (2000) in which it was found that by the year 2010 the number of needed ICT engineers in France would reach 1.1 million.

The Brain Drain

Though brain drain has been a worldwide phenomenon since the 1960s, a UN agency suggested that between 1960 and 1987, 825,000 skilled immigrants entered the U.S. and Canada from developing countries (UNCTAD, 1987).

The UNDP (2003) estimates the number of Arab doctors having emigrated between 1998 and 2000 at more than 15,000 and that 25% of the 300,000 graduates from Arab universities in 1995-1996 migrated. Furthermore, it was estimated that a total of $13 billion was lost to the Arab world and African countries in the 1970s as a result of the brain drain. A further report by the Arab League estimated that the loss to the Arab countries was $200 billion, noting that Western countries are the greatest beneficiaries from hosting a number of 450,000 Arabs of higher scientific qualifications (Arabic News, 2001).

The brain drain is unevenly spread across the Arab region. North Africa and the Middle East suffer higher rates than the Gulf because of difficult socioeconomic conditions, lack of social incentives, limited employment perspectives, shortages in research budgets, and chronic technology underinvestment (Dutta & Coury, 2003).

Losses are not just monetary since this is handled by the amounts of remittances sent home by the diaspora and which, in some cases, more than compensates for the brain drain (Adams, 2003). Adams shows that remittances sent to 24 developing countries (including Egypt, Morocco, and Tunisia) exceeded $500 million dollars in 2000. This is what has been known as the "brain gain" phenomenon whereby a more optimistic light is cast on the brain drain issue. The argument is that remittances, return of skilled migrants and development aid are noted as the ways source countries compensate for the loss of brains (Ramamurthy, 2003).

However, if Arab and developing countries invest heavily in educating young men and women, the consequences are readily felt when a sizeable segment of these migrates and never returns. When the migrants are trained in high-tech know how, the sending countries lose a great advantage because of the migration of labor, with the benefit accruing to the host countries that did not incur the cost of educating them (Asmar, 2003).

The Changing Demographic Scene in the Arab World

One of the most significant factors that developed nations will face in the next two or three decades is the aging of their population. Already between 1998 and 2025 it is estimated that the number of people 65

Figure 6. The age pyramids of the developed world (top) and of the Arab world (bottom) between 2000 and 2030

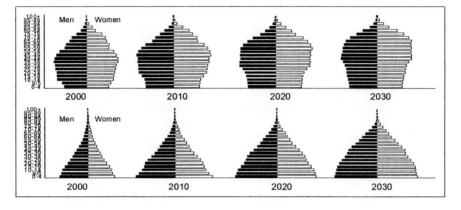

Note: As defined by the United Nations, the developed world is comprised of Europe, North America (U.S. and Canada), Australia, New Zealand and Japan.
Source: United Nations Population Division, World Population Prospects: The 2002 Revision Population Database, Graphics, and computations by the authors

years and older will double, while those less than 15 years old will rise by only 6% (Bureau of the Census, 1999).

One interesting fact is that the Arab world has the most significant proportion (37%) of youths aged 14 years and younger in the world (UNDP, 2002). In the last two decades, the population of the Arab world rose by an average 2.6% relative to the world average of 1.5%. This rate will continue to be above average in 2015 (2.1%) and in 2025 (1.9%) (United Nations, 2003).

North Africa's population grew from 140 million in 1990 to 157 million in 1995 and is likely to reach 211 million in 2010 (Cordesman, 1996). Libya, Morocco, and Tunisia where fertility rates are similar to those in European nations are expected to post populations of 8.7, 35, and 12 million, respectively (Cordesman, 1996). Furthermore, North African women tend to marry later than women from the Gulf countries (Fargues, 2002). However, typically where demographic demands are felt the strongest the greatest constraints are placed on the educational budget, infrastructure, health, water consumption, employment, and even social stability.

A cross-sectional poll of Arab youths conducted by the UNDP (2002) highlights the fact that nearly half (45%) the respondents expressed a desire to emigrate. In fact, close to 100,000 Arabs and Muslims emigrate to the West each year. The aging

populations of Italy, Germany, Great-Britain, France, and especially the United States and Japan create a need for these immigrants (The Economist, 2002).

CONCLUSION

The effect of the demographic developments taking place in various regions of the world on the Arab world begs for our serious attention. In a decade or two, the youths of the Arab world will become even more valued than they have been since 1995. Furthermore, the part of the youths and talented individuals in the Arab world who are most mobile tend to be of the male gender (Hijab, El-Solh, & Ebadi, 2003). As western countries will be seeking new and innovative minds, there is a true possibility that all the investments made by Arab nations in the education and training of Arab engineers will be lost, practically overnight. In addition, the aging populations in the West will represent such a social burden on the governments that the fiscal pie will need to be enlarged and this will be done by rapid and substantial bouts of emigration and naturalization.

Women in the Arab world have an essential role to play in the economic and social development of their nations. If one can use history as a model for the future, the two World Wars correspond to the initial stages of women's emancipation in the West

Figure 7. While net migration rates (the difference between inflow and outflow of human capital) are already negative for North African countries, they will begin to be negative around 2015 for the Middle East

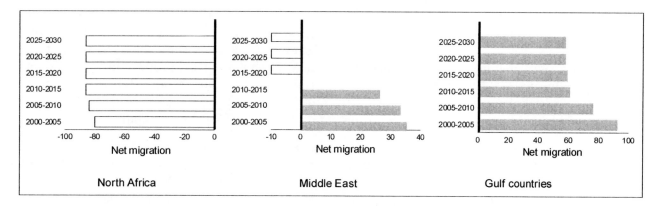

Source: United Nations Population Division, World Population Prospects: The 2002 Revision Population Database, Graphics, and computations by the authors

which was then followed by the 1970s feminist movement in the U.S. When men left for the war front and many did not return, the economic wheels could not afford to stop turning and it was the women who ensured that they did not.

Based on the rate of emigration of Arab youths and the need of western nations, what will be left of the Arab world will essentially be populations that are relatively less educated and less mobile. If we assume that a sizeable portion of the highly qualified and competent men of the Arab world will be lured away by the prosperous and captivating West, the reality we will then face is one where the role of the Arab woman will become much more important and crucial to the survival of the Arab world. Furthermore, this will not only apply to the domain of ICTs. If Gulf nations are assumed to suffer less than those of North Africa and the Middle East, then the women of the former will benefit most from added attention in terms of training and education, especially in the area of ICTs.

Because it would appear that the brain drain situation for the Gulf is not as serious as that of North Africa and the Middle East, it follows that it is women of the Middle East and North Africa (MENA) that need the most urgent attention.

Women should no longer be trained in word processing and other so-called "productivity tools" mostly destined to secretarial jobs but rather in the design of application systems, of Web sites and systems interfaces. Internet technologies might be the feminine "killer application" due to women's hypothesized communicational skills (Taggart & O'Gara, 2000). Training programs such as those launched by Cisco should become commonplace[1]. True technological learning extends beyond word and spreadsheet processing. Arab content owes itself to include women's interests; advances in business and electronic exchanges cannot discount the female gender in the Arab world.

REFERENCES

Adams, R. H., Jr. (2003, June). *International migration, remittances, and the brain drain: A study of 24 labor-exporting countries.* Policy Research Working Paper 3069. Retrieved December 31, 2004, from http://econ.worldbank.org/files/27217_wps3069.pdf

Al-Hamad, A. Y. (2003). *The Arab world: Performance and prospects.* The Per Jacobsson Foundation, Dubai, UAE.

Arabic News, (2001). *$200 billion, Arab loses of brain-drain, regional, economics.* Arab

News.com, February 27. Retrieved December 31, 2004, from http://www.arabicnews.com/ansub/Daily/Day/010227/2001022720.html

Asmar, M. (2003, December 29). Stop the brain drain from the Arab world. *Gulf News*. Gulf News Research Centre. Retrieved December 31, 2004, from http://www.aljazeerah.info/Opinion%20editorials/2003%20Opinion%20Editorials/December/29%20o/Stop%20the%20brain%20drain%20from%20the%20Arab%20world%20By%20Marwan%20Asmar.htm

Bureau of the Census. (1999). *World population at a glance: 1998 and beyond*. U.S. Department of Commerce Economics and Statistics Administration, Bureau of the Census.

Consalvo, M. (2002). Selling the Internet to women: The early years. In M. Consalvo & S. Paasonen (Eds.), *Women and everyday uses of the Internet: Agency and identity*. New York: Peter Lang.

Cordesman, A. H. (1996). *World Population Trends, Regional Issues, and the Middle East as a Case Study, Center for Strategic and International Studies (CSIS)*, 1800 K Street N.W. Washington, DC 20006. Retrieved August 6, 2005, from www.csis.org/mideast/reports/demogra2.pdf

Courbage, Y. (1998, September 3-6). *La pression démographique, passé, présent et futur au Moyen-Orient et en Afrique du Nord, Institut National d'Études Démographiques*, Paris, présenté au Forum sur le Développement Humain au Moyen-Orient et en Afrique du Nord (MENA), Marrakech.

Dos Santos, D. M., & Postel-Vinay, F. (2003). Migration as a source of growth: The perspective of a developing country. *Journal of Population Economics, 16*(1), 161-175.

Dutta, S., & Coury, M. E. (2003). *ICT challenges for the Arab world*. The Global Information Technology Report 2002-2003: Readiness for the Networked World, World Economic Forum, 116-131. Retrieved August 6, 2005, from http://www.developmentgateway.org/download/170136/Chapter_08_ICT_Challenges_for_the_Arab_World.pdf

eMarketer. (2000, September). *The eDemographics & Usage Patterns Report*. eMarketer. Retrieved December 30, 2004, from www.lemoyne.edu/library/ereporter/eDemo.pdf

Fargues, P. (2002, October 16-17). *Dialogue sur la coopération migratoire en Méditerranée occidentale* (5+5), Institut National d'Etudes Démographiques, Paris et Robert Schuman Centre for Advanced Studies, European Insitute, Florence. Ministerial Meeting, Tunis. Retrieved August 6, 2005, from http://www.iom.int/DOCUMENTS/OFFICIALTXT/FR/Fargues.pdf

Herring, S. C. (2001, October). *Gender and power in online communication, center for social informatics* (SLIS, No. WP- 01-05). Indiana University-Bloomington. Retrieved January 4, 2005, from http://www.slis.indiana.edu/CSI/WP/WP01-05B.html

Herring, S. C. (2003). Gender and power in online communication. In J. Holmes & M. Meyerhoff (Eds.), *The handbook of language and gender*. Oxford: Blackwell Publishers. Retrieved January 4, 2004, from http://www.slis.indiana.edu/CSI/WP/WP01-05B.html

Hijab, N., El-Solh, C., & Ebadi, N. (2003). *Social inclusion of women in the Middle East and North Africa*. Washington, DC: The World Bank.

Les Echos. (2000, June 19). *La pénurie d'informaticiens menace les ambitions de la nouvelle économie*, p. 26.

Meyers-Levy, J. (1989). Gender differences in information processing: A selectivity interpretation. In P. Caffarata & A. M. Tybout (Eds.), *Cognitive and affective responses to advertising* (pp. 219-260). Lexington, MA.

Ramamurthy, B. (2003). *International labour migrants: Unsung heroes of globalisation* (Sida Studies No. 8) Prepared for and funded by SIDA Stockholm, 120p. Retrieved January 7, 2005, from www.sida.se/content/1/c6/02/ 20/90/46395 Sida Studies no 8.pdf

Rickert, A., & Sacharow, A. (2000, August 19). *It's a woman's world wide Web: Women's online behavioral patterns across age groups and lifestyles*. Media Metrix, Inc. and Jupiter Communications. Retrieved January 4, 2004, from http://

banners.noticiasdot.com/termometro/boletines/docs/consultoras/jupiter/2000/jupiter_mujeres.pdf

Taggart, N., & O'Gara, C. (2000). *Training women for leadership and success in information technology.* Washington, DC: Academy for Educational Development.

The Economist. (2002, November 2). *The longest journey.*

UNCTAD. (1987, July 13). *United Nations Conference on Trade and Development, Trends and Current Situation in Reverse Transfer of Technology* (TD/B/AC.35/16).

UNDP. (2002). *Arab human development report 2002: Creating opportunities for future generations.* United National Development Program and Arab Fund for Economic and Social Development.

UNDP. (2003). *Arab human development report 2003: Building a knowledge society.* United National Development Program and Arab Fund for Economic and Social Development.

United Nations. (2000). *The world's women 2000: Trends and statistics.* New York: United Nations Statistics Division.

United Nations. (2003). *World population prospects: The 2002 revision population database.* United Nations Population Division, 2003. Retrieved from http://esa.un.org/unpp/

United Nations. (2004). *Where do Arab women stand in the Development process? A gender-based statistical analysis.* New York: Economic and Social Commission for Western Asia. Retrieved January 1, 2005, from http://www.escwa.org.lb/information/publications/sdd/docs/sdd-04-booklet.1.pdf

World Bank. (2004). *Gender and development in the Middle East and North Africa: Women in the public sphere.* Retrieved January 6, 2005, from http://www1.worldbank.org/publications/pdfs/15676frontmat.pdf

KEY TERMS

Arab World: Twenty-two countries and territories, namely: Algeria, Bahrain, Comoros, Djibouti, Egypt, Iraq, Jordan, Kuwait, Lebanon, Libya, Mauritania, Morocco, Oman, Palestine, Qatar, Saudi Arabia, Somalia, Sudan, Syria, Tunisia, United Arab Emirates, and Yemen.

Brain Drain: the migratory movement of skilled labor from one country (the sending country) to another (the host country).

Digital Divide: the difference in access to and usage of information technologies.

Gulf States: Bahrain, Kuwait, Oman, Qatar, Saudi Arabia, and United Arab Emirates.

ICT: Information and communication technologies, including, among others, the PC, the Internet and the Web, mobile and fixed telephony, and their software and applications.

MENA: Middle East and Africa but usually includes the Gulf countries as well.

Middle East: Iraq, Jordan, Lebanon, Palestine, and Syria.

North Africa: Algeria, Egypt, Libya, Mauritania, Morocco, and Tunisia.

ENDNOTE

[1] The scholarship program launched by the Cisco Networking Academy: Women In Technology is sponsored by the United States Agency for International Development (USAID) and managed by the Institute of International Education (IIE). Thanks to a grant from the Internews Network Dot-GOV program, the program grants scholarships to female candidates from Tunisia, Algeria, and Morocco allowing them to be Cisco Certified Network Associate (CCNA). The program trains in Web site design, database design, cabling, Java, Unix, etc. See the Women In Technology Scholarship Program (WIT) Web site (www.iie.org/wcoast/wit.html).

Women and Recruitment to the IT Profession in the UK

Ruth Woodfield
University of Sussex, UK

INTRODUCTION AND BACKGROUND

Early commentary on the development of the field of Computing and its relationship to women was generally optimistic in tone. Many early software workers were female, and the associations of computing with a qualitatively different, and cutting-edge, technological domain, caused projections that women would comfortably enter professional Computing work in a manner unparalleled for scientific and technological occupations (Faulkner, 2002; Woodfield, 2000). The rationalisations shaping the decision of early female entrants to the field often mirrored those buoying up the optimism of commentators. An established female computer professional in the late 1980s, for instance, reported applying for her first job within the IT sector a decade earlier because she had believed the area to be "one of the first businesses with no sex prejudice" (Cowan, cited in *The Guardian*, 1989).

A review of the literature that chronicled the *actual* movement of women into IT work cross-nationally since these early predictions, however, leaves little doubt that women were quickly established as the under represented party within IT roles. As Elizabeth Gerver suggested at the close of the 1980s, Computing effectively became established as a "strangely single-gendered world," and although women's under-representation may have varied "from sector to sector and to some extent from country to country," the evidence of its male-domination and, indeed, its maleness, became so ubiquitous that it tended "to become monotonous" (1989, p. 483). A large body of work has underpinned the ongoing legitimacy of this observation since the 1980s (Faulkner, 2002; Hall, 2004; Millar & Jagger, 2001; Peters, Lane, Rees, & Samuels, 2003; Woodfield, 2000).

The Participation and Progress of Women in UK IT Occupations

Women are under represented within occupational roles within the UK IT sector, or IT roles in other sectors, and especially within the more prestigious and well-rewarded roles.

Their estimated numbers within the broad category of IT workers vary, but available statistics suggest that it is lower in the UK than in other countries, including the U.S. and Canada. Women, it is variously claimed, comprise between 13% and 22% of all IT workers, including those outside of the IT sector itself (see Faulkner, 2002; Millar & Jagger, 2001; Miller, Neathey, Pollard, & Hill, 2004; Sørensen, 2002). This represents a decline since the turn of the millennium (Faulkner, 2002; Millar & Jagger, 2001). Indeed, the president of the British Computer Society recently announced a 3% drop in female participation between 2002 and 2003 alone (Hall, 2004).

Female participation patterns in the UK follow global trends and are unevenly distributed throughout the sector and occupations. Best estimates claim women comprise only 16% of all IT managers, 21% of computer analysts, and 14% of software professionals (although less than 10% of software engineers). They are more substantially represented in the lower echelons of IT work as operating technicians, for instance, where they comprise approximately 30% of the workforce (EOC, 2001b; EOC, 2004; Miller et al., 2004).

Another characteristic of female participation, however, is that women are far more likely to work in an IT role outside of the IT sector itself. Indeed, it is estimated that they comprise only 9% of IT workers within the area (Faulkner, 2002; Millar & Jagger, 2001).

It is worth noting, however, that those women who find themselves joining the IT workforce are likely to experience slightly less severe gender pay differentials than women in other types of work. For instance, computer analysts/programmers can expect to be paid 7.7% less than comparable males, and IT managers can expect a lag of 12.1% behind their male peers (Miller et al., 2004). This compares well against the 18.4% average deficit that women experience more generally. The trend is for pay gaps to increase the further down the occupational scale IT roles are situated, but they are still marginally better than national averages for comparable work (Miller et al., 2004).

This does not, however, help retention rates, which are generally worse for women, including at the more senior levels. The most recent reports suggest that the industry is losing more female staff than it is recruiting (Computer Weekly, 2003; Grey & Healey, 2004; Hall, 2004).

Women's Propensity to Self-Select out of IT

One explanation for low female participation rates in occupational IT is women's propensity to self-select out of the field. The masculine nature of IT's culture and image has, rightly, been cited as a key reason underlying this tendency. Research suggests that the identification of occupations as either gender-appropriate or gender-inappropriate starts early (EOC, 2001a; Miller et al., 2004), and, as females progress through their education, they tend to drop out of subjects allied to gender-atypical careers with increasing frequency. Their over-representation within the lower echelons of the field may also, then, be partly related to their tendency to drop out of educational qualifications that are required for the more professional-level roles within it.

If this were the primary reason for the general picture of female participation and progression rates, however, we would expect women with professional-level IT qualifications to be present within the sector in terms broadly proportionate to comparable men. These women have bucked the gender trend and demonstrated their commitment and interest in the field by choosing to study an IT-related subject to a level that qualifies them for such a role.

The Progress of Women with Computer Science Degrees into Professional-Level IT Work (with Engineering as a Comparator)

An examination of the first occupational destinations of men and women graduating from university with computer science degrees[1] is a useful way of assessing whether women progress into professional IT work with the same frequency as men when they persist with their education in a subject that eminently qualifies them for it. To ascertain whether the conversion rates for Computing are part of a general Science pattern, or more specifically interesting, it is useful to compare them with conversion rates for men and women on a cognate degree programme: engineering. As with computer science, engineering is male-dominated both at the university and occupational level—more so in fact (Scenta, 2005), and also shares the characteristic of having a reasonably clear vocational link between degree-level study and professional work in the field.

Table 1 records the relationship between undergraduate programmes in computer science and engineering, gender and the occupation a graduate is within six months after graduating.

It can be seen that women fail to enter professional IT roles with the same frequency as we would expect, and with about half the frequency that comparably qualified males do. Moreover, over twice as many women as men with a computer science degree end up in an occupation within the non-professional category "administrative & secretarial", and they represent over a quarter of all women with this qualification. Indeed, women with a computer science degree are *more* likely to go into "administrative & secretarial" work than they are to convert their degree into a professional-level job within IT.

Additionally, 10% more women with engineering degrees make the transition to professional-level engineering work than women with computer science degrees make the transition into professional-level IT work[4]. Women with engineering degrees are also more likely to secure a professional level job in any sector. More generally speaking, female Computer Scientists provide the sole exception to the rule that sees women UK graduates more likely than male graduates to secure employment after graduating (Millar & Jagger, 2001).

Table 1. Occupational destination at six months by gender and qualification type

	Computer Science		Engineering	
	Male	**Female**	**Male**	**Female**
Manager & senior officials	10.3%	9%	10.5%	9.1%
Professional:	37%	27.6%	47.2%	36.8%
- ICT professional	30.5%	16%	4.7%	3.4%
- Engineering professional	1.4%	1%	37.8%	26.9%
Associate professional/technical:	22.5%	21.8%	15.8%	21.2%
- IT service delivery occupations	11.3%	6.5%	2%	1.1%
Administrative & secretarial	11.8%	26.3%	7.5%	14.4%
Skilled trades	1.9%	0.1%	3.2%	1.1%
Personal service	0.6%	2.5%	0.9%	2.2%
Sales & customer service	10.3%	10.8%	6.9%	10.6%
Processes, plant, & machine	0.9%	0.2%	2.0%	0.7%
Elementary	4%	1%	5.4%	3.4%
Total number of students	6,685	1,990	6,850	1,315

Source: based on original analysis of data supplied by Higher Educational Statistical Agency, Destinations of Leavers from Higher Education, 2002/03[2] (HESA, 2004)[3]; David Perfect, Equal Opportunities Commission, 2005

Summary of Findings

Although it is widely acknowledged that the IT sector is more graduate-rich than many others, especially at its senior levels, and that employers tend to prefer students graduating in related degrees (e-skills, 2003; Millar & Jagger, 2001; Miller et al., 2004), it would nevertheless seem that many women with directly relevant degrees fail to move into professional-level IT work; they do so at about half the rate that men do. Furthermore, women with degrees from unrelated disciplines are not faring better in relation to securing professional-level IT work—they also lag well behind their male counterparts in managing this transition (HESA, 2004; see also Computer Weekly, 2003).

In some respects, the picture painted in the previous analysis reflects the position of women across all occupational sectors, and within science, engineering, and technology occupations especially (Peters et al., 2003). Generally speaking, women in the UK tend, on average, to hold better educational qualifica-

tions than their male counterparts (EOC, 2004), but be employed in less prestigious and less well remunerated roles and occupations (EOC, 2004), so the rate at which these higher qualifications are converted into professional-level jobs is poor as compared to the conversion rate for men. These tendencies are often amplified when women select to work in gender-atypical sectors. As we have seen, however, there are elements to the story of women's advancement into IT that indicate that it is performing less progressively than some of the more traditional male dominated sectors that the industry was initially predicted to supersede in terms of its gender regime.

IT's Comparative Failure to Recruit Qualified Women: The Limits of the Self-Selection Thesis

The tendency for women to self-select out of IT work may still form part of the explanation of the picture as it is painted here, despite the fact that the women focused on have seemingly demonstrated their commitment to the IT area by opting for, and persisting with, a degree qualification in computer science. Such women still have the duration of their degree courses to be put off from working within the area, and there is evidence that this happens (Margolis & Fisher, 2002; Peters et al., 2003). Relying on the self-selection thesis as a mode of explanation is helpful only to a point here, however, unless we are to accept that the image of IT work and the educational experience of undertaking an IT degree, are *more* off-putting to women than the image and educational experience of engineering, and there is no evidence for this assumption.

The fact that women with computer science degrees do not manage to convert their qualifications to professional-level IT work at the same rate as their male counterparts points to there also being substantial issues in relation to the selection of women on the part of employers, and to the fact that women, consequently, 'encounter significant barriers when they attempt to pursue a professional career' in the sector (Millar & Jagger, 2001, p. A-7). These issues would seem to be independent of the quality and commitment of female applicants. Indeed, given the sometimes problematic experiences women often report having on their computer science degree courses, coupled with the fact that they are still more likely than men to achieve a "good" degree in the subject, it does not seem unreasonable to assume that those who attempt to convert their degrees into professional-level jobs in IT are exceptionally motivated and focused.

One key problem on the recruitment side is the fact that movement into IT work is still often based upon informal contacts and networks, especially in the context of the field's serial contract culture (Clarke, Beck, & Michielsens, 2002; Flood, 2005). Indeed, Roger Ellis, chairman of the UK's IT Directors' Network, has been recently quoted as saying, "professional networking ... is essential ... You can always answer advertisements, but there might be 300 applicants or more. Most senior IT professionals get their jobs through networks" (Flood, 2005). The claim has also been made in terms of promotion (Clarke et al., 2005), and is concerning in both respects because of research evidence suggesting that the more informal the recruitment process, the

more likely it is that the stereotype of an ideal candidate for a job in IT will act as the default template guiding the selection process.

It is suggested that this stereotype reflects the existing composition of the occupational area, as well as elements within the culture's self-image, and operates to produce a "hidden" job specification list: male, white, graduate, with no discernible domestic commitments etc. (Clarke et al., 2002; Faulkner, 2002; Massey, Quintas, & Wield, 1992; Millar & Jagger, 2001; Woodfield, 2000). Evidence suggests that, whilst this profile will remain influential even within formal processes, in the context of informal ones, it can operate far more powerfully and is more likely to do so unchecked. With less formality, individual rapport, often based on personal identification with candidates, coupled with an "I'll know it when I see it" criteria, can more easily supersede the impact of formal qualifications, recommendations, job specifications and appraisals (Woodfield, 2000, 2002).

It is unsurprising, then, that the limited evidence of good levels of job satisfaction (CEL, 2005) amongst professional female IT workers, is more than eclipsed by that which consistently confirms the experience of women feeling like the "odd girl out" (Trauth, 2002). Professional women within the field, and across a diverse range of roles and life-cycle stages, report feeling that they fail to fit the default "ideal" worker profile well enough to experience, or progress within, the occupation as well as men (Computer Weekly, 2003; Faulkner, 2002; Grey & Healy, 2004; Millar & Jagger, 2001).

FUTURE TRENDS

There is increasing recognition that IT's culture, in privileging men over women, as well as failing to provide genuine equality of opportunity for female workers, may also be jeopardising businesses and short-changing the UK's economy (e-skills, 2004; Peters et al., 2003). There is very little evidence, however, that the sector will respond rapidly or effectively to pressures to develop more genuinely open recruitment and promotion policies and practices, either on the basis of business or equity principles (Clarke et al., 2002). This is all the more surprising given its chequered history of skill crises and personnel shortages. Macro-level equality initiatives generated by governments and agencies, and designed to persuade more women and girls to consider IT as a career option and more employers to select them, have been unevenly adopted on the latter's part (Clarke et al., 2002). The field remains more resistant than we might expect to family-friendly policies such as flexible working, return-to-work schemes, and a more general cultural shift towards addressing work/life balance issues (Faulkner, 2002; Grey & Healey, 2004; Millar & Jagger, 2001).

This is also surprising in the context of the initial, and, in some ways, persistent association of IT with the future, even, as the introduction here has indicated, with a less gendered future. IT's highly educated workforce, not unreasonably, can be taken as a superficial indicator of tolerance and broadmindedness. As well as being externally generated, these image elements are actively reproduced within occupational IT cultures, alongside legacy elements from its hobbyist and creative past that signify its workforce as edgy, cynical and creative individualists, too interested in the work itself to be distracted into irrelevancies such as office politics, too excited by, and admiring of, another person's skill to notice their gender or ethnicity (Meiskins & Whalley, 2002; Woodfield 2000). Although in some respects offering a welcome counter to the male, geeky and boring image the sector perennially endures, there is evidence that these more positive aspects of its image can operate discursively within the culture of IT to obscure structural inequalities based on gender differences, and then mute discussions of their implications (Woodfield, 2000). As Faulkner has suggested, the barriers to women within UK "may be the more tenacious precisely because they are not so immediately obvious" (Faulkner, 2002).

CONCLUSION

We need to take note of the demand-side factors when addressing the reasons for the pattern of female progress into IT, as much as those related to the supply of "willing" female applicants. Indeed, the female propensity to select out of IT work, even once within the industry, has also to be understood in

this context, rather than analysed as if it were more an independent trait indicative of proactive "feminine" preferences than a reaction to perceived inequalities.

REFERENCES

CEL. (2005). *The management services company.* Graduate Tracking 2000. Presentation to Department of Trade and Industry, 25/112005.

Clarke, L., Beck, V., & Michielsens, E. (2002). *Gender and ethnic segregations in the British labour market: Mechanisms of marginalisations and inclusion.* University of Westminster, London: Educational Training & the Labour Market Group.

Computer Weekly. (2003). *Stop the female braindrain,* 30/1.

Cowan, J. (1989, February 17). Women into IT. *The Guardian.*

e-Skills UK. (2003). *Quarterly review of the ICT labour market.* Retrieved May 9, 2005, from http://www.e-skills.com

e-Skills UK. (2004). *IT insights: Trends and UK skills implications.* Retrieved May 9, 2005, from http://www.e-skills.com

EOC. (2001a). *Women and men in Britain. Sex stereotyping: From school to work.* Retrieved May 9, 2005, from http://www.eoc.org.uk

EOC (Equal Opportunities Commission). (2001b). *Women and men in Britain: Professional occupations.* Retrieved May 9, 2005, from http://www.eoc.org.uk

EOC. (2004). *Facts about women and men in Great Britain.* Retrieved May 9, 2005, from http://www.eoc.org.uk

Faulkner, W. (2002). *Women, gender in/and ICT: Evidence and reflections from the UK* (Document D02_part 3). SIGIS: Gender, ICTs, and Inclusion.

Flood, S. (2005). Making all the right connections. *Computer Weekly, 18*(2), 2.

Gerver, E. (1989). Computers and gender. In T. Forester (Ed.), *Computers in the human context:*

Information technology, productivity and people (pp. 481-501). Oxford.

Grey S., & Healey, G. (2004). .Women and IT contracting work—A testing process. *New Technology, Work, and Employment, 19*(1), 30-43.

Hall, W. (2004). Women in IT. *Computing, 13*(5).

Higher Education Statistics Agency (HESA). (2004). Destinations of leavers from Higher Education 2002/03.

Margolis, J., & Fisher, A. (2002). *Unlocking the clubhouse: Women in computing.* Cambridge, MA: MIT Press.

Massey, D., Quintas, P., & Wield, D. (1992). *High-tech fantasies. Science parks in society, science, and space.* London; New York: Routledge.

Meiskins P., & Whalley, P. (2002). *Putting work in its place: A quiet revolution.* Ithaca, NY: Cornell.

Millar, J., & Jagger, N. (2001). *Women in ITEC courses and careers.* Brighton, UK: Science and Technology Policy Research Unit & Institute of Employment Studies.

Miller, L., Neathey, F., Pollard, E., & Hill, D. (2004). *Occupational segregation, gender gaps, and skill gaps.* Brighton, UK: EOC Working Paper 15.

Peters, J., Lane, N., Rees, T., & Samuels, G. (2003). *SET FAIR: A report on women in science, engineering, and technology.* HMSO, UK: Office of Science & Technology.

Scenta. (2005). *The online gateway to the best in Science, engineering, and technology.* Retrieved May 9, 2005, from http://www.scenta.co.uk

Sørensen, K. (2002). *Love, duty, and the s-curve—An overview of some current literature on gender and ICT* (Deliverable Number: D02_Part 1). SIGIS: Gender, ICTs, and inclusion.

Trauth, E. (2002), Odd girl out: An individual difference perspective on women in the IT profession. *Information, Technology, and People, 15*(2), 98-119.

Woodfield, R. (2000). *Women, work, and computing.* Cambridge, UK: Cambridge University Press.

Woodfield, R. (2002). Women and information systems development: Not just a pretty (inter)face. *Information Technology and People, 15*(2), 119-138.

KEY TERMS

British Computer Society: The industry body for IT professionals. (http://www.bcs.org/bcs)

E-Skills UK: e-skills UK is a not-for-profit, employer-led organisation, licensed by government as the Sector Skills Council for IT.

Equal Opportunities Commission: The Equal Opportunities Commission is an independent, non-departmental public body, funded primarily by the government. It provides free research reports and statistics chronicling the position of women and men in all aspects of UK life. (http://www.eoc.org.uk)

"Good Degree": A term used in academic literature and common parlance to indicate a UK degree which has been classified as a "Upper Second" or "First," (i.e., in the top two degree classifications). It is normally assumed that those going on to postgraduate study will have such a degree, and they are preferred in many employment contexts.

Higher Educational Statistics Agency (HESA): HESA was set up in 1993 by the UK Government to act as a central source for higher education statistics and has become a respected point of reference. (http://www.hesa.ac.uk)

IT/Computer Professional: An individual working within the UK IT/Computer sector within a complex and skilled role that is classified within the category "Professional occupations" (e.g., software engineers), or sometimes within "Associate professional and technical occupations" (e.g., computer programmers)—in the National Statistics classifications, Standard Occupational Classification Codes (2004). (http://www.statistics.gov.uk)

ENDNOTES

[1]   Defined according to Higher Educational Statistics Agency categories as including: com-

puter science, information systems, software engineering, artificial intelligence, and other programmes within computer science.

2   These data are based upon an 80% response rate from graduating students and their institutions and are therefore not comparable to other HESA data. It is noteworthy, for instance, that more females than males responded here, giving the impression that female computer scien-

tists comprise a third of all graduates in that subject area.

3   Percentages have been stated to the first decimal point, and there are very small numbers of "unknowns" within each column (under 1%)—columns may therefore not total 100%.

4   Also of note is the fact that men within engineering are similarly more likely to make this conversion than men within computer science.

Women and Social Capital Networks in the IT Workforce

Allison J. Morgan
The Pennsylvania State University, USA

Eileen M. Trauth
The Pennsylvania State University, USA

INTRODUCTION

Currently, the IT industry is experiencing explosive growth. As the need for more skilled IT workers increases, the focus on the diversity of individuals participating in IT jobs is highlighted. The under represented populations of women and minorities are being evaluated to determine ways to increase their lasting participation in the technology workforce. Although initiatives and programs have been established to recruit a more diverse labor force, the under representation persists. In an effort to address the problem of under represented populations in the IT workforce, it is necessary to evaluate the situation from a variety of angles and views. Specifically, we seek to better understand the "gender gap" in the IT workforce and the effect of social capital networks in the organization on women.

Social capital can be defined as "an instantiated informal norm that promotes cooperation between two or more individuals" (Fukuyama, 1999, p. 1). Social capital among workers in the organization has been attributed to career success due to increased access to information, resources, and sponsorship (Seibert, Kraimer, & Liden, 2001). One of the ways that social capital can be gained is through participation in networks. Overall, the benefits or advantages gained through the networking process are attributed to an increase in access to and sharing of information.

In this article, we consider social capital networks in the IT workforce and whether the existence of these networks assists in explaining the under representation of women in IT. Our research highlights the experiences of women practitioners and academics currently working in the IT field. Our aim is to uncover the story behind the organizational chart. In doing so, we summarize a study on women's participation with social networks in the IT workforce presented in Morgan, Quesenberry, and Trauth (2004).

BACKGROUND

The notion of an informal social network in the workplace is not a new concept. References have been made to the notion of an "Old Boy's Club," in regard to a network of men in a position of power and privilege in an organization who share resources and information to gain advantage and opportunities. This is particularly germane to the IT field since it is characterized as a male dominated industry.

In this situation, if an "Old Boy's Network" exists, then "women's informal isolation [could] result in men's greater influence and centrality" in networks (Moore, 1988, p. 575). So, the study of informal networks becomes even more important within this context (Morgan et al., 2004). The role of social networks may be playing a critical role in the exclusion of women from opportunities in the field. Social or informal networks have been defined as "the web of relationships that people use to exchange resources and services" (McGuire, 2000, p. 1). Research has pointed to the importance of social networks in areas such as status and power in organizations. Additionally, social networks have been linked to gains in skills, job leads, and mobility in an organization. A popular phrase correlated with social networks is that "it is not what you know, *it is who you know.*"

Social capital, which is gained in these networks, is defined as a virtuous circle of trust, including group membership and informal social ties (Putnam, 1993). The level of trust associated with social capital is critical to access to information. Social capital has a direct relationship to the amount and quality of the information that an individual is privy to. So, it can be inferred that the more social capital one possesses the more advantage they possess in relation to opportunity and resources in the organization.

By the same token, the lack of social capital and access to resources and information may result in a decrease in upward mobility, turnover, and career satisfaction. These factors can be detrimental to maintaining employees and specifically a diverse workforce.

MAIN THRUST OF THE ARTICLE

The literature concerned with social networks highlights the importance of informal and interpersonal relationships with others in an organizational setting. The practice of social networking has been investigated in several industry environments similar to the IT industry. Social networks are often described in terms of strong or weak ties. Strong ties are close personal relationships that are similar in nature to interactions that an individual would carry out with their family or friends. Weak ties reflect more superficial relationships in an organization that a person may have with a co-worker or colleague. The literature on social networks in organizations explains the benefits of them in terms of mentoring, acquisition of information, and sharing of information.

The importance of interpersonal ties in a social network was discussed by Granovetter (1973). According to the author, the "strength of an interpersonal tie" is determined from a "combination of the amount of time, the emotional intensity, the intimacy, and the reciprocal services which characterize each tie" (p. 1361). The importance of "weak ties" highlights an opportunity for community acceptance. This takes place due to ties that extend out of an individual's primary social network that connect him or her to other important social networks. These additional social networks are often valuable information resources. Granovetter showed that weak ties were often the source of job opportunities for the subjects in his study. Lin, Ensel, and Vaughn (1981) discuss the effect of networks in the process of job

seeking. This research suggests that a job seeker's ability to reach a job contact with high-status is influenced by their personal resources and use of their weak ties. It has also been argued by Wegener (1991) that social networks are beneficial to subgroups of job seekers in a variety of ways. The study explains that individuals with previous experience in high status jobs benefit from weak ties, while those from low status jobs do not. According to Brass (1985) being connected informally to the management and supervisors of an organization affects a person's influence in that organization. Those in high-level decision making positions in an organization are deemed as the "dominant coalition" (p. 329). Traditionally, men have occupied these positions, something that has increased the difficulty for women to be a part of informal interactions with people in power in an organization. Mentoring is also an important process that occurs in an organization which is affected by the presence of social networks.

The process of mentoring and its effect on women is discussed by Burke and McKeen (1990). The authors point out that a potential hindrance to women participating in cross gender mentor relationships may be due to their inability to access information networks. This circumstance is the tendency to develop relationships with people with similar characteristics, or male management excluding women. Eby (1997) discusses the benefits of the mentoring process in the organization. Mentoring is described as a medium through which individuals gain specialized knowledge and skills, which then provide people an increased ability to adapt to change in an organization. Participation in traditional mentoring increases the likelihood for an individual to develop peer networks. Peer networks, then, may also result in peer mentoring which is an additional point of leverage for the protégé. The expansion of the peer network increases an individual's opportunity to access resources and information regarding skill sets, career prospects, and strategies. The process of information acquisition is another important process which occurs through formal communication and through social networks.

The notion that a large informal network supports a person's mobility in an organization through the acquisition of resources and information is discussed by Podolny and Baron (1997). The absence of "structural holes" in an individual's network with management and others in the organization with "fate control" increases this mobility. People in an organization with "fate control" have some critical investment in the success and direction of the organization. The types of information that flow through a social network are described as task advice, strategic information, buy-in, and social support. Siebert et al. (2001) provide a framework which details the importance of social capital on career success. It was shown that a person's social network was influenced by access to information, access to resources, and career sponsorship which all may have an impact on career success. In addition, information sharing is an important process which can be made even more beneficial through the use of social networks.

Research has shown that gender inequality in organizations can be reinforced by sex-related differences in social networks (Ibarra, 1992). Homophily was cited as a cause of this gender inequality. Homophily refers to the preference to create same-sex work relationships in networks. An additional factor which has proven problematic in joining social networks is in translating personal characteristics and resources into a means of advantage in a network situation. Mehra, Martin, and Brass (2001) detail how different actions can be enhanced or inhibited by interaction in social networks. The authors report that people in networks may experience a variety of outcomes as a result of the attribute of their network. Often, people who are able to bring together less familiar individuals with one another benefit from an increase in information, resources, mobility, and control. Those individuals who participate in smaller social networks generally do not benefit from the myriad of information that is achieved through participation with expansive networks. The next section will explain how women's participation in the IT workforce was evaluated to identify their relationships with social networks.

Methodology

The methodology guiding the study is based upon an NSF funded study on individual differences in the social shaping of gender and IT. This study is a qualitative research project which seeks to test an empirical theory that addresses the experience of women in IT. This article reports on interviews with women conducted between October 2002 and December 2003 with 44 female practitioners and academics working in the IT field in the United States. These in-depth interviews were held with women in Massachusetts, North Carolina, and Pennsylvania. The women were diverse in their demographic and personal characteristics. The interviews lasted approximately 90 minutes, and the interview items were derived from prior research into gender and IT (Trauth, 1995, 2002). The interviews were coded based on a coding scheme which is informed by the Individual Differences Theory of Gender and IT (Trauth, 2002; Trauth, Quesenberry, & Morgan, 2004). The data being collected reflects information about the participants' personal information, shaping and influencing factors, and environmental context. The theory which is guiding this research investigates the individual attributes, individual influences, and environmental influences of women in IT to determine how these factors influence their participation in IT. Additionally, the theory asserts that women will respond to and experience different socio-cultural elements that affect their participation in the IT workforce in an individual manner. The research reported in this article seeks to provide further evidence in support of individual differences among women working in the IT industry.

Results

Analysis of the women's accounts reveals that informal social networks assist in the flow of information through nontraditional channels in the organization. The type of information that is gained though these social networks can be categorized into four areas: career opportunities, task information, mentoring, and personal advantage.

Career Opportunities

Information regarding job openings, promotion, and opportunities for advancement are largely passed through social networks. In certain situations, network contacts can provide individual access to recruiters, managers, interviewers, or others in charge of hiring. For example, *Allison was offered a job she did not apply for because of network contacts*. In addition, the participation in a social network can provide exposure to decision makers who are in charge of career decisions. *Betty Jean discussed finding her permanent job through friendships gained by personal networking.*

Task Information

Task information that is gained through social networks regarding job specific activities assists in gaining important skills for advancement in the workplace. Often, the sharing of information can provide a way in which to resolve problems or situations, and achieve desired goals or results more quickly. *Julia discussed her experiences of bonding with men socially outside of the workplace, so when she needed help they were more than willing to help her.* In addition, assistance can be granted by other members of the social network who are outside an individual's immediate team members or colleagues. *Donna mentioned that the other developers would help her to understand the technical things she did not know.*

Mentoring

Some forms of mentoring are established at an organization level, but a great deal of personal mentoring goes on with other members of a social network or while participating in networking activities. The gain in knowledge and social capital can be tremendous if a person has both formal mentors, and informal ones as a result of social networks. For example, *Irene learned how to handle career situations though her mentors.*

Personal Advantage

Social capital networks also provide advantage by allowing a person to form a personal bond or level of trust with other members of the organization. This advantage can come from exchanging information about family, hobbies, or interests, as well as by participating in social events such as lunches, happy hours, golfing, or shopping. The bond formed by people in social networks can also lead to a greater appreciation and value of a person and their work. *Jeria discussed her male work environment where she made an effort to fit in with men and became their platonic friends. She discussed going out after work with her male colleagues.*

The analysis of our data also revealed that women participate with social networks in a variety of ways: in the network, outside the network, and in alternative networks.

In the Network

Those women who participate in the established social network in their organization may share some common bond of experience, interest, or likeness with other members of the network. Thus, the women's participation in the network may appear to require less effort and be more meaningful to them. *Joanne discussed her male work environment where she worked to fit in with men and became their platonic friend. She discussed playing video games with her male colleagues.* Among the women who did not share similarities with members of the network, some decided to develop interests in common with those involved in the network. For example, *Sharon took flying lessons to be able to join in lunch conversations with her male colleagues.* These women proactively made a decision to establish a bridge into the network to gain acceptance from the members. It appears that in this type of situation, there is some element of choice which is provided to the women regarding whether or not they are able participate and interact with the members.

Outside the Network

Being outside the network may be based on the group discriminating with respect to admission. The members of this social network may not be open to including individuals who do not share the commonality that links them together. Often, the sole criterion for membership in this group is gender. *Jeanette described her boss who had two guys who acted as his henchmen and who established an inner circle where everybody else was excluded.* When one is outside of the network and excluded, the members of the group often deny these contributions. *Claire explained that she was an experienced programmer who was given clerical work on her project due to her gender and her non-membership in the network.* In addition, due to some other personal responsibilities, such as family duties, many women appear at times to choose not to participate with the network or their activities. *Carol explained that guys would stay at work overnight and come in on weekends, she did not feel that was an option for her because of her responsibility to her family.*

Alternative Networks

The alternative network is a new network. The women who have chosen this option have experienced elements of either a closed or open network, but have nevertheless chosen to develop or participate in a network better suited to them. These women appear to see the value in networking and are motivated to interact with and create other network opportunities. Some alternative networks may be based outside of the workplace, but provide networking opportunities nonetheless. *Emily described being involved with a support group for "design" Web women outside of the workplace.*

The analysis of women's experiences has shown that they respond in a variety of ways to inclusion and exclusion from the network. The women in this study reacted to the situation through a mechanism of their environment, personality, and responsibilities. The interaction with these social networks will continue to be an important area of research with respect to women's participation with technology.

FUTURE TRENDS

A recent study conducted by Forret and Dougherty (2004) found that "the relationship between engaging in professional activities and total compensation for females was negative, while for males the relationship was positive" and that "increasing internal visibility was significantly related to number of promotions and total compensation for men, but not for women" (p. 429). These findings are important because it shows evidence that the professional advancement of women is not occurring in the same manner as it is for their male counterparts. So while networking behavior has been deemed critical to career success, there are still factors which prevent the contribution of women from being realized at their full capacity. Future research into this area

should not only investigate the women in different workplaces, but also the policies, processes, and initiatives that actively or subtly facilitate gender inequality in organizations.

CONCLUSION

This article has examined the presence of social networks in the IT workforce and the subsequent effect on women in the field. The analysis of our data has revealed, through the lens of the Individual Differences Theory of Gender and IT, evidence that this informal network does indeed affect how women interact with others on a daily basis in the IT workforce. Good networking skill is essential for success in most industries, but when obstacles prevent the prospect of networking, a challenging situation becomes even more difficult. The women in this study provided some insight into their strategies and coping mechanisms utilized for continued participation in the workplace. Some women are in opposition to an artificial interaction with people with whom they do not share common interests, while others choose to neglect their own personal interests in order to fit in. Through these responses, women are expressing how they cope with an organizational phenomenon that has no sign of dissipating in the near future.

It is important to point out that a position inside or outside of the network is a dynamic factor. Very possibly, over time, a woman may move among social networking as her career and environment evolve. The contribution of this article is to provide further insight into the gender gap in the IT workforce by addressing a potential barrier to the participation of women. Further research in this area may evaluate the effect of the positioning with regard to networks and how that impacts overall career satisfaction and success.

ACKNOWLEDGMENTS

This article is from a study funded by a National Science Foundation Grant (EIA-0204246).

REFERENCES

Brass, D. (1985). Men's and women's networks: A study of interaction patterns and influence in an organization. *Academy of Management Journal, 28*, 327-343.

Burke, R., & McKeen, C. (1990). Mentoring in organizations: Implications for women. *Journal of Business Ethics, 9*, 317-332.

Eby, L. (1997). Alternative forms of mentoring in changing organizational environments: A conceptual extension of the mentoring literature. *Journal of Vocational Behavior, 51*, 125-144.

Forret, M., & Dougherty, T. (2004). Networking behaviors and career outcomes: Differences for men and women? *Journal of Organizational Behavior, 25*(3), 419-437.

Fukuyama, F. (1999). *Social capital and civil society*. Paper presented at the International Monetary Fund Conference on Second Generation Reforms, Washington, DC. Retrieved January 12, 2006, from https://www.imf.org/external/pubs/ft/seminar/1999/reforms/fukuyama.htm#I

Granovetter, M. S. (1973). The strength of weak ties. *American Journal of Sociology, 78*, 1360-1380.

Ibarra, H. (1992). Differential returns: Gender differences in network structure and access in an advertising firm. *Administrative Science Quarterly, 37*, 442-447.

Lin, N., Ensel, W., & Vaughn, J. (1981). Social resources and strength of ties: Structural factors in occupational status attainment. *American Sociological Review, 46*, 393-405.

McGuire, G. (2000). *Gender, race, and informal networks: A study of network inclusion, exclusion, and resources*. Proposal submitted to Indiana University's Faculty Research Grant.

Mehra, A., Martin, K., & Brass, D. (2001). The social networks of high and low self-monitors: Implications for workplace performance. *Administrative Science Quarterly, 46*, 121-146.

Moore, G. (1988). Women in elite positions: Insiders or outsiders. *Sociological Forum, 3*, 566-585.

Morgan, A. J., Quesenberry, J. L., & Trauth, E. M. (2004). Exploring the importance of social networks in the IT workforce: Experiences with the "Boy's Club". In *Proceedings of the Americas Conference on Information Systems*, New York.

Podolny, J., & Baron, J. (1997). Resources and relationships: Social networks and mobility in the workplace. *American Sociological Review, 62*, 673-693.

Putnam, R. (1993). *Making democracy work: Civic traditions in modern Italy*. Princeton: Princeton University Press.

Seibert, S., Kraimer, M., & Liden, R. (2001). A social capital theory of career success. *Academy of Management Journal, 44*(2), 219-237.

Trauth, E. M. (1995). Women in Ireland's information industry: Voices from inside. *Eire-Ireland, 30*(3), 133-150.

Trauth, E. (2002). Odd girl out: An individual differences perspective on women in the IT profession [Special Issue on Gender and Information Systems]. *Information Technology and People, 15*(2), 98-118.

Trauth, E. M., Quesenberry, J. L., & Morgan, A. J. (2004, April 22-24). Understanding the under representation of women in IT: Toward a theory of individual differences. In *Proceedings of the ACM SIGMIS Computer Personnel Research Conference*, Tucson, AZ (pp. 114-119). New York: ACM Press.

Wegener, B. (1991). Job mobility and social ties: Social resources, prior job, and status attainment. *American Sociological Review, 56*, 60-71.

KEY TERMS

Individual Differences Theory of Gender and IT: A social theory developed by Trauth et al. (2004) that focuses on within-group rather than between-group differences to explain differences in male and female relationships with information technology and IT careers. This theory posits that the under representation of women in IT can best be explained by considering individual characteristics and individual influences that result in individual and varied responses to generalized environmental influences on women.

"Old Boy's Club": An informal network in which men are able to share information in a less formal setting, learn to trust each other, and establish personal relationships which generally provide advantage to those who participate in it.

Social Capital: An instantiated informal norm that promotes cooperation between two or more people.

Social Networks: The web of relationships that people use to exchange resources and services.

Women and the IT Workplace in North West England

Angela Tattersall
University of Salford, UK

Claire Keogh
University of Salford, UK

Helen J. Richardson
University of Salford, UK

Alison Adam
University of Salford, UK

INTRODUCTION

The United Kingdom (UK) information technology (IT) industry is highly male dominated, and women are reported to account for an estimated 15% of the sector's workforce (EOC, 2004). In Spring 2003 it was estimated that there were 151,000 women working in IT occupations compared to 834,000 men (EOC, 2004) Additionally, it has been reported that these numbers are rapidly declining, as women are haemorrhaging from the industry in disproportionate numbers (George, 2003). Although they are making inroads into senior and technical roles, "vertical segregation" is observable. Overall, women tend to be represented in lower-level IT jobs, with the majority, 30%, in operator and clerical roles; and the minority in technical and managerial roles, 15% of ICT management and 11% of IT strategy and planning professionals (EOC, 2004). This renders a "feminisation" of lower-level IT occupations. Educational statistics have also shown that fewer women are enrolling onto computer-related courses; there was a drop from 24% in 2000 to 20% in 2003 (E-Skills, 2004a)

BACKGROUND

Women's exclusion from technology has been attributed to the historical and socio-cultural construction of technology as a "masculine domain" due to the relationship between masculinity and technological skills, rejecting claims that technology is "neutral" (Wajcman, 1991; Woodfield, 2000). Further, Cockburn (1985) argues that power differences between the sexes and their relationship with technology were consolidated in the development of capitalism and the move to manufacturing production. Working-class tradesmen formed into trade unions, which sought to exclude women, therefore denying them access to gain necessary skills to exploit in the labor market.

Wajcman (1991) suggests that "skill" is not some objectively identifiable quality, but rather, is an ideological category, one over which women were and continue to be denied the rights of contestation. Notions of "masculine" and "feminine" skills are problematic in the workplace, as there is more emphasis on the nature of the worker rather than the work. Additionally, categorisations and equations relating to skill are deeply ingrained with sexual bias; for example, men/skilled, women/unskilled (Woodfield, 2000), and further distinctions such as dirty/clean, heavy/light and technical/non-technical, which Game and Pringle (1984) argue have been constructed to preserve the sexual division of labor. Therefore, technological skill has historically been defined as "exclusively male," whereas women's traditional work, such as nursing, is defined as non-technical (Cockburn, 1985). Additionally, as this type of work has been socially constructed as un-skilled and un-technical, it has also been undervalued; consequently, work of women is often deemed as inferior just because "it's work women do" (Wajcman, 1991). Furthermore, this perception has led to the "gender typing" of roles in the workplace.

Gender divisions are actively created and sustained in organisational and domestic life. IT organisational culture is one embedded in a socio-political context of gender discrimination and the masculine domain of technology. If it is credible that (technical) skill and masculinity are so intimately entwined, then it is hardly surprising that women who challenge these masculine skills by gaining them themselves are subject to working under female unfriendly conditions and in hostile environments. It is also not surprising, then, that women must develop a number of coping strategies to deal with these sometimes impossible situations.

THE WINWIT STUDY

The purpose of this article is to present evidence of the working conditions and the cultural barriers women experience when working in the North West IT workplace. The Women in North West IT (WINWIT) project funded by the European Social Fund (ESF) was established to investigate the current regional situation for women in the IT labour market. The study was conducted over a 12-month period (January 2004-December 2004) by researchers at the University of Salford's 6* RAE Information Systems Institute. The aim of the research project was to determine reasons for the under-representation of technical and senior women from public and private North West IT organisations and departments. The study explored issues for women entering, progressing, leaving and returning to the industry, as no specific data was previously available at the regional level. The researchers conducted 11 in-depth interviews with a heterogeneous sample of women at various stages of their careers and hosted an online questionnaire via a project Web site (www.isi.salford.ac.uk/gris/winwit). These women were selected from a variety of sources and were highly technical, with various levels of managerial status. The women were chosen for interview on the basis that they were interested in contributing to encouraging other women into or back to the industry and to support those already participating. The

qualitative data was central to the study, as this provided valuable views and experiences from the labor market sample. The researchers developed trusting relationships with the interviewees to alleviate worries and reluctance to participate. Interviews were conducted in a relaxed neutral setting, away from the workplace. A critical approach was taken when analysing the interview data to gain greater understanding of the under-representation of women in the sector. The main themes to arise from the investigation are barriers to work-life balance, organisational culture, pay and problematic equal opportunities and diversity policies.

Working Conditions in IT

The following section will examine the working conditions in IT and offer examples from the WINWIT study. In the UK, current Equal Opportunities legislation makes it acceptable to offer work only on a full-time basis even though this is difficult for many women to take up (Liff, 1997), particularly in the IT sector, as work is typically full-time, with only 5.3% (Platman & Taylor, 2004) of the total workforce working part-time. A long-hours culture is ubiquitous in today's British companies, where UK employees work some of the highest hours in Europe—more than 48 per week (Rutherford, 2001), often involving unpaid overtime. These practices were reported as problematic for women from both public- and private-sector IT organisations, who gave accounts of long hours being used as a bargaining tool by women with families in return for four-day work weeks and reports of working 50 hours plus; these type of working conditions are indicative of the "hacker culture." Males and females experience conflict between work and family lives differently, with women exhausted trying to maintain domestic and professional roles; conversely, men regret that they couldn't spend more time with their children (Liff, 2001).

The flexibility required by clients often means constant availability (Hoque & Noon, 2004), particularly for those in technical and managerial roles. This can involve time spent away from home, sometimes for long periods. An interviewee explained how she often worked long hours and away from home: "I know I am single and don't have a family to take care of, but I still have priorities and don't just want home to be somewhere I dump my bags at weekends." Further, she commented that this was something that had to be done to get on and the fact that not everyone was in the situation to do this was secondary to the organisation. Other women that did not have children had contemplated that they would have to change careers or leave the industry if they ever did start a family. Those women (10 out of the 11) who were aware of "flexible working practices" within their organisation described how attitudes of resentment were evident as comments were made relating to "unfair policies," as part-time and home working were only offered to women with caring responsibilities. Flexible working models severely limit women's progression opportunities, as employers assume it shows a lack of commitment to their career (Guffens et al., 2004).

Individualized approaches are primarily understood as a way of reducing the influence of traditional industrial relations with Human Resource Management (HRM) policies focusing on individualized pay packages, contracts and development needs (Guffens et al., 2004). Women told how this approach was problematic, as being expected to vie for projects requires a degree of self-advocacy, and the ability to negotiate contracts and pay packages requires a high level of confidence and excellent negotiation skills. Of the 11 women interviewed, 10 conveyed that these were "not natural characteristics"; however, they were aware that this was something "men are good at." A culture of "salary secrets" and pay discrimination was evident from the sample women; figures of up to £17,000 a year were indicated as the difference women experienced compared with that of their male colleagues. Some of the key factors attributed to the gap were said to be returning to work after having a baby and choosing a work-life balance, a lack of confidence to push for higher salaries and informal pay structures. Although some felt that this wasn't deliberate discrimination by their employers, putting it down to "it's just the road women end up on historically" and "its just the way things are," others, however, did express extreme anger.

Where clients require constant availability, working long hours and blatant pay discrimination, it is no wonder that women working in the North West IT sector reported difficulties with confidence, often framing the problem in individualistic terms rather than a structural feature of the IT workplace. "Gen-der blind" organisations that encourage these working practices undermine equality policies, further excluding women (and men) from the technical sphere.

IT a "Fitting" Place for a Woman?

Questions were asked about the working environments and how comfortable women felt with a male-dominated culture. The following points were raised: Five out of the 11 interviewees started their career in a purely technical role and had transcended to a hybrid (technical and interpersonal) role due to the "maleness" of the workplace. The working environment was described in terms of "hostile,"and experiencing "isolation" and "exclusion." One woman explained how she was reduced to feelings of regret with her choice of career because she was so unhappy and lonely working in an all-male environment. Another used the word "painful" when describing her experience of being the only woman out of 80 people in an IT department.

Amplification of the masculine culture is often demonstrated through male camaraderie and perpetuated by sexual discussions and crude humour (Simpson, 2000). One participant described how she perceived that it was a prerequisite to have grown up with lots of brothers if you wanted a career in the IT industry and commented that you have to fit in with the majority of men and not be upset by the banter, adding, "sensitive types wouldn't last 2 minutes." Men were reported to group together in and out of work, forming after-work pub cultures, which for some women was neither desirable nor, in some cases for those with commitments, accessible. Informal networking and leisure pursuits were often male oriented—reported options were mountain biking, go carting and football. Some women expressed anxiety about being excluded from informal networks, as they felt it was imperative for career advancement to be seen. Two participants explained how the centricity of their organizations made it difficult to network, as all the senior managers were in the South of England, one referred to being "out of sight out of mind" and both were aware that had they been able to liaise and socialize with senior managers, their careers would have advanced sooner.

Displays of femininity in the IT workplace were problematic for women; one respondent who was the only women in charge of a group of 15 men didn't want to stand out as a female and felt that high visibility would make it difficult for her to "fit in," be accepted and taken seriously as a manager. In a bid to mask her femininity, she took to wearing gender-neutral clothes, "chinos" and a "polo shirt," and described herself as a genderless "it," only gaining confidence to revert back to feminine attire once she was completely satisfied she had proved herself professionally.

All interviewees expressed high levels of satisfaction, particularly regarding the opportunity to combine technical and communication skills; however, often, highly skilled women played down their technical knowledge when talking about their work in general terms. I'm not in IT but ... "I'm in the people part of the organization," "I'm a manager" and "I'm in sales." As our data reveal, women struggle to get comfortable in technical environments. Women's estrangement from purely technical roles and identifying oneself as having hybrid technical-people skills presents a very positive coping strategy for many women. However, this vehement rejection of any identification with technology does not bode well for future "women into IT" initiatives, and indicates a major cultural barrier that needs to be confronted.

FUTURE TRENDS

The UK has the highest employment rate in the world (Stanfield, Campbell, & Giles, 2004) although this source of labor is declining as the population is aging; individuals more than 60 years old now form a larger part of the population than those younger than 16. The Office for National Statistics (ONS, 2004) predict that by the year 2012 there will be 1.3 million new vacancies in the economy and the vast majority of these will be taken by women. However, 80% of the current workforce will still be employed in the labour force in 2010 (Stanfield et al., 2004). Platman and Taylor (2004) indicate there is a "fundamental cultural problem" with the way the IT sector approaches recruitment and employment; the market is failing to employ the younger generation, to retain sufficient numbers of women and to take advantage of the aging workforce. In addition, part-time and flexible working is rare, and is a reason women and older people are leaving the sector. Labor market intelligence (E-Skills, 2004b) envisage the need for advanced and high-level skills in particular to increase, and a vital need to adequately anticipate skill demand, as many IT professionals require years of training.

Given that women make up an essential element of this personnel and the current acute under-representation of women in IT jobs (George, 2003), women in the IT industry are facing real difficulties working in male environments where long hours are the norm, UK men work the longest hours in Europe and cultures exist where employers reward those who give their time entirely to work. Additionally, women are experiencing added strain from client demands, peer pressure and the tension of having to work harder and longer, coupled with difficulties of negotiating "individual" flexible working conditions, resulting in poor employment conditions that cannot be improved while this norm dominates. Consequently, a substantial proportion of women who wish to work part time or have primary caring responsibilities results in a section of the workforce being excluded, neglected, not encouraged to enter or forced to leave the industry.

Breaking through the glass ceiling (2003) states work-life policies cannot be seen as a perk; attention needs to be focused on the "sticky floor" phenomenon as much as if not more so than the "glass ceiling" syndrome. It is employees predominantly at the lower levels of organisations who experience the least amount of control over their work and time and have less access to work-life policies (Wise, 2003).

However, according to Brandth and Kvande (2001), businesses implementing new working practices without designing or generating new job specifications are unlikely to achieve the best outcome for individuals or businesses. Questioning rigid and unsuccessful working practices in conjunction with the way work is allocated, organised and rewarded has to be fundamentally challenged, as concerns are unlikely to be resolved until men's work-life issues are explored and understood, the results of which ought to produce a more equal share of paid work and caring responsibility between the sexes (Brannen, Lewis, Nilsen, & Smithson, 1997).

CONCLUSION

The points raised by the WINWIT sample were neither surprising nor original, yet they provide a unique view of women's experiences of working in the North West IT industry and provide a sound basis for future study. Challenging women's under-representation in the IT sector is problematic, as this article clearly shows that technology is considered to be a "masculine domain." Despite years of equal opportunities and related policies, many obstacles still remain for women working in the IT sector, chiefly the insidious and embedded organisational culture that maintains workplace discrimination through institutional processes and prevents women from entering male-dominated, higher-paid and higher-status occupations. Women's under-representation in the IT industry should not be looked upon as a "women's problem," but rather a problem for the industry as a whole. But organizational change is not a simple, single process that can come about immediately, as Steel (2000) states: "change is very fragile and can easily be destroyed." Once genuine change seems likely, the organisations' members will attempt to resist new ideas and practices, thus the need to encourage and nurture new behaviours will prove problematic.

REFERENCES

Brandth, B., & Kvande, E. (2001). Flexible work and flexible fathers. *Work, Employment and Society, 15*(2), 251-267.

Brannen, J., Lewis, S., Nilsen, A., & Smithson, J. (Eds). (1997). *Young Europeans, work and family: Futures in transition* (pp. 140-161). London: Routledge.

Cockburn, C. (1985). *Machinery of dominance: Women, men, and technical know-how*. London: Pluto Press.

Equal Opportunities Commission (EOC). (2004). *Occupational segregation, gender gaps and skills gaps*. Occupational Segregation Working Paper Series, No. 15.

e-Skills UK/Gartner. (2004a, November). *IT insights: Trends and UK skills implications*. London. Retrieved November, 2004, from http://www.e-skills.com.register

e-Skills UK/Gartner. (2004b, November). *IT insights: Employment forecasts*. London. Retrieved November, 2004, from http://www.e-skills.com.register

Game, A., & Pringle, R (1984). *Gender at work*. London: Pluto Press.

George, R. (2003). IBM/women in IT champions. *Achieving workforce diversity in the e-business on demand era*. Retrieved November, 2004, from http://www.intellectuk.org/sectors/it/women_it/2003/Achievingworkforcediversity.pdf

Guffens, C., Vendramin, P., Valenduc, G., Ponzellini, A., Lebano, A., Collet, I., et al. (2004). *Widening women's work in information communications and technology*. Work and Technology Research Centre. Retrieved February, 2005, from http://www.ftu-namur.org/www-ict/index.html

Hoque, K., & Noon, M. (2004). Equal opportunities policy and practice in Britain: Evaluating the 'empty shell' hypothesis. *Work Employment and Society, 18*(3), 481-506.

International Labour Office. (n.d.). *BREAKING through the glass ceiling, women in management*. Update 2004. Geneva: International Labour Office, Geneva. Retrieved April, 2005, from http://www.ilo.org/dyn/gender/docs/RES/292/F267981337/Breaking%20Glass%20PDF%20English.pdf

Liff, S. (1997). Two routes to managing diversity: Individual differences or social group characteristics. *Employee Relations, 19*(1), 11-26. MCB University Press.

Liff, S., & Ward, K. (2001). Distorted views through the glass ceiling: The construction of women's understanding of promotion and senior management positions. *Gender, Work and Organisation, 8*(1), 9-36.

Office for National Statistics. (2004, April). Labour market projections from 2002 to 2012. *Labour*

Market Trends (138). London: The Stationary Office.

Platman, K., & Taylor, P. (2004). *Workforce ageing in the new economy: A comparative Study of information technology employment.* Cambridge: University of Cambridge. Retrieved April, 2005, from www.wane.ca/PDF/Platman&TaylorSummary Report2004.pdf

Rutherford, S. (2001). "Are you going home already?" The long hours culture, women managers, and patriarchal closure. *Time and Society, 10*(2-3), 259-276. London; Thousand Oaks; New Delhi: Sage.

Simpson, R. (2000). Gender mix and organisational fit: How gender imbalance at different levels of the organisation impacts on women managers. *Women in Management Review, 15*(1), 5-19.

Stanfield, C., Campbell, M., & Giles, L. (2004). *The UK workforce: Realising our potential* (Research Report No. 7, Sector Skills Development Agency). Retrieved March, 2005, from http://www.ssda.org.uk/pdf/research%207.pdf

Steel, R. (2000). Culture and complexity: New insights on organisational change. *Organisations and People, 7*(2), 2-9.

Wajcman, J. (1991). *Feminism confronts technology.* London: Polity Press.

Wacjamn, J. (1998). *Managing like a man, women and men in corporate management.* London: Polity Press.

Wise, S. (2003). DTI and fair play. *Work-life balance literature and research review.* Retrieved April, 2005, from http://www.napier.ac.uk/depts/eri/Downloads/LitReview.pdf

Woodfield, R. (2000). *WOMEN work and computing.* Cambridge: Cambridge University Press.

KEY TERMS

Feminisation: Where the rise in female labour force participation has been driven largely by an increase in the demand for female labour, pulling women into the labor market.

Flexible Work Practice: Alternative work arrangements for people whose life circumstances make standard employment practices of 9 a.m. to 5 p.m. restrictive and not the best option. Flexible work practices include various types of work arrangements, such as part-time work, job sharing, working from home, flexi-time and term-time working.

Gender: The term gender refers to culturally based expectations of the roles and behaviours of men and women. Sex identifies the biological difference between men and women. Gender identifies the social relations between men and women. It therefore refers not to men and women, but to the relationship between them and the way this is socially constructed.

Gender Blindness: Gender blindness refers to a failure to identify or acknowledge difference on the basis of gender where it is significant. It can be a person, policy or institution that does not recognize that gender is an essential determinant of the life choices available to us in society.

Gender Typing: The process through which occupations come to be seen as appropriate for workers with masculine or feminine characteristics.

Salary Secrets: The hidden truth regarding pay inequality between men and women often being allowed to go unnoticed until women reach senior management levels.

Vertical Segregation: Concentrating individuals in the lower echelons of an organisation. (Term taken from Occupational Segregation (EOC).

Women Embrace Computing in Mauritius

Joel C. Adams
Calvin College, USA

Shakuntala Baichoo
University of Mauritius, Mauritius

Vimala Bauer
Barco Orthogon AG, Germany

INTRODUCTION

Studies like Camp (1997), Gurer and Camp (2002), Sigurdardóttir (2000), and Vegso (2005) have documented the declining percentage of women in computer science (CS) in the United States and other countries. While women are underrepresented in the United States overall, there are cultural pockets within the country that are exceptions to the rule. For example, Lopez and Shultze (2002) note that African-American women earned the majority of CS bachelor's degrees each year from 1989 through 1997 at historically black U.S. colleges and universities. Fisher and Margolis (2002) and Frieze and Blum (2002) report some success in increasing the percentage of women studying computing at Carnegie-Mellon. Camp, Miller, and Davies (2001) point out that the problem is significantly worse for CS departments housed in a school of engineering compared to those housed in a school of arts and sciences, a phenomenon dubbed "the school of engineering effect." So while women are on average underrepresented in CS in the United States, such national averages can hide significant variance within a country's subcultures.

Outside the United States, Schinzel (1999) notes that the situation in Anglo-Saxon, Scandinavian, and German-speaking countries (ASGs) is similar to that in the United States, but female representation in CS is comparatively constant and high (45-50%) in Greece, Turkey, and the Romanic countries (e.g., France, Italy). Schinzel's data are fragmentary, but they offer intriguing hints that culture plays an important role in encouraging or discouraging women from studying CS.

These and reports like Galpin (2002) indicate that there are non-ASG countries where women are equally represented in CS. This in turn suggests that the problem is one of culture: ASG cultures apparently in some way discourage women from choosing IT-related careers, while the cultures of these other countries apparently encourage women to do so. If the root of the problem is the culture in the ASG countries, then that is where we should focus our efforts.

What is it about the culture of the United States and other ASG countries that discourages women from studying CS? Trying to analyze the negative cultural factors from within an ASG country is rather like a fish trying to analyze the water in which it is swimming. A preferable approach is to become a "fish out of water" and visit a non-ASG country where women are studying CS. By identifying those cultural differences in non-ASG countries that are leading women to study CS, we can identify those aspects of ASG culture that are problematic.

In this article, we examine the country of Mauritius, a 25x40-mile island roughly 500 miles east of Madagascar that is home to 1.2 million people. Ethnically, its population is 68% Indo-Mauritian, 27% Creole-African, 3% Sino-Mauritian, and 2% Franco-Mauritian. Religiously, its people are 52% Hindu, 28% Christian, 17% Muslim, and 3% other religions. With this dynamic mix of people, Mauritius is one of the world's most culturally diverse countries.

BACKGROUND

Prior to 2001, the University of Mauritius (UoM) was the sole university in Mauritius, offering bachelor's and some graduate degrees to roughly 4,000 students. The university is free, and admittance is based solely on standardized entrance-exam scores. With roughly 1.2 million people in Mauritius, admission is extremely competitive and the admitted students are highly capable.

Applicants to UoM indicate the program they wish to study, plus alternatives should their first choices be full. Beginning with the top-scoring students on the entrance exam, students are matched to programs using their first choices unless that program is filled, in which case they are matched to their alternative choices. Admission is thus based on merit, plus supply and demand for particular programs; UoM has no special admissions policy to increase underrepresented groups.

UoM's Department of Computer Science and Engineering (CSE) provides the country's primary source of computing-related education. CSE offers bachelor's degrees similar to those of a U.S. technical university, and has periodically updated its programs and curriculum to reflect technological changes. Since 1990, it has offered the following programs.

- **1990-1997:** Bachelor of Technology in CSE (BT-CSE)
- **1997-2000:** Bachelor of Engineering in CSE (BE-CSE)
- **2000-Present:** Bachelor of Science in CSE (BS-CSE)
- **2000-Present:** Bachelor of Science in Information Systems (BS-IS)
- **2001-Present:** Bachelor of Science in CS and Multimedia (BS-CSM)

The BS-CSE and BS-IS programs are very similar to computer-science and information-systems programs in the United States. The BS-CSM blends traditional CS training with training in graphical design and multimedia applications. While the BT-CSE and BE-CSE are 4-year programs, the BS-CSE, BS-IS, and BS-CSM are all 3-year "UK style" bachelor's programs.

GENDER IN MAURITIUS

In this section, we explore the representation of women in the CSE department at UoM. More precisely, we present data showing the rates at which female students enroll in and graduate from CSE programs. These data show that the gender and IT situation in Mauritius is quite different from ASG countries. In an attempt to explain these differences, we conclude this section with some aspects of Mauritian culture that, in our opinion, are responsible for these differences.

Students Entering CSE Programs at UoM

Figure 1 presents the number of students enrolling in CSE programs each year.

Figure 1. First-year CSE students by program

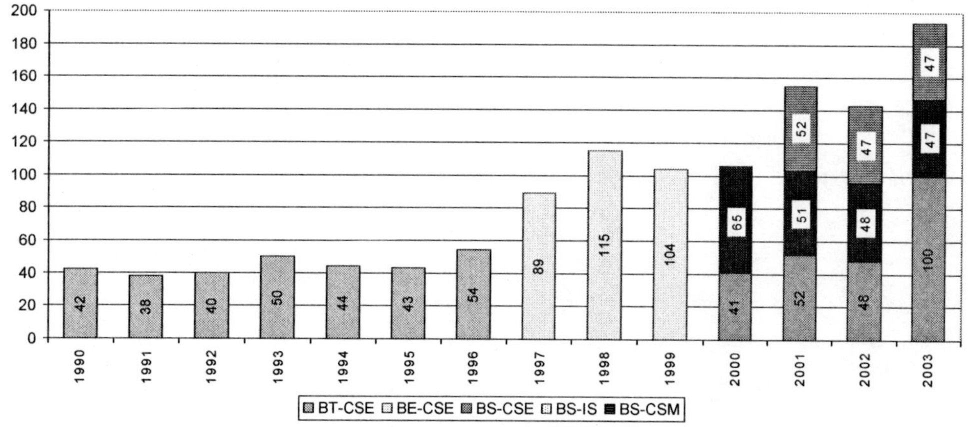

Figure 2. Percent of 1ˢᵗ-year CSE females by program

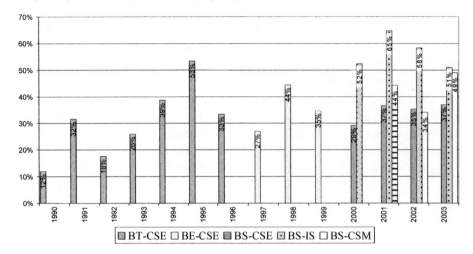

Figure 1 shows rapid increases in CSE enrollments since 1997. Each year, the department has admitted the maximum number of students for which it had staff.

Figure 2 gives the percentages of these students who were women by program.

While the data are noisy, Figure 2 shows increasing representation of women overall. By 2003, women were choosing to enroll in computing-related UoM programs at levels most of us in ASG countries can only dream about: 37% for CSE, 51% for IS, and 49% for CSM. It also indicates that Mauritian women are more attracted to the BS-IS and BS-CSM programs than to the BS-CSE program. One possible explanation is that the BS-CSE has high-school physics as an admissions prerequisite, but the BS-IS and BS-CSM do not, and Mauritian girls are less likely to have this prerequisite. Alternatively, it may be that simply having the word engineering in the name of the program is sufficient to discourage women, similar to the school-of-engineering effect noted in Camp et al. (2001).

Figure 3 presents the percentage of first-year CSE students who were women across all three of the department's programs.

The linear-regression trend line in Figure 3 shows that the percentage of women entering a CSE program at UoM has steadily increased from 12% to well over 40%, even as this percentage has declined in ASG countries.

Students Completing CSE Programs at UoM

Figures 1 through 3 indicate that increasing numbers of women are enrolling in computing-related programs at UoM. What about their graduation rates? Analogous to Figure 2, Figure 4 presents the percentage of female students graduating from each CSE program at UoM.

Aside from the short-lived BE-CSE program, Figure 4 shows the percentage of women graduating from CSE's programs increasing over both the long term and the short term.

Figure 3. Percent of 1st-year CSE females across all programs

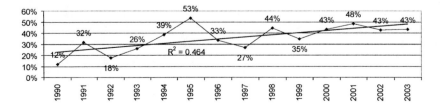

Figure 4. Percent of graduating CSE females by program

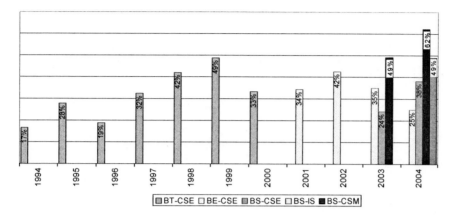

Figure 5. Percent of final-year CSE females across all programs

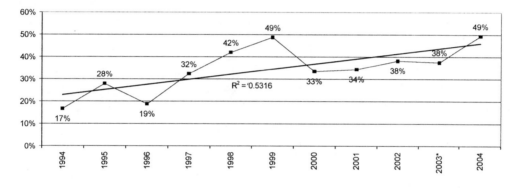

Figure 5 (analogous to Figure 3) presents the percentage of graduating students who were women across all of the CSE department's programs.

Once again, we see steady growth to levels approaching those of the population during the same period of declining representation in ASG countries. When coupled with the increased intake of first-year students shown in Figure 1, Figure 5 implies very rapid growth in the number of Mauritian women choosing a computing-related program.

Are Positive Role Models the Reason?

The increase in the representation of CSE women at UoM described above cannot be attributed to abundant female CSE instructors serving as positive role models or mentors. From 1990 to 2003, the female-to-male CSE instructor ratios were 0:7, 1:5, 1:7, 2:11, 1:8, 1:9, 2:14, 2:13, 2:14, 2:12, 2:14, 2:15, 3:18, and 3:20, respectively. Figure 6 shows the percentage of

female CSE instructors during these years, omitting 1990 (when there were no female CSE instructors).

That is, the representation of women as a percentage of CSE instructors is low relative to the percentage of students who are women, and it has declined slightly as the total number of instructors has increased. During these years, the median percentage of female instructors was 14% (average is 15%).

There has also been significant instability in the set of female CSE instructors.

- In 1992, the sole female CSE instructor was a visitor from the United States; she was also one of the two instructors in 1993, after which she returned to the United States.
- In 1996, one of the female instructors was on leave.
- In 1997, both female instructors were on leave.
- In 1999, one female instructor was on maternity leave.

Figure 6. Percent of female CSE instructors

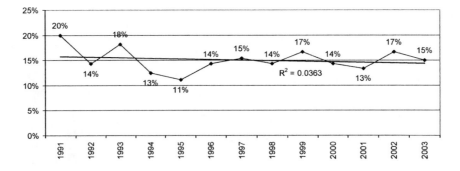

- In 2001, one female instructor was new, and the other was on leave without pay.
- In 2002, one female instructor was new, and another was on maternity leave.
- Under pressure to balance graduate work, family responsibilities, and teaching obligations, these female instructors had very heavy workloads.

This instability in the set of female instructors, coupled with their comparatively low percentage (Figure 6), indicates that abundant, positive female role models are not responsible for the increasing participation and retention of women in computing-related programs at UoM (Figures 3 and 5). While abundant and positive female role models are likely a good thing (for contrasting views, see Bettenger & Long, in press; Pearl, Pollock, Riskin, Thomas, Wolf, & Wu, 1990; Townsend, 2002), the data from Mauritius indicates that they are not a necessary condition for attracting young women to or retaining young women in computing-related programs.

Cultural Factors

We believe that cultural factors are why increasing numbers of women are studying computing in Mauritius. Table 1 lists few that seem especially relevant.

The most obvious difference is that Mauritius is a developing country, whereas most ASG countries are first-world countries. The needs and priorities in a developing country are necessarily different from those of the first world. That is, where a first-world country is maintaining its existing infrastructure, a developing country is building its infrastructure. The cultural imperatives of a developing country are thus different from those of a first-world country. For example, IT is seen in Mauritius as fresh, modern, challenging, and the path to rapid social advancement and national development. Negative words like "geek" and "nerd" are nonexistent in the Mauritian cultural vocabulary.

Relatedly, the Mauritian government is actively and visibly working to build its national IT infrastructure. In the late 1990s, the government began a national IT initiative to turn Mauritius into a "cyber-island" (Ackbarally, 2002). This initiative has further elevated the importance of computing and IT in the national consciousness, making them prestigious (and even patriotic) subjects to study. (The rapid post-1997 increases in CSE admissions visible in Figure 1 are one result of this initiative.)

Table 1. Some Mauritian and ASG cultural differences

Economic level (developing vs. first world)
Governmental promotion of IT
Family emphasis on the importance of education and career
Gender-separate but equal secondary-education system
Limited ability to change major programs in universities

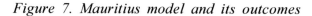

Figure 7. Mauritius model and its outcomes

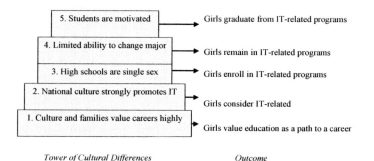

Another related difference is that Mauritian families place a strong emphasis on the importance of education in preparation for a career. From an early age, most Mauritian parents teach their children that education is the path to a good career and a better life. The merit-based admission to the free university creates a highly competitive environment. To improve their children's chances of admission, a high percentage of Mauritian parents send their high-school-age children to after-school tutoring sessions and pressure them to study hard.

Another major cultural difference is that Mauritian boys and girls are taught in separate but equal high schools. Research suggests that girls attending such schools have higher self-confidence and achieve greater career success than those attending coed schools (Coursten & Coleman, 1996). We believe that such separation creates an environment that frees Mauritian girls to discover their academic strengths. The evidence for this is that Mauritian girls do at least as well as boys on the standard entrance exam and have equal chances for admission to the program of their choice at UoM. Moreover, computing-related programs are not perceived as male disciplines because girls have the chance to discover their aptitudes in this separate, nurturing environment.

A final difference is that when a Mauritian student has been admitted to the university, she or he cannot easily change her or his major program. To change one's major program, a student must leave the university and reapply for admission to the new program the following year. This, coupled with UoM's competitive, merit-based admission process, helps the CSE department retain its students (both male and female) because once admitted, most students are loath to jeopardize their prospects by reentering that fierce competition.

The Mauritius Model

We can summarize and enumerate the Mauritian cultural differences as follows.

1. Mauritian families see education as the key to a good career and life. This is an essential foundation for any country seeking to become stronger.
2. The Mauritian government has actively and visibly promoted computing with its IT initiative to become a cyber-island. This has impressed the importance of IT on its national culture, making IT studies and careers honorable, prestigious, and patriotic. It has also encouraged increasing numbers of Mauritian women to consider a computing-related career.
3. Mauritian students are taught in single-sex high schools. This allows Mauritian girls to discover their aptitudes and interests in an environment that is free of gender-based distractions. Because of this and the preceding differences, Mauritian girls are applying to computing-related UoM programs in increasing numbers.
4. Once admitted to a UoM program, students cannot easily change their majors. Because of this, students who are dissatisfied with their CSE programs (e.g., female students discouraged by sexist male peers) are less likely to switch to different programs. This in turn helps the CSE department retain the young women who begin its programs.
5. Access to a university education is not guaranteed in Mauritius. This motivates students admitted to UoM to make the most of the opportunity they have been given. This and the preceding differences are helping Mauritian women graduate from CSE programs in increasing numbers.

We like to visualize these five differences as a tower in which each level builds upon the ones beneath it to produce a positive outcome, as shown in Figure 7.

We conjecture that each of the outcomes in Figure 7 must be achieved to increase the representation of women in IT. For example, if a girl values education but will not consider an IT career, she will not enroll in an IT program. Similarly, if she enrolls in but does not remain in an IT program, she will not graduate from an IT program. Taken together, the achievement of these outcomes forms a pipeline that guides a girl toward the successful completion of an IT program.

We believe that the tower of cultural differences shown in Figure 7 is how Mauritius achieves these outcomes. This tower thus forms a model, the Mauritius model, of one way to increase the representation of women in IT.

Some pieces of this model could be adopted in ASG countries (e.g., national IT initiatives, gender-separate but equal high schools). Other pieces of the Mauritius model seem harder to adopt in ASG countries, at least in the United States (e.g., limiting the ability of a student to change her or his major).

If we cannot use a given piece of the Mauritius model to achieve a given outcome, we in the ASG countries must find an alternative means of achieving that outcome. For example, special IT programs for middle-school girls might provide an alternative means for ASG countries to get girls to consider an IT-related career. Special IT programs for high-school girls might provide a means of getting girls to enroll in an IT-related program. Positive female role models and mentoring relationships might provide a means of attracting young women to and retaining young women in IT-related programs.

FUTURE TRENDS

Since 1990, the University of Mauritius has done very well at increasing the representation of women in its computing-related programs. This achievement is all the more remarkable because admission to UoM is based solely on a student's merit and area of interest. Put differently, there are no special programs at UoM to increase the representation of women or other underrepresented groups. We hope that this happy accident of increasing representation will continue in the future.

Unanswered questions that could benefit from further study in Mauritius include the following.

- UoM students are admitted to the unfilled programs that are highest on their lists of choices. It would be interesting to know the number of CSE women for whom CSE was their first choice, the number for whom it was their second choice, and so on. This would provide a more precise gauge of how interested Mauritian women are in computing-related studies.

- In the late 1990s, the Mauritian government began its IT initiative to make Mauritius a cyber-island. One part of this initiative was the September 2001 opening of the new fee-based University of Technology, Mauritius (UTM), to expand the country's ability to train IT professionals. UTM offers bachelor's degrees in computer applications (CA), software engineering (SE), information-technology-enabled services (ITES), business informatics (BI), and computer science with network security (CSNS). The first students graduated from UTM's SE and BI programs in 2004, with about five students graduating from each program. We have no other data from UTM, so it would be interesting to see how the gender representation at UTM compares with that at UoM.

- The authors are computer scientists, not sociologists, so the preceding cultural-factors and Mauritius-model sections are conjectural. We would welcome a less conjectural analysis of the Mauritius phenomenon by a sociologist.

CONCLUSION

Women in Mauritius are choosing and graduating from computing-related disciplines in increasing numbers, even as those numbers are dropping in ASG countries. Mauritian culture is different from that of ASG countries in several ways, including the following.

1. Mauritian families see education as the key to success.
2. The Mauritian government is aggressively promoting IT.
3. Mauritian students are taught in single-sex high schools.
4. Mauritian students cannot easily change their major programs.
5. Mauritian students are highly motivated and career oriented.

Together, these differences form a model that explains why young Mauritian women consider, enroll in, remain in, and graduate from the CSE programs at UoM at rates far exceeding those of ASG countries.

To increase the representation of women in IT in ASG countries, we must succeed in getting academically serious young women to consider, enroll in, remain in, and graduate from our own IT-related programs.

REFERENCES

Ackbarally, N. (2002). Mauritius: A cyber-island in the making. *The Communication Initiative.* Retrieved July 10, 2005, from http://www.comminit.com/strategicthinking/commentary/sld-2129.html

Bettinger, E., & Long, B. (in press). Do faculty members serve as role models? The impact of faculty gender on female students. *American Economic Review.*

Camp, T. (1997). The incredible shrinking pipeline. *Communications of the ACM, 40*(10), 103-110.

Camp, T., Miller, K., & Davies, V. (2001). *The incredible shrinking pipeline unlikely to reverse.* Retrieved July 10, 2005, from http://www.mines.edu/fs_home/tcamp/new-study/new-study.html

Corston, R., & Colman, A. (1996). Gender and social facilitation effects on computer competence and attitudes toward computers. *Journal of Educational Computing Research, 14*(2), 171-183.

Fisher, A., & Margolis, J. (2002). Unlocking the clubhouse: The Carnegie Mellon experience. *SIGCSE Bulletin, 34*(2), 79-83.

Frieze, C., & Blum, L. (2002). Building an effective computer science student organization: The Carnegie Mellon women@SCS action plan. *SIGCSE Bulletin, 34*(2), 74-78.

Galpin, V. (2002). Women in computing around the world. *SIGCSE Bulletin, 34*(2), 94-99.

Gurer, D., & Camp, T. (2002). An ACM-W literature review on women in computing. *SIGCSE Bulletin, 34*(2), 121-127.

Lopez, A., & Schultze, L. (2002). African American women in the computing sciences: A group to be studied. *SIGCSE Bulletin, 34*(1), 87-90.

Pearl, A., Pollock, M., Riskin, E., Thomas, B., Wolf, E., & Wu, A. (1990). Becoming a computer scientist. *Communications of the ACM, 33*(11), 47-57.

Schinzel, B. (1999). *Contingent construction of the relation between gender and computer science.* Presentation at the International Symposium on Technology and Society, Rutgers, NJ.

Sigurdardóttir, G. (2000). *Women in the Icelandic information society.* Presentation at the Conference about Women and the Information Society, Reykjavík, Iceland. Retrieved July 10, 2005, from http://www.simnet.is/konur/erindi/engish_gudbjorg_erindi.htm

Townsend, G. (2002). People who make a difference: Mentors and role models. *SIGCSE Bulletin, 34*(1), 57-61.

Vegso, J. (2005). Interest in CS as a major drops among incoming freshmen. *Computing Research News, 17*(3), 17.

KEY TERMS

ASG Country: A country whose primary ethnicity or culture is Anglo, Scandinavian, or Germanic, or that has its origin in one of these regions.

CSE: The Department of Computer Science and Engineering at the University of Mauritius.

IT Initiative: A multifaceted national program sponsored by the government of Mauritius to turn the country into a cyber-island.

Mauritius: A 25-mile by 40-mile island nation of 1.2 million people, located about 500 miles east of Madagascar.

Mauritius Model: Five Mauritian cultural differences, the combination of which has increased the percentage of women in computing-related programs at the University of Mauritius.

UoM (University of Mauritius): Until 2001, it was the sole university of Mauritius and the main source of the nation's IT professionals.

UTM (University of Technology, Mauritius): A university opened in September 2001 by the government of Mauritius to help the country expand its population of IT professionals as part of its IT initiative.

Women Entrepreneurs in Finnish ICT Industry

Tarja Pietiläinen
Life Works Consulting Ltd., Finland

Hanna Lehtimäki
Life Works Consulting Ltd. & University of Tampere, Finland

Heidi Keso
Life Works Consulting Ltd. & University of Tampere, Finland

INTRODUCTION

The Nordic countries—Finland, Denmark, Norway, and Sweden—offer interesting material to investigate gendering processes. In these societies, gender equality policy has long traditions and many propagated goals have been researched: women and men participate in paid work almost to the same extent; women make a significant contribution to family income, because it rests on a dual income model; public, low cost day-care is available to all children over one year of age; women's level of education is exceeding that of men's. Yet, the labour market is notoriously segregated both horizontally, meaning that men and women work in different occupations, and vertically meaning that men hold high ranking positions in public and private organisations. The focal phenomenon of this article, entrepreneurship shows even more profound segregation with women and men enterprising in different lines of business and within the same lines of business in different branches (Kovalainen, 1995; Spilling & Berg, 2000).

Nordic experience shows that gaining access to men-only spaces does not bring the same prestige, make women equally influential, and powerful as men. Thus, the question remains what are the processes which hinder women entrepreneurs from achieving a significant position? The article offers one possible answer by reporting a study by Pietiläinen (2002) who set out to investigate what kinds of spaces for entrepreneurial action women business owners are offered in the Finnish information and communication technology (ICT) industry.

BACKGROUND

Since the early 1980s, academic interest in women entrepreneurs has increased steadily due to the increasing impact businesses owned by women have on society and the economy. Presently female entrepreneurship research is a subfield of entrepreneurship studies (Carter, Anderson, & Shaw 2001). In this field, scholars are united in their view that women's unequal access to economic power needs to be changed.

Within the field, researchers differ in their views about the sources of gender inequality and the means to analyse and battle it. This, of course, is based on the presumption that inequality is undesirable. In Pietiläinen's study the different views were categorised into three broad empowerment agendas in female entrepreneurship literature. Table 1 displays the three feminist lines of inquiry. The first two lines of inquiry provided the study with analytical concepts which aided the interpretation of gendering processes in the empirical material. The third approach "doing gender," was followed throughout the research process.

Gender Equality

The overwhelming majority of research is inspired by "gender equality." Researchers following this line of inquiry believe that overt discrimination prevents women from realising their full entrepreneurial potential (Fischer, Reuber, & Dyke, 1993). Over the years they have devoted much research into identifying what kinds of barriers women have to overcome before, during, and after a business start up. Results point to gender differences and cultural prejudices. The latter have been identified as the biggest obstacle for entrepreneurial women. Repeatedly, they encounter gender-based stereotyping and suffer from lack of credibility and support (e.g., Carter & Kolvereid, 1998; Fabowale & Orser, 1995; Kolvereid, Shane, & Westhead, 1993).

Voice to Women

Researchers who are interested in investigating women entrepreneurs' "own voice" and experiences start from the assumption that the source of inequality is women's lifetime experiences of subjugation. These studies represent the second empowerment agenda. Socialization into a woman's position results in uniquely female worldview, and consequently, in different entrepreneurial behaviours than men's (Ahl, 2002). This research has excelled in revealing that women's entrepreneurial choices are greatly shaped by the overall pattern of women's labour market behaviour, life style, and stage of life (e.g., Brush, 1992; Goffee & Scase, 1985; Green & Cohen, 1995; Sundin & Holmquist, 1989).

Doing Gender

Recently, a growing number of European researchers claim that inequality in entrepreneurship is a result of gendering processes which privilege male-typical behaviours and values. Their proposition is to make use of the concept "doing gender" (West &

Zimmerman, 1991). These scholars suggest that researchers of female entrepreneurship need to take a critical look at their own empowerment agendas to move forward. Critical assessment of research inspired by "gender equality" thinking shows that gender equality, in fact, diverts attention away from deeply masculine connotations of entrepreneurship. "Voice to women," in turn, suffers from searching for unique female behaviours, which does more to mystify female experience than to give space for real life women entrepreneurs with their differing aspirations, possibilities, and life strategies. The proponents of "doing gender" approach suggest that effort should be put into exposing the gendered power relations at work in entrepreneurship (Kovalainen, 1995), and move toward change in gendered social and symbolic arrangements governing entrepreneurial activity (Ahl, 2002).

METHODS AND MATERIALS

Pietiläinen's study (ibid.) applied qualitative methodology to identify what kinds of spaces for entrepreneurial action women business owners are offered in ICT industry. Textual analysis was used on a city's strategy documents, media articles about one female-owned new media company Nicefactory Ltd, and transcribed interviews with the company's female owner-entrepreneurs. Empirical material covered the years 1997-2002. As to the theoretical background gender was examined as doing. "Doing gender" approach was chosen because it allowed for studying gender as a process, not as individual

Table 1. Three lines of feminist inquiry in female entrepeneurship literature

Line of Inquiry	Source of Inequality	Analytical Concepts	Empowerment Agenda	Examples of Research
Gender equality	Overt discrimination of women	Gender stereotyping; gender differences	Removal of obstacles; incentives to women; change of attitudes	Kolvereid, Shane, & Westhead, 1993
Voice to women	Women's experiences of subjugation	Female typical qualities and characteristics; Female typical experiences	Creating women-only opportunities and spaces; raising consciousness; change of attitudes	Brush, 1992; Sundin & Holmquist, 1989
Doing gender	Gendering processes which give power to men and male typical behaviours	Gendered social and symbolic arrangements; gendered power relations; hidden masculine connotations	Exposing gendered power relations; change in social + symbolic arrangements; change of attitudes	Ahl, 2002; Kovalainen, 1995

characteristics. In the study, "doing gender" was used both as a theoretical concept and methodological means. Theoretically, "doing gender" allowed focusing away from understanding just the individual entrepreneur to understanding how gendered social and symbolic arrangements of entrepreneurship in ICT industry are shaped. Methodologically, "doing gender" focused qualitative analysis on that continuous meaning making of gender which is embedded in the every-day activities of people and organisations involved.

THE CASE-COMPANY AND CONTEXT OF THE STUDY

The focal company of the study, Tampere based Nicefactory Ltd, was conceived when its owners observed the lack of well-produced Web contents. Specifically they wanted to provide a Web service to female Internet users. The two founding partners of Nicefactory Ltd. worked in radio, TV and print media for an extended period of time before establishing their own company. One of the partners has a university degree in naval engineering and the other holds a master's degree in social psychology. In autumn 1998, they launched their first Web site, www.Nicehouse.fi with content featuring information and discussions about various family-related topics. Their next Web site, www.49er.net, providing real time sailing race information, was launched shortly thereafter. In 2000, a technical university commissioned Nicefactory Ltd. to create a customized learning environment, an opportunity for the company to develop yet another type of Internet content service. The same year, Nicefactory Ltd. launched two new Web sites: www.Addiktio.net, which focuses on "all kinds of addictions," and www.Sooda.com, a Web service designed primarily for teenage girls, but expanded to attract teenage boys, too. Since its launch, Sooda.com has become one of the most popular Finnish Web communities.

At the turn of the century, Finland was well known for its excellence in information and communication technology (ICT) and a highly developed, widespread Internet culture. Nokia had lead the Finnish ICT-sector's growth and rapid internationalization. There was and still is extensive technological knowledge at all levels of society, and Finns are eager to use the latest technological innovations. What was still lacking from this success story, however, was an innovative and user-friendly media culture and business-oriented content production with international potential (e.g., Castells & Himanen, 2001; Tarkka & Mäkelä, 2002). While the ICT-sector had largely developed through technological innovations the content innovations and production had lagged behind.

ICTs—A World without Women

The masculine image of the ICT industry is so strong that it seems paradoxical to link women and entrepreneurship in that industry. Most explicitly, ICT-industry is a world without women: male engineers run the businesses, develop, and market the services/products while women tend to be responsible for office work (Vehviläinen, 1997) and human resources function. Women business leaders are bound to stand out and draw attention as a rarity. A well-known example is Carly Fiorino's position at Hewlett Packard.

When it comes to the definitions that are widely used to describe ICT business, a more, subtle, yet distinctively masculine image emerges (Lie, 1995). It is unanimously recognised that the core competencies of the industry build upon high-tech know how (e.g., Lovio, 1993). From this acknowledgment, it is a small step to define technical accomplishments as groundbreaking innovations and maintain that more is to be expected due to heavy investments in product development (e.g., Schienstock, 2004). Expectations are also supported by the qualities of the people working in the industry. Men who found ICT companies tend to have (technical) degrees. This fact is believed to indicate that the companies are more strategically oriented, more prepared to growth, and more prone to risk taking than the case is in an average start-up.

THREE CONSTRUCTS OF SPACE FOR WOMEN'S ENTREPRENEURSHIP

Based on the empirical analysis, Pietiläinen's (2002) study resulted in three constructs for understanding processes which hinder women entrepreneurs from achieving a more significant position in the Finnish

ICT business. The three constructs are "gender neutral entrepreneurship," "female entrepreneurship," and "powerful women's entrepreneurship."

The first construct, "gender neutral entrepreneurship," gives support to the notion that entrepreneurship, although a human activity is not linked to gender in any fundamental way. The construct is most explicit in strategic management discourse, which becomes important when business opportunities, future growth, and internationalisation or globalisation are visioned. These topics point to the issues which need entrepreneurs' attention in emerging businesses.

The construct provides interpretations to fade out femininities that female entrepreneurs inevitably bring with them to entrepreneurship. The gender neutrality is achieved by meaning making, which first, excludes femininities from the meaning horizon of the category of "entrepreneur" (see also Ahl, 2002) and then, equates selected masculinities of competitiveness, rationality, instrumentality, and control with the ideal entrepreneurial figure.

As such, "gender neutral entrepreneurship" provides a highly appealing space of action to a woman. The idea of that female sex presents an exception can be overcome. This space could secure the position of the "real entrepreneur" for women as well, not only for male business owners. There is, of course, a price attached. It is not easy to find acceptable ways to display femininities as they threaten to expose the otherwise hidden masculinities of the "entrepreneur."

As embodied physical women, female entrepreneurs remind constantly that gender cannot be excluded from entrepreneurship for good. The second construct is labelled "female entrepreneurship." It deals with the femininities and maintains that there is a distinction between "entrepreneurship" and "female entrepreneurship." The distinction rests on gender hierarchy, which makes it rational to relegate gender to women business owners and consider their business actions as gendered exceptions. The construct offers space for action, which many women entrepreneurs are not comfortable with. Experiences of gender stereotyping and (dis)credibility problems give no explanation as to why marginalise women's entrepreneurship by yet another separation. The construct opens up also space for positive action, because it allows for appreciative understanding that living the life of a woman creates a valuable source of business information. Often, women's preferences and ways of doing things are easily bypassed in service and product development as well as marketing, because gender hierarchy prevents from seeing the value of women's activities.

The third construct, "powerful women's entrepreneurship", couples female entrepreneurship with the success stories of start-up businesses in the new economy. It represents selected women entrepreneurs as outstanding representatives of female sex in the male world. The construct dominates media's way of representing female entrepreneurs in the new media business. The positive side of space for action is that media has the power to increase general public's awareness of women's business start-ups. Young men's ICT-business endeavors dominate media publicity and therefore coverage of women's efforts makes a difference. Although women entrepreneurs are treated as a special category, the construct gives meaning making resources to modify particularities of female entrepreneurship into prime means of entrepreneurial advancement.

Ambiguous Spaces for Women's Entrepreneurship

The three constructs offer conflicting and ambiguous spaces for female entrepreneurship in ICT-industry. "Gender neutral entrepreneurship" is a construct which is deeply rooted in strategic management discourse. It is one of the most powerful streams of thinking and acting in the present business world. This connection means that strategic management offers no easy support for entrepreneurial initiative by women, especially if it is innovative in the business field.

"Female entrepreneurship" is more open to women's entrepreneurial behaviour. However, it creates a boundary between entrepreneurship and women's entrepreneurship. The boundary is based on hierarchy, where women's business represents "lesser" form of economic activity. The culturally embedded hierarchy is hard to overcome by individual women.

Yet another type of marginalisation takes place by the construct "powerful women's entrepreneurship." Here, media creates an ideal image of a female entrepreneur and presents it to the public. When a real-life woman compares herself to the ideal, she will most likely find herself lacking the qualities, skills and time to perform within the standards set by the ideal image. Research shows that many business women experience feelings of being inadequate when they think that they are not able to meet the expectations created by the ideal.

Typically, people build their entrepreneurial identities on professional identity which is supported by education and work experience (Hytti, 2003). Ambiguity and conflict stemming from the three constructs refer to the puzzling experience that sex as an anatomical fact presents a more decisive factor to evaluate an entrepreneur's business competence than professional and business experience. In these moments of evaluation, gender is at the same time highly visible and virtually invisible. This is due to fact that entrepreneurship is invisibly male gendered, but thought of as gender neutral and, at the same time, visibly gendered when the actors are women. For individual women entrepreneurs, the dynamic interplay between the visible gendering and the invisible, seemingly gender neutral gendering presents a "rough" field to practise entrepreneurship.

FUTURE TRENDS

The study by Pietiläinen (2002) points out that more research is needed to investigate gender as an ongoing process of meaning making. There is abundant evidence, that to an individual woman entrepreneur, the key question is how to successfully share time between entrepreneurial and personal responsibilities. However, the means by which balancing is achieved by women entrepreneurs, has been overlooked by researcher. The pressure of balancing tends to increase in knowledge intensive business (e.g., ICT-business) where long working hours, project based work, and travel are prevalent practices. What has attracted less attention is that typically, entrepreneurship is associated with for-profit activities in the public sphere and entrepreneurs' needs of nurture and emotional support are seen as private matters, not belonging to for-profit world. The division between work performed in public and work performed in the private reveals that entrepreneurship is considered gender neutral, public activity. Individual entrepreneurs, instead, are gendered and therefore gender "intervenes" to female and male typical ways of coping with the division. Consequently, we need new, critical conceptualisations of entrepreneurship, to overcome deeply held divisions and hierarchies.

CONCLUSION

Female entrepreneurship research shows that economic factors explain only partially women's entrepreneurship. A fairly recent newcomer in this line of inquiry, doing gender studies contribute to understanding that interpretations of entrepreneurship create spaces which men and women are able to occupy differently. These studies alter the theoretical emphasis of entrepreneurship research from examining gender differences per se to analysing how such differences are produced. The theoretical usefulness of examining individual women is questioned. The challenge is to relinquish the stable notions of gender differences and similarities between women's and men's entrepreneurship, popular in the extant literature. Even if there are no clear references to gender, there are seemingly neutral articulations that produce different material outcomes for women. In this sense, meaning making of entrepreneurship rearticulates the power imbalance informing relations of gender.

REFERENCES

Ahl, H. J. (2002). *The making of the female entrepreneur. A discourse analysis of research texts on women's entrepreneurship*. JIBS Dissertation Series No. 015. Jönköping University: Jönköping International Business School.

Brush, C. (1992). Research on women business owners: Past trends, a new perspective and future directions. *Entrepreneurship, Theory, and Practice, 16*(4), 5-30.

Carter, N. M., & Kolvereid, L. (1998). Women starting business: The experience in Norway and the United States. In *Proceedings of Women Entrepreneurs in Small and Medium Enterprises* (pp. 185- 202). Paris: OECD.

Carter, S., & Rosa, P. (1998). The financing of male- and female-owned businesses. *Entrepreneurship and Regional Development, 10*(3), 225-241.

Castells, M., & Himanen, P. (2001). *The Finnish model of the information society.* Helsinki, Finland: Sitra and WSOY.

Fabowale, L., & Orser, B. (1995). Gender, structural factors, and credit terms between Canadian small businesses and financial institutions. *Entrepreneurship: Theory and Practice, 19*(4), 41-65.

Goffee, R., & Scase, R. (1985). *Women in charge: The experiences of female entrepreneurs.* London: George Allen & Unwin.

Green, E., & Cohen, L. (1995). Women's business: Are women entrepreneurs breaking new ground or simply balancing the demands of "women's work" in a new way? *Journal of Gender Studies, 4*(3), 297-314.

Hytti, U. (2003). *Stories of entrepreneurs—Narrative construction of identities.* Turun kauppakorkeakoulun julkaisuja, Sarja A-1:2003. Turku: Turku School of Economics.

Kolvereid, L., Shane, S., & Westhead, P. (1993). Is it equally difficult for female entrepreneurs to start businesses in all countries? *Journal of Small Business Management, 31*(4), 43-51.

Kovalainen, A. (1995). *At the margins of economy. Women's self-employment in Finland, 1960-1990.* Avebury: Adlershot.

Lie, M. (1995). Technology and masculinity. The case of the computer. *The European Journal of Women's Studies, 2,* 379-394.

Lovio, R. (1993). *Evolution of firm communities in new industries. The case of the Finnish electronics industry.* Helsinki School of Economics and Business Administration, A:92. Helsinki, Finland: Helsingin kauppakorkeakoulu.

Pietiläinen, T. (2002). *Moninainen yrittäminen. Sukupuoli ja yrittäjänaisten toimintatila tietoteollisuudessa.* Acta Universitatis Oeconomicae Helsinginensis, A-207. Helsinki, Finland: Helsinki School of Economics.

Schienstock, G. (2004). *Embracing the knowledge economy. The dynamic transformation of the Finnish innovation system.* Cheltenham: Edward Elgar.

Spilling, O. R., & Berg, N. G. (2000). Gender and small business management: The case of Norway in the 1990s. *International Small Business Journal, 18*(2), 38-59.

Sundin, E., & Holmquist, C. (1989). *Kvinnor som företagare—osynlighet, mångfald och anpassning—en studie.* Malmö: Liber.

Tarkka, M., & Mäkelä, T. (2002). Uusi mediakulttuuri innovaatioympäristönä. Kotimainen ja kansainvälinen tutkimus. Mediakulttuuriyhdistys m-cult ry. Sisältötuotanto-työryhmän väliraportti 7.

Vehviläinen, M. (1997). *Gender, expertise, and information technology.* Department of Computer Science, A.1997-1. Tampere: University of Tampere,

West, C., & Zimmerman, D. H. (1991). Doing gender. In J. Lorber & S. A. Farrell (Eds.), *The social construction of gender* (pp. 13-37). Newbury Park: Sage.

KEY TERMS

Doing Gender: A gender theoretical line of reasoning that gender is a process of power which first, creates and maintains differences between the sexes and second, constructs a hierarchy to the benefit of men and masculinities.

Empowerment Agenda: A proposal of actions which are needed to enable discriminated people to act upon their own interests.

Gender-Based Stereotyping: Using common sense criteria to categorise people into males and females. The criteria include assumptions about gender appropriate characteristics, behaviours, and physical qualities.

Gender Equality: A gender theoretical line of reasoning that overt discrimination prevents women from realising their full potential.

Women's Own Voice: A gender theoretical line of reasoning that meanings and interpretations a woman or a group of women invent and use to make sense of their experiences remain silenced due to prevailing gender order.

Women in Computing in the Czech Republic

Eva Turner
University of East London, UK

INTRODUCTION

While the countries of Western Europe and the USA are mostly in control of the design and construction of computing technology, the numbers of women actively involved in this process are very low and decreasing. The Czech Republic is an Eastern European country with highly developed system of tertiary computing education and levels of computer usage comparable to Western Europe. Whereas under capitalist regimes of the period equal opportunities legislation has often been achieved despite Government resistance, Communism built it into its constitution, and professed equality of men and women in every field of human activity. Publicly and in the national subconscious that equality became a reality. However, at a time when Western European governments and European Union (EU) legislators are finally awakening to the unequal position of women in technology, it is a perception that invites closer inspection. This article is based on a set of interviews carried out in the Czech Republic in August 2004 and a collection of official reports and quantitative data published in the Czech Republic between 2002 and 2004. The aim was to find out what has the new Czech regime done about gender equality in the field of computing and what importance the Czech officialdom assigns to the perception of equality. For comparisons this article assumes that the reader is acquainted with gender and computing debates in the "West."

BACKGROUND

Gender Equality Before and After the Velvet Revolution

The Czech Republic (then part of Czechoslovakia) became a "capitalist country" in the change of regime that occurred during the Velvet Revolution of November 1989 and subsequently became a member country of the European Union (EU) in May 2004. In preparation for its entry to the EU and as a reaction to direct criticism by the United Nations, the Czech political leadership was forced to follow gender equality movements, gender mainstreaming, and EU legislation and include gender equality on the country's political agenda.

During the years 1948-1989, the communist regime guaranteed the right to work for every citizen, stated quotas for women's employment and guaranteed women's rights in the constitution. The general population had no opportunity, nor the necessity, to discuss the meaning of gender or women's equal rights. Taking this "guaranteed equality" for granted, the population slipped deeper into accepting biologically and socially deterministic gendered views of the roles of men and women in society. While this supposed equality applied also in further and higher education, views of women not being suitable to study technology, and engineering prevailed and very few women applied. Most women worked in administration, school education, services and caring professions, men worked in fields of engineering, technology and were represented in large numbers in specialist medicine, politics, university education, and research.

When the communist regime was overthrown, gender initiatives developed mainly from the former dissident movement, the Czech Sociological Institute and interested individual academics. Based mainly on The Czech Sociological Institute's findings, the Czech Helsinki Committee reported in 1996 about the human rights situation in the Czech Republic. This very lengthy report contains a section on "Some aspects of women's rights in the Czech Republic" which mentions sexual harassment and unequal treatment in places of employment, advertising, pay, and the existence of glass ceilings. All this was described as almost a "Czech cultural norm" (a term coined by Hana Havelkova).

In 1990, the sociologist Dr. Jirina Siklova founded The Gender Studies Centre, a women's NGO. It is an educational, information and advice centre for equal opportunities. It has its own library and Web site, a number of publications, and participates in national and international research projects. During a brief examination, I found that the library contains little about women in science and technology and nothing on women in computing.

By 1996, the Czech government was forced to begin the debate on equal rights for men and women in Czech society and in April 1998 it agreed its "Action Plan of Priorities and Activities for Enforcement of Equality of Men and Women." The government decreed that each of its departments had to cooperate with women's organisations.

Women in Computing in the Development of Government Gender Policies

The Ministry of Work and Social Affairs has been charged with collecting and publishing gender information and advising the Government. It has established a Department for Equal Opportunities for Women and Men, which publishes (since 1998) an annual "Aggregate Report on Fulfilment of Priorities and Activities of the Government for Enforcement of Equality of Men and Women." In October 2001, the Government established its advisory body The Council for Equal Opportunities of Men and Women with members representing ministries and NGOs (MPSV, 2004). The one ministry, which is *not* represented, is the Ministry of Informatics. The Aggregate Reports are only presented to the government after individual ministerial reports have been debated and accepted by this council.

The Ministry of Work and Social Affairs publishes a range of reports containing gendered statistics and reports on fulfilment of international agreements and recommendations (e.g., Beijing+5). There appears to be no specific interest in women in science and technology in all these publications. Most of them stress that equal rights and opportunities at work are guaranteed by national legislation. By such reports the ministry is trying to demonstrate to the EU that gender inequality at work does not exist. This appears to be in strict contrast with situations reported by the Gender Centre or the Czech Sociological Institute (Study on Paternity Leave, 2003). Using European Union grants, the government has published material intended to educate employees about gender equality. This material has been distributed to government departments and local councils. There seems to be no evaluation of their direct influence on official decision-making and it was not in evidence during my interview at the Ministry of Informatics.

In my search for evidence of understanding of questions of gender and computing, I sent a questionnaire to the Minister for Informatics. His response can be summarised as follows: women in the Czech Republic do have equal opportunities in the workplace and in access to higher education. If they are under-represented in the IT industry, then it is because they do not choose to enter that field. Admittedly a stronger emphasis needs to be put on IT education, but this is the responsibility of the Ministry of Education, and the wider area of equal opportunities is the responsibility of the Ministry of Work and Social Affairs. The role of the Ministry of Informatics seems to be restricted to offering money for training courses in ICT literacy (National Programme).

The Commission for Informatics and Telecommunication—an advisory body on questions of policy in Informatics to the ruling Czech Social Democratic Party—comprises 30 members, of whom just one is a woman. An interview with two male founding members of the Commission confirmed that questions of gender and equal opportunities never reach the agenda. Neither saw any point in trying to persuade women to work in an industry where they "do not want to be." One of them believed that glass ceilings do not exist in the Czech Republic and that women do not wish to be competitive and "collect trophies," and therefore have no need to enter higher management positions. He also "knew" that women do not have the ability to manage technical university education, particularly mathematics, as they possess "different logical skills."

It is not difficult to infer the nature of the political advice that is informed by this kind of gender stereotyping.

Women in Computing in Higher Education

In general, Czech universities are divided into technical and humanities universities. Both types offer some teaching in computing, but it is the technical universities that students would choose to study computer science and informatics. A brief study of the Web pages of computer science/informatics/ engineering faculties of the two main Czech technical universities in Prague and Brno and the two main humanities universities in both cities (accessed 10.9.2004) revealed that no gendered statistics of staff and students have been published. As the Czech language differentiates clearly between male and female names, I was able to do a head count of men and women staff and PhD students. Of the total of 209 full time members of staff only 15 (7%) were women and of the total of 179 PhD students only one was a woman. None of the women academics was in a position of dean (faculty) or professorial (department) management. Only the Masaryk University Brno published their list of students, which showed that of the 1390 undergraduate student only 97 (7%) were women.

The Czech Statistical Office (2003) publishes gendered statistics of the total numbers of men and women students but these are not broken down by individual faculties or subject areas. Table 1 is based on figures available for 2001/02.

There are in total almost equal numbers of women and men entering university education. Some fields, like pedagogy or nursing, are heavily populated by female students. The difference between technical universities (Prague and Brno) and those primarily concerned with humanities (Charles and Masaryk) is striking. The humanities universities have much higher

numbers of women in traditional female fields and thus the statistics hide the low numbers of women in their departments of computing or informatics. Table 2, which shows the numbers of students in the field of "electronics, telecommunication, and computer technologies" for the whole of the Czech Republic, presents a bleak picture.

The Ministry of Work and Social Affairs (December 2003, p. 38) states that "...it is mostly women that study social sciences and services fields, most frequently medicine and pharmacy. We find few women in technical and natural science fields..." It also says (p. 36) that while there are slightly more women in bachelors' degrees, women only account for about one third of all PhD students. Certainly among these there are almost none in computing or informatics. Again while observing the shortages, the ministry has no analytical explanation of these findings, or suggestions for improving the situation.

An interview with an employee responsible for equal opportunities at the Czech Ministry of Education revealed complete inactivity in this matter. There exists in the Czech Republic a Centre for the Study of University Education, established in 1991 as a state-funded institute of the Ministry of Education. Its mission is "to collect, analyze, collate, and disseminate information concerning higher education and research policy." This Centre has published a report evaluating dropout rates from technical universities in the Czech Republic (Menclova, Bastova, Konradova, 2004). There is a section entitled Sex in this article, which translates as follows: "Because we are investigating technical fields of university studies, the substantial majority (73.5%) of men over women (26.5%) is quite natural" (p. 9). The report offers no gendered data, no gender analysis and no proportions of men and women in the 50% dropout rate. As a government advisory body, this institution may be able to influence political decisions, however there appears to be no evidence that its gender reports are being actively used.

Table 1.

University	Men students	Women students
Czech universities in total	114,322	105,192 (48%)
Technical University Prague	17,925	**3,357 (16%)**
Technical University Brno	12,329	**2,765 (18%)**
Charles University Prague (humanities)	17,743	23,180 (57%)
Masaryk University Brno (humanities)	9,678	12,543 (56%)

Table 2.

	Total no. of students	Total women students	Men graduated in 2001/02	Women graduated in 2001/02
2001/02	11,875	458 (3.9%)	1,413	32 (2.2%)
2002/03	14,859	678 (4.6%)	1,619	64 (3.8%)

Women in Computing Employment

While it is difficult to find data on women in computing in the Czech education system, it is almost impossible to find data on the numbers and positions of women in the computing and informatics industry. In 2002 there were 194,000 people who worked in information and communication technologies (ICT). This represents 4% of the total labour force in the Czech Republic (source http://www.czso.cz, provided by Kristova). As elsewhere in the world the definition of "the industry" is not stated and thus some may argue that for 4% the questions of equality are not substantially important.

The Czech Statistical Office has two classification categories, which relate to computing and ICT: "technical, health, and pedagogic workers" and "research and specialist professional workers." Taken together, these categories represent 30% of the total Czech work force, and in both of them women earned almost 30% less than men in 2001. A figure of 30% is not quite so easily dismissed.

An interview with an employee from the Gender Department of the Ministry of Work and Social Affairs confirmed, that while the ministry is responsible for the National Gender Action Plan, the questions of equal opportunities for women and men in employment are of no interest to the ministry. It is predominantly interested in women's unemployment, women's access to employment, and sees no reason for collecting statistical or other data on women employed in any particular field.

An interview with the personnel director of one of the five largest Czech IT companies revealed that of the 350 workers 74% were men, of whom 69% had a university education. Of the 26% who were female, only 33% had a university education, and most did not work in technical jobs. The company employed 241 non-administrative/technical staff for whom gendered statistics are not collected. 2/10 women were employed as top managers, and 4/18 women in middle management.

FUTURE TRENDS

The gender debate and most research in the Czech Republic focuses on issues of domestic violence, trafficking, reproductive health and gender equality education, and has not engaged with ICT issues (Simerska & Fialova, 2004).

There are exceptions to this trend. Sometimes these are committed individuals like the PhD student in Prague whose thesis is on gender and computing, and the Gender Studies Centre researcher who is the ICT & Economy project manager and participates in several international IT and gender related initiatives. The latter is co-author of a report entitled "Bridging the Gender Digital Divide in Central and Eastern Europe" (2004), which concludes that women in this part of the world are under-represented at all levels within ICT initiatives. She also criticises inadequate integration of gender and/or women-specific issues, lack of gendered statistics and unequal access to ICT training. She is dismissive of ministerial initiatives, seeing them as wholly ineffective.

Following the European Technology Assessment Network (ETAN) report on Women and Science in 1999 *Promoting Excellence through Mainstreaming Gender Equality*, the Czech gender specialists have also published an explanatory educational booklet on the glass ceiling for women in science (Linkova, Saldova, Cervinkova, 2002). The workers interviewed at Ministry of Informatics responsible for equality had never seen it.

Women in Science is a "National Contact Centre" with a Web presence, which elects a woman scientist every month, publishes Czech and international news on women and science and a quarterly magazine called Kontext. The April 2004 issue contains an article about Aida Lovelace. I am not clear what political influence this Centre has and how widely it is known among the population.

An important and welcome contribution to the debate on gender politics is the Shadow Report on Equal Opportunities (Bouckova et al., 2004). The Shadow Report is highly critical of the official Aggregate Reports mentioned earlier and accuses the government of inaccuracies, misreporting, little proactivity, misusing statistics to paint a rosier picture of the position of women in the Czech Republic and its own activity in this field, and of not acting on its own recommendations. The report also criticises the Ministry of Education for the lack of gender sensitive education, and for not fulfilling any of the "gender priorities" set for 2003. In the field of science and research, it points out that despite many acknowledged problems like the "brain drain" and lack of money, the accepted norm seems to be no gendered statistics, low gender awareness, lack of support for research in the areas of gender, stereotyping of researchers, financial discrimination and lack of equal opportunities.

The report makes a number of recommendations for improvement, but there has been insufficient time since its publication to assess what impact it may have.

CONCLUSION

Under pressure from the EU, the Czech government appears to expend a great deal of energy on promoting equal rights for women and men. A government committee, educational material, annual reports and a dedicated person at each ministry all now exist. While the government can call on an extensive body of statistical and educational material to back up its claims of success in the field of gender equality, little evidence exists that there has been any real change in public opinion or behaviour. There is an overwhelming impression that gender equality reports are published with the primary objective of satisfying EU requirements and directives.

There is no evidence that political circles have even begun to perceive the shortage of women in technology and ICT as a problem. Although a grant application for a gender and ICT project has been turned down (Simerska), the ministries are "not aware" of any application, and are not issuing any calls for projects of this nature. There appears to be a belief that higher levels of general computer literacy will automatically lead to more women choosing IT as a field to study and work in. There is also no awareness of the need for critical evaluation of the possibility that social conditioning and technological gendered prejudices and stereotypes might be the reason for the unequal position of women in this field.

Communism stifled any debate on gender stereotypes and gender equality and the time that has lapsed since the Velvet Revolution seems to have been too short to reverse this process. There are pockets of refreshing enthusiasm for change among gender researchers of the Gender Studies Centre, but the overwhelming impression is of a Czech officialdom with a bureaucratic attitude to report writing which has little analytical content.

NOTE

This article is based on material published in Turner (2005).

REFERENCE

Bastova, J. (2004). Zpracovani ziskanych dat k ukolu "rovne prilezitosti"—rizene rozhovory. Work in progress.

Bouckova, P. et al. (2004). *Stinova zprava v oblasti rovneho zachazeni a rovnych prilezitosti zen a muzu.* Prague: Gender Studies Centre.

Czech Helsinki Committee. (1996). *Nektere aspekty zenskych prav v Ceske Republice* (Some aspects of women's rights in the Czech Republic). Retrieved August 28, 2004, from http://www.helcom.cz/index.php?p=2&rid=100&cid=373

Czech Statistical Office. (2002, 2003). *Focus on women and men, comprehensive information.* In house publication.

Czech Sociological Institute. (2003). *Study on use of parenting leave by men.* Retrieved September 20, 2004, from http://www.mpsv.cz/files/clanky/712/pruzkum.pdf

Gender Studies, Faculty of Philosophy, Charles University. (n.d.). Retrieved August 28, 2004, from http://soc-prace.ff.cuni.cz/gender/menu.html

Gender Studies Centre. (2004). *An interview with Mrs. Curdova.* Retrieved September 20, 2004, from http://www.feminismus.cz/fulltext.shtml?x=168464

Gender Studies Web site. (n.d.). Retrieved September 20, 2004, from http://www.feminismus.cz/historie.shtml

Linkova, M., Saldova, K., & Cervinkova, K. (2002). *Glass ceiling, position of women in science.* National Contact Centrum—Woman and Science.

Menclova, L., Bastova, J., & Konradova, K. (2004). *Neuspesnost studia posluchacu 1. rocniku technickych studiejnich programu verejnych vysikych skolv CR a jeji priciny, Ministry of*

Education. In house publication, Prague, Brno. Retrieved September 14, 2004, from http://www.csvs.cz/projekty/Neuspesnoststudia.pdf

Ministerstvo Prace a Socialnich Veci. (n.d.). *Website containing reports in the area of equal opportunities of women and men.* Retrieved September 20, 2004, from http://www.mpsv.cz/scripts/clanek.asp?lg=1&id=712

Ministerstvo Prace a Socialnich Veci. (n.d.). *All aggregate reports on equal opportunities of men and women.* Retrieved August 28, 2004, from http://www.mpsv.cz/scripts/clanek.asp?lg=1 &id=696

Ministerstvo Prace a Socialnich Veci. (2004). *Satus Rady vlady pro rovne prilezitosti muzu a zen* (Constitution of the Council for Equal Opportunities of Men and Women). Retrieved August 28, 2004, from http://www.mpsv.cz/scripts/clanek.asp?lg=1 &id=3138. In house publication.

Ministry of Work and Social Affairs and Czech Statistical Office. (December, 2003). *Women and men in data.* In house publication.

Simerska, L., & Fialova, K. (2004). Bridging the gender digital divide. Retrieved September 13, 2004, from http://www.witt-project.net/article46.html

Turner E. (2005). New Europe—New Attitudes? Some initial findings on women in computing in the Czech Republic. In J. Archibald, J. Emms, F. Grundy, J. Payne, & E. Turner (Eds.), *The gender politics of ICT.* London: Middlesex University Press.

Women Networking Support Programme. (n.d.). Retrieved August 28, 2004, from http://www.apc women.org/about/

Women in Science. (n.d.). Retrieved September 14, 2004, from http://www.zenyaveda.cz

KEY TERMS

Czech Government Type: Parliamentary democracy with a president and a prime minister, bicameral Parliament which consists of the Senate and the Chamber of Deputies, regional and local councils and ministries in expected fields.

Czech Republic: A landlocked Central European country, capital Prague (Praha), bordering Germany, Poland, Slovakia and Austria. It has two regions Bohemia and Moravia and 10.5 million inhabitants. As of May 1, 2005 it is a member state of the European Union.

Humanities University: A university delivering degrees in humanities. The Czech Republic also has specialised agricultural, military and economic universities.

Technical University: A university delivering degrees in science and technology subjects.

Velvet Revolution: Term given to a peaceful change of regime from Communism to Capitalism in the Czech Republic in November 1989.

Women in Technology in Sub–Saharan Africa

Vashti Galpin
University of the Witwatersrand, South Africa

INTRODUCTION

International research has shown that in most countries, there are few women studying towards *information technology* (IT) careers (Galpin, 2002), and there is much research, particularly in the United States (U.S.), United Kingdom (UK) and Australia into why this is the case (Gürer & Camp, 2002). This article considers the situation in *sub-Saharan Africa* and focuses on women's involvement in the generation and creation of *information and communication technologies* (ICTs) in sub-Saharan Africa, as opposed to ICT use in sub-Saharan Africa, which is considered elsewhere in this volume. There are a number of aspects to the generation and creation of ICTs: how women are involved in this process as IT professionals and how they are educated for these careers, as well how technology can be used appropriately within the specific conditions of sub-Saharan Africa. ICTs will be considered in the broadest sense of the word, covering all electronic technologies, from computers and networking to radio and television.

Women's participation is important: The World Summit on the Information Society (WSIS) Gender Caucus (www.genderwsis.org) has identified women's involvement in the design and development of technology as well as technology management policy, as key principles for the information society. Marcelle (2001) emphasizes the necessity for African women to become involved in technological and scientific areas, including "computer science, software engineering, network design, network management and related disciplines" (Marcelle, 2001, para. 15) to create an information society appropriate for African women. The diversity of those involved in design leads to higher-quality and more appropriate technological solutions (Borg, 2002; Lazowska, 2002).

BACKGROUND

Sub-Saharan Africa has a population of 641 million, young (almost half under 15) and rural (35% urban). Significant problems are undernourishment, poverty and HIV/AIDS (United Nations Development Programme (UNDP), 2004). All the countries in sub-Saharan Africa are classified as developing countries. Some countries are relatively wealthy, such as Mauritius, South Africa, and Nigeria, but have large wealth disparities within their populations. Women in sub-Saharan Africa are expected to focus on the home, they have less access to education and health, and their contribution to family and community is not valued (Huyer, 1997).

Technology and Infrastructure

Per 1,000 people, there are only 15 landlines, 39 cellular subscribers and 10 Internet users (UNDP, 2004). There is less access to electricity in rural areas, so battery, solar-powered or wind-up radios are important.

The higher rate of cellular subscribers is due to liberalization of telecommunications policies and investment in infrastructure by private companies, but these advances benefit mostly those in urban areas. Mbarika, Jensen, and Meso (2002) identify the urban nature of Internet access, lack of telephone infrastructure, high cost of international links and a lack of technical staff as inhibitors of Internet growth in sub-Saharan Africa. However, growth is being stimulated by growing numbers of Internet cafes; a decrease in charges for access; increased mobile telephony, which releases landlines for Internet access; and deregulation of the Internet and telecommunications sector; as well as reduction of import duties on computer equipment (Mbarika et al., 2002).

Historically, African *non-governmental organizations* (*NGOs*) were pioneers in the use of communicating via computer networks in Africa, convinced of their importance in spite of skepticism from people outside Africa (Esterhuysen, 2002). WorkNet (which became SANGONeT) was founded in 1987 and played an important role in the resistance against apartheid in South Africa internationally. In the early 1990s, NGONET provided e-mail services in East Africa, and ESANet linked five universities in East Africa using *FidoNet*. By 1995, 12 sub-Saharan African countries had full Internet connectivity, and by 2000, all countries had permanent Internet connectivity and dial-up ISPs (Levey & Young, 2002).

IT Education

African women have low participation rates in science and technology education (Hafkin & Taggart, 2001). For undergraduate computer science (CS) students, the following rates of female participation occurred (Bunyi, 2004; Galpin, 2002; Hoffman-Barthes, Nair, & Malpede, 1999; Mariro, 1999; South African Qualifications Authority, 2004):

- Botswana, 10% (1998)
- Eritrea, less than 10% (2001)
- Nigerian polytechnics and some universities, around 30% (mid-1990s)
- University of Nairobi, Kenya, 11% (2001)
- Kenyatta University, Kenya, 14.3% (2002/2003)
- University of Makerere, Uganda, 27% (2000)
- Zimbabwean technical colleges, 41.7% (1996).
- South African universities, 31% (2001); technikons, 37% (2001).

In Tanzania, at the Universities of Dar-es-Salam, Sokoine, and Muhimbili in 1995-1996, only 3% of those receiving the Dip. Sci. Informatic qualification were female (Mariro, 1999).

Odedra-Straub (1995) highlights the lack of female lecturing staff at African universities, and that few universities offered CS degrees in the early 1990s. In Madagascar, 11.1% of CS education teachers were female (Mariro, 1999). In contrast, in South Africa in 2001, 46% of CS instruction staff in higher education institutions were female (South African Reference Group on Women in Science and Technology, 2004), which is unexplainably higher than the percentage of female CS graduates.

An attempt to address the imbalance is the Cisco Networking Academy Program Gender Initiative (www.ciscolearning.org/Initiatives/Gender_Initiative.html) which, with NGOs and governments, provides opportunities for women to study networking. African countries involved include Rwanda, South Africa and Uganda. The Department of Women and Gender Studies at Makerere University, Uganda (www.makerere.ac.ug/womenstudies/ict.html) is an approved Cisco Academy and aims to increase the number of women working with ICTs and to change perceptions of computing as a male domain. To date, the majority of students have been female, and its courses cover basic computer skills, networking skills and the use of ICTs in education. Additionally, the Cisco networking qualification allows entry into the undergraduate CS and IT programs (Bantebya-Kyomuhendo, 2004). A similar program was run at the United Nations Economic Commission for Africa in Addis Ababa, which was exclusively for women and covered gender as well as technical issues (Hafkin, 2002).

IT Professionals

With low participation rates at the tertiary level, there is likely to be a low number of women working as IT professionals. There is little data available on this topic (Hafkin, 2003; Odedra-Straub, 1995). In South Africa, only 27% of IT employees were female, and they were more likely to be in sales and marketing, end-user computing, or education and training than in hardware, management or networking (South African Information Technology Industry Strategy, 2000). In sub-Saharan Africa, it seems that women are unlikely to be found in management or system analysis positions, as well as teachers or lecturers of computing courses, although they can be found as programmers and operators (Odedra-Straub, 1995). Similarly, in developing countries, women are more likely to be in low-level jobs relating to word-processing and data entry than in IT management or in jobs designing or maintaining computer systems and programs (Hafkin & Taggart, 2001). Worldwide, there are few women in senior management positions or on the boards of ICT companies, involved in policy, professional and regulatory organizations or employed by government departments responsible for ICT (Hafkin, 2003).

The International Telecommunication Union (ITU) collects statistics on telecommunications staff. In terms of percentage of female staff in 2001, Benin, Burkina Faso, Central African Republic, Ghana, Malawi, Nigeria, and Togo had less than 20%; Angola, Botswana, Côte d'Ivoire, Kenya, Madagascar, Mali, Senegal, South Africa, and Zambia had between 20% and 29%; Guinea, Lesotho, Seychelles, and Tanzania had between 30% and 49%; Eritrea had 51%; Cape Verde 67%; and São Tomé and Principe 78% (ITU, 2003). Hafkin (2003) notes that the inclusion of telephone operators and data-entry staff explain why some countries have high figures, because these positions are usually held by women.

APPROPRIATE USE OF TECHNOLOGY

Although many people in sub-Saharan Africa, particularly in major cities, have access to ICTs in a way very similar to that of Europe and North America, this is not true throughout. Because of lack of infrastructure and development, a standard model of access cannot be assumed (Kole, 2003), and specific solutions are required for the specific conditions. This section lists some of the issues (for a broad assessment of wireless technologies for Africa, see Jensen, 1996; for other examples of technological solutions, see Holmes, 2004).

Wireless Technology

Two examples that use wireless technology to deal with a lack of telecommunications infrastructure are *high-frequency (HF) radio e-mail* and *digital satellite broadcast*. HF radio e-mail is used to provide infrastructure in remote areas of Guinea (Holmes, 2004; Marshall, 2002). The Arid Lands Information Network (ALIN) in East Africa uses digital satellite broadcasting to transfer content that is downloaded to computer via radio (ALIN, 2002).

Content Delivery

Content appropriate for local conditions is crucial, and more organizations are generating their own

content, both from generally available Internet material and from African women themselves. A reason for the move towards Internet publication is that the barriers to entry are often lower than for other media, such as print, radio and television (Holmes, 2004).

Locally developed CD-ROMs can use audio and video to provide material in local languages that is accessible to people who are not literate. In Uganda, a CD-ROM titled "Rural Women in Africa: Ideas for Earning Money" was successfully developed for use in telecenters (Mijumbi, 2002).

CD-ROMs and other optical media can also be used to provide information that will not change. This avoids the cost and time often necessary for downloading material via telephone modems with slow speeds, unreliable connections and via international links with limited bandwidth and high latency due to distance. Design of Web sites should also take these technical constraints into consideration.

Information brokering is also important; information is obtained via the Internet, modified for local conditions and then redistributed, possibly in hardcopy or by radio, as well as by computers (Morna & Khan, 2000; Pacific Institute for Women's Health, 2002). This process also happens in reverse; for example, during the Fourth World Conference on Women in Beijing in 1995, comments on draft versions of the Platform Actions were obtained from women's groups via e-mail (Hafkin & Taggart, 2001). Additionally, African women have indigenous knowledge that is important to collect and disseminate (Huyer, 1997), although the introduction of ICTs should lead to the preservation of this knowledge and not its appropriation (Karelse & Sylla, 2000).

Radio

Transmission of information by radio plays a large role in development. A participant at a workshop on empowering rural women noted that "radio is the rural Internet of Africa" (Women'sNet/Dimitra, 2004, p.4). Listening clubs meet to listen to radio programs and then discuss their content, record the discussions and pass them onto women in other areas (Kole, 2003). This permits dissemination of indigenous knowledge, as well as permitting access to radios in areas of extreme poverty. Content and timing of broadcasts are important issues (Women'sNet/Dimitra, 2004).

The Women'sNet Community Radio Pilot Project in South Africa inter-phases ICTs with radio. It works with community radio stations on gender sensitivity, repackaging of information from the Internet for broadcast, and training of women at women's NGOs in ICT skills (Morna & Khan, 2000).

Free/Libre/Open Source Software

An approach to cost reduction, non-dependent development and sustainability is to move away from proprietary software and towards the use of *free/libre/open source software* (*FLOSS*) (Waag Society, 2003). This has had limited application in Africa to date (Kagai & Kimolo, 2004). In 2004, Women'sNet held a workshop for women in Southern Africa to learn about FLOSS focusing on educating technicians, decision makers and end-users (www.womensnet.org.za/FOSS/more.shtml).

FUTURE TRENDS

To ensure that technology is designed so it is appropriate for all of its users and their specific conditions, it is essential that women are involved in the generation and creation of ICTs in positions such as programmers, system designers, information technologists, network specialists, software engineers, system analysts and content developers.

Women's involvement in technology policy decisions is also important. The WSIS Gender Caucus has identified equal participation in decision making as a key principle for the information society. Additionally, gender must be part of national and international ICT policy (Hafkin & Taggart, 2001; Huyer, 1997; Marcelle, 2000). This requires the involvement of women who are IT professionals. Hence, it is crucial that there is a concerted effort on behalf of national governments in sub-Saharan Africa to improve the participation of women in science and technology, particularly in IT-related fields. Regional approaches may also be appropriate. Research from elsewhere in the world may be a starting point to understand how to effect change,

but it must be noted that although the problem of insufficient women in IT occurs in most countries worldwide, local, cultural and societal explanatory factors differ from country to country (Galpin, 2002). Solutions must, therefore, be assessed as to whether they are appropriate before they are applied.

CONCLUSION

As this article has shown, there is low participation by women as IT professionals in the generation and creation of ICTs. This is caused in part by lower literacy and education rates of women in sub-Saharan Africa compared to men, as well as by societal and other factors that influence the choices of women who do have the opportunities to study beyond the secondary level. As a result, few women are involved in design, development management or policy. The final outcome is that ICTs in sub-Saharan Africa cannot be as successful in their application as they could and should be. This may negatively affect the implementation of ICTs in sub-Saharan Africa for development. The way forward is to increase the participation of women from all backgrounds in IT careers.

REFERENCES

Arid Lands Information Network. (2002). *Technologies*. East Africa. Retrieved January 23, 2005, from http://www.alin.or.ke/data/technologies.htm

Bantebya-Kyomuhendo, G. (2004). *Bridging the gender Digital Divide through training at the Department of Women and Gender Studies, Makerere University*. World Bank Group. Retrieved January 29, 2005, from http://siteresources.worldbank.org/INTGENDER/Seminar-Series/20211708/GraceBantebya.ppt

Borg, A. (2002). Computing 2002: Democracy, education and the future. *ACM SIGCSE Bulletin, 34*(2), 13-14.

Bunyi, G. (2004). Gender disparities in higher education in Kenya: Nature, extent and the way forward. *The African Symposium, 4*(1). Retrieved January 21, 2005, from http://www2.ncsu.edu/ncsu/aern/gendaedu.htm

Bush, R. (1992). *FidoNet: Technology, use, tools, and history.* Retrieved January 23, 2005, from http://www.fidonet.org/inet92_Randy_Bush.txt

Esterhuysen, A. (2002). Networking for a purpose: African NGOs using ICT. In L. Levey & S. Young (Eds.), *Rowing upstream: Snapshots of pioneers of the information age in Africa* (pp. 3-20). Johannesburg, South Africa: Sharp Sharp Media. Available at http://www.sn.apc.org/Rowing_Upstream/

Galpin, V. (2002). Women in computing around the world. *ACM SIGCSE Bulletin, 34*(2), 94-100.

Gürer, D., & Camp, T. (2002). An ACM-W literature review on women in computing. *ACM SIGCSE Bulletin, 34*(2), 121-127.

Hafkin, N. (2002). Are ICTs gender neutral? A gender analysis of six case studies of multi-donor ICT projects. In *UN/INSTRAW virtual seminar series on gender and ICTs.* Retrieved January 21, 2005, from http://www.un-instraw.org/en/docs/gender_and_ict/Hafkin.pdf

Hafkin, N. (2003). *Some thoughts on gender and telecommunications/ICT statistics and indicators.* International Telecommunication Union. Retrieved June 22, 2005, from http://www.itu.int/ITU-D/ict/wict02/doc/pdf/Doc46_Erev1.pdf

Hafkin, N., & Taggart, N. (2001). *Gender, information technology, and developing countries: An analytic study.* LearnLink/Office of Women in Development, USAID. Retrieved January 21, 2005, from http://learnlink.aed.org/Publications/Gender_Book/Home.htm

Hoffman-Barthes, A., Nair, S., & Malpede, D. (1999). *Scientific, technical and vocational education of girls in Africa: Summary of 21 national reports.* UNESCO working document. Retrieved December 2001 from http://unesdoc.unesco.org/images/0011/001180/118078eo.pdf

Holmes, R. (2004). *Advancing rural women's empowerment: Information and communication technologies (ICTs) in the service of good governance, democratic practice and development for rural women in Africa.* Women'sNet. Retrieved January 21, 2005, from http://womensnet.org.za/dimitra_conference/Empowering_Rural_Women.doc

Huyer, S. (1997). *Supporting women's use of information technologies for sustainable development.* IDRC/Acacia. Retrieved January 21, 2005, from http://web.idrc.ca/en/ev-10939-201-1-DO_TOPIC.html

International Telecommunication Union. (2003). *Female telecom staff, 2001.* Retrieved January 29, 2005, from http://www.itu.int/ITU-D/ict/statistics/at_glance/f_staff.html

Jensen, M. (1996). *A guide to improving Internet access in Africa with wireless technologies.* IDRC/Acacia. Retrieved January 23, 2005, from http://web.idrc.ca/en/ev-11190-201-1-DO_TOPIC.html

Kagai, B., & Kimolo, N. (2004). FOSS usage in Africa: Untapped potential. *i4d/Information for Development, II*(9). Retrieved February 1, 2005, from http://www.i4donline.net/oct04/untapped_full.asp

Karelse, C.-M., & Sylla, F. (2000). Rethinking education for the production, use, and management of ICTs. In E. Rathgeber & E. Adera (Eds.), *Gender and the information revolution in Africa* (Chap. 5). IDRC. Retrieved January 22, 2005, from http://web.idrc.ca/en/ev-9409-201-1-DO_TOPIC.html

Kole, E. (2003). Digital divide or information revolution? *i4d/Information for development.* Retrieved January 21, 2005, from http://www.i4donline.net/issue/sept-oct2003/digital_full.htm

Lazowska, E. (2002). Pale and male: 19th century design in a 21st century world. *ACM SIGCSE Bulletin, 34*(2), 11-12.

Levey, L., & Young, S. (Eds.). (2002). *Rowing upstream: Snapshots of pioneers of the Information Age in Africa.* Johannesburg: Sharp Sharp Media. Available at http://www.sn.apc.org/Rowing_Upstream/

Marcelle, G. (2000). Getting gender into African ICT policy: A strategic view. In E. Rathgeber & E. Adera (Eds.), *Gender and the information revolution in Africa* (Chap. 3). IDRC. Retrieved January

22, 2005, from http://web.idrc.ca/en/ev-9409-201-1-DO_TOPIC.html

Marcelle, G. (2001). Creating an African women's cyberspace. *The Southern African Journal of Information and Communication, 1*(1). Retrieved January 31, 2005, from http://link.wits.ac.za/journal/j-01-gm.htm

Mariro, A. (1999). *Access of girls and women to scientific, technical and vocational education in Africa*. Dakar, Senegal: UNESCO.

Marshall, W. (2002). Radio e-mail in West Africa. *Linux Journal, 2002*(103), 2.

Mbarika, V., Jensen, M., & Meso, P. (2002). Cyberspace across sub-Saharan Africa. *Communications of the ACM, 45*(12), 17-21.

Mijumbi, R. (2002). A case on the use of a locally-developed CD-ROM by rural women in Uganda. In *UN/INSTRAW virtual seminar series on gender and ICTs*. Retrieved January 21, 2005, from http://www.un-instraw.org/en/docs/gender_and_ict/Mijumbi_summary.pdf

Morna, C., & Khan, Z. (2000). *Net gains: African women take stock of information and communication technologies*. APC-Africa-Women/FEMNET. Retrieved January 21, 2005, from http://www.apcafricawomen.org/netgains.htm

Odedra-Straub, M. (1995). Women and information technology in sub-Saharan Africa. In S. Mitter & S. Rowbotham (Eds.), *Women encounter technology: Changing patterns of employment in the third world* (Chap. 12). Routledge/UNU. Retrieved January 21, 2005, from http://www.unu.edu/unupress/unupbooks/uu37we/uu37we00.htm

Pacific Institute for Women's Health. (2002). *Women connect! The power of communications to improve women's lives*. Retrieved January 21, 2005, from http://www.piwh.org/pdfs/wc2002.pdf

South African Information Technology Industry Strategy. (2000). *SAITIS baseline studies: A survey of the IT industry and related jobs and skills in South Africa*. Retrieved January 29, 2005, from http://www.dti.gov.za/saitis/studies/index.html

South African Qualifications Authority. (2004). *Trends in public higher education in South Africa 1991 to 2001*. Pretoria, South Africa: South African Qualifications Authority.

South African Reference Group on Women in Science and Technology. (2004). *Women's participation in science, engineering and technology*. Department of Science and Technology. Retrieved June 22, 2005, from http://www.sarg.org.za/docs/pdf/womens_part_set_2004.pdf

United Nations Development Programme. (2004). *Human development report 2004*. Retrieved January 19, 2005, from http://hdr.undp.org/reports/global/2004/

Waag Society. (2003). *Manifesto on the role of Open Source Software for development cooperation*. The Hague. Retrieved June 28, 2005, from http://sarai.waag.org/display.php?id=28

Women'sNet/Dimitra. (2004). *Report back on the e-consultation: ICTs for the advancement of rural women's empowerment*. Retrieved January 23, 2005, from http://womensnet.org.za/dimitra_conference/e-conference_reportback.doc

KEY TERMS

Digital Satellite Broadcasting: A means of using satellites to broadcast information to computers via radio using a radio interface card (ALIN, 2002). This technology can be used when there is a lack of telecommunications infrastructure.

FidoNet: A point-to-point and store-and-forward e-mail technology using telephone modems as a means of communicating between computers (Bush, 1992). It played an important role in the start of computer communication in the late 1980s and early 1990s in Africa, with usage predominantly by NGOs and universities (Levey & Young, 2002).

Free/Libre/Open Source Software (FLOSS): Software whose source code is available and, hence, different from proprietary software. FLOSS (also called FOSS) "may be used, copied, and distributed with or without modifications, and ... may be offered either with or without a fee" (Waag Society, 2003, para. 3).

High Frequency (HF) Radio E-Mail: A technology for transmitting e-mail using radio modems and a store-and-forward approach (Marshall, 2002), with similarities to FidoNet. It is an appropriate technology when telephone modems cannot be used because of lack of telecommunications infrastructure. Using HF radio allows transmission over long distances but with low bandwidth, hence the necessity to use store-and-forward applications such as e-mail.

Information and Communication Technologies (ICTs): Electronic means of communicating and conveying information, covering media such as radio and television, computer and computer-networking technology and telecommunications.

Information Technology (IT): A term that can be used broadly to cover all computing disciplines, including computer science, IT and software engineering. Computing can also be used as a synonym for IT.

Non-Governmental Organizations (NGOs): An NGO is a non-profit organization focused on particular issues for the public good. It may provide services, lobby government or perform monitoring.

Sub-Saharan Africa: The area of Africa south of the Sahara. In terms of the United Nations definition, this covers all countries on the African continent excluding Algeria, Djibouti, Egypt, Libyan Arab Jamahiriya, Morocco, Somalia, Sudan, and Tunisia, and covers the island states of Comoros, Madagascar, Mauritius, São Tomé and Principe, and Seychelles (UNDP, 2004). All countries in sub-Saharan Africa are developing countries, and 31 are classified as Least Developed Countries by the United Nations because of their very low GDP per capita, low levels of health and education, and economic vulnerability.

Women in the Free/Libre Open Source Software Development

Yuwei Lin
Vrije Universiteit Amsterdam, The Netherlands

INTRODUCTION

Free/libre open source software (FLOSS) has become a prominent phenomenon in the ICT field and the wider public domain for the past years. However, according to a FLOSS survey on FLOSS developers in 2002, "women do not play a role in the [FLOSS] development; only 1.1% of the FLOSS sample is female." (Ghosh, Glott, Krieger, & Robles, 2002). In the mainstream research on FLOSS communities, many researchers also overlook different processes of community-building and diverse experiences of members, and presume a stereotyped male-dominated "hacker community" (e.g., Levy, 1984; Raymond, 2001; Himanen, 2001; Thomas, 2002). Moreover, issues around gender inequality are often ignored and/or muted in the pile of FLOSS studies. Female programmers often are rejected ex/implicitly from the software labour market (Levesque & Wilson 2004). The requirements of female users are not respected and consulted either (European Commission, 2001). This feature is opposite to the FLOSS ideal world where users should be equally treated and embraced (op. cit.). While many researchers endeavour to understand the FLOSS development, few found a gender-biased situation problematic. In short, women are almost invisible in current FLOSS-related literature. Most policies targeting at advocating FLOSS are also gender blind.

Thus, this essay highlights the need for increased action to address imbalances between women's and men's access to and participation in the FLOSS development in cultural (e.g., chauvinistic and/or gender-biased languages in discussions on mailing lists or in documentations), economic (e.g., unequal salary levels for women and men), political (e.g., male-dominated advocacy environment) and technical (e.g., unbalanced students gender in technical tutorials) spheres. On the other hand, it also emphasises the powerful potential of FLOSS as a vehicle for advancing gender equality in software expertise. FLOSS helps transport knowledge and experience of software engineering through distributing source code together with the binary code almost without any limit. Many FLOSS licences such as the General Public Licence (GPL) also facilitates the flow of information and knowledge. In other words, if appropriately harnessed, FLOSS stands to meaningfully contribute to and mutually reinforce the advancement of effective, more expedited solutions to bridging the gender digital divide.

In the end, this article points out that while women in more advanced countries have a better chance of upgrading their ICT skills and knowledge through participating in the FLOSS development, the opportunity is less available for women in the developing world. It is worth noting that although the gender issues raised in this article are widespread, they should not be considered as universally indifferent. Regional specificities in gender agenda in software engineering should be addressed distinctly (UNDP/UNIFEM, 2004).

TOWARD A FEMINIST ANALYTICS[1] ON THE GENDER ISSUES IN THE FLOSS DEVELOPMENT

To a degree, the gender problems in the FLOSS development can be seen as an extension of the ongoing gender issues in new-tech service industries and/or software industry (e.g., Mitter & Rowbotham, 1995). These long-term problems mainly include low-level work content, unequal payment, emotional distress from discrimination and prejudice, physical ache from the long working hour in front of the computer, division of labour within the home (child-rearing), essentialist notions of women's roles, sexism, informal networks, prejudice, lack of role models and support, and "glass ceilings." Generally speaking, women within the software industry have to work harder than men in order to get the same respect and conquer the glass-ceiling problem in this patriarchy world (DeBare, 1996).

Although FLOSS has dramatically changed the way software is produced, distributed, supported, and used, and has a visible social impact enabling a richer digital inclusion, most of the gender problems existing in the software industry have been duplicated in the FLOSS field.

A FLOSS *social world* (Lin, 2004) is different from what Turkle (1984) argues: "computer systems [mainly proprietary] represent a closed, controllable microworld—which appeals to more men than women" (Turkle, 1984). It requires a holistic perspective to capture the complexity and dynamics within and across the social world. While the heterogeneity and the contingency in this social world are not yet fully explored, analysis from a feminist perspective is almost absent. Little attention has been paid to the internal differences and to the private arena linked with the FLOSS innovation system. However, this methodological lack has not stopped us from observing the gender problems within the field. Instead, by means of the FLOSS development, some gender problems in ICT become even more apparent.

Additionally, in a world of volunteers, we clearly see that men and a competitive worldview are more present in all forms of media. Many women participating in the FLOSS development are invisible: their labour in fields such as NGOs that help implement and promote FLOSS, documentation translation, book editing, teaching and tutoring (e.g., E-Riders[2]) are less visible than male-dominated coding work. Indeed, FLOSS advocates have not adequately addressed this critique of gender equality. They tend to treat the FLOSS community as a monolithic culture—to pay more attention to differences between and among groups than to differences within them. They are so eager uniting the voices on freedom of information that they give little or no recognition to the fact that FLOSS groups, "like the societies in which they exist (though to a greater or lesser extent), are themselves *gendered*, with substantial differences of power and advantage between men and women" (Okin, 1999).

A number of key dilemmas that hinder women's participation in the FLOSS development can be summarised:

1. **A Lack of "Mentors" and Role Models:** It is true that there is a very low percentage of female participants in the FLOSS social world. However, we should not overlook the importance and possible future of outstanding female figures in the FLOSS field. It is difficult to make the majority of male peers respect these female figures. I am not suggesting that men all look down on women, but it is more difficult for women to be assertive in front of male-dominated audience. The whole way the world is constructed means there are just men at every level, which makes it really hard for women to get their feet in the door. A way of overcoming this is to establish more female figures in the world. While few in the computer world actually know that Ms. Ada Byron is the first programmer in the world, how could we expect people to recognise women's ability?

2. **Discriminated Languages Online and/or Offline (e.g., Phrases in Documentaries):** Many female FLOSS developers have complained the highly unfriendly atmosphere within the social world, online (e.g., mailing lists, IRC) and/or offline (e.g., documentation). For instance, referred to prospective readers, existed FLOSS documentation usually use single sex term, he, rather than she or they. This kind of gender-biased words subtly exclude women from participating in the FLOSS development. While the online languages are in a direct way full of men's jargon, reading the documentation offline does not make a female developer/user feel more included in the field. If women need to be encouraged to participate in FLOSS-related discussions, a sexist or discriminative surrounding is definitely not attractive.

3. **A Lack of Women-Centred View in the FLOSS Development:** The consequence of the lack of female FLOSS developers is that there is a greater amount of female-unfriendly software in the FLOSS system. Some scholars in science and technology studies (STS) have pointed out that technologies are gendered both in their design and use (e.g., Edwards, 1993;

Wajcman, 2004).The social relations of gender within and across the FLOSS social world are reflected in and shaped by the design of FLOSS. And such a lack of women's perspective on software design and use restricts women's participation in the FLOSS development and, in turn, forms the stereotyped fact that women are almost absent in the FLOSS development because they are less adequate in programming or less likely to be advanced computer users. This absence of female developers would also be a loss of the FLOSS development, and result in inequalities in an ICT-based society as a threat to social cohesion and social order.

4. **A Male-Dominated Competitive Worldview:** "[The FLOSS market] is literally a war for the best and brightest. If we don't get there, somebody else will" (Andrew Clark, director of strategy and market intelligence for the venture capital group at IBM—interviewed with C|Net.com on February 14, 2005).

As Arun and Arun (2001) point out, "The project-based, competitive nature of software development reproduces a masculine culture, which further interacts with the different career patterns of women and social norms and tends to disadvantage women." While languages in a similar tone with Clark's words above repeatedly turn up in the mass media such as advertisements from big computer companies, the male-led competitive worldview is continuously represented and reinforced in the society. Since there are fairly clear disparities of power between the sexes within the FLOSS social world, a gender-imbalanced world is ensued. The more powerful male members are those who are generally in a position to determine and articulate the group's beliefs, practices, and interests. Although not all proposals associated with FLOSS are potentially antifeminist under such conditions, but they somehow duplicate and forward the view that might limit the capacities of women and girls to freely choose lives that they would like to live. It is very alarming that a large amount of perspectives and purposes regarding the FLOSS development is determined by white men. This imbalance might give a distorted world view; it is much better to have views from all people from different social worlds.

5. **No Sympathy from Women Peers:** There are many more spoken or unspoken problems for women to take part in the FLOSS development (e.g., Henson, 2002; Spertus, 1991). However, facing these gender inequalities, many women remain remote and feel no need of tackling these problems. While some women-centred online groups have networked together to address the gender issue in the FLOSS movement, many female programmers still do not share the same view on an ongoing and enlarging gap between men and women software developers. While gender issue in FLOSS is not addressed in most of the literature and also not recognised by female peers, it is difficult to network women to tackle the coherent patriarchal hegemony in the computer world.

HOW CAN FLOSS EMPOWER WOMEN?

Like many other ICTs, FLOSS carries the powerful potential as a vehicle for advancing social equality. It opens up an opportunity for women to learn how to communicate and interact with software designers and speak out what kind of software they want (e.g., file bug report, join the user group and online forum etc.), to have access to source code and fork the software (e.g., to have a female-friendly version of the software), if they are interested and competent.

There are three main ends in current "women movement in the FLOSS community": (1) providing women-friendly software and services; (2) creating a women-friendly environment for developing and using FLOSS; and (3) fostering a gender-balanced ICT innovation system for both competition and collaboration. These three points have close connections with one another. In order to create software that engage and build on women's ideas and visions, we need to create a more women-friendly environment for the purpose of attracting more women to participate in the FLOSS development. Encompassing such a women-centred view of design, which usually resembles a more sympathetic and inclusive way of doing, will possibly foster a gender-balanced ICT innovation system that is not only friendly to women but also to various minorities in our society. This system, unlike the current one based on a highly competition-oriented approach, will draw on aptitudes and competences of diverse actors in the FLOSS social world so as to develop a holistic environment which is based on a collaboration-oriented approach.

Networking is important in democratising the access and dissemination of knowledge and establishing a base for a citizenship defined by gender equity. In order to encourage women's participation and also to explain the operation of FLOSS to women, some female developers/users have started to network and form online groups such as, LinuxChix[3], KDE Women[4], Gnurias[5], GenderChangers[6], and Debian-women[7]. They act to dispel the unfriendly wording in documents and in online peer groups, to report this kind of sexist bug reports to other developers, to give online tutorials. These networking and gathering, online or offline, would serve as a base for gender inclusion.

CONCLUSION AND FUTURE RESEARCH

The essay aims to identify the current challenges of gender politics and help formulate strategies and recommendations in order to advance and to empower women in FLOSS. It is anticipated that through conceptualising and documenting the current gender issues in the FLOSS development, it will help enlarge the knowledge base for gender-sensitive policies on ICTs, and propose a women-centred policy towards developing and implementing FLOSS. While FLOSS denotes a new milestone for software development and knowledge making in a broad sense that might alter the social relations of gender, "in this technoscientific advanced era, feminist politics make wider differences in women-machine relationship than the technologies themselves" (Wajcman, 2004). As such, a gender-sensitive agenda for developing FLOSS is urgently needed.

In terms of future research, in order to get a comprehensive overview of the current gender digital divide in the FLOSS social world, more research, both qualitative and quantitative, needs to be conducted. The former would allow us to understand women's experiences and needs better through ethnographical observation, interviews, and focus groups, while the latter would give a fuller picture of general gender problems. Researchers across disciplines are encouraged to analyse FLOSS activities more critically with regard to gender, and to develop conceptual frameworks and methodologies for better understanding and analysing the relationship between FLOSS and gender. Additionally, in encouraging the FLOSS development, governments and organisations should pay extra attention to gender-related issues as well, and take initiatives to include women in the FLOSS development. Holding training workshops for female developers might be a feasible way of bridging the gender digital divide in the FLOSS social world. Other efforts such as design of products and Web sites for women and girls, supporting networks for female professions in FLOSS shall be encouraged.

However, in speaking of implementing and developing FLOSS, most of the cases are centralised or situated in more developed countries. One should bear in mind that there are many undocumented activities that have happened in the developing world. When strengthening the advantages of FLOSS, we should not overlook many problems emerging from implementing FLOSS in developing countries, such as a lack of sufficient training and support. The digital divide shall be considered as a symptom of inequality, not the cause of it. There is a need of understanding what local people really need: water, food, jobs, decent healthcare and sanitation, or software and ICT infrastructure. The gender issue of ICT might be more complex than we thought as well. Female participants very often suffer from hybrid discriminations, both from the male-dominated FLOSS world and the socio-cultural patriarchy in the society. Although virtual groups such Linuxchix Brazil[8] and Linuxchix Africa[9] have started providing women with help on solving problems in implementing Linux, more efforts need to be spent on documenting, analysing and deconstructing the patriarchal hegemony embedded in the whole ICT infrastructure. As such, like many other fields concerned with gendering, this essay is a mere beginning of a feminist accounts about the FLOSS development—an analytic stage on which "we need to place the details contributed by ethnographic research, cultural critiques, sociological surveys, legal scholarship on men and women in their many specific conditions and subjectivities" (Sassen, 1999, p. 2).

REFERENCES

Arun, S., & Arun, T. G. (2001). Gender at work within the software industry: An Indian perspective. *Journal of Women and Minorities in science and engineering, 7*(3), 42-58.

DeBare, I. (1996, January 21). Women in computing: Logged on or left out? [Special report]. *Sacramento Bee.*

Edwards, P. (1993). *Gender and the cultural construction of computing.* Adapted from "From 'impact' to social process: Case studies of computers in politics, society, and culture (Chapter IV-A)," in Handbook of Science and Technology Studies. Beverly Hills, CA: Sage Press.

European Commission. (2001). *Public report on the consultation meeting on European perspectives for open source software.* Retrieved from ftp://ftp.cordis.lu/pub/ist/docs/ka4/tesss-OSS-report.pdf

Ghosh, R. A., Glott, R., Krieger, B., & Robles, G. (2002). *Free/Libre and open source software: Survey and study* (Deliverable D18: Final Report, Part IV: Survey of Developers). International Institute of Infonomics, University of Maastricht and Berlecon Research GmbH. Retrieved original version of this document from http://www.infonomics.nl/FLOSS/report/

Henson, V. (2002). *How to encourage women in Linux.* Retrieved from http://www.tldp.org/HOWTO/Encourage-Women-Linux-HOWTO/index.html

Himanen, P. (2001). *The hacker ethic and the spirit of the information age.* London: Secker & Warburg.

Levesque M., & Wilson, G. (2004). Women in software: Open source, cold shoulder. *Software Development.* Retrieved February 20, 2005, from http://www.sdmagazine.com/documents/s=9411/sdm0411b/sdm0411b.html?temp=TgtgS9YUY8

Levy, S. (1984). *Hackers: Heroes of the computer revolution.* Garden City, NY: Anchor Press/Doubleday.

Lin, Y. W. (2004). *Hacking practices and software development: A social worlds analysis of ICT innovation and the role of open source software.* PhD thesis, Department of Sociology, University of York, UK.

Mitter S., & Rowbotham, S. (1995). *Women encounter technology: Changing patterns of employment in the third world.* London: Routledge and the United Nations University.

Okin, S. M. (1999) *Is multiculturalism bad for women?* Retrieved from http://www.bostonreview.net/BR22.5/okin.html

Raymond, E. S. (2001). *How to become a hacker.* Retrieved June 23, 2005, from http://www.catb.org/~esr/faqs/hacker-howto.html

Sassen, S. (1999). *Blind spots: Towards a feminist analytics of today's global economy.* Retrieved from http://www.uwm.edu/Dept/IGS/presentation/sassen.pdf

Spertus, E. (1991). *Why are there so few female computer scientists?* (MIT Artificial Intelligence Laboratory Technical Report 1315). Retrieved February 20, 2005, from http://www.mills.edu/ACAD_INFO/MCS/SPERTUS/Gender/pap/pap.html

Thomas, D. (2002). *Hacker culture.* Minneapolis, MN: University of Minnesota Press.

Turkle, S. (1984). *The second self: Computers and the human spirit.* New York: Simon and Schuster.

UNDP Bratislava Regional Center and UNIFEM Central and Eastern Europe. (2004). *Bridging the gender digital divide: A report on gender and information communication technologies (ICT) in Central and Eastern Europe and the commonwealth of independent states (CIS).* UNDP/UNIFEM.

Wajcman, J. (2004). *TechnoFeminism.* Cambridge, UK: Polity Press.

KEY TERMS

Debian GNU/Linux and Debian-Women:
Created by the *Debian Project*, is a widely used free software distribution developed through the collaboration of volunteers from around the world. Since its inception, the released system, The Debian Women project, founded in May 2004, seeks to balance and diversify the Debian Project by actively engaging with interested women and encouraging them to become more involved with Debian.

Ethnography: Refers to the qualitative description of human social phenomena, based on months or years of fieldwork. Ethnography may be "holistic," describing a society as a whole, or it may focus on specific problems or situations within a larger social scene.

FLOSS: Free/libre open source software (FLOSS), generically indicates non-proprietary software that allows users to have freedom to run, copy, distribute, study, change and improve the software.

Focus Group: A focus group is a form of qualitative research in which a group of people are asked about their attitude towards a product, concept, advertisement, idea, or packaging. Questions are asked in an interactive group setting where participants are free to talk with other group members.

GPL: General Public License (GPL) is a free software licence that guarantees the freedom of users to share and change free software. It has been the most popular free software license since its creation in 1991 by Richard Stallman.

Hegemony: Is the dominance of one group over other groups, with or without the threat of force, to the extent that, for instance, the dominant party can dictate the terms of trade to its advantage; or more broadly, that cultural perspectives become skewed to favor the dominant group.

KDE(K Desktop Environment): A free desktop environment and development platform built with Trolltech's Qt toolkit. It runs on most Unix® and Unix®-like systems, such as Linux, BSD, and Solaris.

ENDNOTES

1   By "feminism" I mean the belief that women should not be disadvantaged by their sex, that the moral equality of men and women should be endorsed, and that all forms of oppression should be demolished.
2   http://www.eriders.org
3   http://www.linuxchix.org/
4   http://women.kde.org/
5   http://www.gnurias.org.br/
6   http://www.genderchangers.org/
7   http://women.alioth.debian.org/
8   http://www.linuxchix.org.br
9   http://www.africalinuxchix.org

Women Returners in the UK IT Industry

Niki Panteli
University of Bath, UK

Despina Cochliou
University of Bath, UK

Evangelia Baralou
University of Stirling, UK

INTRODUCTION

IT is a sector that incorporates the newest industries, consisting mainly of young firms and relatively freshly constituted forms of working practices. Despite this, several studies exist to-date that show that opportunities are limited for those women who aspire to have a career in IT. Recent research in the UK has revealed that between 1999 and 2003 the proportion of women in the UK IT workforce fell by almost 50%, from 21% to 12.5%, following steady growth (Platman & Taylor, 2004). The focus of the article is to examine a specific group of female IT staff: women returners. The work presented here explores the factors that often constrain women returners to the IT industry and discusses the findings in relation to the characteristics of the industry; it is part of a bigger study that looks at advancing women in high-tech industries[1]. If women are not found in positions of influence in the IT industry, one of the most growing industries, then what image is being given to prospective students, their parents and careers advisers? What influence will women have on the future developments within the discipline, hence on the industry itself?

BACKGROUND: WOMEN RETURNERS IN IT

IT has been a fast growing industry with an impact on most organizations, large and small, traditional and emergent and vast employment opportunities. Indeed, computing work is characterised by growth in demand but it is also simultaneously characterised by obsolescence of skills (Wright & Jacobs, 1995). That means that computer workers have to go through extensive training most of the time in their career if they want to stay up-to-date with the frequent changes in software, hardware and programming. Also the sector, although a comparatively new one, has been characterised as predominantly white, middle-class and male-dominated (Panteli, Stack, & Ramsay, 1999b). This indicates that men hold high profile posts such as developers and managers of systems whereas women are more likely to be seen as the users of those systems. This phenomenon has also been illustrated by Platman and Taylor (2004) in their recent report. They clearly stated that the UK IT industry is male dominated and full-time, which designates that there are substantial obstacles for women working in IT.

Accordingly, despite the industry growth and increasing job opportunities, several research studies exist to-date that draw attention to the gender inequality in IT employment (Panteli, Stack, Atkinson, & Ramsay, 1999a; Panteli et al., 1999b; Roldan, Soe, & Yakura, 2004). For example, women in IT are increasingly concentrated in areas of work that are low in status, power, and rewards. As women move up the career hierarchy, their representation shrinks, thus the proportion of women in this high tech sector remains underrepresented in top management posts and in key technical jobs (Panteli et al., 1999a; Roldan et al., 2004).

For the purpose of this chapter, we take a focus on a particular group of female employees: women who seek re-entry following a career break, thus women returners. Most of the academic studies on women returners have focused on women's choices and career orientations (Doorewaard, Hendrickx, & Verschuren, 2004) as well as on trying to identify either the demands of women for returning to work or their fears and their differences compared to men (Healy, 2004; Healy & Kraithman, 1991; Shaw, Taylor, & Harris, 1999). A major difference for example between men and women is that women shape their working lives around the competing domestic demands, (e.g., childbirth, household etc.). They link and adjust their work with the different phases their life is going through.

A career break is a period of time where an employee is not working for very specific reasons (Institute of Physics, 2004). The length of career break may vary according to the needs it attempts to cover; for example a maternity leave can be from 26 weeks and over, and it is often the main reason for taking a career break (Rothwell, 1980). Another reason for a career break includes the need to study (e.g., for a further qualification or to follow one's partner on sabbatical leave or on a foreign assignment) (Warrior, 1997). Research has also identified a pattern for women's re-entry to the labour force. According to Rothwell (1980), this pattern includes an "in-and-out" period while children are young, followed by a part-time period, and eventually a return to full-time employment when the children grow up. Rothwell (1980) explained that this pattern depends on women's job ambitions, on family's financial conditions and on the local labour market.

Another barrier on women's return to work is the lack of affordable, accessible, quality childcare. Paul, Taylor, and Duncan (2002) found that 18% of mothers of pre-school children who are working part-time reported that they are prevented from working longer hours by having to look after children, compared with 25% for part-time mothers with school, but not pre-school, children. This proportion according to the authors shows that they could work longer if suitable childcare was available. Other studies have showed that many women with care-taking responsibilities at home tend to accept relatively low-skilled part-time jobs (Doorewaard et al., 2004; Houston & Marks, 2003).

Shackleton and Simm (1998) argued that women returners require training in both "hard" and "soft" skills. Generally, such training takes a focus on the updating of existing skills, confidence building, and training in new technology. Work experience may also contribute towards raising the self-esteem levels of those wishing to reenter the labour market, who often feel that they lack the necessary work-related skills to compete. In particular, in the field of IT the rapid change in technological advancements and IT applications in the recent years has affected a great area in an organization's life including personnel requirements and training. Up-to-date training appears to be an imperative for women returners in particular in coordination with more management support so they would be able to return to their previous job.

EMPIRICAL STUDY

The study presented here, which is part of a bigger project on advancing women in the high-tech industries (i.e., ITEC), was carried out using qualitative information based on in-depth interviews with women returners. These were undertaken during the period December 2004 and January 2005. In particular, women who have returned to work after a career break, are currently on a career break, or are planning one, have been invited to participate in our study. The majority of the interviews (84%) were conducted over the telephone due to the geographical dispersion of the interviewees whilst the rest were face-to-face interviews.

Though the interview questions were pre-determined to ensure that the necessary information was collected, these remained open-ended in an attempt to record the views and perceptions of the respondents. This allowed questions and issues that were revealed during the interview to be explored further. Interviews lasted from 30 to 40 minutes and were tape-recorded when the interviewees granted permission. The interview data was transcribed selectively. The analytical approach adopted was exploratory as the aim of this report is to explore the main issues around women's employment in the IT sector and in particular women returners. A simple way to explore data at this stage of the project is to recast it in a way that counts the frequency (i.e., the number of times that certain things happen), or to find ways of displaying that information.

RESULTS

Our interviewees were categorized into three groups based on their employment status:

- Women currently on a career break (WCB)— 16.7% of the participants
- Women who have had a career break but did not return to IT (WnR)—16.7%
- Women who have had a career break and returned either to their previous job or to another job within the ITEC sector (WR)—66.6%

The main difference between the WnR and WR groups was in the duration of their career break. In the first case, the career break was 1-15 years whereas in the second case their break lasted only 1-2 years. For the women belonging to the WnR group the long time they spent upbringing their family has clearly negatively affected their return to the industry.

In brief, the study has shown that the primary reason for a career break was maternity leave. This has implications on their needs as it will be discussed below, when they decide to return to employment and resume their career. The participants in the study were also found to hold a range of qualifications, from vocational training to PhD. It is very interesting to mention the case of women who have not returned to IT (WnR); although these mainly have high-level academic qualifications they have not been able to find a job in accordance with their qualifications. Furthermore, despite the fact that women in the WnR group used to have highly responsible jobs, they are found in jobs of low status and low payment outside the industry.

Moreover, the women who are currently on a career break have shown a preference for part-time jobs despite their previous successful full-time career, whilst the group of women who have not returned to IT and consequently unable to resume their career, would like to find a full-time job or a "proper" job as characteristically one of them stated.

In their UK-based study, Platman and Taylor (2004) found that the lack of part-time work available for women in the IT profession is one of the main reasons why they are discouraged from the profession. The report also showed that the part-time workers in the UK, at 5.3% are lower than the proportion in Germany (8.6%) and the Netherlands (16%). In particular, only the 16.9% of female IT professionals work part-time in the UK, compared with 37% in the Netherlands and 19% in Germany.

The lack of part-time schemes is not the only hurdle however for women in the UK IT industry. As it was argued in Panteli, Stack, and Ramsey (2001), "the offer of part-time work is also likely to mean a sharp reduction in opportunities for advancement. At the least these factors will delay women's advance; or they may permanently damage their prospects in cultures dominated by masculine conceptions of appropriate career paths and staging" (p. 10). It appears therefore that whereas the availability of a part-time scheme may work to the detriment of women and their career development, the lack of a part-time scheme is also a serious hurdle to women's retention in computing work following a career break.

In addition to the above, the current research study has identified the major factors that constrain women to return to IT after a career break and resume their career. Women who are currently on a career break or have returned to work within only 2 years of their career break are concerned about childcare arrangements and have a preference for flexi-time and part-time schemes, than women who have been on a career break until their children have grown up. These women can more easily return to full time employment, are more in need of re-training, and clearly ask for changes in employers' attitudes. Interestingly, however, all groups including those women who have only stayed for a limited time away from the industry indicated that they experienced loss of confidence during the re-entry process. Women in the WR group explained that his was a result of reasons such as changing from full-time to part-time employment, the industry changes very quickly as well as that they might have stayed long enough out of work which makes them feel less competent.

Overall, the rapid changes that the sector experiences contributes to skills obsolesces and loss of confidence and unless opportunities are available for retraining women either do not return to IT or return to lower-status jobs with low pay. Further obstacles include the inflexibility of the sector to offer part-time jobs and other flexible working patterns.

FUTURE TRENDS

In the light of the literature review and empirical findings, one can tentatively conclude that women in this high-tech field have to overcome more constraints than women employees in any other form of industry. It is also evident that the number of women remaining in the IT sector is decreasing often due to the unfavourable working conditions and consequently the limited career prospects. The image of the industry consists of a competitive, male-dominated along with the fact that this is a fast-growing and continuously changing field; these will remain the main barriers for women's recruitment, retention and advancement in IT.

In this study we focused on women's position in the particular field of IT and more generally on what hinders their return to work after they have had a career break. However, there is little progress connecting these outcomes with the factors that enable women to return to work by studying and comparing women's perceptions on the topic. Future research will seek to provide an opportunity to develop a more carefully grounded link between what deters and what encourages women to return according to their insights and experiences. In addition, further research and analysis is considered necessary in order to evaluate the schemes that already exist to support women returners as well as to recommend additional practices and actions to enhance the current conditions.

CONCLUSION

In this article, we examined the career prospects for women in the UK who seek re-entry to IT following a career break. The chapter is primarily based on women's own views and experiences. Using empirical data, we found evidence of barriers and limited opportunities for those women who want to return to their IT job after a career break. The study has shown that women returners are not a homogeneous group and that the length of their career break appears to play a key role in the re-entry process. Overall, career breaks appear to counteract career development due to the lack of support mechanisms such as flexible working, part-time working, and insufficient training.

It is well documented that the IT sector has so-far failed to show the appropriate commitment towards gender inclusion. Lack of well-established gender strategies within the sector do not allow the implementation of policies favourable to women's employment such as part-time working. Our study on women returners in IT reinforces these views.

REFERENCES

Doorewaard, H., Hendrickx, J., & Verschuren, P. (2004), Work orientations of female returners. *Work, Employment, and Society, 18*(1) 7-27.

Healy, G. (2004). Work-life balance and family friendly policies—In whose interest? *Work, Employment, and Society, 18*(1), 219-223.

Healy, G., & Kraithman, D. (1991). The other side of the equation—The demands of women on re-entering the labour market. *Employee Relations, 13*(3), 17-28.

Houston, D. M., & Marks, G. (2003). The role of planning and workplace support in returning to work after maternity leave. *British Journal of Industrial Relations, 41*(2), 197-214.

Institute of Physics. (2004). *Career breaks.* London: Institute of Physics, Daphne Jackson Trust.

Panteli, A., Stack, J., Atkinson, M., & Ramsay, H. (1999a). The status of women in the UK IT industry: An empirical study. *European Journal of Information Systems, 8*, 170-182.

Panteli, A., Stack, J., & Ramsay, H. (1999b). Gender and professional ethics in IT industry. *Journal of Business Ethics, 22*, 51-61.

Panteli, N., Stack, J., & Ramsay, H. (2001, March). Gendered patterns in computing work in the late 1990s. *New Technology Work, and Employment, 16*(1), 3-17.

Paul, G., Taylor, J., & Duncan, A. (2002). *Mother's employment and childcare use in Britain.* London: The Institute for Fiscal Studies.

Platman, K., & Taylor, P. (2004). *Workforce ageing in the new economy: A comparative study of information technology employment*. A European summary report focusing on the United Kingdom, Germany, Netherlands, University of Cambridge, Cambridge (p. 20).

Roldan, M., Soe, L., & Yakura, E. K. (2004). Perceptions of chilly IT organizational contexts and their effect on the retention and promotion of women in IT. *Communications of the ACM*, 108-113.

Rothwell, S. (1980). United Kingdom. In A. M. Yohalem (Ed.), *Women returning to work: Policies and progress in five countries*. London: Frances Pinter Publishers.

Shackleton, R., & Simm, C. (1998). *Women returners in the south west region: A report to Somerset TEC on behalf of South West TEC*. Warwick: University of Warwick: Institute for Employment Research.

Shaw, S., Taylor, M., & Harris, I. (1999). Jobs for the girls: A study of the careers of professional women returners following participation in a European funded updating programme. *International Journal of Manpower, 20*(3/4), 179-188.

Warrior, J. (1997). *Cracking it! Helping women to succeed in science, engineering, and technology*. Watford: The Engineering Council.

Wright, R., & Jacobs, J. A. (1995) Male flight from computer work: A new look at occupational resegragation and ghettoization. In J. A. Jacobs (Ed.), *Gender inequality at work*. Thousand Oaks, CA: Sage.

KEY TERMS

Career Break: A career break is a period of time where an employee is not working for very specific reasons.

ITEC: Information technology, electronics and communication.

Women Returners: Women who seek re-entry to the labor force following a career break.

ENDNOTE

[1] The research is funded by the DTI (Department of Trade and Industry) ITEC Skills Unit and is part of the Development Partnership funded in part by the European Social Fund under the Equal Community Initiative Programme.

Women, Hi-Tech, and the Family-Career Conflict

Orit Hazzan
Technion – Israel Institute of Technology, Israel

Dalit Levy
UC Berkeley School of Education and The Concord Consortium, USA

INTRODUCTION

This article focuses on female software engineers in the Israeli hi-tech industry. We describe findings of our research that was based on in-depth interviews with 17 female software engineers from four organizations of different kinds and sizes.

BACKGROUND

Software is developed by human beings, usually working in teams. Indeed, the professional literature addresses software teamwork, roles in software teams, and other related topics. For example, van Vliet (2000) discusses general principles for team organization, Hughes and Cotterell (2002) dedicate a full chapter of their book to people management and software team organization, and Humphrey (2000) describes a full software development process from a team perspective.

At the same time, however, our literature review indicates that the role of women in software teams has not been researched extensively. Moreover, although the underrepresentation of women in the IT field in general and as software engineers in particular is well documented (Camp, 1997, 2002; Hazzan & Levy, in press), only few studies deal specifically with women working in the software industry (Sosa, 2005; Turner, Bernt, & Pedora, 2002). Continuing this line, the present article focuses on women participation in software teams.

Women in the Information-Technology and Software Industry

The "shrinking pipeline" (Camp, 1997) describes a phenomenon related to women in computer science, according to which, in addition to the pipeline shrinking from high school to graduate school, the pipeline has been shrinking during the last 20 years at the bachelor's level. For example, in the United States, women currently receive less than 20% of all bachelor's degrees in computer science, compared with 37% in 1984. Camp argues that "[s]ince the number of women at the bachelor's level affects the number of women at levels higher in the pipeline and in the job market, these facts are of great concern" (p. 104). Following Camp's 1997 article, this topic received extensive attention (Camp, 2002; Margolis & Fisher, 2002).

Women in the Israeli Software Industry

Our article focuses on Israeli female software engineers. Thus, characteristics of the Israeli software industry are considered in our study as well. For this purpose, we will first briefly describe the Israeli hi-tech industry.[1]

Israel is a very small country with a population of about 7 million people. Still, at its hi-tech economic peak during the 1990s, Israel was one of the world's leading centers of technology start-ups and innovations. Despite its small population, Israel had at that time about 3,000 start-ups,[2] and it came in third (after the USA and Canada) on the list of countries with the highest number of companies listed on NASDAQ.

This blossoming is explained by two main factors. The first is the national security and military needs that led to the development of cutting-edge technologies. Since its establishment in 1948, Israel has been forced to invest huge budgets and efforts to maintain its military advantage in order to survive. In particular, designated army units exist that specialize in technological innovations. As it turns out, many of Israeli's hi-tech entrepreneurs started their careers in the Israeli army.

The second factor that explains the success of the Israeli hi-tech industry in the 1990s is the massive immigration wave of Russian engineers from the former Soviet Union during the 1990-to-2000 decade. This addition of engineers to the Israeli population led Israel to have the highest number of engineers per capita in the world.[3]

One comprehensive, quantitative research on women in the Israeli hi-tech fields was conducted by Frenkel and Izraeli (in press). Among other things, they found that, on average, incomes of mothers who work in the Israeli hi-tech industry are higher than those of women who also work in this industry but are not mothers, mothers in this industry are more satisfied with their work than their nonmother counterparts, and the price that mothers in this industry pay is expressed mainly by higher levels of stress and the lack of leisure time. However, they concluded that the family-career combination is more rewarding than each one of them alone. It should be emphasized that this research was a quantitative study. The women who participated in it were approached through the Internet, and it included only those women who were still working in the hi-tech industry.

Our research findings, presented in the continuation of the article, show that this picture might be correct for those mothers who are still in the hi-tech industry (i.e., did not leave the hi-tech industry after they became mothers). As we shall see, the women who did leave the hi-tech industry after they became mothers (and therefore did not take part in the above-mentioned research) reveal another aspect of the picture: They left the hi-tech industry in their attempt to solve the family-career conflict.

Another major finding of Frenkel and Izraeli's (in press) research indicates a connection between the division of housework between the two spouses and the woman's success. Specifically, women who share the housework with their spouses earn more and are more satisfied with their jobs than women who "own" the housework. Our research refines this finding. Specifically, it indicates a link between the spouse's occupation, the ability of the two spouses to share the housework, and the female software engineer's course of promotion.

By using a qualitative research approach, as described in the next section, our research enables one to understand subtle details that clarify and shed additional light on the above-mentioned numerical data.

RESEARCH METHOD

Our qualitative research used in-depth interviews for data collection and inductive data analysis. Specifically, over the course of 6 months, between April and October 2004, we interviewed 17 female software engineers who work (or worked) at four hi-tech companies of different areas in the Israeli software industry. The women interviewed were very highly educated (three hold PhDs, six have MScs, and eight have BScs) and were at different stages of life: Two women were unmarried, three had no children, and the rest were married (to their first husband) with 2 to 4 children.

Each interview addressed the following main topics by referring to the following issues and additional relevant subjects that emerged during the interview.

- **Past:** How she became a software engineer
- **Present:**
  - **Current Present:** Daily schedule, tasks she is currently working on, what she might be doing if she were not talking to us at the moment
  - **Continuous Present:** Projects she is part of and her role in these projects, her work style, challenges in her work (professional, personal, social, cultural), conflicts she faces and how she bridges them, how she sees an ideal software development environment
- **Future:** How she envisions her future in the field
- **Teamwork:**
  - **Her Team:** Description of the team, the way it functions and her role in it, how she believes others perceive her, who decides on roles and work assignments in her team and according to what criteria

- **Other Teams She is Familiar with:** Their functionality, tasks they have succeeded or failed in

In order to listen to the women's voices, we let the women lead the interview to the extent that they found relevant. Indeed, this nature of the interview enabled us to open up to new directions and point of views suggested by the interviewees (Harding, 2004).

We restricted the interview to 1 hour, even in cases in which we could have continued. Our message was that we value the time the female software engineers dedicated to the interview. We always invited each interviewee to share further thoughts with us, should she find it appropriate.

MAIN THRUST OF THE ARTICLE

Our data analysis reveals that the perspectives of the Israeli female software engineers are best described in relation to the stages in life and factors that influence their career paths. Thus, we organize this section around six typical profiles that emerge from the analysis of the interviews. For each profile, we describe the factors that influence a particular life story and how the interviewees cope with these factors.

Needless to say, these profiles do not necessarily encompass all female software engineers. However, since our interviewees come from different organizations and are at different stages in life, we believe that the profiles can be regarded as representative of Israeli women software engineers with similar characteristics.

Profile A: "I Guess there are Some Gender-Related Issues Here"

This profile characterizes young women who are not mothers. The typical Profile A software engineer is either a single or just-married woman. She is starting to develop her hi-tech career and seems to be pleased about her job. She is, however, concerned about how the future will look like with children. While some Profile A women consider their spouses to be partners in caring for the children, other Profile A software engineers assume, even at this early stage of

their adult life, that they will bear all responsibility related to child raising. In either case, the woman suspects that her spouse's occupation will influence her career. Often, she derives her model from the way she views her mother.

As has emerged from our interviews, Profile A software engineers devote a considerable amount of attention to social issues, such as who goes with whom for lunch and who lives where. As we shall see in the other profiles, this type of concern is not evident in interviews with women who are mothers. Instead of social-related topics, issues such as time management and factors influencing promotion assume a more central place in their set of concerns.

Profile B: "My Spouse's Work Enables Me to Develop a Career"

Having one child usually enables the female software engineer to go on with her professional life as before. Neither she nor her spouse must give up her or his career. The actual family-career conflict emerges when the second child is born. It seems that in many cases, the way in which this conflict is solved is determined almost solely by the spouse's occupation. If the woman's spouse does not work in the hi-tech sector, it is most likely that she will continue to develop her career as she envisioned it. Profile B addresses this case. If the spouse also works in the hi-tech sector, then it is most likely that she belongs to one of the profiles described next.

Profile C: "It is My Decision to Give Up My Career, Even though I Could Succeed"

This profile illustrates how the fact that the spouse also works in the hi-tech industry has a direct influence on the woman's career. Specifically, this profile describes cases in which female software engineers struggle with the family-career conflict (Ahuja, 2002) and the family prevails. A typical line of thought that characterizes Profile C software engineers is, "I'll do the maximum possible within the restricted time I have at work and will be aware of the fact that I could achieve more. But it was my decision to give up my career." In fact, some of the women explicitly say that they are not prepared to transfer the homemaking responsibilities to their husbands. Declaring that it is their decision to give up their career, no matter what the real reason is, naturally makes the dilemma easier.

Profile D: "I'll Go on Struggling with this Lifestyle"

This profile describes a female software engineer whose husband works in hi-tech, and who has decided, despite the fact that she faces the family-career conflict, to go on struggling with the long-hours system that characterizes the hi-tech industry. As it turns out, no matter how much she struggles and how great her dedication and willingness to juggle family and career, it is to no avail and, according to our interviewees, her promotion, in many cases, is still blocked in different ways.

Profile D women were very emotional throughout the interview. Furthermore, they expressed a very clear approach, trying hard to justify it during the interview. On the surface, it seems that they belong to Profile C since they sometimes express a desire to give up. However, they do not express the compromised approach expressed by Profile C women. The main difference between the two profiles lies in the fact that Profile C software engineers gave their careers up a priori, while Profile D software engineers often have a successful career and their struggle with the conflict is expressed after they having gained some achievements beyond being a software team member. Accordingly, the focus in this case is placed on the factors that prevent their promotions from happening faster.

Profile E: "I'll Leave this Job When I Become a Mother"

This profile describes a woman whose husband also works in the hi-tech industry, and who thinks, "I know what it means to juggle career and family. Therefore, the moment I become a mother, instead of struggling, I will immediately leave the hi-tech industry and will live a peaceful life." Such a woman is usually aware of the fact that mothers in her company have a hard time struggling with the family-career conflict. Unlike women of the next and last profile (Profile F), Profile E women refuse to be

part of this struggle and choose to leave their hi-tech jobs as soon as motherhood becomes reality for them.

Profile F: "I'll Fight a Little and then Leave"

This profile represents women in different stages of leaving the hi-tech industry: Some still work in the industry but with a clear feeling that they are about to leave it, others are in the midst of a leaving process, and yet others have already left. In all cases, the husbands are hi-tech employees and are not involved at all in the home-maintenance chores. This, however, is not the only reason the women give for their desire to leave the hi-tech industry. All of the Profile F women express a desire to change their careers and are happy that their husbands' relatively high incomes enable them to leave this demanding system. Furthermore, some of them declare that many of their colleagues, both men and women, would probably prefer to leave the hi-tech work environment, but do not have the courage to forfeit power and income.

Main Observations

Based on the above analysis, we now highlight three observations.

First, it is not gender in itself that constitutes the family-career conflict; rather, it is motherhood. In other words, as long as a woman is single or married without children, the unlimited-hours barrier does not exist; the moment a woman becomes a mother, barriers are set.

Second, the ability to develop a career in the hi-tech industry is determined almost entirely by the spouse's occupation. If he works in the hi-tech industry, one of the spouses must make a decision to waive his or her career in order to maintain the home; in all cases encountered in our interviews, it was the mother. In other words, one of the main factors that influences female software engineers' courses of promotion in the Israeli hi-tech industry is their ability to work unlimited hours. This ability is directly connected to their spouse's occupation. On the one hand, if he works in hi-tech, it is most likely that she will be the one in charge of maintaining the home, and her ability to dedicate the required long work hours will suffer. Consequently, she will not be given responsibilities and will not attain management roles. On the other hand, if her husband does not work in the hi-tech industry, she will be able to dedicate more time to her work, will be given responsibilities, and will be promoted appropriately.

Third, as it turns out, even if in practice women dedicate to their work the same number of hours (or more) as do their male colleagues, it might be insufficient. More specifically, some of the women we interviewed try to compensate for their need to leave work early (due to their responsibilities at home) by arriving at work earlier than most of their colleagues. Such attempts, however, do not help them shatter the image of not being totally dedicated to work. More explicitly, if one is not available 100% of the time, his or her dedication to work is not perceived as high. As a consequence, this perceived lower evaluation prevents the female software engineers from being given management responsibilities, although their actual workload is the same as that of other software engineers. This becomes extreme even if the woman leaves early only twice a week. Once again, the ability to work unlimited hours is determined by the spouse's occupation. As stated above, if the spouse does not work in hi-tech, there is no problem; if he does, the one to give up a career, at least in the case of the women we interviewed, is the woman.

FUTURE TRENDS

In this section, we question whether the phenomena described in this article are typical to the Israeli hi-tech industry, or whether they also characterize the hi-tech industry in other places in the world. Some of the interviewees, who were relocated to the United States for several years, addressed this issue in their interviews, comparing the atmosphere and working environment in the two countries: the United States and Israel.

We present two of the differences mentioned. First mentioned was the fact that women in the United States start their careers earlier in life since they do not serve in the army, which in Israel is compulsory for men and women. Therefore, female software engineers in the United States usually start

developing a career before they become mothers, and even postpone motherhood in order to make the most of the work world before they become mothers, predicting that they will want to stay at home with their children. Naturally, their future career is determined by this fact. Second, according to our interviewees, it is more acceptable in the United States to leave work at 5:00 p.m. (for both men and women). As a result, the career-family conflict described in this article is reduced.

CONCLUSION

In this article, we suggested a framework that describes the course of life of women in the Israeli hi-tech industry from their own perspective. Based on interviews with 17 female software engineers working in Israeli software organizations, we identified six representative profiles. In addition, we presented three main observations that highlight the career-family conflict. We hope that the framework provides the readers of this article with an inside look through which to ponder gender issues in the hi-tech industry.

ACKNOWLEDGMENT

We would like to thank the Samuel Neaman Institute for Advanced Studies in Science and Technology and the Technion Fund for the Promotion of Research for their generous support in this research.

REFERENCES

Ahuja, M. K. (2002). Women in the information technology profession: A literature review, synthesis and research agenda. *European Journal of Information Systems, 11,* 20-34.

Camp, T. (1997). The incredible shrinking pipeline. *Communications of the ACM, 40*(10), 103-110.

Camp, T. (Ed.). (2002). Women in computing [Special issue]. *SIGCSE Bulletin Inroads, 34*(2).

Frenkel, M., & Izraeli, M. (in press). *Women in the hi-tech industry: Was the "motherhood wall fallen?"* The Women Lobby in Israel, the Center for Policy Research.

Harding, S. (2004). Rethinking standpoint epistemology: What is "strong objectivity"? In S. Hesse-Biber & M. L. Yaiser (Eds.), *Feminist perspectives on social research* (pp. 49-82). New York: Oxford University Press.

Hazzan, O., & Levy, D. (in press). ACM's attention to women in IT. In E. Trauth (Ed.), *Encyclopedia of gender and information technology.* Hershey, PA: Idea Group Inc.

Hughes, B., & Cotterell, M. (2002). *Software project management* (3rd ed.). London: McGraw-Hill.

Humphrey, W. S. (2000). *Introduction to the team software process.* Reading, MA: Addison-Wesley.

Margolis, J., & Fisher, A. (2002). *Unlocking the clubhouse: Women in computing.* Cambridge, MA: MIT Press.

Sosa, T. (2005). *Women programmers: The role of achievement in their careers, education, and families.* Paper presented at the Annual Meeting of AERA, Montreal, Canada.

Tomayko, J., & Hazzan, O. (2004). *Human aspects of software engineering.* Rockland, MA: Charles River Media.

Turner, S., Bernt, P., & Pedora, N. (2002). *Why women choose information technology careers: Educational, social and familial influences.* Paper presented at the Annual Meeting of AERA, New Orleans, LA.

van Vliet, H. (2000). *Software engineering: Principles and practice.* Chichester; New York: Wiley.

KEY TERMS

Shrinking Pipeline: The pipeline represents the ratio of women involved in computer science from high school to graduate school. The pipeline shrinkage problem focuses on several critical junctions: from high school to the bachelor's level, and at the seniority levels both in academia and in industry.

Software Engineering: The application of engineering principles to the construction of software products. Specifically, principles of computer science and mathematics are applied in the development, operation, and maintenance of software. Software engineering addresses not only the technical aspects of building software systems, but also social, management, and cognitive topics.

ENDNOTES

[1]   This description is based largely on Tomayko and Hazzan (2004).

[2]   http://www.ite.poly.edu/htmls/role_israel 0110.htm

[3]   http://www.smartcodecorp.com/about_us/israel_profile.asp

Women, Mathematics, and Computing[1]

Paul De Palma
Gonzaga University, USA

INTRODUCTION

In 1963, Betty Friedan wrote these gloomy words:

The problem lay buried, unspoken, for many years in the minds of American women. ... Each suburban wife struggled with it alone. As she made the beds, shopped for groceries, matched slipcover material, ate peanut butter sandwiches with her children, chauffeured Cub Scouts and Brownies, lay beside her husband at night – she was afraid to ask even of herself the silent question—"Is this all?"

The passage, of course, is from the *The Feminine Mystique* (Friedan, 1983, p. 15). Though it took another decade for the discontent that Friedan described to solidify into a political movement, even in 1963 women were doing more than making peanut butter sandwiches. They also earned 41% of bachelor's degrees. By 1995, the number of degrees conferred had nearly tripled. The fraction going to women more than kept pace, at almost 55%. Put another way, women's share of bachelor's degrees increased by 25% since Betty Friedan first noticed the isolation of housewives. Consider two more sets of numbers: In 1965, 478 women graduated from medical school. These 478 women accounted for only 6.5% of the new physicians. Law was even less hospitable. Only 404 women, or just 3% of the total, received law degrees in 1965. By 1996, however, almost 39% of medical degrees and 43% of law degrees were going to women (Anderson, 1997).

If so many women are studying medicine and law, why are so few studying computer science? That's a good question, and one that has been getting a lot of attention. A search of an important index of computing literature, the *ACM Digital Portal* (ACM, 2005a), using the key words "women" and "computer," produced 2,223 hits. Of the first 200, most are about the underrepresentation of women in information technology. Judging by the volume of research, what we can do to increase the numbers of women studying computer science remains an open question.

While most investigators fall on one side or the other of the essentialist/social constructivist divide (Trauth, Quesenberry & Morgan, 2005), this article sidesteps the issue altogether in favor of offering a testable hypothesis: Girls and young women would be drawn to degree programs in computer science in greater numbers if the field were structured with the precision of mathematics. How we arrived at this hypothesis requires a look at the number of women earning degrees in computer science historically and in relation to other apparently similar fields.

BACKGROUND

In 1997, *The Communications of the ACM* published an article titled "The Incredible Shrinking Pipeline" (Camp, 1997). The article points out that the fraction of computer science degrees going to women decreased from 1986 to 1994. This bucks the trend of women entering male-dominated professions in increasing numbers. The graph below shows the percent of women earning degrees in various scientific disciplines between 1970-'71 and 1994-'95 (National Center for Educational Statistics, 1997).

Figure 1.

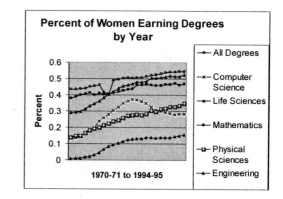

If you did not look at the data over time, you would be justified in concluding that the 13% or so engineering degrees going to women represents a terrible social injustice. Yet the most striking feature of the degrees conferred in engineering and the physical and life sciences is how closely their curves match that of all degrees conferred to women. Stated another way, the fraction of degrees in engineering and the sciences going to women have increased enormously in a single generation. It has, in fact, out-paced the fraction of all degrees going to women. The curves for engineering and the life sciences both have that nice S shape that economists use to describe product acceptance. When a new kind of product comes to market, acceptance is initially slow. When the price comes down and the technology improves, it accelerates. Acceptance finally flattens out as the market becomes saturated. This appears to be exactly what has happened in engineering. Following the growth of the women's movement in the early 1970s, women slowly began to account for a larger share of degrees conferred. By the early 1980s, the fraction grew more rapidly, and then, by the 1990s, the rate of growth began to slow. A parallel situation has occurred in the life sciences, but at a much higher fraction. Women now earn more than 50% of undergraduate degrees in biology.

Computer science is the anomaly. Rapid growth in the mid-1980s was followed by a sharp decline. The fraction of women graduating in computer science flattens out in the 1990s. What is going on here? A study of German women noticed that the sharp increase in the number of degrees in computer science going to women followed the commercial introduction of the microcomputer in the early 1980s (Oechtering, 1993). This is a crucial observation. In a very few years, computers went from something most people were only vaguely aware of to a consumer product. What the graph does not tell you is that great numbers of men also followed the allure of computing in the early and mid-1980s—numbers that declined by the end of the decade.

Despite many earnest attempts to explain why women do not find computer science as appealing as young men (e.g., Bucciarelli, 1997; Wright, 1994), it is important to point out that computer science is not like the other areas we have been considering. Unlike physics, chemistry, mathematics and electri-

cal engineering, there is not an agreed-upon body of knowledge that defines the field. An important textbook in artificial intelligence, for instance, has grown three-fold in 10 years. A common programming language used to teach introductory computing barely existed a decade ago. Noam Chomsky has suggested that the maturity of a scientific discipline is inversely proportional to the amount of material that forms its core. By this measure, computer science is far less mature than other scientific and engineering disciplines.

Many studies have shown that girls are consistently less confident about their abilities in mathematics and science than are boys, even when their test scores show them to be more able (e.g., Mittelberg & Lev-Ari, 1999). Other studies attribute the shortage of women to lack of confidence along with the perception that computing is a male domain (Moorman & Johnson, 2003). Unfortunately, computer science, at least as presently constituted, requires a good bit of confidence. The kinds of problems presented to computer science majors tend to be open-ended. Unlike mathematics, the answers are not in the back of the book—even for introductory courses. There is often not a single best way to come up with a solution and, indeed, the solutions themselves, even for trivial problems, have a stunning complexity to them. The tools that students use to solve these problems tend to be vastly more complex than the problems themselves. The reason for this is that the tools were designed for industrial-scale software development. The move over the last decade to object-oriented languages has only exacerbated an existing problem (Hsia, Simpson, Smith, & Cartwright, 2005). A typical lab assignment to write a program in the C++ or Java language will require that the student have a working knowledge of an operating system, graphical user interface, text editor, debugger and the programming language itself.

One surrogate for complexity is the size of textbooks. Kernighan and Ritchie's classic, *The C Programming Language* (1978) is 228 pages long. The first program in the book, the famous "Hello world," appears on page 6. Deitel, Deitel, Lipari, and Yaeger's (2004) *Visual C++ .NET: How to Program,* on the other hand, weighs in at a hefty four pounds and runs to 1,319 pages. Students have to wade through 52 pages before they reach the book's program equivalent to "Hello, world." The key to successful mastery in this environment is the willingness to tinker and the confidence to press forward with a set of tools that one only partially understands. Although we exhort our students to design a solution before they begin to enter it at the keyboard, in fact, the ready availability of computers has encouraged students to develop a trial-and-error attitude to their work. Those students willing to spend night after night at a computer screen acquire the kind of informal knowledge necessary to write successful programs. This is a world that will welcome only very self-assured young women.

MATHEMATICS, ENGINEERING, AND TINKERING

Recall Chomsky's observation that the most mature disciplines are the most tightly defined. What discipline can boast the tightness and precision of mathematics? As it happens, many reasonable people have attributed at least some of the shortage of women in science and computing as well as the less-than-positive attitudes toward computers to so-called math anxiety among girls (e.g., Chang, 2002; Jennings & Onwuegbuzie, 2001; Mark, 1993). One study says that "The culture of engineering places particular stress on the importance o f mathematical ability. Math is both the most complicated and the purest form of mental activity. It is also the most 'masculine' of subjects" (McIlwee & Robinson, 1992, p. 19, referring to Hacker, 1981). At first glance, the heavier reliance on mathematics might appear to explain why women avoid physics and electrical engineering while embracing biology and oceanography. But this explanation is insufficient for the simple reason that women receive nearly half of the undergraduate degrees in mathematics itself and were receiving almost 40% of them well before the women's movement became a mass phenomenon.

Here, then, is a hypothesis. What if the precision of mathematics is exactly what has appealed to women for so long? And what if the messiness of computing is what has put them off? So far, so good, but we still have to account for electrical engineering and physics. These have a smaller fraction of women than computer science, but are well defined and rely heavily on sophisticated mathematics. What is it about physics and electrical engineering that women find unattractive? The answer is really quite simple. Students drawn to engineering and physics like to tinker with gadgets (e.g., Crawford, Wood, Fowler, & Norell, 1994). That paper describes a grade school curriculum designed to encourage young engineers. It relieves heavily upon "levers, wheels, axles, cams, pulleys, forms of energy to create motion, etc." (p. 173). McIlwee and Robinson (1992) report that 57% of male engineers surveyed chose the field because they like to tinker. Only 16% of women surveyed chose engineering for this reason. It should come as no surprise that the men associated with the microcomputer – Bill Gates, Paul Allen, Jobs and Wozniak—all got their start as tinkerers. And as all parents know but are hard-pressed to explain, their infant sons are drawn to trucks more readily than their infant daughters (Serbin, Poulin-Dubois, Colburne, Sen, & Eichstedt, 2001).

MICROCOMPUTERS, A PROBLEM WITH COMPUTER SCIENCE EDUCATION

Here we find a convergence with computer science and, finally, an explanation for the steep rise in the number of women in the field following the introduction of the microcomputer and its drop a few years later. The development of the microcomputer changed computing enormously. In 1971, a small number of computer science departments awarded fewer than 2,400 degrees. Most people who worked in the thriving data processing industry had received their training in the military, for-profit vocational schools or on the job. By 1986, that number had jumped to nearly 42,000, including almost 15,000 women. Clearly, the microcomputer played a large part in the growth of the academic discipline of computing. Like the dot-com boom, the growth could not be maintained. If the production of computer science degrees had continued to climb at the rate it climbed between 1975 and 1985, by 2001 every American would have had a Bachelor of Science degree in the field. In fact, the number of degrees awarded began to drop sharply in 1987.

We know why both men and women entered IT in the 1980s. Why did the numbers drop by the late 1980s? We cannot really know the answer to this, of course, short of polling those who did not major in computer science during that period; but we can guess. Computer science is hard. What's more, it is not a real profession. There are no licensing barriers to entry, an issue that has been hotly debated in computing literature for at least two decades (ACM, 2005b).Until computing societies agree on licensing and convince state legislatures to go along, students need not earn a degree in computer science to work in the field.

These things are equally true for women, of course, but the tinker factor is an additional burden. Before the mid-1980s and the mass availability of microcomputers, programmers could almost ignore hardware. This article's author wrote programs for a large manufacturer of mainframe computers in the early 1980s without ever having seen the computer he was working on, nor, for that matter, the printer that produced the green bar paper delivered to his cubicle every two hours. There was tinkering going on in those days too, of course. But it was all software tinkering; only computer operators touched the machine. The micro changed all that. Suddenly, those young men who had spent their adolescence installing exotic operating systems and swapping memory chips were in great demand. By adding hardware tinkering to the supposed repertoire of skills necessary to program, the microcomputer reinforced the male-dominated culture of IT (for an account of this very male atmosphere, see De Palma, 2005).

CONCLUSION

Until the day when baby girls like gadgets as much as baby boys, let us look to mathematics itself to see what we can do about attracting young women to computer science. Well before other fields welcomed women, a significant fraction of degrees in mathematics were going to females. Let us assume that the mathematicians have been on the right track all along. A testable hypothesis presents itself. If we make computer science education more like mathematics education, we will make computer science more appealing to women. Computer science grew out of mathematics. How do we get back to basics?

First, teach girls who like to manipulate symbols how to program. Programming is weaving patterns with logic. If girls can do calculus, they can write programs. Second, try not to stray from logic. If we make computer science education less dependent on complex software tools, we remove some of the barriers between the student and logic. Third, minimize the use of microcomputers. Microcomputers, for all their cleverness, misrepresent computer science, the study of algorithms, as hardware tinkering. Fourth, ask students new to computer science to write many small functions, just as students of mathematics work countless short problems. Since there is something about the precision of mathematics that young women seem to like, let us try to make computing more precise. Later, as their confidence grows, they can take on larger projects. Fifth, regard programming languages as notation. It could well be that for complex systems, modern languages will produce a better product in a shorter time. But students do not produce complex systems. They produce relatively simple systems with extraordinarily complex tools. Choose a notation appropriate to the problem and do not introduce another until students become skilled programmers. Taken together, these suggestions outline a program to test the hypothesis.

Suppose we test the hypothesis and it turns out to have been correct. Suppose that, as a result, we give computing a makeover, and it comes out as clearly defined and as appropriate to the job as mathematics. Now imagine that able young women flock to the field. How might this change computing? To begin, students will no longer confuse half-formed ideas about proprietary products with computer science. Nor will they confuse the ability to plug in Ethernet cards with system design. It might mean that with a critical mass of women holding undergraduate degrees in computing, systems will be designed, not by tinkerers, but by women (and men) for whom the needs of computer users are front and center. Since stories of systems that failed through an over fondness for complexity are legion (De Palma, 2005), the makeover might even reduce the number of jerry-rigged systems. Thus, does social justice converge with the market place—a very happy outcome, indeed.

REFERENCES

ACM. (2005a). Search using key words: "Women" and "computer." *The Digital Library*. Retrieved August 23, 2005, from http://www.acm.org

ACM. (2005b). Search using key words: "license" and "profession." *The Digital Library*. Retrieved August 27, 2005, from www.acm.org

Anderson, C. (1998). *Fact book on higher education: 1997 edition*. American Council on Education. Phoenix: Oryx Press.

Bucciarelli, L., & Kuhn, S. (1997). Engineering education and engineering practice: Improving the fit. In S. R. Barley & J. Orr (Eds.), *Between craft and science: Technical work in U.S. settings* (pp. 210-229). Ithaca, NY: Cornell University Press.

Camp, T. (1997). The incredible shrinking pipeline. *Communications of the ACM, 40*(10), 103-110.

Chang, J. (2002). Women and minorities in the sciences, mathematics, and engineering pipeline. *Eric Clearinghouse for Community Colleges*. Retrieved February 23, 2006, from ERIC Digest (ED467855).

Crawford, R., Wood, K., Fowler, M., & Norell, J. (1994, April). An engineering design curriculum for the elementary grades. *Journal of Engineering Education, 83*(2), 172-181.

De Palma, P. (2001). Viewpoint: Why women avoid computer science. *The Communications of the ACM, 44*(6), 27-29.

De Palma, P. (2005). The software wars. *The American Scholar, 74*(1), 69-83.

Deitel, H., Deitel, P., Liperi, J., & Yaeger, C. (2003). *Visual C++ .NET: How to program*. Upper Saddle River: Pearson Education Inc.

Friedan, B. (1983). *The feminine mystique*. New York: W.W. Norton & Co.

Hacker, S. (1981). The culture of engineering: Women, workplace and machine. *Women's Studies International Journal Quarterly, 4*, 341-53.

Hsia, J., Simpson, E., Smith, D., & Cartwright, R. (2005). Taming Java for the classroom. Proceedings of the 36th SIGCSE technical symposium on Computer science education. *ACM SIGCSE Bulletin, 37*(1), 327-331.

Kernighan, B., & Ritchie, D. (1978). *The C programming language*. Englewood Cliffs: Prentice-Hall.

Mark, J. (1993). *Beyond equal access: Gender equity in learning with computers*. The Eisenhower National Clearinghouse for Mathematics and Science Education. *Retrieved August 27, 2005, from www.enc.org/topics/equity/articles*

McIlwee, J., & Robinson, J. G. (1992). *Women in engineering: Gender, power and workplace culture*. Albany: State University of New York Press.

Mittelberg D., & Lev-Ari, L. (1999). Confidence in mathematics and its consequences: Gender differences among Israeli Jewish and Arab youth. *Gender and Education, 11*(1), 75-92.

Moorman, P., & Johnson, E. (2003). Still a stranger here: Attitudes among secondary school students towards computer science. *Proceedings of the 8th Annual Conference on Innovation and Technology in Computer Science Education*, 193-197.

National Center for Educational Statistics. (1997). *Digest of educational statistics NCES 98-015*. Washington, DC: U.S. Government Printing Office.

Oechtering, V., & Behnke, R. (1995). Situations and advancement measures in Germany. *Communications of the ACM, 38*(1), 75-82.

Serbin, L., Poulin-Dubois, D., Colburne, K., Sen, M., & Eichstedt, J. (2001). Gender stereotyping in infancy: Visual preferences for and knowledge of gender-stereotyped *toys* in the second year. *International Journal of Behavioral Development, 25*(1), 7-15.

Trauth, E., Quesenberry, J., & Morgan, A. (2004). *Proceedings of the 2004 SIGMIS Conference on Computer Personnel Research, 114-119*. New York: ACM Press.

Wright, R., & Jacobs, J. (1994). Male flight from computer work. *American Sociological Review, 59*, 511-536.

KEY TERMS

Computer Science: An academic discipline that studies the design and implementation of algorithms. Algorithms are step-by-step procedures for solving well-defined problems. A precise description of a technique for putting words in alphabetical order is an algorithm.

Ethernet Card: Hardware that allows a computer to be attached to a network of computers.

Memory Chip: An informal term for RAM (random access memory) or just plain memory. It is internal to a computer and loses its contents when the power is shut off. Programs must be loaded into RAM to execute.

Microcomputer: Also called a personal computer. The machine on your desk is a microcomputer.

Operating System: The collection of programs that controls all of the computer's hardware and software. Important operating systems are Windows XP and Unix.

Program: A sequence of instructions that tells a computer how to accomplish a well-defined task.

Programming Language: The notational system that a programmer uses to construct a program. This program is transformed by another program, known as a compiler, into the instructions that a computer can execute. Important languages are Java and C++.

ENDNOTE

[1] This article grew out of a shorter opinion piece in the "Viewpoint" column of *Communications of the ACM* (De Palma, 2001).

Women's Role in the Development of the Internet

Shirine M. Repucci
National Coalition of Independent Scholars, USA

INTRODUCTION

The majority of the literature written about the history of the Internet has focused on chronicling only the technical milestones that led to its development. In doing so, most have overlooked a significant period in the Internet's history, the period bounded by the retirement of the United States Department of Defense Advanced Research Projects Agency's network (ARPANET) in the late 1980s and by the commercialization of the network and the excitement over the World Wide Web browser in the mid-1990s. The historical accounts, as a result, include little more than a passing mention of the National Science Foundation Network backbone project (NSFNet) and Merit Network, Inc., which conducted the transfer of this technology to society at large from 1985 to 1995.

Additionally, the literature holds little evidence of women as a force in the Internet's early development. For example, in *Inventing the Internet*, the most thorough book published to date on the history of the Internet, Abbate (2000) mentions more than 60 different men who were involved in the Internet's development but does not recognize a single woman other than to show a female model advertising a computer. The contribution women made to developing the Internet is similarly neglected in Kristula's *The History of the Internet* (2001), Griffiths' *From ARPANET to World Wide Web* (2002), and Castells' *The Internet Galaxy* (2001).

Overall, readers of the Internet's history are left with the impression the Internet was developed solely by men. This impression is incorrect, as is the impression that the Internet's success is solely the result of a series of technical achievements.

This article presents evidence that many women were employed in the Internet industry prior to the mid 1990s, filling in gaps in the literature on this point. In addition, it suggests that, collectively, women may have held a key role in the extraordinarily successful transfer of the internet technology from a small circle of academics and governmental researchers to society at large. The findings presented here are part of a larger body of work examining the role women held in the Internet's development, carried out at Eastern Michigan University for the completion of a master's thesis in interdisciplinary technology (Repucci, 2004).

BACKGROUND

Data for this study came from interviews conducted with eight of the early participants in the NSFNet project,[1] original source documents, and my own observations and knowledge acquired while employed on the NSFNet project. The individuals interviewed for this research were all employed on the NSFNet project, either by Merit Network, Inc. or by the National Science Foundation's division of Networking and Communications Research Infrastructure (NCRI). They included:

- Merit staff
    - Eric Aupperle, President
    - Elise Gerich, Associate Director of National Networking
    - Susan Hares, Internet Engineer, National Networking division
    - Ellen Hoffman, Manager of Network Information Services
    - Jo Ann Ward, User Services Specialist, Network Information Services division
    - Jessica Yu, Internet Engineer, National Networking division
- National Science Foundation (NSF) staff
    - Stephen Wolff, Division Chief NCRI
    - Jane Caviness, Deputy Division Director NCRI

Table 1. Merit's NSFNet Staff, May 1992

Division	Females	Males	Total	Study Participants
Internet Engineering	4	8	12	Gerich, Hares, Repucci,[2] Yu
User Services	6	3	9	Hoffman, Ward
Network Operations	8	10	18	none interviewed

Note: The data are from Merit's 1992 Organizational Chart (Merit, 1992, May 1)

Employee statistics, presented in Table 1, of the Merit staff who were assigned to the NSFNet effort came from Merit's organizational chart dated May 1992, the last such chart found before the transition of the backbone to a commercial provider. Its purpose is simply to show the gender makeup of the staff during this project, and thus the timeframe of this study. These numbers should be viewed in light of the following historical perspective: "Before NSFNet came along and Merit started to grow, Merit only had one female ... outside of the secretarial staff" (E. Hoffman, interview, February 23, 2003).

Note that while the engineering staff had half the number of women as men in 1992, the User Services staff had the opposite proportions: twice as many women as men. While the published history of the Internet to date has focused on the efforts of the engineers and computer scientists, this paper suggests that the contributions of those involved in the user services campaign were equally important to the success of the Internet's early development.

While this research focuses primarily on the gender makeup at Merit, Merit was only one of the many organizations involved with the NSFNet project where high profile women in the field were employed. A sampling of women from other organizations mentioned in the interviews include: Allison Brown at Cornell University, Deborah Estrin at the University of Southern California, Darleen Fisher at the National Science Foundation, Priscilla Jane Houston at Rice University and the National Science Foundation, Radia Perlman at Digital Equipment Corporation, and Lixia Zhang at Xerox Palo Alto Research Center. Clearly, women were involved in the development of the Internet, and the interview data shows that many were considered key contributors as seen from the interview data.

The concept of inclusion served as a central theme in the examination of gender roles within the project. To this end, the study focused on the formal and informal strategies used to make the Internet more socially inclusive. For the purposes of this work, inclusion was defined as satisfying the following three conditions. First, there must be a clearly stated intent to broaden the user base of the technology. Second, a proactive effort to broaden the user base must accompany the stated intent to do so and the techniques to carry this out must include persuasion as well as educational efforts. Finally, there must be evidence that the user base was, in fact, broadened.

MAIN THRUST OF THE CHAPTER

Originally, this work was undertaken to document the participation of individual women in the Internet's early history. However, in examining the role these women played during this period of history, evidence began to emerge that, while they held a variety of positions and titles that in themselves were impressive—management, engineer, and so on—they also collectively played a key role in the transfer of internet technology to society at large. Specifically, this research suggests that the collective action of the women who were involved in the NSFNet project energized the Internet to be more socially inclusive.

Evidence of women's involvement in the Internet's early history was found in both the interview data and in original source materials. A sampling of the evidence suggesting that women were indeed actively engaged in creating an inclusive user community for the emerging Internet is provided below, presented by the qualifying factors of inclusion, as previously defined.

Evidence of a Clearly Stated Intent to Broaden the Community of Users

This factor deals with the origins of the inclusion goals of the Internet, moving the community of users from a small, fairly homogenous, circle of academics, and governmental researchers to a network representing a wide range of society, including business people, K-12 students, and families. Evidence of intent to expand inclusion beyond the boundary of the NSFNet grant was heard in six of the eight interviews. Broken down by gender, this represents five of the six women and one of the two men. It should be noted that the male participant who commented on this topic had a great deal of influence in defining the original NSFNet project and in its ongoing management. His interview made it clear he had actively sought out and hired staff with strong personal values for inclusion. He commented:

I had a personal goal of the Internet [as] a sort of worldwide, in every home, kind of communications system. I'd had that for a long time. I had had a colleague at Aberdeen ... in probably 1980 ... [He] infected me with the notion of the Internet as sort of a universal communication ... [scheme] for every man and I did believe in that. (S. Wolff, interview, October 26, 2003)

Comments on this topic from one of the women included the following:

I was strongly committed to this network being made available to, for instance, all colleges and universities, including not just the major research universities ... And I think that ... commitment to fairness and a commitment to broadening the educational part of it, as well as the research part, were important things to me. (J. Caviness, interview, May 1, 2003)

The research provides significant evidence that women valued and pushed forward the concept of broad social inclusion in the use of this technology. However, there was little in the research findings to suggest that women had a role in setting the initial agenda for inclusion on the NSFNet project. Both Wolff (2003) and Caviness (2003) suggest in their interviews that the agenda of inclusion may have been established in the initial planning stage of the NSFNet project. This planning stage occurred outside of the time frame examined for this research.

Evidence of a Proactive Effort to Broaden the Community of Users

The effort to broaden the community of Internet users was carried out primarily by Merit's User Services group, which, through its contract for the NSFNet project, was specifically tasked with providing information services to the networking community as well as serving in a public relations capacity to aid in the transfer of the technology to the public sector. Women made up the majority of the User Services staff, as seen in Table 1.

Overall, evidence of having an active role in expanding the bounds of the NSFNet project toward more inclusion was heard in five of the eight interviews. By gender, this represents four of the six women and one of the two men.

Persuasion

Merit's technology transfer effort utilized two primary messages in its persuasion campaign. One was directed at society as a whole. The other was directed to colleagues and other potential organizations in an attempt to swell the membership of the Internet community.

The persuasion effort was carried out primarily by Merit's User Services group, again the majority of which were women. The message directed to society was often couched in terms of the excitement of being able to communicate with everyone in the world, any time of day or night. An example of this type of message emanating from Merit can be found in a 1992 article by L. Kelleher of Merit, "The Internet is a vast ocean of data and resources." It is immeasurably large and no one owns it" (1992, p. 460).

The persuasive messages directed to colleagues tended to promote the Internet as a means of increasing productivity. Examples of this type of persuasion can be found in the 1989 issue of Unix World (Fisher, 1989, p. 43) and in the 1990 issue of Supercomputing Review (Turner, 1990, p. 45).

Evidence of a proactive effort to broaden the community of users by employing persuasion was also seen in the interview data from within the User Services group:

A lot of my responsibilities involved helping people who were not even aware of what a network was, understand why they might ever want to use one, and so we did newsletters, trying to persuade universities why this was important, we did seminars for people from universities to K-12 schools, and even from businesses at a point when nobody knew what the Internet was. (E. Hoffman, interview, February 23, 2003)

Educational Efforts

There is strong evidence in the interview data that the NSFNet project focused significant resources on outreach efforts to organizational members of the project, making the user community aware of the availability of this technology. Examples include:

Ultimately we developed a program of NSFNet seminars ... that were held really across the country ... And creating a collaboratory where the topics of discussion would really evolve from the actual development of the technology to how it could be applied. (J. Ward, interview, February 7, 2003)

When I was at Merit I had an award from NASA in the early 90s ... They sent a contingent of about five of us to Kenya for a week and a half in order to help define the requirements for this interconnection, as well as to understand their requirements. (E. Gerich, interview, October 17, 2003)

Original source data was also found, substantiating the educational theme heard in the interviews. Examples of this can be found in the publication of *Zen and the Art of the Internet* (Kehoe, 1992), *The Internet Companion* (LaQuey & Ryer, 1992), and the *Cruise of the Internet* (Kelleher, 1992), a computer disk-based guide to using the Internet which was widely distributed to librarians across the nation.

Evidence of Success in Broadening the Community of Users

In addition, there is evidence that the members of the NSFNet project successfully expanded the community served by the Internet, beyond its initial focus on the research and higher education community in the United States. Two such examples:

Well, I think that NSF became the organization that brought the work that had been done by ARPA and the development of the ARPANET, that technology ... brought it to a broader community ... Up until then ... ARPANET ... had very limited access, [was] very overloaded and it just was pretty restricted. (J. Caviness, interview, May 1, 2003).

I think it's just absolutely remarkable what developed in the period from the mid-80s until the mid-90s; in a decade's time we went from, essentially ... a pure research environment, the ARPANET format, to a worldwide infrastructure that has continued to grow and blossom. (E. Aupperle, interview, February 21, 2003)

Overall, the research highlighted in this article strongly suggests that women did hold a prominent role in the NSFNet's role of migrating the knowledge of the Internet to the larger society. Additionally, significant evidence was found that the individuals on the NSFNet project, although working in an academic environment, were so group oriented that they did not seek individual recognition and thus did not attempt to publish their experiences. Thus, it is conceivable that at the time of the event the individuals did not see, and thus record to memory, their individual actions as their own. Likewise, it is conceivable that as time has passed they have merged the actions of the group into a more singular memory. Either way, this team-focused mindset may well be masking access to the data on the origins of the social inclusion objective of the project. Additional research may uncover this information if a larger study sample is obtained and/or interview questions are included to assist in breaking down the team focus, thus helping to determine if the data is hidden or simply not there.

FUTURE TRENDS

There is some evidence that more interest in the Internet's early history is emerging, as Web sites depicting this history are numerous compared to the number found when the author began research on this topic in early 2003. Additionally, many university courses can now be found that include portions of this history in their Web posted syllabus.

Few sources, however, have yet been found in the academic literature that specifically explore women's role in the historic development of this technology. It is hoped that the work documented in this article will spark an interest not only in documenting women's participation in the Internet's early development but also in the role women held in the social progression of this technology.

CONCLUSION

The broader purpose of this research was to bring to light historical and social factors regarding the Internet's early development and thus to fill an omission in the historical literature regarding women in the development of this technology. As this article documents, women clearly took part in the Internet's development, as managers, engineers, and technology advocates. This work presents a first look at the social factors involved in the Internet's successful transition to the larger society and the role women held in that transition, dispelling the impression that the Internet's success was based exclusively on a series of technical milestones and that it was developed solely by men. Finally, this work has served to both raise the question of how societal factors influenced the transfer of the internet technology and to assert that women were active participants even though the literature fails to note this.

REFERENCES

Abbate, J. (2000). *Inventing the Internet*. Cambridge, MA: The MIT Press.

Aupperle, E. (interviewee), & Repucci, S. (interviewer). (2003). *NSFNet oral history project with Eric Aupperle* [Cassette recording]. Bentley Historical Library, University of Michigan.

Castells, M. (2001). *The Internet galaxy: Reflections on the Internet, business, and society*. Oxford: Oxford University Press.

Caviness, J. (interviewee), & Repucci, S. (interviewer). (2003). *NSFNet oral history project with Jane Caviness* [Cassette recording]. Bentley Historical Library, University of Michigan.

Fisher, S. (1989). Boom times for NSFNet. *UNIX World, Networking,* 43.

Gerich, E. (interviewee), & Repucci, S. (interviewer). (2003). *NSFNet oral history project with Elise Gerich* [Cassette recording]. Bentley Historical Library, University of Michigan.

Griffiths, R. T. (2002, October). From ARPANET to World Wide Web. *The history of the Internet*. Retrieved from http://www.let.leidenuniv.nl/history/ivh/chap2.htm

Hoffman, E. (interviewee), & Repucci, S. (interviewer). (2003). *NSFNet oral history project with Ellen Hoffman* [Cassette recording]. Bentley Historical Library, University of Michigan.

Kehoe, B. P. (1992). *Zen and the art of the Internet: A beginner's guide*. Englewood Cliffs, NJ: Prentice-Hall.

Kelleher, L. A. (1992). *Cruise of the Internet* [computer disk]. Ann Arbor, MI: Merit Network, Inc.

Kristula, D. (2001, August). *The history of the Internet*. Retrieved from http://www.davesite.com/webstation/net-history.html

LaQuey, T., & Ryer, J. C. (1992). *The Internet companion*. Reading, MA: Addison-Wesley.

Merit Network, Inc. (1992, May 1). *University of Michigan, Information Technology Division—Network Systems/Data: Includes Merit Network, Inc. project staff and ANS direct staff*. Ann Arbor, MI: Merit Network, Inc.

Repucci, S. M. (2004). *Women's role in the development of the Internet and the social movement they propelled.* Unpublished master's thesis, Eastern Michigan University, Ypsilanti, Michigan.

Turner, J. A. (1990, August). Washington to Upgrade NSFNET, prepare for gigabit network. *Supercomputing Review, 3*(8), 44-45.

Ward, J. (interviewee), & Repucci, S. (interviewer). (2003). *NSFNet oral history project with Jo Ann Ward* [Cassette recording]. Bentley Historical Library, University of Michigan.

Wolff, S. (interviewee), & Repucci, S. (interviewer). (2003). *NSFNet oral history project with Steve Wolfe* [Cassette recording]. Bentley Historical Library, University of Michigan.

KEY TERMS

Inclusion: Defined as satisfying the following three conditions. There must be a clearly stated intent to broaden the user community, a proactive effort to accomplish this, including both persuasion and educational efforts, and there must be evidence that the community of users was broadened.

internet: Used as a common noun denoting any collection of computer networks that are linked together to form a single, larger, network. Each computer network links to the larger network by both a physical link and a common set of network protocols, or language, through which network communications can occur. An internet may, or may not, be connected to the general purpose Internet.

Internet: Used as a proper noun denoting a collection of wide area computer networks, worldwide, that specifically runs the TCP/IP network protocol suite and is dedicated to general purpose access.

Merit, Merit Network, Inc.: A nonprofit consortium consisting of the four-year public universities within the State of Michigan, established in 1966 to provide wide area networking services to the State's educational institutions.

NSFNet Project: Operating from 1985 to 1995 and funded by the National Science Foundation the project was focused on expanding wide area computer networking technology to a broader segment of the university and research community. The project included both the creation of a physical network infrastructure and an organizational infrastructure to promote and manage the physical infrastructure.

NSFNet Collective: The NSFNet collective was formed by the early researchers in the field of computer networking for the purposes of conceiving of and creating the NSFNet project. The majority of the organizations in this collective were high level research universities or facilities in the United States.

Technology Transfer: The process of migrating a newly developed or under utilized technology to a target population.

ENDNOTES

[1]   All research participants cited in this article have consented to being identified in publications resulting from this research.

[2]   Repucci is included as a participant but due to her role in conducting the study, she was not directly interviewed.

Women's Access to ICT in an Urban Area of Nigeria

Olukunle Babatunde Daramola
Women's Health and Action Research Centre, Nigeria

Bright E. Oniovokukor
Indomitable Youths Association, Nigeria

INTRODUCTION

Despite the rapid and revolutionized development of communication and media around the world in the last few decades, which culminated in the term information communication technology (ICT), most of the developing countries are yet to clearly understand its significance or maximize the use of various forms of ICTs, because of other pressing issues such as access roads, potable drinking water supply, electricity and health facilities. This has greatly caused a wide gap between and within countries in the areas of social, economic, political, health and educational developments.

ICTs encompass all the technologies that facilitate the processing, transfer and exchange of information and communication services. Various forms of ICTs exist, such as radio, television, newspaper, telephone, magazine, billboard, Internet, electronic and print media, and so forth.

In the past few decades, there has been a significant increase in knowledge of the importance and developmental trends of ICTs worldwide. ICTs are very important in analyzing one's existing/potential audience using the most cost-effective way to communicate; evaluating the quality of messages; and making provision for information feedback. ICTs bring about various opportunities, ranging from employment and education to economic, health, social and environmental development.

As a result of the digital divide between and within countries, there are uneven disparities between the economic, social, educational and political status of the international community. This brings about classification of countries into "developed and developing" or "haves and have nots."

Gender disparity has served as a strong barrier to women's use of ICTs, considering the fact that women in most developing countries are still considered unequal in status with their male counterparts. This has, thus, reduced enrollment in sciences and technological fields of study. This also is probably as a result of the limited awareness of the full range of opportunities in ICTs other than access to information.

To achieve the goal of universal access to ICTs, there is a need to bridge the gap between men's and women's access to the use of ICTs. This can be accomplished by making technology accessible, relevant and useful to both women and men. State policies could be made holistic by taking into consideration women's needs as well as addressing related issues, such as the urban-rural bias, promoting enrollment of girls in ICTs programs and empowering women to use ICTs for profitable ventures.

BACKGROUND

This survey is being carried out among in-school and out-of-school youths in Benin City, Edo State, Nigeria. Edo State is one of the 36 states that make up the Nigerian federation. It is located in the South-South geopolitical zone, in the Niger-Delta region of Nigeria. The state is administratively divided into three senatorial districts and 18 local government areas. Edo State had an estimated population of 2.86 million people as of 1999. The major ethnic groups in the state are Bini, Ishan, and Afemai, who are collectively referred to as the Edo-speaking people.

Benin City, the capital of Edo State, is one of the ancient cities in West Africa, with more than 1 million inhabitants and a male-to-female ratio of 1:1. It is divided into three local government councils; namely, Egor, Oredo and Ikpoba-Okha. These represent the grass-roots administrative units and make up the major urban areas of the state[3].

Edo State is one of the poorest and least industrialized states in Nigeria, with more than 60% of its population residing in the rural communities. The majority of them are peasant farmers with no feasible means of livelihood.

Despite its poorly industrialized status, Edo State has one of the highest levels of literacy in the country. Nearly 70% of the inhabitants can read and write, which exceeds the national average of 40%. There are currently four universities, two polytechnics, one college of education and several secondary and primary schools in the state.

Statistics reveal that 32.9% of Edo State populations are young adults aged 10-24 years, with about 1:1 male/female ratio (National Population Census, 1991). Data shows that 83% of the adolescents (10-20 years) are in school, while 17% of them are out of school (Okonofua, Kapiga, & Osuji, 2000). School enrollment for girls is significantly less than for boys, and there is a high dropout rate between the ages of 16 and 19 years, mostly among females (Ministry of Education, Edo State).

This survey was conducted to assess women's access to and beneficial usage of ICTs in Benin City, an urban area of Edo State, Nigeria. The specific objectives are:

1. To clearly document the ratio of male-to-female access to ICTs in Benin City.
2. To identify the age and status of female ICT users in Benin City.
3. To identify the purpose of ICT usage among women in the area.
4. To use the data generated to advocate for women's involvement in ICT.

METHODOLOGY

A total of 3,000 in-school and out-of-school female youth aged 15-30 years will be involved in the survey. Information will be obtained using a questionnaire instrument containing open and close-ended questions.

The questionnaire adopted was divided into three sections—namely, A, B and C—comprising demographic status of the interviewees; previous knowledge about ICTs and perceptions of the interviewees on the significance of ICTs to women and actions to be taken in increasing women's access to ICTs, respectively. Also, a table was developed to determine the ratio of male/female access to Internet service in the urban areas of Edo State. The data generated will be analyzed and a detailed report written.

CONCLUSION

We are currently in the process of administering the questionnaires in the three local government areas of the state. By the end of September, the results will be collated, analyzed and processed for a final report.

At the end of the project, the results obtained from the questionnaire should enable us to identify and document the ratio of male/female access to ICTs in the urban area of the state; the age range and status of ICTs users among women; the common types of ICTs, as well as the purpose of ICT usage among women in an urban area of Edo State, Nigeria.

This, we believe, will build our capacity to advocate for women involvement in ICT policy development in the state, as well as to identify active programs or policies in the promotion of ICTs in Edo State.

In addition, we desire scaling up the project by replicating the research study in other parts of the country, particularly each of the six geopolitical zones of Nigeria, because of the significant regional, cultural and religious diversity that exist in the country. This will empower us to have a clear survey of women's access to ICTs in Nigeria.

However, it will further facilitate to develop strategized programs and policies that will increase women's access to ICTs in Nigeria, thereby reducing all forms of gender discriminations in the areas of educational levels, access to good health services, employment opportunities and economic resources in the country.

Conclusively, ICT has been identified as one important tool that has the potential of catalyzing the attainment of the Millennium Development Goals by the year 2015, provided it is well implemented and funded.

REFERENCES

Africa ICT Monitor (Initiative of Association for Progressive Communications [APC]). (2004). *ICT policy in Nigeria.* (2004). Retrieved from http://www.africa.rights.apc.org

Crowson, K., Sawyer, S., Wigard, R., & Allbritton, M. (2000, December). *How do ICT reshape work?* Retrieved from http://crowston.syr.edu/real-estate/icis2000.pdf

Deaney, R., Ruthven, K., & Hennessy, S. (2000). *Student's perspectives in the use of ICT in subject teaching.* (2000). Retrieved from http://www.educ.cam.ac.uk/istl/WP032.doc

Duncombe, R., & Heeks, R. (2001). *Information and communication technology: A handbook for entrepreneurs in developing countries* (Version 1). Manchester: Institute for Development Policy and Management, University of Manchester, Precinct Centre. Available from http://www.man.ac.uk/idpm/ictsme.htm

Initiative of Policy Project/USAID and Women's Health and Action Research Centre (WHARC). (2000). *Profile of sexual and reproductive health of young adult and adolescents in Edo State: A situation analysis report.*

Literate Futures Project. (2005). *The effective exchange of information.* Retrieved from http://education.qld.gov.au/curriculum/learning/literate-futures/glossary.html

Lord Briggs. (2004). *Technology and the media.* Microsoft Encarta Premium Suite.

Microsoft Encarta Premium Suite. (2004). *Information technology in education; Telecommunication concept; Information revolution; Communication; and Historical perspective of technology.*

National Council on Education. (2002). *Sexuality education curriculum.* Abuja: Federal Ministry of Education.

National Population Commission Edo State Chapter, Nigeria. (2004).

Nzagi, E. (2000, October 16-19). Women in the ICT development within a competitive market environment. *Sixteenth HRM/HRD Regional Meeting for the English Speaking Countries in Africa,* Arusha. Retrieved from http://www.itu.int/ITU-D/hrd/publications/reports/2000/arusha/DOC13-TANZANIA.doc

Okonofua, F. E., Kapiga, S., & Osuji, O. (2000). *Influence of schooling and socio-economic status on sexuality and health seeking behaviour for STDs among Nigerian adolescents.*

Osterwalder, A. ICT in developing countries. *A cross-sectoral snapshot.* Retrieved from http://www.hec.unil.ch/aosterwa/Documents/InternetIn EmergingMarkets/Publications/ISGLOB03

The World Bank Group. (2003). *Global information and communication technologies department.* Retrieved from http://info.worldbank.org/ict

Wikipedia Free Encyclopedia. (2004). Communication *Information technology.* Retrieved from http://en.wikipedia.org/wiki/Information_technology

World Organization of Family Doctors (WONCA). (2005). *Web definition of communication technology.* (2005). Retrieved from http://www.globalfamilydoctor. com/aboutWonca/working_groups/write/itpolicy/ITPoli13.htm

KEY TERMS

Access: Rights or means of approaching or reaching.

Benin City: State capital of Edo State, one of the 36 states that make the nation of Nigeria.

Developing Countries: Poor countries that are building better economic and social conditions.

ICT: Information Communication Technologies.

In-School: Students in secondary or tertiary institutions.

Out-of-School: Students who have completed/dropped out of either secondary or tertiary education.

Urban: Densely populated settlement with large concentrated collection of dwellings.

WSIS Gender and ICT Policy

J. Ann Dumas
The Pennsylvania State University, USA

INTRODUCTION

The World Summit on the Information Society (WSIS) was organized by the United Nations (UN) and the International Telecommunications Union to address the need for international policy and agreement on ICT governance, rights, and responsibilities. It convened in two phases: Geneva in 2003 and Tunis in 2005. International representatives of governments, businesses, and civil society raised issues, and debated and formed policy recommendations. The WSIS Gender Caucus (2003) and other civil-society participants advocated for gender equality to be included as a fundamental principle for action and decision making. The voting plenary session of delegates produced the *WSIS Declaration of Principles* (UN, 2003a) and *WSIS Plan of Action* (UN, 2003b) in Geneva, with gender included in many of the articles.

Two major issues WSIS addressed in Geneva and Tunis were Internet governance and the Digital Solidarity Fund. UN secretary general Kofi Annan established the Working Group on Internet Governance (WGIG) to define Internet and Internet governance to "navigate the complex terrain" (GKP, 2002, p. 6) and to make recommendations for WSIS in Tunis in 2005. WGIG addressed three Internet-governance functions: technical standardization; resources allocation and assignment, such as domain names; and policy formation and enforcement, and dispute resolution. Relevant issues not initially addressed by WGIG included gender, voice, inclusiveness, and other issues rooted in unequal access to ICT and to the decision-making process including governance, now shaping the information society. On February 23, a joint statement on Internet governance was presented in Geneva at the Tunis Prepcom by the Civil Society Internet Governance Caucus, the Gender Caucus, Human Rights Caucus, Privacy Caucus, and Media Caucus on behalf of the Civil Society Content and Themes Group. The statement

asserts, "gender balanced representation in all aspects of Internet Governance is vital for the process and for its outcomes to have legitimacy" (WSIS Gender Caucus, 2005a).

The Digital Solidarity Fund was proposed at WSIS, and the UN Task Force on Financial Mechanisms for ICT for Development was formed. In the 1990s, official development-assistance (ODA) support declined for ICT infrastructure development. In the new millennium, this decline has been offset by funds to integrate ICT programs into development (Hesselbarth & Tambo, 2005). The WSIS Gender Caucus (2003) statement on financing mechanisms affirmed that ICT for development must be framed as a development issue, "encompassing market-led growth but fundamentally a public policy issue." Public finance is central to achieving "equitable and gender just outcomes in ICT for development."

This article examines the WSIS political dynamics over the issue of gender equality as a fundamental principle for action in ICT policy. The WSIS civil-society participants, particularly the Gender Caucus, continued to advocate for gender equality as a fundamental principle for action and decision making within the multiple-stakeholder WSIS process of government delegates and private-sector representatives.

BACKGROUND

The WSIS Gender Caucus was formed at the 2002 WSIS African regional preparatory meeting. The WSIS Gender Caucus presented the following six recommendations for action to the WSIS voting plenary session in the spirit of "creating richness in the information society through inclusion, diversity and gender equality." Gender equality must be a fundamental principle for action. There must be equitable participation in decisions shaping the information society. New and old ICTs must be accessed in a multimodal approach. ICTs must be designed to serve people. ICT empowerment for women and girls is necessary for full participation. Research analysis and evaluation must guide action. These recommendations helped inform the development of the WSIS platform for action.

Many civil-society representatives lobbied the voting delegates for the inclusion of gender equality as a fundamental principle for action in ICT policy. The WSIS Gender Caucus, the NGO Gender Strategies Working Group (NGOGSWG), and other representatives to WSIS advocated for gender equality, basing arguments on the precedent agreements of the *Universal Declaration of Human Rights* (UDHR of 1948), Convention on the Elimination of all Forms of Discrimination against Women (CEDAW; UN Commission on the Status of Women, 1979), the Beijing Platform for Action (BPFA; UN, 1995), and the Millennium Development Goal commitments of 2000.

Global Knowledge Partnership (GKP) helped to structure and coordinate the WSIS civil-society participants at WSIS in the ICT4D civil-society forum. GKP (2002) made recommendations to the digital opportunity task force, an initiative of the G8 nations in 2000, to expand the "digital revolution" to the underserved, particularly women, rural residents, and youth in developing areas. Gender inclusion and mainstreaming were recommended:

GKP experience suggests that gender mainstreaming should be a component of every ICT project to ensure sustainability. A gender perspective must be built into plans, policy and practice, from preliminary project design through implementation and evaluation. The following case studies show that women who are involved in meaningful ICT projects improve their economic and/or social well being in the community. (GKP, 2002, p. 6)

The WSIS Gender Caucus and the NGOGSWG gender advocacy produced results in the documents of the Geneva phase of WSIS. The *WSIS Declaration of Principles* states that the common vision of the information society includes the Millennium Development Goal (MDG) challenges: "promotion of gender equality and empowerment of women;

reduction of child mortality; improvement of maternal health" (ITU, 2005, p. 9).

The documents state that gender equality and sustainable social and economic development are crucial for an equitable information society. Civil-society involvement is acknowledged as key to creating broad-based acceptance and therefore sustainable policy and plans for the information society. Civil society has also developed significant content and provision for a critical perspective. The *WSIS Plan of Action* calls for the removal of gender barriers and the development of gender-sensitive capacity building, e-learning and e-health, and early intervention programmes in science and technology that "target young girls to increase the number of women in ICT careers" (ITU, 2005, p. 45).

The WSIS Gender Caucus contributed to the UN Commission on the Status of Women's Beijing +10 review of the platform for action. Section J: Women and Media called for gender equality in media creation and delivery. The caucus reported on how radio, telecenters, and teleconferencing had enabled experts to share knowledge with rural women in agriculture and to respond to information needs. The caucus stressed the value of these ICTs in helping to deliver services to women in health, education, agricultural extension, law, and social justice. Women's participation helps to ensure that women benefit. Woman can demand more accountable governance with transparent information access, and can participate more fully informed in public discourse. ICT can help empower women as development tools for better business, education, and governance participation. All the new ICTs and media provide connection and networking spaces where women can find voice, own and control information and knowledge, and tackle issues of everyday life, sometimes in new and innovative ways (United Nations Research Institute for Social Development [UNRISD], 2005).

Economic disempowerment and illiteracy are major issues of gender inequality that impact ICT in policy and practice. Two thirds of the world's poor and undereducated are women. Even among the 55 countries with the highest United Nations Development Program (UNDP) Human Development Index (HDI), the ratio of estimated female to male earned income ranged between 34:100 and 74:100. Sweden was the lone exception with a ratio of 82:100. The United States ratio was 62:100 (UNDP, 2004). The income disparities exist despite the fact that gross tertiary enrollment is higher for females than males in all but six of these countries. The equalized education and training of women where it occurs has not yet equalized the ICT-sensitive process of input to the world's knowledge resources. Political representation by women has grown since 1990, but for the 55 top HDI countries, the percentage of positions in government held by women still ranges between 0 and 38%. Sweden again was the exception with 45% women in parliament (UNDP, 2004).

Most development research and policy recommends reduced birthrates, found to promote infant and maternal health. Smaller families and improved health allow women more time for education and employment. Gender-equality advocates lobby for ICT skills training for jobs that offer women opportunities for growth from unskilled labor to small-business entrepreneurial efforts, such as the agriculture and cottage industry.

Gender-equality issues in ICT received mention in the WSIS documents, but only nominal attention in the ongoing WSIS process of policy development for Internet governance and financing mechanisms, key issues of the WSIS Tunis phase. UNIFEM continues to fund the WSIS Gender Caucus and efforts to bring greater gender equality to the information society through research, education, and policy decision making.

The WSIS Gender Caucus identified six policy issues during WSIS Geneva, and some were incorporated in the two WSIS documents. By 2005, WGIG had virtually sidelined the gender-equality issue as a fundamental principle for action without mention in the draft report to the UN secretary general. The caucus issued a statement for the June 2005 WGIG open consultation on Internet governance and the report draft:

We believe that the published WGIG outcome criteria lack the basic and fundamental criterion of gender balance and awareness and suggest that these criteria be amended to become
- *equitable distribution of resources,*
- *access for all,*
- *stable and secure functioning of the Internet,*
- *multilingualism and*

- *gender balance and equity*
...We further request that the WGIG consider gender balance as a fundamental issue in its ongoing assessment of Internet Governance mechanisms (current and future), with the aim of equal representation of women and men at all levels in any and all governance mechanisms proposed by the WGIG. (WSIS Gender Caucus, 2005a)

Sabanes Plou (2003) identified patriarchal structures that perpetuate gender-unequal power relations in the media, in ICT research and development, and in ICT labor and policy. Plou examined gender issues of ICT access, participation, and decision making in the information society. Media are the "vehicle for transmission of ideas, images and information," and new media need "new patterns with a gender perspective" to challenge old patterns of control and decision making on access and content (p. 16).

Though ICT labor has been a source of economic growth for some, overall, women work in less skilled, lower paid, and non-decision-making positions in ICT (UNRISD, 2005). ICT access and participation depend on policy and action to address the primary gendered obstacles. Poverty requires affordable ICT. Illiteracy requires education and capability building, and a gendered approach to ICT integration into human communication systems. Patriarchal institutional structures of political, social, and economic power and relationships will continue in ICT without policy and action initiatives for change toward gender equality at all institutional levels.

FUTURE TRENDS

UNIFEM affirmed and supported the WSIS Gender Caucus' continued efforts to provide policy input for the development of Internet governance and the Digital Solidarity Fund financial mechanism for WSIS Tunis. UNIFEM also supported the WSIS Gender Caucus' cooperative work on ICT gender advocacy with the WSIS NGOGSWG, the Association for Progressive Communications (APC), the World Association of Community Radio Broadcasters (AMARC), the WIN network, which coordinated the WSIS media pool, and the Global Knowledge Partnership, which coordinated civil-society WSIS participation in major WSIS process meetings.

WSIS Gender Caucus achievements during the Geneva phase of WSIS were summarized at the closing plenary session (Marcelle, 2003). The caucus advanced some important research trends including support for gender and ICT research, encouraging increased gender-disaggregated-data collection and analysis. Critical gender analysis of the ICT policies of institutions was also encouraged as another important research area. The idea of developing an archive of gender and ICT research was advanced. The caucus organized events and activities in Geneva "enabling decision makers to interact with gender advocates and scholars on alternative visions of the information society" (Marcelle). Current research and expert perspectives were shared throughout WSIS with plans for a post-WSIS "global platform for reporting back to the development community and women's organizations" (Marcelle) in order for governments to be held accountable for the commitments made during WSIS.

Case studies of ICT for development projects that contribute to poverty elimination for women, men, and children are also part of the ongoing research catalyzed by gender advocacy at WSIS. These include projects on "women's applications of ICTs for mobilization, peace and conflict resolution, enterprise creation, trade, education, and health" (Marcelle, 2003).

During WSIS, UNIFEM announced the Digital Diaspora Initiative set up with an E-Quality Fund for African Women and Innovation to provide flexible funding for African women to have opportunities for capacity development and economic security in the information society.

Beyond the concrete WSIS direct outcomes, civil-society participants developed strong networks for information sharing and knowledge building through face-to- face and ICT communication forums. The WSIS Gender Caucus developed its portal (http://www.genderwsis.org) containing useful resources on gender-equality research and advocacy in ICT, and expanded its communications and outreach network.

The WSIS Gender Caucus, a committed group of gender advocates, continued to work with all stakeholder partners, including "governments, international agencies, the private sector and civil society ... to build an Information Society that benefits all of humanity" (Marcelle, 2003). WSIS Gender Caucus efforts support ongoing policy and action committed to gender equality and ICT applications that further the BPFA and MDGs.

The UN ICT Task Force has been "a global forum for placing ICT at the service of development" (Gilhooly, 2005) since 2001. Its mission and action plan were built on the guidance of the Millennium Declaration. It helped advance the multistakeholder discussion on Internet governance. The ICT Task Force helped to create an enabling environment toward the MDGs and to advance the practice of measuring, monitoring, and analyzing ICT impact on the MDGs. It developed an ICT strategy for knowledge creation and promoted cross-sectoral and cross-regional dialogues and partnerships. This included strengthening relationships with the research scholarship organizations International Communications Association (ICA) and International Association of Media and Communications Researchers (IAMCR). The ICT Task Force mandate was extended to 2005 to facilitate the WSIS process.

The ITU Task Force on Gender Issues (ITU-TFGI) has advanced the issues of gender mainstreaming and gender-disaggregated ICT data collection. ITU-TFGI also developed gender-aware guidelines for policy-making and regulatory agencies. The guidelines promote the establishment of gender units in regulatory bodies, gender-sensitive policy analysis, and gender-disaggregated data collection (Jorge, 2001). Research identified that women are "conspicuously absent from decision making structure in information technology in developed and developing countries" (Hafkin, 2003, p. 4), and recommended action to address the inequity.

The many cooperative gender-equality advocacy efforts have contributed to the WSIS decision making and created a policy for change. The future will unfold how these policy agreements will interact with patriarchal institutions over time to bring about real change in practice and real access to ICT with gender equality.

CONCLUSION

WSIS affirmed the MDGs to end poverty and illiteracy, two main obstacles to ICT access for women. International case studies affirm in practice how ICT applications have improved resource and service distribution for health care, education, economic empowerment, conflict resolution, and political, social, and cultural participation.

Gender equality in ICT access, power sharing, and decision making at all levels will continue as issues for dialogue, debate, and research. Active, perseverant voices for change call for humanizing the technology beyond market-economy indicators and creating an environment of cooperation and sharing rather than competition and commodification.

The argument for gender rights and equality in decision making for the information society continues to be advanced through many advocates and venues beyond WSIS. The importance of decision making was articulated well by Nobel Peace Prize winner and world leader Nelson Mandela (2004):

Our freedom and our rights will only have their full meaning as we succeed together in overcoming our divisions and inequalities of our past and in improving the lives of all, especially the poor. Today we are starting to reap some of the harvest we sowed at the end of a South African famine. Many...have spoken of a miracle. Yet those who have been most closely involved in the transition know it has been the product of human decision.

WSIS, gender, and ICT policy are integrally connected within the process of human decision making, capable of creating a world of gender justice, equality, and peace.

REFERENCES

Gender Working Group-UNCSTD. (1995). *Missing links: Gender equity in science and technology for development*. Ottawa, Ontario, Canada: International Development Research Centre.

Gilhooly, D. (Ed.). (2005). *Creating an enabling environment toward the Millennium Development Goals: Proceedings of the Berlin global forum on the United Nations ICT Task Force* (Document No. E/2005/71). New York: United Nations. Global Knowledge Partnership (GKP). (2002). *GKP recommendations: On issues of bridging the digital divide*. Kuala Lumpur, Malaysia: Author.

Hafkin, N. (2003). *Some thoughts on gender and telecommunications/ICT statistics and indicators* (Document No. WGGI-2/7-E). Retrieved July 11, 2005, from http://www.itu.int/ITU-D/gender

Hamelink, C. (2001). *ICT and social development: The global policy context*. Retrieved April 3, 2005, from http://www.unrisd.org

Haqqani, A. B. (Ed.). (2004). *The role of information and communications technologies in global development: Analysis and policy recommendations* (Series 3). New York: United Nations ICT Task Force.

Hesselbarth, S., & Tambo, I. (2005). Financing ICTs for development: Recent trends of ODA for ICTs. In D. Gilhooly (Ed.), *Creating an enabling environment toward the Millennium Development Goals: Proceedings of the Berlin global forum on the United Nations ICT Task Force* (Document No. E/2005/71, pp. 115-129). New York: United Nations.

International Telecommunications Union (ITU). (2005). *WSIS outcome documents*. Geneva, Switzerland: Author. Retrieved March 3, 2006, from http://www.int.itu/WSIS

Internet Governance Project. (2004). *Internet governance: The state of play*. Retrieved April 3, 2005, from http://www.InternetGovernance.org

Jorge, S. (2001). *Gender-aware guidelines for policy-making and regulatory agencies* (DOC TFGI-4/5 E. ITU-BDT). Retrieved July 11, 2005, from http://www.itu.int/ITU-D/gender

King, E., & Mason, A. (2001). *Engendering development: Through gender equality in rights, resources, and voice*. Washington, DC: World Bank.

Mandela, N. (2004). Diversity: From divisive to inclusive. In *UNDP human development report* (p. 43). New York: United Nations.

Mansell, R., & Wehn, U. (1998). *Knowledge societies: Information technology for sustainable development.* New York: Oxford University Press.

Marcelle, G. (2003). *Statement of the WSIS Gender Caucus to the WSIS plenary.* Retrieved July 8, 2005, from http://www.itu.int

Marcelle, G. (2005). *Technological learning: A strategic imperative for firms in the developing world.* Cheltenham, UK: Edward Elgar.

Marcelle, G., Karelse, C. M., & Goddard, G. (2000). *Engendering ICT policy: Guidelines for action.* Pretoria, Republic of South Africa: African Information Society-Gender Working Group.

Plou, D. S. (2003). What about gender issues in the information society? In B. Girard & S. O'Soichru (Eds.), *Communicating in the information society* (pp. 11-32). Geneva, Switzerland: United Nations Research Institute for Social Development (UNRISD).

United Nations. (1995). *The Beijing declaration and the platform for action: Fourth world conference on women.* New York: United Nations Department of Public Information.

United Nations. (2003a). *WSIS declaration of principles* (WSIS-03/GENEVA/DOC/4). New York: United Nations Department of Public Information.

United Nations. (2003b). *WSIS plan of action* (WSIS-03/GENEVA/DOC/5). New York: United Nations Department of Public Information.

United Nations Commission on the Status of Women. (1979). *Convention on the elimination of all forms of discrimination against women (CEDAW).* New York: UNIFEM.

United Nations Development Program (UNDP). (2004). *Human development report 2004.* New York: UNDP.

United Nations Economic and Social Council. (2005). *Third annual report of the Information and Communications Technologies Task Force* (Document No. E 2005/71). New York: United Nations.

United Nations Research Institute for Social Development (UNRISD). (2005). *Gender equality: Striving for justice in an unequal world.* Geneva, Switzerland: Author.

World Summit on the Information Society (WSIS) Gender Caucus. (2003). *Gender Caucus position on financing mechanisms.* Retrieved July 8, 2005, from http://www.genderwsis.org

World Summit on the Information Society (WSIS) Gender Caucus. (2005a). *Statement by the Civil Society Internet Governance Caucus, the Gender, Human Rights, Privacy and Media Caucuses on behalf of the Civil Society Content and Themes Group.* Retrieved July 8, 2005, from http://www.genderwsis.org

World Summit on the Information Society (WSIS) Gender Caucus. (2005b). *Statement by the Gender Caucus.* Retrieved July 8, 2005, from http://www.wgig.org

KEY TERMS

Beijing Platform for Action (BPFA): Action program produced during the 1995 Beijing Fourth World Conference on Women. The BPFA defined strategic objectives for action in 12 areas identified as essential to developing gender equality and compliance with the 1979 Convention on the Elimination of all Forms of Discrimination against Women. Global progress on the BPFA was reported at the 2000 Beijing +5 United Nations General Assembly Special Session (UNGASS) and the 2005 Beijing +10.

G8: Group of eight nations with the strongest economies, including, Canada, France, Germany, Italy, Japan, Russia, the United Kingdom, and the United States.

Human Development Index (HDI): An index used by the United Nations Development Program to measure development. HDI is composed of health indicators, infant and maternal mortality and life expectancies, education indicators of literacy, and economic indicators of gross domestic product (GDP).

NGO Gender Strategies Working Group: Formed at the first WSIS Prepcom Meeting in

Geneva in July 2002 as one of the subcommittees of the Civil Society Coordinating Group (CSCG). The groups currently involved in this effort are the African Women's Development and Communications Network (FEMNET), Agencia Latino Americana de Informacion, Association for Progressive Communication-Women's Networking Support Programme (APC-WNSP), International Women's Tribune Centre (IWTC), and Isis International-Manila. The working group is open to all NGOs and individuals interested in gender issues and the information society (http://www.wougnet.org/WSIS/wsisgc.html).

WSIS: The World Summit on the Information Society is a two-phase international summit from 2003 to 2005 organized by the United Nations and the International Telecommunications Union to address the need for international policy and agreement on ICT governance, rights, and responsibilities.

WSIS Declaration of Principles: It affirms gender equality and other MDGs:

Our challenge is to harness the potential of information and communication technology to promote the development goals of the Millennium Declaration, namely the eradication of extreme poverty and hunger; achievement of universal primary education; promotion of gender equality and empowerment of women; reduction of child mortality; improvement of maternal health; to combat HIV/AIDS, malaria and other diseases; ensuring environmental sustainability; and development of global partnerships for development for the attainment of a more peaceful, just and prosperous world. (ITU, 2005, p. 9)

WSIS Gender Caucus: Formed during the WSIS African Prepcom in Bamako in 2002 when representatives of organizations responded to an invitation by UNIFEM to contribute to ensuring that gender dimensions are included in the process of defining and creating a global information society that contributes to sustainable development and human security; they issued the WSIS Gender Caucus Bamako Statement (http://www.genderwsis.org).

Young Women and Persistence in Information Technology

Tiffany Barnes
University of North Carolina at Charlotte, USA

Sarah Berenson
North Carolina State University, USA

Mladen A. Vouk
North Carolina State University, USA

INTRODUCTION

The underrepresentation of women in science, technology, and engineering careers is of growing national concern (Vesgo, 2005; National Academy of Engineering, 2002; National Science Foundation, 2004; National Research Council, 2001). While the information technology (IT) workforce appears to be becoming more diverse in terms of race and country of birth, it is becoming less diverse in terms of gender (AAUW, 2000; Malcom, Babco, Teich, Jesse, Campbell, & Bell, 2005; NSF, 2004; Vesgo, 2005). This trend is of particular concern, since women may face unequal access to rewarding IT careers, while society and the IT workforce suffer without the valuable contributions that women might make through the creation of new information technologies (Cohoon, 2005; Freeman & Cuny, 2005).

Past studies have highlighted a tendency of talented young girls to enroll in less rigorous mathematics courses beginning in the middle grades (e.g., Kerr, 1997) and have hypothesized that this lack of preparation creates a barrier to science, technology, and engineering disciplines. In response to the increased under-representation of women in IT, Girls on Track (Got), a year round enrichment program and summer camp, was created in 1998 to encourage talented middle school girls to persist in taking college-bound courses in math, science, and computer science through high school. It was our conjecture that some of these well-prepared girls would later become creative future IT workers.

We have undertaken a longitudinal study of approximately 200 girls who were enrolled in the NSF funded 1999-2001 Girls on Track program, with the goal of creating a model of persistence of these young women into IT careers. This study is now in its seventh year. In this article, we present our somewhat surprising findings. It would appear that talented young women, though prepared and able, are not choosing to pursue IT careers. We suggest some ways the thinking about IT may need to change to encourage broader career-level participation.

BACKGROUND

The demand for information technology workers is projected to surpass demand for all other occupations through 2012 (Sargent, 2004), yet overall enrollments in IT-related fields continue to decline (Zweben, 2005). The percentage of women in IT has also continued to decline (Malcom et al., 2005; Vesgo, 2005). The reasons for this are not well understood, although the "dot-com bubble" deflation in the 2000 may play a part (Malcom et al., 2005). In recent years, the achievement gap in mathematics and science has been closing as more women select advanced courses in high school science and mathematics (National Science Board, 2000). However, enrollments of young women in computer science courses and advanced placement (AP) exams in high school continue to remain low (AAUW, 2000; CCAWM, 2000; Freeman & Cuny, 2005).

Some researchers examine girls' experiences from the middle grades to high school for the root causes of women's underrepresentation in IT. For example, Freeman and Aspray (1999) note that girls have less experience with computers and perceive IT-related work to be solitary and competitive, requiring long hours and unsafe working environments. During this age range, many girls become more involved in extra-curricular activities and take less rigorous courses (Kerr, 1997). At the same time, girls lower their career aspirations between the middle grades and high school (Kerr, 1997), through choosing less competitive careers and post-secondary institutions. Since the rigorous preparatory courses for prestigious fields, including courses in advanced math, science, and technology, are often filters for technical fields, these factors may have a strong influence on women's participation in IT. Our previous findings indicate that parental influence may also be a strong factor in girls' choices (Howe, Berenson, & Vouk, 2005).

Several studies that explore recruiting and retention of women in undergraduate IT curricula have reported factors that positively influence the enrollment and persistence of women in IT-related fields. Margolis and Fisher found that prior class experiences, as well as interest in computers and the promise of the field, were primary motivators for majoring in computer science (2001). In the first national study exploring gendered outcomes in undergraduate computer science programs, Cohoon found that faculty attitudes and behaviors could have a powerful influence on gendered attrition. Factors significantly correlated with higher retention rates for women include: having sufficient faculty, responsiveness to the job market, and faculty who mentored for the purpose of retaining underrepresented minorities (Cohoon, 2005). The availability of same-sex peer support and professional experiences are also important factors in women's retention in computer science programs (Blum & Frieze, 2005; Cohoon, 2005).

Recent results from efforts to gender-balance the undergraduate program at Carnegie Mellon indicate that fundamental misconceptions about computer science, as opposed to gender differences, may be the root cause of the under-representation of women in IT, as well as the declining interest in computer science overall (Blum & Frieze, 2005; Vesgo, 2005; Zweben, 2005). Computer science, Blum and Frieze (2005) argue, is not equal to programming, although the advanced placement exam in computer science reinforces this unfortunate misconception.

In January 2005, Freeman and Cuny identified several areas where efforts could make a difference in broadening participation in computing, including defining computer science to override popular misconceptions, training faculty in cross-cultural mentoring, providing research experiences for undergraduates, and working with K-12 teachers to define computer science curricula.

A recent study sponsored by the American Association for the Advancement of Science and the Commission on Professionals in Science and Technology includes an insightful discussion of the complexities of the IT educational and employment markets, and recommends policies to support the increased diversity of the IT workforce (Malcom et. al, 2005). These recommended policies include a change in admissions criteria—by shifting the emphasis from programming experience to problem solving skills that are relevant to IT/CS. Four year institutions should offer more career guidance and workplace experiences, as well as opportunities for nontraditional students to take courses online or while working full-time. These recommendations align with findings at Carnegie Mellon, where these types of changes have been effective in increasing the participation of women to about a third (Blum & Frieze, 2005), while national averages of participation for women are less than 20% (Vesgo, 2005).

GIRLS ON TRACK: INSIGHTS

Girls on Track, a program funded by the National Science Foundation (NSF #9813902) from 1999-2001, was created in response to the need to increase women's representation in IT-related careers. The Girls on Track program provided 1 year-round enrichment for mathematically high achieving girls in grades 7 and 8. We define high achieving girls as those selected to take Algebra I on the "fast track," thus enabling them to take Advanced Placement Calculus in high school, a necessary preparation for college courses in mathematics, sciences, engineering, or computer science. The age range was 11-13, with about 60% Caucasian, about 30% African American, and about 10% Asian.

Figure 1. GoT participants still "on the fast math track" in 2004

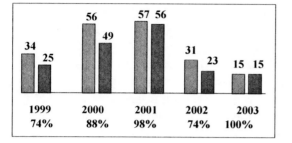

Girls selected for Girls on Track participated in a two-week summer learning experience that included group projects using math and information technologies to address community problems, increasing girls' awareness of gender issues and career planning, along with time for active math-related games. A mentoring program provided continuing support and enrichment activities throughout the school year. Girls on Track also incorporated a professional development component for middle school math teachers, pre-service teachers, and guidance counselors.

In the creation of Girls on Track, we assumed that high achieving young women would be excellent candidates for future IT careers. However, as we have followed these participants through our NSF funded *Women in Technology* (NSF #0204222) longitudinal study, we have discovered that few of these same young women are taking computer science courses in high school, and although one girl has shown interest, so far none of the initial program cohort (1999) have elected IT programs in college. A parallel study at North Carolina State University, *Pair Programming and Agile Software Development*, revealed a theme of the perceptions of the time needed for IT careers, which we found echoed in our interviews of Girls on Track participants.

There are certainly social, personal, and school components that enter into the decisions of the girls to stay involved with IT. In our studies, we are tracking 207 girls using a number of indicators such as proportional reasoning test scores, annual math achievement scores, math attitude (eight factors), courses taken in math, science, and computer science, and PSAT and SAT scores. We have found that a surprisingly large number stayed "on the fast math track" as we have defined it. Approximately

90% of the girls in this study remain on the fast math track until their junior year. At this time, some girls are advised to take a third year of algebra, while others are tracked into advanced mathematics, leaving about 75% of the GoT girls on the fast track. Figure 1 illustrates these percentages for each year of the program; the first bar for each year indicates the total number of students we have been able to follow over time while the second bar shows the number who are taking advanced math courses.

Phone interviews of 39 girls from the first and second cohorts found only one who expressed an interest in an IT career in high school, and only 6 who have elected computer science courses in high school. Five of these six girls noted the low numbers of girls in the courses, and expressed their dissatisfaction with the course content. Only one reported that she liked her CS class and enjoyed solving problems. Only one student took four CS classes in high school, reporting that she enjoyed her CS classes but she is not considering a career in IT.

Reasons a sample of interviewed girls gave for this low CS enrollment in high school include: (a) a lack of interest in computer science, (b) computer science was not an advanced or advanced placement (AP) class (and therefore also provided no extra GPA points), (c) students had no room in their schedules of other advanced classes, (d) computer science is not a college entrance requirement, and (e) some students preferred other electives, such as music or medical academy. These successful young women are willing to work hard but do not wish to be bored by the work, and as these reasons for avoiding CS classes show, they wish to enroll in courses which show clear evidence of advancing them toward college entrance requirements.

These results show both positive and negative aspects. High achieving girls continue to enroll in advanced math courses and are motivated to learn and to find a career that will be both interesting and beneficial to society. They receive strong support from their families and teachers. However, most of these girls do not elect CS courses in high school, do not enjoy CS courses when they do, and few express interest in IT careers, even though 35 to 40% of Girls on Track participants were interested while in middle school.

Although firm data are not available on the effects of the low involvement of girls in high school

CS courses on collegiate IT enrollments, evidence seems to indicate that the limited computer and programming experiences of talented girls can affect admissions into IT/CS departments (e.g., Blum & Frieze, 2005). How to influence the choices of these girls, however, remains an open issue.

In-depth interviews with 30 high school girls revealed that more than half have no image or have incorrect images about "computer science" careers. Some perceive "computer science" as the use of tools such as spreadsheets and databases rather than the design and development (creation) of tools, and the repairing of hardware rather than the engineering of hardware. The creative and inventive benefits of IT careers were hidden from these girls.

Parents are a powerful influence in these girls' lives: two thirds of the young women we interviewed attribute their academic success to their parents. Many of the young women have parents working in the IT industry yet had little to no idea what their parents' work really involved. For those who were more informed about IT careers, more than 20% commented that work with computers required too much time while nearly the same number stated that they did not want to work in a cubicle environment.

One young woman who planned to pursue an IT career spoke positively of her father's IT career in a large company acknowledged for its campus-like, worker-centered environment. This lead us to conjecture that parents' workplace experiences and attitudes may be an important factor in high achieving girls' decisions (not) to pursue IT careers. One possible way to counter these effects would be to provide computer science and engineering experiences that engage girls in inventing, rather than simply using the technology.

The girls in our Girls on Track program exhibited an enthusiastic view towards information technologies while in middle school. However, in our follow-up *Women and Information Technology* study we found few of these same young women interested in high school computer science courses 2-4 years later. None of these high achieving young women have yet elected IT concentrations in college. Yet 75% of the young women in these longitudinal studies take calculus before high school graduation. On the other hand, the study is not over yet. We do not know whether, once they finish the college, these girls who are certainly capable of succeeding in IT careers, may still choose to re-enter IT career paths laterally.

A parallel study at North Carolina State University, *Pair Programming and Agile Software Development*, is a pedagogical intervention to increase face-to-face collaboration in an upper level software engineering course. Interviews with young college women taking this course revealed that the instructional interventions of pair programming and agile development methods saved time, an important value for them (Berenson, Slaten, Williams, & Ho, 2004). Examining the transcripts of the high school girls' interviews for evidence of the "time" theme, we found that some believed that computer science careers demanded too much time, based on their observations of their parents' career experiences in IT and their observations of high school computer "boy geeks."

Having "enough time" for other activities was not an issue for these same girls while in middle school. This difference implies that young women place more value on time and on having freedom to engage in several activities as they mature. This suggests that long hours of university study combined with the long working hours necessary to advance in IT careers may not be attractive to many young women. Margolis and Fisher (2001) speak of the discouragement college women feel when comparing their values with the intensity and focus found in the computer geek culture. To attract women into IT careers, the existing cultures of IT in the university and in the workplace may require a change.

FUTURE TRENDS

A balanced inclusion of women and men into IT should introduce new innovations and improved working conditions for all IT workers (CCAWM, 2000). This balance should also provide the peer support that is particularly needed in recruiting and retaining women in IT (Cohoon, 2005), and should lead to increased opportunities for leadership and full participation (Blum & Frieze, 2005).

However, to achieve this balance, substantial changes may need to be made in the culture and perceptions of IT from within, such as those changes made at Carnegie Mellon University in recent years (Blum & Frieze, 2005). University admissions criteria and advanced placement courses should emphasize the creative and problem-solving aspects of IT, and shift away from a focus on programming (Malcom et. al, 2005). In addition, introductory IT/CS courses should de-emphasize programming and provide a broader view of IT/CS. In order to achieve this goal, it may be necessary to foster IT/CS education programs to prepare teachers and students for new ways of teaching and learning IT/CS concepts and skills, including pair programming and agile software development practices (Berenson, Williams, Michael, & Vouk, 2005). Professional experiences, including mentoring and opportunities for early exposure to real IT applications and careers, should be made available to all students.

CONCLUSION

Providing all U.S. citizens with the "opportunity to gain the skills and knowledge to compete" in the IT workforce will enable the U.S. to remain competitive on a global scale (Malcom et. al, 2005, p. 20). Creativity and innovation, which require a much higher level of preparation, are key in our nation's ability to compete. The participation of women and under-represented minorities will ensure a diversity of perspective that can enrich the quality of innovation in IT.

Findings from our Girls on Track and Women in Technology studies indicate that, although girls are participating to a greater degree in advanced mathematics preparation, high achieving girls are not enrolling in IT/CS courses or choosing IT careers. To recruit these talented girls, we believe it will be necessary to make substantial changes in the culture of IT, including both perceptions and education. Respecting the values of young women and other nontraditional populations, including the need for time for life outside of work, while also maintaining high standards and emphasizing the problem-solving and creative aspects of the field, can make broad improvements in working conditions, diverse participation, and innovation in IT.

REFERENCES

American Association of Undergraduate Women (AAUW)'s Educational Foundation. (2000). *Tech-Savvy: Educating girls in the new computer age.* Washington, DC: American Association of Undergraduate Women.

Berenson, S., Slaten, K., Williams, L., & Ho, C. (2004). Voices of women in a software engineering course: Reflections on collaboration. *Journal on Educational Resources in Computing, 4*(1), 3.

Berenson, S., Williams, L., Michael, J., & Vouk, M. (2005). Examining time as a factor in young women's IT career decisions. In *Proceedings of the Crossing Cultures, Changing Lives International Research Conference*, Oxford, England.

Blum, L., & Frieze, C. (2005, May). In a more balanced computer science environment, similarity is the difference and computer science is the winner. *Computing Research News, 17*(3), 2-16.

Cohoon, J. M. (2005). Just get over it of just get on with it. In *Women and information technology: Research on under-representation* (pp. 205-238). Cambridge, MA: MIT Press.

Congressional Commission on the Advancement of Women and Minorities in Science, Engineering and Technology Development (CCAWM). (2000, September). *Land of plenty: Diversity as America's competitive edge in science, engineering and technology.* Washington, DC: Author.

Freeman, P., & Aspray, W. (1999). *The supply of information technology workers in the United States.* Washington, DC: Computing Research Association.

Freeman, P., & Cuny, J. (2005, January). Common ground: A diverse CS community benefits all of us. *Computing Research News, 17*(1), 1-6.

Kerr, B. S. (1997). *Smart girls: A new psychology of girls, women and giftedness.* Scottsdale, AZ: Gifted Psychology.

Malcom, S., Babco, E., Teich, A., Jesse, J. K., Campbell, L., & Bell, N. (2005). *Bringing women and minorities into the IT workforce: The role of nontraditional educational pathways*. Washington, DC: American Association for the Advancement of Science and the Commission on Professionals in Science and Technology.

Margolis, J., & Fisher, A. (2001). *Unlocking the clubhouse: Women in computing*. Cambridge, MA: MIT Press.

National Academy of Engineering. (2002). *Diversity in engineering: Managing the workforce of the future*. Washington, DC: National Academy Press.

National Research Council. (2001). *Building a workforce for the information economy*. Washington, DC: National Academy Press.

National Science Board. (2000). *Science & engineering indicators*. Arlington, VA: National Science Board (NSB 00-1).

National Science Foundation (NSF). (2004). *Women, minorities, and persons with disabilities in science and engineering*. Arlington, VA: National Science Foundation (NSF 04-317).

Sargent, J. (2004). An overview of past and projected employment changes in the professional IT occupations. *Computing Research News, 16*(3), 1-21.

Vesgo, J. (2005). CRA Taulbee trends: Female students & faculty. *Computing Research Association Taulbee Survey 2005*. Retrieved January 13, 2006, from http://www.cra.org/info/taulbee/women.html

Zweben, S. (2005, May). 2003-2004 Taulbee survey: Record Ph.D. production on the horizon; Undergraduate enrollments continue in decline. *Computing Research News, 17*(3), 7-15.

KEY TERMS

Agile Software Development: A philosophy of software development that values people, collaboration, response to change, and working software More information is available at. http://agilemanifesto.org

Girls on Track: An intervention program designed to keep talented middle school girls on the "fast math track." More information is available at http://ontrack.ncsu.edu.

IT Career: A career requiring an electrical engineering, computer science, or computer engineering degree. Emphasis is placed on technical and creative roles rather than support roles.

IT Education: A proposed new field to prepare teachers to introduce students to IT in more effective ways.

Nontraditional Student: Women and underrepresented minorities often enter the IT workforce through nontraditional pathways, including part-time education, through for-profit trade schools, or by starting a baccalaureate degree after the age of 21.

Pair Programming: An agile software development process wherein two programmers work side-by-side at one computer, collaborating to write software.

Women in Technology (WIT): A longitudinal study of Girls on Track program designed to model the educational persistence of young women in IT-related fields. More information is available at http://wit.ncsu.edu.

ENDNOTE

[1] The program is still successful and continues annually under various sponsorships.

Index of Key Terms

Volume I, pp. 1-744 / Volume II, pp. 745-1330

Index

Volume I, pp. 1-744 / Volume II, pp. 745-1330

W